Student's Solutions Manual for Basic Technical Mathematics and Basic Technical Mathematics with Calculus, 11th Edition

Chapter 1

Basic Algebraic Operations

1.1 Numbers

1. The numbers –7 and 12 are integers. They are also rational numbers since they can be written as $\frac{-7}{1}$ and $\frac{12}{1}$.

5. 3 is an integer, rational $\left(\frac{3}{1}\right)$, and real.

$\sqrt{-4}$ is imaginary.

9. $|3| = 3$

$|-3| = 3$

$\left|-\frac{\pi}{2}\right| = \frac{\pi}{2}$

13. $\pi < 3.1416$; π (3.1415926...) is to the left of 3.1416.

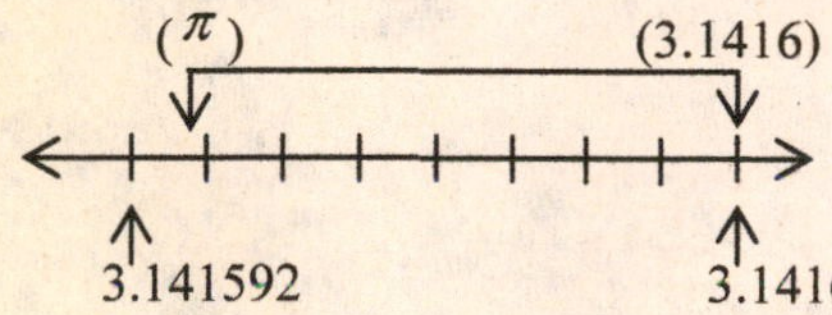

17. $-\frac{2}{3} > -\frac{3}{4}$; $-\frac{2}{3} = -0.666\ldots$ is to the right of $-\frac{3}{4} = -0.75$.

21. Find 2.5, $-\frac{12}{5} = -2.4$; $-\frac{3}{4} = -0.75$; $\sqrt{3} = 1.732...$

25. The reciprocal of the reciprocal of any positive or negative number is the number itself.

The reciprocal of n is $\frac{1}{n}$; the reciprocal of $\frac{1}{n}$ is $\frac{1}{1/n} = 1 \cdot \frac{n}{1} = n$.

29. List these numbers from smallest to largest: -1, 9, $\pi = 3.14$, $\sqrt{5} = 2.236$, $|-8| = 8$, $-|-3| = -3$, -3.1.

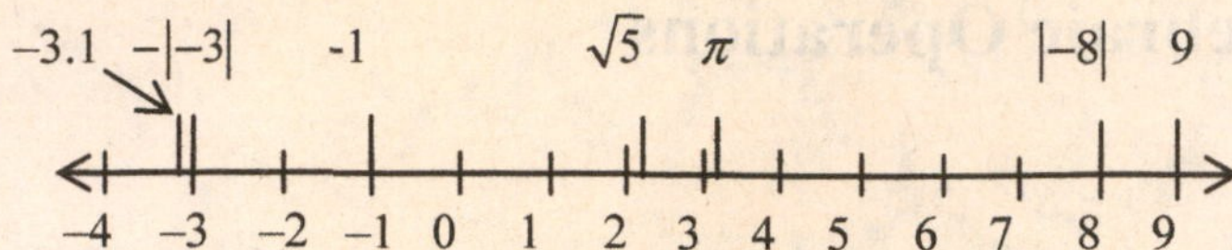

So, from smallest to largest, they are -3.1, $-|-3|$, -1, $\sqrt{5}$, π, $|-8|$, 9.

33. **(a)** Is the absolute value of a positive or a negative integer always an integer?

$|x| = x$, so the absolute value of a positive integer is an integer.

$|-x| = x$, so the absolute value of a negative integer is an integer.

(b) Is the reciprocal of a positive or negative integer always a rational number?

If x is a positive or negative integer, then the reciprocal of x is $\frac{1}{x}$. Since both 1 and x are integers, the reciprocal is a rational number.

37. If $x > 1$, then $\frac{1}{x}$ is a positive number less than 1. Or $0 < \frac{1}{x} < 1$.

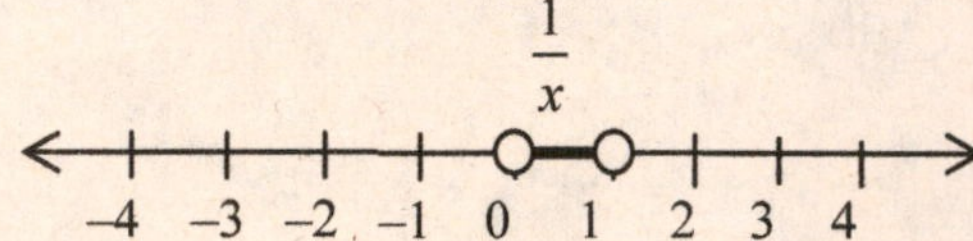

41. $\frac{1}{C_T} = \frac{1}{C_1} + \frac{1}{C_2}$. Find C_T, where $C_1 = 0.0040\text{F}$ and $C_2 = 0.0010\text{F}$.

$$\frac{1}{C_T} = \frac{1}{0.0040} + \frac{1}{0.0010}$$

$$\frac{1}{C_T} = \frac{1(0.0040) + 1(0.0010)}{0.0040 \times 0.0010}$$

$$C_T = \frac{0.0040 \times 0.0010}{0.0040 + 0.0010} = \frac{0.0000040}{0.0050}$$

$$C_T = 0.00080 \text{ F}$$

45. Yes, $-20\,°\text{C} > -30\,°\text{C}$ because $-30\,°\text{C}$ is found to the left of $-20\,°\text{C}$ on the number line.

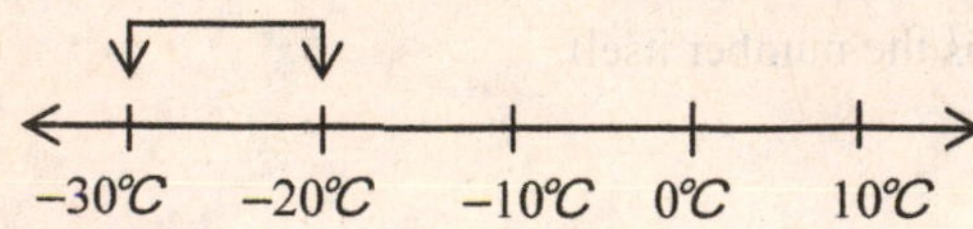

1.2 Fundamental Operations of Algebra

1. $16-2\times(-2)=16-(-4)=16+4=20$

5. $5+(-8)=5-8=-3$

9. $-19-(-16)=-19+16=-3$

13. $-7(-5)=+(7\times5)=35$

17. $-2(4)(-5)=-8(-5)=40$

21. $16\div2(-4)=8(-4)=-32$

25. $\dfrac{17-7}{7-7}=\dfrac{10}{0}$ is undefined

29. $-2(-6)+\left|\dfrac{8}{-2}\right|=12+|-4|=12+4=16$

33. $\dfrac{24}{3+(-5)}-4(-9)=\dfrac{24}{-2}+(4\times9)=-12+36=24$

37.
$$\frac{3|-9-2(-3)|}{1-10}=\frac{3|-9+6|}{-9}=\frac{3|-3|}{-9}=\frac{9}{-9}=-1$$

41. $6(3+1)=6(3)+6(1)$ demonstrates the distributive law.

45. $(\sqrt{5}\times3)\times9=\sqrt{5}\times(3\times9)$ demonstrates the associative law of multiplication.

49. $-b-(-a)=-b+a=a-b$, which is expression (b).

53. **(a)** The sign of a product of an even number of negative numbers is positive.
Example: $-3(-6)=18$

(b) The sign of a product of an odd number of negative numbers is negative.
Example: $-5(-4)(-2)=-40$

57. **(a)** $-xy=1$ is true for values of x and y that are negative reciprocals of each other or $y=-\dfrac{1}{x}$, providing that the number x in the denominator is not zero. So if $x=12$, then $y=-\dfrac{1}{12}$ and $-xy=-(12)\left(-\dfrac{1}{12}\right)=1$.

(b) $\dfrac{x-y}{x-y}=1$ is true for all values of x and y, providing that $x\neq y$ to prevent division by zero.

61. The change in the meter energy reading E would be:
$$E_{change}=E_{used}-E_{generated}$$
$$E_{change}=2.1\text{ kW}\cdot\text{h}-1.5\text{ kW}(3.0\text{ h})$$
$$E_{change}=2.1\text{ kW}\cdot\text{h}-4.5\text{ kW}\cdot\text{h}$$
$$E_{change}=-2.4\text{ kW}\cdot\text{h}$$

65. The sum of the voltages is
$$V_{sum}=6\text{V}+(-2\text{V})+8\text{V}+(-5\text{V})+3\text{V}$$
$$V_{sum}=6\text{V}-2\text{V}+8\text{V}-5\text{V}+3\text{V}$$
$$V_{sum}=10\text{V}$$

69. The total time spent browsing these websites is the total time spent browsing the first site on each day + the total time spent browsing the second site on each day
$$t=7\text{ days}\times25\frac{\text{minutes}}{\text{day}}+7\text{ days}\times15\frac{\text{minutes}}{\text{day}}$$
$$t=175\text{ min}+105\text{ min}$$
$$t=280\text{ min}$$
OR
$$t=7\text{ days}\times(25+15)\frac{\text{minutes}}{\text{day}}$$
$$t=7\text{ days}\times40\frac{\text{minutes}}{\text{day}}$$
$$t=280\text{ min}$$
which illustrates the distributive law.

1.3 Calculators and Approximate Numbers

1. 0.390 has three significant digits since the zero is after the decimal. The zero is not necessary as a placeholder and should not be written unless it is significant.

5. 8 cylinders is exact because they can be counted. 55 km/h is approximate since it is measured.

9. Both 1 cm and 9 g are measured quantities and so they are approximate.

13. 6.80 has 3 significant digits since the zero indicates precision; 6.08 has 3 significant digits; 0.068 has 2 significant digits (the zeros are placeholders.)

17. 5000 has 1 significant digit; 5000.0 has 5 significant digits; $500\bar{0}$ has 4 significant digits since the bar over the final zero indicates that it is significant.

21. **(a)** Both 0.1 and 78.0 have the same precision as they have the same number of decimal places.
(b) 78.0 is more accurate because it has more significant digits (3) than 0.1, which has 1 significant digit.

25. **(a)** 4.936 rounded to 3 significant digits is 4.94.
(b) 4.936 rounded to 2 significant digits is 4.9.

29. **(a)** 5968 rounded to 3 significant digits is 5970.
(b) 5968 rounded to 2 significant digits is $6\bar{0}00$.

33. **(a)** Estimate: $13+1-2=12$
(b) Calculator: $12.78+1.0495-1.633=12.1965$, which is 12.20 to 0.01 precision

37. **(a)** Estimate $9+(1)(4)=9+4=13$
(b) Calculator: $8.75+(1.2)(3.84)=13.358$, which is 13 to 2 significant digits

41. **(a)** Estimate $4.5-\dfrac{2(300)}{400}=3.0$, to 2 significant digits
(b) Calculator: $4.52-\dfrac{2.056(309.6)}{395.2}=2.9093279$, which is 2.91 to 3 significant digits

45. $-3.142(65)=-204.23$, which is -204.2 because the least accurate number has 4 significant digits.

49. The speed of sound is $3.25\text{ mi}\div 15\text{ s}=0.21666...\text{ mi/s}=1144.0...\text{ ft/s}$. However, the least accurate measurement was time since it has only 2 significant digits. The correct answer is 1100 ft/s.

53. **(a)** $2+0=2$
(b) $2-0=2$
(c) $0-2=-2$
(d) $2\times 0=0$
(e) $2\div 0=\text{error}$; from Section 1.2, an equation that has 0 in the denominator is undefined when the numerator is not also 0.

57. **(a)** $1\div 3=0.333...$ It is a rational number since it is a repeating decimal.
(b) $5\div 11=0.454545...$ It is a rational number since it is a repeating decimal.
(c) $2\div 5=0.400...$ It is a rational number since it is a repeating decimal (0 is the repeating part).

61. We would compute $12(129)+16(298.8)=6328.8$ and round to three significant digits for a total weight of 6330 g. The values 12 and 16 are exact.

65. **(a)** Estimate $8\times 5-10=30$, to 1 significant digit.
(b) Calculator: $7.84\times 4.932-11.317=27.34988$ which is 27.3 to 3 significant digits.

1.4 Exponents and Units Conversions

1. $(-x^3)^2=\left[(-1)x^3\right]^2=(-1)^2(x^3)^2=(1)x^6=x^6$

5. $2b^4b^2=2b^{4+2}=2b^6$

9. $\dfrac{-n^5}{7n^9}=-\dfrac{n^{5-9}}{7}=-\dfrac{n^{-4}}{7}=-\dfrac{1}{7n^4}$

13. $\left(aT^2\right)^{30}=a^{30}T^{2(30)}=a^{30}T^{60}$

17. $\left(\frac{x^2}{-2}\right)^4 = \frac{x^{2(4)}}{(-2)^4} = \frac{x^8}{16}$

21. $-3x^0 = -3(1) = -3$

25. $\frac{1}{R^{-2}} = R^2$

29. $-\frac{L^{-3}}{L^{-5}} = -L^{-3-(-5)} = -L^2$

33. $\frac{(n^2)^4}{(n^4)^2} = \frac{n^{2(4)}}{n^{4(2)}} = \frac{n^8}{n^8} = 1$

37. $(-8g^{-1}s^3)^2 = (-8)^2 g^{-1(2)} s^{3(2)} = \frac{64s^6}{g^2}$

41. $\frac{15n^2T^5}{3n^{-1}T^6} = \frac{5n^{2-(-1)}}{T} = \frac{5n^3}{T}$

45. $-(-26.5)^2 - (-9.85)^3 = -(702.25) - (-955.671625) = 253.421625$

which gets rounded to 253 because 702.25 and –955.671625 are both accurate to only 3 significant digits due to the original numbers having only 3 significant digits.

49. $2.38(-60.7)^2 - \frac{254}{1.17^3} = 2.38(3684.49) - \frac{254}{1.601613}$

$= 8769.0862 - 158.5901213339$

$= 8610.4960786661$

which gets rounded to 3 significant digits: 8610.

53. If $a^3 = 5$, then

$a^{12} = a^{3(4)}$

$a^{12} = \left(a^3\right)^4$

$a^{12} = (5)^4$

$a^{12} = 625$

57. ${\frac{kT}{hc}}^{3} (GkThc)^2 c = \frac{k^3T^3}{h^3c^3} \cdot (G^2k^2T^2h^2c^2)c$

$= \frac{k^3T^3}{h^3c^3} \cdot (G^2k^2T^2h^2c^3)$

$= \frac{(G^2k^{2+3}T^{2+3}c^{3-3})}{h^1}$

$= \frac{G^2k^5T^5}{h}$

61. $$2500 \quad 1+\frac{0.042}{4} \quad ^{24} = \$2500(1.0105)^{24}$$
$$= \$2500(1.28490602753)$$
$$= \$3212.26700688$$
$$= \$3212.27$$

65. $\left(28.2\frac{\text{ft}}{\text{s}}\right)(9.81\text{ s}) = 276.642$ ft which is rounded to 277 ft.

69. $15.7\text{ qt} = 15.7\text{ qt}\times\left(\frac{1\text{ L}}{1.057\text{ qt}}\right) = 14.8533586\text{ L}$ which is rounded to 14.9 L.

73. $65.2\frac{\text{m}}{\text{s}} = 65.2\frac{\text{m}}{\text{s}}\times\left(\frac{60\text{ s}}{1\text{ min}}\right)\times\left(\frac{1\text{ ft}}{0.3048\text{ m}}\right) = 12834.6457\ \frac{\text{ft}}{\text{min}}$ which is rounded to $12800\ \frac{\text{ft}}{\text{min}}$.

77. $575{,}000\frac{\text{gal}}{\text{day}} = 575{,}000\frac{\text{gal}}{\text{day}}\times\left(\frac{1\text{ day}}{24\text{ hr}}\right)\times\left(\frac{3.785\text{ L}}{1\text{ gal}}\right) = 90{,}682.2917\ \frac{\text{L}}{\text{hr}}$ which is rounded to $90{,}700\ \frac{\text{L}}{\text{hr}}$.

81. $14.7\frac{\text{lb}}{\text{in}^2} = 14.7\frac{\text{lb}}{\text{in}^2}\times\left(\frac{4.448\text{ N}}{1\text{ lb}}\right)\times\left(\frac{1\text{ in}}{2.54\text{ cm}}\right)^2\times\left(\frac{100\text{ cm}}{1\text{ m}}\right)^2 = 101{,}347.883\ \frac{\text{N}}{\text{m}^2}$ which is rounded to 101,000 Pa.

1.5 Scientific Notation

1. $8.06\times10^3 = 8060$

5. $2.01\times10^{-3} = 0.00201$

9. $1.86\times10 = 18.6$

13. $0.0087 = 8.7\times10^{-3}$

17. $0.0528 = 5.28\times10^{-2}$

21. $\frac{88{,}000}{0.0004} = \frac{8.8\times10^4}{4\times10^{-4}} = 2.2\times10^8$

25. $0.0973 = 97.3\times10^{-3}$

29. $2\times10^{-35} + 3\times10^{-34} = 0.2\times10^{-34} + 3\times10^{-34} = 3.2\times10^{-34}$

33. $1320(649{,}000)(85.3) = 7.3074804\times10^{10}$
which gets rounded to 7.31×10^{10}.

37. $(3.642\times10^{-8})(2.736\times10^5) = 9.964512\times10^{-3}$
which gets rounded to 9.965×10^{-3}.

41. 500,000,000 tweets = 5×10^8 tweets

45. 1,200,000,000 Hz = 1.2×10^9 Hz

49. 1.6×10^{-12} W = 0.0000000000016 W

53. **(a)** googol $= 1\times10^{100} = 10^{100}$

(b) googolplex $= 10^{\text{googol}} = 10^{10^{100}}$

57. $\dfrac{7.5\times10^{-15}\text{ s}}{\text{addition}}\times5.6\times10^6\text{ additions} = 4.2\times10^{-8}\text{ s}$

61. $\dfrac{1.66\times10^{-27}\text{ kg}}{\text{amu}}\times\dfrac{1.6\times10^1\text{ amu}}{\text{oxygen atoms}}\times1.25\times10^8\text{ oxygen atoms} = 3.32\times10^{-18}\text{ kg}$

1.6 Roots and Radicals

1. $-\sqrt[3]{64} = -\sqrt[3]{(4)^3} = -4$

5. $\sqrt{49} = \sqrt{7^2} = 7$

9. $-\sqrt{64} = -\sqrt{8^2} = -8$

13. $\sqrt[3]{125} = \sqrt[3]{5^3} = 5$

17. $\left(\sqrt{5}\right)^2 = \sqrt{5}\times\sqrt{5} = 5$

21. $\left(-\sqrt[4]{53}\right)^4 = (-1)^4\left(\sqrt[4]{53}\right)^4 = (1)(53) = 53$

25. $\sqrt{1200} = \sqrt{(100)(4)(3)} = \sqrt{100}\times\sqrt{4}\times\sqrt{3} = 10\times2\times\sqrt{3} = 20\sqrt{3}$

29. $\sqrt{\dfrac{80}{|3-7|}} = \sqrt{\dfrac{80}{4}} = \sqrt{20} = \sqrt{4\times5} = \sqrt{4}\times\sqrt{5} = 2\times\sqrt{5} = 2\sqrt{5}$

33. $\dfrac{7^2\sqrt{81}}{(-3)^2\sqrt{49}} = \dfrac{(49)(9)}{(9)(7)} = \dfrac{(49)\cancel{(9)}}{\cancel{(9)}(7)} = 7$

37. $\sqrt{3^2+9^2} = \sqrt{9+81} = \sqrt{90} = \sqrt{(9)(10)} = \sqrt{9}\times\sqrt{10} = 3\sqrt{10}$

41. $\sqrt{0.8152} = 0.9028842672$, which is rounded to 0.9029

45. **(a)** $\sqrt{0.0429^2 - 0.0183^2} = \sqrt{0.00184041 - 0.00033489}$
$= \sqrt{0.00150552}$
$= 0.03880103091$
$= 0.0388$

(b) $\sqrt{0.0429^2} - \sqrt{0.0183^2} = 0.0429 - 0.0183$
$= 0.0246$

49. $\sqrt{\frac{B}{d}} = \sqrt{\frac{2.18\times10^9 \text{ Pa}}{1.03\times10^3 \text{ kg/m}^3}}$
$= \sqrt{2116504.85436 \frac{\text{N/m}^2}{\text{kg/m}^3}}$
$= \sqrt{2116504.85436 \frac{(\text{kg}\cdot\text{m/s}^2)/\text{m}^2}{\text{kg/m}^3}}$
$= \sqrt{2116504.85436 \text{ m}^2/\text{s}^2}$
$= 1454.82124481 \text{ m/s}$
$= 1450 \text{ m/s}$

53. $\sqrt{gd} = \sqrt{(9.8)(3500)}$
$= \sqrt{34300}$
$= 185.20259$
$= 190 \text{ m/s}$

57. **(a)** $\sqrt[3]{2140} = 12.8865874254$, which is rounded to 12.9
(b) $\sqrt[3]{-0.214} = -0.59814240297$, which is rounded to –0.598

```
³√(2140)
         12.88658743
³√(-0.214
        -.598142403
```

```
7ˣ√0.382
        .8715549346
7ˣ√-382
        -2.338116758
```

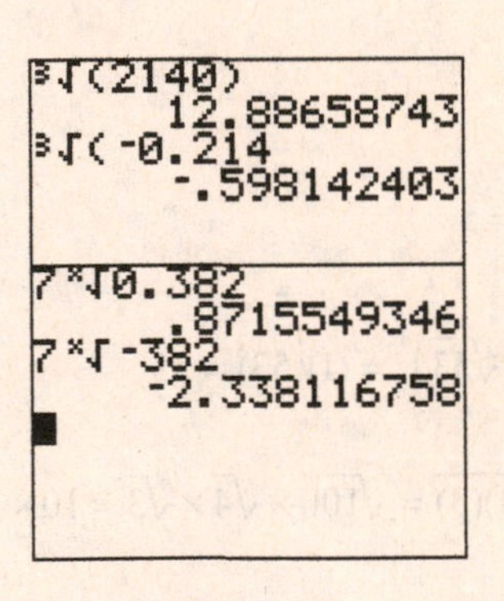

1.7 Addition and Subtraction of Algebraic Expressions

1. $3x + 2y - 5y = 3x - 3y$

5. $5x + 7x - 4x = 8x$

9. $3t - 4s - 3t - s = 0t - 5s = -5s$

13. $a^2b - a^2b^2 - 2a^2b = -a^2b - a^2b^2$

17. $v - (7 - 9x + 2v) = v - 7 + 9x - 2v = -v + 9x - 7$

21. $(a-3)+(5-6a)=a-3+5-6a=-5a+2$

25. $3(2r+s)-(-5s-r)=6r+3s+5s+r=7r+8s$

29. $-[(4-6n)-(n-3)]=-[4-6n-n+3]$
$=-[-7n+7]$
$=7n-7$

33. $-2[-x-2a-(a-x)]=-2[-x-2a-a+x]$
$=-2[-3a]$
$=6a$

37. $5z-\{8-[4-(2z+1)]\}=5z-\{8-[4-2z-1]\}$
$=5z-\{8-4+2z+1\}$
$=5z-\{5+2z\}$
$=5z-5-2z$
$=3z-5$

41. $-2\{-(4-x^2)-[3+(4-x^2)]\}=-2\{-4+x^2-[3+4-x^2]\}$
$=-2\{-4+x^2-3-4+x^2\}$
$=-2\{2x^2-11\}$
$=-4x^2+22$

45. $-(3t-(7+2t-(5t-6)))=-(3t-(7+2t-5t+6))$
$=-(3t-(-3t+13))$
$=-(3t+3t-13)$
$=-(6t-13)$
$=-6t+13$

49. $3D-(D-d)=3D-D+d=2D+d$

53. Memory = x(4 terabytes) + $(x+25)$(8 terabytes)
= $(4x+8x+200)$ terabytes
= $(12x+200)$ terabytes

1.8 Multiplication of Algebraic Expressions

1. $2s^3(-st^4)^3(4s^2t)=2s^3(-1)^3s^3t^{12}(4s^2t)$
$=-2s^6t^{12}(4s^2t)$
$=-8s^8t^{13}$

5. $(a^2)(ax)=a^3x$

9. $(2ax^2)^2(-2ax) = (2ax^2)(2ax^2)(-2ax)$
$= (4a^2x^4)(-2ax)$
$= -8a^3x^5$

13. $-3s(s^2 - 5t) = (-3s)(s^2) + (-3s)(-5t)$
$= -3s^3 + 15st$

17. $3M(-M - N + 2) = (3M)(-M) + (3M)(-N) + (3M)(2)$
$= -3M^2 - 3MN + 6M$

21. $(x-3)(x+5) = (x)(x) + (x)(5) + (-3)(x) + (-3)(5)$
$= x^2 + 5x - 3x - 15$
$= x^2 + 2x - 15$

25. $(y+8)(y-8) = (y)(y) + (y)(-8) + (8)(y) + (8)(-8)$
$= y^2 - 8y + 8y - 64$
$= y^2 - 64$

29. $(2s+7t)(3s-5t) = (2s)(3s) + (2s)(-5t) + (7t)(3s) + (7t)(-5t)$
$= 6s^2 - 10st + 21st - 35t^2$
$= 6s^2 + 11st - 35t^2$

33. $(x-2y-4)(x-2y+4)$
$= (x)(x) + (x)(-2y) + (x)(4) + (-2y)(x) + (-2y)(-2y) + (-2y)(4) + (-4)(x) + (-4)(-2y) + (-4)(4)$
$= x^2 - 2xy + 4x - 2xy + 4y^2 - 8y - 4x + 8y - 16$
$= x^2 + 4y^2 - 4xy - 16$

37. $-3(3-2T)(3T+2) = -3[(3)(3T) + (3)(2) + (-2T)(3T) + (-2T)(2)]$
$= -3[-6T^2 + 9T - 4T + 6]$
$= -3[-6T^2 + 5T + 6]$
$= 18T^2 - 15T - 18$

41. $(3x-7)^2 = (3x-7)(3x-7)$
$= (3x)(3x) + (3x)(-7) + (-7)(3x) + (-7)(-7)$
$= 9x^2 - 21x - 21x + 49$
$= 9x^2 - 42x + 49$

45. $(xyz-2)^2 = (xyz-2)(xyz-2)$
$= (xyz)(xyz) + (xyz)(-2) + (-2)(xyz) + (-2)(-2)$
$= x^2y^2z^2 - 2xyz - 2xyz + 4$
$= x^2y^2z^2 - 4xyz + 4$

49. $(2+x)(3-x)(x-1)=[(6-2x+3x-x^2)](x-1)$
$$=(x-1)[-x^2+x+6]$$
$$=(x)(-x^2)+(x)(x)+(6)(x)+(-1)(-x^2)+(-1)(x)+(-1)(6)$$
$$=-x^3+x^2+6x+x^2-x-6$$
$$=-x^3+2x^2+5x-6$$

53. **(a)** $(x+y)^2=(3+4)^2=7^2=49$
$x^2+y^2=3^2+4^2=9+16=25$
$(x+y)^2\neq x^2+y^2$
$49\neq 25$
(b) $(x-y)^2=(3-4)^2=(-1)^2=1$
$x^2-y^2=3^2-4^2=9-16=-7$
$(x-y)^2\neq x^2-y^2$
$1\neq -7$

57. $P(1+0.01r)^2=P(1+0.01r)(1+0.01r)$
$$=P[(1)(1)+(1)(0.01r)+(0.01r)(1)+(0.01r)(0.01r)]$$
$$=P[1+0.01r+0.01r+0.0001r^2]$$
$$=0.0001r^2P+0.02rP+P$$

61. $(2R-X)^2-(R^2+X^2)=(2R-X)(2R-X)-(R^2+X^2)$
$$=[(2R)(2R)+(2R)(-X)+(2R)(-X)+(-X)(-X)]-(R^2+X^2)$$
$$=4R^2-2RX-2RX+X^2-R^2-X^2$$
$$=3R^2-4RX$$

65. $(R_1+R_2)^2-2R_2(R_1+R_2)=\left[(R_1+R_2)(R_1+R_2)\right]-2R_2(R_1+R_2)$
$$=\left[(R_1)(R_1)+(R_1)(R_2)+(R_2)(R_1)+(R_2)(R_2)\right]-2R_1R_2-2R_2^2$$
$$=R_1^2+R_1R_2+R_1R_2+R_2^2-2R_1R_2-2R_2^2$$
$$=R_1^2-R_2^2$$

1.9 Division of Algebraic Expressions

1. $\dfrac{-6a^2xy^2}{-2a^2xy^5}=\left(\dfrac{-6}{-2}\right)\dfrac{a^{2-2}x^{1-1}}{y^{5-2}}=\dfrac{3}{y^3}$

5. $\dfrac{8x^3y^2}{-2xy}=-4x^{3-1}y^{2-1}=-4x^2y$

9. $\dfrac{(15x^2y)(2xz)}{10xy}=\dfrac{30x^3yz}{10xy}=3x^{3-1}y^{1-1}z=3x^2z$

13. $\dfrac{3a^2x+6xy}{3x}=\dfrac{3a^2x}{3x}+\dfrac{6xy}{3x}=\dfrac{3a^2x^{1-1}}{3}+\dfrac{6x^{1-1}y}{3}=a^2+2y$

17.
$$\begin{aligned}\frac{4pq^3+8p^2q^2-16pq^5}{4pq^2}&=\frac{4pq^3}{4pq^2}+\frac{8p^2q^2}{4pq^2}-\frac{16pq^5}{4pq^2}\\&=p^{1-1}q^{3-2}+2p^{2-1}q^{2-2}-4p^{1-1}q^{5-2}\\&=-4q^3+2p+q\end{aligned}$$

21.
$$\begin{aligned}\frac{-7a^2b+14ab^2-21a^3}{14a^2b^2}&=-\frac{7a^2b}{14a^2b^2}+\frac{14ab^2}{14a^2b^2}-\frac{21a^3}{14a^2b^2}\\&=-\frac{a^{2-2}}{2b^{2-1}}+\frac{b^{2-2}}{a^{2-1}}-\frac{3}{2}a^{3-2}b^{-2}\\&=-\frac{1}{2b}+\frac{1}{a}-\frac{3a}{2b^2}\end{aligned}$$

25.
$$\begin{array}{r}x+5\\ x+4\overline{)x^2+9x+20}\\ \underline{x^2+4x}\\ 5x+20\\ \underline{5x+20}\\ 0\end{array}$$

$\dfrac{x^2+9x+20}{x+4}=x+5$

29.
$$\begin{array}{r}x-1\\ x-2\overline{)x^2-3x+2}\\ \underline{x^2-2x}\\ -x+2\\ \underline{-x+2}\\ 0\end{array}$$

$\dfrac{x^2-3x+2}{x-2}=x-1$

33.
$$\begin{array}{r}Z-2\\ 4Z+3\overline{)4Z^2-5Z-7}\\ \underline{4Z^2+3Z}\\ -8Z-7\\ \underline{-8Z-6}\\ -1\end{array}$$

$\dfrac{4Z^2-5Z-7}{4Z+3}=Z-2-\dfrac{1}{4Z+3}$

37.
$$\begin{array}{r}
2a^2 + \qquad 8 \\
a^2-2\overline{)2a^4+0a^3+4a^2+0a-16} \\
\underline{2a^4 \qquad -4a^2} \qquad\qquad \\
8a^2 \qquad -16 \\
\underline{8a^2 \qquad -16} \\
0
\end{array}$$

$$\frac{2a^4+4a^2-16}{a^2-2}=2a^2+8$$

41.
$$\begin{array}{r}
x-y \\
x-y\overline{)x^2-2xy+y^2} \\
\underline{x^2-xy} \qquad \\
-xy+y^2 \\
\underline{-xy+y^2} \\
0
\end{array}$$

$$\frac{x^2-2xy+y^2}{x-y}=x-y$$

45. We know that $2x+1$ multiplied by $x+c$ will give us $2x^2-9x-5$, so $2x^2-9x-5$ divided by $2x+1$ will give us $x+c$:

$$\begin{array}{r}
x-5 \\
2x+1\overline{)2x^2-9x-5} \\
\underline{2x^2+x} \qquad \\
-10x-5 \\
\underline{-10x-5} \\
0
\end{array}$$

$$x+c=x-5$$
$$c=-5$$

49.
$$V_1 \; 1+\frac{T_2-T_1}{T_1} = V_1 \; 1+\frac{T_2}{T_1}-\frac{T_1}{T_1}$$
$$=V_1 \; 1+\frac{T_2}{T_1}-1$$
$$=V_1\frac{T_2}{T_1}$$
$$=\frac{V_1T_2}{T_1}$$

53.
$$\frac{GMm[(R+r)-(R-r)]}{2rR}=\frac{GMm[R+r-R+r]}{2rR}$$
$$=\frac{GMm[2r]}{2rR}$$
$$=\frac{GMm\cancel{[2r]}}{\cancel{2r}R}$$
$$=\frac{GMm}{R}$$

1.10 Solving Equations

1. **(a)**
$$x-3=-12$$
$$x-3+3=-12+3$$
$$x=-9$$

(b)
$$x+3=-12$$
$$x+3-3=-12-3$$
$$x=-15$$

(c)
$$\frac{x}{3}=-12$$
$$3\;\frac{x}{3}=3(-12)$$
$$x=-36$$

(d)
$$3x=-12$$
$$\frac{3x}{3}=\frac{-12}{3}$$
$$x=-4$$

5.
$$x-2=7$$
$$x=7+2$$
$$x=9$$

9.
$$\frac{t}{2}=-5$$
$$t=2(-5)$$
$$t=-10$$

13.
$$4E=-20$$
$$E=\frac{-20}{4}$$
$$E=-5$$

17. $5-2y=-3$

$$-2y=-3-5$$
$$y=\frac{-8}{-2}$$
$$y=4$$

21. $2(3q+4)=5q$

$$6q+8=5q$$
$$6q-5q=-8$$
$$q=-8$$

25. $8(y-5)=-2y$

$$8y-40=-2y$$
$$8y+2y=40$$
$$10y=40$$
$$y=\frac{40}{10}$$
$$y=4$$

29. $-4-3(1-2p)=-7+2p$

$$-4-3+6p=-7+2p$$
$$-7+6p-2p=-7$$
$$4p=-7+7$$
$$p=\frac{0}{4}$$
$$p=0$$

33. $|x|-9=2$

$$|x|=2+9=11$$
$$x=11 \text{ or } x=-11$$

37. $5.8-0.3(x-6.0)=0.5x$

$$0.5x=5.8-0.3x+1.8$$
$$0.5x+0.3x=7.6$$
$$0.8x=7.6$$
$$x=\frac{7.6}{0.8}$$
$$x=9.5$$

41. $\frac{x}{2.0}=\frac{17}{6.0}$

$$x=2.0\ \frac{17}{6.0}$$
$$x=5.6666666...$$
$$x=5.7$$

45. **(a)** $2x+3=3+2x$

$$2x+3=2x+3$$

Is an identity, since it is true for all values of x.

(b) $2x-3=3-2x$

$$4x=6$$
$$x=\frac{6}{4}=\frac{3}{2}$$

Is conditional as x has one answer only.

49. $0.03x+0.06(2000-x)=96$

$$0.03x+120-0.06x=96$$
$$-0.03x=96-120$$
$$-0.03x=-24$$
$$x=\frac{-24}{-0.03}$$
$$x=\$800$$

53. $0.14n+0.06(2000-n)=0.09(2000)$

$$0.14n+120-0.06n=180$$
$$0.14n-0.06n=180-120$$
$$0.08n=60$$
$$n=\frac{60}{0.08}$$
$$n=750 \text{ L}$$

1.11 Formulas and Literal Equations

1.
$$v = v_0 + at$$
$$v - v_0 = at$$
$$a = \frac{v - v_0}{t}$$

5.
$$E = IR$$
$$R = \frac{E}{I}$$

9.
$$B = \frac{nTWL}{12}$$
$$12B = nTWL$$
$$n = \frac{12B}{TWL}$$

13.
$$F_c = \frac{mv^2}{r}$$
$$rF_c = mv^2$$
$$r = \frac{mv^2}{F_c}$$

17.
$$ct^2 = 0.3t - ac$$
$$ac + ct^2 = 0.3t$$
$$ac = 0.3t - ct^2$$
$$a = \frac{-ct^2 + 0.3t}{c}$$

21.
$$\frac{K_1}{K_2} = \frac{m_1 + m_2}{m_1}$$
$$K_2(m_1 + m_2) = K_1 m_1$$
$$K_2 m_1 + K_2 m_2 = K_1 m_1$$
$$K_2 m_2 = K_1 m_1 - K_2 m_1$$
$$m_2 = \frac{K_1 m_1 - K_2 m_1}{K_2}$$

25.
$$C_0^2 = C_1^2(1 + 2V)$$
$$C_0^2 = C_1^2 + 2C_1^2 V$$
$$2C_1^2 V = C_0^2 - C_1^2$$
$$V = \frac{C_0^2 - C_1^2}{2C_1^2}$$

29.
$$T_2 = T_1 - \frac{h}{100}$$
$$100T_2 = 100T_1 - h$$
$$h + 100T_2 = 100T_1$$
$$h = 100T_1 - 100T_2$$

33.
$$N = N_1 T - N_2(1 - T)$$
$$N_1 T = N + N_2(1 - T)$$
$$N_1 = \frac{N + N_2 - N_2 T}{T}$$

37.
$$P = \frac{V_1(V_2 - V_1)}{gJ}$$
$$gJP = V_1 V_2 - V_1^2$$
$$V_1 V_2 = V_1^2 + gJP$$
$$V_2 = \frac{V_1^2 + gJP}{V_1}$$

41.
$$V = C \;\; 1 - \frac{n}{N}$$
$$V = C - \frac{Cn}{N}$$
$$V + \frac{Cn}{N} = C$$
$$\frac{Cn}{N} = C - V$$
$$Cn = CN - NV$$
$$n = \frac{CN - NV}{C}$$

45.
$$F = \frac{9}{5}C + 32$$
$$90.2 = \frac{9}{5}C + 32$$
$$\frac{5}{9}(90.2 - 32) = C$$
$$C = \frac{5}{9} \times 58.2$$
$$C = 32.3^\circ \mathrm{C}$$

49.
$$d = v_2 t_2 + v_1 t_1$$
$$d = v_2(4\text{ h}) + v_1(t + 2\text{ h})$$
$$tv_1 + v_1(2\text{ h}) = d - v_2(4\text{ h})$$
$$tv_1 = d - v_2(4\text{ h}) - v_1(2\text{ h})$$
$$t = \frac{d - v_2(4\text{ h}) - v_1(2\text{ h})}{v_1}$$

1.12 Applied Word Problems

1. Let x = the number of 25 W lights.
Let $31-x$ = the number of 40 W lights.

$$25x+40(31-x)=1000$$
$$25x+1240-40x=1000$$
$$-15x=1000-1240$$
$$-15x=-240$$
$$x=16$$

There are 16 of the 25 W lights and(31– 16)=15 of the 40 W lights.
Check:

$$25\cdot 16+40(31-16)=1000$$
$$400+40(15)=1000$$
$$400+600=1000$$
$$1000=1000$$

5. Let x = the cost of the car 6 years ago.
Let x + \$5000 = the cost of the car model today.

$$x+(x+\$5000)=\$49\ 000$$
$$2x=\$44\ 000$$
$$x=\frac{\$44\ 000}{2}$$
$$x=\$22\ 000$$

The cost of the car 6 years ago was \$22 000, and the cost of the today's model is (\$22 000 + 5000) = \$27 000.
Check:

$$\$22\ 000+(\$22\ 000+\$5000)=\$49\ 000$$
$$\$22\ 000+\$27\ 000=\$49\ 000$$
$$\$49\ 000=\$49\ 000$$

9. Let x = the number acres of land leased for \$200 per acre.
Let $140-x$ = the number of acres of land leased for \$300 per acre.

$$\$200\ /\ \text{acre}\ x+\$300\ /\ \text{acre}(140\ \text{acre}-x)=\$37\ 000$$
$$-\$100\ /\ \text{acre}\ (x)=-\$5\ 000$$
$$x=\frac{-\$5000}{-\$100\ /\ \text{acre}}$$
$$x=50\ \text{acres}$$

There are 50 acres leased at \$200 per acre and (140 acres – 50 acres) = 90 hectares leased for \$300 per hectare.
Check:

$$\$200\ /\ \text{acre}\ (50\ \text{acres})+\$300\ /\ \text{acre}(140\ \text{acres}-50\ \text{acres})=\$37\ 000$$
$$\$10\ 000+\$27\ 000=\$37\ 000$$
$$\$37\ 000=\$37\ 000$$

13. Let x = amount paid per month for first six months.
Let $x + 10$ = amount paid per month for final four months.

$$(6\text{ mo})x+(4\text{ mo})(x+\$10/\text{mo})=\$890$$
$$(10\text{ mo})x+\$40=\$890$$
$$(10\text{ mo})x=\$850$$
$$x=\frac{\$850}{10\text{ mo}}$$
$$x=\$85/\text{mo}$$

The bill was \$85/mo for the first six months and \$95/mo for the next four months.
Check:

$$(6\text{ mo})\$85/\text{mo}+(4\text{ mo})(\$85/\text{mo}+\$10/\text{mo})=\$890$$
$$\$510+(4\text{ mo})(\$95/\text{mo})=\$890$$
$$\$510+\$380=\$890$$
$$\$890=\$890$$

17. Let x = the length of the first pipeline in km.
Let $x + 2.6$ km = the length of the 3 other pipelines.

$$x+3(x+2.6\text{ km})=35.4\text{ km}$$
$$x+3x+7.8\text{ km}=35.4\text{ km}$$
$$4x=27.6\text{ km}$$
$$x=\frac{27.6\text{ km}}{4}$$
$$x=6.9\text{ km}$$

The first pipeline is 6.9 km long, and the other three pipelines are each (6.9 km + 2.6 km) = 9.5 km long.
Check:

$$6.9\text{ km}+3(6.9\text{ km}+2.6\text{ km})=35.4\text{ km}$$
$$6.9\text{ km}+3(9.5\text{ km})=35.4\text{ km}$$
$$6.9\text{ km}+28.5\text{ km}=35.4\text{ km}$$
$$35.4\text{ km}=35.4\text{ km}$$

21. Let x = the amount of time in seconds between when the start of the trains pass each other to when the end of the trains pass each other.
The total distance the ends must travel in this time is 960 feet. We first convert mi/hr into ft/sec.

$$1\text{ mi/hr }=\frac{5280\text{ ft}}{3600\text{ s}}=\frac{22\text{ ft}}{15\text{ s}}=\frac{22}{15}\text{ ft/s}$$

Therefore, train A travels at 60(22/15)=88 ft/s and train B travels at 40(22/15)=176/3 ft/s.

$$(88\text{ ft/s})x+(176/3\text{ ft/s})x=960\text{ ft}$$
$$(440/3\text{ ft/s})x=960\text{ ft}$$
$$x=\frac{960\text{ ft}}{440/3\text{ ft/s}}$$
$$x=\frac{72}{11}\text{ s}$$

The trains completely pass each other in about 6.55 seconds.
Check:

$$(88\text{ ft/s})\frac{72}{11}\text{ s}+(176/3\text{ ft/s})\frac{72}{11}\text{ s}=960\text{ ft}$$
$$576\text{ ft}+384\text{ ft}=960\text{ ft}$$
$$960\text{ ft}=960\text{ ft}$$

25. Let x = the speed the train leaving England in km/h.
Let x + 8 km/h = speed of the train leaving France in km/h.
The distance travelled by each train is speed ×time.

$$x\left(\frac{17\text{ min}}{60\text{ min/ h}}\right)+(x+8\text{ km/h})\left(\frac{17\text{ min}}{60\text{ min/ h}}\right)=50\text{ km}$$

$$(0.28333\text{ h})x+(x+8\text{ km/h})(0.28333\text{ h})=50\text{ km}$$

$$(0.28333\text{ h})x+(0.28333\text{ h})x+2.26667\text{ km}=50\text{ km}$$

$$(0.56666\text{ h})x=47.73333\text{ km}$$

$$x=\frac{47.73333\text{ km}}{0.56666\text{ h}}$$

$$x=84.23529421\text{ km/h}$$

$$x=84.2\text{ km/h}$$

The train leaving England was travelling at 84.2 km/h, and the train leaving France was travelling at (84.2 km/h + 8 km/h) = 92.2 km/h.
Check:

$$84.23529421\text{ km/h}\left(\frac{17\text{ min}}{60\text{ min/ h}}\right)+(84.23529421\text{ km/h}+8\text{ km/h})\left(\frac{17\text{ min}}{60\text{ min/ h}}\right)=50\text{ km}$$

$$23.86666\text{ km}+(92.23529421\text{ km/h})\left(\frac{17\text{ min}}{60\text{ min/ h}}\right)=50\text{ km}$$

$$23.86666\text{ km}+26.13333\text{ km}=50\text{ km}$$

$$50\text{ km}=50\text{ km}$$

29.

x mi (228 – x) mi

\$2.90/gal 228 mi \$2.70/gal

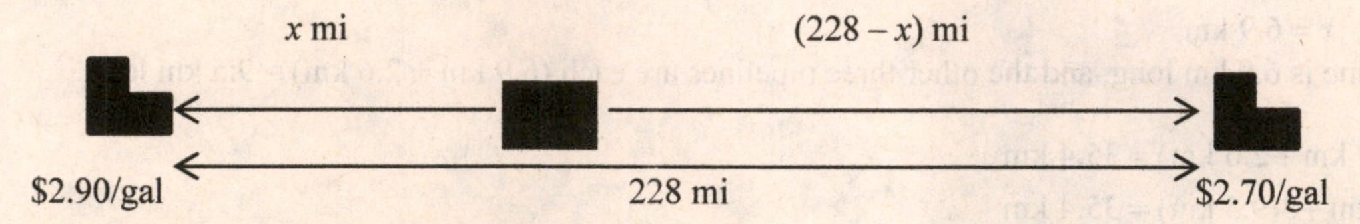

Assuming that the customer is located between the two gasoline distributors:
Let x = the distance in km to the first gasoline distributor that costs \$2.90/gal.
Let 228 mi− x= the distance in km to the second gasoline distributor that costs \$2.70/gal.

$$\$2.90+\$0.002(x)=\$2.70+\$0.002(228-x)$$

$$\$2.90+\$0.002(x)=\$2.70+\$0.456-\$0.002(x)$$

$$\$0.004(x)=\$0.256$$

$$x=\frac{\$0.256}{\$0.004}$$

$$x=64\text{ mi}$$

The customer is 64 mi away from the first gas distributor (\$2.90/gal) and (228 mi – 64 mi)= 164 mi away from the second gas distributor (\$2.70).
Check:

$$\$2.90+\$0.002(64)=\$2.70+\$0.002(228-64)$$

$$\$2.90+\$0.128=\$2.70+\$0.002(164)$$

$$\$3.028=\$2.70+\$0.328$$

$$\$3.028=\$3.028$$

33.

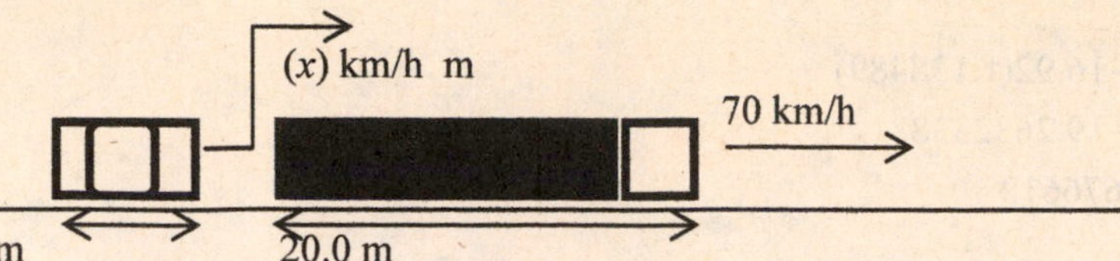

Let x = the speed the car needs to travel in km/h to pass the semi in 10 s.
Speed = distance/time. 10 s is 10s/3600 s/h = 0.002777777 h.

$$x = \frac{\text{distance needed to pass truck + distance travelled by truck in 10s}}{10s}$$

$$x = \frac{0.025 \text{ km} + 70 \text{ km/h}(0.0027777 \text{ h})}{0.0027777 \text{ h}}$$

$$x = \frac{0.025 \text{ km} + 0.19444 \text{ km}}{0.0027777 \text{ h}}$$

$$x = \frac{2.19444 \text{ km}}{0.0027777 \text{ h}}$$

$$x = 79 \text{ km/h}$$

The car needs to travel at a speed of 79 km/h to pass the semitrailer in 10s.
Check:

$$79 \text{ km/h} = \frac{0.025 \text{ km} + 70 \text{ km/h}(0.0027777 \text{ h})}{0.0027777 \text{ h}}$$

$$79 \text{ km/h} = \frac{0.025 \text{ km} + 0.19444 \text{ km}}{0.0027777 \text{ h}}$$

$$79 \text{ km/h} = 79 \text{ km/h}$$

Review Exercises

1. False, because $|0| = 0$ which is not a positive value.

5. True.

9. False. The left-hand side simplifies to $\frac{6x+2}{2} = \frac{6x}{2} + \frac{2}{2} = 3x + 1$.

13. $(-2)+(-5)-3 = -2-5-3 = -10$

17. $-5-|2(-6)| + \frac{-15}{3} = -5-|-12|+(-5) = -5-12-5 = -22$

21. $\sqrt{16}-\sqrt{64} = \sqrt{(4)(4)} - \sqrt{(8)(8)} = 4-8 = -4$

25. $(-2rt^2)^2 = (-2)^2 r^2 t^{2\times 2} = 4r^2t^4$

29. $\frac{-16N^{-2}(NT^2)}{-2N^0T^{-1}} = \frac{8N^{-2+1}T^{2+1}}{(1)} = \frac{8N^{-1}T^3}{(1)} = \frac{8T^3}{N}$

33. 8000 has 1 significant digit. Rounded to 2 significant digits, it is $8\bar{0}00$.

37. $$\begin{aligned} 37.3-16.92(1.067)^2 &= 37.3-16.92(1.138489) \\ &= 37.3-19.26323388 \\ &= 18.03676612 \end{aligned}$$
which rounds to 18.0.

41. $$875\text{ Btu} = 875\text{ Btu}\times\frac{778.2\text{ ft}\cdot\text{lb}}{1\text{ Btu}}\times\frac{1.356\text{ J}}{1\text{ ft}\cdot\text{lb}}$$
$$= 923{,}334.3\text{ J}$$
which rounds to 923,000 J.

45. $$225\text{ hp} = 225\text{ hp}\times\frac{550\text{ ft}\cdot\text{lb/s}}{1\text{ hp}}\times\frac{1.356\text{ J}}{1\text{ ft}\cdot\text{lb}}\times\frac{60\text{ s}}{1\text{ min}}$$
$$= 10068300\ \frac{\text{J}}{\text{min}}$$
which rounds to $10100000 = 1.01\times10^7$ J/min.

49. $6LC-(3-LC) = 6LC-3+LC = 7LC-3$

53. $$\begin{aligned} (x+8)^2 &= (x+8)(x+8) \\ &= (x)(x)+(x)(8)+(8)(x)+(8)(8) \\ &= x^2+8x+8x+64 \\ &= x^2+16x+64 \end{aligned}$$

57. $$\begin{aligned} 4R-[2r-(3R-4r)] &= 4R-[2r-3R+4r] \\ &= 4R-[6r-3R] \\ &= 4R-6r+3R \\ &= 7R-6r \end{aligned}$$

61. $$\begin{aligned} (2x+1)(x^2-x-3) &= (2x)(x^2)+(2x)(-x)+(2x)(-3)+(1)(x^2)+(1)(-x)+(1)(-3) \\ &= 2x^3-2x^2-6x+x^2-x-3 \\ &= 2x^3-x^2-7x-3 \end{aligned}$$

65. $$\begin{aligned} 3p[(q-p)-2p(1-3q)] &= 3p[q-p-2p+6pq] \\ &= 3p[q-3p+6pq] \\ &= 18p^2q-9p^2+3pq \end{aligned}$$

69. $$\begin{array}{r} 2x-5 \\ x+6\overline{\smash{)}2x^2+7x-30} \\ \underline{2x^2+12x} \\ -5x-30 \\ \underline{-5x-30} \\ 0 \end{array}$$

73.

$$\begin{array}{r} 4x^3-2x^2+6x \\ x+3\overline{)4x^4+10x^3+0x^2+18x-1} \\ \underline{4x^4+12x^3} \\ -2x^3+0x^2 \\ \underline{-2x^3-6x^2} \\ 6x^2+18x \\ \underline{6x^2+18x} \\ 0x-1 \end{array}$$

$$\frac{4x^4+10x^3+18x-1}{x+3}=4x^3-2x^2+6x-\frac{1}{x+3}$$

77.

$$\begin{array}{r} y^2+5y-1 \\ 2y-1\overline{)2y^3+9y^2-7y+5} \\ \underline{2y^3-1y^2} \\ 10y^2-7y \\ \underline{10y^2-5y} \\ -2y+5 \\ \underline{-2y+1} \\ 4 \end{array}$$

$$\frac{2y^3+9y^2-7y+5}{2y-1}=y^2+5y-1+\frac{4}{2y-1}$$

81.

$$\frac{5x}{7}=\frac{3}{2}$$
$$2(5x)=3(7)$$
$$10x=21$$
$$x=\frac{21}{10}$$

85.

$$2s+4(3-s)=6$$
$$2s+12-4s=6$$
$$-2s=-6$$
$$s=\frac{-6}{-2}$$
$$s=3$$

89.

$$2.7+2.0(2.1x-3.4)=0.1$$
$$2.7+4.2x-6.8=0.1$$
$$4.2x-4.1=0.1$$
$$4.2x=4.2$$
$$x=\frac{4.2}{4.2}$$
$$x=1.0$$

93. 15,400,000,000 km = 1.54×10^{10} km

97. 10^{-12} W/m^2 = 0.000000000001 W/m^2

101.

$$V=\pi r^2L$$
$$L=\frac{V}{\pi r^2}$$

105.

$$Pp+Qq=Rr$$
$$Qq=Rr-Pp$$
$$q=\frac{Rr-Pp}{Q}$$

109.

$$N_1=T(N_2-N_3)+N_3$$
$$N_1-N_3=N_2T-N_3T$$
$$N_2T=N_1-N_3+N_3T$$
$$N_2=\frac{N_1-N_3+N_3T}{T}$$

113.

$$d=kx^2[3(a+b)-x]$$
$$d=kx^2[3a+3b-x]$$
$$d=3akx^2+3bkx^2-kx^3$$
$$3akx^2=d-3bkx^2+kx^3$$
$$a=\frac{d-3bkx^2+kx^3}{3kx^2}$$

117. $\dfrac{0.553\text{ km}}{0.442\text{ km}}=1.25113122$
which rounds to 1.25. The CN Tower is 1.25 times taller than the Sears tower.

121.

$$\begin{aligned}(x-2a)+3\text{ ft/yd}\cdot(x+2a)&=x-2a+3x+3(2a)\\&=x-2a+3x+6a\\&=4x+4a\end{aligned}$$

The sum of their length is $4x+4a$ ft.

125. $3\times18\div(9-6)=54\div(3)=18$
$3\times18\div9-6=54\div9-6=6-6=0$
Yes, the removal of the parentheses does affect the answer.

129. **(a)** $2|2|-2|4|=4-8$
$=-4$

(b) $2|2-(-4)|=2|6|$
$=12$

133. $(x-y)^3=(x-y)(x-y)(x-y)$
$=(-(y-x))(-(y-x))(-(y-x))$
$=-(y-x)(y-x)(y-x)$
$=-(y-x)^3$

137. $250 \text{ hp} = 250 \text{ hp} \times \dfrac{746.0 \text{ W}}{1 \text{ hp}} \times \dfrac{1 \text{ kW}}{1000 \text{ W}}$

$= 186.5 \text{ kW}$
This is rounded to 190 kW.

141. Let x = the cost of the first computer program.
Let x + \$72 = the cost of the second computer program.

$$x+(x+\$72)=\$190$$
$$2x+\$72=\$190$$
$$2x=\$118$$
$$x=\frac{\$118}{2}$$
$$x=\$59$$

The cost of the first computer program is \$59, and the other program costs (\$59 + \$72) = \$131.
Check: \$59 + \$131 = 190

145. Let x = the resistance in the first resistor in Ω .
Let $x + 1200\,\Omega$ = the resistance in the second resistor in Ω .
Voltage = current × resistance. 2.4 μA = 2.3 × 10^{-6} A. 12 mV = 0.0120 V

$$(2.4\times10^{-6}\text{A})(x)+(2.4\times10^{-6}\text{A})(x+1200\ \Omega)=0.0120\text{ V}$$
$$(2.4\times10^{-6}\text{A})(x)+(2.4\times10^{-6}\text{A})(x)+(2.4\times10^{-6}\text{A})(1200\ \Omega)=0.0120\text{ V}$$
$$(4.8\ \times10^{-6}\text{A})(x)+(0.00288\text{V})=0.0120\text{ V}$$
$$(4.0\times10^{-6}\text{A})(x)=0.00912\text{ V}$$
$$x=\frac{0.00912\text{ V}}{4.8\times10^{-6}\text{A}}$$
$$x=1900\ \Omega$$

The first resistor's resistance is 1900 Ω and the second resistor's is (1900 Ω + 1200 Ω) = 3100 Ω .
Check:

$$(2.4\times10^{-6}\text{A})(1900\ \Omega)+(2.4\times10^{-6}\text{A})(1900\ \Omega+1200\ \Omega)=0.0120\text{ V}$$
$$0.00456\text{ V}+0.00744\text{ V}=0.0120\text{ V}$$
$$0.0120\text{ V}=0.0120\text{ V}$$

149. 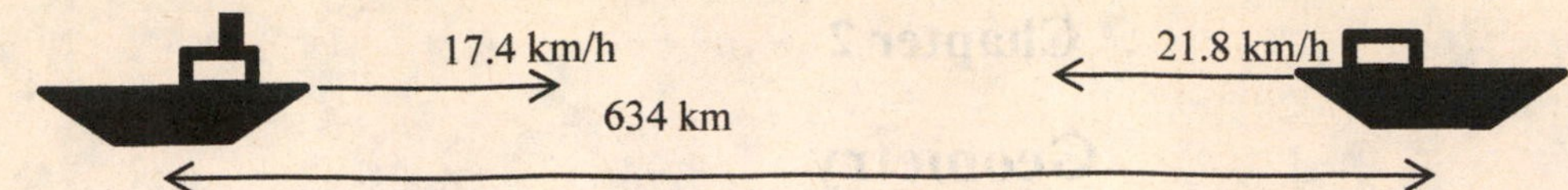

Let x = the time taken by the second ship in hours.
Let $x + 2$ h = the amount time taken by the first ship in hours.
The distance travelled adds up to 634km. Distance = speed × time.

$$21.8\text{ km/h}(x)+17.4\text{ km/h}(x+2\text{ h})=634\text{ km}$$
$$21.8\text{ km/h}(x)+17.4\text{ km/h}(x)+17.4\text{ km/h}(2\text{ h})=634\text{ km}$$
$$39.2\text{ km/h}(x)+34.8\text{ km}=634\text{ km}$$
$$39.2\text{ km/h}(x)=599.2\text{ km}$$
$$x=\frac{599.2\text{ km}}{39.2\text{ km/h}}$$
$$x=15.2857\text{ h}$$

which rounds to 15.2 h. The ships will pass 15.2 h after the second ship enters the canal.
Check:

$$21.8\text{ km/h}(15.2857\text{ h})+17.4\text{ km/h}(15.2857\text{ h}+2\text{ h})=634\text{ km}$$
$$333.23\text{ km}+300.77\text{ km}=634\text{ km}$$
$$634\text{ km}=634\text{ km}$$

153. Let x = the area of space in ft^2 in the kitchen and bath.

$$\frac{\text{ft}^2\text{ of tile in the house}}{\text{ft}^2\text{ in the house}}=0.25$$
$$\frac{x+0.15(2200\text{ ft}^2)}{(x+2200\text{ ft}^2)}=0.25$$
$$x+330\text{ ft}^2=0.25(x)+(0.25)(2200\text{ ft}^2)$$
$$x+330\text{ ft}^2=0.25(x)+550\text{ ft}^2$$
$$0.75x=220\text{ ft}^2$$
$$x=\frac{220\text{ ft}^2}{0.75}$$
$$x=293.33333333\text{ ft}^2$$

which rounds to 290 ft^2. The kitchen and bath area is 290 ft^2.
Check:

$$\frac{293.33333333\text{ ft}^2+0.15(2200\text{ ft}^2)}{(293.33333333\text{ ft}^2+2200\text{ ft}^2)}=0.25$$
$$\frac{623.3333333\text{ ft}^2}{2493.3333333\text{ ft}^2}=0.25$$
$$0.25=0.25$$

Chapter 2

Geometry

2.1 Lines and Angles

1. $\angle ABE = 90^\circ$ because it is a vertically opposite angle to $\angle CBD$ which is also a right angle.

5. $\angle EBD$ and $\angle DBC$ are acute angles (i.e., $< 90°$).

9. The complement of $\angle CBD = 65^\circ$ is $\angle DBE$

$$\angle CBD + \angle DBE = 90^\circ$$
$$65^\circ + \angle DBE = 90^\circ$$
$$\angle DBE = 90^\circ - 65^\circ$$
$$\angle DBE = 25^\circ$$

13. $\angle AOB = \angle AOE + \angle EOB$

but $\angle AOE = 90^\circ$ because it is vertically opposite to $\angle DOF$ a given right angle,

and $\angle EOB = 50^\circ$ because it is vertically opposite to $\angle COF$ a given angle of 50°,

so $\angle AOB = 90^\circ + 50^\circ = 140^\circ$

17. $\angle 1$ is supplementary to 150°, so

$$\angle 1 = 180^\circ - 150^\circ = 30^\circ$$
$$\angle 2 = \angle 1 = 30^\circ$$

$\angle 4$ is vertically opposite to $\angle 2$, so

$$\angle 4 = \angle 2$$
$$\angle 4 = 30^\circ$$

21. $\angle 6 = 90 - 62^\circ$ since they are complementary angles

$$\angle 6 = 28^\circ$$

$\angle 3$ is an alternate-interior angle to $\angle 6$, so

$$\angle 3 = \angle 6$$
$$\angle 3 = 28^\circ$$

25. $\angle EDF = \angle BAD = 44^\circ$ because they are corresponding angles

$$\angle BDE = 90^\circ$$
$$\angle BDF = \angle BDE + \angle EDF$$
$$\angle BDF = 90^\circ + 44^\circ$$
$$\angle BDF = 134^\circ$$

29. $\angle EDF = \angle BAD = 44^\circ$ because they are corresponding angles

Angles $\angle ADB$, $\angle BDE$, and $\angle EDF$ make a straight angle

$$\angle ADB + \angle BDE + \angle EDF = 180^\circ$$
$$\angle ADB = 180^\circ - \angle BDE - \angle EDF$$
$$\angle ADB = 180^\circ - 90^\circ - 44^\circ$$
$$\angle ADB = 46^\circ$$

$\angle DFE = \angle ADB$ because they are corresponding angles

$$\angle DFE = 46^\circ$$

33. Using Eq. (2.1),

$$\frac{c}{3.20} = \frac{5.05}{4.75}$$
$$c = \frac{(3.20)(5.05)}{4.75}$$
$$c = 3.40 \text{ m}$$

37. $\angle BCH = \angle DCG$ is given and $\angle BCH, 50^\circ$, and $\angle DCG$ together form a straight angle.

$$\angle BCH + 50^\circ + \angle DCG = 180^\circ$$
$$2\angle BCH = 130^\circ$$
$$\angle BCH = 65^\circ$$

Since $\angle BCH$ and $\angle GHC$ are alternate-interior angles,

$$\angle BCH = \angle GHC.$$

Therefore, $\angle GHC = 65^\circ$.

Similarly, $\angle HGC = \angle GCD = 65^\circ$.

41.

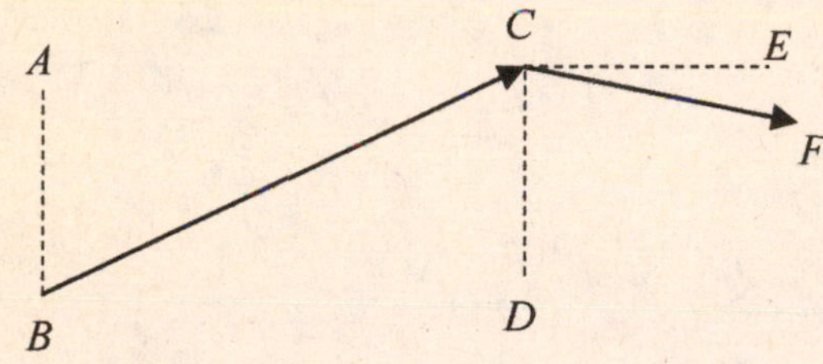

We are given $\angle ABC = 58°$ and $\angle ECF = 18°$. We draw CD parallel to AB.
Since $\angle ABC$ and $\angle BCD$ are alternate-interior angles, they are equal and so $\angle BCD = 58°$.
Since $\angle ECF$ and $\angle DCF$ are complementary angles,
$\angle ECF + \angle DCF = 90°$. Therefore, $\angle DCF = 90° - 18° = 72°$
Thus, $\angle BCF = \angle BCD + \angle DCF = 58° + 72° = 130°$.

45. Using Eq. (2.1),

$$\frac{AB}{3}=\frac{BC}{2}=$$

$$AB=\frac{3(2.15)}{2}$$

$$AB=3.225 \text{ cm}$$

$$AC=AB+BC$$

$$AC=3.225 \text{ cm}+2.15 \text{ cm}$$

$$AC=5.375 \text{ cm}$$

$$AC=5.38 \text{ cm}$$

49. The sum of the angles with vertices at *A*, *B*, and *D* is 180°. Since those angles are unknown quantities, the sum of interior angles in a closed triangle is 180°.

2.2 Triangles

1. $\angle 5=45^\circ$

$\angle 3=45^\circ$ since $\angle 3$ and $\angle 5$ are alternate interior angles.

$\angle 1$, $\angle 2$, and $\angle 3$ make a stright angle, so

$$\angle 1+\angle 2+\angle 3=180^\circ$$

$$70^\circ+\angle 2+45^\circ=180^\circ$$

$$\angle 2=65^\circ$$

5.

$$\angle A+\angle B+\angle C=180^\circ$$

$$\angle A+40^\circ+84^\circ=180^\circ$$

$$\angle A=56^\circ$$

9.

$$A=\frac{1}{2}bh$$

$$A=\frac{1}{2}(6.3)(2.2)$$

$$A=6.9 \text{ ft}^2$$

13. One leg can represent the base, the other leg the height.

$$A=\frac{1}{2}bh$$

$$A=\frac{1}{2}(3.46)(2.55)$$

$$A=4.41 \text{ ft}^2$$

17. We add the lengths of the sides to get

$$p=205+322+415$$

$$p=942 \text{ cm}$$

21. $c^2 = a^2 + b^2$

$c = \sqrt{a^2 + b^2}$

$c = \sqrt{3^2 + 4^2}$

$c = 5 \text{ in}$

25. $c^2 = a^2 + b^2$

$b = \sqrt{c^2 - a^2}$

$b = \sqrt{551^2 - 175^2}$

$b = 522 \text{ cm}$

29. Length c is found in Question 26, $c = 98.309\ 77$ cm

$p = 98.309\ 77 + 90.5 + 38.4 = 227.2$ cm

33. An equilateral triangle.

37. $\angle LMK$ and $\angle OMN$ are vertically opposite angles and thus equal.
Since each triangle has a right angle, the remaining angle in each triangle must be the same.

$\angle KLM = \angle MON$.

The triangles ΔMKL and ΔMNO have all the same angles, so therefore the triangles are similar:

$\Delta MKL \sim \Delta MNO$

41. $s = \frac{p}{2} = \frac{6+25+29}{2} = 30$

By Hero's formula,

$A = \sqrt{s(s-a)(s-b)(s-c)}$

$A = \sqrt{30(30-6)(30-25)(30-29)}$

$A = \sqrt{3600} = 60$

$p = 6 + 25 + 29 = 60$

and so a triangle with sides 6, 25, 29 is perfect.

45.

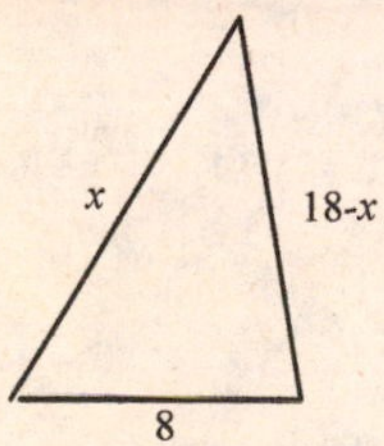

The tree could break at any point between 5 and 13 feet to have the top land 8 feet away from the base.

49.

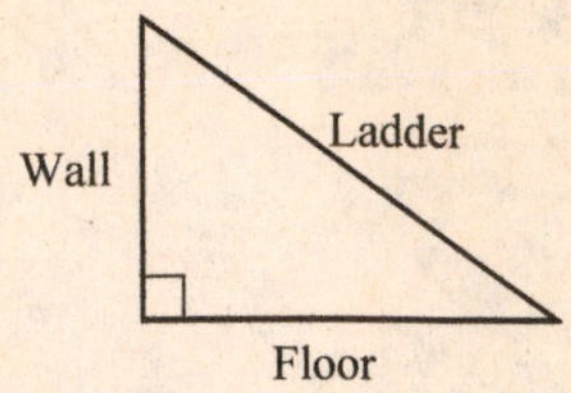

The base of the window is leg b satisfying

$b = \sqrt{c^2 - a^2}, c = 10, a = 6$

$b = \sqrt{100 - 36} = 8$

The top of the window is $b + 2.5 = 10.5$ ft.

and so the ladder's new length must be

$l = \sqrt{6^2 + 10.5^2} = 12.1$ ft.

53.

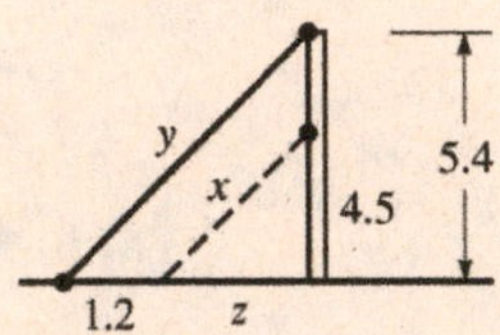

By Eq. (2.1),

$$\frac{z}{4.5} = \frac{1.2}{0.9}$$

$$z = \frac{(4.5)(1.2)}{0.9}$$

$$z = 6.0 \text{ m}$$

$$x^2 = z^2 + 4.5^2$$

$$x = \sqrt{56.25} \text{ m}$$

$$x = 7.5 \text{ m}$$

$$y^2 = (1.2 + 6)^2 + 5.4^2$$

$$x = \sqrt{81.0} \text{ m}$$

$$y = 9.0 \text{ m}$$

57. Redraw ΔBCP as

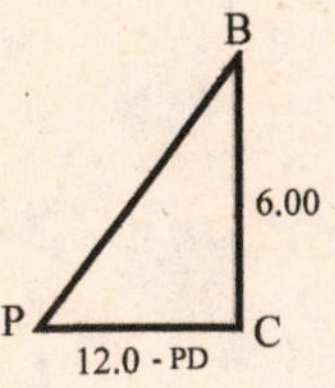

ΔAPD is

A

10.0

P D

since $\Delta BCP \sim \Delta ADP$

$$\frac{6.00}{12.0-PD}=\frac{10.0}{PD}$$

$$6PD = 120 - 10PD$$

$$16PD = 120$$

$$PD = 7.50 \text{ km}$$

$$PC = 12.0 - PD$$

$$PC = 4.50 \text{ km}$$

$$l = PB + PA$$

$$l = \sqrt{4.50^2 + 6.00^2} + \sqrt{7.50^2 + 10.0^2}$$

$$l = 7.50 + 12.5$$

$$l = 20.0 \text{ km}$$

2.3 Quadrilaterals

1.

b_2

b_1

trapezoid

5. $p = 4s = 4(85) = 340$ m

9. $p = 2l + 2w = 2(3.7) + 2(2.7) = 12.8$ m

13. $A = s^2 = 6.4^2 = 41 \text{ mm}^2$

17. $A = bh = 3.7(2.5) = 9.2 \text{ m}^2$

21. $p = 2b + 4a$

25. The parallelogram is a rectangle.

29. The diagonal divides the quadrilateral into two triangles. The sum of the interior angles in each triangle is 180° and so the sum of the interior angles in the quadrilateral must be 360°.

33. The diagonals of a rhombus bisect the angles formed by pairs of adjacent sides of the rhombus. This also implies that if one diagonal is horizontal, then the sides on either side of this diagonal form the same angle above and below the diagonal. This corresponds perfectly to the handle of the jack being horizontal and the sides of the jack being positioned as they appear in figure 2.70.

37.

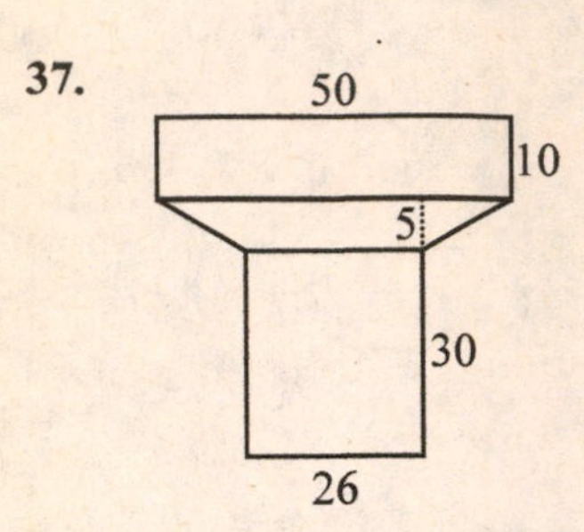

The total area is

$26(30)+\frac{1}{2}(5)(26+50)+50(10)=780+190+500=1470 \text{ ft}^2$

41. Let h be the height. Then the width is $w=1.60h$.

The diagonal is 43.3 in and so

$43.3^2=h^2+(1.60h)^2$

$1874.89=3.56h^2$

$h^2=526.6545$

$h=22.94895$ which rounds to 22.9 in

$w=36.71833$ which rounds to 36.7 in

45. 360°. A diagonal divides a quadrilateral into two triangles, and the sum of the interior angles of each triangle is 180°.

2.4 Circles

1.
$$\begin{aligned} \angle OAB+OBA+\angle AOB &= 180^\circ \\ \angle OAB+90^\circ+72^\circ &= 180^\circ \\ \angle OAB &= 18^\circ \end{aligned}$$

5. (a) AD is a secant line.

(b) AF is a tangent line.

9. $c=2\pi r=2\pi(275)=1730 \text{ ft}$

13. $A=\pi r^2=\pi(0.0952)^2=0.0285 \text{ yd}^2$

17. $r = \dfrac{c}{2\pi} = \dfrac{40.1}{2\pi} = 6.38211 \text{ cm}$

$A = \pi r^2 = \pi(6.38211)^2 = 127.96 \text{ cm}^2$

which rounds to 128 cm^2.

21. A tangent to a circle is perpendicular to the radius drawn to the point of contact. Therefore,

$\angle ABT = 90^\circ$

$\angle CBT = \angle ABT - \angle ABC = 90^\circ - 65^\circ = 25^\circ;$

$\angle CAB = 25^\circ$

25. $\angle ABC = (1/2)(80^\circ) = 40^\circ$ since the measure of an inscribed angle is one-half its intercepted arc.

29. $125.2^\circ = 125.2\left(\dfrac{\pi \text{ rad}}{180^\circ}\right) = 2.185 \text{ rad}$

33. $\text{Perimeter} = \dfrac{\pi r}{2} + \sqrt{r^2 + r^2} = \dfrac{\pi r}{2} + r\sqrt{2}$

37.

6.00

6.00

A of sector = A of quarter circle − A of triangle

$A = \dfrac{1}{4} \cdot \pi(6.00)^2 - \dfrac{1}{2}(6.00)(6.00)$

$A = 10.3 \text{ in}^2$

41. A of sector = A of quarter circle − A of triangle

$A = \dfrac{1}{4} \cdot \pi r^2 - \dfrac{1}{2} r^2$

$A = \dfrac{\pi}{4} - \dfrac{1}{2} \; r^2$

45. We calculate $\sqrt{68^2 + 58^2} = 89.3756$ km which is farther than the 85 km range of the radio station. A clear signal cannot be received at this distance.

49. $\dfrac{A_{\text{basketball}}}{A_{\text{hoop}}} = \dfrac{\pi\left(\dfrac{12.0}{2}\right)^2}{\pi\left(\dfrac{18.0}{2}\right)^2} = \dfrac{0.444}{1}$

53. Let D = diameter of large conduit, then

$D = 3d$, where d = diameter of smaller conduit

$$F = \frac{\text{area large conduit}}{\text{area 7 small conduits}}$$

$$= \frac{7\pi\frac{d^2}{4}}{\pi\frac{D^2}{4}}$$

$$= \frac{7d^2}{D^2} = \frac{7d^2}{(3d)^2} = \frac{7d^2}{9d^2}$$

$$F = \frac{7}{9}$$

The smaller conduits occupy $\frac{7}{9}$ of the larger conduits.

57. Horizontally and opposite to original direction

2.5 Measurement of Irregular Areas

1. The use of smaller intervals improves the approximation since the total omitted area or the total extra area is smaller. Also, since the number of intervals would be 10 (an even number) Simpson's Rule could be employed to achieve a more accurate estimate.

5. The area is exact for areas whose boundaries are line segments.

9. $A_{\text{simp}} = \frac{h}{3}(y_0 + 4y_1 + 2y_2 + 4y_3 + \cdots + 2y_{n-2} + 4y_{n-1} + y_n)$

$A_{\text{simp}} = \frac{1.00}{3}\left[0 + 4(0.52) + 2(0.75) + 4(1.05) + 2(1.15) + 4(1.00) + 0.62\right]$

$A_{\text{simp}} = 4.9\text{ ft}^2$

13. We can use Simpson's rule (which is usually more accurate) because we have an even number of intervals.

$A_{\text{Simp}} = \frac{h}{3}[y_0 + 4y_1 + 2y_2 + \ldots + 4y_{n-1} + y_n]$

$A_{\text{Simp}} = \frac{10.0}{2}\left[38 + 4(24) + 2(25) + 4(17) + 2(34) + 4(29) + 2(36) + 4(34) + 30\right]$

$A_{\text{Simp}} = \left(3370\text{ mm}^2\right)\left(\frac{23\text{ km}}{10\text{ mm}}\right)^2$

$A_{\text{Simp}} = 17{,}827\text{ km}^2$

We round this to 18,000 km^2 (two significant digits.)

17. $A_{\text{simp}} = \frac{h}{3}(y_0 + 4y_1 + 2y_2 + 4y_3 + \cdots + 2y_{n-2} + 4y_{n-1} + y_n)$

$A_{\text{simp}} = \frac{50}{3}[5 + 4(12) + 2(17) + 4(21) + 2(22) + 4(25) + 2(26) + 4(16) + 2(10) + 4(8) + 0]$

$A_{\text{simp}} = 8050 \text{ ft}^2 = 8.0 \times 10^3 \text{ ft}^2$

21. $A_{\text{simp}} = \frac{h}{3}(y_0 + 4y_1 + 2y_2 + 4y_3 + \cdots + 2y_{n-2} + 4y_{n-1} + y_n)$

$A_{\text{simp}} = \frac{0.500}{3}[0.000 + 4(1.732) + 2(2.000) + 4(1.732) + 0.000]$

$A_{\text{simp}} = 2.98 \text{ cm}^2$

The ends of the areas are curved so they can get closer to the boundary, including more area in the calculation.

2.6 Solid Geometric Figures

1. $V_1 = lwh$

$V_2 = (2l)(w)(3h)$

$V_2 = 6lwh$

$V_2 = 6V_1$

The volume increases by a factor of 6.

5. $V = s^3$

$V = (6.95 \text{ ft})^3$

$V = 336 \text{ ft}^3$

9. $V = \frac{4}{3}\pi r^3$

$V = \frac{4}{3}\pi(1.037 \text{ yd})^3$

$V = 4.671 \text{ yd}^3$

13. $V = \frac{1}{3}Bh$

$V = \frac{1}{3}(0.76 \text{ in})^2(1.30 \text{ in})$

$V = .250\ 293 \text{ in}^3$

$V = 2.50 \times 10^{-1} \text{ in}^3$

17. $S = ph$

$S = (3 \times 1.092 \text{ m})(1.025 \text{ m})$

$S = 3.358 \text{ m}^2$

21. $s^2 = h^2 + r^2$

$s = \sqrt{h^2 + r^2}$

$s = \sqrt{0.274^2 + 3.39^2}$

$s = 3.401055 \text{ cm}$

$A = \pi r^2 + \pi r s$

$A = \pi(3.39 \text{ cm})^2 + \pi(3.39 \text{ cm})(3.401055 \text{ cm})$

$A = 72.3 \text{ cm}^2$

25. Let r = radius of cone,

Let h = height of the cone

$$\frac{V_{\text{cylinder}}}{V_{\text{cone}}} = \frac{\pi(2r)^2 \frac{h}{2}}{\frac{1}{3}\pi r^2 h}$$

$$\frac{V_{\text{cylinder}}}{V_{\text{cone}}} = \frac{2\pi r^2 h}{\frac{1}{3}\pi r^2 h}$$

$$\frac{V_{\text{cylinder}}}{V_{\text{cone}}} = 6$$

29. $A = A_{base} + A_{ends} + A_{sides}$

$A = 2lw + 2wh + 2lh$

$A = 2(12.0)(9.50) + 2(9.50)(8.75) + 2(12.0)(8.75)$

$A = 604 \text{ in}^2$

33. $V = \frac{1}{3}\pi h(R^2 + Rr + r^2)$

$h = 62.5 \text{ m}, R = \frac{3.88}{2} = 1.94 \text{ m}, r = \frac{1.90}{2} = 0.95 \text{ m}$

$V = \frac{1}{3}\pi(62.5)(1.94^2 + (1.94)(0.95) + 0.95^2)$

$V = 426.02 \text{ m}^3$

$V = 426 \text{ m}^3$

37. $V = \frac{4}{3}\pi r^3$

$V = \frac{4}{3}\pi\left(\frac{d}{2}\right)^3$

$V = \frac{4}{3}\pi\left(\frac{165}{2}\right)^3$

$V = 2{,}352{,}071 \text{ ft}^3$

$V = 2.35 \times 10^6 \text{ ft}^3$

41. $c = 2\pi r$

$29.8 = 2\pi r$

$r = \frac{29.8}{2\pi}$

$V = \frac{4}{3}\pi r^3$

$V = \frac{4}{3}\pi\left(\frac{29.8}{2\pi}\right)^3$

$V = 447 \text{ in}^3$

45. $V_{\text{new}} = V_{\text{old}} - 0.08\ V_{\text{old}} = 0.92\ V_{\text{old}}$

$V_{\text{new}} = \frac{4}{3}\pi r_{\text{new}}^3; V_{\text{old}} = \frac{4}{3}\pi r_{\text{old}}^3$

$r_{\text{new}}^3 = \frac{3V_{\text{new}}}{4\pi}$

$= \frac{3(0.92\ V_{\text{old}})}{4\pi}$

$= 0.92\frac{3V_{\text{old}}}{4\pi}$

$r_{\text{new}}^3 = 0.92\ r_{\text{old}}^3$

$r_{\text{new}} = \sqrt[3]{0.92}\ r_{\text{old}} = 0.9726\ r_{\text{old}} = (1-0.0274)\ r_{\text{old}}$

and so the radius has decreased by 2.7%
(rounded to two significant digits.)

Review Exercises

1. This is false; the angles are supplementary, not complementary.

5. This is true. There is an even number of intervals, making Simpson's rule applicable.

9. $\angle CGE = 32^\circ$ from Question 1

$\angle CGE$ and $\angle DGH$ are vertically opposite angles

$\angle DGH = \angle CGE$

$\angle DGH = 32^\circ$

13. $c^2 = a^2 + b^2$

$c = \sqrt{400^2 + 580^2}$

$c = \sqrt{496\ 400}$

$c = 704.55659815$

$c = 700$

17. $c^2 = a^2 + b^2$

$a^2 = c^2 - b^2$

$a = \sqrt{52.9^2 - 38.3^2}$

$a = \sqrt{1331.52}$

$a = 36.4899986$

$a = 36.5$

21. $A = \frac{1}{2}bh$

$A = \frac{1}{2}(0.125 \text{ ft})(0.188 \text{ ft})$

$A = 0.0118 \text{ ft}^2$

25. $A = \frac{1}{2}h(b_1 + b_2)$

$A = \frac{1}{2}(34.2 \text{ in})(67.2 \text{ in} + 126.7 \text{ in})$

$A = 3315.69 \text{ in}^2$

$A = 3320 \text{ in}^2$

29. $V = \frac{1}{3}Bh$

$V = \frac{1}{3}(3850 \text{ ft}^2)(125 \text{ ft})$

$V = 160416.6667 \text{ ft}^3$

$V = 1.60 \times 10^5 \text{ ft}^3$

33. $A = 6s^2$

$A = 6(0.520 \text{ m})^2$

$A = 1.6224 \text{ m}^2$

$A = 1.62 \text{ m}^2$

37. $\angle BTA = \frac{50^\circ}{2} = 25^\circ$

41. $\angle ABE$ and $\angle ADC$ are corresponding angles since $\Delta ABE \sim \Delta ADC$

$\angle ABE = \angle ADC$

$\angle ABE = 53^\circ$

45. p = base of triangle + hypotenuse of triangle + semicircle perimeter

$p = b + \sqrt{b^2 + (2a)^2} + \frac{1}{2}\pi(2a)$

$p = b + \sqrt{b^2 + 4a^2} + \pi a$

49. A square is a rectangle with four equal sides.
A rectangle is a parallelogram with perpendicular intersecting sides so a square is a parallelogram.
A rhombus is a parallelogram with four equal sides and since a square is a parallelogram, a square is a rhombus.

53.

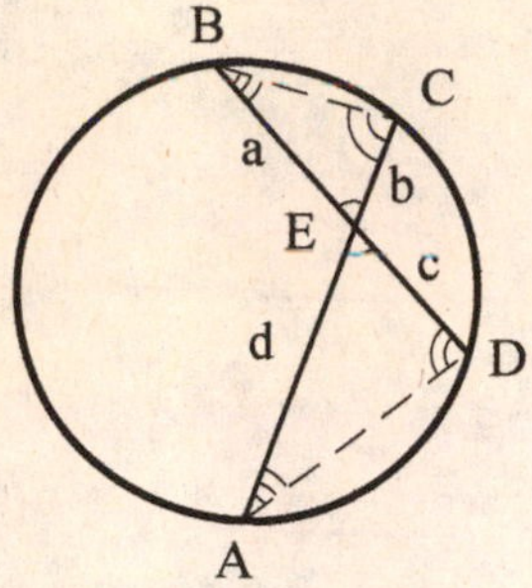

$\angle BEC = \angle AED$, since they are vertically opposite angles

$\angle BCA = \angle ADB$, both are inscribed in $\overset{\frown}{AB}$

$\angle CBD = \angle CAD$, both are inscribed in $\overset{\frown}{CD}$

which shows $\Delta AED \sim \Delta BEC$

$$\frac{a}{d} = \frac{b}{c}$$

57.

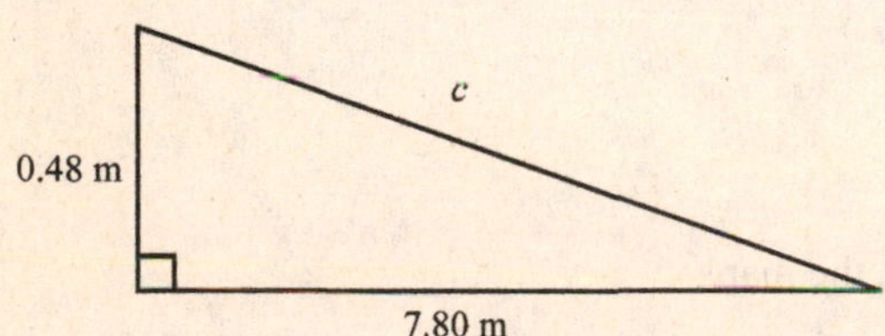

$$c^2 = a^2 + b^2$$
$$c = \sqrt{(0.48\text{ m})^2 + (7.80\text{ m})^2}$$
$$c = \sqrt{61.0704\text{ m}^2}$$
$$c = 7.814755274\text{ m}$$
$$c = 7.81\text{ m}$$

61.

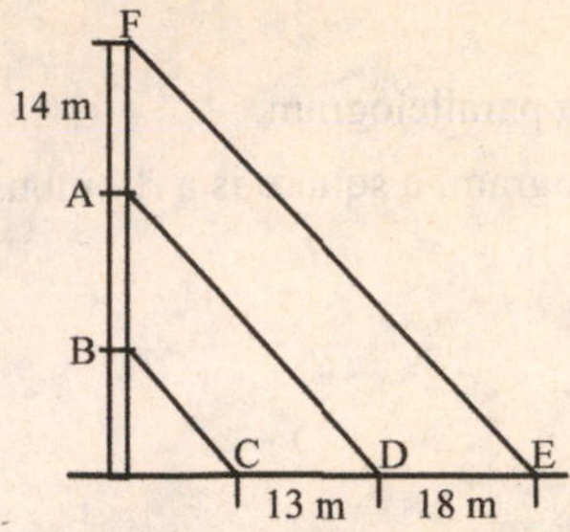

Since line segments BC, AD, and EF are parallel, the segments AB and CD are proportional to AF and DE

$$\frac{AB}{CD} = \frac{AF}{DE}$$

$$\frac{AB}{13\text{ m}} = \frac{14\text{ m}}{18\text{ m}}$$

$$AB = \frac{13\text{ m}(14\text{ m})}{18\text{ m}}$$

$$AB = 10.111111\text{ m}$$

$$AB = 10\text{ m}$$

65. The longest distance between points on the photograph is

$$c^2 = a^2 + b^2$$

$$c = \sqrt{(8.00\text{ in})^2 + (10.00\text{ in})^2}$$

$$c = \sqrt{164\text{ in}^2}$$

$$c = 12.806248\text{ in}$$

Find the distance in km represented by the longest measure on the map

$$x = (12.806248\text{ in})\ \frac{18\ 450}{1}\ \frac{1\text{ ft}}{12\text{ in}}\ \frac{1\text{ mi}}{5280\text{ ft}}$$

$$x = 3.7290922\text{ mi}$$

$$x = 3.73\text{ mi}$$

69. Area of the drywall is the area of the rectangle subtract the two circular cutouts.

$$A = lw - 2(\pi r^2)$$

$$A = lw - 2\ \frac{\pi d^2}{4}$$

$$A = lw - \frac{\pi d^2}{2}$$

$$A = (4.0\text{ ft})(8.0\text{ ft}) - \frac{\pi(1.0\text{ ft})^2}{2}$$

$$A = 30.42920367\text{ ft}^2$$

$$A = 3.0\times 10^1\text{ ft}^2$$

73. $V = \pi r^2 h$

$$V = \frac{\pi d^2}{4} h$$

$$V = \frac{\pi (4.3\text{ m})^2}{4}(13\text{ m})$$

$$V = 188.7861565\text{ m}^3$$

$$V = 189\text{ m}^3$$

77.

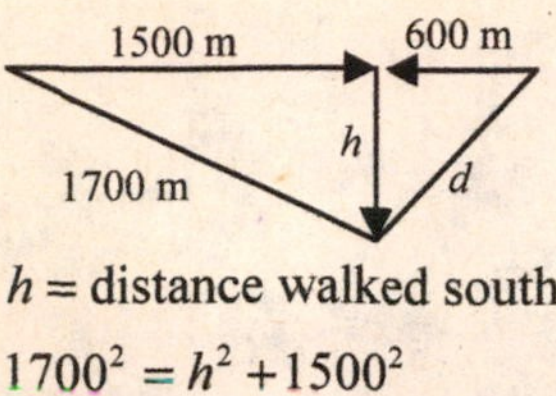

h = distance walked south

$$1700^2 = h^2 + 1500^2$$

$$h = \sqrt{1700^2 - 1500^2}$$

$$h = 800\text{ m}$$

$$d^2 = 600^2 + h^2$$

$$d = \sqrt{600^2 + 800^2}$$

$$d = 1000\text{ m}$$

81.

$$\frac{w}{h} = \frac{16}{9}$$

$$w = \frac{16h}{9}$$

$$152^2 = w^2 + h^2$$

$$23104 = \left(\frac{16h}{9}\right)^2 + h^2$$

$$23104 = \frac{256}{81}h^2 + h^2$$

$$23104 = \frac{337}{81}h^2$$

$$h^2 = \frac{81(23104)}{337}$$

$$h = \sqrt{5553.18694\text{ cm}^2}$$

$$h = 74.519708\text{ cm}$$

$$h = 74.5\text{ cm}$$

$$w = \frac{16h}{9}$$

$$w = \frac{16(74.519708\text{ cm})}{9}$$

$$w = 132.4794816\text{ cm}$$

$$w = 132\text{ cm}$$

85.

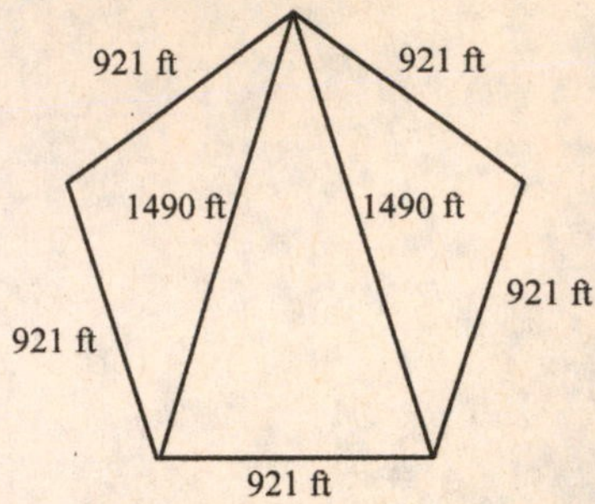

The area is the sum of the areas of three triangles, one with sides 921, 1490, and 1490 and two with sides 921, 921, and 1490. The semi-perimeters are given by

$$s_1 = \frac{921+921+1490}{2} = 1666$$

$$s_2 = \frac{1490+1490+921}{2} = 1950.5$$

$$A = 2\sqrt{1666(1666-921)(1666-921)(1666-1490)} + \sqrt{1950.5(1950.5-1490)(1950.5-1490)(1950.5-921)}$$

$$A = 806826 \text{ ft}^2 + 652{,}553 \text{ ft}^2 = 1{,}459{,}379 \text{ ft}^2$$

$$A = 1.46\times10^6 \text{ ft}^2$$

Chapter 3

Functions and Graphs

3.1 Introduction to Functions

1. $f(x) = 3x - 7$

$f(-2) = 3(-2) - 7 = -13$

5. (a) $A(r) = \pi r^2$

(b) $A(d) = \pi\left(\frac{d}{2}\right)^2 = \frac{\pi d^2}{4}$

9. To get A as a function of the diagonal d of a square, we note that d satisfies

$s^2 + s^2 = d^2$

$2s^2 = d^2$

$d = s\sqrt{2}$

where s is its side length

and so $s = \frac{d}{\sqrt{2}}$

$A = s^2 = \left(\frac{d}{\sqrt{2}}\right)^2 = \frac{d^2}{2}$

$A(d) = \frac{d^2}{2}$

To get d as a function of A, starting with $A = \frac{d^2}{2}$,

$d^2 = 2A$

$d = \sqrt{2A}$

$d(A) = \sqrt{2A}$

13. $f(x) = 2x + 1$

$f(3) = 2(3) + 1 = 7$

$f(-5) = 2(-5) + 1 = -9$

17. $\phi(x) = \frac{6 - x^2}{2x}$

$\phi(2\pi) = \frac{6 - (2\pi)^2}{2(2\pi)} = \frac{6 - 4\pi^2}{4\pi} = \frac{3 - 2\pi^2}{2\pi} = -2.66$

$\phi(-2) = \frac{6 - (-2)^2}{2(-2)} = \frac{2}{-4} = -\frac{1}{2}$

21. $K(s) = 3s^2 - s + 6$

$K(-s+2) = 3(-s+2)^2 - (-s+2) + 6 = 3s^2 - 12s + 12 + s - 6 = 3s^2 - 11s + 6$

$K(-s) + 2 = \left(3(-s)^2 - (-s) + 6\right) + 2 = 3s^2 + s + 8$

25. $f(x) = 5x^2 - 3x$

$f(3.86) = 5(3.86)^2 - 3(3.86) = 62.918 = 62.9$

$f(-6.92) = 5(-6.92)^2 - 3(-6.92) = 260.192 = 2.60 \times 10^2$

29. $f(x) = 3x^2 - 4$

33. $f(x) = x^2 + 2$

Square the value of the input and add 2 to the result.

37. $R(r) = 3(2r + 5)$

Double the value of the input and then add 5.

Multiply this result by 3.

41. $A = 5200 - 120t$

$f(t) = 5200 - 120t$

45. $d = f(v) = v + 0.05v^2$

$f(30) = 30 + 0.05(30)^2 = 75$ ft

$f(2v) = 2v + 0.05(2v)^2 = 2v + 0.2v^2$

$f(60) = 60 + 0.05(60)^2 = 240$ ft

$f(60) = f(2(30)) = 2(30) + 0.2(30)^2 = 240$ ft

49. distance = rate × time

For the first leg,

$d_1 = 55t$

For the second leg,

$d_2 = 65(t + 1)$

The total distance traveled is

$d_1 + d_2 = 55t + 65(t + 1)$

$f(t) = 120t + 65$

3.2 More about Functions

1. $f(x) = -x^2 + 2$ is defined for all real values of x.
Domain: all real numbers $\mathbb{R}$ or $(-\infty, \infty)$
Since x^2 cannot be negative, the maximum value of $f(x)$ is 2.
Range: all real numbers $f(x) \le 2$, or $(-\infty, 2]$

5. $f(x) = x + 5$
Domain: all real numbers $\mathbb{R}$ or $(-\infty, \infty)$
Range: all real numbers $\mathbb{R}$ or $(-\infty, \infty)$

9. $f(s) = \sqrt{s-2}$
$f(s)$ is not defined for $s < 2$.
Domain: all real values $s \ge 2$, or $[2, \infty)$.
Since $\sqrt{s-2}$ is never negative
Range: all real numbers $f(s) \ge 0$, or $[0, \infty)$

13. $y = |x - 3|$
Domain: all real numbers $\mathbb{R}$ or $(-\infty, \infty)$
Absolute value is never negative, so
Range: all real numbers $y \ge 0$, or $[0, \infty)$

17. $f(D) = \sqrt{D} + \dfrac{1}{D-2}$
Division by zero is undefined, so the domain must be restricted to exclude any value for which $D - 2$ is equal to zero. In this case, $D \ne 2$.
Also, $\sqrt{D}$ is undefined for $D < 0$ and so negative numbers are excluded from the domain.
Domain: all non-negative real numbers $D \ne 2$ or $[0, 2)$ and $(2, \infty)$.

21. $F(t) = 3t - t^2$ for $t \le 2$
$F(1) = 3(1) - 1^2 = 2$
$F(2) = 3(2) - 2^2 = 6 - 4 = 2$
$F(3)$ does not exist since function is not defined at that location.

25. Distance d is expressed in mi.
$d(t) = 40t + 55 \cdot 2$
$d(t) = 40t + 110$

29. For every metre over 1000 m in altitude, an additional 0.5 kg is added to the mass, if mass is expressed in kilograms, and the altitude is restricted to be greater than 1000 m,

$m(h) = 110 + 0.5(h - 1000)$

$m(h) = 0.5h - 390 \quad$ for $h > 1000$ m

33. (a)

$x(0.10) + y(0.40) = 1200$

$$y(x) = \frac{1200 - 0.1x}{0.4}$$

(b)

$$y(400) = \frac{1200 - 0.1(400)}{0.4}$$

$y(400) = 2900$ L

37. For the square, with the length of a side s,

$p = 4s$

$$s = \frac{p}{4}$$

$A_{\text{square}} = (s)^2$

$$A_{\text{square}} = \left(\frac{p}{4}\right)^2$$

$$A_{\text{square}} = \frac{p^2}{16}$$

For the circle, the circumference will be whatever is left over after the square perimeter is cut from the wire, $60 - p$

$c = 60 - p$

$2\pi r = 60 - p$

$$r = \frac{60 - p}{2\pi}$$

$A_{\text{circle}} = \pi r^2$

$$A_{\text{circle}} = \pi\left(\frac{60 - p}{2\pi}\right)^2$$

$$A_{\text{circle}} = \frac{(60 - p)^2}{4\pi}$$

Thus, the total area if both figures A will be a function of square perimeter p,

$A(p) = A_{\text{square}} + A_{\text{circle}}$

$$A(p) = \frac{p^2}{16} + \frac{(60 - p)^2}{4\pi}$$

41. distance = velocity × time

$300 = v \cdot (t-3)$

$v(t) = \dfrac{300}{t-3}$

In this example, negative time and speed have no meaning.

Cannot divide by zero, so $t \neq 3$.

Domain: all real values $t > 3$, or $(3, \infty)$

Since all $t > 3$, this forces v to always be positive.

No object with mass can reach the speed of light c $(3 \times 10^8$ m/s), but in practical terms the upper limit for speed is probably less than that of sound (around 330 m/s), and most real trucks would have a limit much lower than that.

Range: all real numbers $0 < v < c$, or $(0, c)$

45. For every metre over 1000 m in altitude, an additional 0.5 kg is added to the mass, but up to that altitude, mass is a constant 110 kg. If mass is expressed in kilograms, and from Exercise 29, when altitude is restricted to be greater than 1000 m, $m(h) = 0.5h - 390$

$$m(h) = \begin{cases} 110 & \text{for } 0 \leq h \leq 1000 \text{ m} \\ 0.5h - 390 & \text{for } h > 1000 \text{ m} \end{cases}$$

49. $f(x+2) = |x|$

In order to evaluate $f(0)$, the value of x must be -2.

$f(-2+2) = f(0)$

So, when $x = -2$ the value of the function will be:

$f(0) = |-2|$

$f(0) = 2$

3.3 Rectangular Coordinates

1. $A(0, -2)$, $B(4, -2)$, $C(4, 1)$

The vertices of base AB both have y-coordinates of -2, which means the base CD which must be parallel, has the same y-coordinates for its vertices. Since at point C the y-coordinate is 1, then at D, y must also be 1. In the same way, the x-coordinates of the left side must both be -1. Therefore the fourth vertex is $D(0, 1)$.

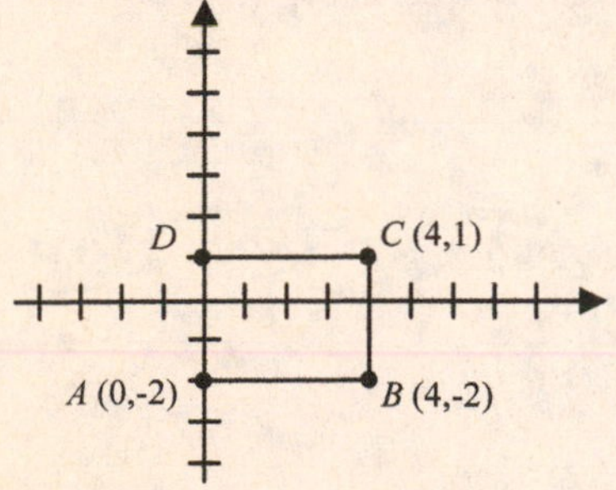

5.

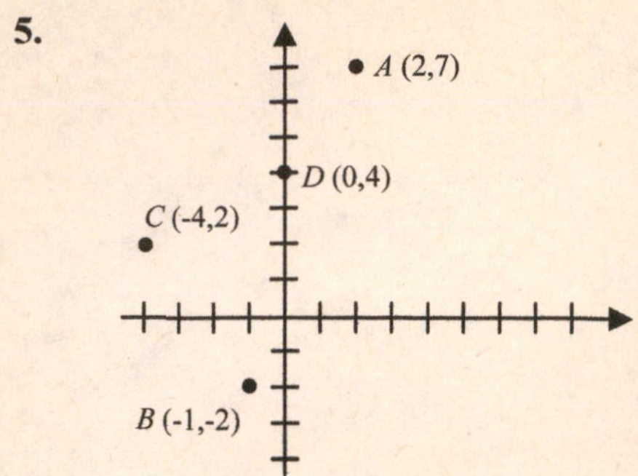

9. Figure $ABCD$ forms a rectangle.

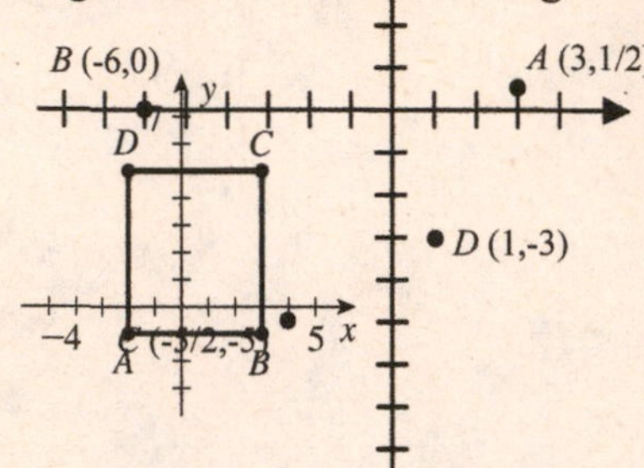

13. In order for the x-axis to be the perpendicular bisector of the line segment joining $P(3,\ 6)$ and $Q(x,\ y)$, Q must be located the same distance on the opposite side of the x-axis from P, and at the same x-coordinate so that the segment PQ is vertical (perpendicular to the x-axis). Therefore the point is $Q(3,\ -6)$.

17. If $x < 0$ then the point lies in either the second or third quadrant. If $y < 0$ then the point lies in either the third or fourth quadrant. For both of these to occur, the point must lie in the third quadrant.

21. All points with y-coordinate of 3 are on a horizontal line 3 units above the x-axis, passing through $(0,\ 3)$. The equation of this horizontal line is $y = 3$.

25. The x-coordinate of all points on the y-axis is zero, since all points on that line have form $(0,\ y)$, where y can be any real number.

29. All points for which $x < -1$ are in Quadrant II and Quadrant III to the left of the line $x = -1$, which is parallel to the y-axis, one unit to the left of the y-axis.

33. If $xy = 0$ then the product of the coordinates x and y must be zero.
Therefore, either coordinate may be zero.
For $x = 0$ points lie on the y-axis , and
for $y = 0$ points lie on the x-axis.
So, all points where $xy = 0$ lie on the x- or y-axis.

37.

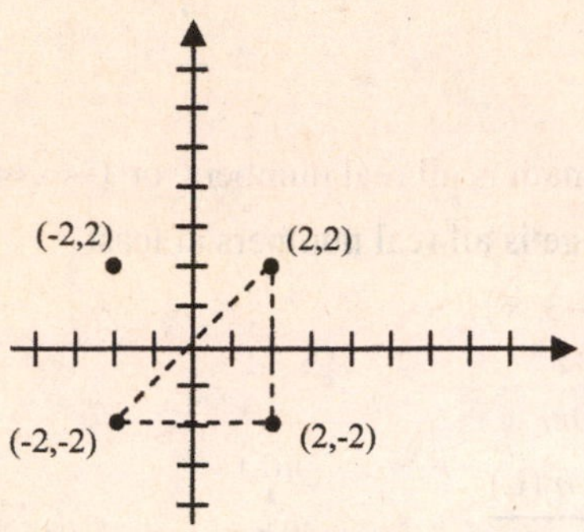

The distance between (2, 2) and (−2, −2) can be found using the Pythagorean theorem applied to the right triangle whose third vertex is at (2,−2). This triangle has legs of length 4 and 4 and so its hypotenuse has length $\sqrt{4^2+4^2}=\sqrt{32}=4\sqrt{2}$.

3.4 The Graph of a Function

1. $f(x) = 3x + 5$

x	y
−3	−4
−2	−1
−1	2
0	5
1	8

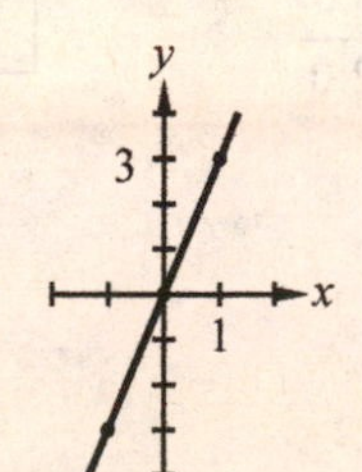

5. $y = 3x$

x	y
−1	−3
0	0
1	3

9. $s = 7 - 2t$

s	t
−1	9
0	7
1	5

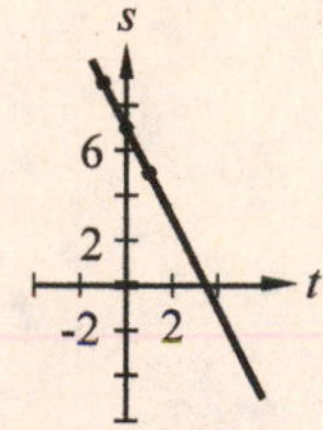

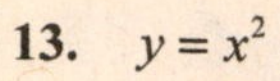

13. $y = x^2$

x	y
−2	4
−1	1
0	0
1	1
2	4

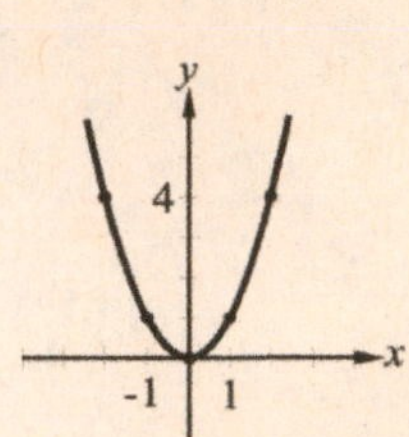

17. $y = \frac{1}{2}x^2 + 2$

x	y
−4	10
−2	4
0	2
2	4
4	10

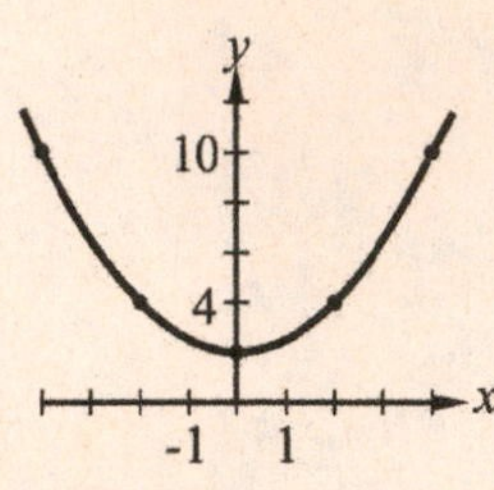

21. $y = x^2 - 3x + 1$

x	y
3	1
2	−1
1.5	−1.25
1	−1
0	1

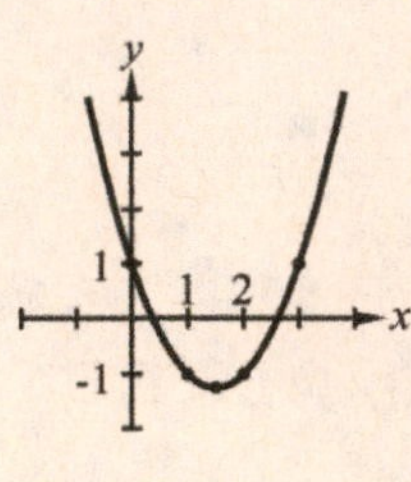

25. $y = x^3 - x^2$

x	y
−2	−12
−1	−2
0	0
2/3	−4/27
1	0
2	4

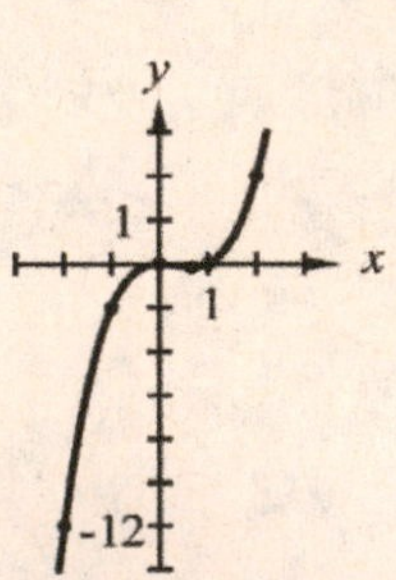

29. $P = \frac{8}{V} + 3$

V	P
−4	1
−2	−1
−1	−5
2	7
4	5

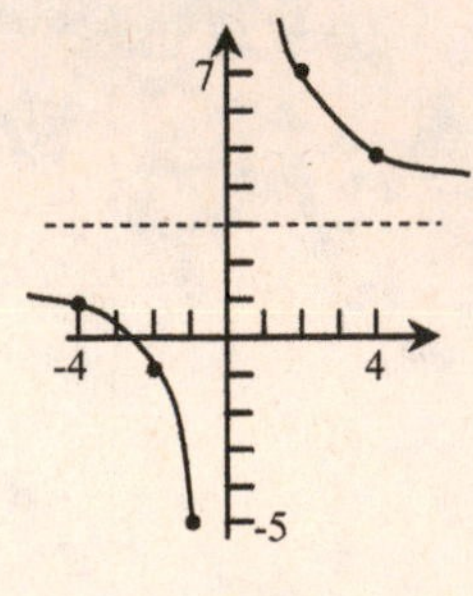

33. $y = \sqrt{9x}$

x	y
0	0
1	3
4	6
9	9
16	12

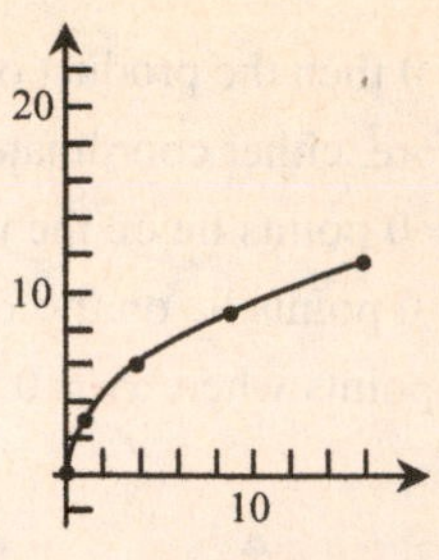

37. The domain is all real numbers, or $(-\infty, \infty)$. The range is all real numbers at least −3, or $[-3, \infty)$.

41. $n = 0.40m$

m (L)	n (L)
10	4
50	20
80	32

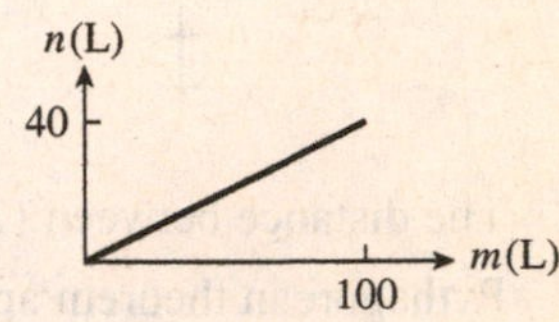

45. $H = 240I^2$

I (A)	H (W)
0	0
0.2	9.6
0.4	38.4
0.6	86.4
0.8	153.6

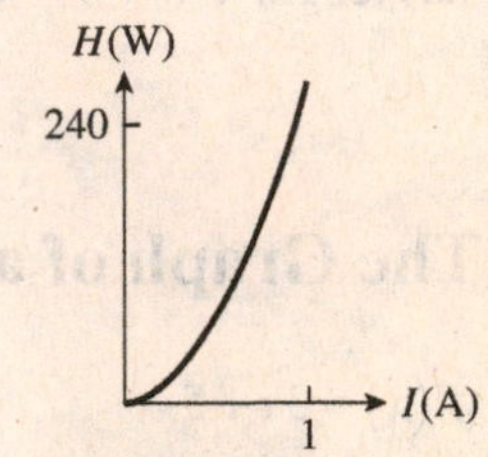

49. $P = 0.004v^3$

v (km/h)	P (W)
0	0
5	0.5
10	4.0
15	13.5
20	32.0

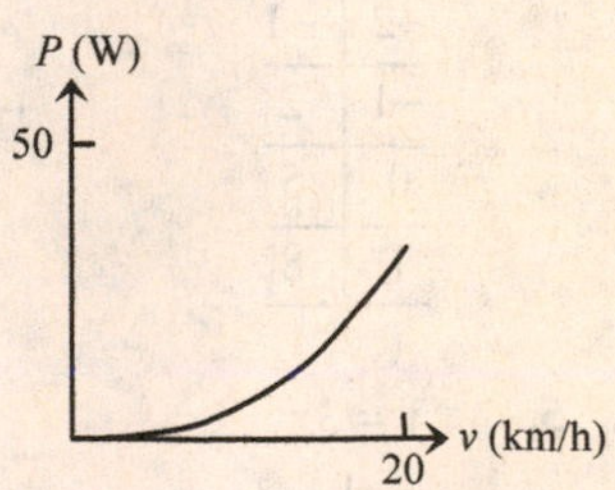

53. Since $l + w + h = 45$ and $l = 2w$ are given,
$h = 45 - w - l = 45 - 32$ and
$V = lwh = 2w^2(45 - 3w) = 90w^2 - 6w^3$.

$V(w) = 90w^2 - 6w^3$

w (in)	V (in^3)
3	648
6	1944
9	2916
12	2592

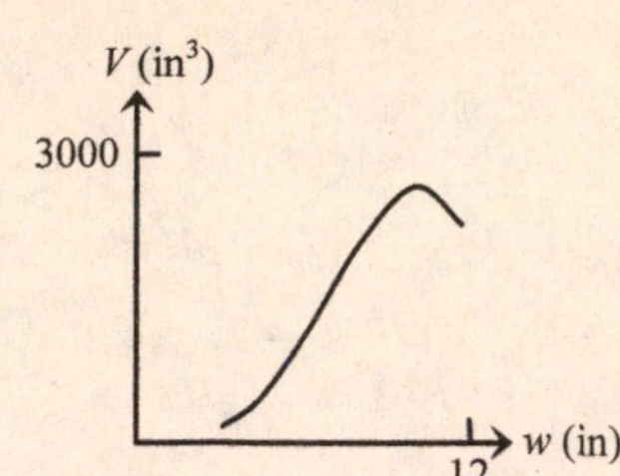

57. $N = \sqrt{n^2 - 1.69}$

n	N
1.3	0
1.5	0.748
1.7	1.095
2.0	1.520

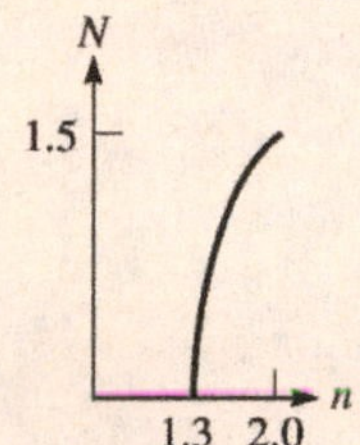

61. (1, 2) on the graph means if we evaluate the function at $x = 1$ it gives $f(1) = 2$ which says nothing about $f(2)$, the function evaluated at $x = 2$, which may or may not equal 1.

65.

x	y
−2	2
−1	1
0	0
1	1
2	2

$y = x$ is the same as $y = |x|$ for $x \ge 0$.
$y = |x|$ is the same as $y = -x$ for $x < 0$.

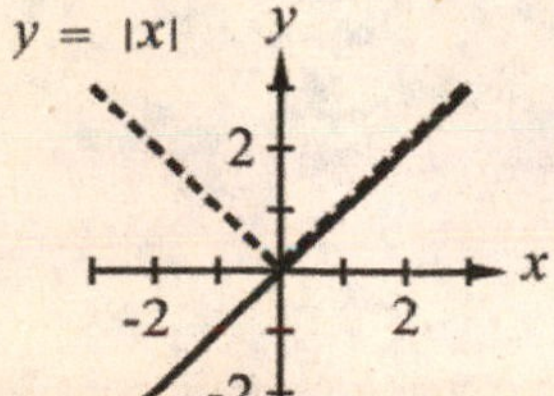

For negative values of x, $y = |x|$ becomes $y = -x$.

69. **(a)** $y = x + 2$

x	y
−3	−1
−2	0
−1	1
0	2
1	3
2	4

(b) $y = \dfrac{x^2 - 4}{x - 2}$

x	y
−3	−1
−2	0
−1	1
0	2
1	3
2	undefined

The graphs of $y = x + 2$ and $y = \dfrac{x^2 - 4}{x - 2}$ are identical except the second curve is undefined at a single point $x = 2$. It turns out that if you factor the second function (see Chapter 6), it will reduce to the first function everywhere except at $x = 2$ where it is undefined.

73. No. Some vertical lines will intercept the graph at multiple points when $x > 0$. The graph is that of a relation.

3.5 Graphs on the Graphing Calculator

1. $x^2 + 2x = 1$

$x^2 + 2x - 1 = 0$

Graph the following function, and estimate solutions.

$y = x^2 + 2x - 1$

x	y
−4	7
−3	2
−2	−1
−1	−2
0	−1
1	2
2	7

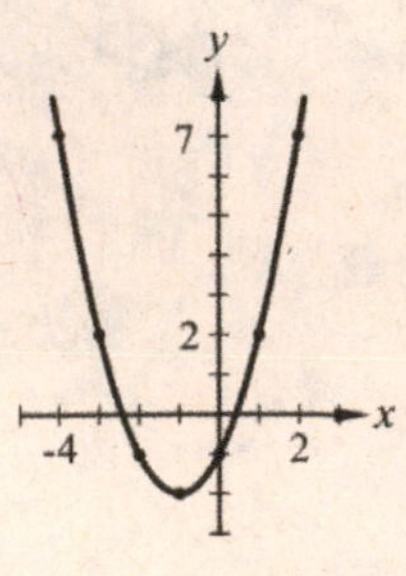

$x = -2.4,\ x = 0.4$

5. $y = x^2 - 4x$

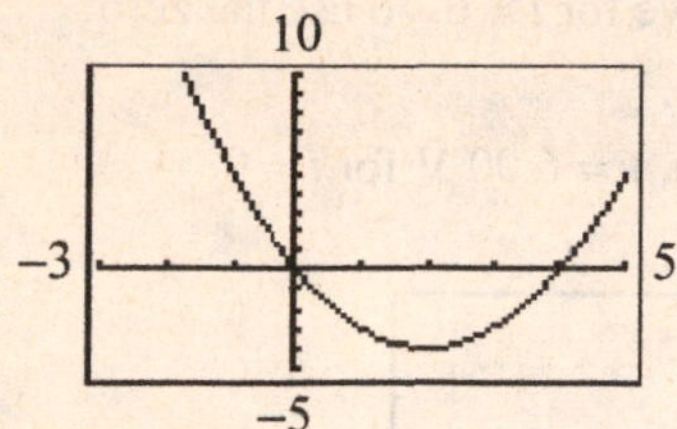

9. $y = x^4 - 2x^3 - 5$

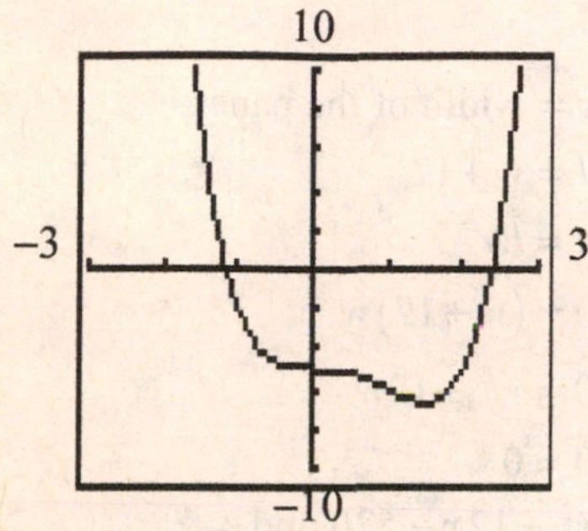

13. $y = x + \sqrt{x+3}$

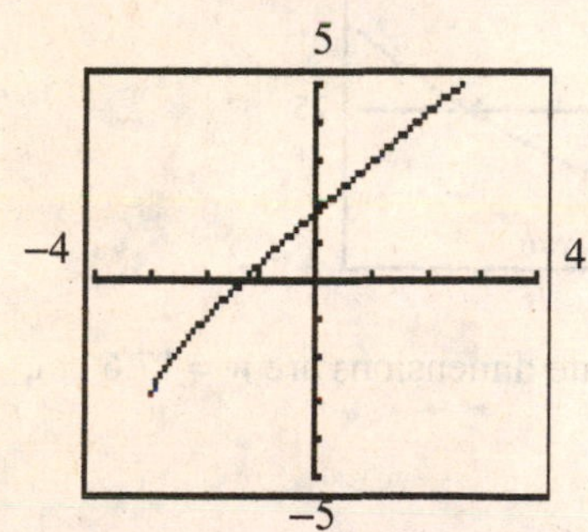

17. $y = 2 - |x^2 - 4|$

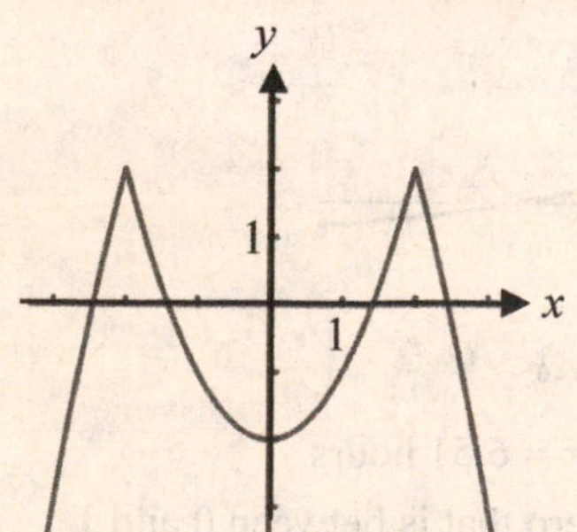

21. To solve $x^3 - 3 = 3x$ or $x^3 - 3x - 3 = 0$,
graph $y = x^3 - 3x - 3$
and use the zero feature to solve.

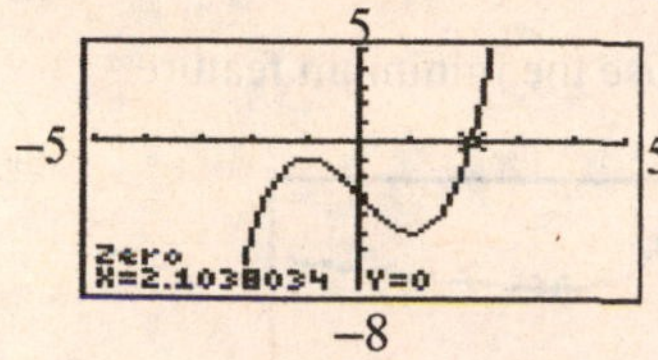

Solution $x = 2.104$

25. To solve $\sqrt{5R+2} = 3$ or $\sqrt{5R+2} - 3 = 0$
graph $y = \sqrt{5x+2} - 3$ and use the zero feature to solve.

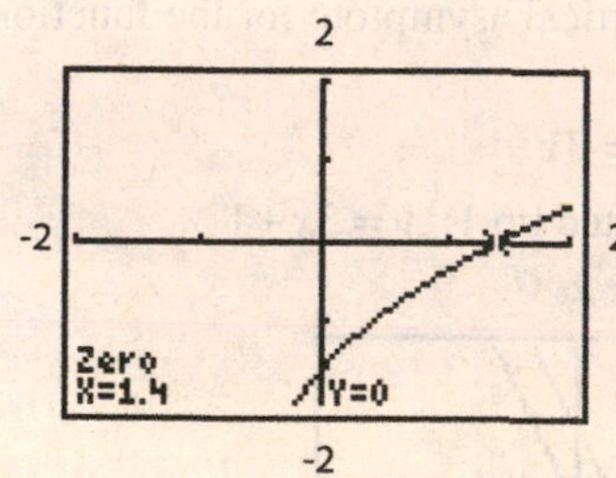

Solution $R = 1.400$

29. From the graph, $y = -4x^2 + 8x + 3$ has
Range: all real values $y \le 7$.

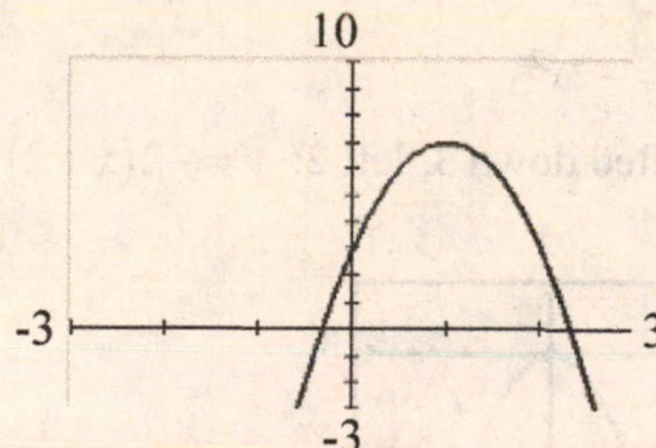

33. From the graph, $y = \dfrac{4}{x^2 - 4}$ has
Range: all real values $y > 0$ when $x < -2$ or $x > 2$
Range: all real values $y \le -1$ when $-2 < x < 2$
$x = \pm 2$ are vertical asymptotes for the function.

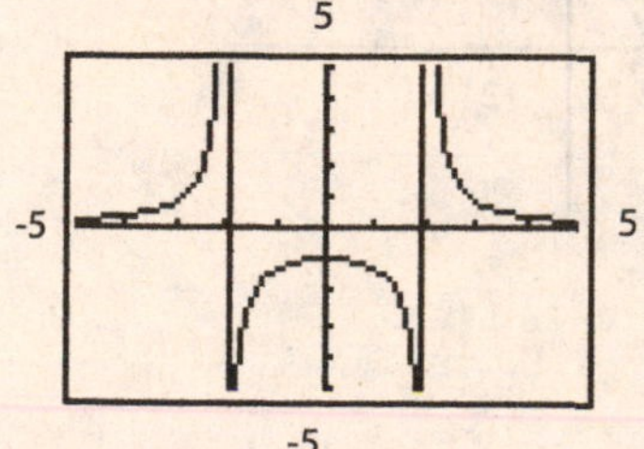

37. To find range of $Y(y)=\dfrac{y+1}{\sqrt{y-2}}$

graph $y=\dfrac{x+1}{\sqrt{x-2}}$ on the graphing calculator and use the minimum feature.

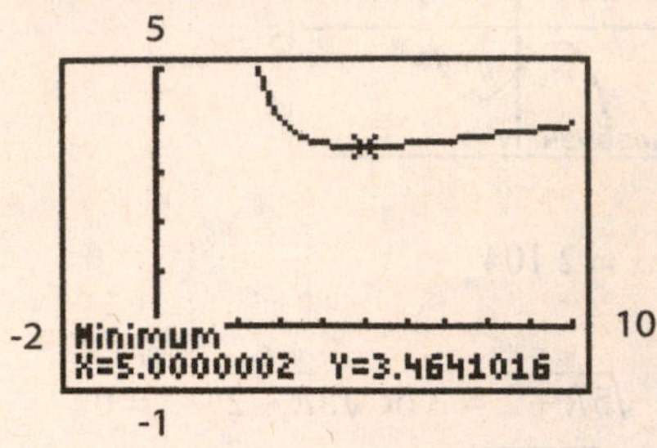

Range: all real values $Y(y)\geq 3.464$

$x=2$ is a vertical asymptote for the function.

41. function: $y=3x$

function shifted up 1: $y=3x+1$

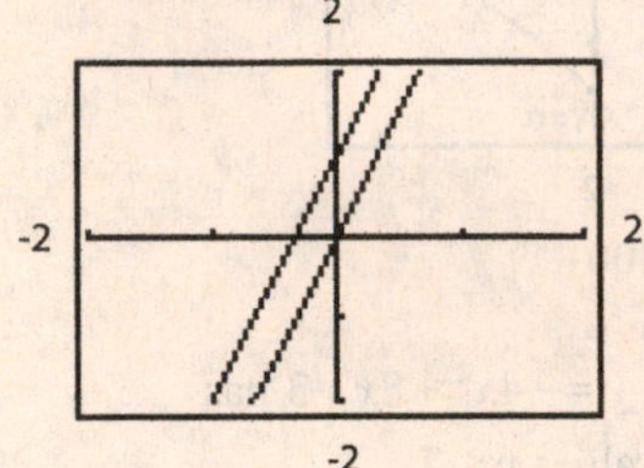

45. function: $y=-2x^2$

function shifted down 3, left 2: $y=-2(x+2)^2-3$

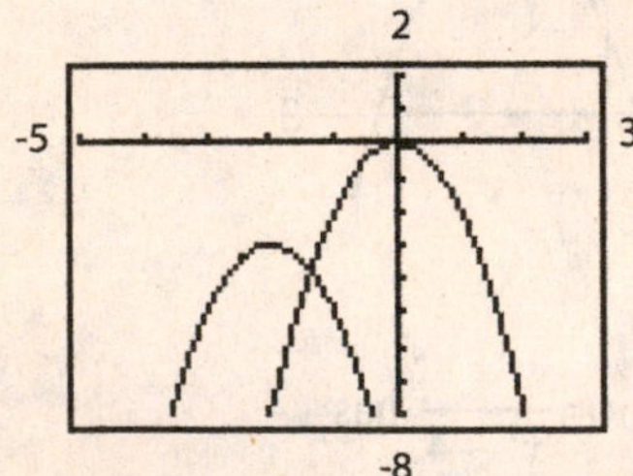

49.

-2 2

53. $i=0.01v-0.06$, graph $y=0.01x-0.06$

We need to solve for $i=0$, so use the zero feature to solve.

From the graph, $v=6.00$ V for $i=0$.

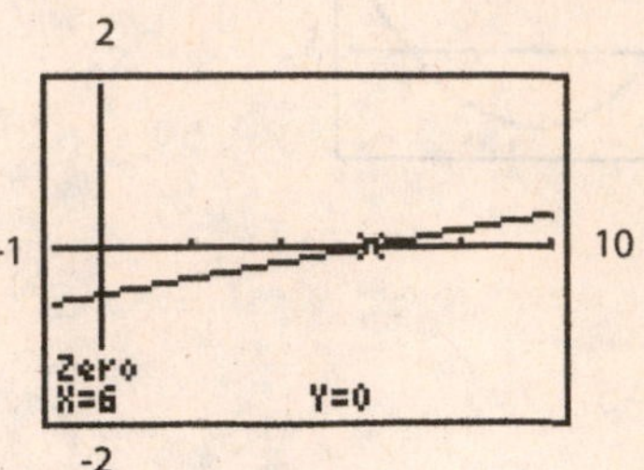

57. Let w = width of the panel

Let $l=w+12$

$$A=lw$$

$$520=(w+12)w$$

$$520=w^2+12w$$

$$w^2+12w-520=0$$

So, graph $y=x^2+12x-520$ and use the zero feature to solve

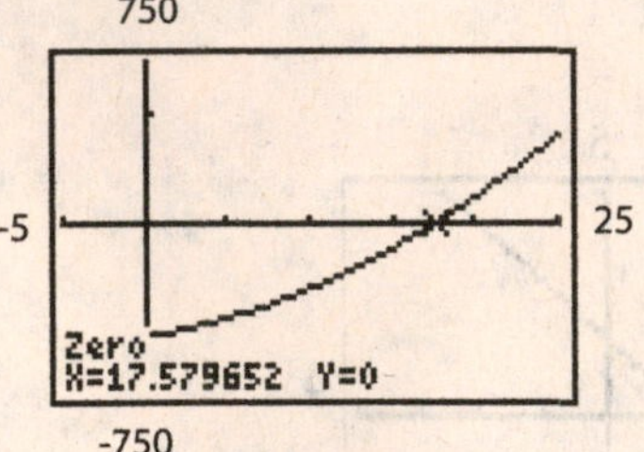

The approximate dimensions are $w=17.6$ cm, $l=29.6$ cm.

61. Graph $y=\dfrac{10.0x}{x^2+1.00}-1.50$

and use the zero feature to solve.

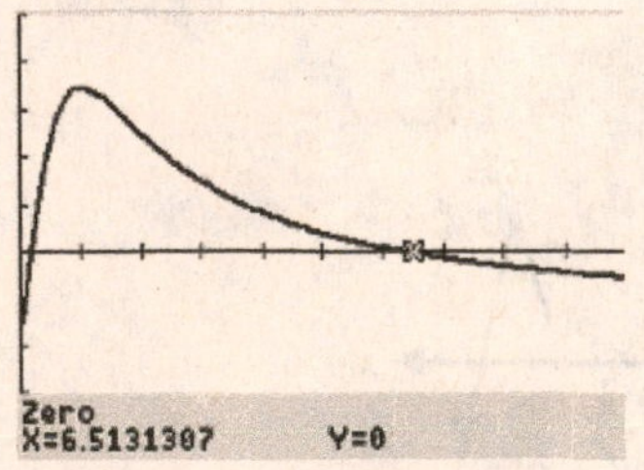

From the graph $x=6.51$ hours.

We ignore the zero that is between 0 and 1 since that corresponds to when the concentration first increases above 1.50 mg/L.

65. **(a)**

Since V is increasing at a constant rate and $V = \frac{4}{3}\pi r^3$ it seems reasonable that r^3 should be increasing at a constant rate. Thus, r should be proportional to the cube root of time, $r = k\sqrt[3]{t}$. The graph will look similar to the graph shown below.

(b) $r = \sqrt[3]{3t}$ would be a typical situation.

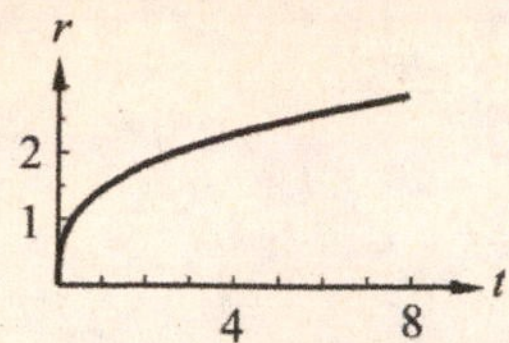

3.6 Graphs of Functions Defined by Tables of Data

1.

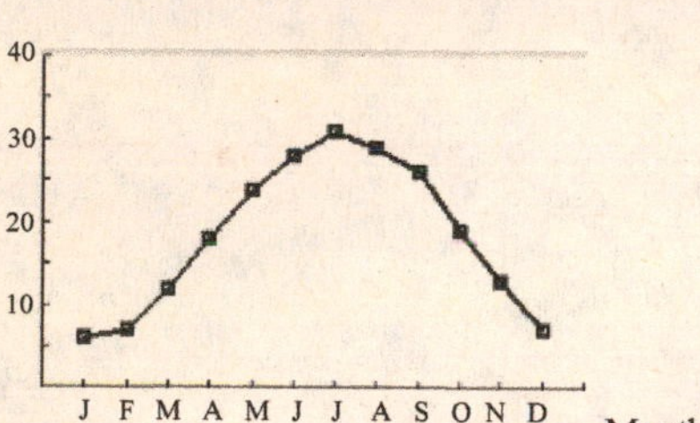

5.

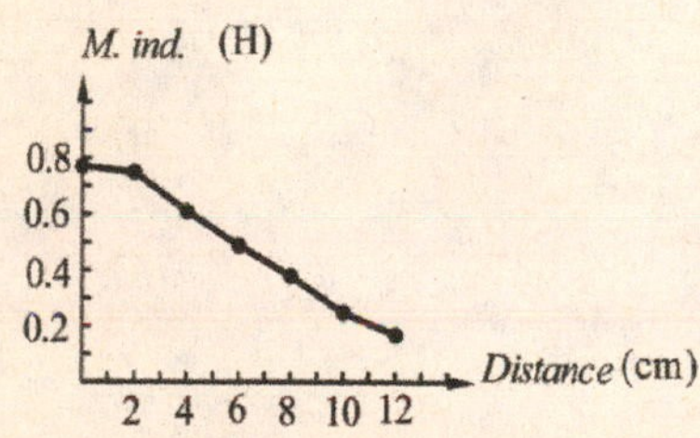

9. Estimate 0.3 of the interval between 4 and 5 on the t-axis and mark it. Draw a line vertically from this point to the graph. From the point where it intersects the graph, draw a horizontal line to the T-axis which it crosses at 132°C. Therefore for $t = 4.3$ min, $T = 132°$C

13.

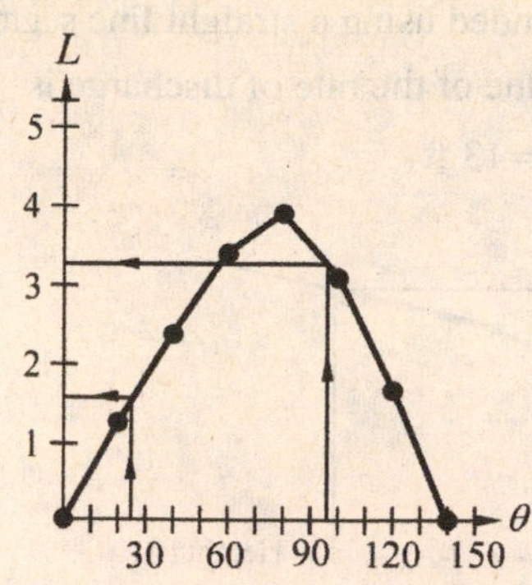

(a) For $\theta = 25°$, $L = 1.5$ mm

(b) For $\theta = 96°$, $L = 3.2$ mm

17. $1.2\left[\begin{array}{cc} 0.2\left[\begin{array}{cc} 10.2 & 7 \\ 10.0 & ? \end{array}\right]x \\ 9.0 \quad 8 \end{array}\right]1$

$$\frac{0.2}{1.2} = \frac{x}{1}$$

$$x = \frac{0.2}{1.2}$$

$$x = 0.1666$$

$$r = 7 + 0.1666$$

$$r = 7.2\%$$

21. $1.0\left[\begin{array}{cc} 0.7\left[\begin{array}{cc} 1.0 & 10 \\ 1.7 & ? \end{array}\right]x \\ 2.0 \quad 15 \end{array}\right]5$

$$\frac{0.7}{1.0} = \frac{x}{5}$$

$$x = 3.5$$

$$R = 10 + 3.5$$

$$R = 13.5 \text{ ft}^3/\text{s}$$

25. $0.05\left[\begin{array}{cc} 0.03\left[\begin{array}{cc} 0.56 & 70 \\ 0.59 & ? \end{array}\right]x \\ 0.61 \quad 80 \end{array}\right]10$

$$\frac{0.03}{0.05} = \frac{x}{10}$$

$$x = \frac{0.03(10)}{0.05}$$

$$x = 6$$

$$A = 70 + 6$$

$$A = 76 \text{ m}^2$$

29. The graph is extended using a straight line segment. The estimated value of the rate of discharge is $35.5\ \text{ft}^3/\text{s}$ for $H = 13$ ft

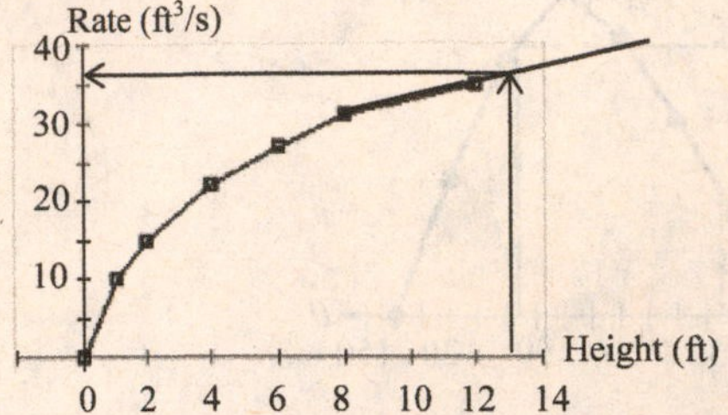

Review Exercises

1. It is false that for any function $f(x)$, $f(-x) = -f(x)$. Consider $f(x) = x^2$. Then $f(-x) = (-x)^2 = x^2$ and $-f(x) = -x^2$ which is not equal to x^2 for any $x \neq 0$.

5. The statement is true. If $(2,0)$ is an x-intercept, then $f(2) = 0$ and so $2f(2) = 2(0) = 0$ as well.

9.
$$0.05x + 0.04y = 2000$$
$$0.04y = 2000 - 0.05x$$
$$y(x) = \frac{2000 - 0.05x}{0.04}$$

13.
$$H(h) = \sqrt{1-2h}$$
$$H(-4) = \sqrt{1-2(-4)} = \sqrt{1+8} = \sqrt{9} = 3$$
$$H(2h) + 2 = \sqrt{1-2(2h)} + 2 = \sqrt{1-4h} + 2$$

17.
$$f(x) = 4 - 5x$$
$$f(2x) - 2f(x) = 4 - 5(2x) - 2(4 - 5x)$$
$$f(2x) - 2f(x) = 4 - 10x - 8 + 10x$$
$$f(2x) - 2f(x) = -4$$

21.
$$G(S) = \frac{S - 0.087629}{(3.0125)S}$$
$$G(0.17427) = \frac{0.17427 - 0.087629}{(3.0125)(0.17427)} = \frac{0.086641}{0.524988375} = 0.16503413 = 0.16503$$
$$G(0.053206) = \frac{0.053206 - 0.087629}{3.0125(0.053206)} = \frac{-0.034423}{0.160283075} = -0.214763785 = -0.21476$$

25. $g(t) = \dfrac{8}{\sqrt{t+4}}$

To avoid a division by zero error $t \neq -4$

To avoid a negative root, $t \geq -4$, so

Domain: all real numbers $t > -4$, or $(-4, \infty)$

No value of t will produce $g(t) = 0$, and if we consider only principal roots,

Range: all real numbers $g(t) > 0$, or $(0, \infty)$

29. The graph of $y = 4x + 2$ is

x	y
−2	−6
−1	−2
0	2
1	6
2	10

33. The graph of $y = 3 - x - 2x^2$ is

x	y
−3	−12
−2	−3
−1	2
0	3
1	0
2	−7

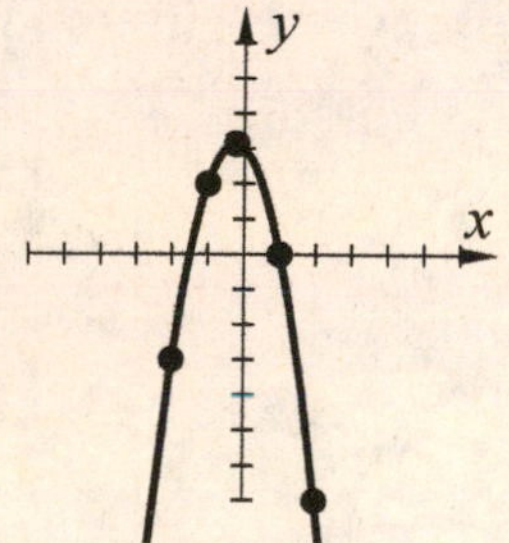

37. The graph of $y = \dfrac{x}{x+1}$ is

x	y
−3	1.5
−2	2
0	0
1	1/2
2	2/3
3	3/4

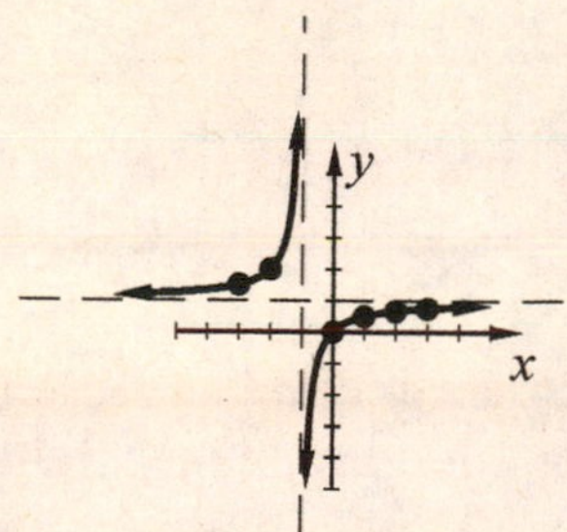

41. $x^2+1=6x$

$x^2-6x+1=0$

Graph $y=x^2-6x+1$ and use the zero feature to solve.

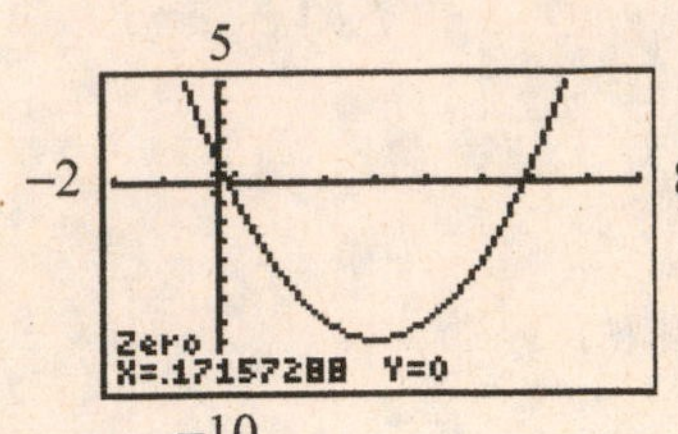

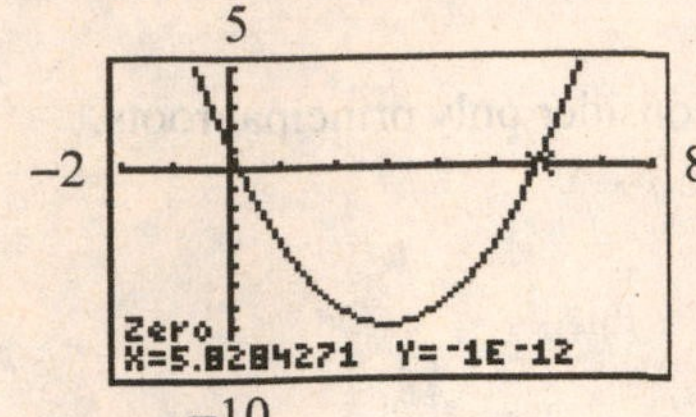

Solutions $x=0.17$ and $x=5.83$

45. $\dfrac{5}{v^3}=v+4$

$v+4-\dfrac{5}{v^3}=0$

Graph $y=x+4-\dfrac{5}{x^3}$ and use the zero feature to solve.

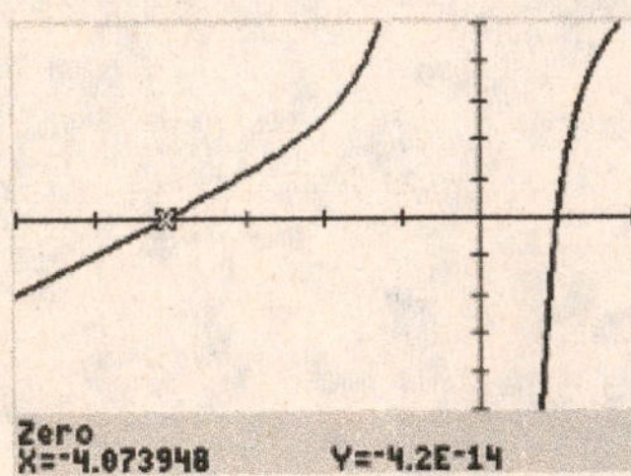

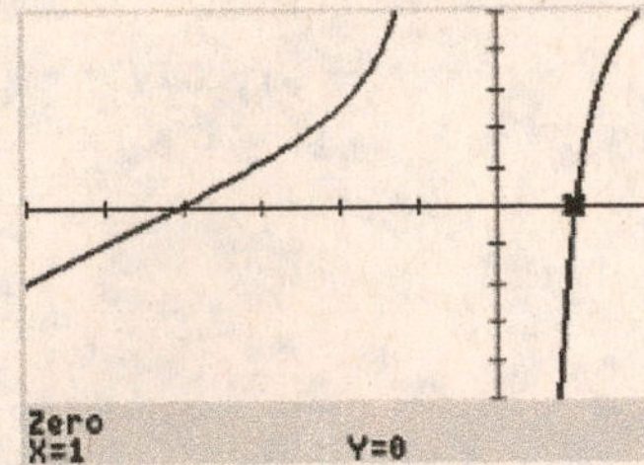

Solutions $x=-4.07$ and $x=1.00$

49. If $A=w+\dfrac{2}{w}$, then graph $y=x+\dfrac{2}{x}$ and use the minimum and maximum feature.

Range: all real values $A\le -2.83$ when $w<0$ or $(-\infty,-2.83]$

Range: all real values $A\ge 2.83$ when $w>0$ or $[2.83,\infty)$

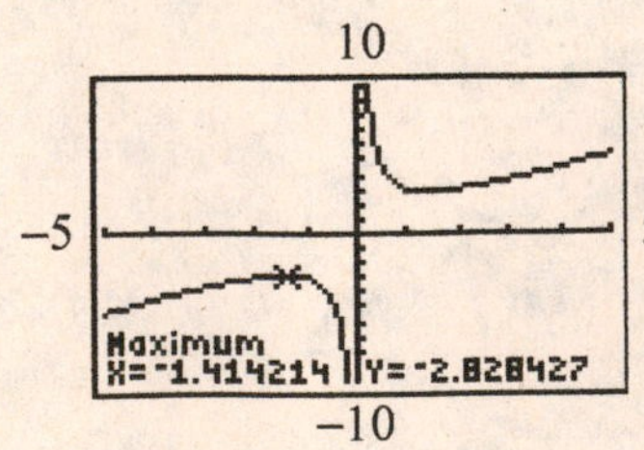

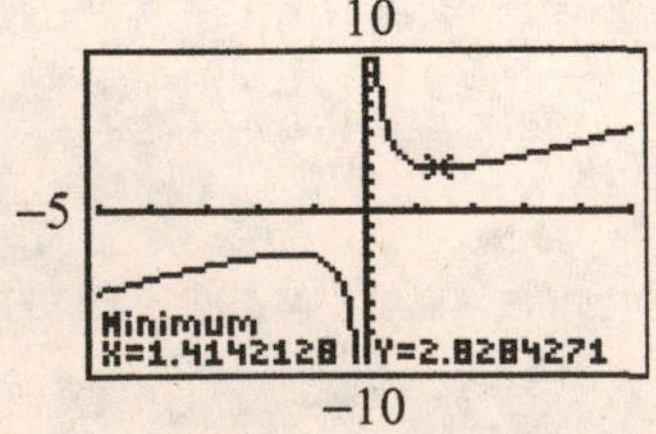

53.

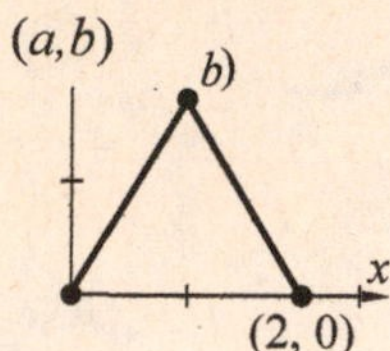

We know the x-component of the third vertex is halfway between the other two points because it is an equilateral triangle, and the base is horizontal.

so $a = \frac{2-0}{2} = 1$

The y-component of the vertex is the b of the Pythagorean theorem, and we know $c = 2$ since all sides of the triangle are the same length.

$b^2 = c^2 - a^2$

$b^2 = 2^2 - 1^2$

$b^2 = 4 - 1$

$b^2 = 3$

$b = \sqrt{3}$

$b = \pm\sqrt{3}$

$\left(1, \pm\sqrt{3}\right)$ are the coordinates of the third vertex

57. There are many possibilities. Two are shown.

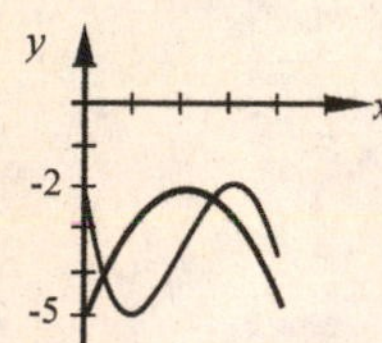

61. $f(x) = 2x^3 - 3,$

$f(-x) = 2(-x)^3 - 3 = -2x^3 - 3.$

The graphs are reflections of each other across the y-axis.

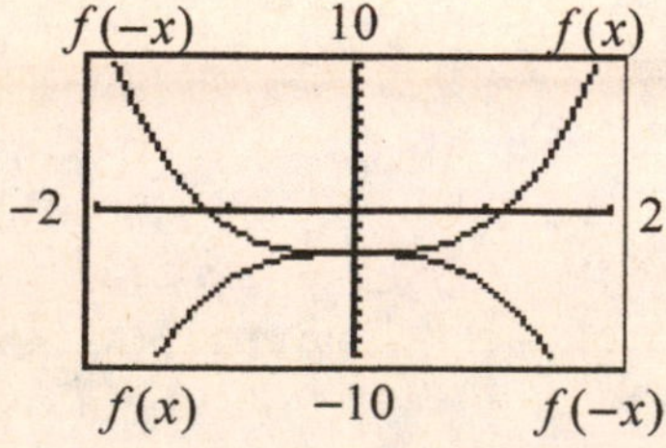

65. Let r be the radius of the circle. Then since the triangle formed by two radii and the side of the square is a right isosceles triangle, we have $r^2 + r^2 = s^2$ by the Pythagorean theorem. Therefore, $2r^2 = s^2$, or $r^2 = \frac{1}{2}s^2$. The area of the circle is $A = \pi r^2 = \frac{1}{2}\pi s^2$.

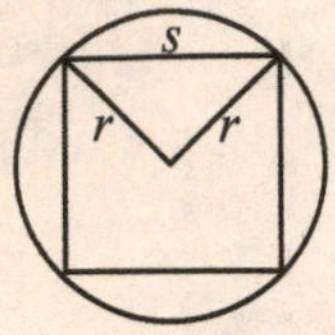

69. $A = 8.0 + 12t^2 - 2t^3$

Graph $y = 8.0 + 12x^2 - 2x^3$ for $0 \le x \le 6$ and use the maximum feature.

The maximum angle is 72°.

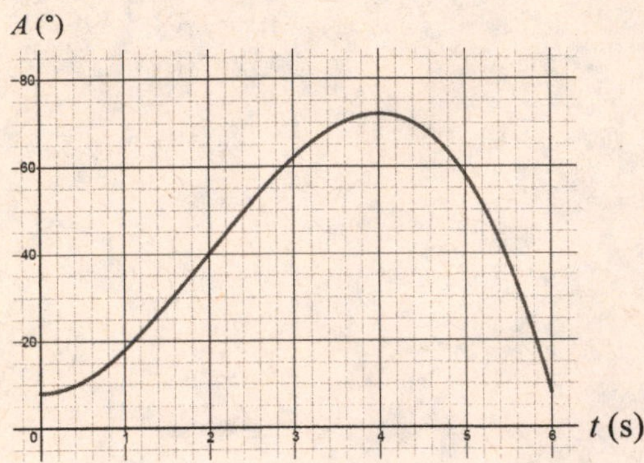

73. $P = f(p) = 50(p-50) = 50p - 2500, \quad \$30 \le p \le \$150$

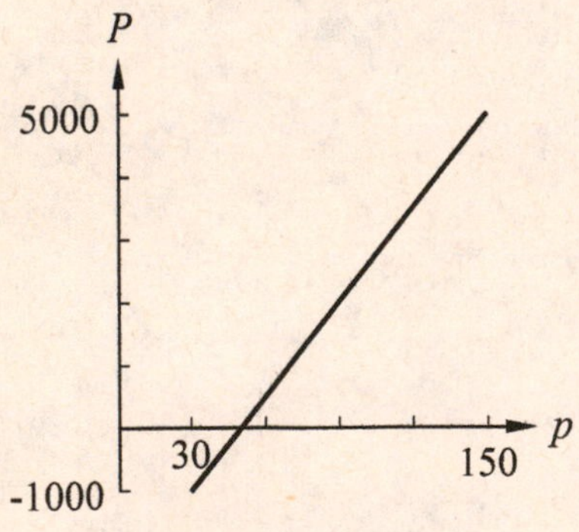

77. $e = f(T) = \dfrac{100\left(T^4 - 307^4\right)}{307^4}$

$f(309) = \dfrac{100\left(309^4 - 307^4\right)}{307^4}$

$f(309) = \dfrac{100(233\ 747\ 360)}{8\ 882\ 874\ 001}$

$f(309) = 2.631438428$

$f(309) = 2.63\%$

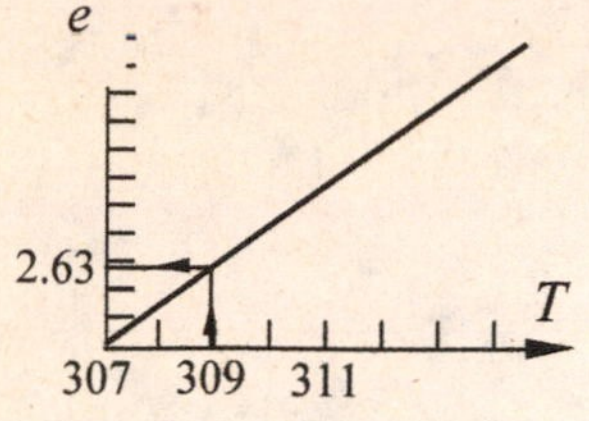

81. 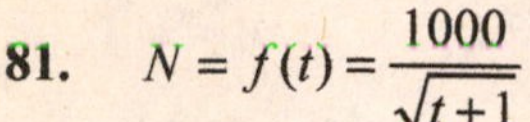

$N = f(t) = \dfrac{1000}{\sqrt{t+1}}$

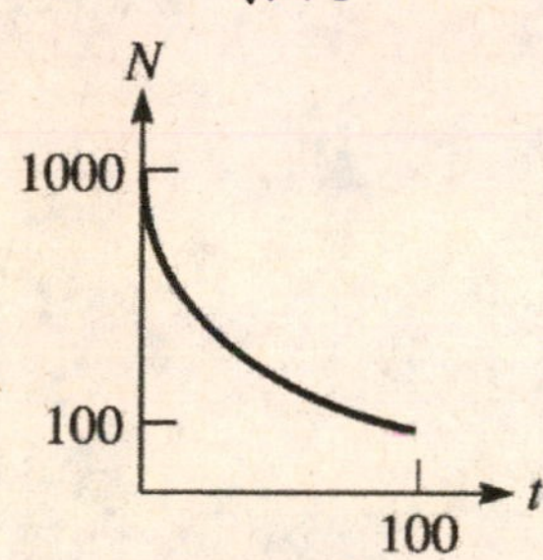

85. Since $f(47)$ lies between $f(40) = 11.1$ ft and $f(60) = 11.7$ ft, we interpolate between 40 and 60.

The distance between 40 and 60 is 20. The distance between 40 and 47 is 7.

The distance between the corresponding distances is $11.7 - 11.1 = 0.6$

40 11.1

20 7 47 ? x 0.6

60 11.7

so, $\dfrac{x}{7} = \dfrac{0.6}{20}$

$x = \dfrac{(0.6)(7)}{20}$

$x = 0.21$

$f(47) = 11.1 + 0.21 = 11.31$ ft.

89.
$$s = 135 + 4.9T + 0.19T^2$$
$$500 = 135 + 4.9T + 0.19T^2$$

$0.19T^2 + 4.9T - 365 = 0$

Graph $y = 0.19x^2 + 4.9x - 365$ and use the zero feature to solve.

Then $T = 32.8$ °C, the negative temperature is likely too low to be physical.

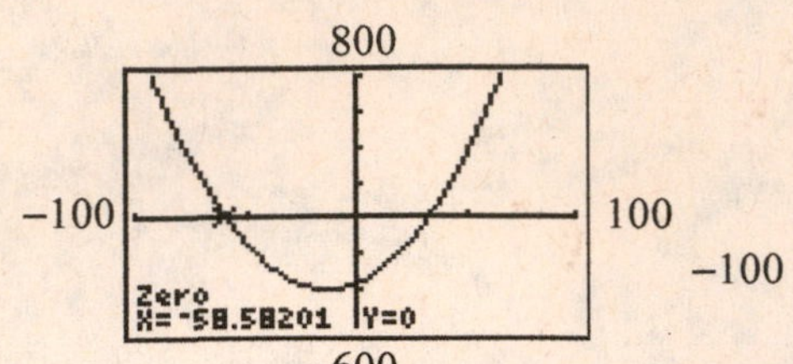

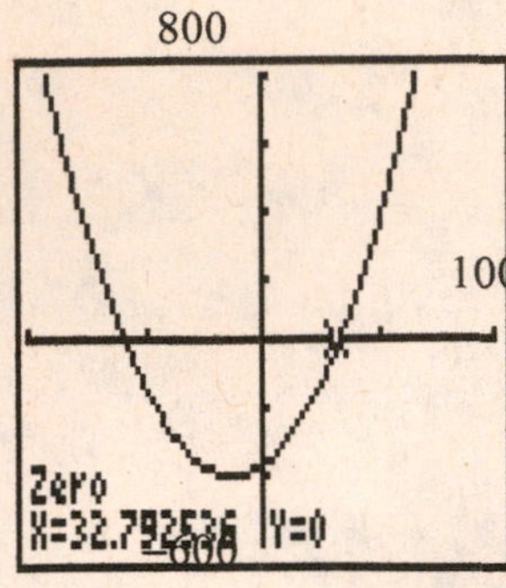

93. $A(r) = \pi(r + 45.0)^2$

$11500 = \pi(r + 45.0)^2$

$\pi(r + 45.0)^2 - 11500 = 0$

We solve for r using the zero feature on the graphing calculator.

We find $x = 15.5$ ft.

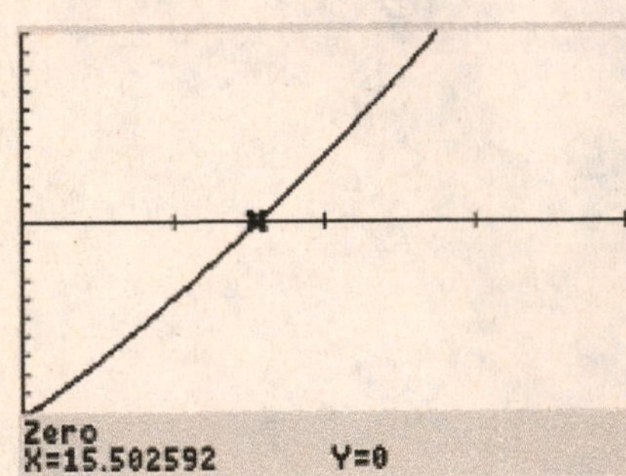

97. Volume is a function of radius and height, so we can write the volume equation

$V = \pi r^2 h = 250.0 \text{ cm}^3$

$h = \dfrac{250.0}{\pi r^2}$

The area is the surface area of the bottom and sides of the cup (no top)

$A = A_{\text{base}} + A_{\text{side}}$

$A = \pi r^2 + 2\pi rh$

$A = \pi r^2 + 2\pi r \cdot \dfrac{250.0}{\pi r^2}$

$A = \pi r^2 + \dfrac{500.0}{r}$

If $A = 175.0 \text{ cm}^2$, then solve for r:

Graph $y = \pi r^2 + \dfrac{500.0}{x}$, for $x > 0$, and $y_2 = 175.0$, for $x > 0$ and use the intersect feature to solve.

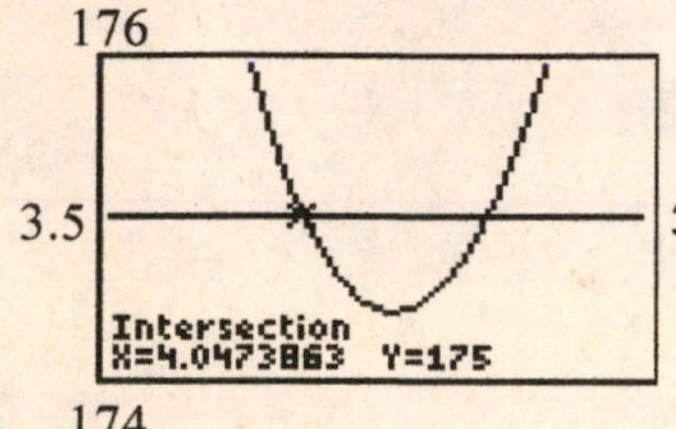

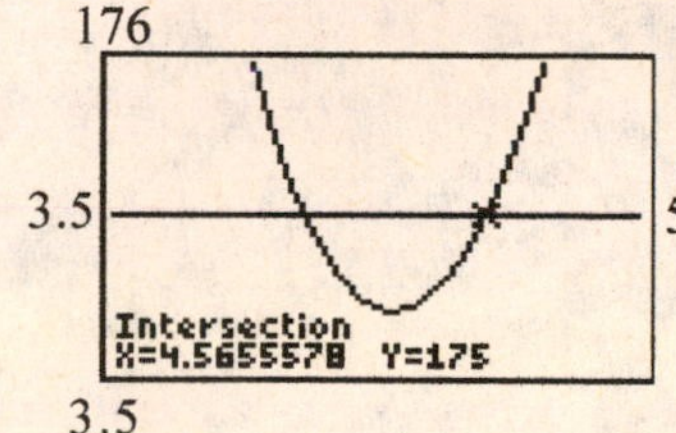

Solutions: $x = 4.047$ cm or $x = 4.566$ cm

Chapter 4

Trigonometric Functions

4.1 Angles

1. $145.6^\circ + 2(360^\circ) = 865.6^\circ$, or
$145.6^\circ - 2(360^\circ) = -574.4^\circ$

5. $60^\circ, 120^\circ, -90^\circ$

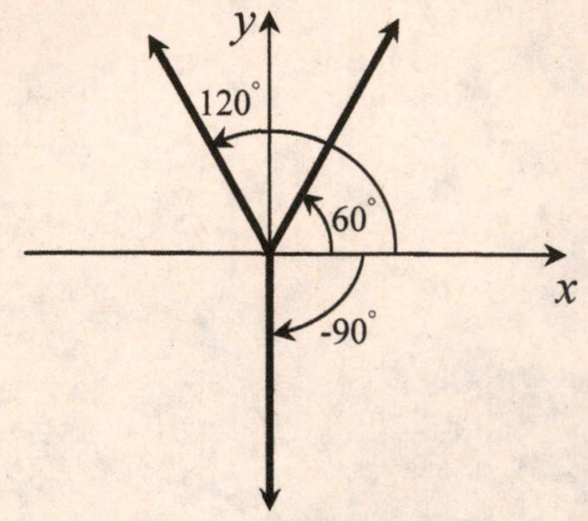

9. positive: $125^\circ + 360^\circ = 485^\circ$
negative: $125^\circ - 360^\circ = -235^\circ$

13. positive: $278.1^\circ + 360^\circ = 638.1^\circ$
negative: $278.1^\circ - 360^\circ = -81.9^\circ$

17. $4.447 \text{ rad} = 4.447 \text{ rad}\left(\dfrac{180^\circ}{\pi \text{ rad}}\right) = 254.79^\circ$

21. With calculator in Degree mode, enter -4.110^r (ANGLE menu #3 on TI-83+)
$-4.110 \text{ rad} = -235.49^\circ$

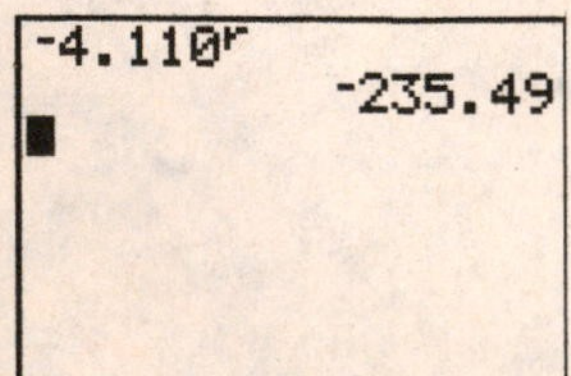

25. With calculator in Radian mode, enter 384.8° (ANGLE menu #1 on TI-83+)

$384.8^\circ = 6.72$ rad

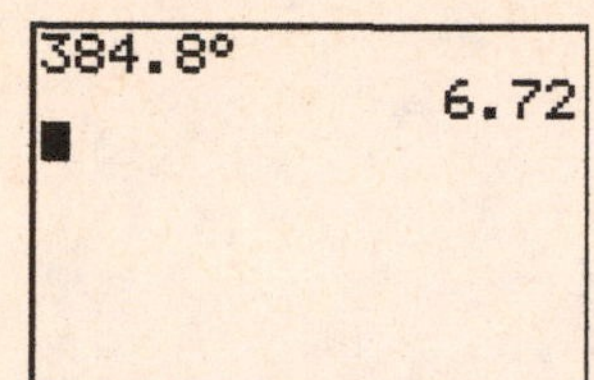

29. $-5.62^\circ = -5^\circ - 0.62^\circ\left(\frac{60'}{1^\circ}\right) = -5^\circ - 37' = -5^\circ 37'$

33. $301^\circ 16' = 301^\circ + 16'\left(\frac{1^\circ}{60'}\right) = 301^\circ + 0.27^\circ = 301.27^\circ$

37. Angle in standard position terminal side passing through $(-3, -5)$.

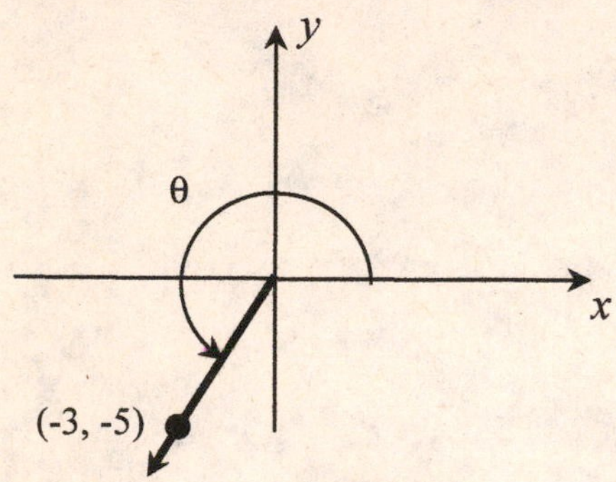

41. Angle in standard position terminal side passing through $(-2, 0)$.

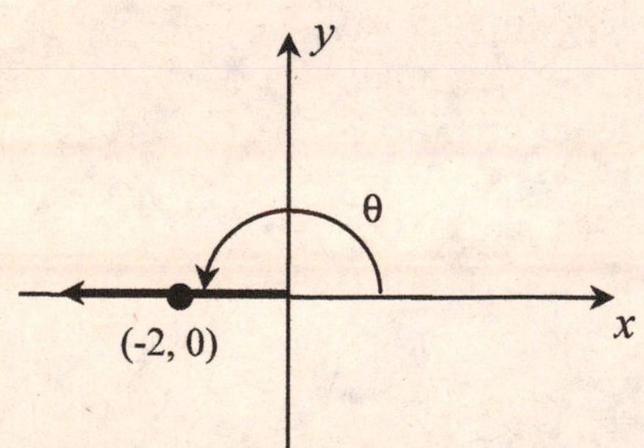

45. 435°, coterminal with 75°, lies in Quadrant I, so it is a first-quadrant angle.

-270° lies between Quandrant III and Quandrant IV, so it is a quadrantal angle.

49. 4 rad = 229.2° lies in Quadrant III, so it is a third-quadrant angle.

$\frac{\pi}{3}$ rad = 60° lies in Quadrant I, so it is a first-quadrant angle.

53. Convert 86.274° to dms format

$$0.274^\circ \; \frac{60'}{1^\circ} = 16.44'$$

$$0.44' \; \frac{60''}{1'} = 26''$$

so $86.274^\circ = 86^\circ 16' 26''$

4.2 Defining the Trigonometric Functions

1. For a point (4, 3) on the terminal side,

$x = 4,\ y = 3$

$r = \sqrt{4^2 + 3^2} = \sqrt{25} = 5$

$\sin\theta = \frac{y}{r} = \frac{3}{5}$ $\qquad \csc\theta = \frac{r}{y} = \frac{5}{3}$

$\cos\theta = \frac{x}{r} = \frac{4}{5}$ $\qquad \sec\theta = \frac{r}{x} = \frac{5}{4}$

$\tan\theta = \frac{y}{x} = \frac{3}{4}$ $\qquad \cot\theta = \frac{x}{y} = \frac{4}{3}$

5. For a point (15, 8) on the terminal side,

$x = 15,\ y = 8$

$r = \sqrt{15^2 + 8^2} = \sqrt{289} = 17$

$\sin\theta = \frac{y}{r} = \frac{8}{17}$ $\qquad \csc\theta = \frac{r}{y} = \frac{17}{8}$

$\cos\theta = \frac{x}{r} = \frac{15}{17}$ $\qquad \sec\theta = \frac{r}{x} = \frac{17}{15}$

$\tan\theta = \frac{y}{x} = \frac{8}{15}$ $\qquad \cot\theta = \frac{x}{y} = \frac{15}{8}$

9. For a point (1.2, 3.5) on the terminal side,

$x = 1.2,\ y = 3.5$

$r = \sqrt{1.3^2 + 3.5^2} = \sqrt{13.69} = 3.7$

$\sin\theta = \frac{y}{r} = \frac{3.5}{3.7} = \frac{35}{37}$ $\qquad \csc\theta = \frac{r}{y} = \frac{37}{35}$

$\cos\theta = \frac{x}{r} = \frac{1.2}{3.7} = \frac{12}{37}$ $\qquad \sec\theta = \frac{r}{x} = \frac{37}{12}$

$\tan\theta = \frac{y}{x} = \frac{3.5}{1.2} = \frac{35}{12}$ $\qquad \cot\theta = \frac{x}{y} = \frac{12}{35}$

13. For a point (7, 7) on the terminal side,

$x = 7,\ y = 7$

$r = \sqrt{7^2 + 7^2} = \sqrt{98} = 7\sqrt{2}$

$\sin\theta = \dfrac{y}{r} = \dfrac{7}{7\sqrt{2}} = \dfrac{1}{\sqrt{2}}$ $\csc\theta = \dfrac{r}{y} = \sqrt{2}$

$\cos\theta = \dfrac{x}{r} = \dfrac{7}{7\sqrt{2}} = \dfrac{1}{\sqrt{2}}$ $\sec\theta = \dfrac{r}{x} = \sqrt{2}$

$\tan\theta = \dfrac{y}{x} = \dfrac{7\sqrt{2}}{7\sqrt{2}} = 1$ $\cot\theta = \dfrac{x}{y} = 1$

17. For a point (0.687, 0.943) on the terminal side,

$x = 0.687,\ y = 0.943$

$r = \sqrt{0.687^2 + 0.943^2} = \sqrt{1.361218} = 1.17$

$\sin\theta = \dfrac{y}{r} = \dfrac{0.943}{1.17} = 0.808$ $\csc\theta = \dfrac{r}{y} = \dfrac{1.17}{0.943} = 1.24$

$\cos\theta = \dfrac{x}{r} = \dfrac{0.687}{1.17} = 0.589$ $\sec\theta = \dfrac{r}{x} = \dfrac{1.17}{0.687} = 1.70$

$\tan\theta = \dfrac{y}{x} = \dfrac{0.943}{0.687} = 1.37$ $\cot\theta = \dfrac{x}{y} = \dfrac{0.687}{0.943} = 0.729$

21. $\tan\theta = \dfrac{y}{x} = \dfrac{2}{1}$

Use $y = 2,\ x = 1$, so

$r = \sqrt{x^2 + y^2} = \sqrt{1^2 + 2^2} = \sqrt{5}$

$\sin\theta = \dfrac{y}{r} = \dfrac{2}{\sqrt{5}}$

$\sec\theta = \dfrac{r}{x} = \dfrac{\sqrt{5}}{1} = \sqrt{5}$

25. $\cot\theta = \dfrac{x}{y} = \dfrac{0.254}{1}$

Use $x = 0.254,\ y = 1$

$r = \sqrt{x^2 + y^2} = \sqrt{0.254^2 + 1^2} = \sqrt{1.064516}$

$\cos\theta = \dfrac{x}{r} = \dfrac{0.254}{\sqrt{1.064516}} = 0.246$

$\tan\theta = \dfrac{y}{x} = \dfrac{1}{0.254} = 3.94$

29. For $(0.3, 0.1)$, $x = 0.3$, $y = 0.1$, $r = \sqrt{0.3^2 + 0.1^2} = \sqrt{0.1}$

$$\tan\theta = \frac{y}{x} = \frac{0.1}{0.3} = \frac{1}{3} \text{ and}$$

$$\sec\theta = \frac{r}{x} = \frac{\sqrt{0.1}}{0.3} = \frac{\sqrt{10}}{3}$$

For $(9, 3)$, $x = 9$, $y = 3$, $r = \sqrt{9^2 + 3^2} = \sqrt{90} = \sqrt{9(10)} = 3\sqrt{10}$

$$\tan\theta = \frac{y}{x} = \frac{3}{9} = \frac{1}{3} \text{ and}$$

$$\sec\theta = \frac{r}{x} = \frac{3\sqrt{10}}{9} = \frac{\sqrt{10}}{3}$$

For $(33, 11)$, $x = 33$, $y = 11$, $r = \sqrt{33^2 + 11} = \sqrt{1210} = \sqrt{121(10)} = 11\sqrt{10}$

$$\tan\theta = \frac{y}{x} = \frac{11}{33} = \frac{1}{3} \text{ and}$$

$$\sec\theta = \frac{r}{x} = \frac{11\sqrt{10}}{33} = \frac{\sqrt{10}}{3}$$

33. We use $x = 0.28, y = 0.96$ to obtain

$$\sin\theta = y = 0.96$$

$$\csc\theta = \frac{1}{y} = \frac{1}{0.96} = 1.04$$

37. If $\tan\theta = \frac{y}{x} = \frac{3}{4}$

Use $y = 3$, $x = 4$, $r = \sqrt{x^2 + y^2} = \sqrt{4^2 + 3^2} = \sqrt{25} = 5$

$$\sin\theta = \frac{y}{r} = \frac{3}{5} \quad \text{and} \quad \cos\theta = \frac{x}{r} = \frac{4}{5}$$

$$\sin^2\theta + \cos^2\theta = \frac{3}{5}^2 + \frac{4}{5}^2$$

$$\sin^2\theta + \cos^2\theta = \frac{9}{25} + \frac{16}{25}$$

$$\sin^2\theta + \cos^2\theta = \frac{25}{25}$$

$$\sin^2\theta + \cos^2\theta = 1$$

41. $\cos\theta = \frac{x}{r}$

$$\frac{1}{\cos\theta} = \frac{r}{x} = \sec\theta$$

4.3 Values of the Trigonometric Functions

1. $\sin\theta = 0.3527$

$\theta = \sin^{-1}(0.3527)$

$\theta = 20.65^\circ$

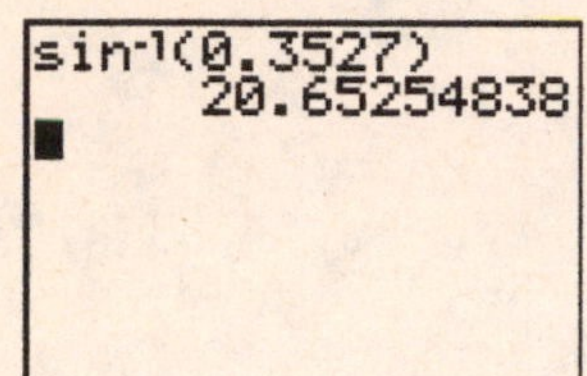

5. $\sin 34.9^\circ = 0.572$

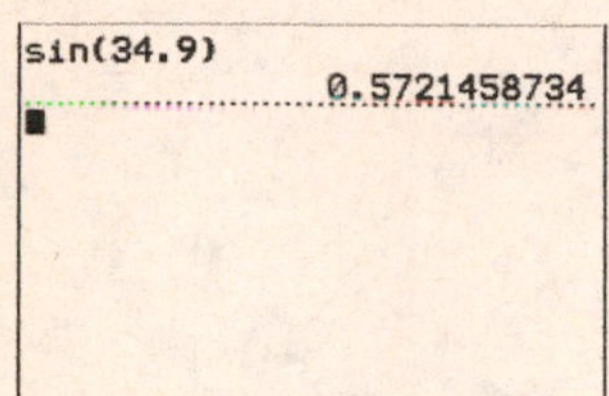

9. $\cos 15.71^\circ = 0.9626$

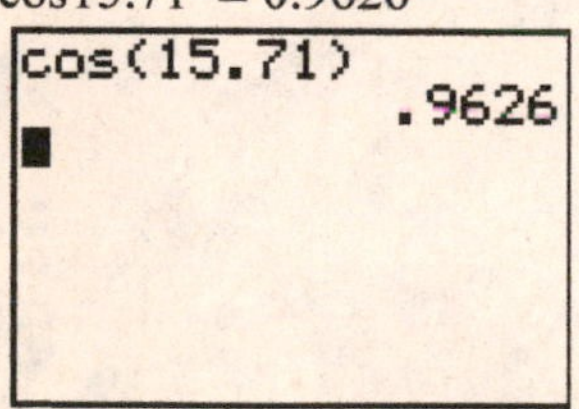

13. $\cot 57.86^\circ = 0.6283$

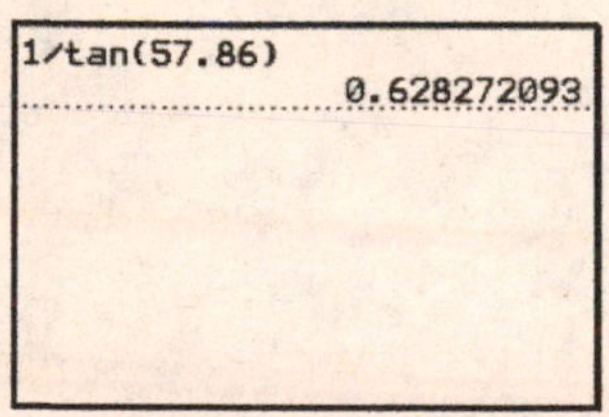

17. $\csc 0.49^\circ = 116.9$

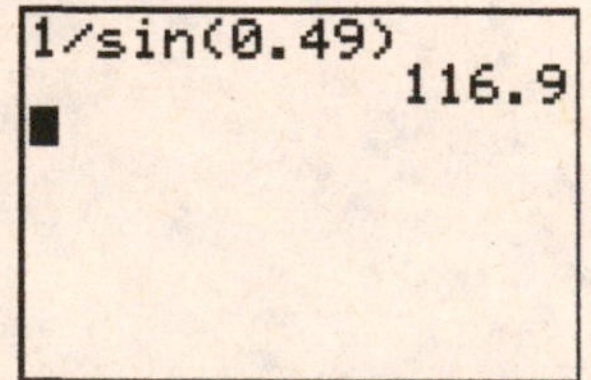

21. $\cos\theta = 0.3261$

$\theta = \cos^{-1}(0.3261)$

$\theta = 70.97^\circ$

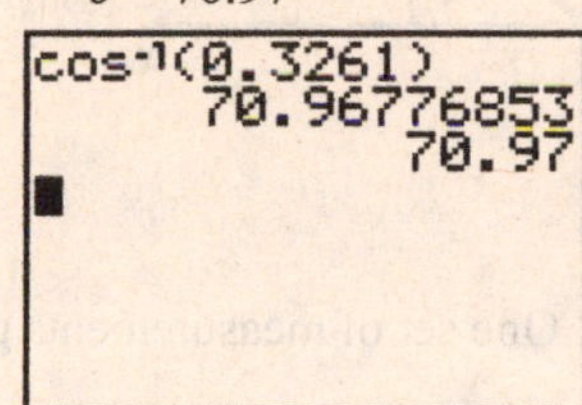

25. $\tan\theta = 0.317$

$\theta = \tan^{-1}(0.317)$

$\theta = 17.6^\circ$

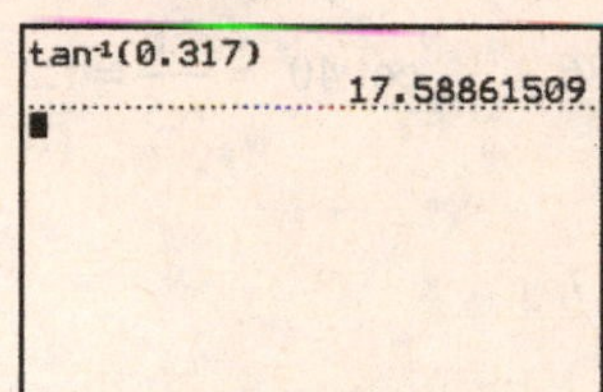

29. $\csc\theta = 2.574$

$\sin\theta = 1/2.574$

$\theta = \sin^{-1}(1/2.574)$

$\theta = 22.86^\circ$

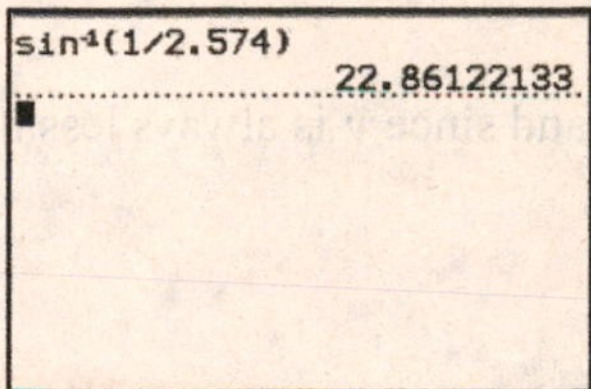

33. $\sec\theta = 0.305$

$\cos\theta = 1/0.305$

This has no solution. There is no angle whose cosine is greater than 1.

37.

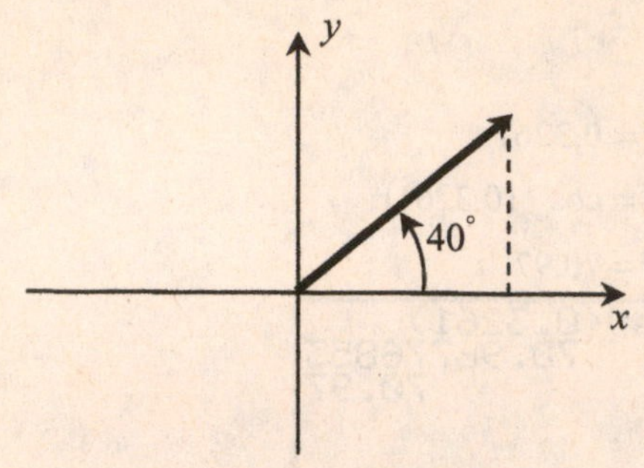

Answers may vary. One set of measurements gives $x = 7.6$ and $y = 6.5$.

$\sin 40^\circ = \frac{6.5}{10} = 0.65 \qquad \csc 40^\circ = \frac{10}{6.5} = 1.5$

$\cos 40^\circ = \frac{7.6}{10} = 0.76 \qquad \sec 40^\circ = \frac{10}{7.6} = 1.3$

$\tan 40^\circ = \frac{6.5}{7.6} = 0.86 \qquad \cot 40^\circ = \frac{7.6}{6.5} = 1.2$

41. $\frac{\sin 43.7^\circ}{\cos 43.7^\circ} = \tan 43.7^\circ$

$0.956 = 0.956$

```
sin(43.7)/cos(43
.7)
      .9556207567
tan(43.7)
      .9556207567
```

45. Since $\sin\theta = \frac{y}{r}$, and since y is always less than or equal to r,

$y \le r$

$\frac{y}{r} \le 1$

$\sin\theta \le 1$

The minimum value of y is 0, so

$\sin\theta \ge 0$

Together,

$0 \le \sin\theta \le 1$

49. Since $\sin\theta = \frac{y}{r}$ and $\cos\theta = \frac{x}{r}$ and $x, y \geq 0$

$$\sin\theta + \cos\theta = \frac{y+x}{r}$$

$$(\sin\theta + \cos\theta)^2 = \left(\frac{y+x}{r}\right)^2$$

$$= \frac{x^2 + 2xy + y^2}{r^2}$$

$$= \frac{x^2 + 2xy + y^2}{x^2 + y^2}$$

$$= 1 + \frac{2xy}{x^2 + y^2} \geq 1$$

implying $\sin\theta + \cos\theta \geq 1$.

53. $\sec\theta = 1.3698$

$\cos\theta = 1.3698^{-1}$

$\theta = \cos^{-1}\left(1.3698^{-1}\right)$

$\tan\theta = \tan\left(\cos^{-1}\left(1.3698^{-1}\right)\right)$

$\tan\theta = 0.93614$

57. $l = a(\sec\theta + \csc\theta)$

$l = 28.0\text{ cm}\left(\sec 34.5^\circ + \csc 34.5^\circ\right)$

$l = 28.0\text{ cm}\left(\frac{1}{\cos 34.5^\circ} + \frac{1}{\sin 34.5^\circ}\right)$

$l = 83.4\text{ cm}$

4.4 The Right Triangle

1.

B

$c = \sqrt{65}$ $a = 7$

A C

$b = 4$

$\sin A = \frac{7}{\sqrt{65}} = 0.868$

$\cos A = \frac{4}{\sqrt{65}} = 0.496$

$\tan A = \frac{7}{4} = 1.75$

$\sin B = \frac{4}{\sqrt{65}} = 0.496$

$\cos B = \frac{4}{\sqrt{65}} = 0.868$

$\tan B = \frac{4}{7} = 0.571$

5. Once the given parts are in place one side remains to be included. Only one length and one direction of the line will close up the space to form a triangle.

9. Given $a = 150,\ c = 345$

$$b = \sqrt{c^2 - a^2}$$
$$b = \sqrt{345^2 - 150^2}$$
$$b = 311$$
$$\sin A = \frac{a}{c}$$
$$A = \sin^{-1}\frac{a}{c}$$
$$A = \sin^{-1}\frac{150}{345}$$
$$A = 25.8^\circ$$
$$\cos B = \frac{a}{c}$$
$$B = \cos^{-1}\left(\frac{a}{c}\right)$$
$$B = \cos^{-1}\left(\frac{150}{345}\right)$$
$$B = 64.2^\circ$$

13. Given $b = 82.0,\ c = 881$

$$a = \sqrt{c^2 - b^2}$$
$$a = \sqrt{881^2 - 82.0^2}$$

$a = 877$ (rounded to three significant digits.)

$$\cos A = \frac{b}{c}$$
$$A = \cos^{-1}\frac{82.0}{881}$$
$$A = 84.7^\circ$$
$$\sin B = \frac{b}{c}$$
$$B = \sin^{-1}\frac{82.0}{881}$$
$$B = 5.34^\circ$$

17. Given $a = 56.73,\ b = 44.09$

$$c = \sqrt{a^2 + b^2}$$
$$c = \sqrt{56.73^2 + 44.09^2}$$
$$c = 71.85$$
$$\tan A = \frac{a}{b}$$
$$A = \tan^{-1}\frac{56.73}{44.09}$$
$$A = 52.15^\circ$$
$$\tan B = \frac{b}{a}$$
$$B = \tan^{-1}\frac{44.09}{56.73}$$
$$B = 37.85^\circ$$

21. Given $B = 74.18^\circ,\ b = 1.849$

$$\sin B = \frac{b}{c}$$
$$c = \frac{b}{\sin B}$$
$$c = \frac{1.849}{\sin 74.18^\circ}$$
$$c = 1.922$$
$$\tan B = \frac{b}{a}$$
$$a = \frac{b}{\tan B}$$
$$a = \frac{1.849}{\tan 74.18^\circ}$$
$$a = 0.5239$$
$$A = 90^\circ - 74.18^\circ$$
$$A = 15.82^\circ$$

25. Given $A = 2.975^\circ,\ b = 14.592$

$$\cos A = \frac{b}{c}$$
$$c = \frac{b}{\cos A}$$
$$c = \frac{14.592}{\cos 2.975^\circ}$$
$$c = 14.61$$
$$\tan A = \frac{a}{b}$$
$$a = b\tan A$$
$$a = 14.592\tan 2.975^\circ$$
$$a = 0.7584$$
$$B = 90^\circ - 2.975^\circ$$
$$B = 87.025^\circ$$

29. The given information does not determine a unique triangle. An infinte number of solutions exist for b, c, A, and B.

33. $\cos A = \dfrac{0.6673}{0.8742}$

$$A = \cos^{-1}\frac{0.6673}{0.8742}$$

$$A = 40.24^\circ$$

37.

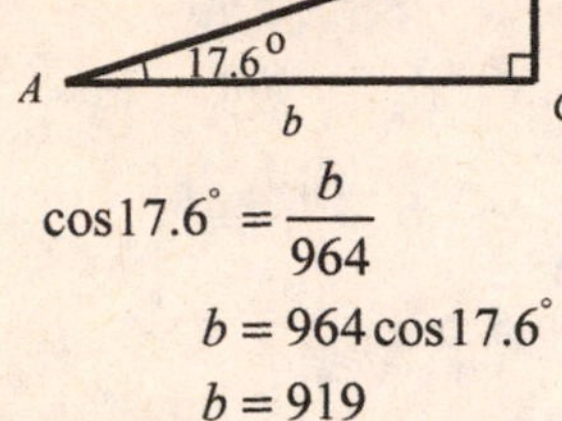

$$\cos 17.6^\circ = \frac{b}{964}$$

$$b = 964\cos 17.6^\circ$$

$$b = 919$$

41. If h is the height, then

$$350 = h\cos 78.85°$$

$$h = \frac{350}{\cos 78.85°}$$

$$h = 1810 \text{ ft.}$$

4.5 Applications of Right Triangles

1. $\dfrac{1850 \text{ ft}}{d} = \cos 27.9^\circ$

$$d = \frac{1850 \text{ ft}}{\cos 27.9^\circ}$$

$$d = 2090 \text{ ft}$$

5.

θ
12.0 m
85.0 m

$$\tan\theta = \frac{12.0}{85.0} = 0.1411765$$

$$\theta = \tan^{-1}(0.1411765)$$

$$\theta = 8.04°$$

9. $Z = \sqrt{12.0^2 + 15.0^2} = 19.2$

$$\tan\phi = \frac{15.0}{12.0}$$

$$\phi = \tan^{-1}\left(\frac{15.0}{12.0}\right)$$

$$\phi = 51.34^\circ$$

13.

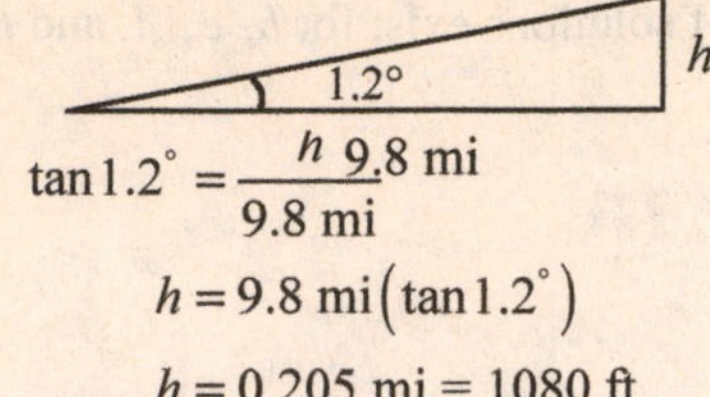

$\tan 1.2^\circ = \dfrac{h\ 9.8\text{ mi}}{9.8\text{ mi}}$

$h = 9.8\text{ mi}\left(\tan 1.2^\circ\right)$

$h = 0.205\text{ mi} = 1080\text{ ft}$

17.

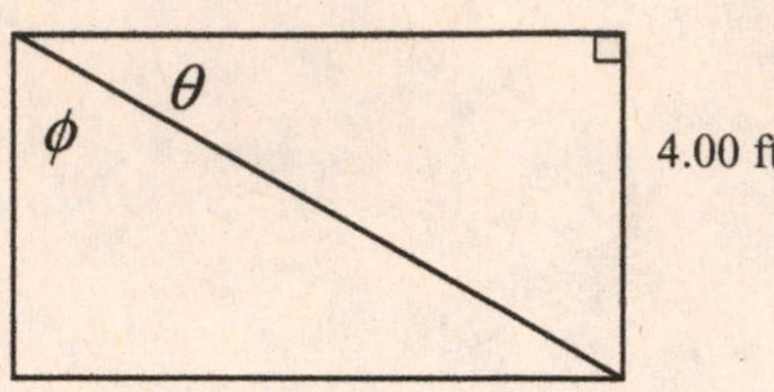

8.00 ft

$\tan\phi = \dfrac{8.00}{4.00}$

$\phi = \tan^{-1}\left(\dfrac{8.00}{4.00}\right) = 63.4^\circ$

$\tan\theta = \dfrac{4.00}{8.00}$

$\theta = \tan^{-1}\left(\dfrac{4.00}{8.00}\right) = 26.6^\circ$

The angles are 26.6°, 63.4° between the two pieces.

21.

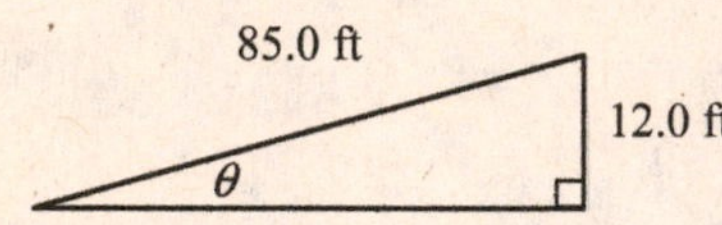

$\sin\theta = \dfrac{12.0\text{ ft}}{85.0\text{ ft}}$

$\theta = \sin^{-1}\dfrac{12.0}{85.0}$

$\theta = 8.12^\circ$

The octagon has eight sides, each side has $\dfrac{360^\circ}{8\text{ sides}}$ or $\dfrac{45^\circ}{\text{side}}$,

where each piece makes two right triangles, of angle $\dfrac{45^\circ}{2}$ or 22.5°

with the hypotenuse being the radius of the octagon, and

opposite side being half the octagon side length.

$\sin 22.5^\circ = \dfrac{s/2}{0.6\text{ m}}$

$s = 2(\sin 22.5^\circ)(0.6\text{ m})$

$s = 0.45922\text{ m}$

$p = 8s = 3.67\text{ m}$

25.

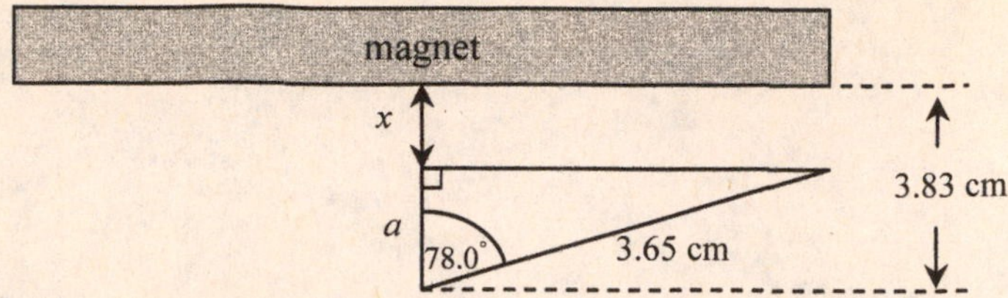

$$\cos 78.0^\circ = \frac{a}{3.65\text{ cm}}$$
$$a = 3.65\text{ cm}\left(\cos 78.0^\circ\right)$$
$$a = 0.759\text{ cm}$$
$$x = 3.83\text{ cm} - 0.759\text{ cm}$$
$$x = 3.07\text{ cm}$$

29.

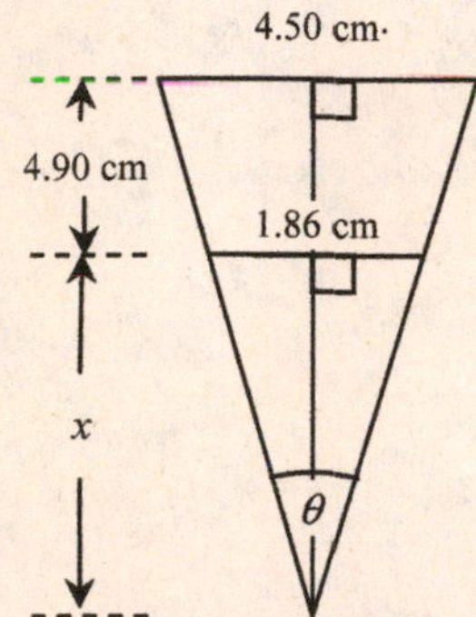

Let x be the vertical distance from the bottom of the taper to the vertex of angle θ. These are similar triangles.

$$\frac{x}{1.86} = \frac{x + 4.90}{4.50}$$
$$4.50x = 1.86(x + 4.90)$$
$$4.50x = 1.86x + 9.114$$
$$2.64x = 9.114$$
$$x = 3.45\text{ cm}$$
$$\tan\tfrac{\theta}{2} = \frac{1.86/2}{x}$$
$$\tfrac{\theta}{2} = \tan^{-1}\frac{0.93}{3.45227}$$
$$\tfrac{\theta}{2} = 15.077^\circ$$
$$\theta = 30.2^\circ$$

33. The angles α and β satisfy

$$\tan\alpha = \frac{65.3}{224}; \tan\beta = \frac{65.3}{302}$$

and so

$$\alpha - \beta = \tan^{-1}\frac{65.3}{224} - \tan^{-1}\frac{65.3}{302}$$

$$= 16.25234669° - 12.20095859°$$

$$= 4.05°$$

37.

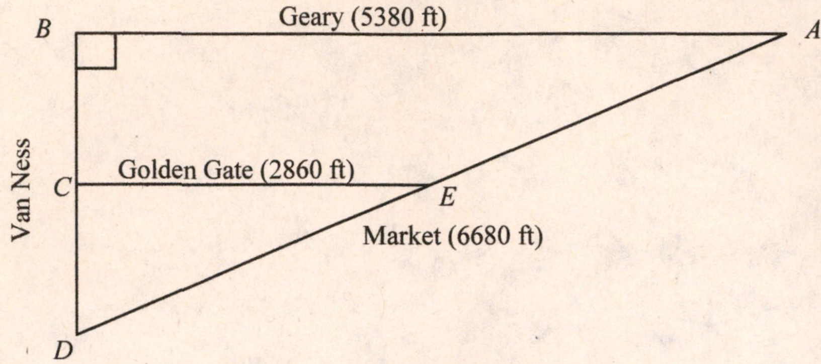

$$BD = \sqrt{6680^2 - 5380^2} = 4000$$

Triangles ΔABD and ΔECD are similar.

$$\frac{2860}{5380} = \frac{CD}{BD}$$

$$CD = 4000 \times \frac{2860}{5380} = 2126.387 \text{ ft}$$

Intersections C and D are 2130 ft apart.

41.

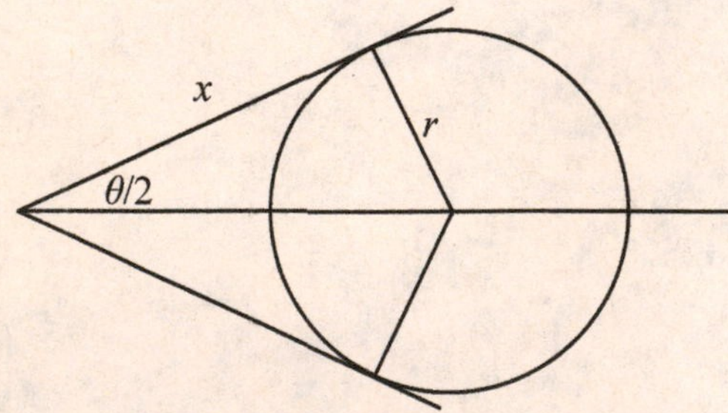

$$\tan\frac{\theta}{2} = \frac{r}{x}$$

$$r = x\tan\frac{\theta}{2}$$

And so the diameter d is

$$d = 2r = 2x\tan\frac{\theta}{2}$$

Review Exercises

1. This is false. A standard position angle of 205° is a third-quadrant angle.

5. This is true.

9. positive: $-217.5^\circ + 360^\circ = 142.5^\circ$
negative: $-217.5^\circ - 360^\circ = -577.5^\circ$

13. $-83^\circ 21' = -\left(83^\circ + 21'\left(\frac{1^\circ}{60'}\right)\right) = -83^\circ - 0.35^\circ = -83.35^\circ$

17. $749.75^\circ = 749^\circ + 0.75^\circ\left(\frac{60'}{1^\circ}\right) = 749^\circ 45'$

21. $x = 48,\ \ y = 48$

$r = \sqrt{x^2 + y^2} = \sqrt{48^2 + 48^2} = \sqrt{2(48^2)} = 48\sqrt{2}$

$\sin\theta = \frac{y}{r} = \frac{48}{48\sqrt{2}} = \frac{1}{\sqrt{2}}$ $\qquad \csc\theta = \frac{r}{y} = \sqrt{2}$

$\cos\theta = \frac{x}{r} = \frac{48}{48\sqrt{2}} = \frac{1}{\sqrt{2}}$ $\qquad \sec\theta = \frac{r}{x} = \sqrt{2}$

$\tan\theta = \frac{y}{x} = \frac{48}{48} = 1$ $\qquad \cot\theta = \frac{x}{y} = 1$

25. $\tan\theta = \frac{y}{x} = \frac{2}{1}$, so $y = 2,\ \ x = 1$

$r = \sqrt{x^2 + y^2} = \sqrt{2^2 + 1^2} = \sqrt{4+1} = \sqrt{5}$

$\cos\theta = \frac{x}{r} = \frac{1}{\sqrt{5}} = 0.447$

$\csc\theta = \frac{r}{y} = \frac{\sqrt{5}}{2} = 1.12$

29. $\sin 72.1^\circ = 0.952594403 = 0.952$

33. $\sec 36.2^\circ = \frac{1}{\cos 36.2^\circ} = 1.2392183 = 1.24$

37. $\cos\theta = 0.850$

$\theta = \cos^{-1}(0.850)$

$\theta = 31.8^\circ$

41. $\csc\theta = \dfrac{1}{\sin\theta} = 4.713$

$\sin\theta = \dfrac{1}{4.713}$

$\theta = \sin^{-1}\left(\dfrac{1}{4.713}\right)$

$\theta = 12.25^\circ$

45. $\cot\theta = \dfrac{1}{\tan\theta} = 7.117$

$\tan\theta = \dfrac{1}{7.117}$

$\theta = \tan^{-1}\left(\dfrac{1}{7.117}\right)$

$\theta = 7.998^\circ$

49. $\sin\theta = \dfrac{2}{5} = \dfrac{y}{r}$

$y = 2, r = 5, x = \sqrt{r^2 - y^2} = \sqrt{25-4} = \sqrt{21}$

The point where the terminal side intersects

is $\dfrac{x}{r}, \dfrac{y}{r}$

or $\dfrac{\sqrt{21}}{5}, \dfrac{2}{5}$.

53. Given $a = 81.0$, $b = 64.5$

$c = \sqrt{a^2 + b^2}$

$c = \sqrt{81.0^2 + 64.5^2}$

$c = 104$

$\tan A = \dfrac{a}{b}$

$A = \tan^{-1}\dfrac{81.0}{64.5}$

$A = 51.5^\circ$

$\tan B = \dfrac{b}{a}$

$B = \tan^{-1}\dfrac{64.5}{81.0}$

$B = 38.5^\circ$

57. Given $b = 6.508$, $c = 7.642$

$a = \sqrt{c^2 - b^2}$

$a = \sqrt{7.642^2 - 6.508^2}$

$a = 4.006$

$\cos A = \dfrac{b}{c}$

$A = \cos^{-1}\dfrac{6.508}{7.642}$

$A = 31.61^\circ$

$\sin B = \dfrac{b}{c}$

$B = \sin^{-1}\dfrac{6.508}{7.642}$

$B = 58.39^\circ$

61.

$\sin 25^\circ = \dfrac{d}{12}$

$d = 12\sin 25^\circ$

$d = 5.0714$

$90^\circ - 25^\circ = 65^\circ$

$\sin 65^\circ = \dfrac{x}{d}$

$x = d\sin 65^\circ$

$x = 5.0714\sin 65^\circ$

$x = 4.6$

65.

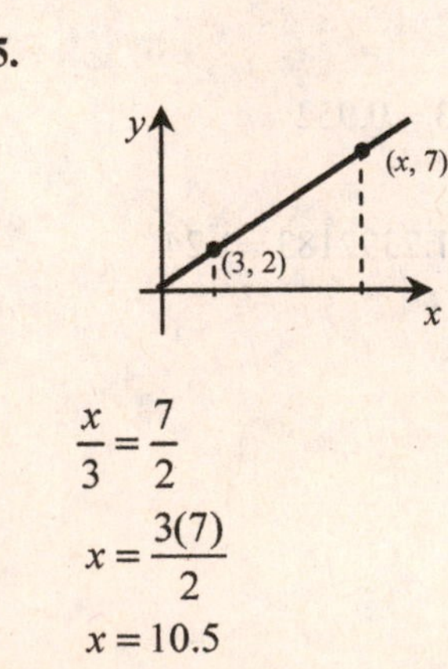

$\dfrac{x}{3} = \dfrac{7}{2}$

$x = \dfrac{3(7)}{2}$

$x = 10.5$

69.

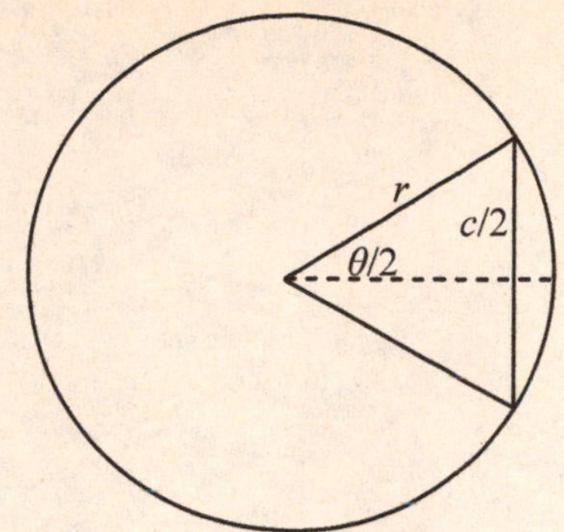

$$\sin\frac{\theta}{2}=\frac{c/2}{r}$$

$$\frac{c}{2}=r\sin\frac{\theta}{2}$$

$$c=2r\sin\frac{\theta}{2}$$

73.

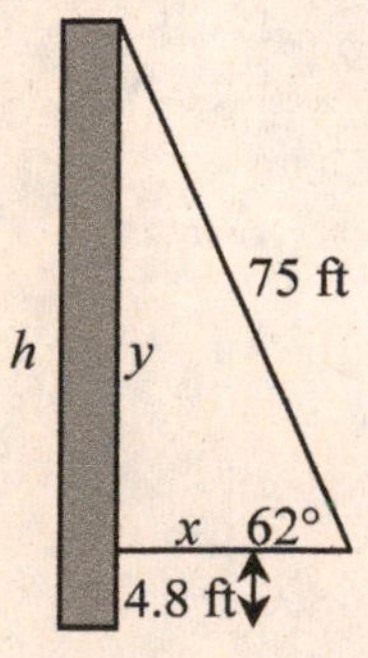

$$y=(75\text{ ft})(\sin 62^\circ)=66.2\text{ ft}$$

$$h=y+4.8\text{ ft}=71\text{ ft}$$

$$x=(75\text{ ft})(\cos 62^\circ)=35.2\text{ ft}$$

The ladder reaches 71 feet high up the building and the fire truck must be 35.2 feet away from the building.

77. $\tan\theta=\frac{v^2}{gr}$

$$\theta=\tan^{-1}\frac{v^2}{gr}$$

$$\theta=\tan^{-1}\frac{(80.7\text{ ft/s})^2}{(950\text{ ft})(32.2\text{ ft/s}^2)}$$

$$\theta=12.0^\circ$$

81.

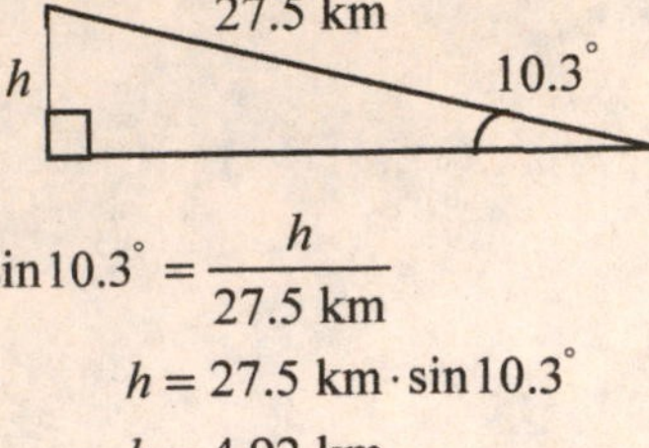

$$\sin 10.3^\circ = \frac{h}{27.5\text{ km}}$$
$$h = 27.5\text{ km}\cdot\sin 10.3^\circ$$
$$h = 4.92\text{ km}$$

85.

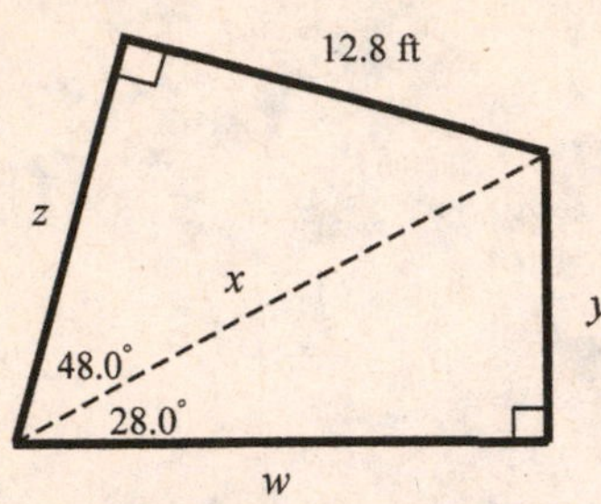

$$76.0^\circ - 28.0^\circ = 48.0^\circ$$
$$\tan 48.0^\circ = \frac{12.8}{z}$$
$$z = \frac{12.8}{\tan 48.0^\circ}$$
$$z = 11.525\text{ ft}$$
$$\sin 48.0^\circ = \frac{12.8}{x}$$
$$x = \frac{12.8}{\sin 48.0^\circ}$$
$$x = 17.224\text{ ft}$$
$$\sin 28.0^\circ = \frac{y}{x}$$
$$y = x\sin 28.0^\circ$$
$$y = 17.224\text{ ft}\cdot\sin 28.0^\circ$$
$$y = 8.086\text{ ft}$$
$$\cos 28.0^\circ = \frac{w}{x}$$
$$w = x\cos 28.0^\circ$$
$$w = 17.224\text{ ft}\cdot\cos 28.0^\circ$$
$$w = 15.208\text{ ft}$$
$$A = \frac{1}{2}(12.8\text{ ft})z + \frac{1}{2}wy$$
$$A = \frac{1}{2}(12.8\text{ ft})(11.525\text{ ft}) + \frac{1}{2}(15.208\text{ ft})(8.086\text{ ft})$$
$$A = 135\text{ ft}^2$$

89.

source
65 km
h
0.030°
city

$$\sin 0.030^\circ = \frac{h}{65\text{ km}}$$
$$h = (65\text{ km})\sin 0.030^\circ$$
$$h = 0.034033918\text{ km}\left(\frac{1000\text{ m}}{1\text{ km}}\right)$$
$$h = 34\text{ m}$$

93. $$\sin 21.8^\circ = \frac{d}{14.2\text{ in}}$$
$$d = (14.2\text{ in})\sin 21.8^\circ$$
$$d = 5.2734\text{ in}$$
$$\sin 31.0^\circ = \frac{d}{x}$$
$$x = \frac{d}{\sin 31.0^\circ}$$
$$x = \frac{5.2734\text{ in}}{\sin 31.0^\circ}$$
$$x = 10.2\text{ in}$$

97. $$\sin\frac{0.00200^\circ}{2} = \frac{d/2}{52\,500\text{ km}}$$
$$d = 2(52500\text{ km})(\sin 0.00100^\circ)$$
$$d = 1.83\text{ km}$$

101.

Glider
y
Observer
42.0°
d
25.0°
x
375 m
Shore

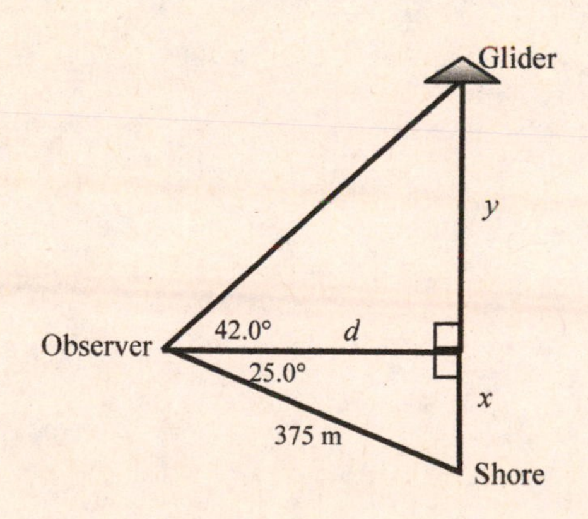

$$\cos 25.0^\circ = \frac{d}{375}$$
$$d = 375\cos 25.0^\circ$$
$$d = 339.87\text{ m}$$

$$\sin 25.0^\circ = \frac{x}{375}$$
$$x = 375\sin 25.0^\circ$$
$$x = 158.48\text{ m}$$
$$\tan 42.0^\circ = \frac{y}{d}$$
$$y = (339.87\text{ m})\tan 42.0^\circ$$
$$x = 306.02\text{ m}$$
$$x + y = 306.02\text{ m} + 158.48\text{ m}$$
$$x + y = 464\text{ m}$$

105.

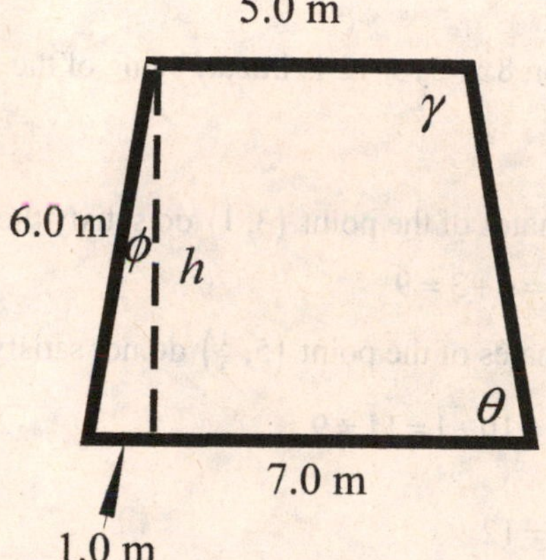

base angle
$$\cos\theta = \frac{1.0}{6.0}$$
$$\theta = \cos^{-1}\tfrac{1}{6} = 80.4^\circ$$
upper angle
$$\sin\phi = \frac{1.0}{6.0}$$
$$\phi = \sin^{-1}\tfrac{1}{6} = 9.59^\circ$$
$$\gamma = 90^\circ + 9.59^\circ$$
$$\gamma = 99.6^\circ$$
$$\tan\phi = \frac{1.0}{h}$$
$$h = \frac{1.0}{\tan 9.59^\circ}$$
$$h = 5.9161\text{ m}$$
$$A = \tfrac{1}{2}h(b_1 + b_2)$$
$$A = \tfrac{1}{2}(5.9161\text{ m})(7.0\text{ m} + 5.0\text{ m})$$
$$A = 35\text{ m}^2$$

Chapter 5

Systems of Linear Equations; Determinants

5.1 Linear Equations and Graphs of Linear Functions

1. $2x-y-4=0$

For $x=-3$, the solution is

$2(-3)-y-4=0$

$-6-y-4=0$

$-10=y$

The point $(-3,-10)$ must lie on the graph of the line.

5. The equation $8x-3y=12$ is linear, being of the form $ax+by=c$ where $a=8, b=-3$ and $c=12$.

9. $2x+3y=9$

The coordinates of the point $(3, 1)$ do satisfy the equation since

$2(3)+3(1)=6+3=9$

The coordinates of the point $\left(5, \frac{1}{3}\right)$ do not satisfy the equation since

$2(5)+3\left(\frac{1}{3}\right)=10+1=11\neq 9.$

13. $3x-2y=12$

If $x=2$,

$3(2)-2y=12$

$2y=6-12$

$2y=-6$

$y=-3$

If $x=-3$,

$3(-3)-2y=12$

$2y=-9-12$

$2y=-21$

$y=-\frac{21}{2}$

17. $m=\dfrac{8-0}{3-1}=\dfrac{8}{2}=4$

21. $m=\dfrac{(-5)-(-3)}{(-2)-5}=\dfrac{-2}{-7}=\dfrac{2}{7}$

25. The line has slope 2 and passes through $(0,-1)$.
Since the slope is 2, the line rises 2 units for each unit of run going from left to right.
Thus, $(1,1)$ is another point on the line.

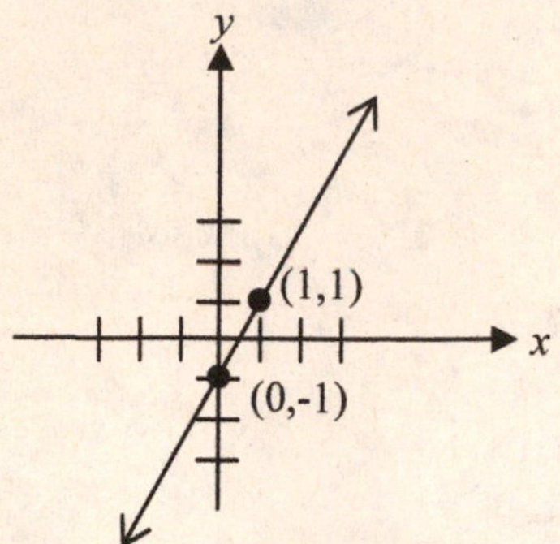

29. The line has slope $\frac{1}{2}$ and passes through $(0,0)$.
Since the slope is $\frac{1}{2}$, the line rises 1 unit for each 2 units of run going from left to right.
Thus, $(2,1)$ is another point on the line.

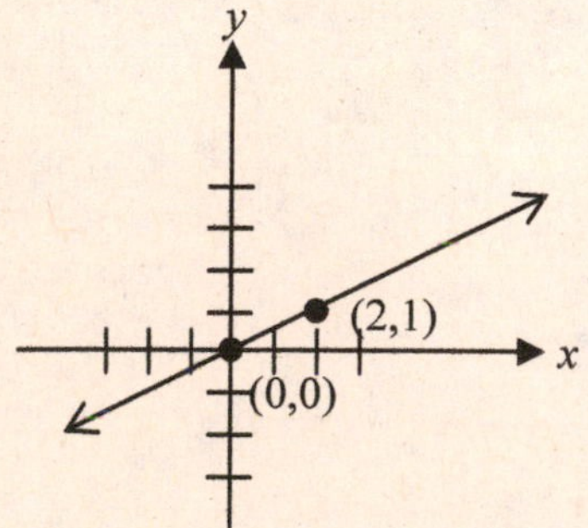

33. $y=-2x+1$, compare to $y=mx+b$
$m=-2$, $b=1$
Plot the y-intercept point $(0, 1)$. Since the slope is $-2/1$, from this point go right 1 unit and down 2 units, and plot a second point. Sketch a line passing through these two points.

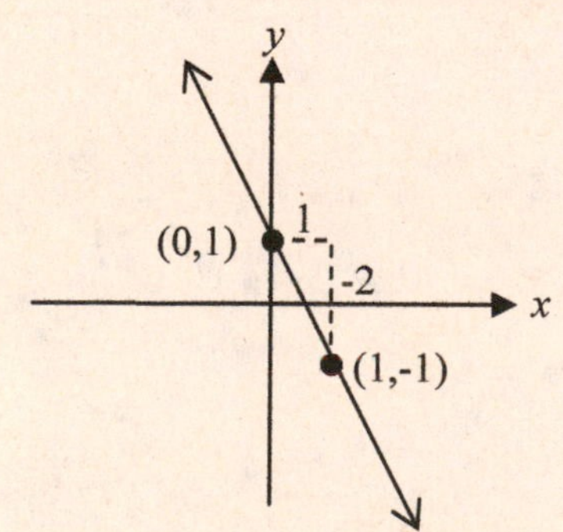

37. $5x-2y=40$

$2y=5x-40$

$y=\frac{5}{2}x-20$, compare to $y=mx+b$

$m=\frac{5}{2}$, $b=-20$

Plot the y-intercept point $(0,-20)$. Since the slope is 5/2, from this point go right 2 units and up 5 units, and plot a second point. Sketch a line passing through these two points.

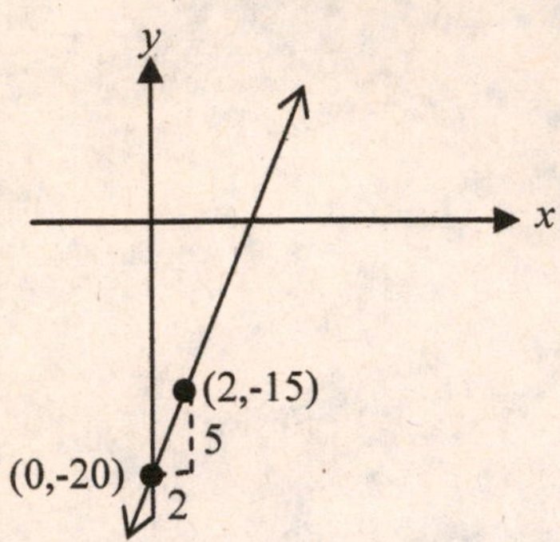

41. $x+2y=4$

For y-int, set $x=0$

$0+2y=4$

$2y=4$

$y=2$ y-int is (0, 2)

For x-int, set $y=0$

$x+0=4$

$x=4$ x-int is (4, 0)

Plot the x-intercept point $(4, 0)$ and the y-intercept point $(0, 2)$.
Sketch a line passing through these two points.
A third point is found as a check.

Let $x=2$

$2+2y=4$

$2y=2$

$y=1$. Therefore the point $(2, 1)$ should lie on the line.

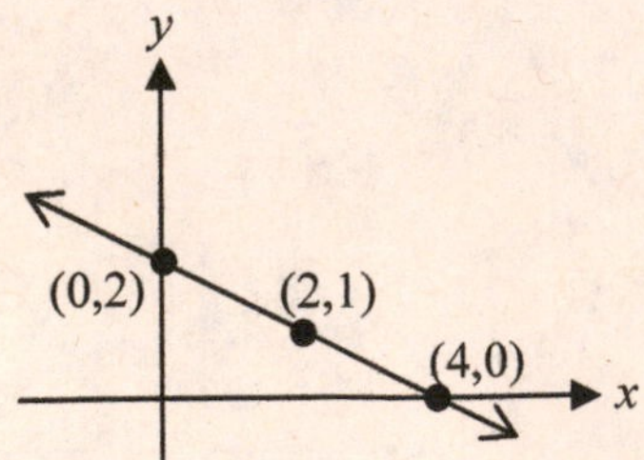

45. $y = 3x + 6$

For y-int, set $x = 0$

$y = 0 + 6$

$y = 6$ y-int is (0, 6)

For x-int, set $y = 0$

$0 = 3x + 6$

$x = -2$ x-int is $(-2, 0)$

Plot the x-intercept point $(-2, 0)$ and the y-intercept point $(0, 6)$.

Sketch a line passing through these two points.

A third point is found as a check.

Let $x = -1$

$y = 3(-1) + 6$

$y = 3$

Therefore the point $(-1, 3)$ should lie on the line.

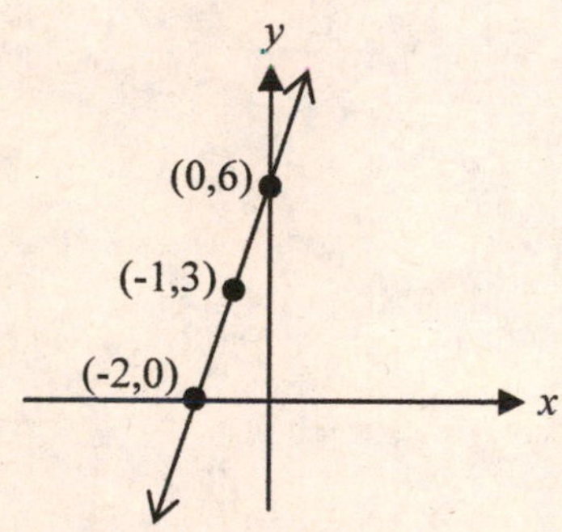

49. Using the LinReg(ax+b) feature of the TI-84, we find the regression line is $y = 32.500x - 84.714$. Substituting $x = 4.5$ yields $y = 61.536$ which is rounded to a prediction of $s = 62$ kN/mm.

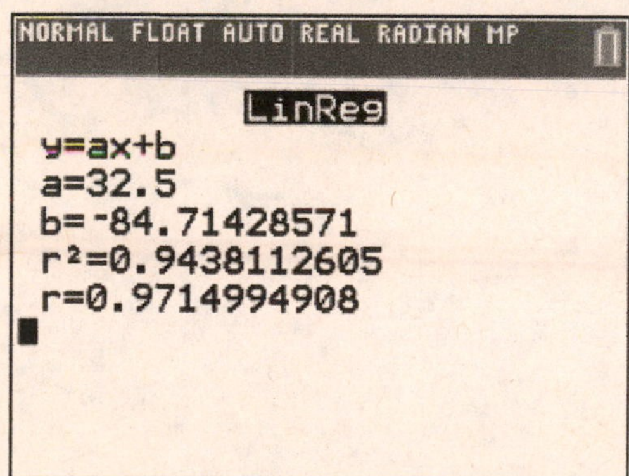

53. Given the line $\frac{x}{a}+\frac{y}{b}=1,$

For y-int, set $x=0$

$$\frac{0}{a}+\frac{y}{b}=1$$

$$\frac{y}{b}=1$$

$y=b$ The y-int is $(0, b)$

For x-int, set $y=0$

$$\frac{x}{a}+\frac{0}{b}=1$$

$$\frac{x}{a}=1$$

$x=a$ The x-int is $(a, 0)$

57. For the line $d=0.2l+1.2$, we have a slope of 0.2 and an intercept of $(0,1.2)$. When $l=10$ we have $d=0.2\cdot 10+1.2=3.2$ and so $(10,3.2)$ is another point on the line.

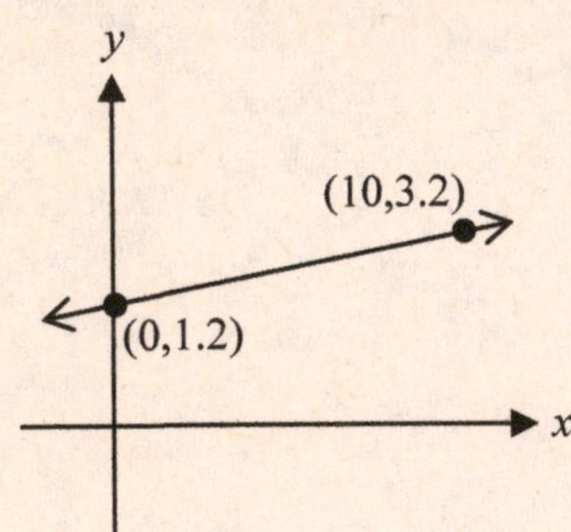

61. Letting t be time in minutes and h be altitude in kilometers, we are told $h=2.5$ at $t=0$ and so $(0,2.5)$ is the y – intercept.
After 10 minutes, the plane has descended $150\cdot 10=1500\text{m}=1.5$ km.
Thus, when $t=10, h=2.5-1.5=1.0$ and so $(10,1.0)$ is another point on the line.

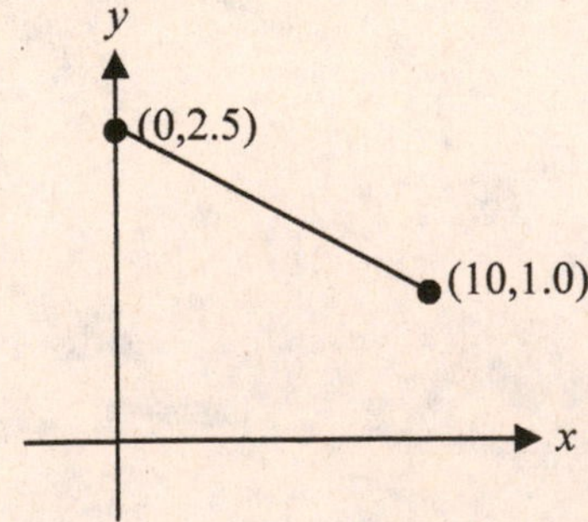

5.2 Systems of Equations and Graphical Solutions

1. In Example 2, we check to see if $F_1=20$, $F_2=40$ is a solution to

$$2F_1+4F_2=200$$
$$F_2=2F_1$$

$$2(20)+4(40)\stackrel{?}{=}200 \qquad 40\stackrel{?}{=}2(20)$$
$$200=200 \qquad 40=40$$

and we do in fact have a solution to the system of equations.

5. $x-y=5$

$2x+y=7$

If the values $x=4$ and $y=-1$ satisfy both equations, they are a solution.

$$4-(-1)=4+1=5$$
$$2(4)+(-1)=8-1=7$$

Therefore the given values are a solution.

9. $2x-5y=0$

$4x+10y=4$

If the values $x=\frac{1}{2}$ and $y=-\frac{1}{5}$ satisfy both equations, they are a solution.

$$2\left(\tfrac{1}{2}\right)-5\left(-\tfrac{1}{5}\right)=1+1=2\neq 0$$
$$4\left(\tfrac{1}{2}\right)+10\left(-\tfrac{1}{5}\right)=2-2=0\neq 4$$

Neither equation is satisfied, therefore the given values are not a solution.

13. $y=-x+4$ and $y=x-2$

The slope of the first line is -1, and the y-intercept is 4.

The slope of the second line is 1 and the y- intercept is -2.

From the graph, the point of intersection is approximately at $(3, 1)$.

Therefore, the solution of the system of equations is

$x=3.0,\ y=1.0.$

Check:

$$y=-x+4 \qquad y=x-2$$
$$1.0=-3.0+4 \qquad 1.0=3.0-2$$
$$1.0=1.0 \qquad 1.0=1.0$$

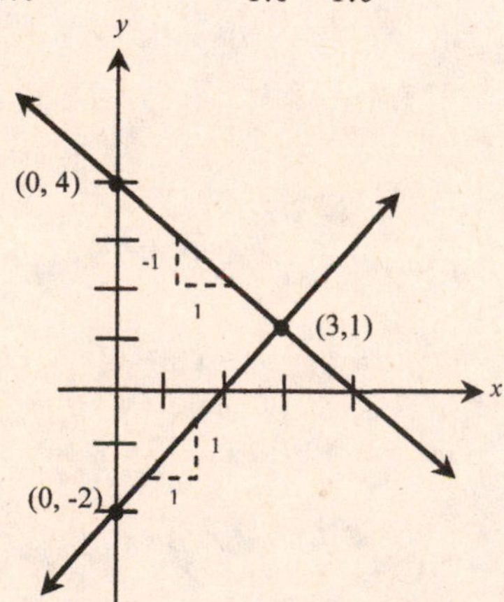

17. For line $3x+2y=6$

Let $x=0$ to find the y-int

$2y=6$

$y=3$ y-int is $(0, 3)$

Let $y=0$ to find the x-int

$3x=6$

$x=2$ x-int is $(2, 0)$

Let $x=4$ to find a third point

$3(4)+2y=6$

$2y=-6$

$y=-3$ A third point is $(4,-3)$

For line $x-3y=3$

Let $x=0$ to find the y-int

$-3y=3$

$y=-1$ y-int is $(0,-1)$

Let $y=0$ to find the x-int

$x=3$ x-int is $(3, 0)$

Let $x=1$ to find a third point

$1-3y=3$

$-3y=2$

$y=-\frac{2}{3}$ A third point is $\left(1,-\frac{2}{3}\right)$

From the graph the solution is approximately $(2.2,-0.3)$

$x=2.2,\ y=-0.3$

Checking both equations,

$3x+2y=6$	$x-3y=3$
$3(2.2)+2(-0.3)=6$	$2.2-3(-0.3)=3$
$6.0=6$	$3.1\approx 3$

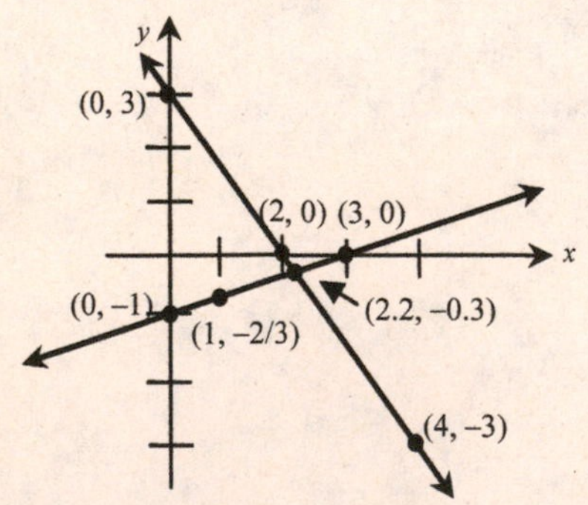

21. $s-4t=8$ and $2s=t+4$

$s=4t+8$ $\quad s=\frac{1}{2}t+2$

The slope of the first line is 4, and the s-intercept is 8.

The slope of the second line is 1/2 and the s-intercept is 2.

From the graph, the point of intersection is $(-1.7, 1.1)$.

Therefore, the solution of the system of equations is

$t = -1.7,\ s = 1.1.$

Check:

$s = 4t + 8 \qquad s = \frac{1}{2}t + 2$

$1.1 = 4(-1.7) + 8 \qquad 1.1 = \frac{1}{2}(-1.7) + 2$

$1.1 \approx 1.2 \qquad 1.1 \approx 1.2$

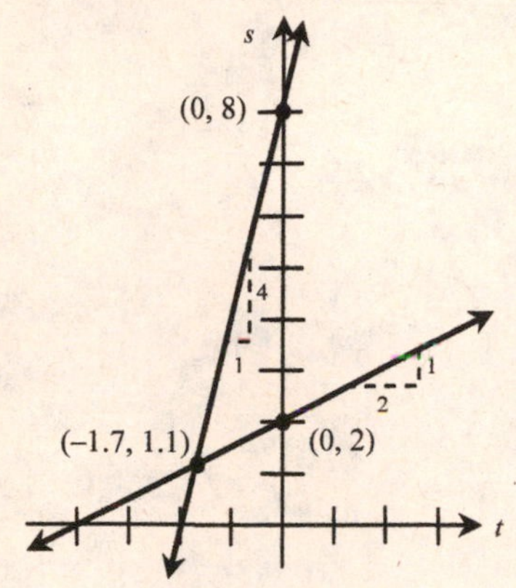

25. $x - 4y = 6$ and $2y = x + 4$

$4y = x - 6 \qquad y = \frac{1}{2}x + 2$

$y = \frac{1}{4}x - \frac{3}{2}$

The slope of the first line is 1/4, and the y-intercept is $-3/2$.

The slope of the second line is 1/2 and the y-intercept is 2.

From the graph, the point of intersection is $(-14, -5)$.

Therefore, the solution of the system of equations is

$x = -14.0,\ y = -5.0$

Check:

$x - 4y = 6 \qquad 2y = x + 4$

$-14 - 4(-5) = 6 \qquad 2(-5) = -14 + 4$

$6 = 6 \qquad -10 = -10$

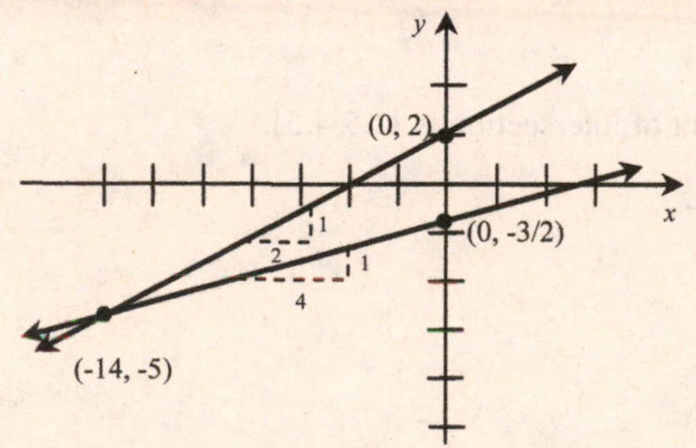

29. $x = 4y + 2$ and $3y = 2x + 3$

$y = \frac{x-2}{4}$ $\quad$ $y = \frac{2x+3}{3}$

On a graphing calculator let

$y_1 = \frac{x-2}{4}$ and $y_2 = \frac{2x+3}{3}$

Using the intersect feature, the point of intersection is $(-3.6, -1.4)$, and the solution of the system of equations is

$x = -3.600$

$y = -1.400$

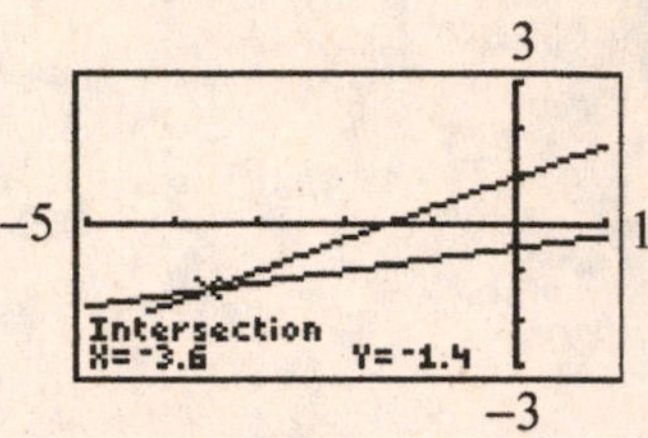

33. $x - 5y = 10$ and $2x - 10y = 20$

$y = \frac{x-10}{5}$ $\quad$ $y = \frac{x-10}{5}$

On a graphing calculator let

$y_1 = \frac{x-10}{5}$ and $y_2 = \frac{x-10}{5}$

From the graph the lines are the same. The system is dependent.

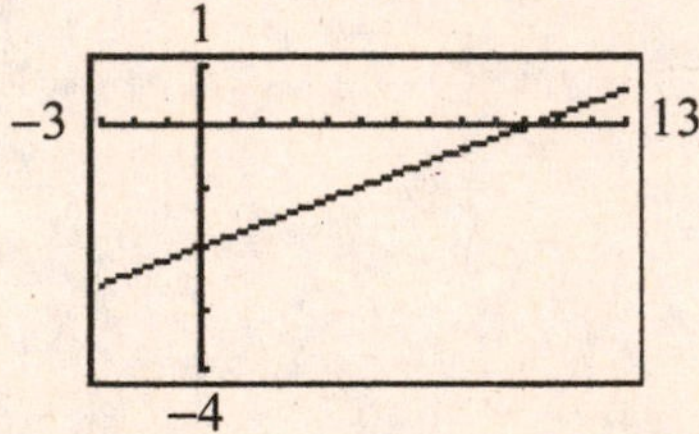

37. $5x = y + 3$ and $4x = 2y - 3$

$y = 5x - 3$ $\quad$ $y = 2x + \frac{3}{2}$

On a graphing calculator using the intersect feature, the point of intersection is $(1.5, 4.5)$, and the solution to the system of equations is

$x = 1.500$

$y = 4.500$

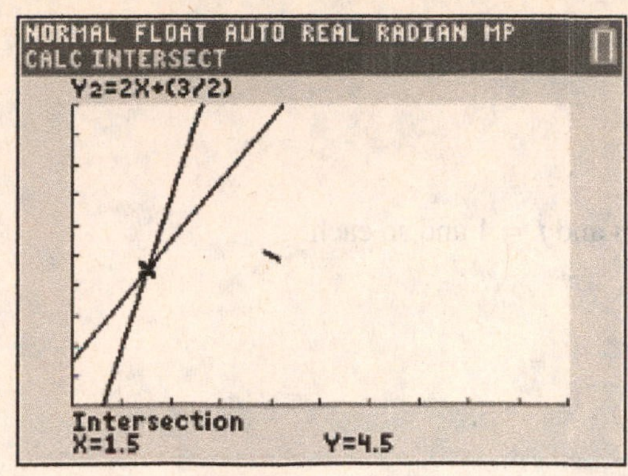

41. We verify whether $p = 260$ and $w = 40$ satisfy the equations

$p + w = 300$ and $p - w = 220$:

$260 + 40 = 300$ and $260 - 40 = 220$

and so the two speeds are in fact $p = 260$ and $w = 40$.

45. $0.8T_1 - 0.6T_2 = 12$ and $0.6T_1 + 0.8T_2 = 68$

$$T_2 = \frac{0.8T_1 - 12}{0.6} \qquad T_2 = \frac{68 - 0.6T_1}{0.8}$$

On a graphing calculator use

$x = T_1$ and $y = T_2$ and

let $y_1 = \frac{0.8x - 12}{6}$ and $y_2 = \frac{68 - 0.6x}{0.8}$.

Using the intersect feature, the point of intersection is $(50, 47)$.

The tensions (to the nearest 1 N) are

$T_1 = 50$ N

$T_2 = 47$ N

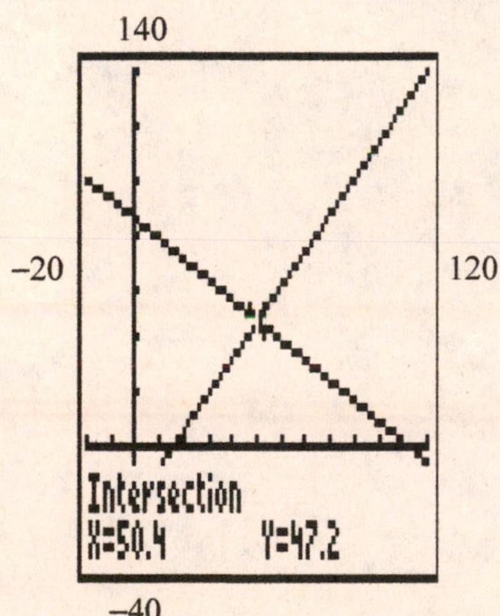

49. We let p be the cost per sheet of plywood and f be the cost of one framing stud. The equations are $304 = 8p + 40f$ and $498 = 25p + 12f$.
Solving these for p, we have
$p = 38 - 5f$ and $p = 19.92 - 0.48f$
which we plot on a graphing calculator. The intersection is at $p = 18$ and $f = 4$ and so each sheet of plywood costs \$18.00 and each framing stud costs \$4.00.

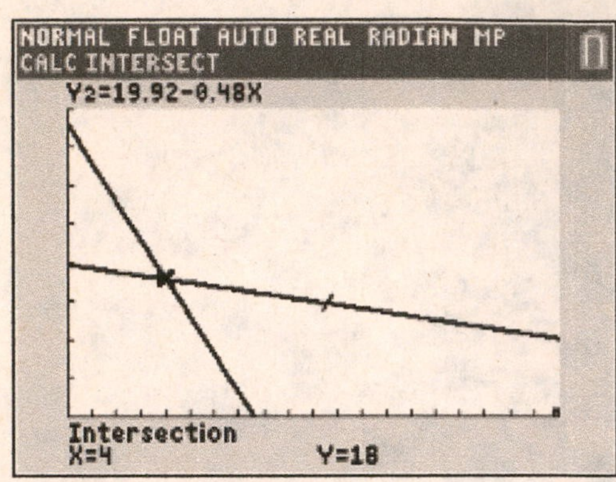

5.3 Solving Systems of Two Linear Equations in Two Unknowns Algebraically

1.

$$
\begin{aligned}
x - 3y &= 6 \\
x &= 3y + 6 && \text{(A)} \\
2x - 3y &= 3 && \text{(B)} \\
2(3y + 6) - 3y &= 3 && \text{substitute } x \text{ from (A) into (B)} \\
6y + 12 - 3y &= 3 \\
3y &= -9 \\
y &= -3 \\
x &= 3(-3) + 6 && \text{substitute } -3 \text{ for } y \text{ into (A)} \\
x &= -3
\end{aligned}
$$

The solution to the system is
$x = -3,\ y = -3$

5.

$$
\begin{aligned}
x &= y + 3 && \text{Equation (A)} \\
x - 2y &= 5 && \text{Equation (B)} \\
(y + 3) - 2y &= 5 && \text{substitute } x \text{ from (A) into (B)} \\
-y &= 2 \\
y &= -2 \\
x &= -2 + 3 && \text{substitute } -2 \text{ for } y \text{ into (A)} \\
x &= 1
\end{aligned}
$$

The solution to the system is
$x = 1,\ y = -2$

9.

$x+y=-5$

$y=-x-5$ Equation (A)

$2x-y=2$ Equation (B)

$2x-(-x-5)=2$ substitute y from (A) into (B)

$3x=-3$

$x=-1$

$y=-(-1)-5$ substitute -1 for x into (A)

$y=-4$

The solution to the system is

$x=-1,\ y=-4$

13.

$33x+2y=34$

$y=\dfrac{34-33x}{2}$ Equation (A)

$40y=9x+11$

$40y-9x=11$ Equation (B)

$40\ \dfrac{34-33x}{2}-9x=11$ substitute y from (A) into (B)

$680-660x-9x=11$

$-669x=-669$

$x=1$

$y=\dfrac{34-33(1)}{2}$ substitute 1 for x into (A)

$y=\frac{1}{2}$

The solution to the system is

$x=1,\ y=\frac{1}{2}$

17.

$2x-3y=4$ Equation (A)

$2x+y=-4$ Equation (B)

If we subtract Eq. (A) – Eq. (B)

$$\begin{array}{r} 2x-3y=\ \ 4 \\ -\ \ 2x+\ \ y=-4 \\ \hline -4y=8 \end{array}$$

$y=-2$

$2x+(-2)=-4$ substitute -2 for y into (B)

$2x=-2$

$x=-1$

The solution to the system is

$x=-1,\ y=-2$

21. $v + 2t = 7$ Equation (A)

$2v + 4t = 9$ Equation (B)

If we subtract $2 \times$ Eq. (A) – Eq. (B)

$$\begin{array}{r} 2v + 4t = 14 \\ -\quad 2v + 4t = 9 \\ \hline 0 = 5 \end{array}$$

The system of equations is inconsistent.

25. $2x - y = 5$ Equation (A)

$6x + 2y = -5$ Equation (B)

If we add $2 \times$ Eq. (A) + Eq. (B)

$$\begin{array}{r} 4x - 2y = 10 \\ +\quad 6x + 2y = -5 \\ \hline 10x \quad = 5 \end{array}$$

$x = \frac{1}{2}$

$2\left(\frac{1}{2}\right) - y = 5$ substitute $\frac{1}{2}$ for x into (A)

$y = 1 - 5$

$y = -4$

The solution to the system is

$x = \frac{1}{2},\ y = -4$

29. $15x + 10y = 11$ Equation (A)

$20x - 25y = 7$ Equation (B)

If we add $5 \times$ Eq. (A) + $2 \times$ Eq. (B)

$$\begin{array}{r} 75x + 50y = 55 \\ +\quad 40x - 50y = 14 \\ \hline 115x \quad = 69 \end{array}$$

$x = \frac{69}{115} = \frac{3}{5}$

$15\left(\frac{3}{5}\right) + 10y = 11$ substitute $\frac{3}{5}$ for x into (A)

$10y = 2$

$y = \frac{1}{5}$

The solution to the system is

$x = \frac{3}{5},\ y = \frac{1}{5}$

33. $44A = 1 - 15B$

$A = \dfrac{1 - 15B}{44}$ Equation (A)

$5B = 22 + 7A$ Equation (B)

$5B = 22 + 7\left(\dfrac{1 - 15B}{44}\right)$ substitute A from (A) into (B)

$5B = 22 + \dfrac{7}{44} - \dfrac{105}{44}B$

$$\frac{220+105}{44}B=\frac{968+7}{44}$$
$$325B=975$$
$$B=3$$
$$A=\frac{1-15(3)}{44} \quad \text{substitute 3 for } B \text{ into (A)}$$
$$A=-1$$

The solution to the system is

$A=-1,\ B=3$

37. $0.3x-0.7y=0.4$ multiply every term by 10

$3x-7y=4$ Equation (A)

$0.2x+0.5y=0.7$ multiply every term by 10

$2x+5y=7$ Equation (B)

If we subtract $2\times$ Eq. (A) $-\ 3\times$ Eq. (B)

$$\begin{array}{r} 6x-14y=\ 8 \\ -\quad 6x+15y=\ 21 \\ \hline -29y=-13 \end{array}$$

$y=\frac{13}{29}$

$3x-7\ \frac{13}{29}\ =4$ substitute $\frac{13}{29}$ for y into (A)

$3x=\frac{116+91}{29}=\frac{207}{29}$

$x=\frac{69}{29}$

The solution to the system is

$x=\frac{69}{29},\ y=\frac{13}{29}$

41. $\frac{x}{3}+\frac{2y}{3}=2$ Multiply by 3

$x+2y=6$ Equation (A)

$\frac{x}{2}-2y=\frac{5}{2}$ Multiply by 2

$x-4y=5$ Equation (B)

If we subtract Eq. (A) – Eq. (B)

$$\begin{array}{r} x+2y=6 \\ -\quad x-4y=5 \\ \hline 6y=1 \end{array}$$

and so $y=\frac{1}{6}$.

$x+2(\frac{1}{6})=6$ Substitute $\frac{1}{6}$ for y into (A)

$x=6-2(\frac{1}{6})$

$x=\frac{17}{3}$

The solution to the system is

$x=\frac{17}{3},\ y=\frac{1}{6}$

45. **(a)** Solving for x and substituting:

$2x + y = 4$

$2x = 4 - y$

$x = 2 - \frac{1}{2}y$ Equation (A)

$3x - 4y = -5$ Equation (B)

$3(2 - \frac{1}{2}y) - 4y = -5$ substitute for x into (B)

$6 - \frac{11}{2}y = -5$

$-\frac{11}{2}y = -11$

$y = 2$

$x = 2 - \frac{1}{2}(2)$ substitute for y into (A)

$x = 1$

The solution to the system is

$x = 1, y = 2$

(b) Solving for y and substituting:

$2x + y = 4$

$y = 4 - 2x$ Equation (A)

$3x - 4y = -5$ Equation (B)

$3x - 4(4 - 2x) = -5$ substitute for x into (B)

$11x - 16 = -5$

$11x = 11$

$x = 1$

$y = 4 - 2(1)$ substitute for x into (A)

$y = 2$

The solution to the system is

$x = 1, y = 2$

49.

$x + y = 10\ 000$

$y = 10\ 000 - x$ Equation (A)

$0.0180x + 0.0100y = 0.0150(10\ 000)$

$0.0180x + 0.0100y = 150$ Equation (B)

$0.0180x + 0.0100(10\ 000 - x) = 150$ substitute y from (A) into (B)

$0.0080x = 50$

$x = 6250$

$y = 10\ 000 - 6250$ substitute 6250 for x into (A)

$y = 3750$

The solution to the system is

$x = 6250$ L, $y = 3750$ L

53. Let W_f = weight supported by front wheels

Let W_r = weight supported by rear wheels

$W_r + W_f = 17\ 700$ Equation (A)

$\frac{W_r}{W_f} = 0.847$

$W_r = 0.847W_f$ Equation (B)

$0.847\ W_f + W_f = 17\ 700$ substitute W_r from (B) into (A)

$1.847W_f = 17\ 700$

$W_f = 9583.108$ N

$W_f = 9580$ N weight supported by front wheels

$W_r = 0.847(9583.108)$ substitute 9580 for W_f into (B)

$W_r = 8116.892$ N

$W_r = 8120$ N weight supported by rear wheels

57. Let x = number of offices renting at \$900 per month

Let y = number of offices renting at \$1250 per month

$x + y = 54$

$y = 54 - x$ Equation (A)

$900x + 1250y = 55\ 600$ Equation (B)

$900x + 1250(54 - x) = 55\ 600$ substitute y from (A) into (B)

$900x + 67\ 500 - 1250x = 55\ 600$

$-350x = -11\ 900$

$x = 34$

$y = 54 - 34$ substitute 34 for x into (A)

$y = 20$

There are 34 offices renting at \$900 per month
and 20 offices renting at \$1250 per month.

61. Let x = windmill power capacity (in kW)

Let y = gas generator power capacity (in kW)

Energy produced = power × time

In a 10-day period, there are 240 h.

For the first 10-day period:

$(0.450x)240 + y(240) = 3010$

$108x + 240y = 3010$

$y = \frac{3010 - 108x}{240}$ Equation (A)

For the second 10-day period:

$$(0.720x)240 + y(240-60) = 2900$$

$$172.8x + 180y = 2900 \qquad \text{Equation (B)}$$

$$172.8x + 180\,\frac{3010-108x}{240} = 2900 \qquad \text{substitute } y \text{ from (A) into (B)}$$

$$172.8x + 2257.5 - 81x = 2900$$

$$91.8x = 642.5$$

$$x = 7.00 \text{ kW} \qquad \text{substitute 7.00 for } x \text{ into (A)}$$

$$y = \frac{3010-108(7.00)}{240}$$

$$y = 9.39 \text{ kW}$$

The windmill power capacity is 7.00 kW, and the gas generator capacity is 9.39 kW.

65. $ax + y = c$

$bx + y = d$

In order to create a unique solution, the lines must have different slopes. Putting both equations into slope-intercept form,

$y = -ax + c$ has slope $-a$

$y = -bx + d$ has slope $-b$

To make the slopes different, $a \neq b$.

5.4 Solving Systems of Two Linear Equations in Two Unknowns by Determinants

Note to students: In all questions where solving a system of equations is required, substitution of the solutions into *both* equations serves as a check.

1. $\begin{vmatrix} 4 & -6 \\ 3 & 17 \end{vmatrix} = 4(17) - 3(-6) = 68 + 18 = 86$

5. $\begin{vmatrix} 8 & 3 \\ 4 & 1 \end{vmatrix} = (8)(1) - (3)(4) = 8 - 12 = -4$

9. $\begin{vmatrix} 15 & -9 \\ 12 & 0 \end{vmatrix} = (15)(0) - (12)(-9) = 0 - (-108) = 108$

13. $\begin{vmatrix} 0.75 & -1.32 \\ 0.15 & 1.18 \end{vmatrix} = 0.75(1.18) - (0.15)(-1.32) = 0.885 + 0.198 = 1.083$

17. $\begin{vmatrix} 2 & a-1 \\ a+2 & a \end{vmatrix} = 2(a) - (a+2)(a-1) = 2a - (a^2 + a - 2) = -a^2 + a + 2$

21. $2x - 3y = 4$

$2x + y = -4$

$$x = \frac{\begin{vmatrix} 4 & -3 \\ -4 & 1 \end{vmatrix}}{\begin{vmatrix} 2 & -3 \\ 2 & 1 \end{vmatrix}} = \frac{4(1) - (-4)(-3)}{2(1) - 2(-3)} = \frac{-8}{8} = -1$$

$$y = \frac{\begin{vmatrix} 2 & 4 \\ 2 & -4 \end{vmatrix}}{\begin{vmatrix} 2 & -3 \\ 2 & 1 \end{vmatrix}} = \frac{2(-4) - 2(4)}{8} = \frac{-16}{8} = -2$$

25. $v + 2t = 7$

$2v + 4t = 9$

$$v = \frac{\begin{vmatrix} 7 & 2 \\ 9 & 4 \end{vmatrix}}{\begin{vmatrix} 1 & 2 \\ 2 & 4 \end{vmatrix}} = \frac{7(4) - 9(2)}{1(4) - 2(2)} = \frac{28 - 18}{4 - 4} = \frac{10}{0} = \text{inconsistent}$$

29. $0.3x - 0.7y = 0.4$

$0.2x + 0.5y = 0.7$

$$x = \frac{\begin{vmatrix} 0.4 & -0.7 \\ 0.7 & 0.5 \end{vmatrix}}{\begin{vmatrix} 0.3 & -0.7 \\ 0.2 & 0.5 \end{vmatrix}} = \frac{0.4(0.5) - 0.7(-0.7)}{0.3(0.5) - 0.2(-0.7)} = \frac{0.69}{0.29} = \frac{69}{29}$$

$$y = \frac{\begin{vmatrix} 0.3 & 0.4 \\ 0.2 & 0.7 \end{vmatrix}}{\begin{vmatrix} 0.3 & -0.7 \\ 0.2 & 0.5 \end{vmatrix}} = \frac{0.3(0.7) - 0.2(0.4)}{0.3(0.5) - 0.2(-0.7)} = \frac{0.13}{0.29} = \frac{13}{29}$$

33. $301x - 529y = 1520$

$385x - 741y = 2540$

$$x = \frac{\begin{vmatrix} 1520 & -529 \\ 2540 & -741 \end{vmatrix}}{\begin{vmatrix} 301 & -529 \\ 385 & -741 \end{vmatrix}} = \frac{217340}{-19376} = -11.2$$

$$y = \frac{\begin{vmatrix} 301 & 1520 \\ 385 & 2540 \end{vmatrix}}{\begin{vmatrix} 301 & -529 \\ 385 & -741 \end{vmatrix}} = \frac{179340}{-19376} = -9.26$$

37. For $c = d = 0$

$$\begin{vmatrix} a & b \\ c & d \end{vmatrix} = \begin{vmatrix} a & b \\ 0 & 0 \end{vmatrix} = a(0) - 0(b) = 0$$

41. Rewrite both equations in standard form.

$$F_1 + F_2 = 21$$
$$2F_1 - 5F_2 = 0$$

$$F_1 = \frac{\begin{vmatrix} 21 & 1 \\ 0 & -5 \end{vmatrix}}{\begin{vmatrix} 1 & 1 \\ 2 & -5 \end{vmatrix}} = \frac{21(-5)-0(1)}{1(-5)-2(1)} = \frac{-105}{-7} = 15 \text{ N}$$

$$F_2 = \frac{\begin{vmatrix} 1 & 21 \\ 2 & 0 \end{vmatrix}}{\begin{vmatrix} 1 & 1 \\ 2 & -5 \end{vmatrix}} = \frac{1(0)-2(21)}{-7} = \frac{-42}{-7} = 6.0 \text{ N}$$

45. Let x = number of three bedroom homes

Let y = number of four bedroom homes

$$2\ 000x + 3\ 000y = 560\ 000$$
$$25\ 000x + 35\ 000y = 6\ 800\ 000$$

$$x = \frac{\begin{vmatrix} 560\ 000 & 3\ 000 \\ 6\ 800\ 000 & 35\ 000 \end{vmatrix}}{\begin{vmatrix} 2\ 000 & 3\ 000 \\ 25\ 000 & 35\ 000 \end{vmatrix}} = \frac{560\ 000(35\ 000) - 6\ 800\ 000(3\ 000)}{35\ 000(2\ 000) - (25\ 000)(3\ 000)} = \frac{-800\ 000\ 000}{-5\ 000\ 000} = 160$$

$$y = \frac{\begin{vmatrix} 2\ 000 & 560\ 000 \\ 25\ 000 & 6\ 800\ 000 \end{vmatrix}}{\begin{vmatrix} 2\ 000 & 3\ 000 \\ 25\ 000 & 35\ 000 \end{vmatrix}} = \frac{6\ 800\ 000(2\ 000) - 560\ 000(25\ 000)}{-5\ 000\ 000} = 80$$

49. Let L = the length of the rectangle in meters

Let w = the width of the rectangle in meters

$$L = 1.62w$$
$$2L + 2w = 4.20$$

Rewrite these in standard form

$$L - 1.62w = 0$$
$$2L + 2w = 4.20$$

$$L = \frac{\begin{vmatrix} 0 & -1.62 \\ 4.20 & 2 \end{vmatrix}}{\begin{vmatrix} 1 & -1.62 \\ 2 & 2 \end{vmatrix}} = \frac{0 - 4.20(-1.62)}{1(2) - 2(-1.62)} = \frac{6.804}{5.24} = 1.30 \text{ m}$$

$$w = \frac{\begin{vmatrix} 1 & 0 \\ 2 & 4.20 \end{vmatrix}}{\begin{vmatrix} 1 & -1.62 \\ 2 & 2 \end{vmatrix}} = \frac{4.20 - 0}{5.24} == 0.80 \text{ m}$$

53. Let t_1 = time taken by drug boat

Let t_2 = time taken by Coast Guard

$$24\text{ min} = 24\text{ min}\ \frac{1\text{ h}}{60\text{ min}} = 0.40\text{ h}$$

We know the drug boat had a 0.40 h head start

$$t_1 = t_2 + 0.40$$
$$t_1 - t_2 = 0.40$$

Remember $d = vt$, and the total distance travelled by each boat is the same

$$42t_1 = 50t_2$$
$$42t_1 - 50t_2 = 0$$

$$t_1 = \frac{\begin{vmatrix} 0.4 & -1 \\ 0 & -50 \end{vmatrix}}{\begin{vmatrix} 1 & -1 \\ 42 & -50 \end{vmatrix}} = \frac{0.4(-50) - 0}{-50 - (-42)} = \frac{-20}{-8} = 2.5\text{ h}$$

$$t_2 = \frac{\begin{vmatrix} 1 & 0.4 \\ 42 & 0 \end{vmatrix}}{\begin{vmatrix} 1 & -1 \\ 42 & -50 \end{vmatrix}} = \frac{0 - 42(0.4)}{-8} = \frac{-16.8}{-12} = 2.1\text{ h}$$

5.5 Solving Systems of Three Linear Equations in Three Unknowns Algebraically

Note to students: In all questions where solving a system of equations is required, substitution of the solutions into *all three* equations serves as a check.

1.

(1)	$4x + y + 3z = 1$	
(2)	$2x - 2y + 6z = 12$	
(3)	$-6x + 3y + 12z = -14$	
(4)	$8x + 2y + 6z = 2$	(1) multiplied by 2
(2)	$2x - 2y + 6z = 12$	add
(5)	$10x + 12z = 14$	
(6)	$12x + 3y + 9z = 3$	(1) multiplied by 3
(3)	$-6x + 3y + 12z = -14$	subtract
(7)	$18x - 3z = 17$	

$$
\begin{array}{lrl}
(8) & 72x - 12z = 68 & \text{(7) multiplied by 4} \\
(5) & 10x + 12z = 14 & \text{add} \\
\hline
(9) & 82x = 82 & \\
 & x = 1 & \\
(11) & 18(1) - 3z = 17 & \text{substituting } x = 1 \text{ into (7)} \\
 & -3z = -1 & \\
 & z = \frac{1}{3} & \\
(12) & 4(1) + y + 3\frac{1}{3} = 1 & \text{substitute } x = 1 \text{ and } z = \frac{1}{3} \text{ into (1)} \\
 & 4 + y + 1 = 1 & \\
 & y = -4 &
\end{array}
$$

The solution is $x = 1$, $y = -4$, $z = \frac{1}{3}$

5.

$$
\begin{array}{lrl}
(1) & 2x+3y+z=2 & \\
(2) & -x+2y+3z=-1 & \\
(3) & -3x-3y+z=0 & \\
\hline
(4) & 5x+6y=2 & \text{Subtract (1) } - \text{ (3)}
\end{array}
$$

$$
\begin{array}{lrl}
(5) & 6x+9y+3z=6 & \text{multiply (1) by 3} \\
(2) & -x+2y+3z=-1 & \text{subtract} \\
\hline
(6) & 7x+7y=7 &
\end{array}
$$

$$
\begin{array}{lrl}
(7) & 5x+5y=5 & \text{multiply (6) by 5/7} \\
(4) & 5x+6y=2 & \text{subtract} \\
\hline
 & -y=3 & \\
 & y=-3 & \\
(8) & 7x+7(-3)=7 & \text{substitute } -3 \text{ for } y \text{ into (6)} \\
 & 7x=28 & \\
 & x=4 & \\
(9) & -3(4)-3(-3)+z=0 & \text{substitute 4 for } x, \text{ and } -3 \text{ for } y \text{ into (3)} \\
 & z=3 &
\end{array}
$$

The solution is $x = 4$, $y = -3$, $z = 3$.

9.

(1)	$2x-2y+3z=5$	
(2)	$2x+y-2z=-1$	
(3)	$4x-y-3z=0$	
(4)	$6x-5z=-1$	add (2) and (3)
(5)	$4x+2y-4z=-2$	multiply (2) by 2
(1)	$2x-2y+3z=5$	add
(6)	$6x-z=3$	
(4)	$6x-5z=-1$	subtract
	$4z=4$	
	$z=1$	
(7)	$6x-1=3$	substitute 1 for z into (6)
	$x=\frac{4}{6}=\frac{2}{3}$	
(8)	$2\left(\frac{2}{3}\right)+y-2(1)=-1$	substitute $\frac{2}{3}$ for x, and 1 for z into (2)
	$\frac{4}{3}+y-\frac{6}{3}=-\frac{3}{3}$	
	$y=-\frac{1}{3}$	

The solution is $x=\frac{2}{3}$, $y=-\frac{1}{3}$, $z=1$.

13.

(1)	$10x+15y-25z=35$	
(2)	$40x-30y-20z=10$	
(3)	$16x-2y+8z=6$	
(4)	$20x+30y-50z=70$	multiply (1) by 2
(2)	$40x-30y-20z=10$	add
(5)	$60x-70z=80$	
(6)	$240x-30y+120z=90$	multiply (3) by 15
(2)	$40x-30y-20z=10$	subtract
(7)	$200x+140z=80$	
(8)	$120x-140z=160$	multiply (5) by 2, then add to (7)
(9)	$320x=240$	
	$x=\frac{3}{4}$	
(10)	$60\left(\frac{3}{4}\right)-70z=80$	substitute $\frac{3}{4}$ for x into (5)
	$-70z=35$	
	$z=-\frac{1}{2}$	
(11)	$16\left(\frac{3}{4}\right)-2y+8\left(-\frac{1}{2}\right)=6$	substitute $\frac{3}{4}$ for x, and $-\frac{1}{2}$ for z into (3)
	$12-2y-4=6$	
	$-2y=-2$	
	$y=1$	

The solution is $x=\frac{3}{4}$, $y=1$, $z=-\frac{1}{2}$.

17. In standard form, the equations are

$-I_1 + I_2 - I_3 = 0$

$8.00I_1 + 10.0I_2 + 0.00I_3 = 80.0$

$0.00I_1 + 10.0I_2 + 6.00I_3 = 60.0$

which are encoded in the augmented matrix

−1	1	−1	0
8.00	10.0	0	80.0
0	10.0	6.00	60.0

Using the rref feature on a calculator, we find $I_1 = 3.62, I_2 = 5.11, I_3 = 1.49$ to three significant figures.

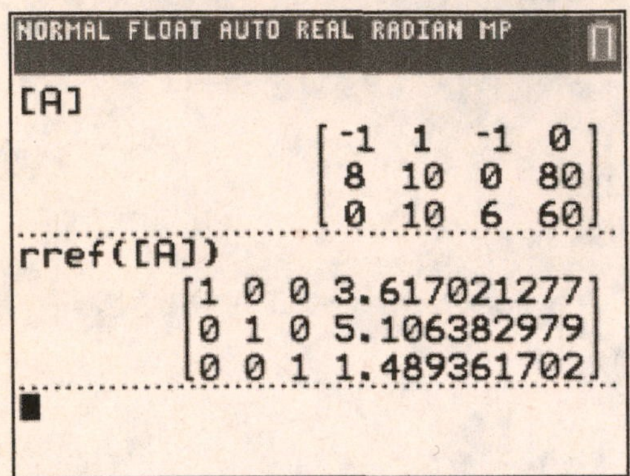

21.

(1)	$P + M + I = 1150$	
(2)	$P - 4I = -100$	
(3)	$P - 6M = 50$	
(4)	$4P + 4M + 4I = 4600$	multiply (1) by 4
(2)	$P - 4I = -100$	add
(5)	$5P + 4M = 4500$	
(6)	$5P - 30M = 250$	multiply (3) by 5, then subtract from (5)
(7)	$34M = 4250$	
	$M = 125$	
(8)	$P - 6(125) = 50$	substitute 125 for M into (3)
	$P = 800$	
(9)	$800 + 125 + I = 1150$	substitute 125 for M and 800 for P into (1)
	$I = 225$	

The solution is $P = 800$ h, $M = 125$ h, $I = 225$ h.

25.

(1*a*)	$A+(A+B)=90$	from left triangle
(1*b*)	$2A+B=90$	
(2*a*)	$A+2B=C$	from angle C
(2*b*)	$A+2B-C=0$	
(3)	$A+B+C=180$	from large (outside) triangle
(4)	$2A+3B=180$	add (2b) and (3)
(1*b*)	$2A+B=90$	subtract
(5)	$2B=90$	
	$B=45.0$	
(6)	$2A+45=90$	substitute 45 for *B* into (1b)
	$2A=45$	
	$A=22.5$	
(7)	$22.5+45+C=180$	substitute 22.5 for *A*, 45 for *B* into (3)
	$C=112.5$	

The solution is $A=22.5^\circ$, $B=45.0^\circ$, $C=112.5^\circ$.

29. Let x, y, z represent numbers of MA, MS, and PhD degrees, respectively.

(1)	$x+y+z=420$	
(2)	$x-y-z=100$	
(3)	$y-3z=0$	
(4)	$2x=520$	add (1) and (2)
(5)	$x=260$	solve for x
	$2y+2z=320$	subtract (2) from (1)
(6)	$y+z=160$	divide by 2
	$4z=160$	subtract (3) from 6
(7)	$z=40$	solve for z
(8)	$y+40=160$	substitute (7) into (6)
	$y=120$	solve for y

Thus, 260 MA degrees, 120 MS degrees, and 40 PhD degrees were awarded.

33.

$$\begin{array}{lrl}
(1) & x-2y-\ 3z=2 & \\
(2) & x-4y-13z=14 & \\
(3) & -3x+5y+\ 4z=0 & \\
\hline
(4) & 2y+10z=-12 & \text{subtract } (1)-(2) \\
(5) & y+5z=-6 & \text{divide (4) by 2}
\end{array}$$

$$\begin{array}{lrl}
(6) & 3x-6y-9z=6 & \text{multiply (1) by 3} \\
(3) & -3x+5y+4z=0 & \text{add} \\
\hline
(7) & -y-5z=6 & \\
(8) & y+5z=-6 & \text{divide (7) by } -1 \\
(5) & y+5z=-6 & \text{subtract} \\
\hline
 & 0=0 &
\end{array}$$

The system is dependent, there are an infinite number of solutions.

One possible solution, if we let $z = 0, y = -6$ from Eq. (8) or Eq. (5).

Substituting into Eq. (1),

$x-2(-6)-0=2$

$x=-10$

A possible solution is $x=-10,\ y=-6,\ z=0$

5.6 Solving Systems of Three Linear Equations in Three Unknowns by Determinants

Note to students: In all questions where solving a system of equations is required, substitution of the solutions into *all* equations serves as a check.

1.

$$\begin{vmatrix} -2 & 3 & -1 \\ 1 & 5 & 4 \\ 2 & -1 & 5 \end{vmatrix} \begin{matrix} -2 & 3 \\ 1 & 5 \\ 2 & -1 \end{matrix}$$

$=-2(5)(5)+3(4)(2)+(-1)(1)(-1)-2(5)(-1)-(-1)(4)(-2)-5(1)(3)$

$=-50+24+1+10-8-15$

$=-38$

This is the same determinant as that of Example 1 except the sign has changed.

Interchanging a single pair of rows alters the determinant in sign only.

5.

$$\begin{vmatrix} 8 & 9 & -6 \\ -3 & 7 & 2 \\ 4 & -2 & 5 \end{vmatrix} \begin{matrix} 8 & 9 \\ -3 & 7 \\ 4 & -2 \end{matrix} = 280+72+(-36)-(-168)-(-32)-(-135)=651$$

9.

$$\begin{vmatrix} 4 & -3 & -11 \\ -9 & 2 & -2 \\ 0 & 1 & -5 \end{vmatrix} \begin{matrix} 4 & -3 \\ -9 & 2 \\ 0 & 1 \end{matrix} = -40+0+99-0-(-8)-(-135)=202$$

13. $\left|\begin{array}{ccc|cc} 0.1 & -0.2 & 0 & 0.1 & -0.2 \\ -0.5 & 1 & 0.4 & -0.5 & 1 \\ -2 & 0.8 & 2 & -2 & 0.8 \end{array}\right. = 0.2 + 0.16 + 0 - 0 - 0.032 - 0.2 = 0.128$

17. $x + y + z = 2$
$x \quad - z = 1$
$x + y \quad = 1$

$$x = \frac{\left|\begin{array}{ccc|cc} 2 & 1 & 1 & 2 & 1 \\ 1 & 0 & -1 & 1 & 0 \\ 1 & 1 & 0 & 1 & 1 \end{array}\right.}{\left|\begin{array}{ccc|cc} 1 & 1 & 1 & 1 & 1 \\ 1 & 0 & -1 & 1 & 0 \\ 1 & 1 & 0 & 1 & 1 \end{array}\right.} = \frac{0 + (-1) + 1 - 0 - (-2) - 0}{0 + (-1) + 1 - 0 - (-1) - 0} = \frac{2}{1} = 2$$

$$y = \frac{\left|\begin{array}{ccc|cc} 1 & 2 & 1 & 1 & 2 \\ 1 & 1 & -1 & 1 & 1 \\ 1 & 1 & 0 & 1 & 1 \end{array}\right.}{1} = \frac{0 + (-2) + 1 - 1 - (-1) - 0}{1} = \frac{-1}{1} = -1$$

$$z = \frac{\left|\begin{array}{ccc|cc} 1 & 1 & 2 & 1 & 1 \\ 1 & 0 & 1 & 1 & 0 \\ 1 & 1 & 1 & 1 & 1 \end{array}\right.}{1} = \frac{0 + 1 + 2 - 0 - 1 - 1}{1} = \frac{1}{1} = 1$$

Solution: $x = 2,\ y = -1,\ z = 1.$

21. $5l + 6w - 3h = 6$
$4l - 7w - 2h = -3$
$3l + w - 7h = -1$

$$l = \frac{\left|\begin{array}{ccc|cc} 6 & 6 & -3 & 6 & 6 \\ -3 & -7 & -2 & -3 & -7 \\ 1 & 1 & -7 & 1 & 1 \end{array}\right.}{\left|\begin{array}{ccc|cc} 5 & 6 & -3 & 5 & 6 \\ 4 & -7 & -2 & 4 & -7 \\ 3 & 1 & -7 & 3 & 1 \end{array}\right.} = \frac{294 + (-12) + 9 - 21 - (-12) - 126}{245 + (-36) + (-12) - 63 - (-10) - (-168)} = \frac{156}{312} = \frac{1}{2}$$

$$w = \frac{\left|\begin{array}{ccc|cc} 5 & 6 & -3 & 5 & 6 \\ 4 & -3 & -2 & 4 & -3 \\ 3 & 1 & -7 & 3 & 1 \end{array}\right.}{312} = \frac{105 + (-36) + (-12) - 27 - (-10) - (-168)}{312} = \frac{208}{312} = \frac{2}{3}$$

$$h = \frac{\left|\begin{array}{ccc|cc} 5 & 6 & 6 & 5 & 6 \\ 4 & -7 & -3 & 4 & -7 \\ 3 & 1 & 1 & 3 & 1 \end{array}\right.}{312} = \frac{-35 + (-54) + 24 - (-126) - (-15) - 24}{312} = \frac{52}{312} = \frac{1}{6}$$

Solution: $l = \frac{1}{2},\ w = \frac{2}{3},\ h = \frac{1}{6}$

25. $3x-7y+3z=6$
$3x+3y+6z=1$
$5x-5y+2z=5$

$$x=\frac{\left|\begin{array}{ccc|cc}6&-7&3&6&-7\\1&3&6&1&3\\5&-5&2&5&-5\end{array}\right|}{\left|\begin{array}{ccc|cc}3&-7&3&3&-7\\3&3&6&3&3\\5&-5&2&5&-5\end{array}\right|}=\frac{36+(-210)+(-15)-45-(-180)-(-14)}{18+(-210)+(-45)-45-(-90)-(-42)}=\frac{-40}{-150}=\frac{4}{15}$$

$$y=\frac{\left|\begin{array}{ccc|cc}3&6&3&3&6\\3&1&6&3&1\\5&5&2&5&5\end{array}\right|}{-150}=\frac{6+180+45-15-90-36}{-150}=\frac{90}{-150}=-\frac{3}{5}$$

$$z=\frac{\left|\begin{array}{ccc|cc}3&-7&6&3&-7\\3&3&1&3&3\\5&-5&5&5&-5\end{array}\right|}{-150}=\frac{45+(-35)+(-90)-90-(-15)-(-105)}{-150}=\frac{-50}{-150}=\frac{1}{3}$$

Solution: $x=\frac{4}{15}$, $y=-\frac{3}{5}$, $z=\frac{1}{3}$

29. $\left|\begin{array}{ccc|cc}-2&1&1&-2&1\\2&3&1&2&3\\6&5&1&6&5\end{array}\right|=-6+6+10-18-(-10)-2=0$

and so the points (–2,1), (2,3), and (6,5) are collinear.

33. $\begin{vmatrix}2&4&1\\5&10&6\\7&9&8\end{vmatrix}=160+168+45-70-108-160=35$

Adding a multiple of one row to another does not change the value of the determinant.

37. $s_0 + 2v_0 + 2a = 20$

$s_0 + 4v_0 + 8a = 54$

$s_0 + 6v_0 + 18a = 104$

$$s_0 = \frac{\begin{vmatrix} 20 & 2 & 2 \\ 54 & 4 & 8 \\ 104 & 6 & 18 \end{vmatrix}}{\begin{vmatrix} 1 & 2 & 2 \\ 1 & 4 & 8 \\ 1 & 6 & 18 \end{vmatrix}} = \frac{1440 + 1664 + 648 - 832 - 960 - 1944}{72 + 16 + 12 - 8 - 48 - 36} = \frac{16}{8} = 2$$

$$v_0 = \frac{\begin{vmatrix} 1 & 20 & 2 \\ 1 & 54 & 8 \\ 1 & 104 & 18 \end{vmatrix}}{8} = \frac{972 + 160 + 208 - 108 - 832 - 360}{8} = \frac{40}{8} = 5$$

$$a = \frac{\begin{vmatrix} 1 & 2 & 20 \\ 1 & 4 & 54 \\ 1 & 6 & 104 \end{vmatrix}}{8} = \frac{416 + 108 + 120 - 80 - 324 - 208}{8} = \frac{32}{8} = 4$$

Solution: $s_0 = 2.00$ m, $v_0 = 5.00$ m/s, $a = 4.00$ m/s^2

41. $V = f(T) = a + bT + cT^2$

$a + 2.0b + 4.0c = 6.4$

$a + 4.0b + 16.0c = 8.6$

$a + 6.0b + 36.0c = 11.6$

$$a = \frac{\begin{vmatrix} 6.4 & 2.0 & 4.0 \\ 8.6 & 4.0 & 16.0 \\ 11.6 & 6.0 & 36.0 \end{vmatrix}}{\begin{vmatrix} 1 & 2.0 & 4.0 \\ 1 & 4.0 & 16.0 \\ 1 & 6.0 & 36.0 \end{vmatrix}} = \frac{921.6 + 371.2 + 206.4 - 185.6 - 614.4 - 619.2}{144 + 32 + 24 - 16 - 96 - 72} = \frac{80}{16} = 5$$

$$b = \frac{\begin{vmatrix} 1 & 6.4 & 4.0 \\ 1 & 8.6 & 16.0 \\ 1 & 11.6 & 36.0 \end{vmatrix}}{16} = \frac{309.6 + 102.4 + 46.4 - 34.4 - 185.6 - 230.4}{16} = \frac{8}{16} = \frac{1}{2}$$

$$c = \frac{\begin{vmatrix} 1 & 2.0 & 6.4 \\ 1 & 4.0 & 8.6 \\ 1 & 6.0 & 11.6 \end{vmatrix}}{16} = \frac{46.4 + 17.2 + 38.4 - 25.6 - 51.6 - 23.2}{16} = \frac{1.6}{16} = \frac{1}{10}$$

$V = f(T) = 5.00 + 0.500T + 0.100T^2$

45. Let v_c = velocity of the car (in mi/h)
Let v_j = velocity of the jet (in mi/h)
Let v_t = velocity of the taxi (in mi/h)
Remeber $d = vt$

$$\begin{aligned} v_j &= 12v_c & 12v_c - v_j &= 0 \\ v_c &= v_t + 15 & v_c - v_t &= 15 \\ 1.10v_c + 1.95v_j + 0.52v_t &= 1140 & 1.10v_c + 1.95v_j + 0.52v_t &= 1140 \end{aligned}$$

$$v_c = \frac{\begin{vmatrix} 0 & -1 & 0 \\ 15 & 0 & -1 \\ 1140 & 1.95 & 0.52 \end{vmatrix}}{\begin{vmatrix} 12 & -1 & 0 \\ 1 & 0 & -1 \\ 1.10 & 1.95 & 0.52 \end{vmatrix}} = \frac{0+1140+0-0-0-(-7.8)}{0+1.10+0-0-(-23.4)-(-0.52)} = \frac{1147.8}{25.02} = 45.9$$

$$v_j = \frac{\begin{vmatrix} 12 & 0 & 0 \\ 1 & 15 & -1 \\ 1.10 & 1140 & 0.52 \end{vmatrix}}{25.02} = \frac{93.6+0+0-0-(-13680)-0}{25.02} = \frac{13773.6}{25.02} = 551$$

$$v_t = \frac{\begin{vmatrix} 12 & -1 & 0 \\ 1 & 0 & 15 \\ 1.10 & 1.95 & 1140 \end{vmatrix}}{25.02} = \frac{0+(-16.5)+0-0-351-(-1140)}{25.02} = \frac{772.5}{25.02} = 30.9$$

The average speeds for the trip were
$v_c = 45.9$ mi/h, $v_j = 551$ mi/h, $v_t = 30.9$ mi/h.

Chapter 5 Review Exercises

1. This is false.
We substitute $x = 2, y = -3$ into $4x - 3y = -1$, obtaining

$$\begin{aligned} 4(2) - 3(-3) &= -1 \\ 8 + 9 &= -1 \\ 17 &= -1 \end{aligned}$$

which is a false statement.

5. This is false.
We compute the determinant:

$$\begin{vmatrix} 2 & 4 \\ -1 & 3 \end{vmatrix} = (2)(3) - (4)(-1) = 10.$$

9. $\begin{vmatrix} -2 & 5 \\ 3 & 1 \end{vmatrix} = (-2)(1) - (5)(3) = -17.$

13. $m = \dfrac{0-(-8)}{2-4} = \dfrac{8}{-2} = -4$

17. $m = -2, b = 4$

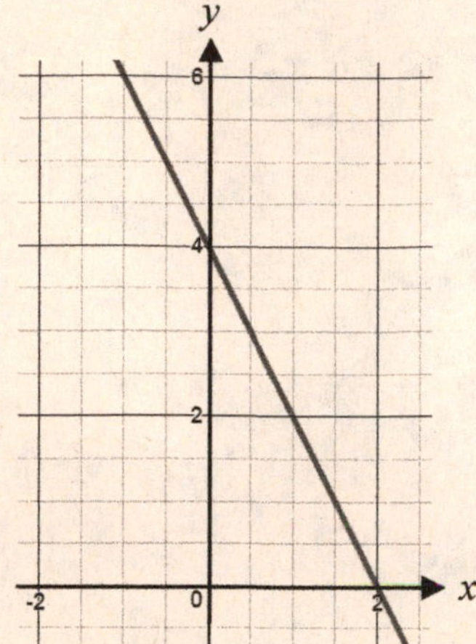

21. $x = 2, y = 0$

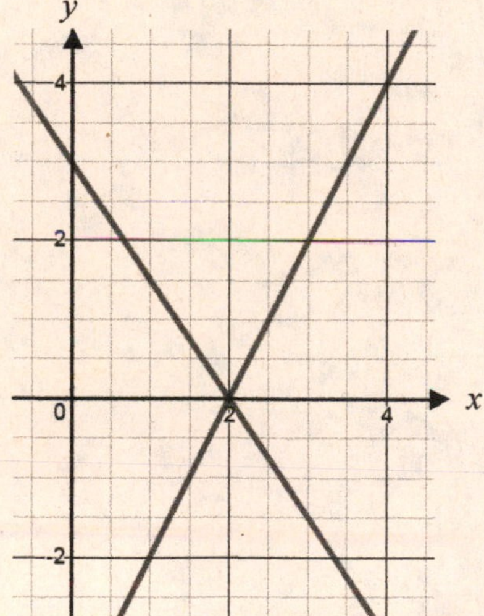

25. $x = 1.467, y = -1.867$

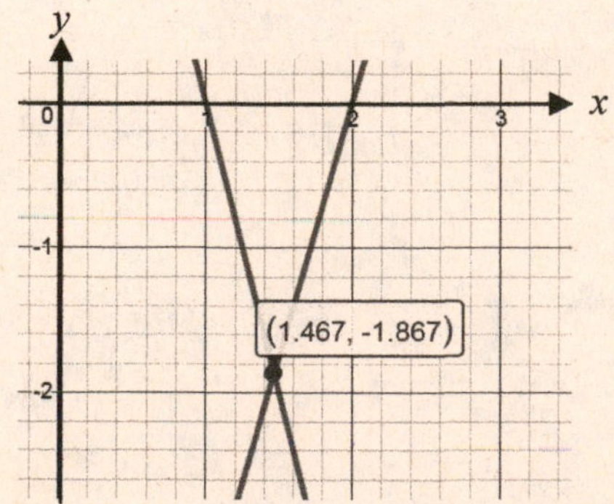

29. $x + 2y = 5 \quad \text{(I)}$

$x + 3y = 7 \quad \text{(II)}$

Subtract (I) from (II) to eliminate x :

$y = 2$

Substitute $y = 2$ into (I):

$x + 4 = 5 \rightarrow x = 1$

Solution: $x = 1, y = 2$

33. $10i - 27v = 29 \quad \text{(I)}$

$40i + 33v = 69 \quad \text{(II)}$

Multiply (I) by 4:

$40i - 108v = 116 \quad \text{(III)}$

Subtract (III) from (II) to eliminate i :

$141v = -37 \rightarrow v = -\frac{1}{3}$

Substitute $v = -\frac{1}{3}$ into (I):

$10i + 9 = 29 \rightarrow i = 2$

Solution: $i = 2, v = -\frac{1}{3}$

37. $90x - 110y = 40 \quad \text{(I)}$

$30x - 15y = 25 \quad \text{(II)}$

Multiply (II) by 3:

$90x - 45y = 75 \quad \text{(III)}$

Subtract (I) from (III) to eliminate x :

$65y = 35 \rightarrow y = \frac{7}{13}$

Substitute $y = \frac{7}{13}$ into (II):

$30x - \frac{105}{13} = 25$

$30x = \frac{430}{13} \rightarrow x = \frac{43}{39}$

Solution: $x = \frac{43}{39}, y = \frac{7}{13}$

41. $4x + 3y = -4 \quad \text{(I)}$

$-2x + y = -3 \quad \text{(II,rearranged)}$

Using Cramer's rule:

$$x = \frac{\begin{vmatrix} -4 & 3 \\ -3 & 1 \end{vmatrix}}{\begin{vmatrix} 4 & 3 \\ -2 & 1 \end{vmatrix}} = \frac{(-4)(1) - (3)(-3)}{(4)(1) - (3)(-2)} = \frac{5}{10} = \frac{1}{2}$$

$$y = \frac{\begin{vmatrix} 4 & -4 \\ -2 & -3 \end{vmatrix}}{\begin{vmatrix} 4 & 3 \\ -2 & 1 \end{vmatrix}} = \frac{(4)(-3) - (-4)(-2)}{(4)(1) - (3)(-2)} = \frac{-20}{10} = -2$$

45. $7x - 2y = -6$ (I, rearranged)

$4x + 7y = 12$ (II, rearranged)

Using Cramer's rule:

$$x = \frac{\begin{vmatrix} -6 & -2 \\ 12 & 7 \end{vmatrix}}{\begin{vmatrix} 7 & -2 \\ 4 & 7 \end{vmatrix}} = \frac{(-6)(7)-(-2)(12)}{(7)(7)-(-2)(4)} = \frac{-18}{57} = -\frac{6}{19}$$

$$y = \frac{\begin{vmatrix} 7 & -6 \\ 4 & 12 \end{vmatrix}}{\begin{vmatrix} 7 & -2 \\ 4 & 7 \end{vmatrix}} = \frac{(7)(12)-(-6)(4)}{(7)(7)-(-2)(4)} = \frac{108}{57} = \frac{36}{19}$$

49. Exercise 41 is well-suited for substitution since the variable y is already isolated in the second equation.

53. $\begin{vmatrix} 4 & -1 & 8 \\ -1 & 6 & -2 \\ 2 & 1 & -1 \end{vmatrix} = (4)(6)(-1) - (4)(-2)(1) + (-1)(-2)(2) - (-1)(-1)(-1) + (8)(-1)(1) - (8)(6)(2)$

$= -24 + 8 + 4 + 1 - 8 - 96$

$= -115$

57. $2x + y + z = 4$ (I)

$x - 2y - z = 3$ (II)

$3x + 3y - 2z = 1$ (III)

Add (I) and (II) to produce (IV); Add 2(I) and (III) to produce (V):

$3x - y = 7$ (IV)

$7x + 5y = 9$ (V)

Add 5(IV) and (V):

$22x = 44 \rightarrow x = 2$

Substitute $x = 2$ into (IV):

$6 - y = 7 \rightarrow y = -1$

Substitute $x = 2$ and $y = -1$ into (I):

$4 - 1 + z = 4 \rightarrow z = 1$

The solution is $x = 2, y = -1, z = 1$.

61. $3.6x + 5.2y - z = -2.2$ (I)
$3.2x - 4.8y + 3.9z = 8.1$ (II)
$6.4x + 4.1y + 2.3z = 5.1$ (III)
Add 3.9(I) and (II) to produce (IV); Add 2.3(I) and (III) to produce (V):
$17.24x + 15.48y = -0.48$ (IV)
$14.68x + 16.06y = 0.04$ (V)
Add 16.06(IV) and -15.48(V):
$49.628x = -8.328 \to x = -0.1678085$
Substitute $x = -0.1678085$ into (V):
$-2.46342878 + 16.06y = 0.04 \to y = 0.1558798$
Substitute $x = -0.1678085$ and $y = 0.1558798$ into (I):
$-0.6041106 + 0.8105747 - z = -2.2 \to z = 2.4064641$
The solution is $x = -0.17, y = 0.16, z = 2.4$.

65. $2x + y + z = 4$ (I)
$x - 2y - z = 3$ (II)
$3x + 3y - 2z = 1$ (III)

$$x = \frac{\begin{vmatrix} 4 & 1 & 1 \\ 3 & -2 & -1 \\ 1 & 3 & -2 \end{vmatrix}}{\begin{vmatrix} 2 & 1 & 1 \\ 1 & -2 & -1 \\ 3 & 3 & -2 \end{vmatrix}} = \frac{44}{22} = 2;\ y = \frac{\begin{vmatrix} 2 & 4 & 1 \\ 1 & 3 & -1 \\ 3 & 1 & -2 \end{vmatrix}}{\begin{vmatrix} 2 & 1 & 1 \\ 1 & -2 & -1 \\ 3 & 3 & -2 \end{vmatrix}} = \frac{-22}{22} = -1;\ z = \frac{\begin{vmatrix} 2 & 1 & 4 \\ 1 & -2 & 3 \\ 3 & 3 & 1 \end{vmatrix}}{\begin{vmatrix} 2 & 1 & 1 \\ 1 & -2 & -1 \\ 3 & 3 & -2 \end{vmatrix}} = \frac{22}{22} = 1$$

The solution is $x = 2, y = -1, z = 1$.

69. $3.6x + 5.2y - z = -2.2$ (I)
$3.2x - 4.8y + 3.9z = 8.1$ (II)
$6.4x + 4.1y + 2.3z = 5.1$ (III)

$$x = \frac{\begin{vmatrix} -2.2 & 5.2 & -1.0 \\ 8.1 & -4.8 & 3.9 \\ 5.1 & 4.1 & 2.3 \end{vmatrix}}{\begin{vmatrix} 3.6 & 5.2 & -1.0 \\ 3.2 & -4.8 & 3.9 \\ 6.4 & 4.1 & 2.3 \end{vmatrix}} = \frac{8.328}{-49.628};\ y = \frac{\begin{vmatrix} 3.6 & -2.2 & -1.0 \\ 3.2 & 8.1 & 3.9 \\ 6.4 & 5.1 & 2.3 \end{vmatrix}}{\begin{vmatrix} 3.6 & 5.2 & -1.0 \\ 3.2 & -4.8 & 3.9 \\ 6.4 & 4.1 & 2.3 \end{vmatrix}} = \frac{-7.736}{-49.628};\ z = \frac{\begin{vmatrix} 3.6 & 5.2 & -2.2 \\ 3.2 & -4.8 & 8.1 \\ 6.4 & 4.1 & 5.1 \end{vmatrix}}{\begin{vmatrix} 3.6 & 5.2 & -1.0 \\ 3.2 & -4.8 & 3.9 \\ 6.4 & 4.1 & 2.3 \end{vmatrix}} = \frac{-119.428}{-49.628}$$

The solution is $x = -0.17, y = 0.16, z = 2.4$

73. $$5 = \begin{vmatrix} x & 1 & 2 \\ 0 & -1 & 3 \\ -2 & 2 & 1 \end{vmatrix} = x(-1-6) + 1(-6-0) + 2(0-2) = -7x - 10$$

$x = -\frac{15}{7}$

77. The system is dependent or inconsistent (here, it is dependent) if

$0 = \begin{vmatrix} 3 & -k \\ 1 & 2 \end{vmatrix} = 6 + k$, or if $k = -6$.

81. $F_1 + 2.0F_2 = 280 \quad$ (I)

$0.87F_1 - F_3 = 0 \quad$ (II)

$3.0F_1 - 4.0F_2 = 600 \quad$ (III)

$$F_1 = \frac{\begin{vmatrix} 280 & 2 & 0 \\ 0 & 0 & -1 \\ 600 & 4 & 0 \end{vmatrix}}{\begin{vmatrix} 1 & 2 & 0 \\ 0.87 & 0 & -1 \\ 3 & 4 & 0 \end{vmatrix}} = \frac{-80}{-2}; F_2 = \frac{\begin{vmatrix} 1 & 280 & 0 \\ 0.87 & 0 & -1 \\ 3 & 600 & 0 \end{vmatrix}}{\begin{vmatrix} 1 & 2 & 0 \\ 0.87 & 0 & -1 \\ 3 & 4 & 0 \end{vmatrix}} = \frac{-240}{-2}; F_3 = \frac{\begin{vmatrix} 1 & 2 & 280 \\ 0.87 & 0 & 0 \\ 3 & 4 & 600 \end{vmatrix}}{\begin{vmatrix} 1 & 2 & 0 \\ 0.87 & 0 & -1 \\ 3 & 4 & 0 \end{vmatrix}} = \frac{-69.6}{-2};$$

The solution is $F_1 = 40, F_2 = 120, F_3 = 35$.

85. Let x be the sales in a month. At parity, we have

$0.10x = 2400 + 0.04x$

$0.06x = 2400$

$x = 40000$

The monthly sales at parity is \$40000

and the sales reps would make 10% of this amount, or \$4000.

89. At $x = 0, T = 14 = \frac{a}{100} + b$, or $a + 100b = 1400$ (I)

At $x = 900, T = 10 = \frac{a}{1000} + b$, or $a + 1000b = 10000$ (II)

Subtracting (I) from (II),

$900b = 8600 \rightarrow b = 9.556$

and $a = 1400 - 100(9.556) = 444.4$

We have $a = 440$ and $b = 9.6$.

93. Let x and y denote the numbers of 800 ft^2 and 1100 ft^2 offices, respectively.

The system of equations becomes

$800x + 1100y = 49200$

$900x + 1250y = 55600$

Using Cramer's rule,

$$x = \frac{\begin{vmatrix} 49200 & 1100 \\ 55600 & 1250 \end{vmatrix}}{\begin{vmatrix} 800 & 1100 \\ 900 & 1250 \end{vmatrix}} = \frac{340000}{10000}; y = \frac{\begin{vmatrix} 800 & 49200 \\ 900 & 55600 \end{vmatrix}}{\begin{vmatrix} 800 & 1100 \\ 900 & 1250 \end{vmatrix}} = \frac{200000}{10000}$$

$x = 34$ and $y = 20$

There are 34 smaller and 20 larger offices.

97. The system of equations becomes

$R_1 + 9R_2 = 14$ (I)

$9R_1 + R_2 = 6$ (II)

Using Cramer's rule,

$$R_1 = \frac{\begin{vmatrix} 14 & 9 \\ 6 & 1 \end{vmatrix}}{\begin{vmatrix} 1 & 9 \\ 9 & 1 \end{vmatrix}} = \frac{-40}{-80}; R_2 = \frac{\begin{vmatrix} 1 & 14 \\ 9 & 6 \end{vmatrix}}{\begin{vmatrix} 1 & 9 \\ 9 & 1 \end{vmatrix}} = \frac{-120}{-80}$$

$R_1 = 0.5\Omega$ and $R_2 = 1.5\Omega$

101. Let a, b, c be the measures of angles A, B, C respectively.

We have

$a + b + c = 180$

$a = 2b - 55$

$c = b - 25$

Substitute the latter two equations into the first:

$(2b - 55) + b + (b - 25) = 180$

$4b - 80 = 180$

$b = 65$

$c = 65 - 25 = 40$

$a = 130 - 55 = 75$

The three angles have measures $a = 75°, b = 65°, c = 40°$.

105. Let a and b represent the weights of gold and silver in air, respectively.

The system of equations becomes

$a + \quad b = 6.0$ (I)

$0.947a + 0.9b = 5.6$ (II)

Subtracting 0.9(I) from (II) yields

$0.047a = 0.2 \rightarrow a = 4.255$

$b = 6.0 - a = 1.745$

The gold weighs 4.3 N and the silver weighs 1.7 N.

Chapter 6

Factoring and Fractions

6.1 Factoring: Greatest Common Factor and Difference of Squares

1. $$\begin{aligned} 4ax^2 - 2ax &= 2ax(2x) - 2ax(1) \\ &= 2ax(2x-1) \end{aligned}$$

5. $$\begin{aligned} (T+6)(T-6) &= T^2 - 6^2 \\ &= T^2 - 36 \end{aligned}$$

9. $7x + 7y = 7(x+y)$

13. $3x^2 - 9x = 3x(x-3)$

17. $72n^3 + 24n = 24n(3n^2+1)$

21. $3ab^2 - 6ab + 12ab^3 = 3ab(b - 2 + 4b^2)$

25. $2a^2 - 2b^2 + 4c^2 - 6d^2 = 2(a^2 - b^2 + 2c^2 - 3d^2)$

29. $$\begin{aligned} 100 - 4A^2 &= (10)^2 - (2A)^2 \\ 100 - 4A^2 &= (10+2A)(10-2A) \end{aligned}$$

33. $$\begin{aligned} 162s^2 - 50t^2 &= 2(81s^2 - 25t^2) \\ &= 2((9s)^2 - (5t)^2) \\ &= 2(9s+5t)(9s-5t) \end{aligned}$$

37. $$\begin{aligned} (x+y)^2 - 9 &= ((x+y)+3)((x+y)-3) \\ &= (x+y+3)(x+y-3) \end{aligned}$$

41. $$\begin{aligned} 300x^2 - 2700z^2 &= 300(x^2 - 9z^2) \\ &= 300(x+3z)(x-3z) \end{aligned}$$

45. $$\begin{aligned} x^4 - 16 &= (x^2+4)(x^2-4) \\ &= (x^2+4)(x+2)(x-2) \end{aligned}$$

49. Solve $2a - b = ab + 3$ for a.

$$\begin{aligned} 2a - ab &= b + 3 \\ a(2-b) &= b + 3 \\ a &= \frac{b+3}{2-b} \end{aligned}$$

53. Solve $(x+2k)(x-2)=x^2+3x-4k$ for k.

$$\begin{aligned}(x+2k)(x-2)&=x^2+3x-4k\\ x^2-2x+2kx-4k&=x^2+3x-4k\\ 2kx&=5x\\ k&=\frac{5}{2}\end{aligned}$$

57. $$\begin{aligned}a^2+ax-ab-bx&=\left(a^2+ax\right)-(ab+bx)\\ &=a(a+x)-b(a+x)\\ &=(a+x)(a-b)\end{aligned}$$

61. $$\begin{aligned}x^2-y^2+x-y&=\left(x^2-y^2\right)+(x-y)\\ &=(x+y)(x-y)+(x-y)\\ &=(x-y)(x+y+1)\end{aligned}$$

65. Since $n^2+n=n(n+1)$, this is the product of two consecutive integers of which one must be even. Therefore, the product is even.

69. $$\begin{aligned}81s-s^3&=s(81-s^2)\\ &=s(9+s)(9-s)\end{aligned}$$

73.

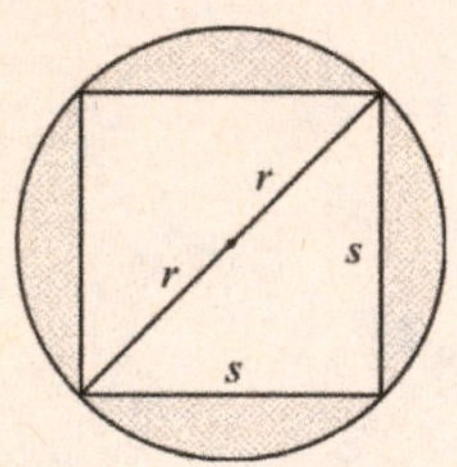

Using Pythagoras' Theorem

$$\begin{aligned}s^2+s^2&=(2r)^2\\ 2s^2&=4r^2\\ s^2&=2r^2\end{aligned}$$

Area of square $= s^2 = 2r^2$

Area of circle $= \pi r^2$

Area left = Area of circle − Area of square

Area left $=\pi r^2-2r^2$

Area left $= r^2(\pi-2)$

77. Solve $i_1R_1 = (i_2 - i_1)R_2$ for i_1.

$$i_1R_1 = (i_2 - i_1)R_2$$
$$i_1R_1 = i_2R_2 - i_1R_2$$
$$i_1R_1 + i_1R_2 = i_2R_2$$
$$i_1(R_1 + R_2) = i_2R_2$$
$$i_1 = \frac{i_2R_2}{R_1 + R_2}$$

81. Solve $ER = AtT_0 - AtT_1$ for t

$$ER = AtT_0 - AtT_1$$
$$ER = At(T_0 - T_1)$$
$$t = \frac{ER}{A(T_0 - T_1)}$$

6.2 Factoring Trinomials

1. $x^2 + 4x + 3 = (x+3)(x+1)$

5. $(x-7)^2 = x^2 - 2(7)x + 7^2$
$= x^2 - 14x + 49$

9. $x^2 + 4x + 3 = (x+1)(x+3)$

13. $t^2 + 5t - 24 = (t+8)(t-3)$

17. $a^2 - 6ab + 9b^2 = (a-3b)(a-3b)$
$= (a-3b)^2$

21. $12y^2 - 32y - 12 = 4(3y^2 - 8y - 3)$
$= 4(3y+1)(y-3)$
because $-8y = y - 9y$

25. $3z^2 - 19z + 6 = (3z-1)(z-6)$
(because $-19z = -18z - z$)

29. $3t^2 - 7tu + 4u^2 = (3t-4u)(t-u)$
(because $-7tu = -4tu - 3tu$)

33. $9x^2 + 7xy - 2y^2 = (x+y)(9x-2y)$
(because $7xy = 9xy - 2xy$)

37. $8x^2 - 24x + 18 = 2(4x^2 - 12x + 9)$
$= 2(2x - 3)(2x - 3)$
$= 2(2x - 3)^2$
(because $-12x = -6x - 6x$)

41. $8b^6 + 31b^3 - 4 = (8b^3 - 1)(b^3 + 4)$
(because $31b^3 = -b^3 + 32b^3$)
But the first factor is actually a difference of cubes
(discussed in Section 6.4)
So this could be further factored to
$8b^6 + 31b^3 - 4 = ((2b)^3 - 1^3)(b^3 + 4)$
$= (2b - 1)(4b^2 + 2b + 1)(b^3 + 4)$

45. $12x^2 + 47xy - 4y^2 = (12x - y)(x + 4y)$
(because $47xy = -xy + 48xy$)

49. $4x^5 + 14x^3 - 8x = 2x(2x^4 + 7x^2 - 4)$
$= 2x(2x^2 - 1)(x^2 + 4)$
because $8x^2 - x^2 = 7x^2$

53. $4x^{2n} + 13x^n - 12 = (4x^n - 3)(x^n + 4)$
(because $16x^n - 3x^n = 13x^n$)

57. $d^4 - 10d^2 + 16^2 = (d^2 - 8)(d^2 - 2)$

61. $V^2 - 2nBV + n^2B^2 = (V - nB)^2$

65. $3Adu^2 - 4Aduv + Adv^2 = Ad(3u^2 - 4uv + v^2)$
$= Ad(3u - v)(u - v)$

69. Find six values of k such that $x^2 + kx + 18$ can be factored
Since $18 = (1)(18) = (2)(9) = (3)(6)$, we obtain
$(x + 1)(x + 18) = x^2 + 19x + 18$
$(x + 2)(x + 9) = x^2 + 11x + 18$
$(x + 3)(x + 6) = x^2 + 9x + 18$
Also, $18 = (-1)(-18) = (-2)(-9) = (-3)(-6)$ and so
$(x - 1)(x - 18) = x^2 - 19x + 18$
$(x - 2)(x - 9) = x^2 - 11x + 18$
$(x - 3)(x - 6) = x^2 - 9x + 18$
and so six values of k are -19, -11, -9, 9, 11, or 19.

6.3 The Sum and Difference of Cubes

1. $x^3-8=x^3-2^3$
$=(x-2)(x^2+2x+2^2)$
$=(x-2)(x^2+2x+4)$

5. $y^3-125=y^3-5^3$
$=(y-5)(y^2+5y+25)$

9. $8a^3-27b^3=(2a)^3-(3b)^3$
$=(2a-3b)(4a^2+6ab+9b^2)$

13. $7n^5-7n^2=7n^2(n^3-1)$
$=7n^2(n-1)(n^2+n+1)$

17. $x^6y^3+x^3y^6=x^3y^3(x^3+y^3)$
$=x^3y^3(x+y)(x^2-xy+y^2)$

21. $\frac{4}{3}\pi R^3-\frac{4}{3}\pi r^3=\frac{4}{3}\pi(R^3-r^3)$
$=\frac{4}{3}\pi(R-r)(R^2+Rr+r^2)$

25. $(a+b)^3+64=(a+b)^3+4^3$
$=(a+b+4)((a+b)^2-4(a+b)+4^2)$
$=(a+b+4)(a^2+2ab+b^2-4a-4b+16)$

29. $x^6-2x^3+1=(x^3-1)^2$
$=((x-1)(x^2+x+1))^2$
$=(x-1)^2(x^2+x+1)^2$

33. $D^4-d^3D=D(D^3-d^3)$
$=D(D-d)(D^2+Dd+d^2)$

37. $(a+b)(a^2-ab+b^2)=a(a^2-ab+b^2)+b(a^2-ab+b^2)$
$=(a^3-a^2b+ab^2)+(a^2b-ab^2+b^3)$
$=a^3+b^3$

41. $n^3+1=(n+1)(n^2-n+1)$
which, for $n>1$, is the product of two integers each greater than 1.
Thus, this product cannot be prime.

6.4 Equivalent Fractions

1. $$\frac{x^2-4x-12}{x^2-4}=\frac{(x+2)(x-6)}{(x+2)(x-2)}$$
$$=\frac{x-6}{x-2} \text{ where } x\neq 2, x\neq -2$$

5. $$\frac{ax}{y}=\frac{ax(3a)}{y(3a)}=\frac{3a^2x}{3ay}$$

9. $$\frac{a(x-y)}{x-2y}=\frac{a(x-y)(x+y)}{(x-2y)(x+y)}$$
$$=\frac{a(x^2-y^2)}{x^2-xy-2y^2}$$
$$=\frac{ax^2-ay^2}{x^2-xy-2y^2}$$

13. $$\frac{4x^2y}{8xy^2}=\frac{\frac{4x^2y}{2x}}{\frac{8xy^2}{2x}}=\frac{2xy}{4y^2} \text{ where } x,y\neq 0$$

17. $$\frac{s^2-3s-10}{2s^2-3s-2}=\frac{\frac{(s+2)(s-5)}{(s+2)}}{\frac{(s+2)(2s-1)}{(s+2)}}=\frac{s-5}{2s-1} \text{ where } s\neq -2,\tfrac{1}{2}$$

21. $$\frac{6a-24}{A}=\frac{6}{(a+4)}\cdot\frac{(a-4)}{(a-4)}$$
$$\frac{6a-24}{A}=\frac{6a-24}{a^2-16}$$
$$A=a^2-16$$

25. $$\frac{x+4b}{A}=\frac{x^2+3bx-4b^2}{x-b}$$
$$\frac{x+4b}{A}=\frac{(x+4b)(x-b)}{(x-b)}$$
$$\frac{x+4b}{A}=\frac{x+4b}{1} \text{ where } x\neq b$$
$$A=1$$

29. $$\frac{18x^2y}{24xy}=\frac{6xy(3x)}{6xy(4)}=\frac{3x}{4} \text{ where } x,y\neq 0$$

33. $$\frac{4a-4b}{4a-2b}=\frac{4(a-b)}{2(2a-b)}=\frac{2(a-b)}{2a-b}$$

37. $$\frac{3x^2-6x}{x-2}=\frac{3x(x-2)}{(x-2)}=3x \text{ where } x\neq 2$$

41. $\dfrac{x^2-10x+25}{x^2-25}=\dfrac{(x-5)(x-5)}{(x+5)(x-5)}=\dfrac{x-5}{x+5}$ where $x \neq -5, 5$

45. $\dfrac{5x^2-6x-8}{x^3+x^2-6x}=\dfrac{(5x+4)(x-2)}{x(x^2+x-6)}$

$=\dfrac{(5x+4)(x-2)}{x(x+3)(x-2)}$

$=\dfrac{5x+4}{x(x+3)}$ where $x \neq -3, 0, 2$

49. $\dfrac{t+4}{(2t+9)t+4}=\dfrac{t+4}{2t^2+9t+4}$

$=\dfrac{(t+4)}{(2t+1)(t+4)}$

$=\dfrac{1}{2t+1}$ where $t \neq -4, -\frac{1}{2}$

53. $\dfrac{y^2-x^2}{2x-2y}=\dfrac{(y+x)(y-x)}{2(x-y)}$

$=\dfrac{(y-x)(y+x)}{-2(y-x)}$

$=-\dfrac{y+x}{2}$ where $x \neq y$

57. $\dfrac{(x+5)(x-2)(x+2)(3-x)}{(2-x)(5-x)(3+x)(2+x)}=\dfrac{-(x+5)(x-2)(x+2)(x-3)}{-(5-x)(x-2)(x+2)(x+3)}$

$=\dfrac{(x+5)(x-3)}{(5-x)(x+3)}$ where $x \neq \pm 2, -3, 5$

61. $\dfrac{6x^2+2x}{27x^3+1}=\dfrac{2x(3x+1)}{(3x+1)(9x^2-3x+1)}$

$=\dfrac{2x}{9x^2-3x+1}$ where $x \neq -\frac{1}{3}$

65. **(a)** $\dfrac{x^2-x-2}{x^2-x}=\dfrac{(x-2)(x+1)}{x(x-1)}$

Numerator and denominator have no common factor.

(b) $\dfrac{x^2-x-2}{x^2+x}=\dfrac{(x-2)(x+1)}{x(x+1)}=\dfrac{x-2}{x}$ where $x \neq 0, -1$

Numerator and denominator have no common factor.

69. $\dfrac{v^2-v_0^2}{vt-v_0t}=\dfrac{(v+v_0)(v-v_0)}{t(v-v_0)}$

$=\dfrac{v+v_0}{t}$ where $v\neq v_0, t\neq 0$

73. $\dfrac{E^2R^2-E^2r^2}{\left(R^2+2Rr+r^2\right)^2}=\dfrac{E^2\left(R^2-r^2\right)}{\left[\left[(R+r)(R+r)\right]^2\right]^2}$

$=\dfrac{E^2(R+r)(R-r)}{(R+r)^4}$

$=\dfrac{E^2(R-r)}{(R+r)^3}$

where $R\neq -r$ (shouldn't be an issue since resistance is always positive)

6.5 Multiplication and Division of Fractions

1. $\dfrac{2x-4}{4x-10}\times\dfrac{2x^2+x-15}{3x-1}=\dfrac{2(x-2)(2x-5)(x+3)}{2(2x-5)(3x-1)}$

$=\dfrac{(x-2)(x+3)}{3x-1}$ where $x\neq\frac{5}{2},\frac{1}{3}$

5. $\dfrac{4x}{3y}\times\dfrac{9y^2}{2}=\dfrac{2(2x)(3y)(3y)}{(3y)(2)}$

$=\dfrac{(2x)(3y)}{(1)(1)}$

$=6xy$ where $y\neq 0$

(divide out a common factor of 2 and $3y$)

9. $\dfrac{yz}{az}\div\dfrac{bz}{ay}=\dfrac{yz}{az}\times\dfrac{ay}{bz}=\dfrac{yz(ay)}{az(bz)}=\dfrac{y^2}{bz}$ where $a,b,y,z\neq 0$

(divide out common factors of a and z)

13. $\dfrac{u^2-v^2}{u+2v}(3u+6v)=\dfrac{(u+v)(u-v)(3)(u+2v)}{u+2v}$

$=3(u+v)(u-v)$ where $u\neq -2v$

(divide out a common factor of $(u+2v)$)

17. $\dfrac{x^4-9}{x^2}\div\left(x^2+3\right)^2=\dfrac{x^4-9}{x^2}\times\dfrac{1}{\left(x^2+3\right)^2}$

$$=\frac{\left(x^2+3\right)\left(x^2-3\right)}{x^2\left(x^2+3\right)^2}$$

$$=\frac{x^2-3}{x^2\left(x^2+3\right)} \text{ where } x\neq 0$$

(divide out a common factor of $\left(x^2+3\right)$)

21. $\dfrac{x^4-1}{8x+16}\quad\dfrac{2x^2-8x}{x^3+x}=\dfrac{\left(x^2-1\right)\left(x^2+1\right)(2x)(x-4)}{8(x+2)(x)\left(x^2+1\right)}$

$$=\frac{2x(x+1)(x-1)(x-4)}{8x(x+2)}$$

$$=\frac{(x+1)(x-1)(x-4)}{4(x+2)} \quad \text{where } x\neq 0,-2$$

(divide out common factors of 2, x, and $\left(x^2+1\right)$)

25. $\dfrac{35a+25}{12a+33}\div\dfrac{28a+20}{36a+99}=\dfrac{5(7a+5)}{3(4a+11)}\times\dfrac{9(4a+11)}{4(7a+5)}$

$$=\frac{15}{4} \quad \text{where } a\neq -\tfrac{5}{7},-\tfrac{11}{4}$$

(divide out common factors of 3, $(7a+5)$, and $(4a+11)$)

29. $\dfrac{\frac{6T^2-NT-N^2}{2V^2-9V-35}}{\frac{8T^2-2NT-N^2}{20V^2+26V-60}}=\dfrac{6T^2-NT-N^2}{2V^2-9V-35}\times\dfrac{20V^2+26V-60}{8T^2-2NT-N^2}$

$$=\frac{(2T-N)(3T+N)}{(2V+5)(V-7)}\times\frac{2\left(10V^2+13V-30\right)}{(4T+N)(2T-N)}$$

$$=\frac{2(2T-N)(3T+N)(5V-6)(2V+5)}{(2V+5)(V-7)(4T+N)(2T-N)}$$

$$=\frac{2(3T+N)(5V-6)}{(V-7)(4T+N)} \quad \text{where } T\neq\frac{N}{2} \text{ and } V\neq -\frac{5}{2}$$

(divide out a common factor of $(2V+5)$ and $(2T-N)$

33. $\dfrac{4t^2-1}{t-5}\div\dfrac{2t+1}{2t}\times\dfrac{2t^2-50}{1+4t+4t^2}=\dfrac{4t^2-1}{t-5}\times\dfrac{2t}{2t+1}\times\dfrac{2t^2-50}{4t^2+4t+1}$

$$=\frac{(2t+1)(2t-1)(2t)(2)\left(t^2-25\right)}{(t-5)(2t+1)(2t+1)(2t+1)}$$

$$=\frac{4t(2t+1)(2t-1)(t+5)(t-5)}{(t-5)(2t+1)(2t+1)(2t+1)}$$

$$=\frac{4t(2t-1)(t+5)}{(2t+1)^2} \quad \text{where } t\neq 5,0,-\frac{1}{2}$$

(divide out common factors $(2t+1)$ and $(t-5)$)

37. $$\left(\frac{ax+bx+ay+by}{p-q}\right)\left(\frac{3p^2+4pq-7q^2}{a+b}\right)=\frac{x(a+b)+y(a+b)}{p-q}\times\frac{(3p+7q)(p-q)}{a+b}$$
$$=\frac{(a+b)(x+y)(3p+7q)(p-q)}{(p-q)(a+b)}$$
$$=(x+y)(3p+7q) \text{ where } a\neq -b \text{ and } p\neq q$$
(divide out common factors of $(a+b)$ and $(p-q)$)

41. $$\frac{2x^2+3x-2}{2+3x-2x^2}\div\frac{5x+10}{4x+2}=\frac{2x^2+3x-2}{-(2x^2-3x-2)}\times\frac{4x+2}{5x+10}$$
$$=\frac{(2x-1)(x+2)}{-(2x+1)(x-2)}\times\frac{2(2x+1)}{5(x+2)}$$
$$=-\frac{2(2x-1)}{5(x-2)} \text{ where } x\neq -\frac{1}{2},\pm 2$$

45. $$\frac{c\lambda^2-c\lambda_0^2}{\lambda_0^2}\div\frac{\lambda^2+\lambda_0^2}{\lambda_0^2}=\frac{c(\lambda^2-\lambda_0^2)}{\lambda_0^2}\times\frac{\lambda_0^2}{\lambda^2+\lambda_0^2}$$
$$=\frac{c(\lambda+\lambda_0)(\lambda-\lambda_0)}{\lambda_0^2}\times\frac{\lambda_0^2}{\lambda^2+\lambda_0^2}$$
$$=\frac{c(\lambda+\lambda_0)(\lambda-\lambda_0)}{\lambda^2+\lambda_0^2} \text{ where } \lambda_0\neq 0$$

6.6 Addition and Subtraction of Fractions

1. $4a^2b=2\cdot 2\cdot a\cdot a\cdot b$

$6ab^3=2\cdot 3\cdot a\cdot b\cdot b\cdot b$

$4a^2b^2=2\cdot 2\cdot a\cdot a\cdot b\cdot b$

$L.C.D.=2^2\cdot 3\cdot a^2\cdot b^3=12a^2b^3$

5. $$\frac{5}{6}+\frac{7}{6}=\frac{5+7}{6}=\frac{12}{6}=2$$

9. $$\frac{1}{2}+\frac{3}{4}=\frac{1(2)+3}{4}=\frac{2+3}{4}=\frac{5}{4}$$

13. $$\frac{a}{x}-\frac{b}{x^2}=\frac{a(x)-b}{x^2}=\frac{ax-b}{x^2}$$

17. $$\frac{2}{5a}+\frac{1}{a}-\frac{a}{10}=\frac{2(2)+1(10)-a(a)}{10a}=\frac{14-a^2}{10a}$$

21.
$$\frac{y^2}{y+3}-\frac{2y+15}{y+3}=\frac{y^2-2y-15}{y+3}$$
$$=\frac{(y+3)(y-5)}{y+3}$$
$$=y-5$$

25.
$$\frac{4}{x(x+1)}-\frac{3}{2x}=\frac{4(2)-3(x+1)}{2x(x+1)}$$
$$=\frac{8-3x-3}{2x(x+1)}$$
$$=\frac{5-3x}{2x(x+1)}$$

29.
$$\frac{3R}{R^2-9}-\frac{2}{3R+9}=\frac{3R}{(R+3)(R-3)}-\frac{2}{3(R+3)}$$
$$=\frac{3R(3)-2(R-3)}{3(R+3)(R-3)}$$
$$=\frac{9R-2R+6}{3(R+3)(R-3)}$$
$$=\frac{7R+6}{3(R+3)(R-3)}$$

33.
$$\frac{v+4}{v^2+5v+4}-\frac{v-2}{v^2-5v+6}=\frac{v+4}{(v+4)(v+1)}-\frac{v-2}{(v-3)(v-2)}$$
$$=\frac{1}{v+1}-\frac{1}{v-3} \quad \text{where } v\neq -4, 2$$
$$=\frac{1(v-3)-1(v+1)}{(v+1)(v-3)}$$
$$=\frac{-4}{(v+1)(v-3)}$$

37.
$$\frac{t}{t^2-t-6}-\frac{2t}{t^2+6t+9}+\frac{t}{9-t^2}=\frac{t}{(t-3)(t+2)}-\frac{2t}{(t+3)(t+3)}-\frac{t}{(t+3)(t-3)}$$
$$=\frac{t(t+3)^2-2t(t-3)(t+2)-t(t+3)(t+2)}{(t-3)(t+2)(t+3)^2}$$
$$=\frac{t(t^2+6t+9)-2t(t^2-t-6)-t(t^2+5t+6)}{(t+3)^2(t-3)(t+2)}$$
$$=\frac{t^3+6t^2+9t-2t^3+2t^2+12t-t^3-5t^2-6t}{(t+3)^2(t-3)(t+2)}$$
$$=\frac{-2t^3+3t^2+15t}{(t+3)^2(t-3)(t+2)}$$
$$=\frac{-t(2t^2-3t-15)}{(t+3)^2(t-3)(t+2)}$$

41. $$\frac{\frac{3}{x}}{\frac{1}{x}-1}=\frac{\frac{3}{x}}{\frac{1-x}{x}}$$

$$=\frac{3}{x}\times\frac{x}{1-x}$$

$$=\frac{3}{1-x},\ x\neq 0,1$$

45. $$\frac{\frac{3}{x}+\frac{1}{x^2+x}}{\frac{1}{x+1}-\frac{1}{x-1}}=\frac{\frac{3}{x}+\frac{1}{x(x+1)}}{\frac{1}{x+1}-\frac{1}{x-1}}$$

$$=\frac{\frac{3(x+1)+1}{x(x+1)}}{\frac{1(x-1)-1(x+1)}{(x+1)(x-1)}}$$

$$=\frac{3x+3+1}{x(x+1)}\times\frac{(x+1)(x-1)}{x-1-x-1}$$

$$=\frac{3x+4}{x(x+1)}\times\frac{(x+1)(x-1)}{-2}$$

$$=-\frac{(3x+4)(x-1)}{2x}\quad ,x\neq 0,1,-1$$

49. $$f(x)=\frac{1}{x^2}$$

$$f(x+h)-f(x)=\frac{1}{(x+h)^2}-\frac{1}{x^2}$$

$$=\frac{1(x^2)-1(x^2+2xh+h^2)}{x^2(x+h)^2}$$

$$=\frac{x^2-x^2-2xh-h^2}{x^2(x+h)^2}$$

$$=\frac{-2xh-h^2}{x^2(x+h)^2}$$

$$=\frac{-h(2x+h)}{x^2(x+h)^2}$$

53. $$f(x)=2x-x^2$$

$$f\left(\frac{1}{a}\right)=2\left(\frac{1}{a}\right)-\left(\frac{1}{a}\right)^2$$

$$f\left(\frac{1}{a}\right)=\frac{2}{a}-\frac{1}{a^2}$$

$$f\left(\frac{1}{a}\right)=\frac{2a-1}{a^2}$$

57. $\dfrac{a+b}{\frac{1}{a}+\frac{1}{b}}=\dfrac{a+b}{\frac{b+a}{ab}}$

$\dfrac{a+b}{\frac{1}{a}+\frac{1}{b}}=(a+b)\times\dfrac{ab}{(a+b)}$

$\dfrac{a+b}{\frac{1}{a}+\frac{1}{b}}=ab\quad\text{where } a\neq -b$

61. $\dfrac{3}{4\pi}-\dfrac{3H_0}{4\pi H}=\dfrac{3(H)-3H_0}{4\pi H}=\dfrac{3(H-H_0)}{4\pi H}$

65. $\left(\dfrac{3Px}{2L^2}\right)^2+\left(\dfrac{P}{2L}\right)^2=\dfrac{9P^2x^2}{4L^4}+\dfrac{P^2}{4L^2}$

$=\dfrac{9P^2x^2+P^2(L^2)}{4L^4}$

$=\dfrac{P^2\left(9x^2+L^2\right)}{4L^4}$

69. $\dfrac{1}{R^2}+\quad\omega c-\dfrac{1}{\omega L}\quad^2=\dfrac{1}{R^2}+\quad\dfrac{\omega^2Lc-1}{\omega L}\quad^2$

$=\dfrac{1}{R^2}+\dfrac{\omega^4L^2c^2-2\omega^2Lc+1}{\omega^2L^2}$

$=\dfrac{\omega^2L^2+\omega^4L^2c^2R^2-2\omega^2LcR^2+R^2}{R^2\omega^2L^2}$

6.7 Equations Involving Fractions

1. $\dfrac{x}{2}-\dfrac{1}{b}=\dfrac{x}{2b}$

$\dfrac{x(2b)}{2}-\dfrac{1(2b)}{b}=\dfrac{x(2b)}{2b}$

$xb-2=x$

$x(b-1)=2$

$x=\dfrac{2}{b-1}$

which gives division by zero in the second fraction.
Therefore, no solution.

5. $\dfrac{x}{2}+6=2x$

$\dfrac{x(2)}{2}+6(2)=2x(2)$

$x+12=4x$

$-3x=-12$

$x=4$

9.

$$\frac{1}{4}-\frac{t-3}{8}=\frac{1}{6}$$
$$\frac{1(24)}{4}-\frac{(t-3)(24)}{8}=\frac{1(24)}{6}$$
$$6-3(t-3)=4$$
$$6-3t+9=4$$
$$-3t=-11$$
$$t=\frac{11}{3}$$

13.

$$\frac{3}{T}+2=\frac{5}{3}$$
$$\frac{3(3T)}{T}+2(3T)=\frac{5(3T)}{3}$$
$$9+6T=5T$$
$$T=-9$$

17.

$$3-\frac{x-2}{5x}=\frac{1}{5}$$
$$3(5x)-\frac{(x-2)(5x)}{5x}=\frac{1(5x)}{5}$$
$$15x-x+2=x$$
$$13x=-2$$
$$x=-\frac{2}{13}$$

21.

$$\frac{2}{s}=\frac{3}{s-1}$$
$$\frac{2s(s-1)}{s}=\frac{3s(s-1)}{s-1}$$
$$2(s-1)=3s$$
$$2s-2=3s$$
$$s=-2$$

25.

$$\frac{2}{z-5}-\frac{7}{10-2z}=3$$
$$\frac{2}{Z-5}+\frac{7}{2(Z-5)}=3$$
$$\frac{2(2)(z-5)}{z-5}+\frac{7(2)(z-5)}{2(z-5)}=3(2)(z-5)$$
$$4+7=6z-30$$
$$41=6z$$
$$z=\frac{41}{6}$$

29.

$$\frac{5}{y}=\frac{2}{y-3}+\frac{7}{2y^2-6y}$$

$$\frac{5}{y}=\frac{2}{y-3}+\frac{7}{2y(y-3)}$$

$$\frac{5(2y)(y-3)}{y}=\frac{2(2y)(y-3)}{y-3}+\frac{7(2y)(y-3)}{2y(y-3)}$$

$$10(y-3)=4y+7$$

$$10y-30=4y+7$$

$$6y=37$$

$$y=\frac{37}{6}$$

33.

$$\frac{2}{B^2-4}-\frac{1}{B-2}=\frac{1}{2B+4}$$

$$\frac{2}{(B+2)(B-2)}-\frac{1}{B-2}=\frac{1}{2(B+2)}$$

$$\frac{2(2)(B+2)(B-2)}{(B+2)(B-2)}-\frac{1(2)(B+2)(B-2)}{B-2}=\frac{1(2)(B+2)(B-2)}{2(B+2)}$$

$$4-2(B+2)=B-2$$

$$4-2B-4=B-2$$

$$-3B=-2$$

$$B=\frac{2}{3}$$

37.

$$\frac{t-3}{b}-\frac{t}{2b-1}=\frac{1}{2}, \text{ for } t$$

$$\frac{(t-3)(2b)(2b-1)}{b}-\frac{t(2b)(2b-1)}{2b-1}=\frac{1(2b)(2b-1)}{2}$$

$$(t-3)(2b-1)(2)-t(2b)=b(2b-1)$$

$$(t-3)(4b-2)-2bt=2b^2-b$$

$$4bt-2t-12b+6-2bt=2b^2-b$$

$$2bt-2t=2b^2+11b-6$$

$$2t(b-1)=(2b-1)(b+6)$$

$$t=\frac{(2b-1)(b+6)}{2(b-1)}$$

41.

$$V = 1.2\ \ 5.0 + \frac{8.0R}{8.0+R}$$
$$V = 6.0 + \frac{9.6R}{8.0+R}$$
$$V(8.0+R) = 6.0(8.0+R) + \frac{9.6R(8.0+R)}{8.0+R}$$
$$V(8.0+R) = 48 + 6.0R + 9.6R$$
$$8.0V + RV = 48 + 15.6R$$
$$8.0V - 48 = R(15.6 - V)$$
$$R = \frac{8.0V - 48}{15.6 - V}$$
$$R = \frac{8.0(V-6.0)}{15.6 - V}$$

45.

$$p = \frac{RT}{V-b} - \frac{a}{V^2}$$
$$pV^2(V-b) = \frac{RTV^2(V-b)}{V-b} - \frac{aV^2(V-b)}{V^2}$$
$$pV^2(V-b) = RTV^2 - a(V-b)$$
$$pV^3 - pV^2b = RTV^2 - aV + ab$$
$$RTV^2 = pV^3 - pV^2b + aV - ab$$
$$T = \frac{pV^3 - pV^2b + aV - ab}{RV^2}$$

49.

$$f(n-1) = \frac{1}{\frac{1}{R_1} + \frac{1}{R_2}}$$
$$f(n-1) = \frac{1}{\frac{1R_2 + 1R_1}{R_1R_2}}$$
$$f(n-1) = \frac{R_1R_2}{R_1 + R_2}$$
$$f(n-1)(R_1 + R_2) = \frac{R_1R_2}{R_1 + R_2}(R_1 + R_2)$$
$$f(n-1)(R_1 + R_2) = R_1R_2$$
$$f(n-1)(R_1) - R_1R_2 = -f(n-1)(R_2)$$
$$R_1(nf - f - R_2) = f(1-n)(R_2)$$
$$R_1 = \frac{f(1-n)(R_2)}{nf - f - R_2}$$

53.

$$\text{rate} = \frac{\text{work}}{\text{time}}$$

$$\text{work} = \text{rate} \times \text{time}$$

For machine 1: $r_1 = \frac{100}{12}$ boxes/min

For machine 2: $r_2 = \frac{100}{10}$ boxes/min

For machine 3: $r_3 = \frac{100}{8}$ boxes/min

To complete the whole job of 100 boxes

$$r_1t + r_2t + r_3t = 100 \text{ boxes (complete job)}$$

$$\frac{100}{12}t + \frac{100}{10}t + \frac{100}{8}t = 100$$

$$\frac{100t(120)}{12} + \frac{100t(120)}{10} + \frac{100t(120)}{8} = 100(120)$$

$$1000t + 1200t + 1500t = 12000$$

$$3700t = 12000$$

$$t = \frac{12000}{3700} = 3.24 \text{ min}$$

57.

$d = rt$; d = amount of water

Fast rate is $r_{\text{fast}} = 60.0 \text{ m}^3/\text{h}$

Slow rate is $r_{\text{slow}} = 18.0 \text{ m}^3/\text{h}$

$$d_{\text{fast}} + d_{\text{slow}} = 276 \text{ m}^3$$

$$d_{\text{fast}} = 276 - d_{\text{slow}}$$

$$t_{\text{fast}} + t_{\text{slow}} = 4.83 \text{ h}$$

$$\frac{d_{\text{fast}}}{r_{\text{fast}}} + \frac{d_{\text{slow}}}{r_{\text{slow}}} = 4.83$$

$$\frac{276 - d_{\text{slow}}}{60} + \frac{d_{\text{slow}}}{18} = 4.83$$

$$828 - 3\,d_{\text{slow}} + 10\,d_{\text{slow}} = 870$$

$$7\,d_{\text{slow}} = 42$$

$$d_{\text{slow}} = 6.00 \text{ m}^3$$

61. Use $d = rt$

$$(450+v)\times t = 2580 \text{ with wind}$$
$$(450-v)\times t = 1800 \text{ against wind}$$
$$t = \frac{2580}{450+v}$$
$$\text{and } t = \frac{1800}{450-v}$$
$$\frac{2580}{450+v} = \frac{1800}{450-v}$$
$$\frac{2580(450+v)(450-v)}{450+v} = \frac{1800(450+v)(450-v)}{450-v}$$
$$2580(450-v) = 1800(450+v)$$
$$1\,161\,000 - 2580v = 810\,000 + 1800v$$
$$4380v = 351\,000$$
$$v = 80.1 \text{ km/h, wind speed}$$

65.
$$\frac{x-12}{x^2+x-6} = \frac{A}{x+3} + \frac{B}{x-2}$$
$$\frac{x-12}{(x+3)(x-2)} = \frac{A(x-2)+B(x+3)}{(x+3)(x-2)}$$
$$A(x-2)+B(x+3) = x-12$$
$$Ax - 2A + Bx + 3B = x - 12$$
$$(A+B)x - 2A + 3B = x - 12$$

Comparing the x-term

$$1 = A + B \qquad \text{(Equation 1)}$$

Comparing the constant term

$$-12 = -2A + 3B \qquad \text{(Equation 2)}$$

This is a system of equations (see Chapter 5)

Solving the first equation above

$$B = 1 - A$$

Substituting into the second equation above

$$-12 = -2A + 3(1-A)$$
$$-12 = -2A + 3 - 3A$$
$$5A = 15$$
$$A = 3$$
$$B = 1 - 3$$
$$B = -2$$

Review Exercises

1. False. $(2a-3)^2 = 4a^2 - 12a + 9$

5. False. $\frac{x^2+x-2}{x^2+3x-4} = \frac{(x+2)(x-1)}{(x+4)(x-1)} = \frac{x+2}{x+4}$. One cannot cancel the x^2 terms from the original numerator and denominator in order to simplify the fraction.

9. $3a(4x+5a) = 12ax + 15a^2$

13. $(b-2)(b+8) = b^2 + 6b - 16$

17. $a^2x^2 + a^2 = a^2(x^2+1)$

21.
$$\begin{aligned} 16(x+2)^2 - t^4 &= (4(x+2)+t^2)(4(x+2)-t^2) \\ &= (4x+8+t^2)(4x+8-t^2) \end{aligned}$$

25.
$$\begin{aligned} 25t^2 + 10t + 1 &= (5t)^2 + 2(5t)(1) + 1^2 \\ &= (5t+1)^2 \end{aligned}$$

29.
$$\begin{aligned} t^4 - 5t^2 - 36 &= (t^2-9)(t^2+4) \\ &= (t+3)(t-3)(t^2+4) \end{aligned}$$

33.
$$\begin{aligned} 8x^2 - 8x - 70 &= 2(4x^2 - 4x - 35) \\ &= 2(2x+5)(2x-7) \end{aligned}$$

37.
$$\begin{aligned} 4x^2 - 16y^2 &= 4(x^2 - 4y^2) \\ &= 4(x+2y)(x-2y) \end{aligned}$$

41.
$$\begin{aligned} 8x^3 + 27 &= (2x)^3 + 3^3 \\ &= (2x+3)(4x^2 - 6x + 9) \end{aligned}$$

45.
$$\begin{aligned} nx + 5n - x^2 + 25 &= n(x+5) - (x^2 - 25) \\ &= n(x+5) - (x+5)(x-5) \\ &= (x+5)(n-(x-5)) \\ &= (x+5)(n-x+5) \end{aligned}$$

49. $$\frac{6x^2-7x-3}{2x^2-5x+3}=\frac{(3x+1)(2x-3)}{(2x-3)(x-1)}$$
$$=\frac{3x+1}{x-1} \quad \text{where } x\neq 1,\frac{3}{2}$$

53. $$\frac{18-6L}{L^2-6L+9}\div\frac{L^2-2L-15}{L^2-9}=\frac{18-6L}{L^2-6L+9}\times\frac{L^2-9}{L^2-2L-15}$$
$$=\frac{-6(L-3)}{(L-3)(L-3)}\times\frac{(L+3)(L-3)}{(L-5)(L+3)}$$
$$=\frac{-6}{L-5} \quad \text{where } L\neq \pm 3$$

57. $$\frac{x+\frac{1}{x}+1}{x^2-\frac{1}{x}}=\frac{\frac{x^2+x+1}{x}}{\frac{x^3-1}{x}}$$
$$=\frac{x^2+x+1}{x}\cdot\frac{x}{x^3-1}$$
$$=\frac{x(x^2+x+1)}{x(x-1)(x^2+x+1)}$$
$$=\frac{1}{x-1} \quad \text{where } x\neq 0,1$$

61. $$\frac{6}{x}-\frac{7}{2x}+\frac{3}{xy}=\frac{6(2y)-7(y)+3(2)}{2xy}$$
$$=\frac{12y-7y+6}{2xy}$$
$$=\frac{5y+6}{2xy}$$

65. $$\frac{2x}{x^2+2x-3}-\frac{1}{6x+2x^2}=\frac{2x}{(x+3)(x-1)}-\frac{1}{2x(x+3)}$$
$$=\frac{2x(2x)-1(x-1)}{2x(x+3)(x-1)}$$
$$=\frac{4x^2-x+1}{2x(x+3)(x-1)}$$

69. $$\frac{3x}{x^2+2x-3}-\frac{2}{x^2+3x}+\frac{x}{x-1}=\frac{3x}{(x+3)(x-1)}-\frac{2}{x(x+3)}+\frac{x}{x-1}$$
$$=\frac{3x(x)-2(x-1)+x(x)(x+3)}{x(x+3)(x-1)}$$
$$=\frac{3x^2-2x+2+x^3+3x^2}{x(x+3)(x-1)}$$
$$=\frac{x^3+6x^2-2x+2}{x(x+3)(x-1)}$$

73. The two functions both have the same graph that looks like the following output.

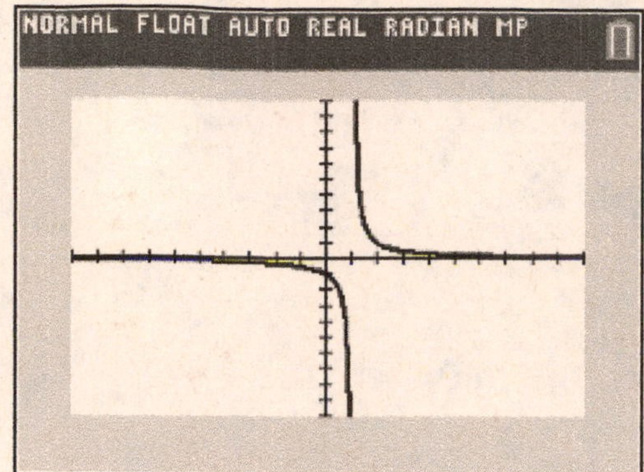

77.

$$\frac{2}{t}-\frac{1}{at}=2+\frac{a}{t}$$

$$\frac{2}{t}-\frac{1}{at}\cdot at=2+\frac{a}{t}\cdot at$$

$$2a-1=2at+a^2$$

$$2at=-a^2+2a-1$$

$$2at=-(a^2-2a+1)$$

$$2at=-(a-1)(a-1)$$

$$t=\frac{-(a-1)^2}{2a}$$

81.

$$f(x)=\frac{1}{x+2}$$

$$f(x+2)=\frac{1}{x+2+2}$$

$$2f(x)=\frac{2}{x+2}$$

But if $f(x+2)=2f(x)$

then $\frac{1}{x+2+2}=\frac{2}{x+2}$

$$\frac{1(x+2)(x+4)}{x+4}=\frac{2(x+2)(x+4)}{x+2}$$

$$x+2=2x+8$$

$$x=-6$$

85. $$xy=\frac{1}{4}\left[(x+y)^2-(x-y)^2\right]$$
$$xy=\frac{1}{4}\left[x^2+2xy+y^2-(x^2-2xy+y^2)\right]$$
$$xy=\frac{1}{4}\left[x^2+2xy+y^2-x^2+2xy-y^2\right]$$
$$xy=\frac{1}{4}[4xy]$$
$$xy=xy$$

89. $$\begin{aligned}\left[2b+(n-1)\lambda\right]^2&=4b^2+4b(n-1)\lambda+(n-1)^2\lambda^2\\&=4b^2+4b\lambda(n-1)+\lambda^2(n^2-2n+1)\\&=4b^2+4bn\lambda-4b\lambda+n^2\lambda^2-2n\lambda^2+\lambda^2\end{aligned}$$

93. $$\begin{aligned}(2R-r)^2-(r^2+R^2)&=4R^2-4Rr+r^2-r^2-R^2\\&=3R^2-4Rr\\&=R(3R-4r)\end{aligned}$$

97. $$\begin{aligned}10a(T-t)+a(T-t)^2&=10aT-10at+a(T^2-2Tt+t^2)\\&=10aT-10at+aT^2-2aTt+at^2\end{aligned}$$

101. $$\begin{aligned}\frac{2wtv^2}{Dg}\quad\frac{b\pi^2D^2}{n^2}\quad\frac{6}{bt^2}&=\frac{12wtv^2b\pi^2D^2}{Dgn^2bt^2}\\&=\frac{bDt(12wv^2\pi^2D)}{bDt(gn^2t)}\\&=\frac{12wv^2\pi^2D}{gn^2t}\end{aligned}$$

105. $$1-\frac{d^2}{2}+\frac{d^4}{24}-\frac{d^6}{120}=\frac{120-60d^2+5d^4-d^6}{120}$$

109. $$1-\frac{3a}{4r}-\frac{a^3}{4r^3}=\frac{4r^3-3ar^2-a^3}{4r^3}$$

113. $$W=mgh_2-mgh_1$$
$$W=mg(h_2-h_1)$$
$$m=\frac{W}{g(h_2-h_1)}$$

117.

$$E = V_0 + \frac{(m+M)V^2}{2} + \frac{p^2}{2I}$$

$$E \cdot 2I = V_0 + \frac{(m+M)V^2}{2} + \frac{p^2}{2I} \cdot 2I$$

$$2EI = 2V_0 I + (m+M)IV^2 + p^2$$

$$2EI - 2V_0 I - (m+M)IV^2 = p^2$$

$$I\left(2E - 2V_0 - (m+M)V^2\right) = p^2$$

$$I = \frac{p^2}{2E - 2V_0 - (m+M)V^2}$$

121.

$$F = \frac{m}{sC} + \frac{F_0}{s}$$

$$F \cdot sC = \left(\frac{m}{sC} + \frac{F_0}{s}\right) \cdot sC$$

$$FsC = m + F_0 C$$

$$FsC - F_0 C = m$$

$$C = \frac{m}{Fs - F_0}$$

125.

$$\frac{1}{m} = \frac{\frac{1}{x} + \frac{1}{y}}{2}$$

$$\frac{1}{m} = \frac{\frac{1}{400} + \frac{1}{1200}}{2}$$

$$\frac{1}{m} = \frac{\frac{1(3)+1}{1200}}{2}$$

$$\frac{1}{m} = \frac{\frac{4}{1200}}{2}$$

$$\frac{1}{m} = \frac{1}{300} \cdot \frac{1}{2}$$

$$\frac{1}{m} = \frac{1}{600}$$

$m = 600$ Hz

The harmonic mean of the two musical notes is 600 Hz.

129.

$$\frac{1}{\mathrm{R}_{eq}} = \frac{1}{R_1} + \frac{1}{R_2} + \frac{1}{R_3}$$

$$\frac{1}{6} = \frac{1}{12} + \frac{1}{R} + \frac{1}{2R}$$

$$\frac{1}{6} 12R = \left(\frac{1}{12} + \frac{1}{R} + \frac{1}{2R}\right) \cdot 12R$$

$$2R = (R + 12 + 6)$$

$$2R = R + 18$$

$$R = 18.0\ \Omega$$

Chapter 7

Quadratic Equations

7.1 Quadratic Equations; Solution by Factoring

1.
$$3x^2+7x+2=0$$
$$(3x+1)(x+2)=0 \quad \text{factor}$$
$$3x+1=0 \quad \text{or} \quad x+2=0$$
$$3x=-1 \qquad x=-2$$
$$x=-\frac{1}{3}$$

The roots are $x=-\frac{1}{3}$ and $x=-2$.

Checking in the original equation:

$$3\left(-\frac{1}{3}\right)^2+7\left(-\frac{1}{3}\right)+2=0 \qquad 3(-2)^2+7(-2)+2=0$$
$$\frac{1}{3}-\frac{7}{3}+2=0 \qquad 12-14+2=0$$
$$0=0 \qquad 0=0$$

The roots are $-\frac{1}{3}, -2$.

5.
$$x^2=(x+2)^2$$
$$x^2=x^2+4x+4$$

$4x+4=0$, no x^2 term so it is not quadratic

9.
$$y^2(y-2)=3(y-2)$$
$$y^3-2y^2=3y-6$$

Not quadratic since there is a y^3 term.

13.
$$4y^2=9$$
$$4y^2-9=0$$
$$(2y+3)(2y-3)=0$$
$$2y+3=0 \quad \text{or} \quad 2y-3=0$$
$$2y=-3 \qquad 2y=3$$
$$y=-\frac{3}{2} \qquad y=\frac{3}{2}$$

17.
$$R^2+12=7R$$
$$R^2-7R+12=0$$
$$(R-4)(R-3)=0$$
$$R-4=0 \quad \text{or} \quad R-3=0$$
$$R=4 \qquad R=3$$

21.
$$12m^2 = 3$$
$$12m^2 - 3 = 0$$
$$3(4m^2 - 1) = 0$$
$$3(2m-1)(2m+1) = 0$$
$$2m - 1 = 0 \quad \text{or} \quad 2m + 1 = 0$$
$$2m = 1 \qquad 2m = -1$$
$$m = \frac{1}{2} \qquad m = -\frac{1}{2}$$

25.
$$4x = 3 - 7x^2$$
$$7x^2 + 4x - 3 = 0$$
$$(7x-3)(x+1) = 0$$
$$7x - 3 = 0 \quad \text{or} \quad x + 1 = 0$$
$$7x = 3 \qquad x = -1$$
$$x = \frac{3}{7}$$

29.
$$4x(x+1) = 3$$
$$4x^2 + 4x - 3 = 0$$
$$(2x-1)(2x+3) = 0$$
$$2x - 1 = 0 \quad \text{or} \quad 2x + 3 = 0$$
$$2x = 1 \qquad 2x = -3$$
$$x = \frac{1}{2} \qquad x = -\frac{3}{2}$$

33.
$$8s^2 + 16s = 90$$
$$8s^2 + 16s - 90 = 0$$
$$(4s-10)(2s+9) = 0$$
$$4s - 10 = 0 \quad \text{or} \quad 2s + 9 = 0$$
$$4s = 10 \qquad 2s = -9$$
$$s = \frac{5}{2} \qquad s = -\frac{9}{2}$$

37.
$$(x+a)^2 - b^2 = 0$$
$$(x+a-b)(x+a+b) = 0$$
$$x + a - b = 0 \quad \text{or} \quad x + a + b = 0$$
$$x = b - a \qquad x = -b - a$$

41. For $a = 2$, $b = -7$, $c = 3$

Equation 7.1 ($ax^2 + bx + c = 0$) becomes

$$2x^2 - 7x + 3 = 0$$
$$(2x-1)(x-3) = 0$$
$$x = \frac{1}{2} \text{ or } x = 3$$

The sum of the roots is

$$\frac{1}{2} + 3 = \frac{7}{2} = -\frac{-7}{2} = -\frac{b}{a}$$

45.

$$V = \alpha I + \beta I^2$$
$$2I + 0.5I^2 = 6$$
$$I^2 + 4I - 12 = 0$$
$$(I+6)(I-2) = 0$$
$$I + 6 = 0 \quad \text{or} \quad I - 2 = 0$$
$$I = -6 \qquad I = 2$$

The current is -6.00 A or 2.00 A.

49. If the solutions are to be $x = 0.5$ and $x = 2$, one such equation is

$(x - 0.5)(x - 2) = 0$

$x^2 - 2.5x + 1 = 0$

53.

$$\frac{1}{x-3} + \frac{4}{x} = 2$$
$$\frac{1(x)(x-3)}{x-3} + \frac{4x(x-3)}{x} = 2x(x-3) \quad \text{multiply by the LCD}$$
$$x + 4x - 12 = 2x^2 - 6x$$
$$-2x^2 + 11x - 12 = 0$$
$$2x^2 - 11x + 12 = 0$$
$$(x-4)(2x-3) = 0$$
$$2x - 3 = 0 \quad \text{or} \quad x - 4 = 0$$
$$2x = 3 \qquad x = 4$$
$$x = \frac{3}{2}$$

57. $\frac{1}{k_c}=\frac{1}{k_1}+\frac{1}{k_2}$

Let k = the spring constant of the first spring in N/cm
Let $k + 3$ N/cm = the spring constant of the second spring in N/cm

$$\frac{1}{2}=\frac{1}{k}+\frac{1}{k+3}$$

$$\frac{1\cdot 2k(k+3)}{2}=\frac{1\cdot 2k(k+3)}{k}+\frac{1\cdot 2k(k+3)}{k+3} \quad \text{multiply by LCD}$$

$$k^2+3k=2k+6+2k$$

$$k^2-k-6=0$$

$$(k-3)(k+2)=0$$

$$k-3=0 \quad \text{or} \quad k+2=0$$

$$k=3 \qquad k=-2 \quad \text{reject this solution since } k>0$$

The one spring constant is 3 N/cm and the other spring constant is (3N/cm + 3 N/cm) = 6N/cm

7.2 Completing the Square

1. $x^2+6x-8=0$

$$x^2+6x=8$$

$$x^2+6x+9=8+9$$

$$(x+3)^2=17$$

$$x+3=\pm\sqrt{17}$$

$$x=-3\pm\sqrt{17}$$

5. $x^2=7$

$$x=\pm\sqrt{7}$$

9. $(x-2)^2=25$

$$x-2=\pm\sqrt{25}$$

$$x-2=\pm 5$$

$$x=2\pm 5$$

$$x=-3 \text{ or } x=7$$

13. $x^2+2x-15=0$

$$x^2+2x=15$$

$$x^2+2x+1=15+1$$

$$(x+1)^2=16$$

$$x+1=\pm\sqrt{16}$$

$$x=-1\pm 4$$

$$x=3 \text{ or } x=-5$$

17.

$$n^2 = 6n - 4$$
$$n^2 - 6n = -4$$
$$n^2 - 6n + 9 = -4 + 9$$
$$(n-3)^2 = 5$$
$$n - 3 = \pm\sqrt{5}$$
$$n = 3 \pm \sqrt{5}$$

21.

$$2s^2 + 5s = 3$$
$$s^2 + \frac{5}{2}s = \frac{3}{2}$$
$$s^2 + \frac{5}{2}s + \frac{25}{16} = \frac{3}{2} + \frac{25}{16}$$
$$\left(s + \frac{5}{4}\right)^2 = \frac{49}{16}$$
$$s + \frac{5}{4} = \pm\frac{7}{4}$$
$$s = -\frac{5}{4} \pm \frac{7}{4}$$
$$s = -3 \text{ or } s = \frac{1}{2}$$

25.

$$2y^2 - y - 2 = 0$$
$$2y^2 - y = 2$$
$$y^2 - \frac{1}{2}y + \frac{1}{16} = 1 + \frac{1}{16}$$
$$\left(y - \frac{1}{4}\right)^2 = \frac{17}{16}$$
$$y - \frac{1}{4} = \pm\frac{\sqrt{17}}{4}$$
$$y = \frac{1}{4} \pm \frac{\sqrt{17}}{4}$$

29.

$$9x^2 + 6x + 1 = 0$$
$$9\left(x^2 + \frac{2}{3}x + \frac{1}{9}\right) = 0$$
$$\left(x + \frac{1}{3}\right)^2 = 0$$
$$x + \frac{1}{3} = 0$$
$$x = -\frac{1}{3} \text{ (double root)}$$

33.

$$V = 4.0T - 0.2T^2 = 15$$
$$0.2T^2 - 4.0T + 15 = 0$$
$$0.2\left(T^2 - 20T + 75\right) = 0$$
$$T^2 - 20T = -75$$
$$T^2 - 20T + 100 = -75 + 100$$
$$(T-10)^2 = 25$$
$$T - 10 = \pm 5$$
$$T = 10 \pm 5$$
$$T = 5 \text{ or } T = 15$$

The voltage is 15.0 V when the temperature is 5.0°C or 15°C.

7.3 The Quadratic Formula

1. $x^2 + 5x + 6 = 0$; $a = 1$, $b = 5$, $c = 6$

$$x = \frac{-5 \pm \sqrt{5^2 - 4(1)(6)}}{2(1)}$$
$$x = \frac{-5 \pm \sqrt{1}}{2}$$
$$x = \frac{-5 \pm 1}{2}$$
$$x = \frac{-5+1}{2} \text{ or } x = \frac{-5-1}{2}$$
$$x = -2 \qquad x = -3$$

5. $D^2 + 3D + 2 = 0$; $a = 1$, $b = 3$, $c = 2$

$$D = \frac{-3 \pm \sqrt{(3)^2 - 4(1)(2)}}{2(1)}$$
$$D = \frac{-3 \pm \sqrt{9-8}}{2}$$
$$D = \frac{-3 \pm \sqrt{1}}{2}$$
$$D = \frac{-3 \pm 1}{2}$$
$$D = -2 \text{ or } D = -1$$

9. $v^2 = 15 - 2v$

$v^2 + 2v - 15 = 0;\ a = 1,\ b = 2,\ c = -15$

$$v = \frac{-2 \pm \sqrt{(2)^2 - 4(1)(-15)}}{2(1)}$$

$$v = \frac{-2 \pm \sqrt{4 - (-60)}}{2}$$

$$v = \frac{-2 \pm \sqrt{64}}{2}$$

$$v = \frac{-2 \pm 8}{2} =$$

$v = -5$ or $v = 3$

13. $3y^2 = 3y + 2$

$3y^2 - 3y - 2 = 0;\ a = 3,\ b = -3,\ c = -2$

$$y = \frac{-(-3) \pm \sqrt{(-3)^2 - 4(3)(-2)}}{2(3)}$$

$$y = \frac{3 \pm \sqrt{9 - (-24)}}{6}$$

$$y = \frac{3 \pm \sqrt{33}}{6}$$

$$y = \frac{1}{6}\left(3 \pm \sqrt{33}\right)$$

17. $30y^2 + 23y - 40 = 0;\ a = 30,\ b = 23,\ c = -40$

$$y = \frac{-23 \pm \sqrt{(23)^2 - 4(30)(-40)}}{2(30)}$$

$$y = \frac{-23 \pm \sqrt{529 - (-4800)}}{60}$$

$$y = \frac{-23 \pm \sqrt{5329}}{60}$$

$$y = \frac{-23 \pm 73}{60}$$

$y = -\frac{8}{5}$ or $y = \frac{5}{6}$

21.
$$s^2 = 9 + s(1-2s)$$
$$s^2 = 9 + s - 2s^2$$
$$3s^2 - s - 9 = 0;\ a = 3,\ b = -1,\ c = -9$$
$$s = \frac{-(-1) \pm \sqrt{(-1)^2 - 4(3)(-9)}}{2(3)}$$
$$s = \frac{1 \pm \sqrt{1-(-108)}}{6}$$
$$s = \frac{1 \pm \sqrt{109}}{6}$$
$$s = \frac{1}{6}\left(1 \pm \sqrt{109}\right)$$

25.
$$15 + 4z = 32z^2$$
$$32z^2 - 4z - 15 = 0;\ a = 32,\ b = -4,\ c = -15$$
$$z = \frac{-(-4) \pm \sqrt{(-4)^2 - 4(32)(-15)}}{2(32)}$$
$$z = \frac{4 \pm \sqrt{16-(-1920)}}{64}$$
$$z = \frac{4 \pm \sqrt{1936}}{64}$$
$$z = \frac{4 \pm 44}{64}$$
$$z = \frac{3}{4} \text{ or } z = -\frac{5}{8}$$

29.
$$0.29Z^2 - 0.18 = 0.63Z$$
$$0.29Z^2 - 0.63Z - 0.18 = 0; a = 0.29,\ b = -0.63,\ c = -0.18$$
$$Z = \frac{-(-0.63) \pm \sqrt{(-0.63)^2 - 4(0.29)(-0.18)}}{2(0.29)}$$
$$Z = \frac{0.63 \pm \sqrt{0.3969-(-0.2088)}}{0.58}$$
$$Z = \frac{0.63 \pm \sqrt{0.6057}}{0.58}$$
$$Z = -0.256 \text{ or } Z = 2.43$$

33.

$$b^2x^2+1-a=(b+1)x$$

$$b^2x^2-(b+1)x+(1-a)=0;\ a=b^2,\ b=-(b+1),\ c=1-a$$

$$x=\frac{-[-(b+1)]\pm\sqrt{[-(b+1)]^2-4(b^2)(1-a)}}{2(b^2)}$$

$$x=\frac{b+1\pm\sqrt{b^2+2b+1-4b^2+4ab^2}}{2b^2}$$

$$x=\frac{b+1\pm\sqrt{4ab^2-3b^2+2b+1}}{2b^2}$$

37.

$$3.6t^2+2.1=7.7t$$

$$b^2-4ac=(-7.7)^2-4(3.6)(2.1)=29.05$$

Since $b^2-4ac>0$ and not a perfect square, the roots are real, irrational, and unequal.

3 is the smallest positive integer value of k for which the roots are imaginary.

41.

$$x^4-5x^2+4=0$$

$$\left(x^2\right)^2-5x^2+4=0$$

$$\left(x^2-4\right)\left(x^2-1\right)=0$$

$$x^2-4=0 \quad \text{or} \quad x^2-1=0$$

$$x^2=4 \qquad x^2=1$$

$$x=\pm2 \qquad x=\pm1$$

45.

For $D=3.625$

$$D_0^2-DD_0-0.250D^2=0$$

$$D_0^2-3.625D_0-0.25(3.625)^2=0$$

$$D_0^2-3.625D_0-3.28515625=0$$

$$a=1,\ b=-3.625,\ c=-3.28515625$$

$$D_0=\frac{-(-3.625)\pm\sqrt{(-3.625)^2-4(1)(-3.28515625)}}{2}$$

$D_0=4.38$ cm or $D_0=-0.751$ cm, reject since $D_0>0$.

49. The cars have traveled x km and $x+2.0$ km, forming the legs of a right triangle. The hypotenuse is 6.0 km. Using the Pythagorean theorem,

$$6.0^2 = x^2 + (x+2.0)^2$$

$$36.0 = 2x^2 + 4.0x + 4.0$$

$$2x^2 + 4.0x - 32.0 = 0$$

$$x = \frac{-4.0 \pm \sqrt{(4.0)^2 - 4(2)(-32.0)}}{2(2)}$$

$x = 3.1231$ or $x = -5.1231$ (negative distance discarded)

The first car traveled 3.1 km and the second car traveled 5.1 km.

53. $Lm^2 + Rm + \frac{1}{C} = 0$; $a = L$, $b = R$, $c = \frac{1}{C}$

$$m = \frac{-R \pm \sqrt{R^2 - 4(L)\left(\frac{1}{C}\right)}}{2(L)}$$

$$m = \frac{-R \pm \sqrt{R^2 - \frac{4L}{C}}}{2L}$$

57. Let w be the added width to each dimension

$$(12+w)\times(16+w) = (12\times16)+80 = 272$$

$$w^2 + 28w + 192 = 272$$

$$w^2 + 28w - 80 = 0$$

$a = 1$, $b = 28$, $c = -80$

$$r = \frac{-28 \pm \sqrt{28^2 - 4(1)(-80)}}{2(1)}$$

$w = 2.61324$ or $w = -30.61324$

$w = 2.6$ ft

61. We have $l = w + 12.8$, $A = lw$ and so

$$262 = (w+12.8)w$$

$$262 = w^2 + 12.8w$$

$$w^2 + 12.8w - 262 = 0$$

$$w = \frac{-12.8 \pm \sqrt{12.8^2 - 4(1)(-262)}}{2(1)}$$

$w = 11.0$ or $w = -23.8$ (negative width discarded)

Therefore, the tennis court is 11.0 m wide and 23.8 m long.

7.4 The Graph of the Quadratic Function

1. $y = 2x^2 + 8x + 6;\ a = 2,\ b = 8,\ c = 6$

$$x\text{-coordinate of vertex} = \frac{-b}{2a} = \frac{-8}{2(2)} = -2$$

$$y\text{-coordinate of vertex} = 2(-2)^2 + 8(-2) + 6 = -2$$

The vertex is $(-2, -2)$ and since $a > 0$, it is a minimum. Since $c = 6$, the y-intercept is (0, 6) and the check is:

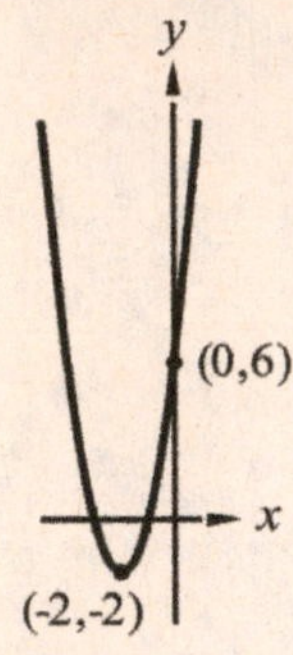

5. $y = -3x^2 + 10x - 4$, with $a = -3$, $b = 10$, $c = -4$. This means that the x-coordinate of the extreme is

$$\frac{-b}{2a} = \frac{-10}{2(-3)} = \frac{10}{6} = \frac{5}{3}$$

and the y-coordinate is

$$y = -3\left(\frac{5}{3}\right)^2 + 10\left(\frac{5}{3}\right) - 4 = \frac{13}{3}.$$

Thus the extreme point is $\left(\frac{5}{3}, \frac{13}{3}\right)$.

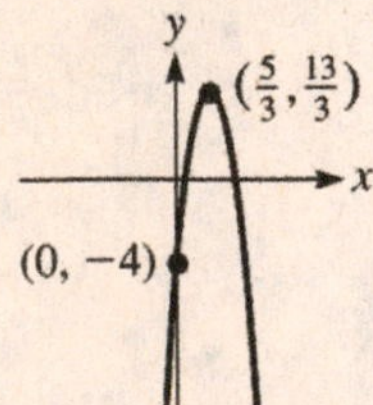

Since $a < 0$, it is a maximum point.
Since $c = -4$, the y-intercept is $(0, -4)$. Use the maximum point $\left(\frac{5}{3}, \frac{13}{3}\right)$, and the y-intercept $(0, -4)$, and the fact that the graph is a parabola, to sketch the graph.

9. $y = x^2 - 4 = x^2 + 0x - 4$; $a = 1$, $b = 0$, $c = -4$
The x-coordinate of the extreme point is
$\frac{-b}{2a} = \frac{-0}{2(1)} = 0$, and the y-coordinate is
$y = 0^2 - 4 = -4$.
The extreme point is $(0, -4)$.

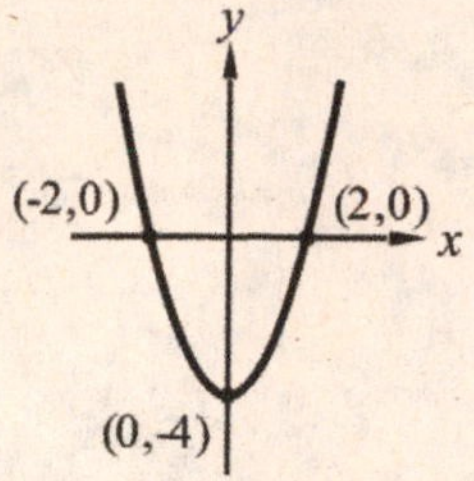

Since $a > 0$, it is a minimum point.
Since $c = -4$, the y-intercept is $(0, -4)$.
$x^2 - 4 = 0$, $x^2 = 4$, $x = \pm 2$ are the x-intercepts.
Use the minimum points and intercepts to sketch the graph.

13. $y = 2x^2 + 3 = 2x^2 + 0x + 3$; $a = 2$, $b = 0$, $c = 3$
The x-coordinate of the extreme point is
$\frac{-b}{2a} = \frac{-0}{2(2)} = 0$, and the y-coordinate is
$y = 2(0)^2 + 3 = 3$.
The extreme point is $(0, 3)$. Since $a > 0$ it is a minimum point.

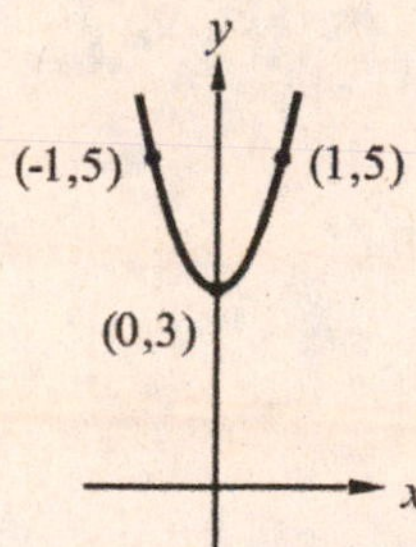

Since $c = 3$, the y-intercept is $(0, 3)$ there are no x-intercepts, $b^2 - 4ac = -24$. $(-1, 5)$ and $(1, 5)$ are on the graph. Use the three points to sketch the graph.

17. $2x^2 - 7 = 0.$

Graph $y = 2x^2 - 7$ and use the zero feature to find the roots.

$x = -1.87$ and $x = 1.87.$

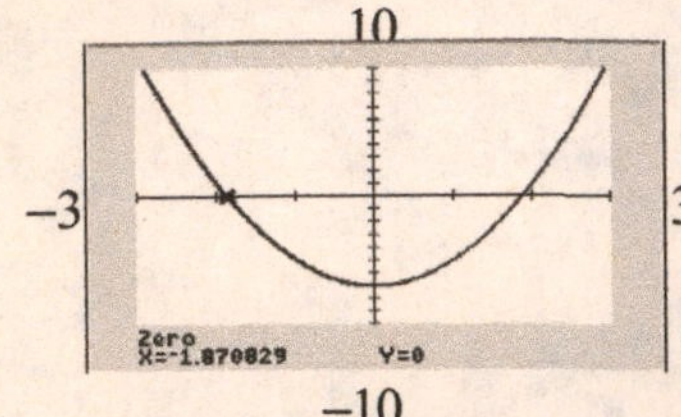

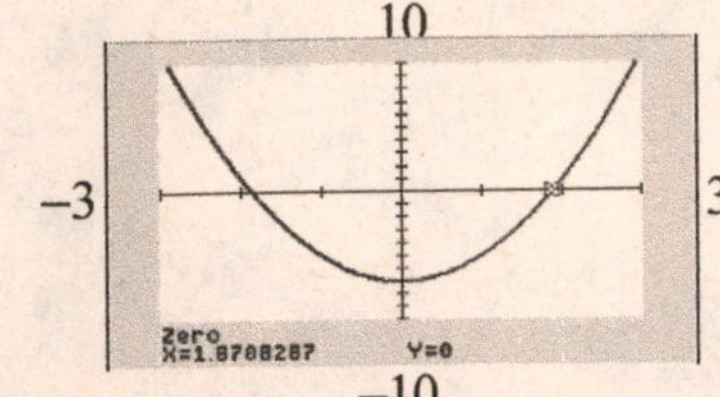

21. $x(2x-1) = -3$

Graph $y_1 = x(2x-1)+3$ and use the zero feature.

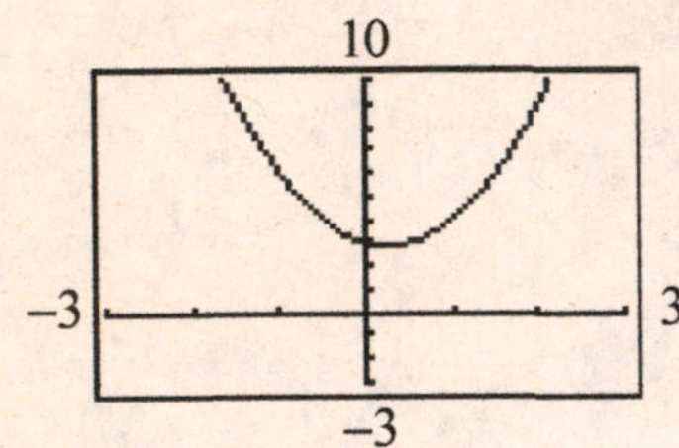

As the graph shows, there are no real solutions.

25. **(a)** $y = x^2$ **(b)** $y = x^2 + 3$ **(c)** $y = x^2 - 3$

The parabola $y = x^2 + 3$ is shifted up +3 units (minimum point $(0,\ 3)$).

The parabola $y = x^2 - 3$ is shifted down -3 units (minimum point $(0,\ -3)$).

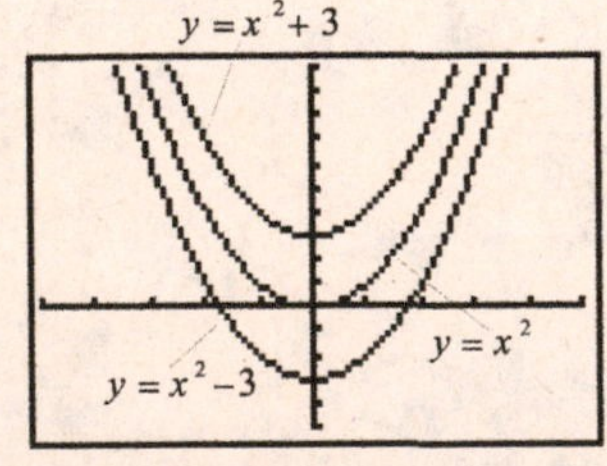

29. **(a)** $y = x^2$ **(b)** $y = 3x^2$ **(c)** $y = \frac{1}{3}x^2$

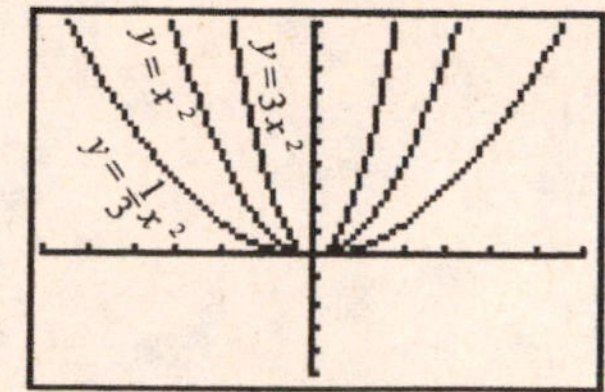

The graph of $y = 3x^2$ is the graph of $y = x^2$ narrowed. The graph of $y = \frac{1}{3}x^2$ is the graph of $y = x^2$ broadened.

33. Since the vertex has x-coordinate equal to 0, $b = 0$. Since it also has y-coordinate equal to 0, $c = 0$ as well. It follows that the equation takes the form $y = ax^2$. By substituting $x = 25$ and $y = 125$, we find $125 = a(25)^2$, or $a = \frac{125}{(25)^2} = \frac{1}{5}$.

The equation is $y = \frac{1}{5}x^2$.

37. Assume the equation is $y = ax^2 + bx + c$. Since $(0, -3)$ is on the parabola, we know $c = -3$.
Substituting $x = 2, y = 5$ yields $5 = 4a + 2b - 3$, or $4a + 2b = 8$ and so $b = 4 - 2a$.
Substituting $x = -2, y = -3$ yields $-3 = 4a - 2b - 3$, or $4a - 2b = 0$ and so $b = 2a$. Thus, $2a = 4 - 2a$, implying $a = 1$ and then $b = 2$.
The desired equation is $y = x^2 + 2x - 3$.

41. If the Arch is modeled by $y = 192 - 0.0208x^2$, then the range must be $0 \le y \le 192$ (we disregard negative values of y.) Thus, the Arch must be 192 meters high.
We determine its width by finding the x-intercepts, solving

$$0 = 192 - 0.0208x^2$$

$$x^2 = \frac{192}{0.0208}$$

$$x = \pm\sqrt{\frac{192}{0.0208}} = \pm 96.07689$$

The distance between the two x-intercepts is $2(96.07689) = 192.154$ and so the Arch is 192 meters wide.

45. Graph $h = -4.9t^2 + 68t + 2$ and use the maximum feature, finding the maximum occurs when $t = 6.9$ s.

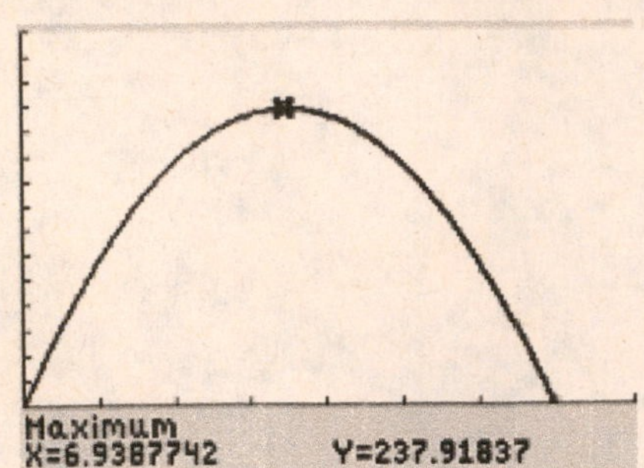

49. We let x represent the length of \$20/ft fence.

$$A = xy = 20000,$$

$$y = \frac{20000}{x}$$

$$\text{cost} = 20x + 30y = 7500$$

$$20x + 30\left(\frac{20000}{x}\right) = 7500$$

$$20x^2 - 7500x + 600000 = 0$$

$$x^2 - 375x + 30000 = 0$$

Using the zero finder, there are two solutions,

$x = 116$ ft. or $x = 259$ ft.

If $x = 116$ ft., then $y = 173$ ft.

If $x = 259$ ft., then $y = 77$ ft.

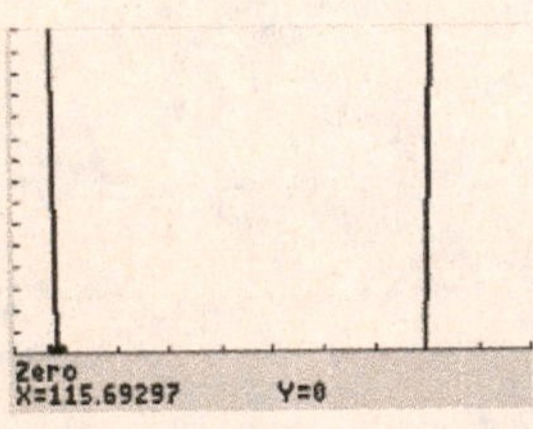

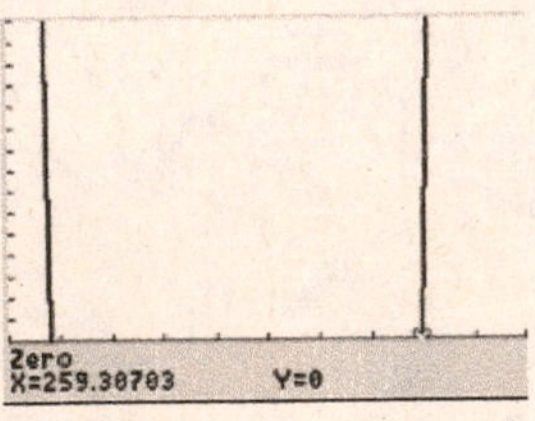

Review Exercises

1. The solution $x = 2$ is not the only solution to $x^2 - 2x = 0$. The other is $x = 0$.

5.
$$x^2 + 3x - 4 = 0$$
$$(x+4)(x-1) = 0$$
$$x+4 = 0 \quad \text{or} \quad x-1 = 0$$
$$x = -4 \qquad\qquad x = 1$$

9.
$$3x^2 + 11x = 4$$
$$3x^2 + 11x - 4 = 0$$
$$(3x-1)(x+4) = 0$$
$$3x - 1 = 0 \quad \text{or} \quad x+4 = 0$$
$$3x = 1 \qquad\qquad x = -4$$
$$x = \frac{1}{3}$$

13.
$$4s^2 = 18s$$
$$4s^2 - 18s = 0$$
$$s(4s-18) = 0$$
$$4s - 18 = 0 \quad \text{or} \quad s = 0$$
$$4s = 18$$
$$s = \frac{9}{2}$$

17. $x^2 - 4x - 96 = 0$; $a = 1$; $b = -4$; $c = -96$
$$x = \frac{-(-4) \pm \sqrt{(-4)^2 - 4(1)(-96)}}{2(1)}$$
$$x = \frac{4 \pm \sqrt{16 - (-384)}}{2}$$
$$x = \frac{4 \pm \sqrt{400}}{2}$$
$$x = \frac{4 \pm 20}{2}$$
$$x = -8 \quad \text{or} \quad x = 12$$

21. $2x^2 - x = 36$

$2x^2 - x - 36 = 0$; $a = 2$; $b = -1$; $c = -36$

$$x = \frac{-(-1) \pm \sqrt{(-1)^2 - 4(2)(-36)}}{2(2)}$$

$$x = \frac{1 \pm \sqrt{1 - (-288)}}{4}$$

$$x = \frac{1 \pm \sqrt{289}}{4}$$

$$x = \frac{1 \pm 17}{4}$$

$$x = \frac{9}{2} \text{ or } x = -4$$

25. $2.1x^2 + 2.3x + 5.5 = 0$; $a = 2.1$; $b = 2.3$; $c = 5.5$

$$x = \frac{-2.3 \pm \sqrt{(2.3)^2 - 4(2.1)(5.5)}}{2(2.1)}$$

$$x = \frac{-2.3 \pm \sqrt{5.29 - 46.2}}{4.2}$$

$$x = \frac{-2.3 \pm \sqrt{-40.91}}{4.2} \text{ (imaginary roots)}$$

29. $4x^2 - 5 = 15$

$4x^2 = 20$

$x^2 = 5$

$x = \pm\sqrt{5}$

33. $3x^2 + 8x + 2 = 0$; $a = 3$; $b = 8$; $c = 2$

$$x = \frac{-8 \pm \sqrt{(8)^2 - 4(3)(2)}}{2(3)}$$

$$x = \frac{-8 \pm \sqrt{64 - 24}}{6}$$

$$x = \frac{-8 \pm \sqrt{40}}{6}$$

$$x = \frac{-8 \pm \sqrt{4 \times 10}}{6}$$

$$x = \frac{-8 \pm 2\sqrt{10}}{6}$$

$$x = \frac{-4 \pm \sqrt{10}}{3}$$

37.

$$7+3C=-2C^2$$

$$2C^2+3C+7=0;\ a=2;\ b=3;\ c=7$$

$$C=\frac{-3\pm\sqrt{(3)^2-4(2)(7)}}{2(2)}$$

$$C=\frac{-3\pm\sqrt{9-56}}{4}$$

$$C=\frac{-3\pm\sqrt{-47}}{4}\text{ (imaginary roots)}$$

41.

$$ay^2=a-3y$$

$$ay^2+3y-a=0;\ a=a;\ b=3;\ c=-a$$

$$y=\frac{-3\pm\sqrt{(3)^2-4(a)(-a)}}{2a}$$

$$y=\frac{-3\pm\sqrt{9+4a^2}}{2a}$$

45.

$$2t^2=t+4$$

$$2t^2-t=4$$

$$t^2-\frac{1}{2}t=2$$

$$t^2-\frac{1}{2}t+\frac{1}{16}=2+\frac{1}{16}$$

$$\left(t-\frac{1}{4}\right)^2=\frac{33}{16}$$

$$t-\frac{1}{4}=\pm\sqrt{\frac{33}{16}}$$

$$t-\frac{1}{4}=\pm\frac{\sqrt{33}}{4}$$

$$t=\frac{1\pm\sqrt{33}}{4}$$

49.

$$\frac{x^2-3x}{x-3}=\frac{x^2}{x+2},\ (x\neq-2,\ 3)$$

$$\frac{x\cancel{(x-3)}}{\cancel{(x-3)}}=\frac{x^2}{x+2}$$

$$x(x+2)=x^2$$

$$x^2+2x=x^2$$

$$2x=0$$

$$x=0$$

53. $y = x - 3x^2; a = -3,\ b = 1,\ c = 0$

$y\text{-int} = (0,\ 0)$

$-3x^2 + x = 0$

$x(-3x + 1) = 0$

$x = 0 \quad \text{or} \quad x = -\frac{1}{3}$ are the x-intercepts

$x \text{ vertex} = \frac{-b}{2a} = \frac{-1}{2(-3)} = \frac{1}{6}$

$y \text{ vertex} = \frac{1}{6} - 3\left(\frac{1}{6}\right)^2 = \frac{1}{12}$

y (1/6, 1/12) (0, 0) (1/3, 0) x (1, -2) (-1, -4)

57. Graph $y_1 = 3x^2 + x + 2$ and use the zero feature to solve.

No real roots.

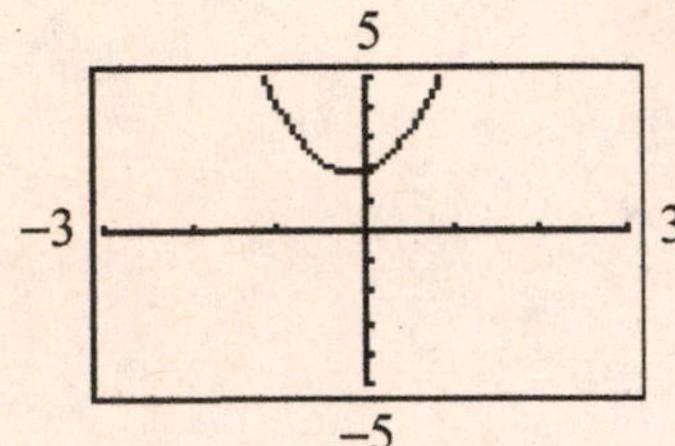

61. $M = 0.5wLx - 0.5wx^2$

$M = 0.5wx(L - x)$

$M = 0$ for $x = 0$ and $x = L$

65. $0.1x^2 + 0.8x + 7 = 50$

$0.1x^2 + 0.8x - 43 = 0;\ a = 0.1, b = 0.8, c = -43$

$$x = \frac{-b \pm \sqrt{b^2 - 4ac}}{2a}$$

$$x = \frac{-(0.8) \pm \sqrt{(0.8)^2 - 4(0.1)(-43)}}{2(0.1)}$$

$$x = \frac{-0.8 \pm \sqrt{17.84}}{0.2}$$

$x = 17.1$ units or -25.1 units

17 units can be made for \$50

69.
$$h = vt\sin\theta - 16t^2$$
$$18.00 = 44.0t\ \sin 65.0^\circ - 16t^2$$
$16t^2 - 39.87754t + 18.00 = 0$ from which
$$a = 16, b = -39.87754, c = 18.00$$
$$t = \frac{-b \pm \sqrt{b^2 - 4ac}}{2a}$$
$$t = \frac{-(-39.87754) \pm \sqrt{(-39.87754)^2 - 4(16)(18.00)}}{2(16)}$$
$$t = \frac{39.87754 \pm \sqrt{438.2182}}{32}$$
$$t = 0.59 \quad \text{or} \quad t = 1.90$$
The height $h = 18.0$ ft is reached when $t = 0.59$ s and 1.90 s.
It reaches that height twice (once on the way up, and once on the way down).

73.
$$A = 2\pi r^2 + 2\pi rh$$
$$2\pi r^2 + 2\pi hr - A = 0$$
$$a = 2\pi, b = 2\pi h, c = -A$$
$$r = \frac{-b \pm \sqrt{b^2 - 4ac}}{2a}$$
$$r = \frac{-2\pi h \pm \sqrt{(2\pi h)^2 - 4(2\pi)(-A)}}{2(2\pi)}$$
$$r = \frac{-2\pi h + \sqrt{4\pi^2h^2 + 8\pi A}}{4\pi}$$
$$r = \frac{-2\pi h + \sqrt{4(\pi^2h^2 + 2\pi A)}}{4\pi}$$
$$r = \frac{-2\pi h + 2\sqrt{\pi^2h^2 + 2\pi A}}{4\pi}$$
$$r = \frac{-\pi h + \sqrt{\pi^2h^2 + 2\pi A}}{2\pi}$$

77.
$$p = 0.090t - 0.015t^2$$
$$p = -0.015t^2 + 0.090t$$
$$a = -0.015,\ b = 0.090,\ c = 0$$
$$y\text{-int} = (0, 0)$$
$$-0.015t^2 + 0.090t = 0$$
$$0.015t(-t + 6) = 0$$
$t = 0$ or $t = 6$ are the x-intercepts
$$t \text{ vertex} = \frac{-b}{2a} = \frac{-0.090}{2(-0.015)} = 3$$
$$p \text{ vertex} = 0.090(3) - 0.015(3)^2 = 0.135$$

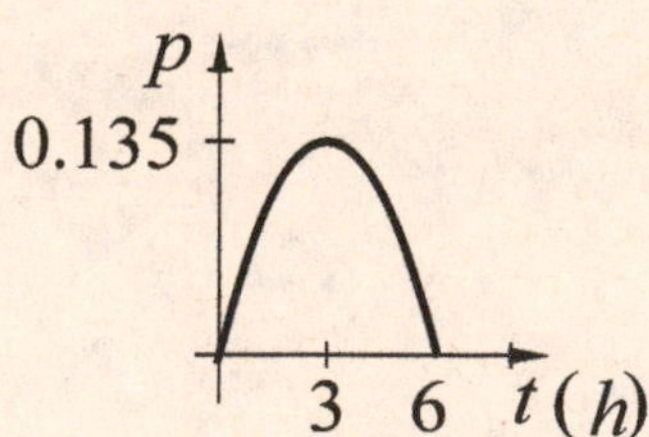

81.

$$\text{Original area } A = l \cdot w$$
$$A = (100 - x) \cdot (80 - x)$$
$$\text{New area } A_{new} = (100) \cdot (80) = 8000$$
$$\text{Difference in area } A_{new} - A = 3000$$
$$8000 - (100 - x) \cdot (80 - x) = 3000$$
$$8000 - 8000 + 100x + 80x - x^2 = 3000$$
$$x^2 - 180x + 3000 = 0$$
$$a = 1,\ b = -180,\ c = 3000$$
$$x = \frac{-b \pm \sqrt{b^2 - 4ac}}{2a}$$
$$x = \frac{-(-180) \pm \sqrt{(-180)^2 - 4(1)(3000)}}{2(1)}$$
$$x = \frac{180 \pm \sqrt{20400}}{2}$$
$$x = 18.59 \quad \text{or } x = 161.41 \text{ (reject since } x < 80)$$
$$x = 18.6 \text{ m}$$

The original dimensions are

$l = 100 - 18.6 = 81.4$ m and

$w = 80 - 18.6 = 61.4$ m

85.

$$(h + 22.9)^2 + h^2 = 60.0^2$$
$$h^2 + 2(22.9)h + 22.9^2 + h^2 = 60.0^2$$
$$2h^2 + 45.8h - 3075.59 = 0$$
$$a = 2,\ b = 45.8,\ c = -3075.59$$
$$h = \frac{-b \pm \sqrt{b^2 - 4ac}}{2a}$$
$$h = \frac{-(45.8) \pm \sqrt{(45.8)^2 - 4(2)(-3075.59)}}{2(2)}$$
$$h = \frac{-45.8 \pm \sqrt{26702.36}}{4}$$
$$h = 29.4 \quad \text{or } h = -52{,}3 \text{ (reject since } h > 0)$$
$$h = 29.4 \text{ in}$$
$$h + 22.9 = 52.3 \text{ in}$$

The dimensions of the screen are 29.4 in × 52.3 in.

89.

$$p = 0.001\,74(10 + 24h - h^2)$$
$$p = -0.00174h^2 + 0.04176h + 0.0174$$
If $p = 0.205$ ppm
$$0.205 = -0.00174h^2 + 0.04176h + 0.0174$$
$$0 = -0.00174h^2 + 0.04176h - 0.1876$$
$$a = -0.00174,\ b = 0.04176,\ c = -0.1876$$
$$h = \frac{-b \pm \sqrt{b^2 - 4ac}}{2a}$$
$$h = \frac{-0.04176 \pm \sqrt{0.04176^2 - 4(-0.00174)(-0.1876)}}{2(-0.00174)}$$
$$h = 5.98 \text{ and } h = 18.0$$

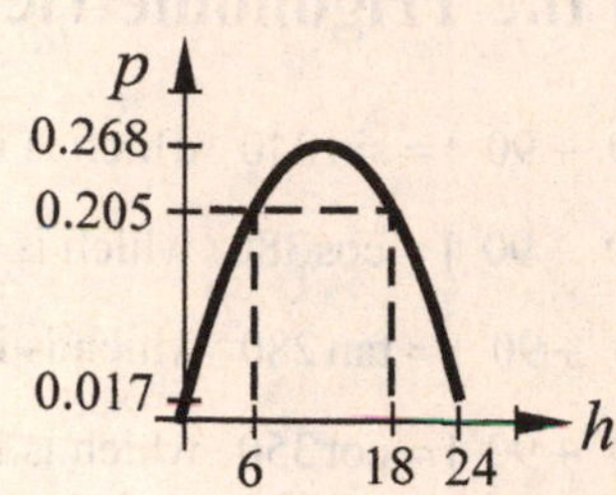

From the graph $p = 0.205$ at 6 h and 18 h.

Chapter 8

Trigonometric Functions of Any Angle

8.1 Signs of the Trigonometric Functions

1. **(a)** $\sin(150^\circ + 90^\circ) = \sin 240^\circ$ which is in Quandrant III is $-$

$\cos(290^\circ + 90^\circ) = \cos 380^\circ$ which is in Quandrant I is $+$

$\tan(190^\circ + 90^\circ) = \tan 280^\circ$ which is in Quandrant IV is $-$

$\cot(260^\circ + 90^\circ) = \cot 350^\circ$ which is in Quandrant IV is $-$

$\sec(350^\circ + 90^\circ) = \sec 440^\circ$ which is in Quandrant I is $+$

$\csc(100^\circ + 90^\circ) = \csc 190^\circ$ which is in Quandrant III is $-$

(b) $\sin(300^\circ + 90^\circ) = \sin 390^\circ$ which is in Quandrant I is $+$

$\cos(150^\circ + 90^\circ) = \cos 240^\circ$ which is in Quandrant III is $-$

$\tan(100^\circ + 90^\circ) = \tan 190^\circ$ which is in Quandrant III is $+$

$\cot(300^\circ + 90^\circ) = \cot 390^\circ$ which is in Quandrant I is $+$

$\sec(200^\circ + 90^\circ) = \sec 290^\circ$ which is in Quandrant IV is $+$

$\csc(250^\circ + 90^\circ) = \csc 340^\circ$ which is in Quandrant IV is $-$

5. $\sin 290^\circ$ is negative since 290° is in Quadrant IV where $\sin\theta$ is negative.
$\cos 200^\circ$ is negative since 200° is in Quadrant III where $\cos\theta$ is negative.

9. $\sec 150^\circ$ is negative since 150° is in Quadrant II, where $\sec\theta$ is negative.
$\tan 220^\circ$ is positive since 220° is in Quadrant III, where $\tan\theta$ is positive.

13. $\tan 460^\circ$ is negative since 460° is in Quadrant II, where $\tan\theta$ is negative.
$\sin(-185^\circ)$ is negative since -185° is in Quadrant III, where $\sin\theta$ is negative.

17. Point $(2, 1), x = 2, y = 1$

$$r = \sqrt{x^2 + y^2}$$
$$r = \sqrt{2^2 + 1^2}$$
$$r = \sqrt{5}$$
$$\sin\theta = \frac{y}{r} = \frac{1}{\sqrt{5}}$$
$$\cos\theta = \frac{x}{r} = \frac{2}{\sqrt{5}}$$
$$\tan\theta = \frac{y}{x} = \frac{1}{2}$$
$$\csc\theta = \frac{r}{y} = \sqrt{5}$$
$$\sec\theta = \frac{r}{x} = \frac{\sqrt{5}}{2}$$
$$\cot\theta = \frac{x}{y} = 2$$

21. Point $(-0.5, 1.2), x = -0.5, y = 1.2$

$$r = \sqrt{x^2 + y^2}$$
$$r = \sqrt{(-0.5)^2 + (1.2)^2}$$
$$r = 1.3$$
$$\sin\theta = \frac{y}{r} = \frac{1.2}{1.3} = \frac{12}{13}$$
$$\cos\theta = \frac{x}{r} = \frac{-0.5}{1.3} = -\frac{5}{13}$$
$$\tan\theta = \frac{y}{x} = \frac{1.2}{-0.5} = -\frac{12}{5}$$
$$\csc\theta = \frac{r}{y} = \frac{1.3}{1.2} = \frac{13}{12}$$
$$\sec\theta = \frac{r}{x} = \frac{1.3}{-0.5} = -\frac{13}{5}$$
$$\cot\theta = \frac{x}{y} = \frac{-0.5}{1.2} = -\frac{5}{12}$$

25. $\sin\theta = 0.500$

$\sin\theta = \frac{y}{r}$. Since $\sin\theta$ is positive, y must be positive and the terminal side of the angle must lie in either Quadrant I or Quadrant II.

29. $\cos\theta = -0.500$

$\cos\theta = \frac{x}{r}$. Since $\cos\theta$ is negative, x must be negative, and the terminal side of the angle must lie in either Quadrant II or Quadrant III.

33. $\sec\theta$ is negative and $\cot\theta$ is negative
$\sec\theta$ is negative in Quadrant II and Quadrant III.
$\cot\theta$ is negative in Quadrant II and Quadrant IV.
The terminal side of θ must lie in Quadrant II to meet both conditions.

37. $\sin\theta$ is positive and $\tan\theta$ is positive
$\sin\theta$ is positive in Quadrant I and Quadrant II.
$\tan\theta$ is positive in Quadrant I and Quadrant III.
The terminal side of θ must lie in Quadrant I to meet both conditions.

41. For (x, y) in Quadrant III,
x is$(-)$ and y is$(-)$

$$\frac{x}{r} = \frac{(-)}{(+)} = (-)$$

8.2 Trigonometric Functions of Any Angle

1. $\sin 200^\circ = -\sin(200^\circ - 180^\circ) = -\sin 20^\circ = -0.342$

$\tan 150^\circ = -\tan(180^\circ - 150^\circ) = -\tan 30^\circ = -0.577$

$\cos 265^\circ = -\cos(265^\circ - 180^\circ) = -\cos 85^\circ = -0.0872$

$\cot 300^\circ = -\cot(360^\circ - 300^\circ) = -\cot 60^\circ = -\frac{1}{\tan 60^\circ} = -0.577$

$\sec 344^\circ = \sec(360^\circ - 344^\circ) = \sec 16^\circ = \frac{1}{\cos 16^\circ} = 1.04$

$\sin 397^\circ = \sin(397^\circ - 360^\circ) = \sin 37^\circ = 0.602$

5. $\tan 105^\circ = -\tan(180^\circ - 105^\circ) = -\tan 75^\circ$

$\csc 328^\circ = -\csc(360^\circ - 328^\circ) = -\csc 32^\circ$

9. $\sin 195^\circ = -\sin(195^\circ - 180^\circ) = -\sin 15^\circ = -0.259$

13. $\sec 328.33^\circ = \sec(360^\circ - 328.33^\circ) = \sec 31.67^\circ = 1.1750$

17. $\cos(-62.7)^\circ = 0.4586$

21. $\csc 194.82^\circ = -3.9096$

25.
$$\sin\theta = -0.8480$$
$$\theta_{\text{ref}} = \sin^{-1} 0.8480$$
$$\theta_{\text{ref}} = 57.99^\circ$$

Since $\sin\theta$ is negative, θ must lie in Quadrant III or Quadrant IV.

Therefore, $\theta_3 = 180^\circ + 57.99^\circ$

$\theta_3 = 237.99^\circ$

or $\theta_4 = 360^\circ - 57.99^\circ$

$\theta_4 = 302.01^\circ$

29.
$$\cot\theta = -0.0122$$
$$\theta_{\text{ref}} = \tan^{-1}(1/0.0122)$$
$$\theta_{\text{ref}} = 89.3^\circ$$

Since $\cot\theta$ is negative, θ must lie in Quadrant II or Quadrant IV.

Therefore, $\theta_2 = 180^\circ - 89.3^\circ$

$\theta_2 = 90.7^\circ$

or $\theta_4 = 360^\circ - 89.3^\circ$

$\theta_4 = 270.7^\circ$

33. $\cos\theta = -0.12$

$\theta_{\text{ref}} = \cos^{-1}(0.12)$

$\theta_{\text{ref}} = 83^\circ$

Since $\cos\theta$ is negative, θ must lie in Quadrant II or Quadrant III.

If $\tan\theta$ is positive, θ must lie in Quadrant I or Quadrant III.

To satisfy both conditions, θ must lie in Quadrant III.

$\theta_3 = 180^\circ + 83^\circ$

$\theta_3 = 263^\circ$

37. $\sec\theta = 2.047$

$\theta_{\text{ref}} = \cos^{-1}(1/2.047)$

$\theta_{\text{ref}} = 60.76^\circ$

Since $\sec\theta$ is positive, θ must lie in Quadrant I or Quadrant IV.

If $\cot\theta$ is negative, θ must lie in Quadrant II or Quadrant IV.

To satisfy both conditions, θ must lie in Quadrant IV.

$\theta_4 = 360^\circ - 60.76^\circ$

$\theta_4 = 299.24^\circ$

41. $\tan 40^\circ + \tan 135^\circ - \tan 220^\circ = \tan 40^\circ - \tan\left(180^\circ - 135^\circ\right) - \tan\left(220^\circ - 180^\circ\right)$

$= \tan 40^\circ - \tan 45^\circ - \tan 40^\circ$

$= -\tan 45^\circ$

$= -1.000$

45. $\tan\theta = -0.809$

$\theta_{ref} = \tan^{-1} 0.809$

$\theta_{ref} = 39.0^\circ$

Since $\tan\theta$ is negative, θ must lie in Quadrant II or Quadrant IV.

If $\csc\theta$ is positive, θ must lie in Quadrant I or Quadrant II.

To satisfy both conditions, θ must lie in Quadrant II.

$\theta_2 = 180^\circ - 39.0^\circ$

$\theta_2 = 141.0^\circ$

In general, the angle could be any solution

$\theta = 141.0^\circ + k \times 360^\circ$ where $k = 0, \pm 1, \pm 2, \cdots$

but all evaluations of the trigonometric functions will be identical

for any integer number of rotations from the Quadrant II solution.

$\cos\theta = \cos 141.0^\circ$

$\cos\theta = -0.777$

49. $\tan 180^\circ = 0$, and

$\tan 0^\circ = 0$

$0 = 0$

$\tan 180^\circ = \tan 0^\circ$

53. For $0^\circ < \theta < 90^\circ$, $270^\circ - \theta$ is in Quadrant III, where tangent is positive.

$\tan\left(270^\circ - \theta\right) = +\tan\left(270^\circ - \theta - 180^\circ\right)$

$\tan\left(270^\circ - \theta\right) = \tan\left(90^\circ - \theta\right)$

$90^\circ - \theta$ and θ are complementary, so their cofunctions are equivalent

$\tan\left(270^\circ - \theta\right) = \cot\ \theta$

57. $i = i_m \sin\theta$

$i = (0.0259\text{ A})\sin 495.2^\circ$

$i = 0.0183$ A

8.3 Radians

1. $2.80 = (2.80)\ \dfrac{180^\circ}{\pi} = 160^\circ$

5. $75^\circ = 75^\circ\ \dfrac{\pi}{180^\circ} = \dfrac{5\pi}{12}$

$330^\circ = 330^\circ\ \dfrac{\pi}{180^\circ} = \dfrac{11\pi}{6}$

9. $720^\circ = 720^\circ\left(\dfrac{\pi}{180^\circ}\right) = 4\pi$

$-9^\circ = -9^\circ\left(\dfrac{\pi}{180^\circ}\right) = -\dfrac{\pi}{20}$

13. $\dfrac{5\pi}{9} = \dfrac{5\pi}{9}\left(\dfrac{180^\circ}{\pi}\right) = 100^\circ$

$\dfrac{7\pi}{4} = \dfrac{7\pi}{4}\left(\dfrac{180^\circ}{\pi}\right) = 315^\circ$

17. $-\dfrac{\pi}{15} = -\dfrac{\pi}{15}\left(\dfrac{180^\circ}{\pi}\right) = -12^\circ$

$\dfrac{3\pi}{20} = \dfrac{3\pi}{20}\left(\dfrac{180^\circ}{\pi}\right) = 27^\circ$

21. $252^\circ = 252^\circ\ \dfrac{\pi \text{ rad}}{180^\circ} = 4.40 \text{ rad}$

25. $478.5^\circ = 478.5^\circ\left(\dfrac{\pi \text{ rad}}{180^\circ}\right) = 8.351 \text{ rad}$

29. $3.407 = 3.407\left(\dfrac{180^\circ}{\pi}\right) = 195.2^\circ$

33. $-16.42 = -16.42\left(\dfrac{180^\circ}{\pi}\right) = -940.8^\circ$

37. $\tan\dfrac{5\pi}{12} = \tan\left[\left(\dfrac{5\pi}{12}\right)\left(\dfrac{180^\circ}{\pi}\right)\right] = \tan 75^\circ = 3.732$

41. $\sec 4.5920 = \sec\ 4.5920\ \dfrac{180^\circ}{\pi}$

$= \sec 263.10^\circ$

$= \dfrac{1}{\cos 263.10^\circ}$

$= -8.3265$

45. $\sin 4.24 = -0.890$

49. $\cot(-4.86) = \dfrac{1}{\tan(-4.86)} = 0.149$

53. $\tan\theta = -0.2126$

$\theta_{ref} = \tan^{-1} 0.2126$

$\theta_{ref} = 0.2095$

Since $\tan\theta$ is negative, θ must lie in Quadrant II or Quadrant IV.

$\theta_2 = \pi - 0.2095$

$\theta_2 = 2.932$

$\theta_4 = 2\pi - 0.2095$

$\theta_4 = 6.074$

57. $\sec\theta = -1.307$

$\theta_{ref} = \cos^{-1}(1/1.307)$

$\theta_{ref} = 0.6996$

Since $\sec\theta$ is negative, θ must lie in Quadrant II or Quadrant III.

$\theta_2 = \pi - 0.6966$

$\theta_2 = 2.442$

$\theta_3 = \pi + 0.6966$

$\theta_3 = 3.841$

61. $\dfrac{5\pi}{8}$ has $\theta_{\text{ref}} = \pi - \dfrac{5\pi}{8} = \dfrac{3\pi}{8}$

$\dfrac{5\pi}{8}$ is in Quadrant II where $\cos\theta$ is negative.

$\cos\dfrac{5\pi}{8} = -\cos\dfrac{3\pi}{8}$

$\dfrac{3\pi}{8}$ and $\dfrac{\pi}{2} - \dfrac{3\pi}{8}$ are complementary, so their cofunctions are equivalent.

$\cos\dfrac{5\pi}{8} = -\sin\ \dfrac{\pi}{2} - \dfrac{3\pi}{8}$

$\cos\dfrac{5\pi}{8} = -\sin\dfrac{\pi}{8}$

$\cos\dfrac{5\pi}{8} = -0.3827$

65. $34.4^\circ = 34.4^\circ\left(\frac{1 \text{ circumference}}{360^\circ}\right)\left(\frac{1 \text{ mil}}{\frac{1}{6400} \text{ circumference}}\right) = 612 \text{ mil}$

69.

$$V = \frac{1}{2}Wb\theta^2$$

$$V = \frac{1}{2}(8.75 \text{ lb})(0.75 \text{ ft})\left(5.5^\circ \cdot \frac{\pi}{180^\circ}\right)^2$$

$$V = 0.030 \text{ ft} \cdot \text{lb}$$

8.4 Applications of Radian Measure

1. $s = \theta r$

$$s = \left(\frac{\pi}{4}\right)(3.00 \text{ cm})$$

$$s = 2.36 \text{ cm}$$

5. $s = \theta r = \left(\frac{\pi}{4}\right)(5.70 \text{ in}) = 4.48 \text{ in}$

9. $\theta = \frac{s}{r} = \frac{0.3913 \text{ mi}}{0.9449 \text{ mi}} = 0.4141 \text{ rad}$

$$A = \frac{1}{2}\theta r^2 = \frac{1}{2}(0.4141)(0.9449 \text{ mi})^2 = 0.1849 \text{ mi}^2$$

13. $\theta = 326.0^\circ\left(\frac{\pi}{180^\circ}\right) = 5.690 \text{ rad}$

$$A = \frac{1}{2}\theta r^2$$

$$r = \sqrt{\frac{2A}{\theta}}$$

$$r = \sqrt{\frac{2\left(0.0119 \text{ ft}^2\right)}{5.690}}$$

$$r = 0.0647 \text{ ft}$$

17. $r = \frac{s}{\theta} = \frac{0.203 \text{ mi}}{\frac{3}{4}(2\pi)}$

$$r = 0.0431 \text{ mi} \cdot \frac{5280 \text{ ft}}{1 \text{ mi}}$$

$$r = 228 \text{ ft}$$

21. If time t is measured in hours from noon,
The hour hand rotates at the rate of a full revolution (2π rad) every 12 hours,

$$\text{Hour hand } \omega = \frac{2\pi \text{ rad}}{12 \text{ h}}$$

$$\theta_H = \omega t$$

$$\theta_H = \frac{2\pi}{12}t$$

The minute hand rotates at the rate of a full revolution (2π rad) every 1 hour,

$$\text{Minute hand } \omega = \frac{2\pi \text{ rad}}{1 \text{ h}}$$

$$\theta_M = 2\pi t$$

If the angle between the minute hand and hour hand is π rad,

$$\theta_H + \pi = \theta_M$$

$$\frac{2\pi}{12}t + \pi = 2\pi t$$

$$\pi = \left(\frac{12\pi}{6} - \frac{\pi}{6}\right)t$$

$$t = \frac{\pi}{11\pi/6}$$

$$t = \frac{6}{11} \text{ h} \left(\frac{60 \text{ min}}{1 \text{ h}}\right) = 32.727 \text{ min}$$

$$t = 32 \text{ min} + 0.727 \text{ min}\left(\frac{60 \text{ s}}{1 \text{ min}}\right)$$

$$t = 32 \text{ min} + 44 \text{ s}$$

The clock will read 12:32:44 when the hour and minute hands will be at 180° apart.

25. $\omega = \frac{\theta}{t} = \frac{\pi \text{ rad}}{6.0 \text{ s}} = 0.52 \text{ rad/s}$

29. $$\theta = 28.0^\circ\left(\frac{\pi \text{ rad}}{180^\circ}\right) = 0.489 \text{ rad}$$

From $s = \theta r$

$$s_1 = 0.489(93.67 \text{ ft}) = 45.80 \text{ ft}$$

$$s_2 = 0.489(93.67 \text{ ft} + 4.71 \text{ ft}) = 48.11 \text{ ft}$$

$$s_2 - s_1 = 2.31 \text{ ft}$$

Outer rail is 2.31 ft longer.

33. $$\theta = 15.6^\circ \frac{\pi}{180^\circ} = 0.272 \text{ rad}$$

$$A = \frac{1}{2}\theta r^2$$

$$r_1 = 285.0 \text{ m}$$

$$r_2 = 285.0 \text{ m} + 15.2 \text{ m} = 300.2 \text{ m}$$

$$A_{road} = A_2 - A_1$$

$$A_{road} = \frac{1}{2}(0.272)\left(300.2^2 - 285.0^2\right)\text{m}^2$$

$$A_{road} = 1210 \text{ m}^2$$

Volume = Area × thickness

$$V = At$$

$$V = 1210 \text{ m}^2 \left(0.305 \text{ m}\right)$$

$$V = 369 \text{ m}^3$$

37. $$\omega = 20.0 \text{ r/min}\left(\frac{2\pi \text{ rad}}{1 \text{ r}}\right)\left(\frac{1 \text{ min}}{60 \text{ s}}\right) = 2.09 \text{ rad/s}$$

$$v = \omega r = \left(2.09 \text{ rad/s}\right)\left(8.50 \text{ ft}\right) = 17.8 \text{ ft/s}$$

41. $$\omega = \frac{2 \text{ r}}{1 \text{ d}}\left(\frac{2\pi \text{ rad}}{1 \text{ r}}\right)\left(\frac{1 \text{ d}}{24 \text{ h}}\right)\left(\frac{1 \text{ h}}{3600 \text{ s}}\right) = 0.0001454 \text{ rad/s}$$

$$v = \omega r = \left(0.0001454 \text{ rad/s}\right)\left(26600 \text{ km}\right) = 3.87 \text{ km/s}$$

45. $$\theta = 82.0^\circ \frac{\pi \text{ rad}}{180^\circ} = 1.43 \text{ rad}$$

$$s = \theta r$$

$$s = 1.43(15.0 \text{ ft})$$

$$s = 21.5 \text{ ft}$$

49. At 2.25 cm from the center, 1590 r/min corresponds to a linear velocity of

$$v = \omega r = \frac{1590 \text{ r}}{1 \text{ min}} \times \frac{2\pi \text{ rad}}{1 \text{ r}} \times \frac{2.25 \text{ cm}}{1 \text{ rad}} = 22478 \text{ cm/min}$$

At this linear velocity, the angular velocity at 5.51 cm is

$$\omega = \frac{v}{r} = \frac{22478 \text{ cm}}{1 \text{ min}} \times \frac{1 \text{ rad}}{5.51 \text{ cm}} \times \frac{1 \text{ r}}{2\pi \text{ rad}} = 649 \text{ r/min}$$

53. $$\omega = 2400\frac{\text{r}}{\text{min}} \frac{2\pi \text{ rad}}{1 \text{ r}} \frac{1 \text{ min}}{60 \text{ s}} = 80\pi \text{ rad/s}$$

$$\theta = \omega t = \left(80\pi \text{ rad/s}\right)\left(1.0 \text{ s}\right) = 250 \text{ rad}$$

57.

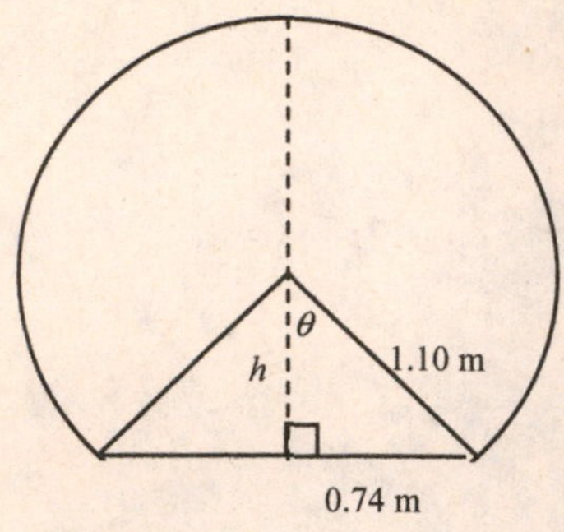

$h = \sqrt{1.10^2 - 0.74^2} = 0.8138$ m;

$\theta = \sin^{-1} \frac{0.74}{1.10} = 42.278^\circ$

$2\theta = 84.556^\circ \frac{\pi \text{ rad}}{180^\circ} = 1.476 \text{ rad}$

$A_{\text{sector}} = \frac{1}{2}(2\theta)r^2 = \frac{1}{2}(1.476)(1.10)^2 = 0.8929 \text{ m}^2$

$A_{\text{triangle}} = \frac{1}{2}bh = \frac{1}{2}(1.48)(0.8138) = 0.6023 \text{ m}^2$

$A_{\text{segment}} = 0.8929 - 0.6023 = 0.2906 \text{ m}^2$

$A_{end} = A_{circle} - A_{segment}$

$A_{end} = \pi(1.10 \text{ m})^2 - 0.2906 \text{ m}^2 = 3.5107 \text{ m}^2$

$V_{\text{tank}} = A_{end}L$

$V_{\text{tank}} = 3.5107 \text{ m}^2(4.25 \text{ m}) = 14.9 \text{ m}^3$

61.

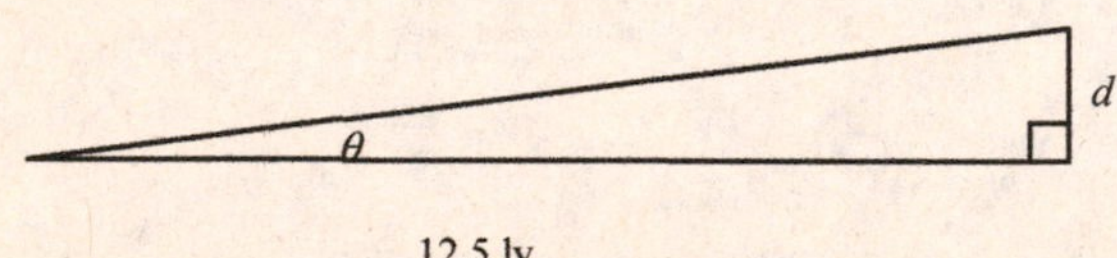

$\theta = 0.2'' \frac{1^\circ}{3600''} \frac{\pi \text{ rad}}{180^\circ} = 9.696 \times 10^{-7} \text{ rad}$

12.5 light years $= (12.5)(9.46 \times 10^{15})$m

$\tan\theta = \frac{d}{12.5 \text{ ly}}$

Using Eq. (8.17), $\tan\theta = \theta$ for small angles, so

$\theta = \frac{d}{12.5 \text{ ly}}$

$d = (12.5 \text{ ly})(9.696 \times 10^{-7})$

$d = (1.212 \times 10^{-7} \text{ ly}) \frac{9.46 \times 10^{12} \text{ km}}{1 \text{ ly}}$

$d = 1.15 \times 10^8 \text{ km}$

Review Exercises

1. This is true. If θ is an angle in quadrant I or quadrant IV, then $\cos\theta$ is positive.

5. This is false. The correct arc length formula is $s = \theta r$.

9. Point $(42,-12)$, $x = 42$, $y = -12$

$$r = \sqrt{42^2 + (-12)^2} = \sqrt{1908} = \sqrt{36(53)} = 6\sqrt{53}$$

$$\sin\theta = \frac{y}{r} = \frac{-12}{6\sqrt{53}} = -\frac{2}{\sqrt{53}}$$

$$\cos\theta = \frac{x}{r} = \frac{42}{6\sqrt{53}} = \frac{7}{\sqrt{53}}$$

$$\tan\theta = \frac{y}{x} = \frac{-12}{42} = -\frac{2}{7}$$

$$\csc\theta = \frac{r}{y} = -\frac{\sqrt{53}}{2}$$

$$\sec\theta = \frac{r}{x} = \frac{\sqrt{53}}{7}$$

$$\cot\theta = \frac{x}{y} = -\frac{7}{2}$$

13. 289° is in Quadrant IV, where $\sin\theta$ is negative.

$\sin 289^\circ = -\sin(360^\circ - 289^\circ) = -\sin 71^\circ$

-15° is in Quadrant IV, where $\sec\theta$ is positive.

$\sec(-15^\circ) = \sec 15^\circ$

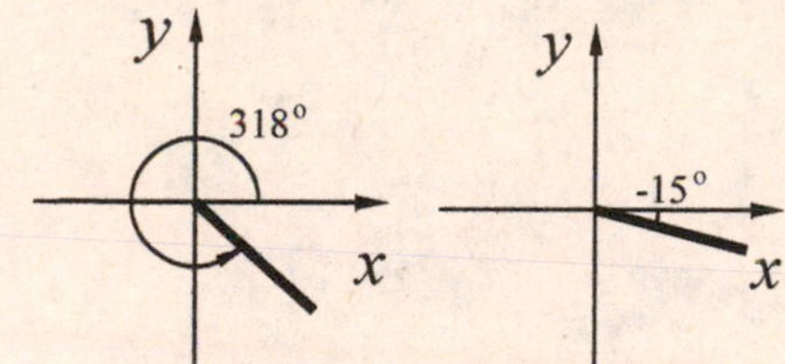

17. $408^\circ = 408^\circ\left(\frac{\pi}{180^\circ}\right) = \frac{34\pi}{15}$

$202.5^\circ = 202.5^\circ\left(\frac{\pi}{180^\circ}\right) = \frac{9\pi}{8}$

21. $\frac{\pi}{15} = \frac{\pi}{15}\left(\frac{180^\circ}{\pi}\right) = 12^\circ$

$\frac{11\pi}{6} = \frac{11\pi}{6}\left(\frac{180^\circ}{\pi}\right) = 330^\circ$

25. $-36.07 = -36.07\left(\frac{180^\circ}{\pi}\right) = -2067^\circ$

29. $20.25^\circ = 20.25^\circ\left(\frac{\pi \text{ rad}}{180^\circ}\right) = 0.3534 \text{ rad}$

33. $270^\circ = 270^\circ\left(\frac{\pi}{180^\circ}\right) = \frac{3}{2}\pi$

37. $\cos 237.4^\circ = -0.5388$

41. $\csc 247.82^\circ = -1.0799$

45. $\tan 301.4^\circ = -1.638$

49. $\sin\frac{9\pi}{5} = -0.5878$

53. $\sin 0.5906 = 0.5569$

57. $\tan\theta = 0.1817,\ 0 \le \theta < 360^\circ$

$\theta_{ref} = \tan^{-1}(0.1817) = 10.30^\circ$

Since $\tan\theta$ is positive, θ must lie in Quadrant I or Quadrant III.

$\theta_1 = 10.30^\circ$

$\theta_3 = 180^\circ + 10.30^\circ = 190.30^\circ$

61. $\cos\theta = 0.8387,\ 0 \le \theta < 2\pi$

$\theta_{ref} = \cos^{-1}(0.8387) = 0.5759$

Since $\cos\theta$ is positive, θ must lie in Quadrant I or Quadrant IV.

$\theta_1 = 0.5759$

$\theta_4 = 2\pi - 0.5759 = 5.707$

65. $\cos\theta = -0.672,\ 0^\circ \le \theta < 360^\circ$

$\theta_{ref} = \cos^{-1}(0.672) = 47.78^\circ$

Since $\cos\theta$ is negative, θ must lie in Quadrant II or Quadrant III.

Since $\sin\theta$ is negative, θ must lie in Quadrant III or Quadrant IV.

To satisfy both conditions, θ must lie in Quadrant III.

$\theta_3 = 180^\circ + 47.78^\circ = 227.78^\circ$

69. $\theta = 107.5^\circ\left(\frac{\pi \text{ rad}}{180^\circ}\right) = 1.876 \text{ rad}$

$r = \frac{s}{\theta} = \frac{20.3 \text{ in}}{1.876} = 10.8 \text{ in}$

73. $A=\frac{1}{2}\theta r^2$

$$\theta=\frac{2A}{r^2}=\frac{2\left(32.8\text{ m}^2\right)}{(4.62\text{ m})^2}=3.07\text{ rad}$$

$s=\theta r=3.07(4.62\text{ m})=14.2\text{ m}$

77. We wish to compute $\tan 200°+2\cot 110°+\tan(-160°)$.

We note $\tan 200°=\tan(-160°)=\tan 20°$ and

$$\cot 110°=\frac{\cos 110°}{\sin 110°}=\frac{-\sin 20°}{\cos 20°}=-\tan 20°.$$

Therefore,

$$\begin{aligned}\tan 200°+2\cot 110°+\tan(-160°)&=\tan 20°-2\tan 20°+\tan 20°\\&=0\end{aligned}$$

81. **(a)** $\theta=20.0^\circ\ \frac{\pi\text{ rad}}{180^\circ}\ =\frac{\pi}{9.00}\text{ rad}$

$r=\sqrt{40.0^2+30.0^2}=50.0\text{ m}$

$A=A_{\text{triangle}}+A_{\text{sector}}$

$A=\frac{1}{2}bh+\frac{1}{2}\theta r^2$

$A=\frac{1}{2}(40.0\text{ m})(30.0\text{ m})+\frac{1}{2}\ \frac{\pi}{9.00}\ (50.0\text{ m})^2$

$A=1040\text{ m}^2$

(b) The perimeter is the total of the lengths of the two legs, a radius of the circle, and the length of the circular arc.

$$\begin{aligned}P&=30.0+40.0+50.0+50.0\cdot 20°\cdot\frac{\pi}{180°}\\&=120.0+17.4533\\&=137\text{ m}\end{aligned}$$

85. $s=\theta r$

$$\theta=\frac{s}{r}=\frac{6.60\text{ in}}{8.25\text{ in}}=0.800\text{ rad}$$

$$\theta=0.800\text{ rad}\ \frac{180^\circ}{\pi\text{ rad}}\ =45.8^\circ$$

89. $\omega=\frac{1\text{ r}}{28\text{ d}}\ \frac{2\pi\text{ rad}}{1\text{ r}}\ \frac{1\text{ d}}{24\text{ h}}\ =0.0093\text{ rad/h}$

$v=\omega r=(0.0093\text{ rad/h})(240000\text{ mi})$

$v=2230\text{ mi/h}$

93. $\omega = 60.0 \text{ r/s} \left(\dfrac{2\pi \text{ rad}}{1 \text{ r}}\right) = 120.0\pi \text{ rad/s}$

$v = \omega r = (120.0\pi \text{ rad/s})\left(\dfrac{0.250 \text{ m}}{2}\right) = 47.1 \text{ m/s}$

97. $\omega = 80\ 000 \text{ r/min} \left(\dfrac{2\pi \text{ rad}}{1 \text{ r}}\right)\left(\dfrac{1 \text{ min}}{60 \text{ s}}\right) = 8377.6 \text{ rad/s}$

$v = r\omega = (3.60 \text{ cm})(8377.6 \text{ rad/s}) = 30159 \text{ cm/s}$

$v = 30159 \text{ cm/s} \left(\dfrac{1 \text{ m}}{100 \text{ cm}}\right) = 302 \text{ m/s}$

101. $\theta = \dfrac{0.0008^\circ}{2}\left(\dfrac{\pi \text{ rad}}{180^\circ}\right) = 6.9813 \times 10^{-6} \text{ rad}$

$\tan\theta = \dfrac{y}{x}$

$x = \dfrac{y}{\tan\theta}$

Using Eq. (8.17),

$\tan\theta = \theta$ for small θ.

$x = \dfrac{y}{\theta} = \dfrac{2.50 \text{ km}}{6.9813 \times 10^{-6}}$

$x = 3.58 \times 10^5 \text{ km}$

Chapter 9

Vectors and Oblique Triangles

9.1 Introduction to Vectors

1.

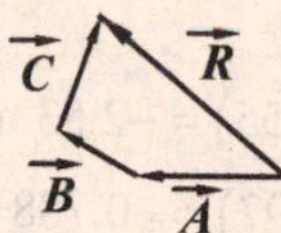

5. **(a)** scalar, no direction given

(b) vector, magnitude and direction both given

9.

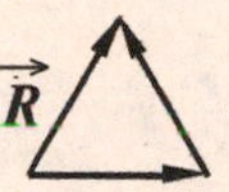

13.

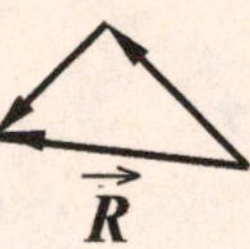

17. 4.3 cm, 156°

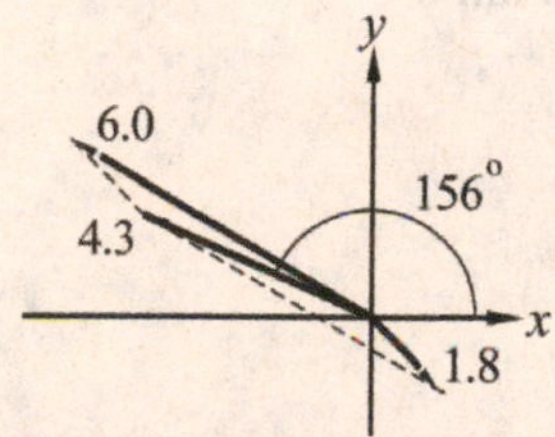

21.

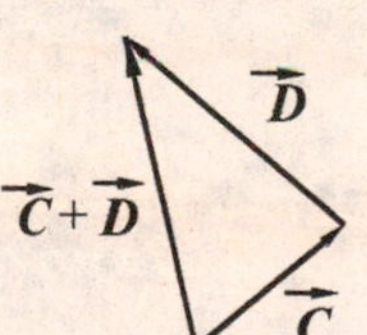

25.

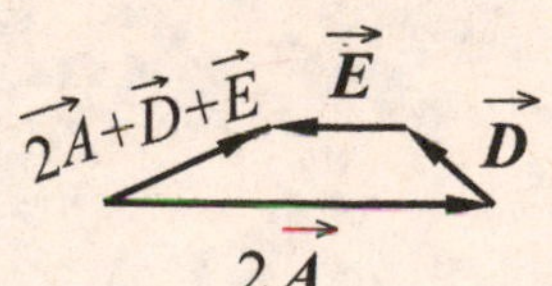

29.

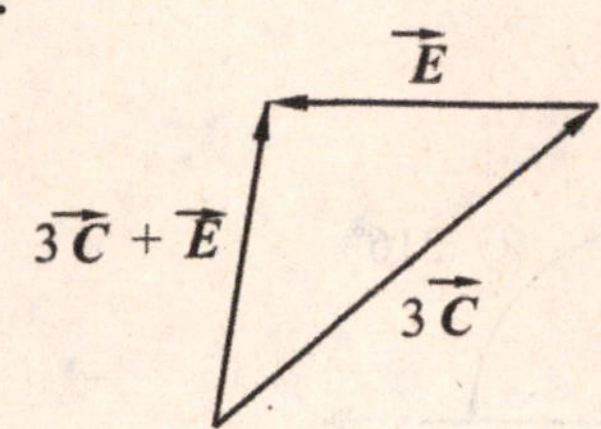

33.

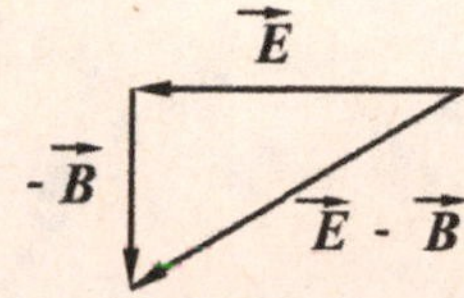

37.

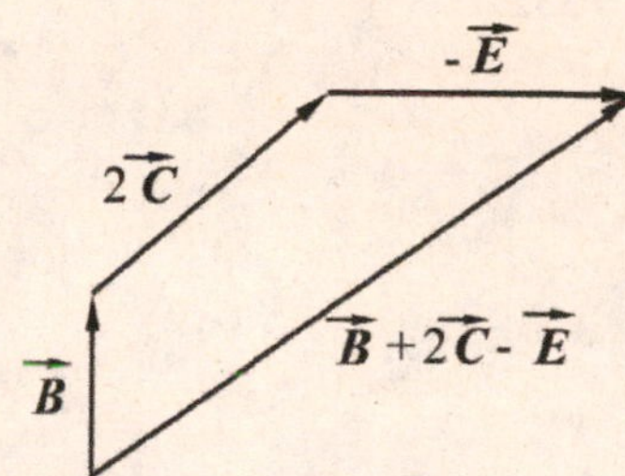

41.

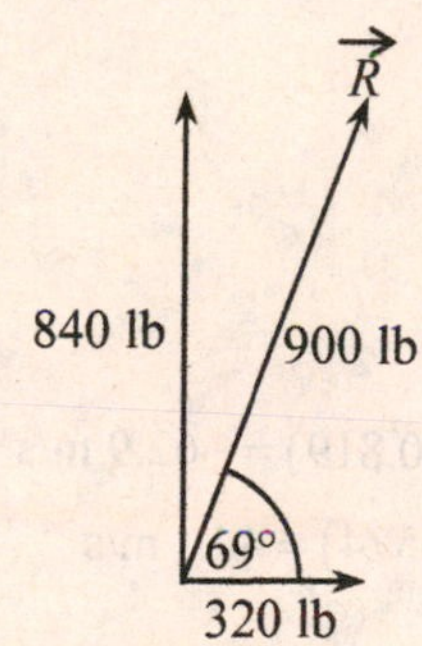

From the drawing, $\vec{R}$ is approximately 900 lb at 69°.

45.

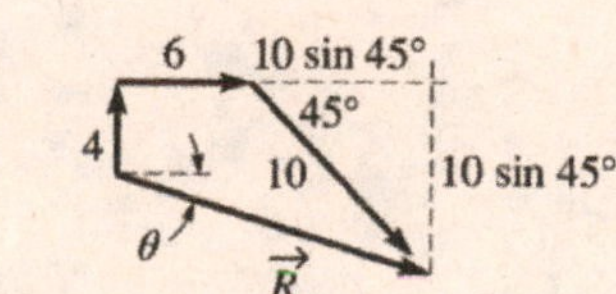

From the drawing,

$R = 13$ km

$\theta = 13^\circ$

9.2 Components of Vectors

1. $d_x = d\cos\theta$

$d_x = 14.4\cos 216^\circ = -11.6 \text{ m}$

$d_y = d\sin\theta$

$d_y = 14.4\sin 216^\circ = -8.46 \text{ m}$

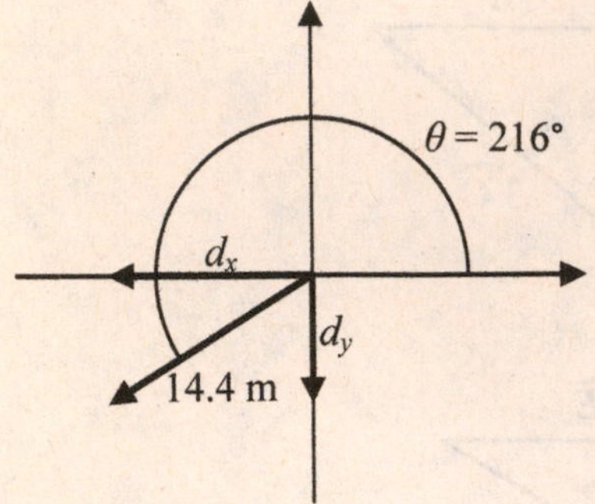

5. $V_x = 750\cos 28^\circ = 662$

$V_y = 750\sin 28^\circ = 352$

9. $V_x = -750$

$V_y = 0$

13. Let $V = 76.8$

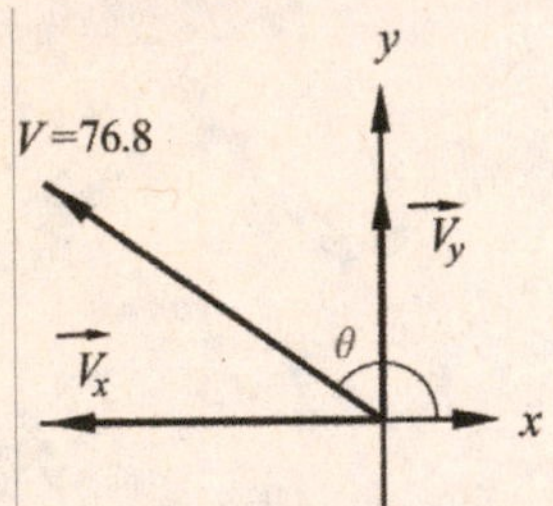

$V_x = V\cos 145.0^\circ = 76.8(-0.819) = -62.9 \text{ m/s}$

$V_y = V\sin 145.0^\circ = 76.8(0.574) = 44.1 \text{ m/s}$

$V_x = V\cos 156.5^\circ = 16.4(-0.917) = -15.0 \text{ cm/s}^2$

$V_y = V\sin 156.5^\circ = 16.4(0.399) = 6.54 \text{ cm/s}^2$

17. Let $V = 2.65$

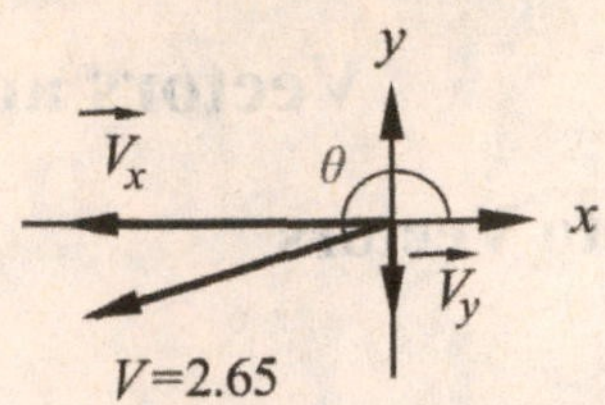

$V_x = V\cos 197.3^\circ = 2.65(-0.955) = -2.53 \text{ mN}$

$V_y = V\sin 197.3^\circ = 2.65(-0.297) = -0.788 \text{ m/N}$

21.

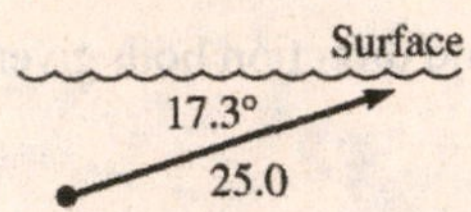

$V_x = 25.0\cos 17.3^\circ = 23.9 \text{ km/h}$

$V_y = 25.0\sin 17.3^\circ = 7.43 \text{ km/h}$

25.

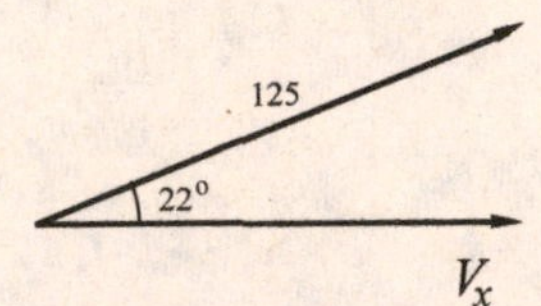

$V_x = 125\cos 22.0^\circ = 116 \text{ km/h}$

29.

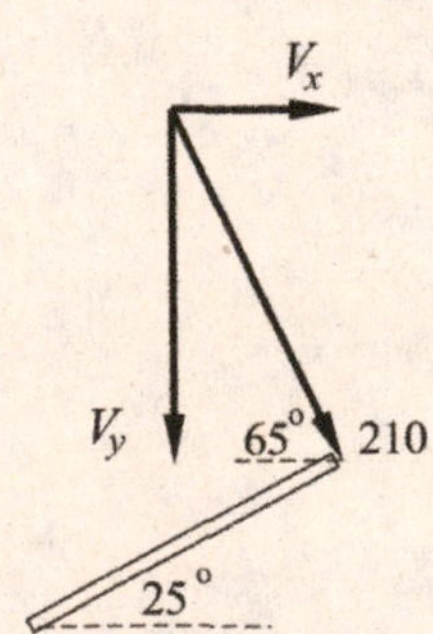

$V_x = 210\cos 65^\circ = 89 \text{ N}$

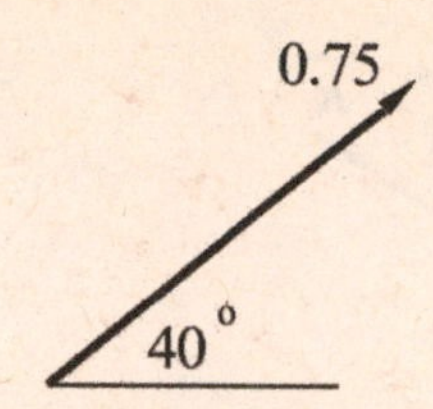

$V_y = -210\sin 65^\circ = -190 \text{ N}$

33.

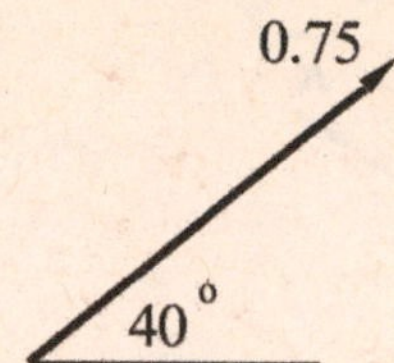

horizontal component = $0.75\cos 40^\circ$

$= 0.57$ (km/h)/m

vertical component = $0.75\sin 40^\circ$

$= 0.48$ (km/h)/m

9.3 Vector Addition by Components

1. $A = 1200 = A_x$, $B = 1750$

$Ay = 0$

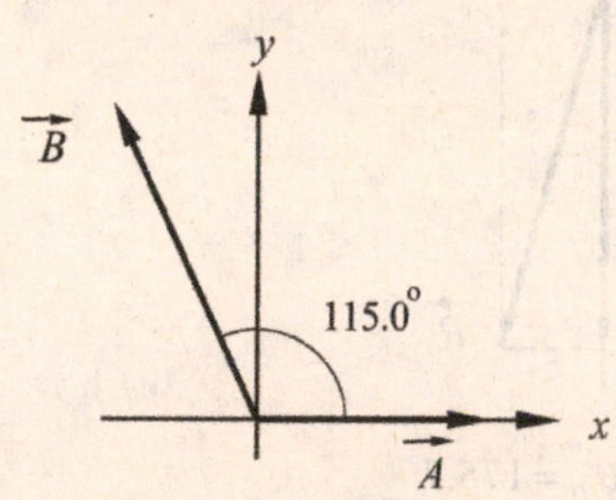

$$R_x = A_x + B_x = 1200 + 1750\cos 115^\circ = 460.4$$
$$R_y = A_y + B_y = 0 + 1750\sin 115^\circ = 1586$$
$$R = \sqrt{R_x^2 + R_y^2} = \sqrt{460.4^2 + 1586^2} = 1650$$
$$\theta = \tan^{-1}\frac{R_y}{R_x} = \tan^{-1}\frac{1586}{460.4} = 73.8^\circ$$

5.

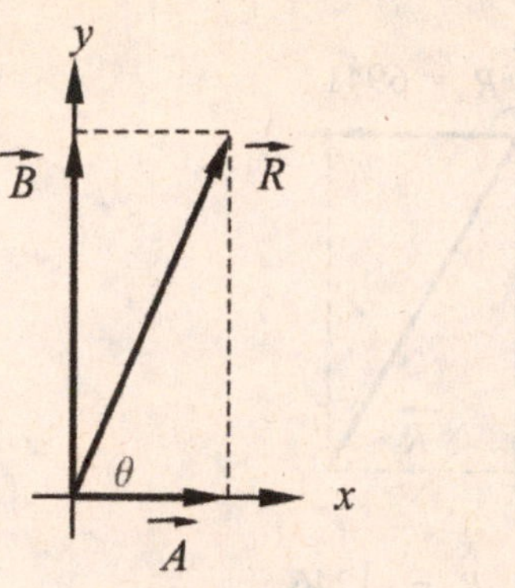

$$R = \sqrt{3.086^2 + 7.143^2} = \sqrt{60.54} = 7.781$$
$$\tan\theta = \frac{7.143}{3.086} = 2.315$$
$$\theta = 66.63^\circ \text{ (with } \vec{A}\text{)}$$

9.

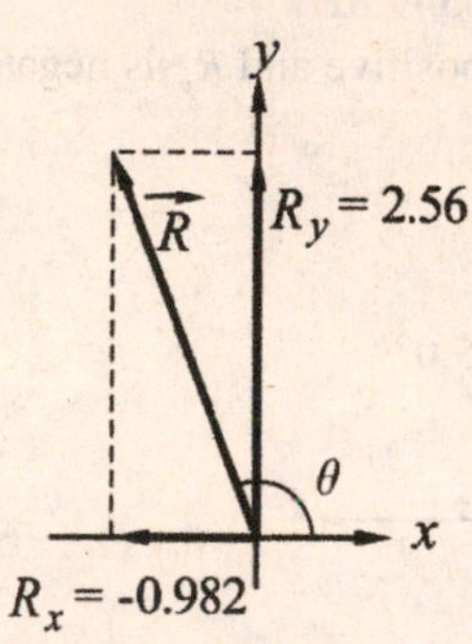

$$R_x = -0.982,\ R_y = 2.56$$
$$R = \sqrt{R_x^2 + R_y^2} = \sqrt{(-0.982)^2 + 2.56^2} = 2.74$$
$$\tan\theta_{\text{ref}} = \left|\frac{2.56}{-0.982}\right| = 2.61$$
$$\theta_{\text{ref}} = 69.0^\circ$$
$$\theta = 180^\circ - 69.0^\circ = 111.0^\circ$$

(θ is in Quad II since R_x is negative and R_y is positive)

13.

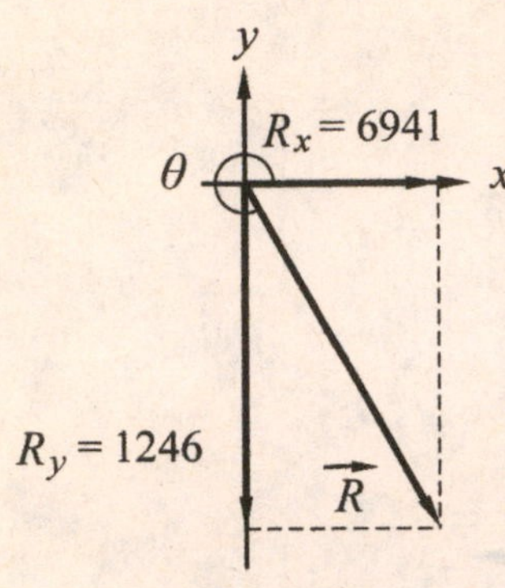

$$R_x = 6941,\ R_y = -1246$$
$$R = \sqrt{6941^2 + (-1246)^2} = 7052$$
$$\tan\theta_{ref} = \left|\frac{-1246}{6941}\right| = 0.1795$$
$$\theta_{ref} = 10.18°$$
$$\theta = 360° - 10.18° = 349.82°$$
(θ is in QIV since R_x is positive and R_y is negative)

17.

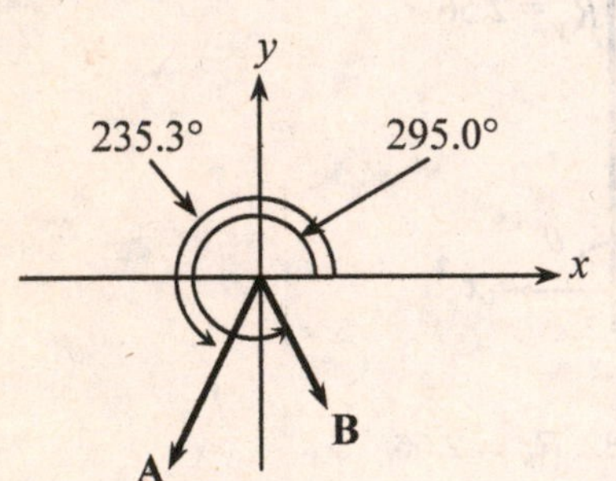

$$A_x = 368\cos 235.3° = -209.4948646$$
$$A_y = 368\sin 235.3° = -302.5490071$$
$$B_x = 227\cos 295.0° = 95.93434542$$
$$B_y = 227\sin 295.0° = -205.7318677$$
$$R_x = A_x + B_x = -113.5605192$$
$$R_y = A_y + B_y = -508.2808748$$
$$R = \sqrt{R_x^2 + R_y^2}$$
$$= 521$$
$$\theta_{ref} = \tan^{-1}\left(\frac{R_y}{R_x}\right) = 77.40576722°$$
$$\theta = 180° + 77.4° = 257.4°$$

21.

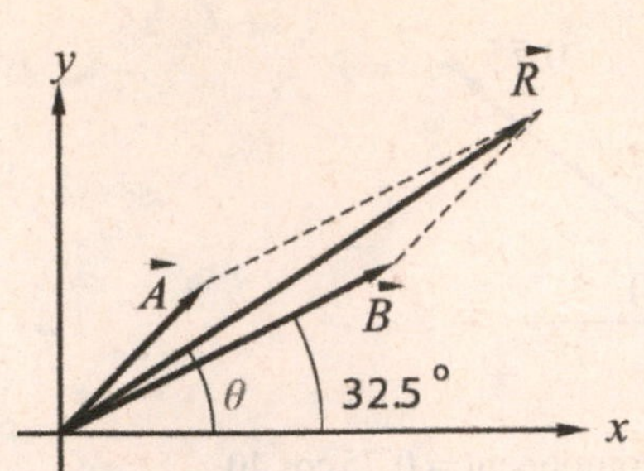

$$R_x = A_x + B_x$$
$$= 9.821\cos 34.27° + 17.45\cos 752.5°$$
$$R_y = 9.821\sin 34.27° + 17.45\sin 752.5°$$
$$R = \sqrt{R_x^2 + R_y^2} = 27.27$$
$$\theta = \tan^{-1}\frac{R_y}{R_x} = 33.14°$$

25.

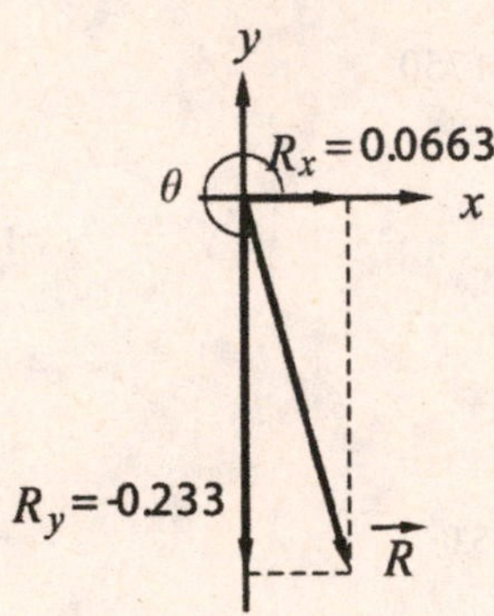

$$U = 0.364, \theta_U = 175.7°$$
$$U_x = 0.364\cos 175.7° = -0.363$$
$$U_y = 0.364\sin 175.7° = 0.0273$$
$$V = 0.596, \theta_V = 319.5°$$
$$V_x = 0.596\cos 319.5° = 0.453$$
$$V_y = 0.596\sin 319.5° = -0.387$$
$$W = 0.129, \theta_W = 100.6$$
$$W_x = 0.129\cos 100.6° = -0.0237$$
$$W_y = 0.139\sin 100.6° = 0.137$$
Using stored values (without rounding),
$$R_x = 0.0665$$
$$R_y = -0.223$$
$$R = \sqrt{R_x^2 + R_y^2} = 0.242$$
$$\theta_{ref} = \tan^{-1}\frac{|-0.223|}{|0.0665|} = 74.1°$$
$$\theta = 360° - 74.1° = 285.9°$$
(θ is in Quad IV since R_x is positive and R_y is negative.)

29.

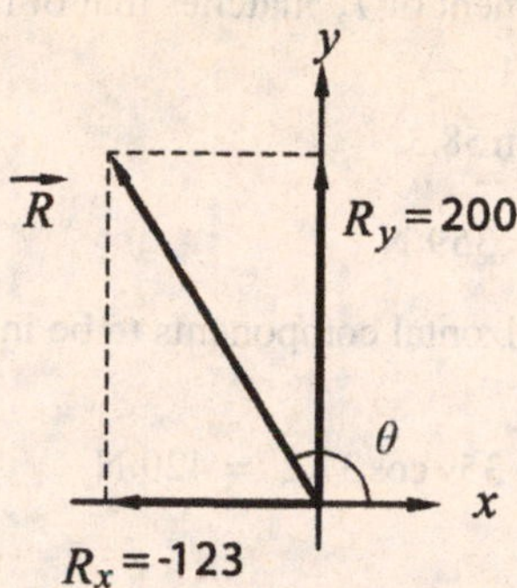

$$R_x = 302\cos(180° - 45.4°) + 155\cos(180° + 53.0°) + 212\cos 30.8° = -123$$

$$R_y = 302\sin(180° - 45.4°) + 155\sin(180° + 53.0°) + 212\sin 30.8° = 200$$

$$R = \sqrt{R_x^{\,2} + R_y^{\,2}} = 235$$

$$\theta_{\text{ref}} = \tan^{-1}\frac{|R_y|}{|R_x|} = 58.4°$$

$$\theta = 180° - 58.4° = 121.6°$$

33.

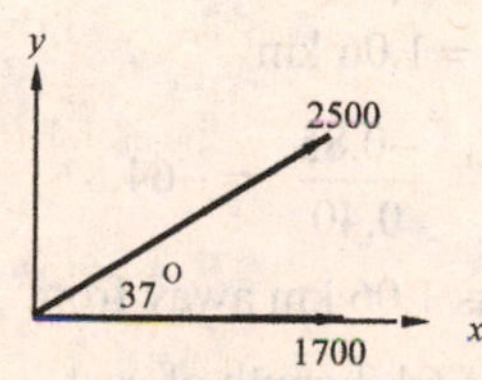

$$R = \sqrt{R_x^{\,2} + R_y^{\,2}}$$

$$R = \sqrt{(1700 + 2500\cos 37°)^2 + (2500\sin 37°)^2}$$

$$R = 4000 \text{ N}$$

9.4 Applications of Vectors

1.

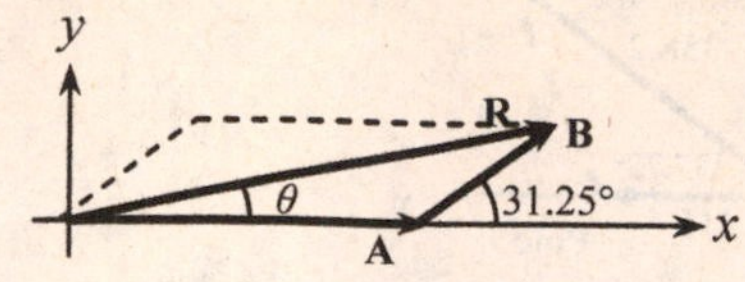

$$R_x = A_x + B_x$$
$$= 32.50 + 16.18\cos 31.25°$$
$$= 46.33$$
$$R_y = 16.18\sin 31.25°$$
$$= 8.394$$
$$R = \sqrt{46.33^2 + 8.394^2}$$
$$= 47.08 \text{ km}$$
$$\theta = \tan^{-1}\frac{8.394}{46.33} = 10.27°$$

The ship is 47.08 km from start in direction 10.27° N of E.

5.

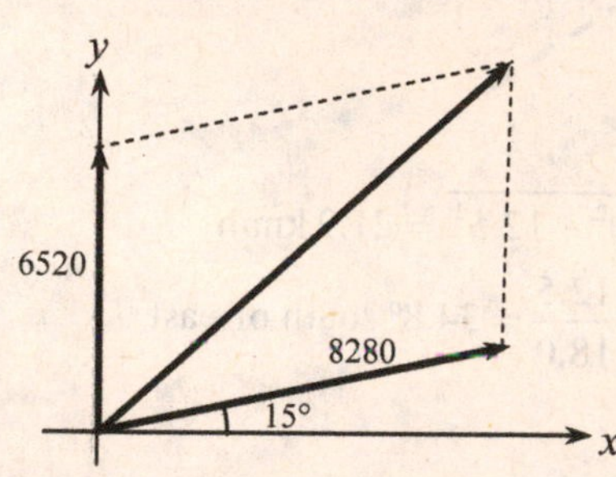

$$R_x = 8280\cos 15.0°$$
$$= 7998 \text{ N}$$
$$R_y = 6520 + 8280\sin 15.0°$$
$$= 8663 \text{ N}$$
$$R = \sqrt{7998^2 + 8663^2}$$
$$= 11800 \text{ N}$$
$$\theta = \tan^{-1}\frac{8663}{7998} = 47.3°$$

9.

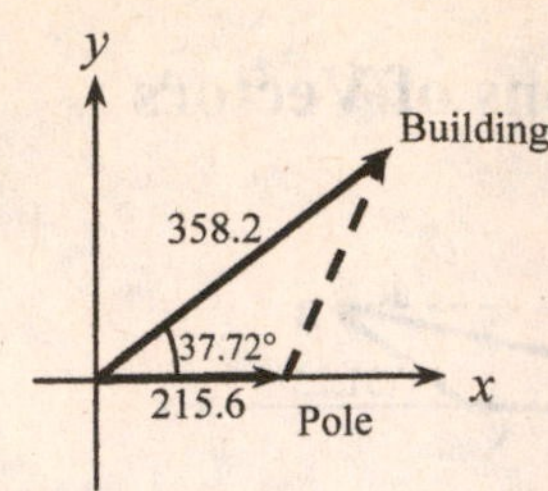

$$358.2\cos(37.72^\circ) = 215.6 + R_x$$

$$R_x = 358.2\cos(37.72^\circ) - 215.6 = 67.74$$

$$R_y = 358.2\sin(37.72^\circ) = 219.15$$

$$R = \sqrt{R_x^2 + R_y^2} = 229.4 \text{ ft}$$

$$\theta = \tan^{-1}\frac{219.15}{67.74} = 72.82^\circ$$

13.

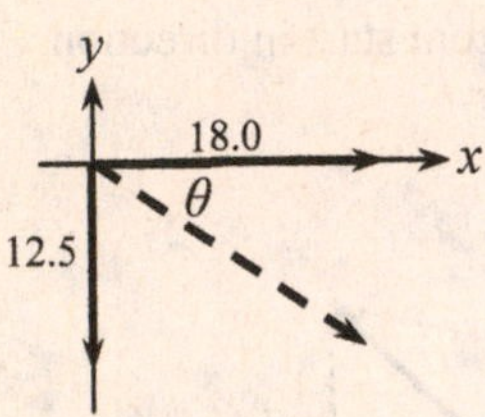

$$R = \sqrt{18.0^2 + 12.5^2} = 21.9 \text{ km/h}$$

$$\theta = \tan^{-1}\frac{12.5}{18.0} = 34.8^\circ \text{ south of east}$$

17.

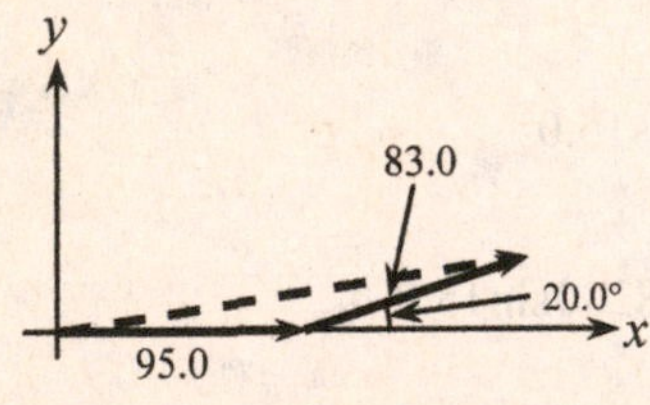

$$F_x = 95.0 + 83.0\cos 20.0^\circ = 173$$

$$F_y = 83.0\sin 20.0^\circ = 28.4$$

$$F = \sqrt{173^2 + 28.4^2} = 175 \text{ lb}$$

$$\tan\theta = \frac{28.4}{173}$$

$\theta = 9.32^\circ$, above horizontal and to the right.

21. The vertical component of T_2 matches that of the vector in Quadrant I:

$$T_2 \sin 37.2^\circ = 255\sin 58.3^\circ$$

$$T_2 = \frac{255\sin 58.3^\circ}{\sin 37.2^\circ} = 359 \text{ N}$$

In order for the horizontal components to be in equilibrium,

$$T_1 = 255\cos 58.3^\circ + 359\cos 37.2^\circ = 420 \text{ N}$$

25.

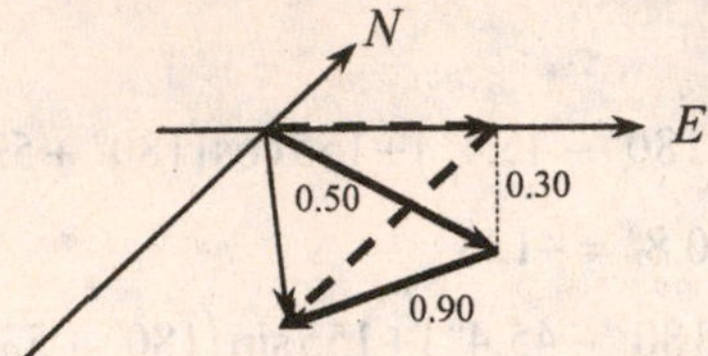

The component along the surface for the eastward portion of the journey is $\sqrt{0.50^2 - 0.30^2} = 0.40$ km. The component along the surface for the southward portion of the journey is $\sqrt{0.90^2 - 0.30^2} = 0.85$ km. The displacement on the surface is

$$s = \sqrt{0.40^2 + (-0.85)^2} = 1.06 \text{ km}$$

with an angle $\theta = \tan^{-1}\dfrac{-0.85}{0.40} = -64.8^\circ$

and so the submarine is 1.06 km away from its origin at an angle of 64.8° south of east.

29.

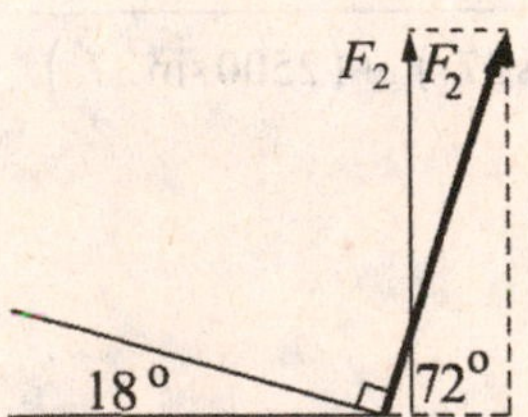

Using the fact that $1.2F_1 = 0.3F_2$, we have

$$F_2 = \frac{1.2F_1}{0.3}$$

$$F_{2v} = F_2 \sin 72^\circ$$

$$= \frac{1.2(240)}{0.3}\sin 72^\circ$$

$$F_{2v} = 910 \text{ N}$$

33.

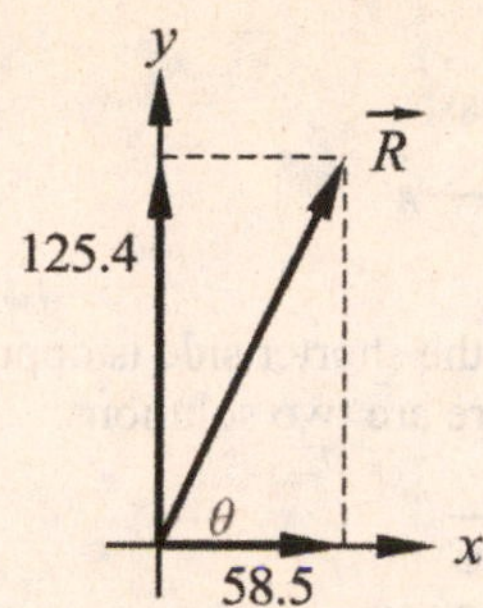

$R_x = 75.0 + 75.0\cos 65.0^\circ + 75.0\cos 130.0^\circ = 58.5$

$R_y = 75.0\sin 65.0^\circ + 75.0\sin 130.0^\circ = 125.4$

$R = \sqrt{58.5^2 + 125.4^2} = 138 \text{ km}$

$\theta = \tan^{-1}\dfrac{125.4}{58.5} = 65.0^\circ \text{ N of E}$

37. The horizontal components satisfy

$T_1 \cos 51.0^\circ + T_2 \cos 119.6^\circ = 0$

$T_2 = -\dfrac{\cos 51.0^\circ}{\cos 119.6^\circ} T_1 = 1.274\ T_1$

The vertical components satisfy

$T_1 \sin 51.0^\circ + T_2 \sin 119.6^\circ = 975$

or

$0.777146 T_1 + 0.869495(1.274\ T_1) = 975$

or

$T_1 = \dfrac{975}{0.777146 + 0.869495(1.274)} = 517 \text{ lb.}$

and

$T_2 = 1.274\ T_1 = 659 \text{ lb.}$

9.5 Oblique Triangles, the Law of Sines

1.

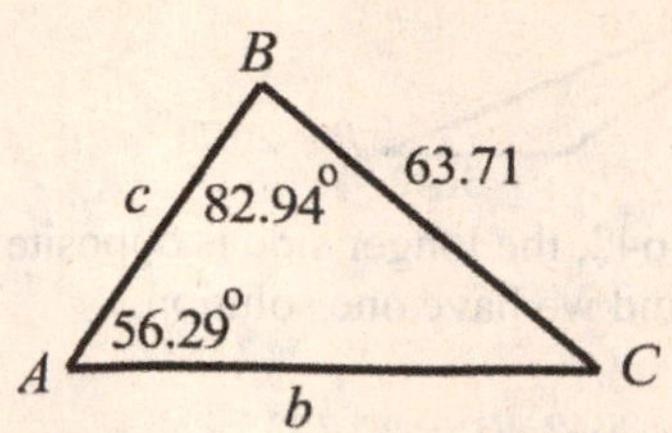

$C = 180° - (56.29° - 82.94°) = 40.77°$

$\dfrac{b}{\sin 82.94°} = \dfrac{63.71}{\sin 56.29°} = \dfrac{c}{\sin 40.77°}$

$b = \dfrac{63.71 \sin 82.94°}{\sin 56.29°} = 76.01$

$c = \dfrac{63.71 \sin 40.77°}{\sin 56.29°} = 50.01$

5.

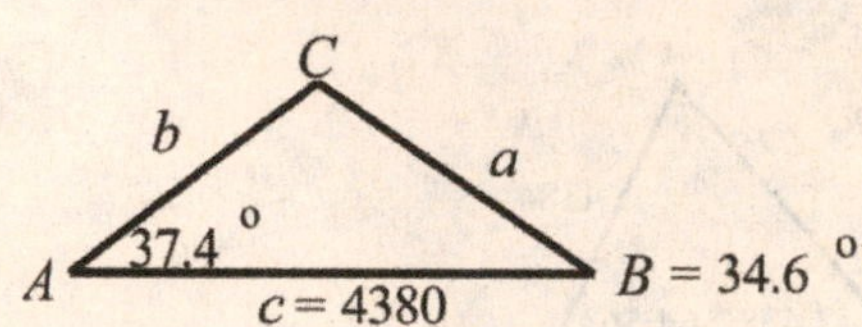

$c = 4380,\ A = 37.4°,\ B = 34.6°$

$C = 180.0° - (37.4° + 34.6°) = 108.0°$

$\dfrac{b}{\sin B} = \dfrac{c}{\sin C};\ \dfrac{b}{\sin 34.6°} = \dfrac{4380}{\sin 108.0°}$

$b = \dfrac{4380 \sin 34.6°}{\sin 108.0°} = 2620$

$\dfrac{a}{\sin A} = \dfrac{c}{\sin C};\ \dfrac{a}{\sin 37.4°} = \dfrac{4380}{\sin 108.0°}$

$a = \dfrac{4380 \sin 37.4°}{\sin 108.0°} = 2800$

9.

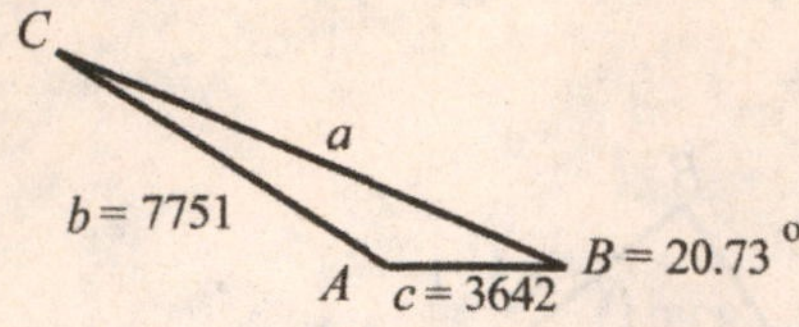

Since 7751 > 3642, the longer side is opposite the known angle, and we have one solution.

$b = 7751, c = 3642, B = 20.73°$

$$\frac{b}{\sin B} = \frac{c}{\sin C}; \frac{7751}{\sin 20.73°} = \frac{3642}{\sin C}$$

$$\sin C = \frac{3642 \sin 20.73°}{7751} = 0.1663$$

$$C = 9.574°$$

$$A = 180.0° - (20.73° + 9.574°) = 149.7°$$

$$\frac{a}{\sin A} = \frac{b}{\sin B}; \frac{a}{\sin 149.7°} = \frac{7751}{\sin 20.73°}$$

$$a = \frac{7751 \sin 149.7°}{\sin 20.73°} = 11\,050$$

13.

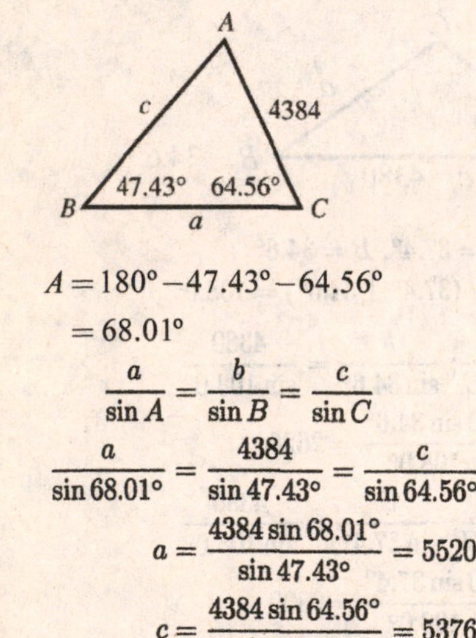

$$A = 180° - 47.43° - 64.56°$$
$$= 68.01°$$

$$\frac{a}{\sin A} = \frac{b}{\sin B} = \frac{c}{\sin C}$$

$$\frac{a}{\sin 68.01°} = \frac{4384}{\sin 47.43°} = \frac{c}{\sin 64.56°}$$

$$a = \frac{4384 \sin 68.01°}{\sin 47.43°} = 5520$$

$$c = \frac{4384 \sin 64.56°}{\sin 47.43°} = 5376$$

17.

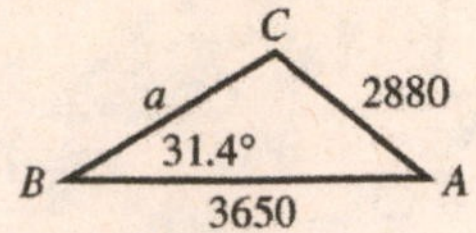

Since 2880 < 3650, the shorter side is opposite the known angle, so there are two solutions.

$$\frac{a}{\sin A} = \frac{b}{\sin B} = \frac{c}{\sin C}$$

$$\frac{a}{\sin A} = \frac{2880}{\sin 31.4°} = \frac{3650}{\sin C}$$

$$\sin C = \frac{3650 \sin 31.4°}{2880}$$

$$C = 41.3° \text{ or } 138.7°$$

Case I.

$C = 41.3°,$

$A = 180° - 31.4° - 41.3° = 107.3°$

$$\frac{a}{\sin 107.3°} = \frac{2880}{\sin 31.4°}$$

$$a = 5280$$

Case II.

$C = 138.7°,$

$A = 180° - 31.4° - 138.7° = 9.9°$

$$\frac{a}{\sin 9.9°} = \frac{2880}{\sin 31.4°}$$

$$a = 950$$

21. The third angle is $180° - (45.0° + 55.0°) = 80.0°$

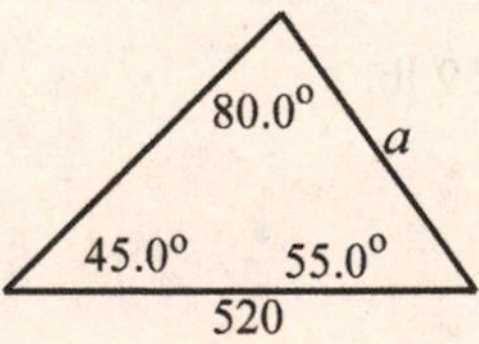

The shortest side is opposite the 45.0° angle.
We use the law of sines with the other side we know.

$$\frac{520}{\sin 80.0°} = \frac{a}{\sin 45.0°}$$

$$a = 373 \text{ m}$$

25.

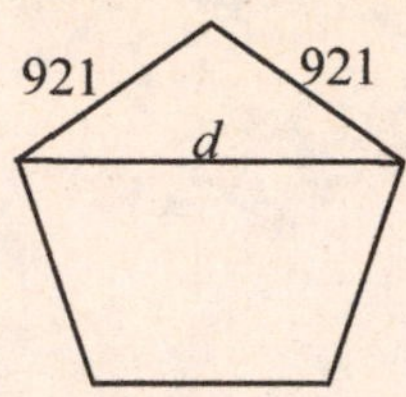

The angles of a regular pentagon are the same and sum to 540°, implying each angle is 540°/5=108°. The remaining angles of the triangle formed by two sides and a diagonal have measure $\frac{180° - 108°}{2} = 36°$

Using the law of sines, the diagonal d satisfies

$$\frac{921}{\sin 36°} = \frac{d}{\sin 108°}$$

$$d = \frac{921 \sin 108°}{\sin 36°} = 1490 \text{ ft}$$

29. $C = 180° - (29.0° + 35.0°) = 114°$

$$\frac{32.5}{\sin 114°} = \frac{c}{\sin 35.0°}$$

$$c = \frac{32.5 \sin 35.0°}{\sin 114°} = 20.4 \text{ m}$$

33. $A = 180° - 89.2° = 90.8°$

$C = 180° - 86.5 - 90.8° = 2.7°$

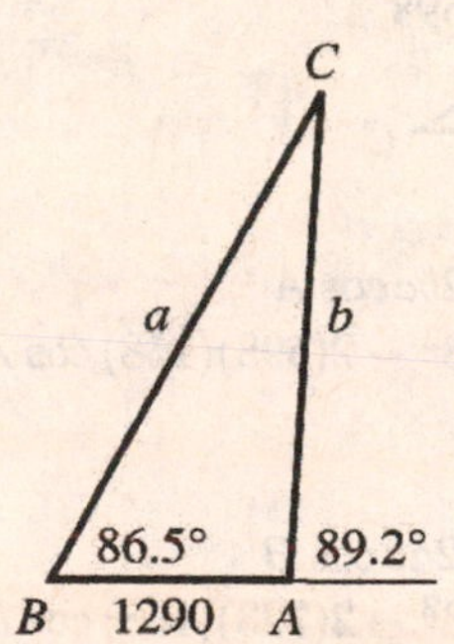

$$\frac{b}{\sin 86.5°} = \frac{1290}{\sin 2.7°}$$

$$b = 27\,300 \text{ km}$$

37.

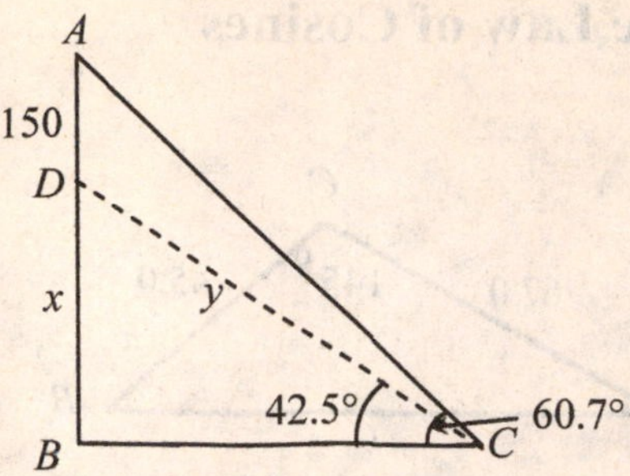

We let x represent the height of the elevator and y represent the distance from the observer to the elevator at the first instant.

We have from the right triangle BDC

$x = y \sin 42.5°$

Also, we know angle A has measure 29.3° and in triangle ACD, angle C has measure 18.2°.

From the law of sines,

$$\frac{y}{\sin 29.3°} = \frac{150}{\sin 18.2°}$$

$$y = \frac{150 \sin 29.3°}{\sin 18.2°} = 235.03 \text{ ft.}$$

$x = y \sin 42.5° = 158.78$ ft.

and so the elevator is $x + 150 = 309$ ft. high at the second instant

9.6 The Law of Cosines

1.

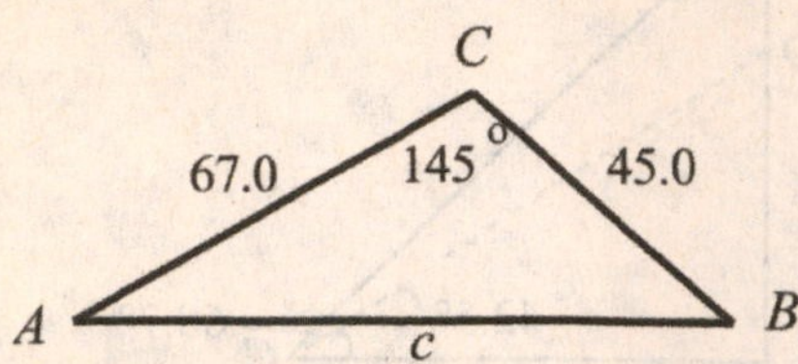

$$c = \sqrt{45.0^2 + 67.0^2 - 2(45.0)(67.0)\cos 145°}$$
$$= 107$$

$$\frac{45.0}{\sin A} = \frac{c}{\sin 145°} = \frac{67.0}{\sin B}$$

$$A = \sin^{-1}\frac{45.0\sin 145°}{c} = 14.0°$$

$$B = \sin^{-1}\frac{67.0\sin 145°}{c} = 21.0°$$

5.

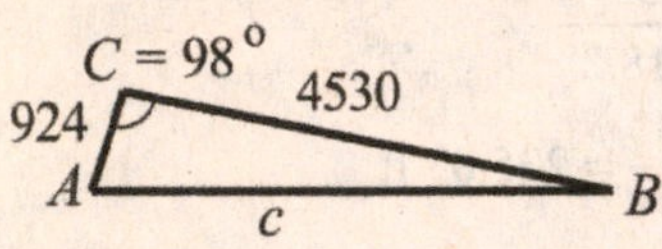

$a = 4530, b = 924, C = 98.0°$

$$c = \sqrt{4530^2 + 924^2 - 2(4530)(924)(\cos 98.0°)}$$
$$= 4750$$

$$\frac{c}{\sin C} = \frac{b}{\sin B}; \frac{4750}{\sin 98.0°} = \frac{924}{\sin B}$$

$$\sin B = \frac{924\sin 98.0°}{4750} = -0.193$$

$$B = 11.1°$$
$$A = 180° - 98.0° - 11.1° = 70.9°$$

9.

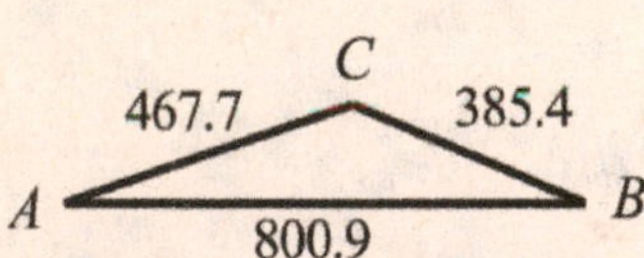

$a = 385.4, b = 467.7, c = 800.9$

$$\cos A = \frac{467.7^2 + 800.9^2 - 385.4^2}{2(467.7)(800.9)} = 0.9499$$

$$A = 18.21°$$

$$\cos B = \frac{385.4^2 + 800.9^2 - 467.7^2}{2(385.4)(800.9)} = 0.9253$$

$$B = 22.28°$$
$$C = 180° - 18.21° - 22.28° = 139.51°$$

13.

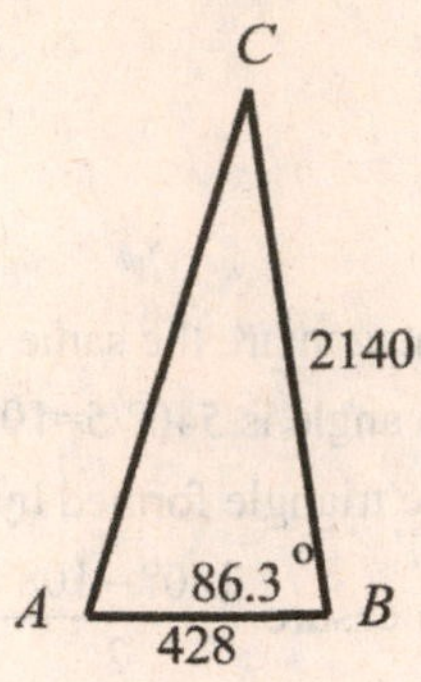

$a = 2140, c = 428, B = 86.3°$

$$b = \sqrt{2140^2 + 428^2 - 2(2140)(428)(\cos 86.3°)}$$
$$= 2160$$

$$\frac{b}{\sin B} = \frac{c}{\sin C}; \frac{2160}{\sin 86.3°} = \frac{428}{\sin C}$$

$$\sin C = \frac{428\sin 86.3°}{2160} = 0.198$$

$$C = 11.4°$$
$$A = 180° - 86.3° - 11.4° = 82.3°$$

17.

A, $c = 158$, $b = 598$, B, C, $a = 723$

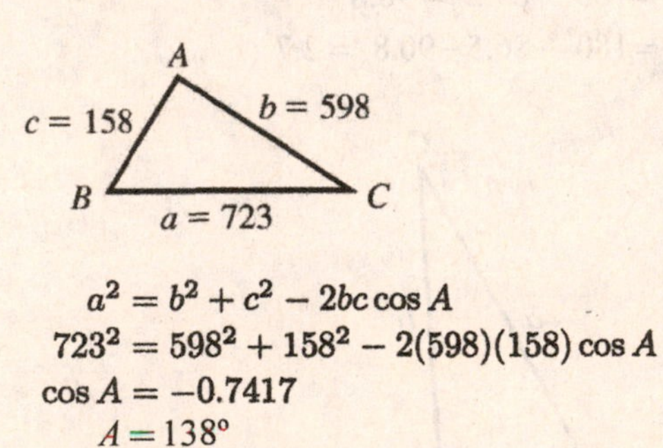

$$a^2 = b^2 + c^2 - 2bc\cos A$$
$$723^2 = 598^2 + 158^2 - 2(598)(158)\cos A$$
$$\cos A = -0.7417$$
$$A = 138°$$
$$b^2 = a^2 + c^2 - 2ac\cos B$$
$$598^2 = 723^2 + 158^2 - 2(723)(158)\cos B$$
$$\cos B = 0.8320$$
$$B = 33.7°$$
$$C = 180° - A - B$$
$$= 180° - 138° - 33.7° = 8.3°$$

21.

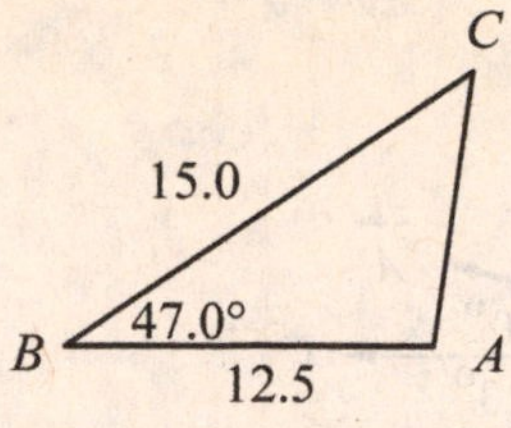

By the law of cosines,

$c^2 = a^2 + b^2 - 2ab\cos C$

$c^2 = 12.5^2 + 15.0^2 - 2(12.5)(15.0)\cos 47.0°$

$c^2 = 125.5$

$c = \sqrt{125.5} = 11.2$ ft

25.

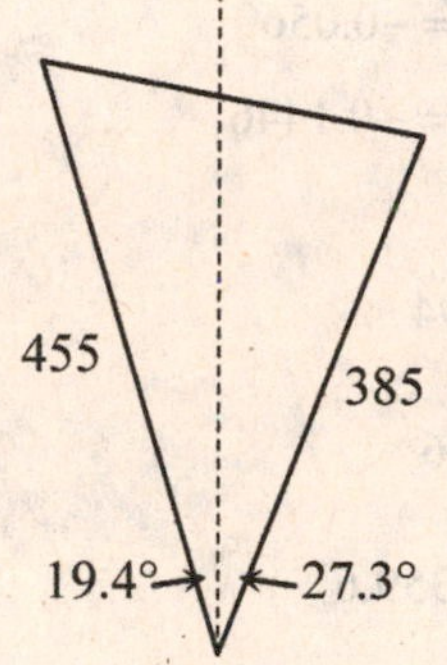

By the law of cosines,

$c^2 = a^2 + b^2 - 2ab\cos C$

where $a = 385, b = 455, C = 19.4° + 27.3° = 46.7°$

$c^2 = 385^2 + 455^2 - 2(385)(455)\cos 46.7°$

$c^2 = 114973$

$c = 339$ mi

29. $19^2 = 21^2 + 25^2 - 2(21)(25)\cos\theta$

$\cos\theta = \dfrac{21^2 + 25^2 - 19^2}{2(21)(25)} = 0.6714286$

$\theta = 48^0$

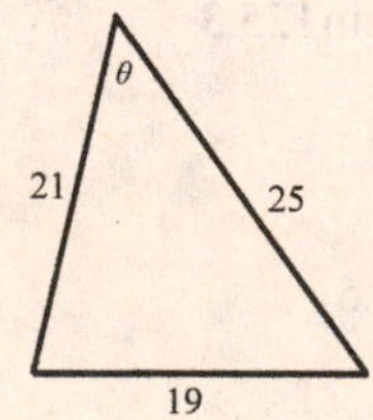

33. $x = \sqrt{9.53^2 + 9.53^2 - 2(9.53)(9.53)(\cos 120^\circ)}$

$= 16.5$ mm

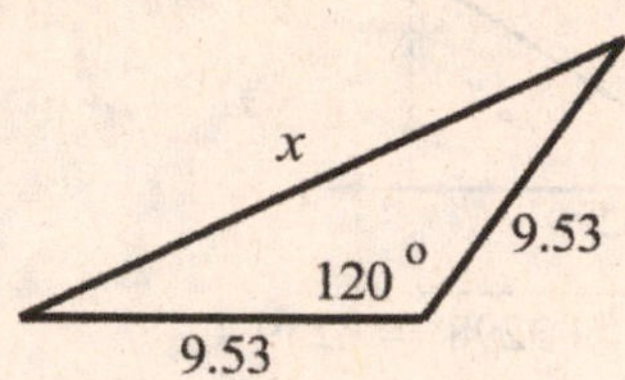

37. $A = 26.4^\circ - 12.4^\circ = 14.0^\circ$

$a = \sqrt{15.8^2 + 32.7^2 - 2(15.8)(32.7)\cos 14.0^\circ}$

$= 17.8$ mi

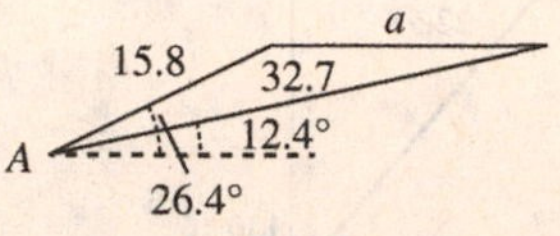

Review Exercises

1. True

5. False. One should use the law of sines to determine a second angle.
It is possible for there to be two solutions. This is Case 2 from the Summary of Solving Oblique Triangles.

9.

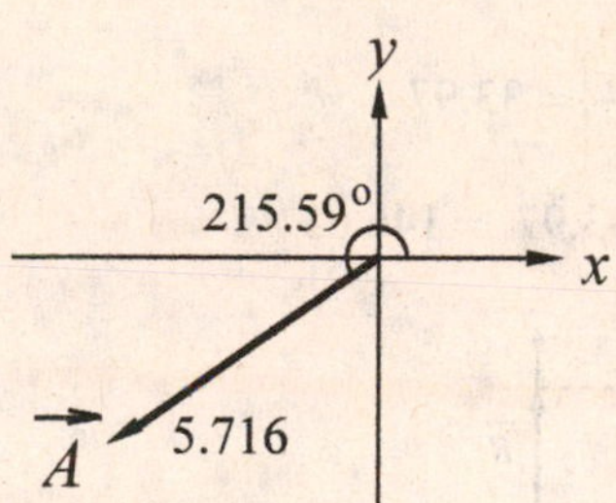

$A_x = A\cos\theta_A = 5.716\cos 215.59^\circ = -4.648$

$A_y = A\sin\theta_A = 5.716\sin 215.59^\circ = -3.327$

13.

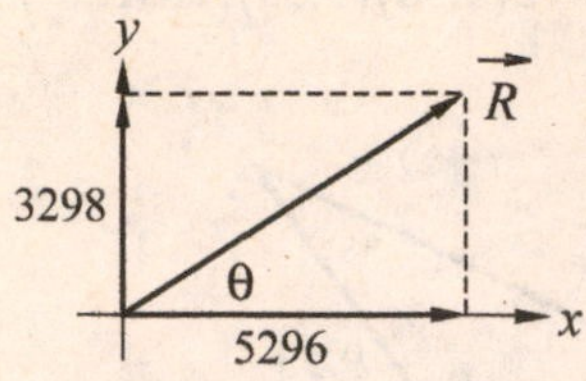

$R = \sqrt{5296^2 + 3298^2} = 6239$

$\theta = \tan^{-1}\dfrac{3298}{5296} = 31.91^\circ$

17.

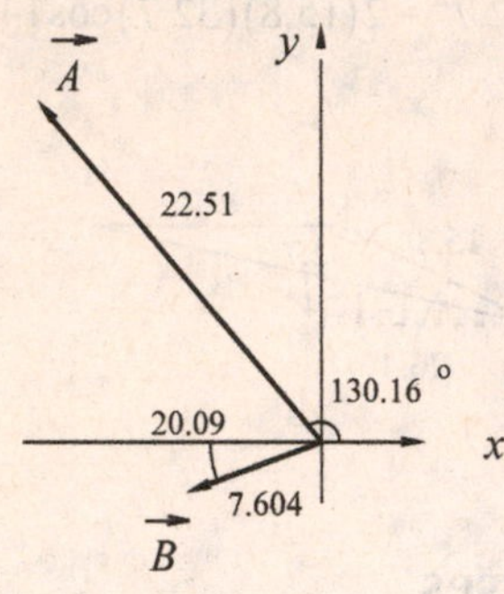

$A_x = 22.51\cos 130.16^\circ = -14.52$

$B_x = 7.604\cos 200.09^\circ = -7.141$

$R_x = A_x + B_x = -21.66$

$A_y = 22.51\sin 130.16^\circ = 17.20$

$B_y = 7.604\sin 200.09^\circ = -2.612$

$R_y = A_y + B_y = 14.59$

$R = \sqrt{R_x^2 + R_y^2} = 26.12$

$\theta_{\text{ref}} = \tan^{-1}\left|\dfrac{R_y}{R_x}\right| = 33.97^\circ$

$\theta_R = 180^\circ - 33.97^\circ = 146.03^\circ$

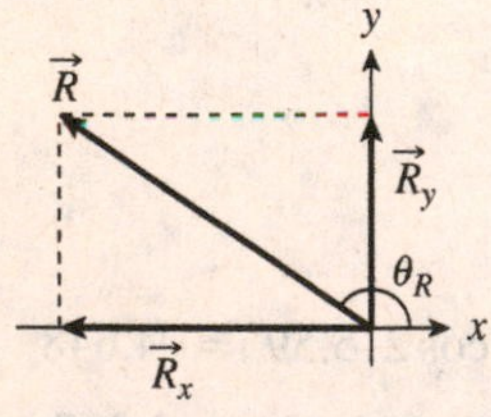

21.

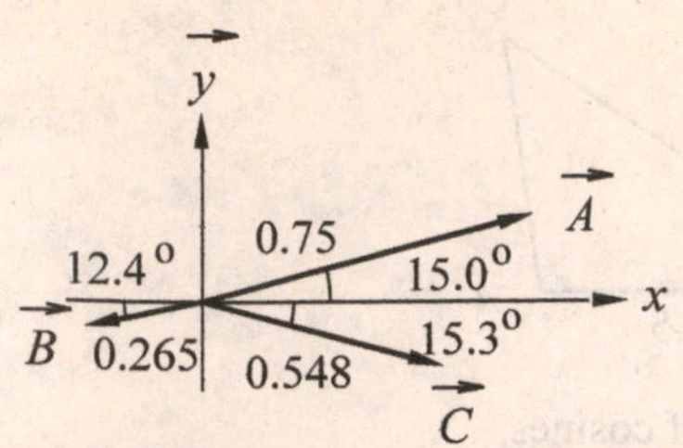

$A_x = 0.750\cos 15.0^\circ = 0.724$

$B_x = 0.265\cos 192.4^\circ = -0.259$

$C_x = 0.548\cos 344.7^\circ = 0.529$

$R_x = 0.994$

$A_y = 0.750\sin 15.0^\circ = 0.1941$

$B_y = 0.265\sin 192.4^\circ = -0.0569$

$C_y = 0.548\sin 344.7^\circ = -0.1446$

$R_y = -0.00740$

$R = \sqrt{R_x^2 + R_y^2} = 0.994$

$\theta_{ref} = \tan^{-1}\left|\dfrac{R_y}{R_x}\right| = 0.426^\circ$

$\theta_R = 360^\circ - 0.426^\circ = 359.6^\circ$

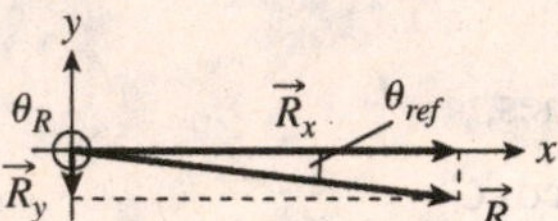

25.

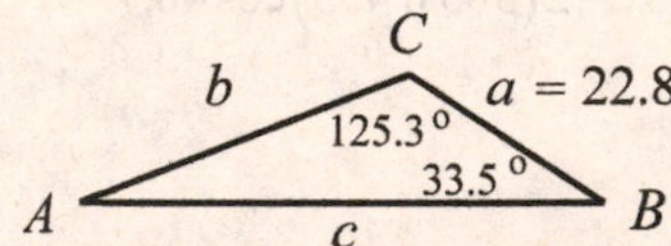

Given: two angles and one side (Case 1).

$A = 180.0^\circ - \left(125.3^\circ + 33.5^\circ\right) = 21.2^\circ$

$\dfrac{22.8}{\sin 21.2^\circ} = \dfrac{b}{\sin 33.5^\circ} = \dfrac{c}{\sin 125.3^\circ}$

$b = \dfrac{22.8\sin 33.5^\circ}{\sin 21.2^\circ} = 34.8$

$c = \dfrac{22.8\sin 125.3^\circ}{\sin 21.2^\circ} = 51.5$

29.

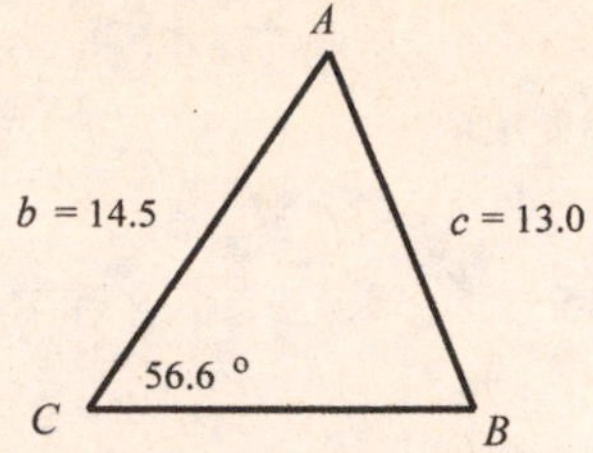

Given: two sides and the angle opposite the shorter side (Case 2, two solutions).

$$\frac{a}{\sin A}=\frac{14.5}{\sin B}=\frac{13.0}{\sin 56.6^\circ}$$

$$\sin B=\frac{14.5\sin 56.6^\circ}{13.0}$$

$B=68.6^\circ$ or 111.4°

Solution 1:

$B=68.6^\circ$

$A=180^\circ-68.6^\circ-56.6^\circ=54.8^\circ$

$$\frac{a}{\sin 54.8^\circ}=\frac{13.0}{\sin 56.6^\circ}$$

$$a=\frac{13\sin 54.8^\circ}{\sin 56.6^\circ}=12.7$$

Solution 2:

$B=111.4^\circ$

$A=180^\circ-111.4^\circ-56.6^\circ=12.0^\circ$

$$\frac{a}{\sin 12.0^\circ}=\frac{13.0}{\sin 56.6^\circ}$$

$$a=\frac{13\sin 12.0^\circ}{\sin 56.6^\circ}=3.24$$

33.

C $a = 7.86$ B

(2.5°) $b = 2.45$ A

Given: two sides and the included angle (Case 3).

$c^2=7.86^2+2.45^2-2(7.86)(2.45)\cos 2.5^\circ$

$c=5.413386814$

$c=5.41$

$7.86^2=2.45^2+c^2-2(2.45)(c)\cos A$

(We use c without rounding for more precision since the angle is so close to 180°).

$A=176.4^\circ$

$B=180^\circ-2.5^\circ-176.4^\circ$

$=1.1^\circ$

37.

C

$b = 12$ $a = 17$

A   B

$c = 25$

Given: three sides (Case 4).

$17^2=12^2+25^2-2(12)(25)\cos A$

$A=37^\circ$

$12^2=17^2+25^2-2(17)(25)\cos B$

$B=25^\circ$

$C=180^\circ-37^\circ-25^\circ=118^\circ$

41.

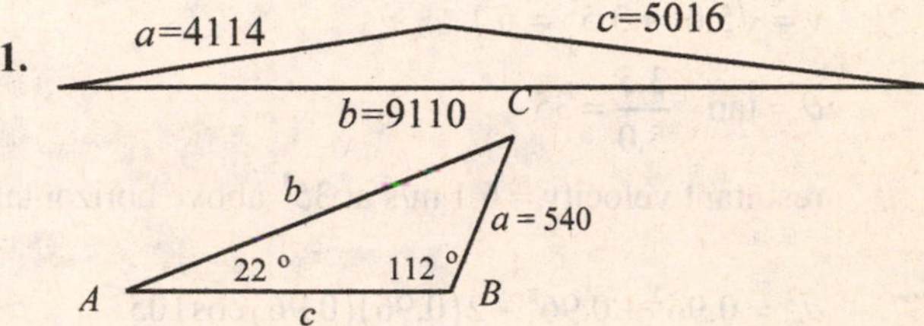

$C=180^\circ-(22^\circ+112^\circ)=46^\circ$

The shortest side is opposite the smallest angle.

$a=540$ m

The longest side is opposite the largest angle.

$b=$ longest side

$$\frac{b}{\sin 112^\circ}=\frac{540}{\sin 22^\circ}$$

$$b=\frac{54\sin 112^\circ}{\sin 22^\circ}$$

$=1300$ m

45.

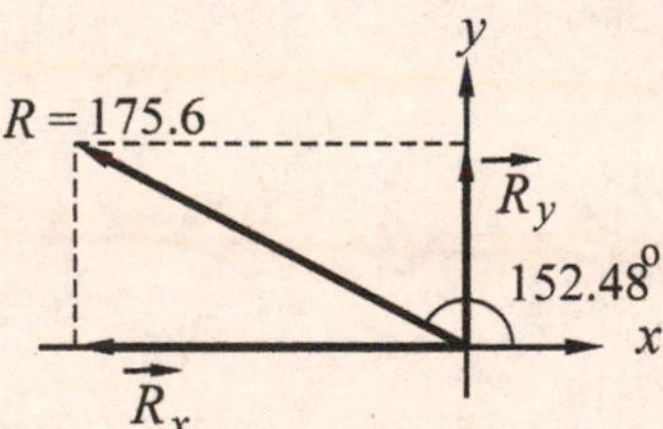

$F_x=175.6\cos 152.48^\circ=-155.7$ lb

$F_y=175.6\sin 152.48^\circ=81.14$ lb

49. The horizontal components must sum to zero:

$T_1 \cos 22.6° + 427 \cos 162.5° = 0$

$T_1 = -\dfrac{427 \cos 162.5°}{\cos 22.6°} = 441 \text{ lb}$

In order for the vertical components to balance,

$T_2 = T_1 \sin 22.6° + 427 \sin 162.5° = 298 \text{ lb}$

53.

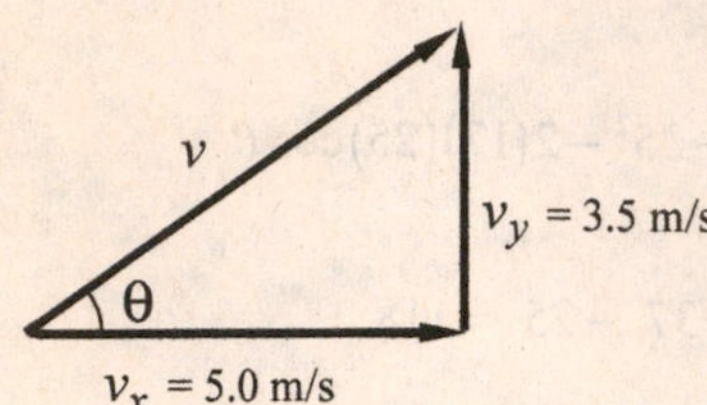

$v = \sqrt{5.0^2 + 3.5^2} = 6.1$

$\theta = \tan^{-1} \dfrac{3.5}{5.0} = 35^\circ$

resultant velocity = 6.1 m/s at 35° above horizontal

57. $d^2 = 0.96^2 + 0.96^2 - 2(0.96)(0.96)\cos 105^\circ$

$d = 1.5$ pm

61.

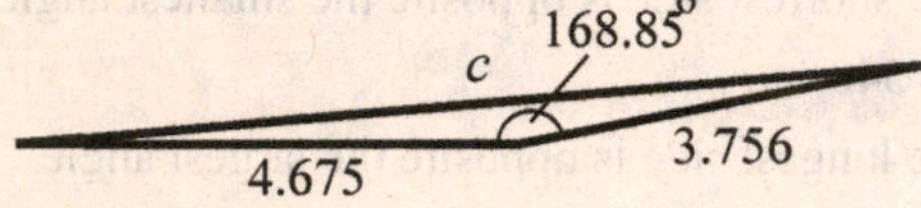

extra pipeline

$= 3.756 + 4.675 - c$

$= 3.756 + 4.675$

$- \sqrt{3.756^2 + 4.675^2 - 2(3.756)(4.675) \cdot \cos 168.85^\circ}$

$= 0.039\ 40$ km

65.

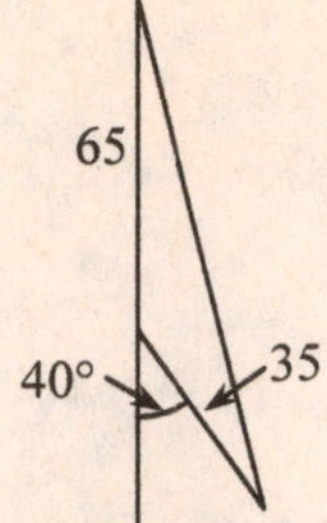

Components:

$D_x = 35 \sin 40° = 22.5 \text{ m}$

$D_y = -65 - 35 \cos 40° = -91.8 \text{ m}$

$D = \sqrt{22.5^2 + (-91.8)^2} = 94.5 \text{ m}$

$\theta = \tan^{-1} \dfrac{-91.8}{22.5} = -76°$

The skier is 94.5 m away at an angle of 14° east of south.

69.

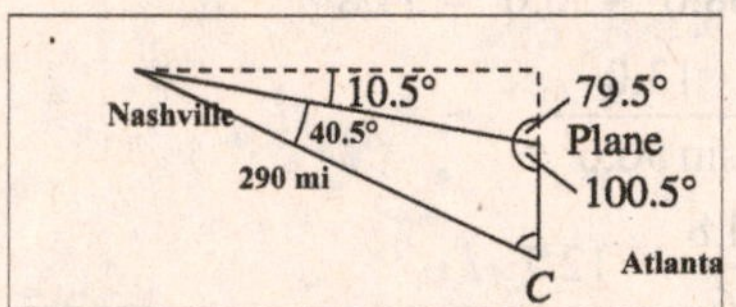

$C = 180^\circ - \left(\left(51.0^\circ - 10.5^\circ\right) - 100.5\right)$

$= 39^\circ$

$\dfrac{290}{\sin 100.5^\circ} = \dfrac{c}{\sin 39^\circ}$

$c = \dfrac{290 \sin 39^\circ}{\sin 100.5^\circ}$

$= 186$ mi

73.

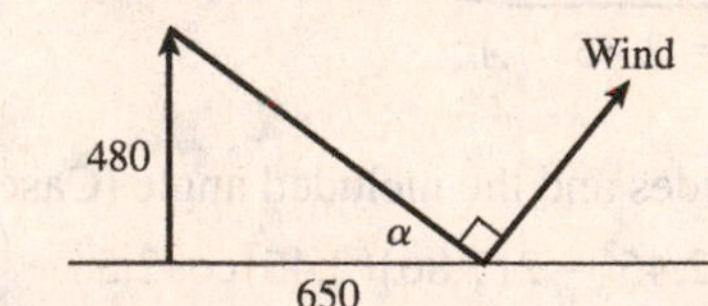

$\tan \alpha = \dfrac{480}{650}$

$\alpha = 36^\circ$ N of E

$F = \sqrt{F_x^2 + F_y^2} = \sqrt{650^2 + 480^2}$

$= 810$ N

Chapter 10

Graphs of The Trigonometric Functions

10.1 Graphs of $y = a \sin x$ and $y = a \cos x$

1. $y = 3 \cos x$

x	0	$\frac{\pi}{6}$	$\frac{\pi}{3}$	$\frac{\pi}{2}$	$\frac{2\pi}{3}$	$\frac{5\pi}{6}$
y	3	2.6	1.5	0	−1.5	−2.6

x	π	$\frac{7\pi}{6}$	$\frac{4\pi}{3}$	$\frac{3\pi}{2}$	$\frac{5\pi}{3}$	$\frac{11\pi}{6}$	2π
y	−3	−2.6	−1.5	0	1.5	2.6	3

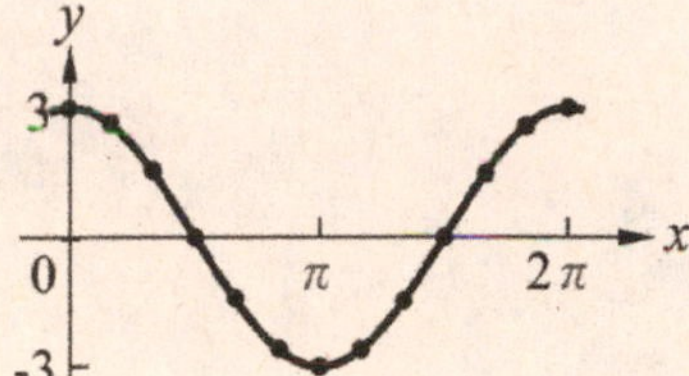

5. $y = -3 \cos x$

x	$-\pi$	$-\frac{3\pi}{4}$	$-\frac{\pi}{2}$	$-\frac{\pi}{4}$	0	$\frac{\pi}{4}$	$\frac{\pi}{2}$	$\frac{3\pi}{4}$	π
y	3	2.1	0	−2.1	−3	−2.1	0	2.1	3

x	$\frac{5\pi}{4}$	$\frac{3\pi}{4}$	$\frac{7\pi}{4}$	2π	$\frac{9\pi}{4}$	$\frac{5\pi}{2}$	$\frac{11\pi}{4}$	3π
y	2.1	0	−2.1	−3	−2.1	0	2.1	3

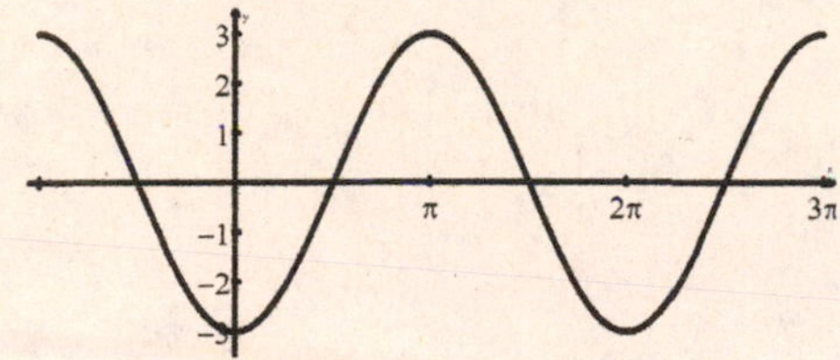

9. $y = \frac{5}{2} \sin x$ has amplitude $\frac{5}{2}$.

The table for key values between 0 and 2π is

x	0	$\frac{\pi}{2}$	π	$\frac{3\pi}{2}$	2π
y	0	$\frac{5}{2}$	0	$-\frac{5}{2}$	0

max min

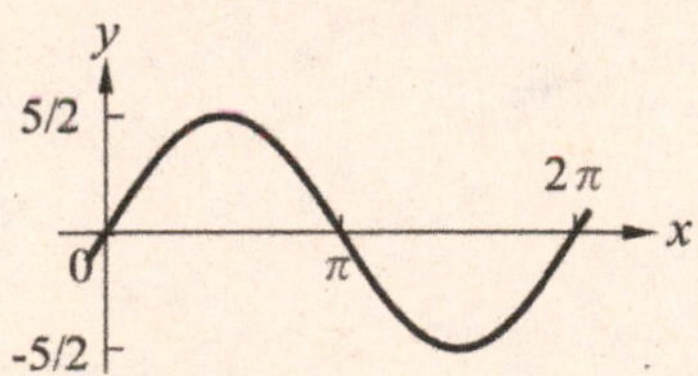

13. $y = 0.8 \cos x$ has amplitude 0.8.

The table for key values between 0 and 2π is

x	0	$\frac{\pi}{2}$	π	$\frac{3\pi}{2}$	2π
y	0.8	0	−0.8	0	0.8

max min max

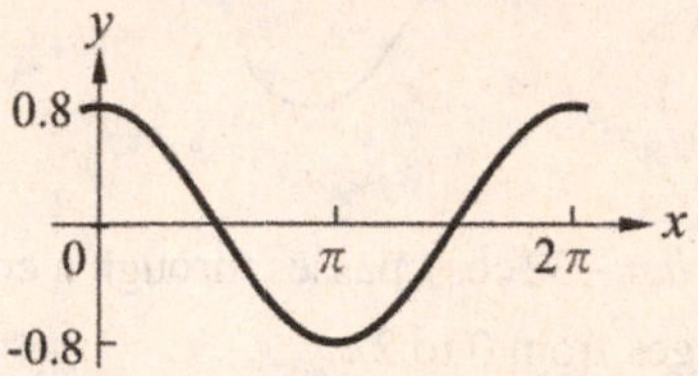

17. $y = -1500 \sin x$ has amplitude 1500.

The negative sign will invert the graph, so the table for key values between 0 and 2π is

x	0	$\frac{\pi}{2}$	π	$\frac{3\pi}{2}$	2π
y	0	−1500	0	1500	0

min max

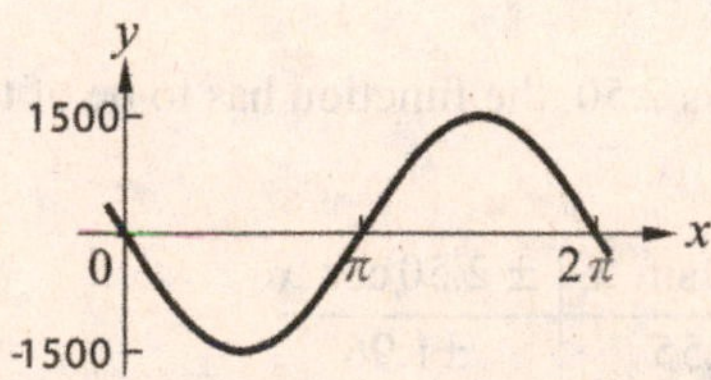

21. $y = -50 \cos x$ has amplitude 50.

The negative sign inverts the graph, so the table for key values between 0 and 2π is

x	0	$\frac{\pi}{2}$	π	$\frac{3\pi}{2}$	2π
y	−50	0	50	0	−50

min max min

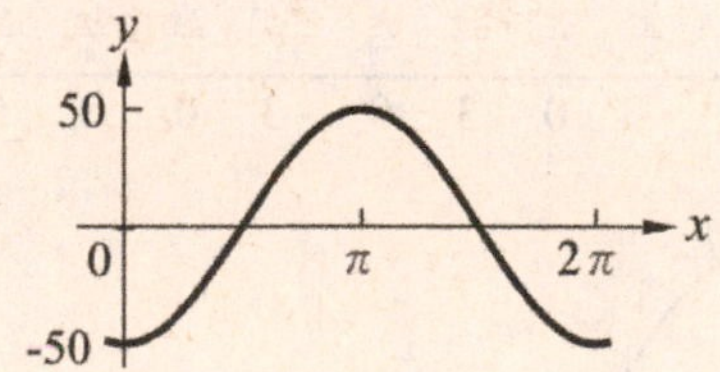

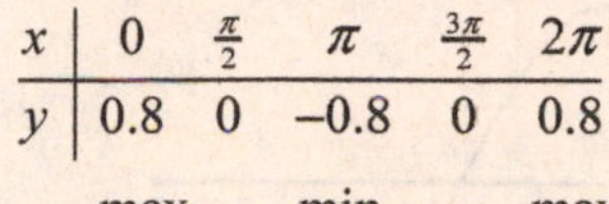

25. Sketch $y = -12\cos x$ for $x = 0, 1, 2, 3, 4, 5, 6, 7$

x	0	1	2	3	4
-12 cos x	−12	−6.48	4.99	11.9	7.84

x	5	6	7
-12 cos x	−3.40	−11.5	−9.05

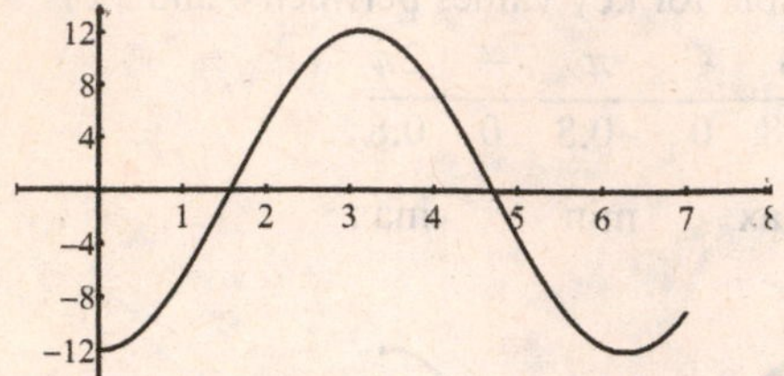

29. The graph of $h = -32\cos t$ passes through a complete cycle as t ranges from 0 to 2π.

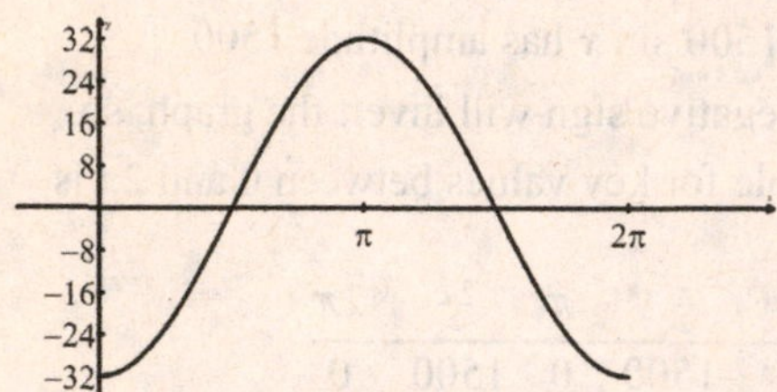

33. The graph has zeros at $x = 0$, π, and, 2π, so it is a sine function. Its amplitude is 4, and it has not been inverted. Hence the function is $y = 4\sin x$.

37. If amplitude is 2.50, the function has to be of the form $y = \pm 2.50\sin x$ or $y = \pm 2.50\cos x$. We evaluate each one at $x = 0.67$:

x	$\pm 2.50\sin x$	$\pm 2.50\cos x$
0.67	±1.55	±1.96

The function is $y = -2.50\sin x$.

10.2 Graphs of $y = a\sin bx$ and $y = a\cos bx$

1. $y = 3\sin 6x$, amplitude = 3, period = $\frac{2\pi}{6} = \frac{\pi}{3}$, key values at multiples of $\frac{1}{4}\left(\frac{\pi}{3}\right) = \frac{\pi}{12}$.

x	0	$\frac{\pi}{12}$	$\frac{\pi}{6}$	$\frac{\pi}{4}$	$\frac{\pi}{3}$	$\frac{5\pi}{12}$	$\frac{\pi}{2}$	$\frac{7\pi}{12}$	$\frac{2\pi}{3}$	$\frac{3\pi}{4}$	$\frac{5\pi}{6}$	$\frac{11\pi}{12}$	π
y	0	3	0	−3	0	3	0	−3	0	3	0	−3	0

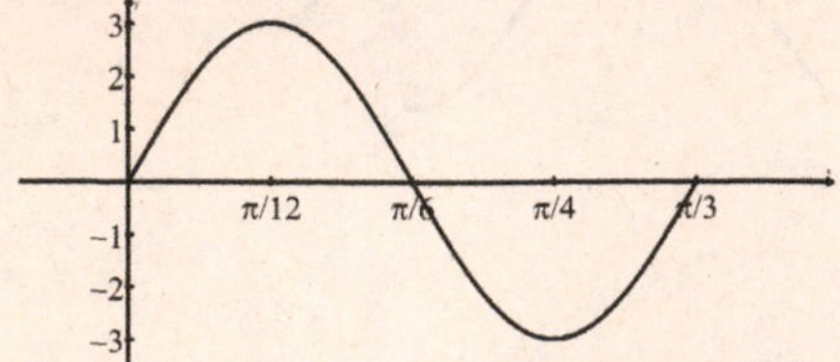

5. $y = 3\cos 8x$ has amplitude of 3 and period $\frac{\pi}{4}$, with key values at multiples of $\frac{1}{4}\left(\frac{\pi}{4}\right) = \frac{\pi}{16}$.

x	0	$\frac{\pi}{16}$	$\frac{\pi}{8}$	$\frac{3\pi}{16}$	$\frac{\pi}{4}$
y	3	0	−3	0	3

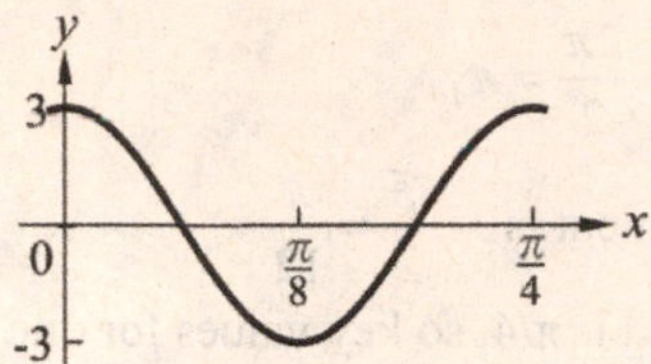

9. $y = -\cos 16x$ has amplitude of $|-1| = 1$, and period of $\frac{\pi}{8}$, with key values at multiples of $\frac{1}{4}\left(\frac{\pi}{8}\right) = \frac{\pi}{32}$.

x	0	$\frac{\pi}{32}$	$\frac{\pi}{16}$	$\frac{3\pi}{32}$	$\frac{\pi}{8}$
y	−1	0	1	0	−1

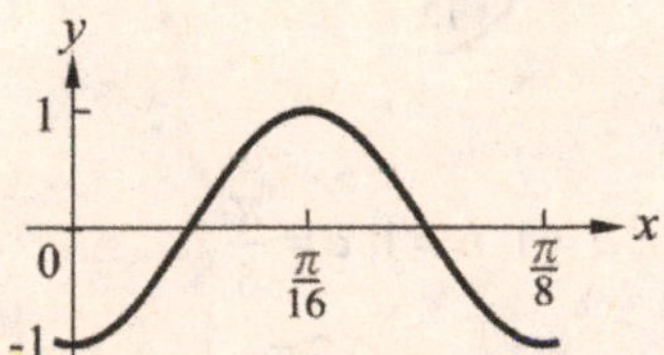

13. $y = 3\cos 4\pi x$ has amplitude of 3 and period of $\frac{1}{2}$, with key values at multiples of $\frac{1}{4}\left(\frac{1}{2}\right) = \frac{1}{8}$.

x	0	$\frac{1}{8}$	$\frac{1}{4}$	$\frac{3}{8}$	$\frac{1}{2}$
y	3	0	−3	0	3

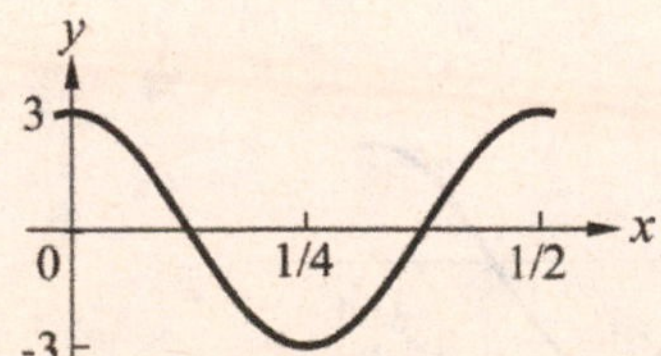

17. $y = -\frac{1}{2}\cos\frac{2}{3}x$ has amplitude of $\left|-\frac{1}{2}\right| = \frac{1}{2}$, and period of 3π, with key values at multiples of $\frac{1}{4}\cdot 3\pi = \frac{3\pi}{4}$.

x	0	$\frac{3\pi}{4}$	$\frac{3\pi}{2}$	$\frac{9\pi}{4}$	3π
y	$-\frac{1}{2}$	0	$\frac{1}{2}$	0	$-\frac{1}{2}$

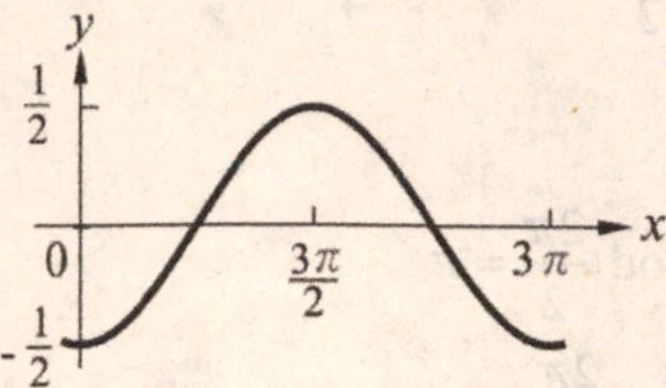

21. $y = 3.3\cos \pi^2 x$ has amplitude of 3.3 and period of $\frac{2}{\pi}$, with key values at multiples of $\frac{1}{4}\left(\frac{2}{\pi}\right) = \frac{1}{2\pi}$.

x	0	$\frac{1}{2\pi}$	$\frac{1}{\pi}$	$\frac{3}{2\pi}$	$\frac{2}{\pi}$
$\pi^2 x$	0	$\frac{\pi}{2}$	π	$\frac{3\pi}{2}$	2π
$\cos \pi^2 x$	1	0	−1	0	1
$3.3\cos \pi^2 x$	3.3	0	−3.3	0	3.3

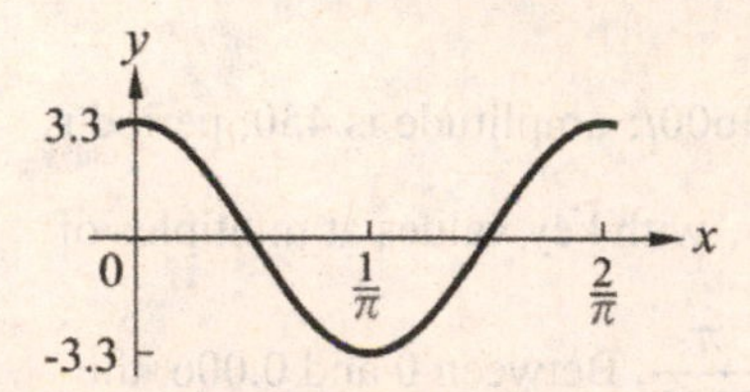

25. $b = \frac{2\pi}{1/3} = 6\pi$; $y = \sin 6\pi x$

29. $y = 8\left|\cos\frac{\pi}{2}x\right|$

This function has amplitude 8. Moreover, $\cos(\pi x/2)$ has period $\frac{2\pi}{\frac{\pi}{2}} = 4$ so that key values are at multiples of 4/4=1.

Also, the absolute value makes the y values positive whenever they are negative and leaves positive values intact, so that the negative parts of the curve are reflected with respect to the x axis. Note that the function repeats itself every 2 units, so that the absolute value changes the period from 4 to 2.

x	0	1	2	3	4
y	8	0	8	0	8

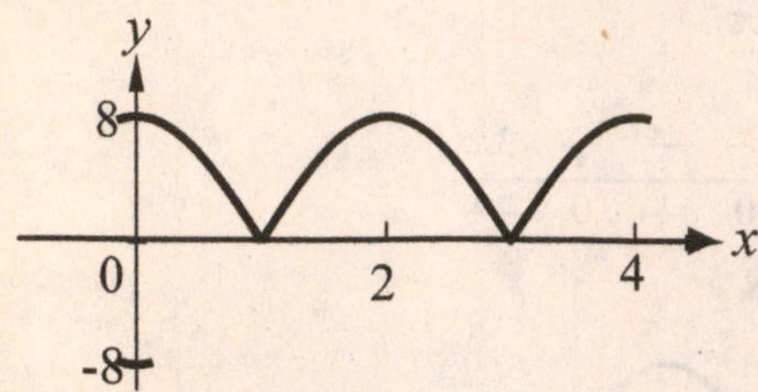

33. $\sin 2x$ has period $\frac{2\pi}{2} = \pi$

$\sin 3x$ has period $\frac{2\pi}{3}$

The period of $\sin 2x + \sin 3x$ is the least common multiple of π, and $\frac{2\pi}{3}$, which is 2π.

37. The period is $\frac{2\pi}{6.60\times10^8}$ and so the frequency is

$\frac{6.60\times10^8}{2\pi} = 1.05\times10^8 \text{ Hz} = 1.05\times10^2 \text{ MHz}.$

41. $v = 450 \cos 3600t$; amplitude is 450; period is $\frac{2\pi}{3600} = \frac{\pi}{1800}$, with key values at multiples of $\frac{1}{4}\frac{\pi}{1800} = \frac{\pi}{7200}$. Between 0 and 0.006, the function completes 3.4 cycles.

$0.006 \div \frac{\pi}{1800} = 3.4$ cycles

t	0	$\frac{\pi}{7200}$	$\frac{\pi}{3600}$	$\frac{\pi}{2400}$	$\frac{\pi}{1800}$
$3600t$	0	$\frac{\pi}{2}$	π	$\frac{3\pi}{2}$	2π
$\cos 3600t$	1	0	−1	0	1
$450 \cos 3600t$	450	0	−450	0	450

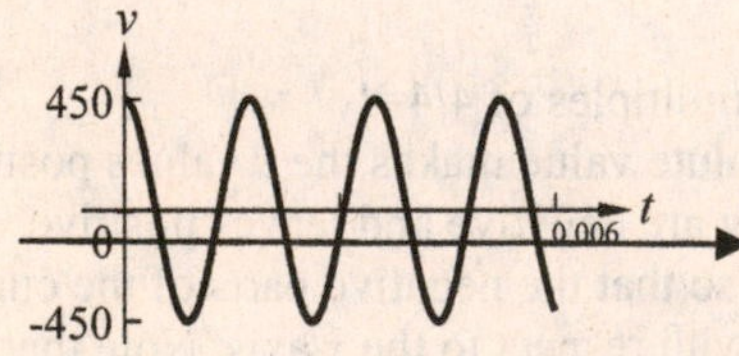

45. The function has amplitude 4, period 2, with a zero at $x = 0$ and a minimum after one-fourth of a period (so it is an inverted sine function).
$b = 2\pi/2$
$b = \pi$
Hence the function is $y = -4 \sin \pi x$

10.3 Graphs of $y = a \sin(bx + c)$ and $y = a \cos(bx + c)$

1. $y = -\cos\left(2x - \frac{\pi}{6}\right)$

(1) the amplitude is 1

(2) the period is $\frac{2\pi}{2} = \pi$

(3) the displacement is $-\frac{-\frac{\pi}{6}}{2} = \frac{\pi}{12}$

One-fourth period is $\pi/4$, so key values for one full cycle start at $\pi/12$, end at $13\pi/12$, and are found $\pi/4$ units apart. The table of key values is

x	$\frac{\pi}{12}$	$\frac{\pi}{3}$	$\frac{7\pi}{12}$	$\frac{5\pi}{6}$	$\frac{13\pi}{12}$
y	−1	0	1	0	−1

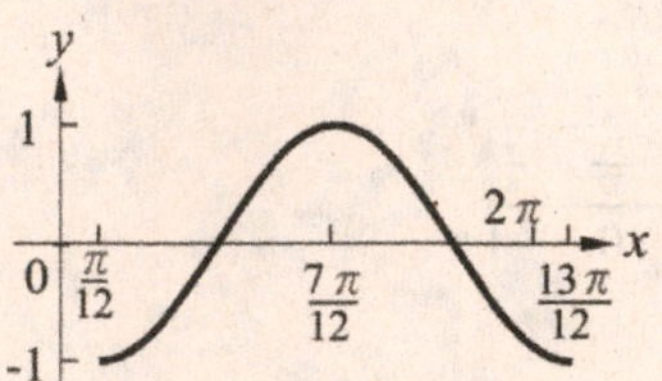

5. $y = \cos\left(x + \frac{\pi}{6}\right)$; $a = 1, b = 1, c = \frac{\pi}{6}$

Amplitude is $|a| = 1$; period is $\frac{2\pi}{b} = 2\pi$;

displacement is $-\frac{c}{b} = -\frac{\pi}{6}$

One-fourth period is $\pi/2$, so key values for one full cycle start at $-\pi/6$, end at $11\pi/6$, and are found $\pi/2$ units apart. The table of key values is

x	$-\frac{\pi}{6}$	$\frac{\pi}{3}$	$\frac{5\pi}{6}$	$\frac{4\pi}{3}$	$\frac{11\pi}{6}$
y	0	1	0	−1	0

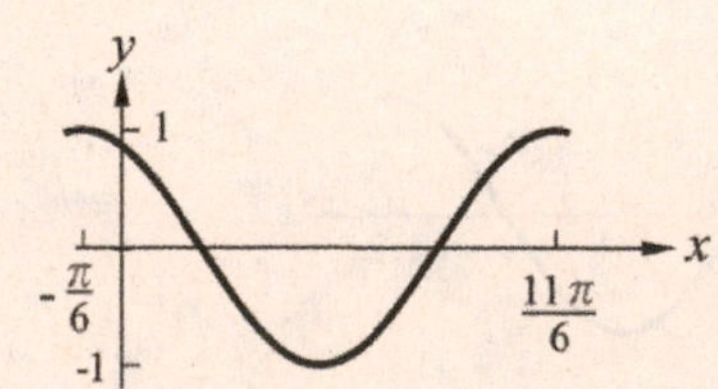

9. $y = -\cos(2x-\pi)$; $a=-1$, $b=2$, $c=-\pi$

Amplitude is $|a|=1$; period is $\dfrac{2\pi}{b}=\dfrac{2\pi}{2}=\pi$;

displacement is $-\dfrac{c}{b}=-\left(\dfrac{-\pi}{2}\right)=\dfrac{\pi}{2}$

One-fourth period is π/4, so key values for one full cycle start at π/2, end at 3π/2, and are found π/4 units apart. The table of key values is

x	$\frac{\pi}{2}$	$\frac{3\pi}{4}$	π	$\frac{5\pi}{4}$	$\frac{3\pi}{2}$
y	−1	0	1	0	−1

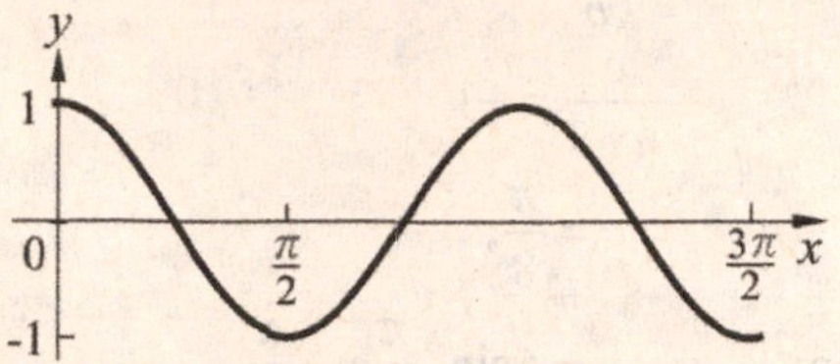

13. $y = 30\cos\left(\dfrac{1}{3}x+\dfrac{\pi}{3}\right)$; $a=30$, $b=\dfrac{1}{3}$, $c=\dfrac{\pi}{3}$

Amplitude is $|a|=30$; period is $\dfrac{2\pi}{b}=\dfrac{2\pi}{1/3}=6\pi$;

displacement is $-\dfrac{c}{b}=\dfrac{-\pi/3}{1/3}=-\pi$

One-fourth period is 3π/2, so key values for one full cycle start at −π, end at 5π, and are found 3π/2 units apart. The table of key values is

x	$-\pi$	$\frac{\pi}{2}$	2π	$\frac{7\pi}{2}$	5π
y	30	0	−30	0	30

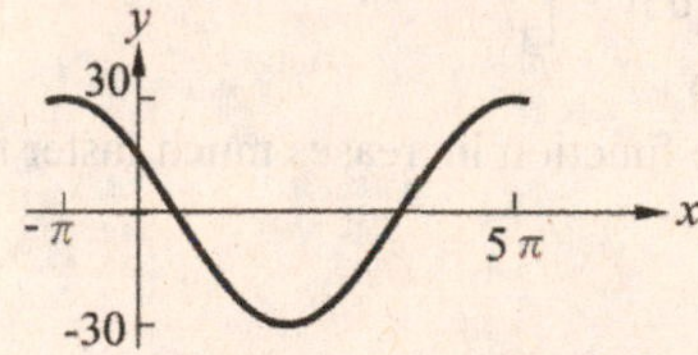

17. $y = 0.08\cos\left(4\pi x-\dfrac{\pi}{5}\right)$; $a=0.08$, $b=4\pi$, $c=-\dfrac{\pi}{5}$

Amplitude is $|a|=0.08$; period is $\dfrac{2\pi}{b}=\dfrac{2\pi}{4\pi}=\dfrac{1}{2}$;

displacement is $-\dfrac{c}{b}=-\left(-\dfrac{\pi/5}{4\pi}\right)=\dfrac{1}{20}$

One-fourth period is 1/8, so key values for one full cycle start at 1/20, end at 11/20, and are found 1/8 units apart. The table of key values is

x	$\frac{1}{20}$	$\frac{7}{40}$	$\frac{3}{10}$	$\frac{17}{40}$	$\frac{11}{20}$
y	0.08	0	−0.08	0	0.08

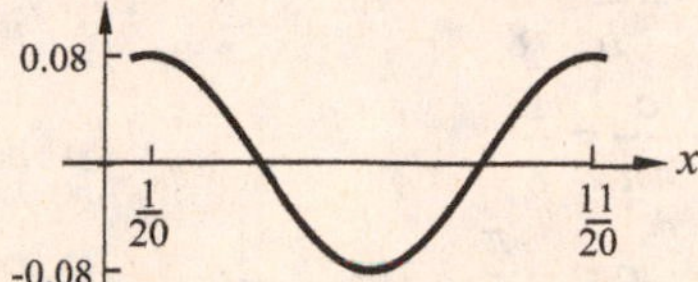

21. $y = 40\cos(3\pi x+1)$; $a=40$, $b=3\pi$, $c=1$

Amplitude is $|a|=40$; period is $\dfrac{2\pi}{b}=\dfrac{2\pi}{3\pi}=\dfrac{2}{3}$;

displacement is $-\dfrac{1}{b}=-\dfrac{1}{3\pi}$

One-fourth period is 1/6, so key values for one full cycle start at −1/(3π), end at 2/3−1/(3π), and are found 1/6 units apart. The table of key values is

x	$-\frac{1}{3\pi}$	$\frac{1}{6}-\frac{1}{3\pi}$	$\frac{1}{3}-\frac{1}{3\pi}$	$\frac{1}{2}-\frac{1}{3\pi}$	$\frac{2}{3}-\frac{1}{3\pi}$
y	40	0	−40	0	40

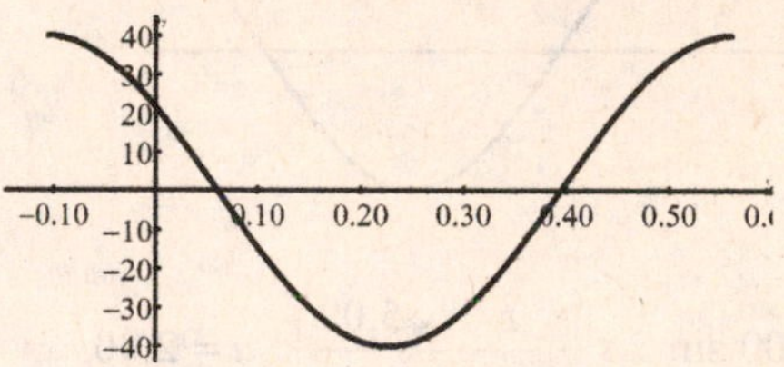

25. $y=-\dfrac{3}{2}\cos\left(\pi x+\dfrac{\pi^2}{6}\right)$; $a=-\dfrac{3}{2}$, $b=\pi$, $c=\dfrac{\pi^2}{6}$

Amplitude is $|a|=\dfrac{3}{2}$; period is $\dfrac{2\pi}{b}=\dfrac{2\pi}{\pi}=2$;

displacement is $-\dfrac{c}{b}=-\dfrac{\pi^2/6}{\pi}=-\dfrac{\pi}{6}$

One-fourth period is 1/2, so key values for one full cycle start at −π/6, end at 2−π/6, and are found 1/2 units apart. The table of key values is

x	$-\frac{\pi}{6}$	$\frac{1}{2}-\frac{\pi}{6}$	$1-\frac{\pi}{6}$	$\frac{3}{2}-\frac{\pi}{6}$	$2-\frac{\pi}{6}$
y	$-\frac{3}{2}$	0	$\frac{3}{2}$	0	$-\frac{3}{2}$

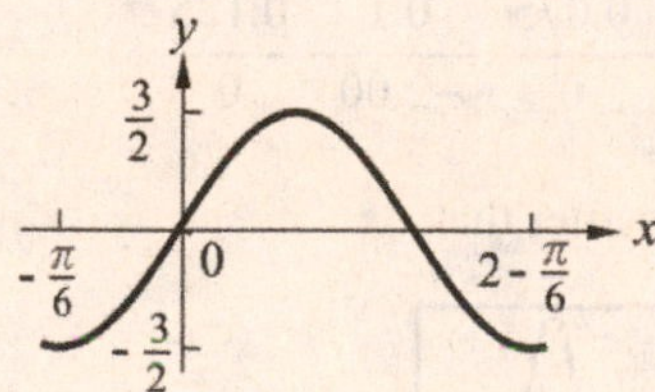

29. amplitude: $a = 12$;

period: $\frac{2\pi}{b} = \frac{1}{2}$

$b = 4\pi$

displacement: $-\frac{c}{b} = \frac{1}{8}$

$-\frac{c}{4\pi} = \frac{1}{8}$

$c = -\frac{\pi}{2}$

Solution: $y = 12\cos\left(4\pi x - \frac{\pi}{2}\right)$

33. Graph $y_1 = 2\sin\left(3x + \frac{\pi}{6}\right)$

and $y_2 = -2\sin\left(-3x + \frac{\pi}{6}\right)$.

The graphs are the same.

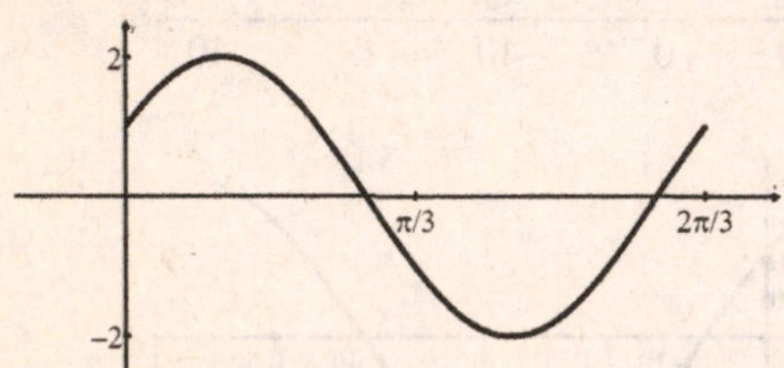

37. $y = 2.00\sin 2\pi\left(\frac{t}{0.100} - \frac{5.00}{20.0}\right)$; $a = 2.00$,

$b = \frac{2\pi}{0.100}$, $c = \frac{-5.00(2\pi)}{20.0}$

Amplitude $= |a| = 2.00$,

period $= \frac{2\pi}{b} = 0.100$,

displacement $= -\frac{c}{b} = 0.025$

One-fourth period is 0.025, so key values for one full cycle start at 0.025, end at 0.125, and are found 0.025 units apart. Three cycles end at 0.325. The table of key values in the first cycle is

t	0.025	0.05	0.075	0.1	0.125
i	0	2.00	0	–2.00	0

The curve repeats after that.

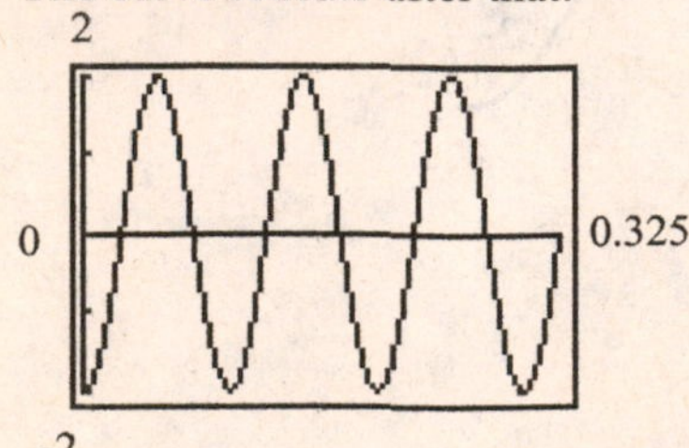

41. The maximum is at $y = 5$, so amplitude is 5. A full cycle takes place between –1 and 15, so the period is 16. The graph has a zero at $x = -1$, so displacement of the sine function is –1. We thus have

amplitude: $a = 5$;

period: $\frac{2\pi}{b} = 16$

$b = \frac{\pi}{8}$;

displacement: $-\frac{c}{b} = -1$

$-\frac{c}{\pi/8} = -1$

$c = \frac{\pi}{8}$;

Solution: $y = 5\sin\left(\frac{\pi}{8}x + \frac{\pi}{8}\right)$

10.4 Graphs of $y = \tan x$, $y = \cot x$, $y = \sec x$, $y = \csc x$

1. $y = 5\cot 2x$.

Since the period of $y = \cot x$ is π, the period of this function is $\pi/2$. We have the following table of key values:

x	0	$\frac{\pi}{4}$	$\frac{\pi}{2}$	$\frac{3\pi}{4}$	π
y	*	0	*	0	*

(* = asymptote)

Since $a = 5$, the function increases much faster than $y = \cot x$.

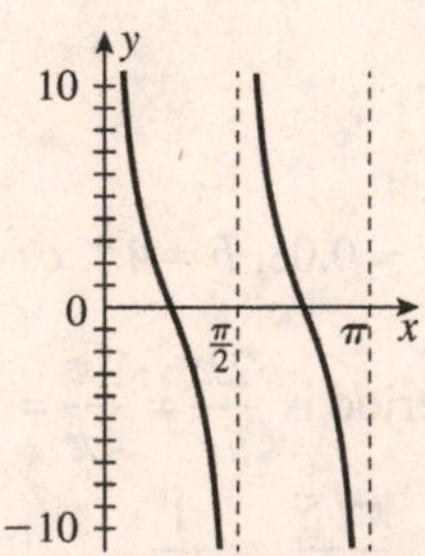

5.

x	$-\frac{\pi}{2}$	$-\frac{\pi}{3}$	$-\frac{\pi}{4}$	$-\frac{\pi}{6}$	0	$\frac{\pi}{6}$	$\frac{\pi}{4}$
$\sec x$	*	2	1.4	1.2	1	1.2	1.4

x	$\frac{\pi}{3}$	$\frac{\pi}{2}$	$\frac{2\pi}{3}$	$\frac{3\pi}{4}$	$\frac{5\pi}{6}$	π
$\sec x$	2	*	−2	−1.4	−1.2	−1

(* = asymptote)

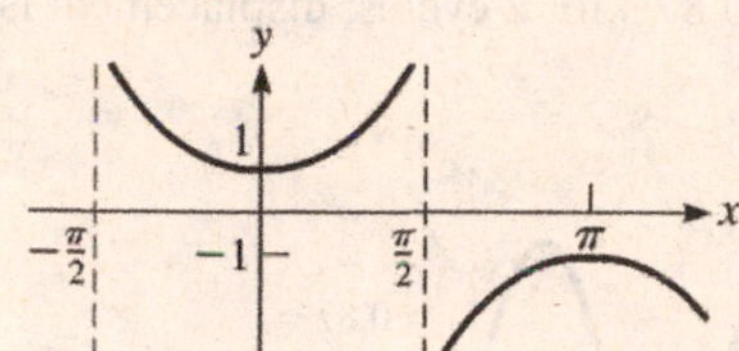

9. For $y = \frac{1}{2}\sec x$, first sketch the graph of $y = \sec x$, then multiply the y – values of the secant function by $\frac{1}{2}$ and graph.

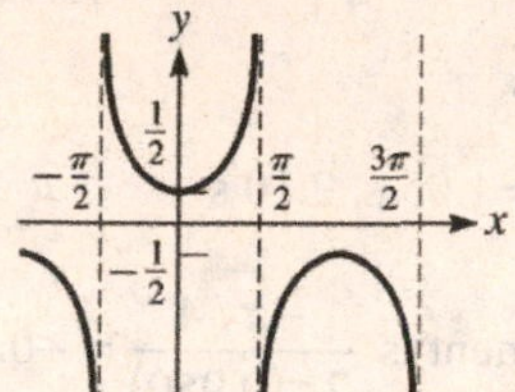

13. For $y = -3\csc x$, sketch the graph of $y = \csc x$, then multiply the y – values by -3, and resketch the graph. It will be inverted.

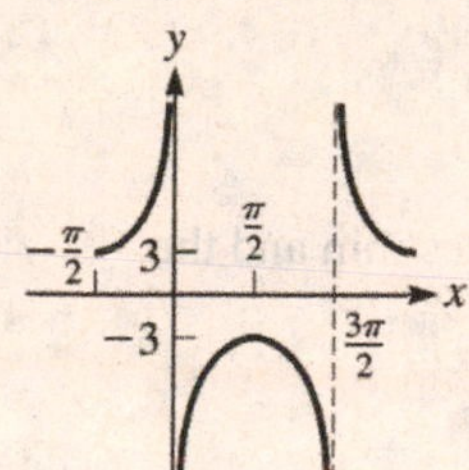

17. Since the period of $\sec x$ is 2π, the period of $y = \frac{1}{2}\sec 3x$ is $\frac{2\pi}{3}$. Graph $y_1 = 0.5(\cos 3x)^{-1}$.

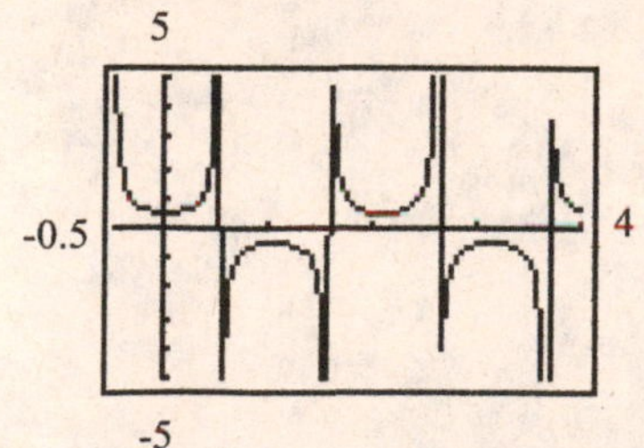

21. Since the period of $\csc x$ is 2π, the period of $y = 18\csc\left(3x - \frac{\pi}{3}\right)$ is $\frac{2\pi}{3}$. The displacement is $-\left(-\frac{\pi/3}{3}\right) = \frac{\pi}{9}$. Graph $y_1 = 18\left(\sin\left(3x - \frac{\pi}{3}\right)\right)^{-1}$.

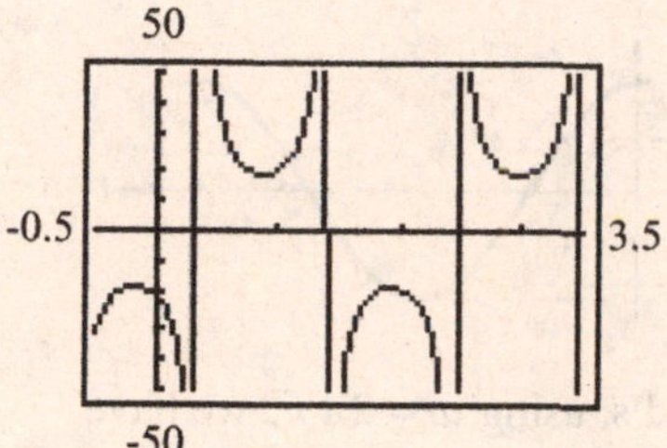

25. Using the graph of $y = \tan x$, we see that $\tan x$ becomes very large and positive as x approaches $\frac{\pi}{2}$ from the left and $\tan x$ becomes very large and negative as x approaches $\frac{\pi}{2}$ from the right.

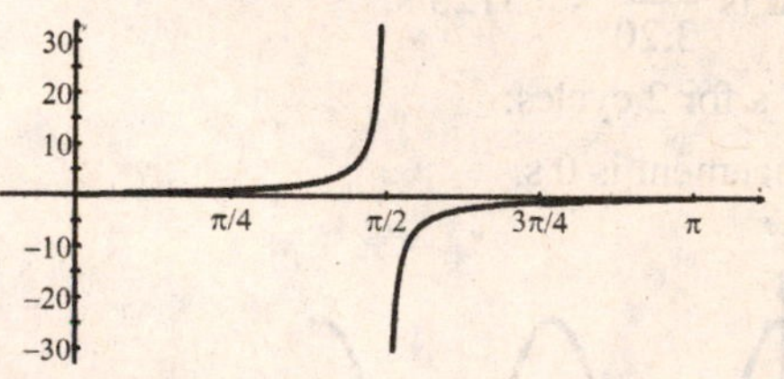

29. $x = 200\cot\theta$

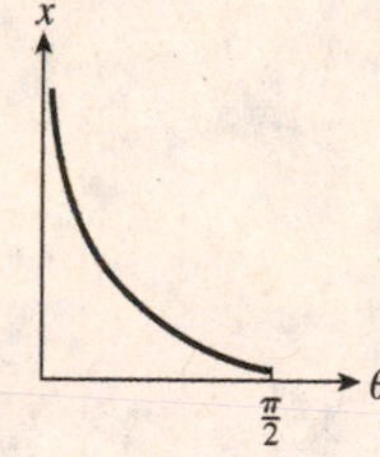

10.5 Applications of the Trigonometric Graphs

1. The displacement of the projection on the y-axis is d and is given by $d = R\cos\omega t$.

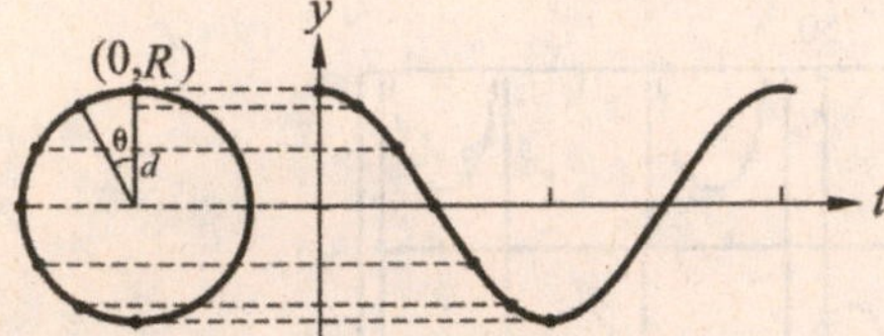

5. If $\omega = 40\pi$ rad/s, using $\omega = 2\pi f$, we have

$f = \dfrac{\omega}{2\pi} = \dfrac{40\pi}{2\pi} = 20$ cycles/s and the period is $\dfrac{1}{20}$ s.

9. $y = R\cos\omega t$

$= 8.30\cos\left[(3.20)(2\pi)\right]t$

Amplitude is 8.30 cm;

period is $\dfrac{1}{3.20} = 0.3125$ s,

0.625 s for 2 cycles;

displacement is 0 s.

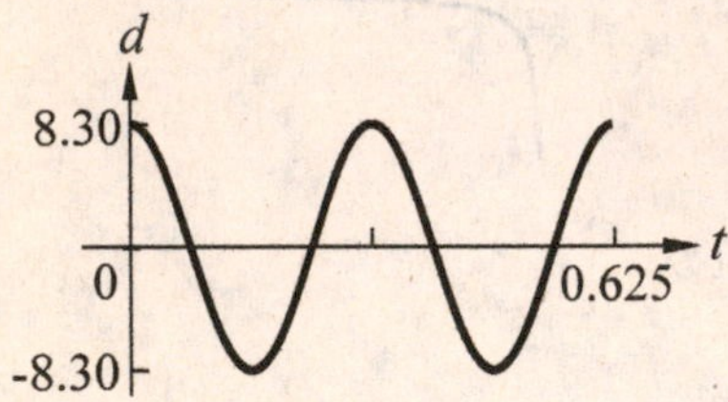

13. $V = E\cos(\omega t + \alpha)$

$= 170\cos\ 2\pi(60.0)t - \dfrac{\pi}{3}$

Amplitude is 170 V,

period is $\dfrac{2\pi}{2\pi(60.0)} = 0.016$ s, 0.033 s

for 2 cycles; displacement is $\dfrac{\pi/3}{2\pi(60.0)} = \dfrac{1}{360}$ s

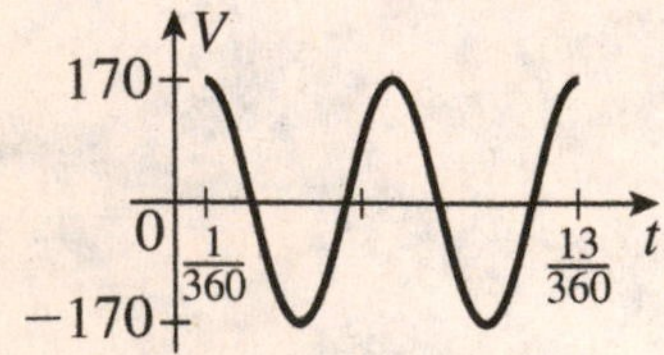

17. $p = p_0 \sin 2\pi f t$

$= 2.80\sin\left[2\pi(2.30)\right]t$

$= 2.80\sin 14.45t$

Amplitude is 2.80 lb/in^2, period is $\dfrac{2\pi}{14.45} = 0.435$ s for 1 cycle, 0.87 s for 2 cycles; displacement is 0 s

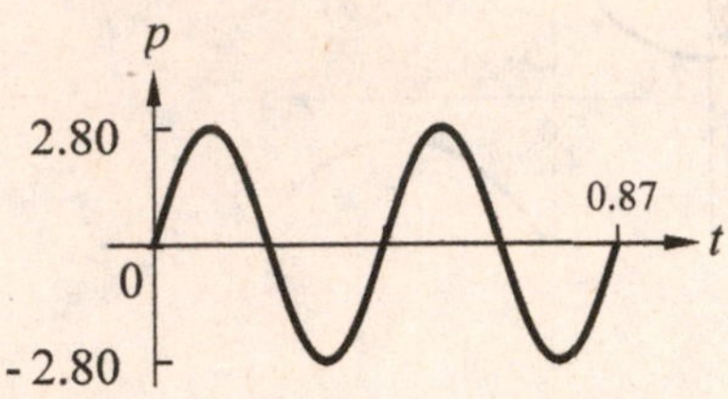

21. $V = 0.014\cos\left(2\pi f t + \dfrac{\pi}{4}\right)$

$= 0.014\cos\left[2\pi(0.950)t + \dfrac{\pi}{4}\right]$

Amplitude is 0.014 V,

period is $\dfrac{2\pi}{2\pi(0.950)} = 1.05$ s, 2.10 s

for 2 cycles; displacement is $\dfrac{-\pi/4}{2\pi(0.950)} = -0.13$ s

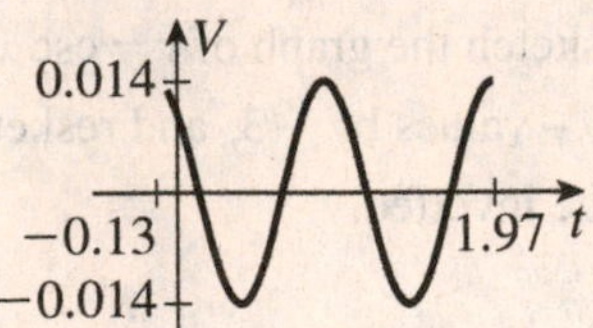

25. The frequency is 18 cycles/min and the amplitude is 12 ft, so

$y = 12\sin(18\cdot 2\pi t)$

$y = 12\sin(36\pi t)$

with y in ft and t in min.

One cycles takes 1/18 min, so two cycles end at t=1/9 min.

10.6 Composite Trigonometric Curves

1. $y = 1 + \sin x$

x	-2π	$-\frac{3\pi}{2}$	$-\pi$	$-\frac{\pi}{2}$	0	$\frac{\pi}{2}$	π	$\frac{3\pi}{2}$	2π
y	1	2	1	0	1	2	1	0	1

This is a vertical shift of 1 unit of the graph of $y = \sin x$.

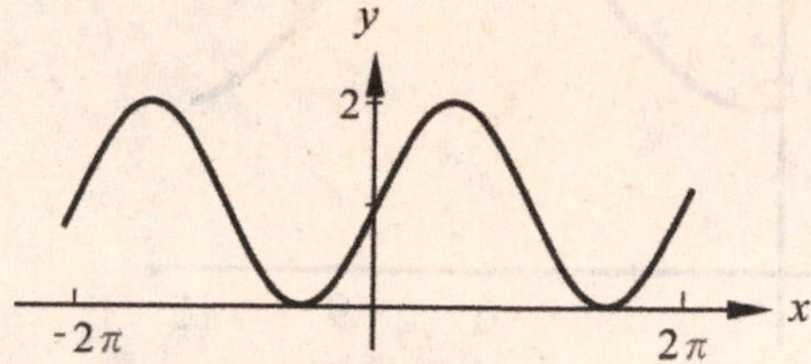

5. $y = \frac{1}{10}x^2 - \sin \pi x$

x	-4	-3.43	-2.55	-1.88	-1.47
y	1.60	0.20	1.64	0	-0.78

x	-1.03	-0.51	0	0.49	0.97	1.53
y	0	1.03	0	-0.98	0	1.23

x	2.15	2.45	2.73	0
y	0	-0.39	4	1.6

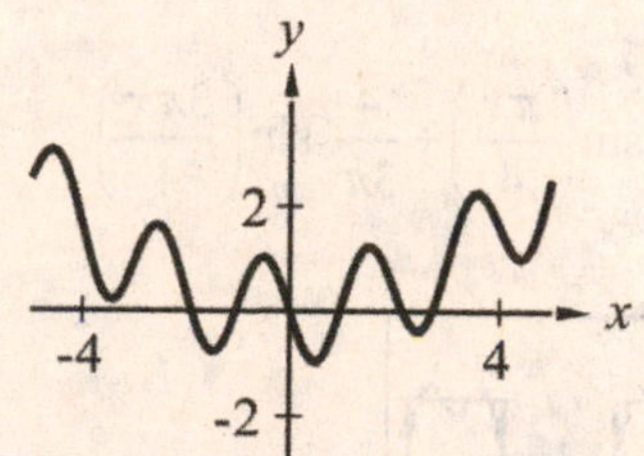

9. Graph $y_1 = x^3 + 10 \sin 2x$.

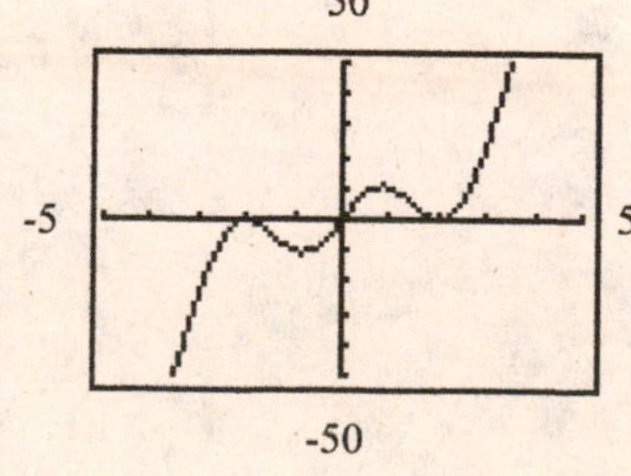

13. Graph $y_1 = 20 \cos 2x + 30 \sin x$.

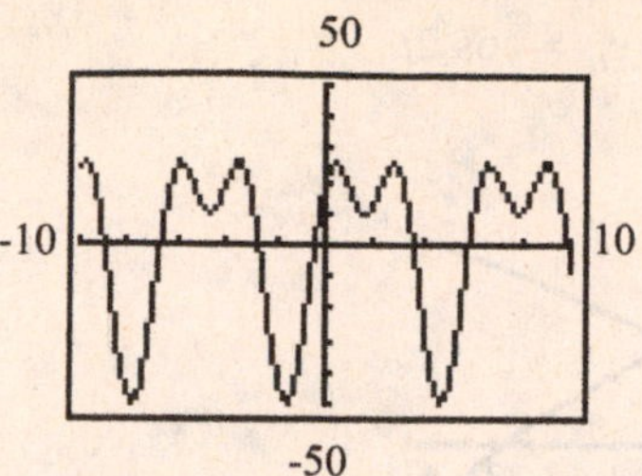

17. Graph $y_1 = \sin \pi x - \cos 2x$.

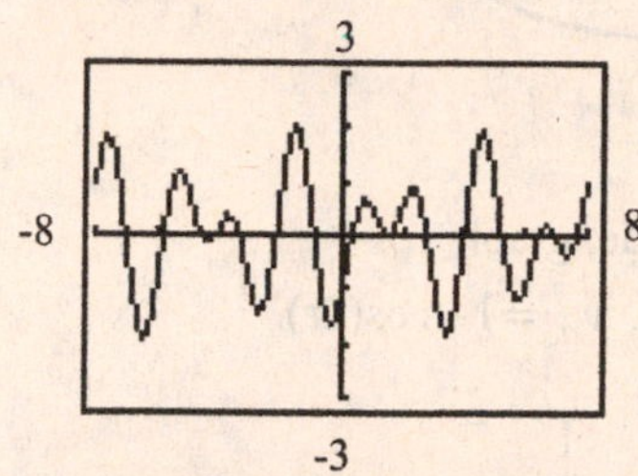

21. In parametric mode graph

$x_{1T} = 8 \cos t,\ y_{1T} = 5 \sin t$

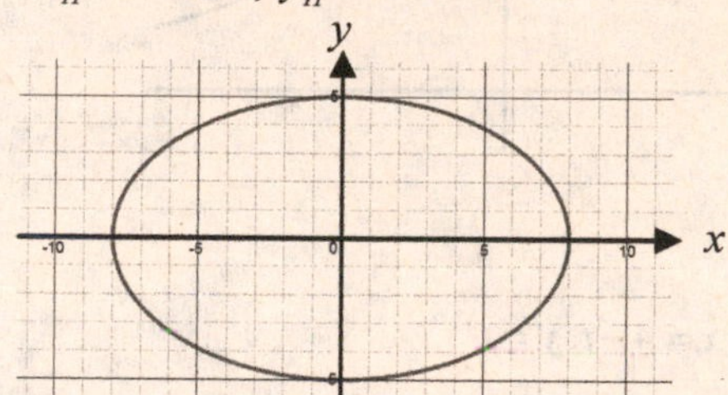

25. In parametric mode graph

$x = \sin \pi t,\ y = 2\cos 2\pi t$

t	0	$\frac{1}{4}$	$\frac{1}{2}$	$\frac{3}{4}$	1
x	0	0.707	1	0.707	0
y	2	0	-2	0	2

t	$\frac{5}{4}$	$\frac{3}{2}$	$\frac{7}{4}$	2
x	-0.707	-1	-0.707	0
y	0	-2	0	2

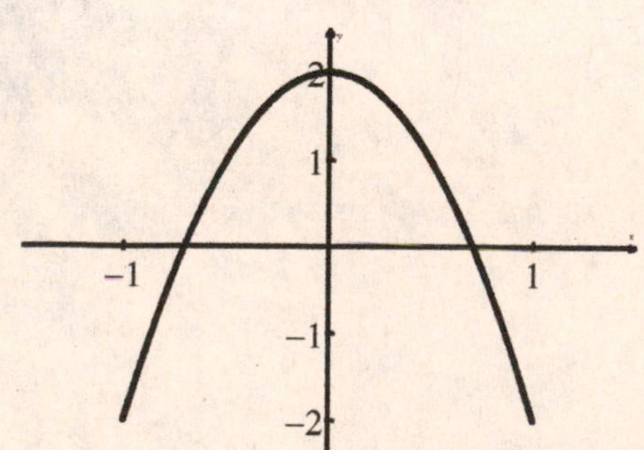

29. In parametric mode, graph
$x_{1T} = 4\cos 3t,\ y_{1T} = \cos 2t.$

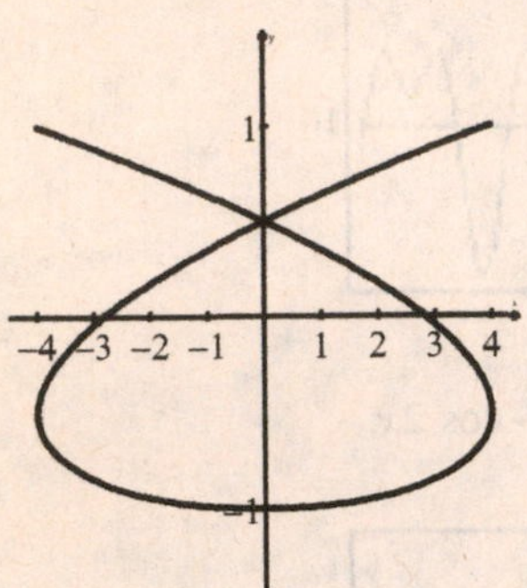

33. In parametric mode, graph
$x_{1T} = 2\cos t/\sin t,\ y_{1T} = 1 - \cos(2t).$

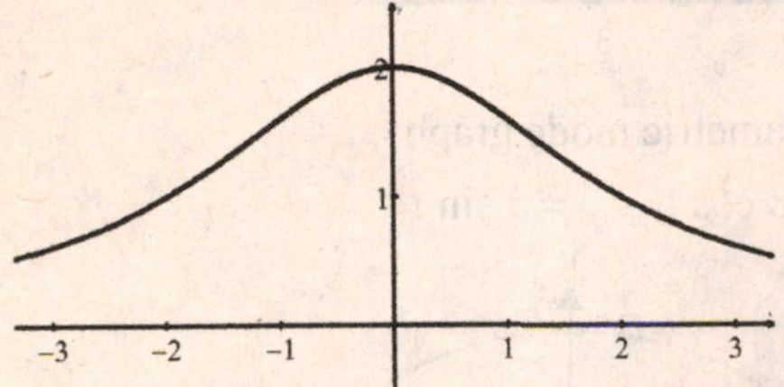

37. Graph $x = 4 - t, y = t^3$

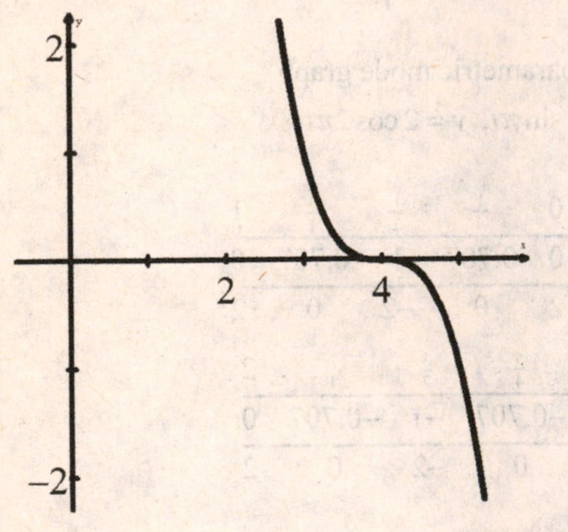

41. $T = 56 - 22\cos\left[\frac{\pi}{6}(x - 0.5)\right]$

Graph $y_1 = 56 - 22\cos\left[\frac{\pi}{6}(x - 0.5)\right]$

for x between 0 and 12

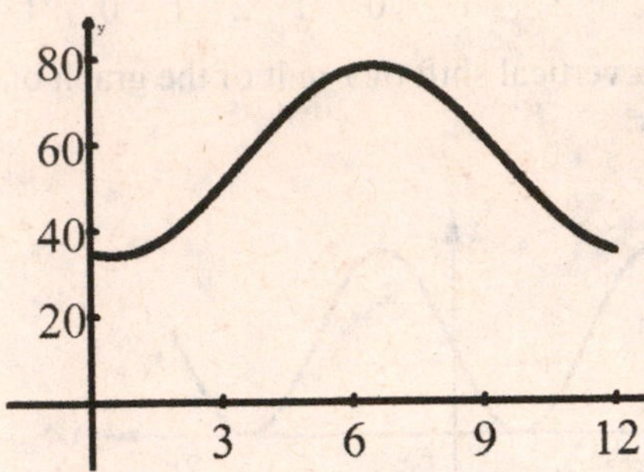

45. To graph $I = 40 + 50\sin t - 20\cos 2t$,
graph $y_1 = 40 + 50\sin x - 20\cos 2x$.

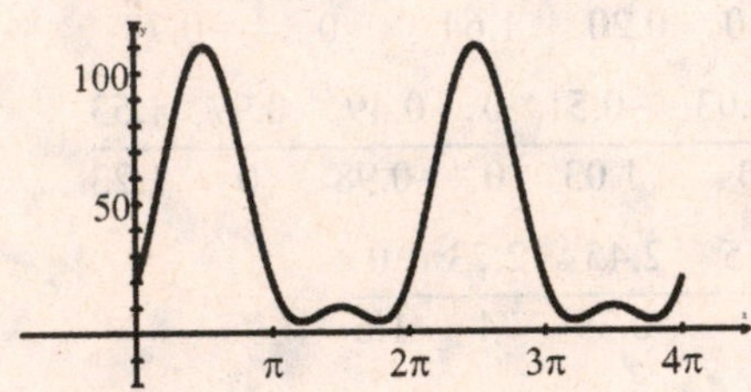

49. Graph $y_1 = 1 + \frac{4}{\pi}\sin\left(\frac{\pi x}{4}\right) + \frac{4}{3\pi}\sin\left(\frac{3\pi x}{4}\right)$.

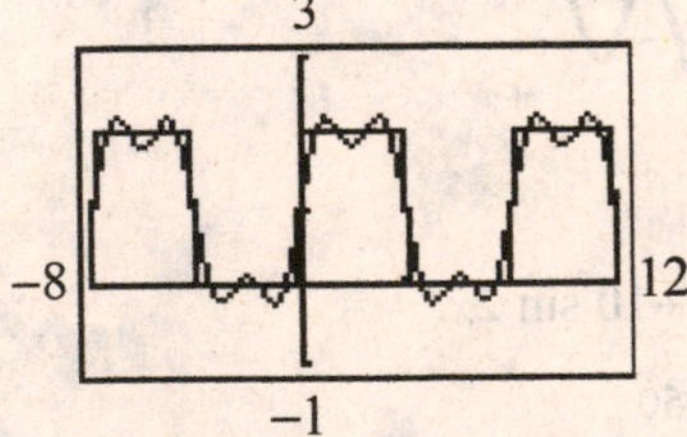

Review Exercises

1. This is true. $-2 = 2\cos\pi$

5. This is false.

For $b = 60\pi$, period $= \frac{2\pi}{60\pi} = \frac{1}{30}$ s.

The frequency is the reciprocal of the period, or 30 Hz.

9. $y = -2\cos x$ has amplitude 2 and has been inverted. The table of key values between 0 and 2π is:

x	0	π/2	π	3π/2	2π
y	-2	0	2	0	-2

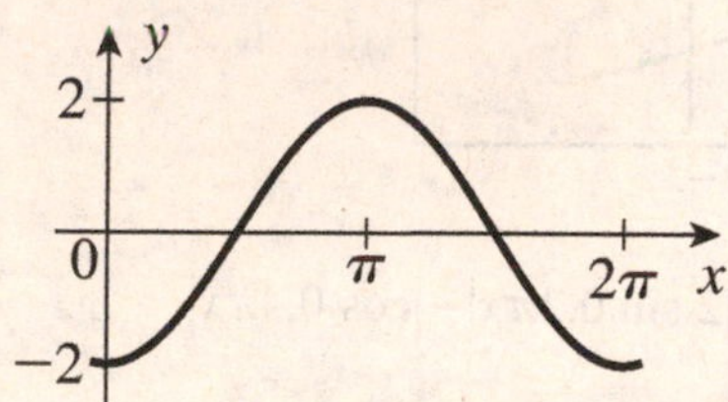

13. $y = 0.4\cos 4x$ has amplitude 0.4 and period 2π/4=π/2, with key values at multiples of $\frac{1}{4}\left(\frac{\pi}{2}\right) = \frac{\pi}{8}$. The table of key values between 0 and π/2 is:

x	0	π/8	π/4	3π/8	π/2
y	0.4	0	-0.4	0	0.4

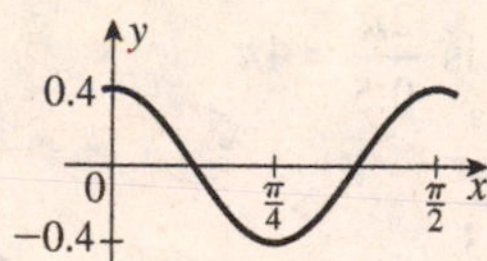

17. $y = \sin\pi x$ has amplitude 1 and period 2π/π=2, with key values at multiples of $\frac{1}{4}(2) = \frac{1}{2}$. The table of key values between 0 and 2 is:

x	0	0.5	1	1.5	2
y	0	1	0	-1	0

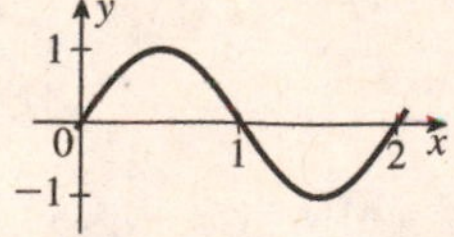

21. $y = -0.5\sin\left(\frac{-\pi x}{6}\right)$ is the same as $y = 0.5\sin\left(\frac{\pi x}{6}\right)$ (see Eq. (8.7)). It has amplitude 0.5 and period 2π/(π/6)=12, with key values at multiples of $\frac{1}{4}(12) = 3$. The table of key values between 0 and 12 is:

x	0	3	6	9	12
y	0	0.5	0	-0.5	0

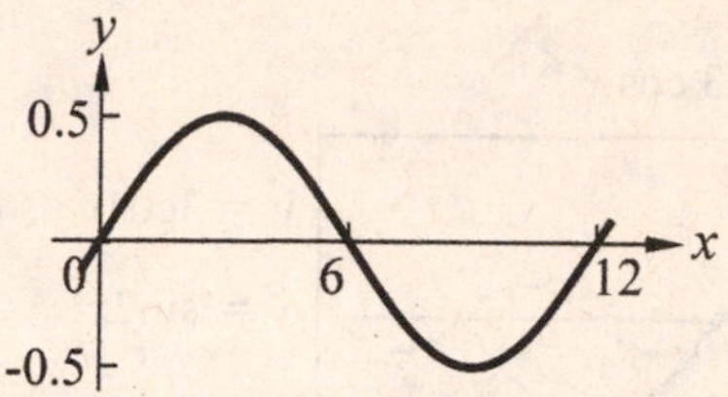

25. $y = -2\cos(4x + \pi)$ has amplitude 2 (inverted), period 2π/4=π/2, and displacement $-\frac{\pi}{4}$. One-fourth period is π/8, so key values for one full cycle start at –π/4, end at π/4, and are found π/8 units apart. The table of key values is:

x	-π/4	-π/8	0	π/8	π/4
y	-2	0	2	0	-2

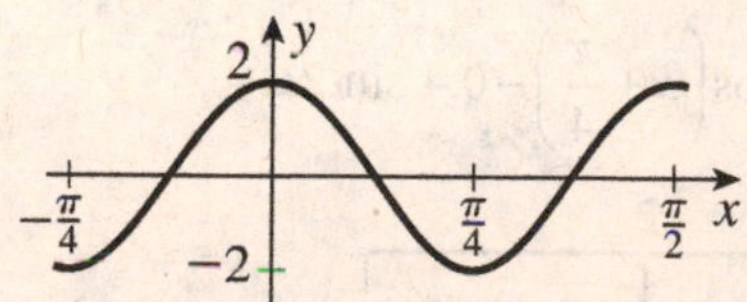

29. $y = 8\cos\left(4\pi x - \frac{\pi}{2}\right)$ has amplitude 8, period 2π/4π=1/2, and displacement $-\left(-\frac{\pi}{2}\right)\cdot\frac{1}{4\pi} = \frac{1}{8}$. One-fourth period is 1/8, so key values for one full cycle start at 1/8, end at 5/8, and are found 1/8 units apart. The table of key values is:

x	1/8	1/4	3/8	1/2	5/8
y	8	0	-8	0	8

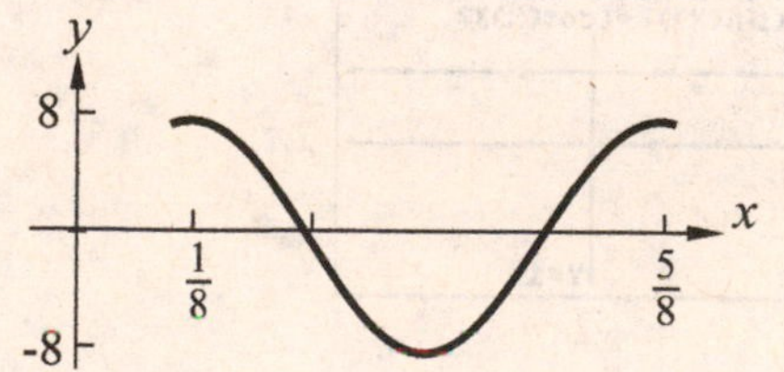

33. $y = -3 \csc x$ has the same period as the cosecant function, with values multiplied by 1/3 and inverted.

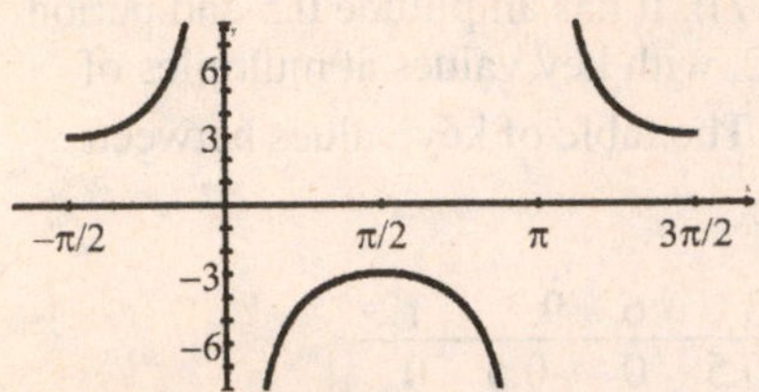

37. $y = \sin 2x + 3\cos x$

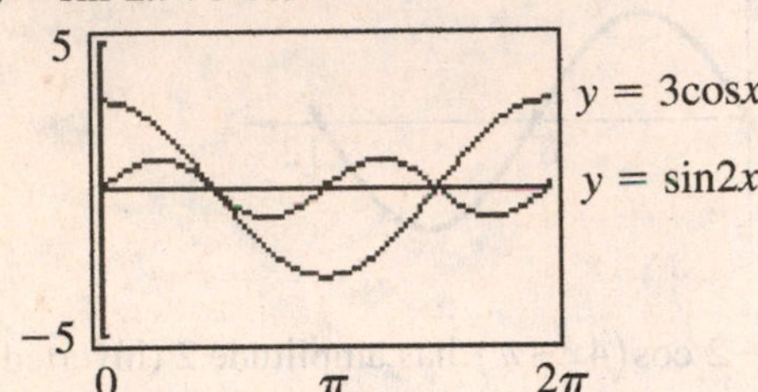

Adding y values of these curves gives

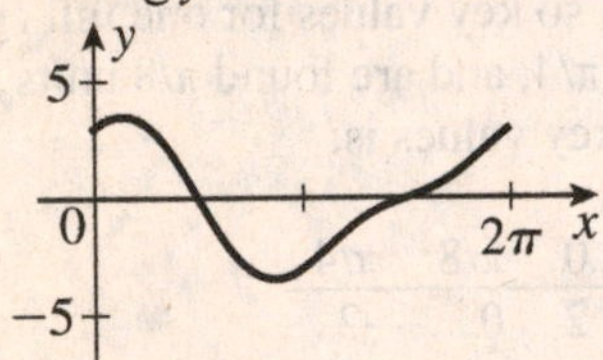

41. Graph $y_1 = \cos\left(x + \frac{\pi}{4}\right) - 0.4 \sin 2x$.

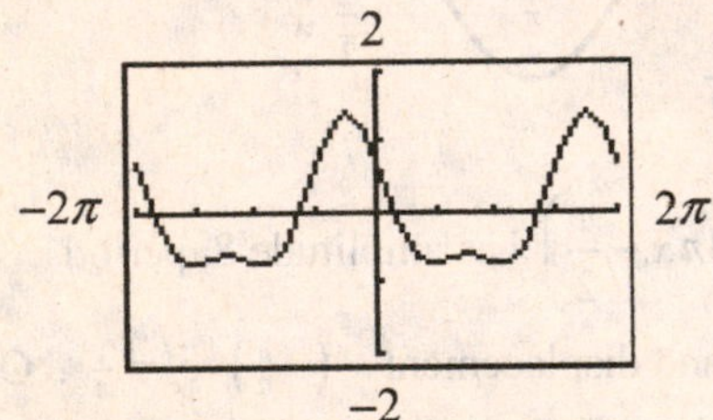

45. Graph $y_1 = \sin^2 x + \cos^2 x$. The graph is the horizontal line $y = 1$.

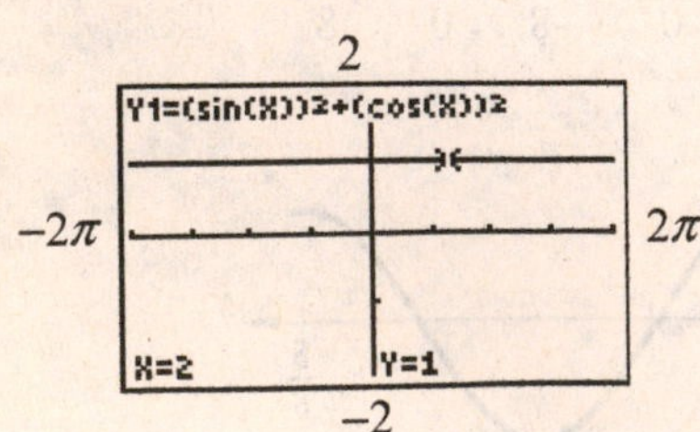

We conclude that the following identity holds:
$\sin^2 x + \cos^2 x = 1$
(We will discuss this identity in Chapter **20**.)

49. amplitude: $a = 1$

period: $\frac{2\pi}{b} = 8$

$b = \frac{\pi}{4}$;

displacement: $-\frac{c}{b} = 3$

$-\frac{c}{\pi/4} = 3$

$c = -\frac{3\pi}{4}$

Solution: $y = \cos\left(\frac{\pi}{4}x - \frac{3\pi}{4}\right)$

53. In parametric mode graph $x_1 = -\cos 2\pi t$, $y_1 = 2\sin \pi t$

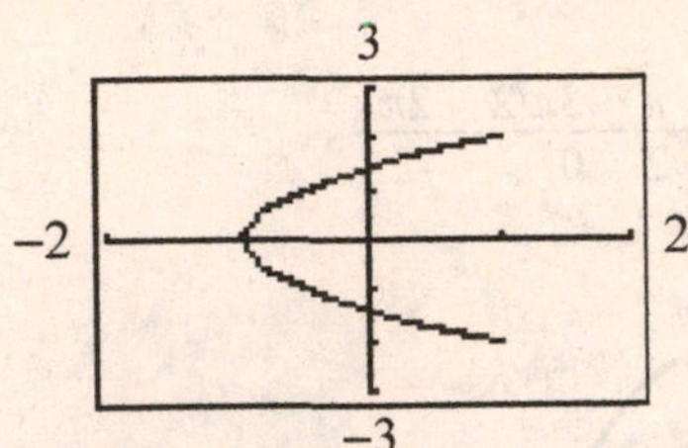

57. Graph $y_1 = 2|2 \sin 0.2\pi x| - |\cos 0.4\pi x|$

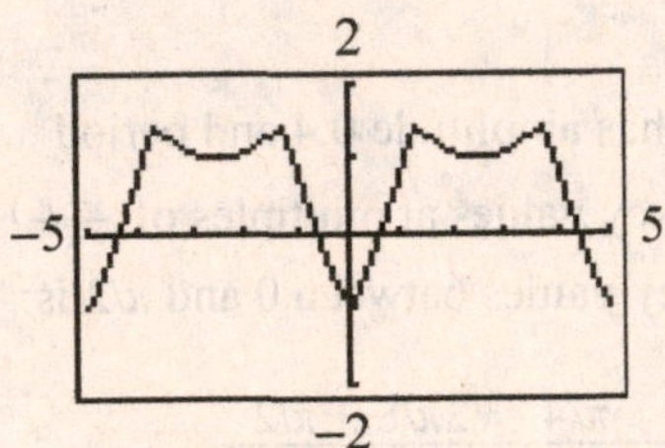

61. The period of $2\cos 0.5x$ is $\frac{2\pi}{0.5} = 4\pi$.

The period of $\sin 3x$ is $\frac{2\pi}{3}$.

The period of $y = 2\cos 0.5x + \sin 3x$ is the least common multiple of 4π and $\frac{2\pi}{3}$, which is 4π.

65. Substitute $x = \pi/3$ and $y = -3$ into $y = 3\cos bx$ to get

$$3\cos\tfrac{\pi}{3}b = -3$$
$$\tfrac{\pi}{3}b = \pi + 2\pi n$$

from where the smallest positive b is $b = 3$.
Solution: $y = 3\cos 3x$

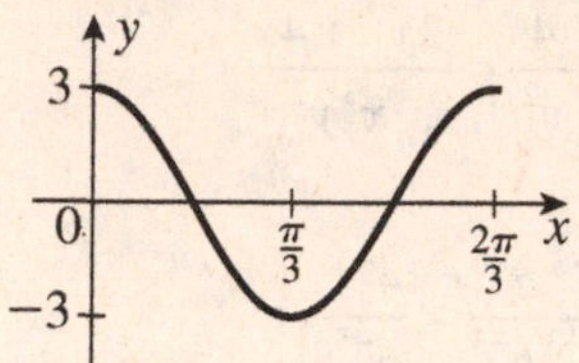

69.

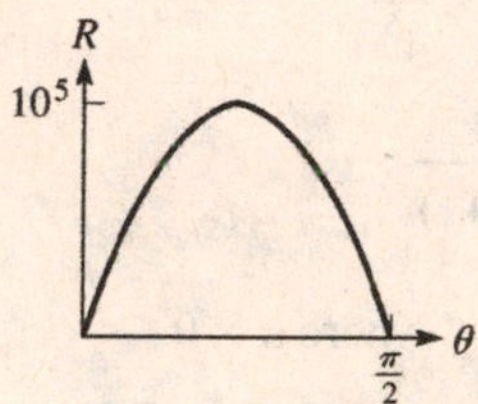

73. The period of $y = \sin 120\pi t$ is $2\pi/120\pi = 1/60$ s. With the absolute value, every negative part of the function is reflected with respect to the x axis to become positive, so the period is reduced by half. The period is 1/120 s. There are six complete cycles between 0 and 0.05 s.

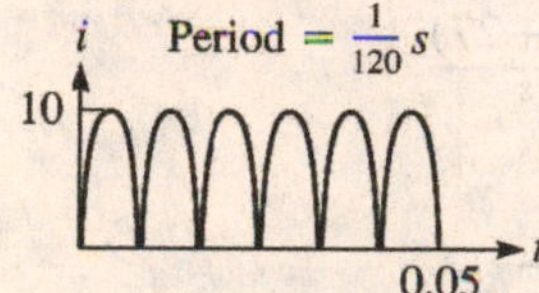

77. Graph $y = 3.0\cos 0.2x + 1.0\sin 0.4x$

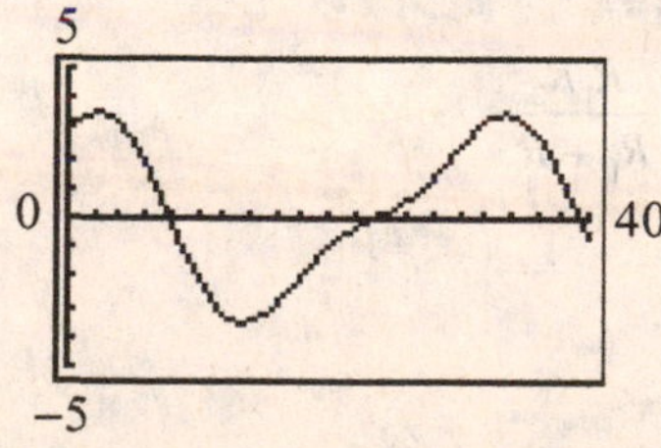

81. $d = a\sec\theta$, $a = 3.00$
$d = 3.00\sec\theta$
We graph the function from 0 to $\pi/2$.

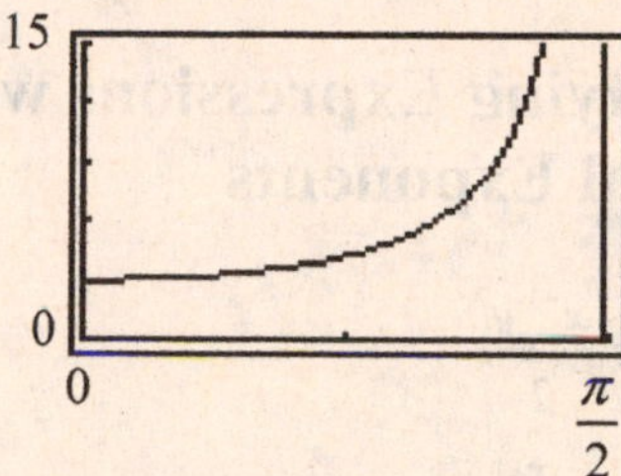

85. $y = 4\sin 2t - 2\cos 2t$

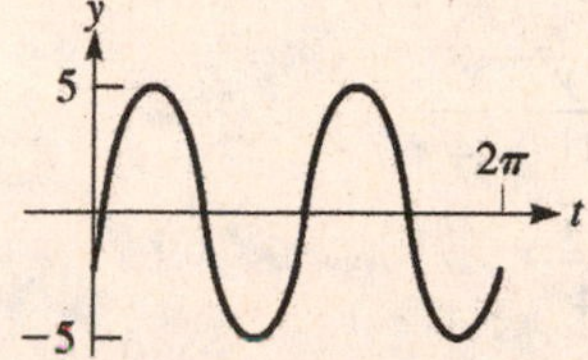

89.

θ	0	$\pi/2$	π	$3\pi/2$	2π
x	0	$\pi/2-1$	π	$3\pi/2+1$	2π
y	0	1	2	1	0

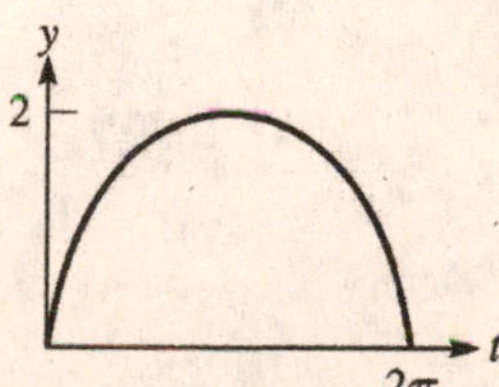

93.

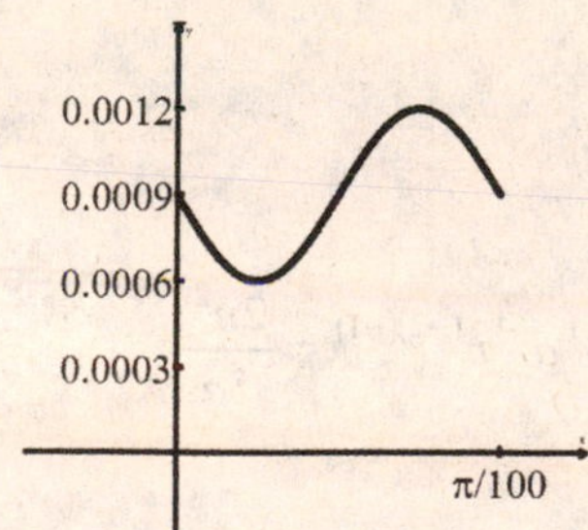

Chapter 11

Exponents and Radicals

11.1 Simplifying Expressions with Integral Exponents

1. $$\left(x^{-2}y\right)^2 \ \frac{2}{x}^{-2} = \frac{x^{-4}y^2}{\frac{2}{x}^{2}}$$
$$= \frac{x^{-4}y^2}{\frac{4}{x^2}}$$
$$= \frac{x^{-4}y^2}{1}\cdot\frac{x^2}{4}$$
$$= \frac{x^{-2}y^2}{4}$$
$$= \frac{y^2}{4x^2}$$

5. $x^8 \cdot x^{-3} = x^{8+(-3)} = x^5$

9. $$\frac{c^7}{c^{-2}} = c^{7-(-2)}$$
$$= c^9$$

13. $$5^0 \times 5^{-3} = 5^{0+(-3)}$$
$$= 5^{-3}$$
$$= \frac{1}{5^3}$$
$$= \frac{1}{125}$$

17. $2\left(5an^{-2}\right)^{-1} = 2\times 5^{-1}a^{-1}n^{(-2)(-1)} = \dfrac{2n^2}{5a}$

21. $-9x^0 = -9(1) = -9$

25. $$\left(7a^{-1}x\right)^{-3} = 7^{-3}a^3x^{-3}$$
$$= \frac{a^3}{7^3x^3}$$
$$= \frac{a^3}{343x^3}$$

29. $3\left(\dfrac{a}{b^{-2}}\right)^{-3} = \dfrac{3a^{-3}}{b^6} = \dfrac{3}{a^3b^6}$

33. $2x^{-3}+4y^{-2} = \dfrac{2}{x^3}+\dfrac{4}{y^2} = \dfrac{2y^2+4x^3}{x^3y^2}$

37. $$\left(\frac{3a^2}{4b}\right)^{-3}\left(\frac{4}{a}\right)^{-5} = \frac{3^{-3}a^{-6}}{4^{-3}b^{-3}}\times\frac{4^{-5}}{a^{-5}}$$
$$= \frac{4^3b^3}{3^3a^6}\times\frac{a^5}{4^5}$$
$$= \frac{b^3}{27a(4^2)}$$
$$= \frac{b^3}{432a} \quad \text{where } a \neq 0$$

41. $$3a^{-2}+\left(3a^{-2}\right)^4 = 3a^{-2}+3^4a^{-8}$$
$$= \frac{3}{a^2}+\frac{81}{a^8}$$
$$= \frac{3a^6+81}{a^8}$$
$$= \frac{3(a^6+27)}{a^8}$$

45. $$\left(R_1^{-1}+R_2^{-1}\right)^{-1} = \frac{1}{\frac{1}{R_1}+\frac{1}{R_2}}$$
$$= \frac{1}{\frac{R_2+R_1}{R_1R_2}}$$
$$= \frac{R_1R_2}{R_1+R_2}$$

49. $$\frac{6^{-1}}{4^{-2}+2} = \frac{\frac{1}{6}}{\frac{1}{4^2}+2}$$
$$= \frac{\frac{1}{6}}{\frac{1}{16}+2}$$
$$= \frac{\frac{1}{6}}{\frac{1+32}{16}}$$
$$= \frac{1}{6}\times\frac{16}{33}$$
$$= \frac{8}{99}$$

53. $2t^{-2}+t^{-1}(t+1)=\frac{2}{t^2}+t^0+t^{-1}$

$=\frac{2}{t^2}+1+\frac{1}{t}$

$=\frac{2+t^2+t}{t^2}$

$=\frac{t^2+t+2}{t^2}$

57. If $x<0$, then $x^2>0$ and $x^{-2}>0$

If $x<0$, then $x^1<0$ and $x^{-1}<0$.

$x^{-2}>0>x^{-1}$ so $x^{-2}>x^{-1}$.

No, there are no values for x where $x^{-2}<x^{-1}$

61. **(a)** $\frac{a}{b}^{-n}=\frac{1}{\frac{a}{b}^{n}}=\frac{1}{\frac{a^n}{b^n}}=1\times\frac{b^n}{a^n}=\frac{b}{a}^{n}$

(b) $\frac{a}{b}^{-n}=\frac{b}{a}^{n}$

$\frac{3.576}{8.091}^{-7}=(0.4419726)^{-7}$

$=303.55182$

$\frac{8.091}{3.576}^{7}=(2.262584)^{7}$

$=303.55182$

Therefore,

$\frac{3.576}{8.091}^{-7}=\frac{8.091}{3.576}^{7}$

65. This is true:

$-2^0-(-1)^0\ ^0=[-1-1]^0=(-2)^0=1$

69. $2^{5x}=2^7\left(2^{2x}\right)^2$

$2^{5x}=2^7\left(2^{4x}\right)$

$2^{5x}=2^{7+4x}$

$5x=7+4x$

$x=7$

73. $1\text{ J}=1\text{ kg}\cdot\text{m}^2\cdot\text{s}^{-2}$, so

$\text{kg}\cdot\text{s}^{-1}\left(\text{m}\cdot\text{s}^{-2}\right)^2=\text{kg}\cdot\text{s}^{-1}\cdot\text{m}^2\cdot\text{s}^{-4}$

$=\text{kg}\cdot\text{m}^2\cdot\text{s}^{-2}\left(\text{s}^{-3}\right)$

$=\text{J/s}^3$

77. $P=\frac{A\ \frac{r}{12}}{1-\ 1+\frac{r}{12}^{-12t}}$; $A=20000, t=5, r=0.04$

$P=\frac{20000\ \frac{0.04}{12}}{1-\ 1+\frac{0.04}{12}^{-60}}=\frac{66.67}{0.181}=368.33$

The monthly payment is \$368.33

11.2 Fractional Exponents

1. $8^{4/3}=\left(8^{1/3}\right)^4=\left(\sqrt[3]{8}\right)^4=2^4=16$

5. $36^{1/2}=\sqrt{36}=6$

9. $100^{3/2}=\left(100^{1/2}\right)^3=\left(\sqrt{100}\right)^3=10^3=1000$

13. $64^{-2/3}=\frac{1}{64^{2/3}}$

$=\frac{1}{\left(\sqrt[3]{64}\right)^2}$

$=\frac{1}{(4)^2}$

$=\frac{1}{16}$

17. $\left(3^6\right)^{2/3}=(3)^{12/3}=3^4=81$

21. $\frac{15^{2/3}}{5^2\cdot 15^{-1/3}}=\frac{15^{2/3+1/3}}{5^2}=\frac{15^1}{25}=\frac{3}{5}$

25. $125^{-2/3} - 100^{-3/2} = \dfrac{1}{125^{2/3}} - \dfrac{1}{100^{3/2}}$

$= \dfrac{1}{\left(\sqrt[3]{125}\right)^2} - \dfrac{1}{\left(\sqrt{100}\right)^3}$

$= \dfrac{1}{5^2} - \dfrac{1}{10^3}$

$= \dfrac{1}{25} - \dfrac{1}{1000}$

$= \dfrac{40-1}{1000}$

$= \dfrac{39}{1000}$

29. $(4.0187)^{-4/9} = 0.53891$

33. $\dfrac{4y^{-1/2}}{-y^{2/5}} = -4y^{-1/2-2/5}$

$= -4y^{-5/10-4/10}$

$= -4y^{-9/10}$

$= -\dfrac{4}{y^{9/10}}$

37. $\left(8a^3b^6\right)^{1/3} = 8^{1/3}a^{3(1/3)}b^{6(1/3)}$

$= \sqrt[3]{8}ab^2$

$= 2ab^2$

41. $\left(\dfrac{a^{5/7}}{a^{2/3}}\right)^{7/4} = \left(a^{5/7-2/3}\right)^{7/4}$

$= \left(a^{15/21-14/21}\right)^{7/4}$

$= \left(a^{1/21}\right)^{7/4}$

$= a^{7/84}$

$= a^{1/12}$

45. $\dfrac{y^{3/8}\left(y^{5/8} - y^{13/8}\right)}{y^{1/2}\left(y^{1/2} - y^{-1/2}\right)} = \dfrac{y^{8/8} - y^{16/8}}{y^1 - y^0}$

$= \dfrac{y - y^2}{y-1}$

$= \dfrac{-y(y-1)}{(y-1)}$

$= -y \quad \text{where} \quad y \neq 1$

49. $\left(a^3\right)^{-4/3} + a^{-2} = a^{-4} + \dfrac{1}{a^2}$

$= \dfrac{1}{a^4} + \dfrac{1}{a^2}$

$= \dfrac{1+a^2}{a^4}$

53. $x^2(2x-1)^{-1/2} + 2x(2x-1)^{1/2}$

$= \dfrac{x^2}{(2x-1)^{1/2}} + 2x(2x-1)^{1/2}$

$= \dfrac{x^2 + 2x(2x-1)^{1/2}(2x-1)^{1/2}}{(2x-1)^{1/2}}$

$= \dfrac{x^2 + 2x(2x-1)}{(2x-1)^{1/2}}$

$= \dfrac{x^2 + 4x^2 - 2x}{(2x-1)^{1/2}}$

$= \dfrac{5x^2 - 2x}{(2x-1)^{1/2}}$

$= \dfrac{x(5x-2)}{(2x-1)^{1/2}}$

57. $y = f(t) = t^{-4/5}$

t	0	1	−1	2	−2	4	−4
y	0	1	1	1.74	1.74	3.03	3.03

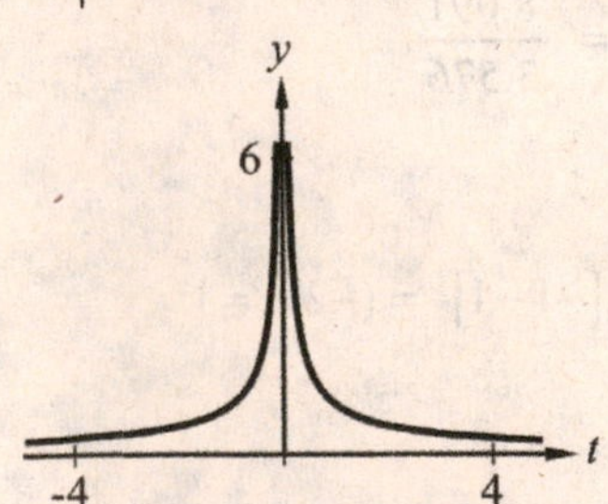

61. $\left(x^{n-1} \div x^{n-3}\right)^{1/3} = \left(x^{n-1-(n-3)}\right)^{1/3}$

$= \left(x^{n-1-n+3}\right)^{1/3}$

$= \sqrt[3]{x^2}$

65. $T^2 = kR^3 \quad 1+\dfrac{d}{R} \quad ^{3}$

$$T^2 = kR^3 \quad \frac{R+d}{R} \quad ^{3}$$

$$T^2 = kR^3 \frac{(R+d)^3}{R^3}$$

$$T^2 = k(R+d)^3$$

$$(R+d)^3 = \frac{T^2}{k}$$

$$R+d = \quad \frac{T^2}{k} \quad ^{1/3}$$

$$R = \frac{T^{2/3}}{k^{1/3}} - d$$

69. $A(t) = A_0 \cdot 2^{-t/5730}; A_0 = 55.0 \text{ mg}$

$$A(3250) = 55.0\left(2^{-3250/5730}\right)$$

$$= 55.0(0.67493)$$

$$= 37.1 \text{ mg}$$

11.3 Simplest Radical Form

1. $\sqrt{a^3b^4} = \sqrt{a^2\left(b^2\right)^2 \cdot a} = ab^2\sqrt{a}$

5. $\sqrt{75} = \sqrt{25\cdot 3} = \sqrt{25}\cdot\sqrt{3} = 5\sqrt{3}$

9. $\sqrt{x^2y^5} = \sqrt{x^2y^4y} = \sqrt{x^2}\sqrt{y^4}\sqrt{y} = xy^2\sqrt{y}$

13. $\sqrt{80R^5TV^4} = \sqrt{16R^4V^4(5RT)}$

$$= \sqrt{16R^4V^4}\cdot\sqrt{5RT}$$

$$= 4R^2V^2\sqrt{5RT}$$

17. $\sqrt[5]{96} = \sqrt[5]{32\cdot 3} = \sqrt[5]{2^5}\cdot\sqrt[5]{3} = 2\sqrt[5]{3}$

21. $\sqrt[4]{64r^3s^4t^5} = \sqrt[4]{16s^4t^4(4r^3t)}$

$$= \sqrt[4]{2^4}\cdot\sqrt[4]{s^4}\cdot\sqrt[4]{t^4}\cdot\sqrt[4]{4r^3t}$$

$$= 2st\sqrt[4]{4r^3t}$$

25. $\sqrt[3]{P}\sqrt[3]{P^2V} = \sqrt[3]{P^3V} = P\sqrt[3]{V}$

29. $\sqrt{\dfrac{3}{2}} = \sqrt{\dfrac{3}{2}}\times\dfrac{\sqrt{2}}{\sqrt{2}}$

$$= \frac{\sqrt{6}}{\sqrt{4}}$$

$$= \frac{\sqrt{6}}{2}$$

33. $\sqrt[5]{\dfrac{1}{9}} = \sqrt[5]{\dfrac{1}{3^2}}\times\dfrac{\sqrt[5]{3^3}}{\sqrt[5]{3^3}}$

$$= \frac{\sqrt[5]{27}}{\sqrt[5]{3^5}}$$

$$= \frac{\sqrt[5]{27}}{3}$$

37. $\sqrt[6]{64} = \sqrt[6]{2^6} = 2$

41. $\sqrt{4\times 10^6} = \sqrt{2^2 10^6} = 2\left(10^3\right) = 2000$

45. $\sqrt[4]{\dfrac{1}{4}} = \sqrt[4]{\dfrac{1}{2^2}}$

$$= \frac{1}{2^{2/4}}$$

$$= \frac{1}{2^{1/2}}$$

$$= \frac{1}{\sqrt{2}}\times\frac{\sqrt{2}}{\sqrt{2}}$$

$$= \frac{\sqrt{2}}{2}$$

49. $\sqrt{\sqrt{\sqrt{n}}} = \left(\left(n^{1/2}\right)^{1/2}\right)^{1/2} = n^{1/8} = \sqrt[8]{n}$

53. $\sqrt{64+144} = \sqrt{2^6 + 2^4 3^2}$

$$= \sqrt{2^4(2^2+3^2)}$$

$$= 2^{4/2}\sqrt{9+4}$$

$$= 4\sqrt{13}$$

57. $\sqrt{\frac{5}{4}-\frac{1}{8}}=\sqrt{\frac{5(2)-1}{8}}$
$=\sqrt{\frac{9}{8}}$
$=\sqrt{\frac{3^2}{2^2(2)}}$
$=\frac{3}{2\sqrt{2}}\times\frac{\sqrt{2}}{\sqrt{2}}$
$=\frac{3\sqrt{2}}{4}$

61. $\sqrt{\frac{C-2}{C+2}}=\frac{\sqrt{C-2}}{\sqrt{C+2}}\times\frac{\sqrt{C+2}}{\sqrt{C+2}}$
$=\frac{\sqrt{(C-2)(C+2)}}{C+2}$

65. $\sqrt{9x^2-6x+1}=\sqrt{(3x-1)^2}=3x-1$

69.

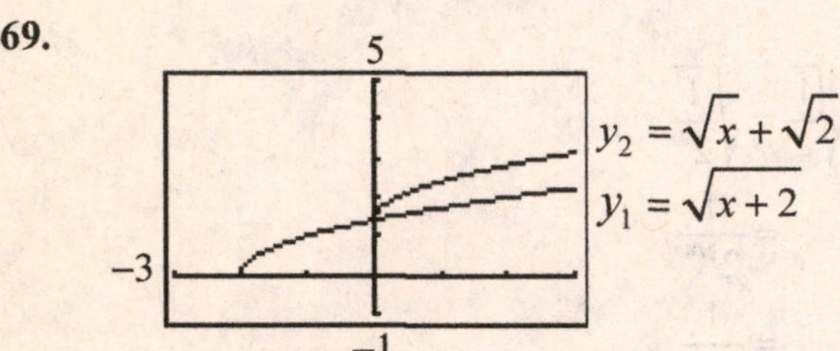

73. $a\sqrt{\frac{2g}{a}}=a\sqrt{\frac{2g}{a}}\times\frac{\sqrt{a}}{\sqrt{a}}$
$=a\sqrt{\frac{2ag}{a^2}}$
$=\frac{a}{a}\sqrt{2ag}$
$=\sqrt{2ag}$

11.4 Addition and Subtraction of Radicals

1. $3\sqrt{125}-\sqrt{20}+\sqrt{45}=3\sqrt{25(5)}-\sqrt{4(5)}+\sqrt{9(5)}$
$=3(5)\sqrt{5}-2\sqrt{5}+3\sqrt{5}$
$=15\sqrt{5}-2\sqrt{5}+3\sqrt{5}$
$=16\sqrt{5}$

5. $\sqrt{28}+\sqrt{5}-3\sqrt{7}=\sqrt{(4)7}+\sqrt{5}-3\sqrt{7}$
$=2\sqrt{7}+\sqrt{5}-3\sqrt{7}$
$=\sqrt{5}-\sqrt{7}$

9. $\sqrt{5}+\sqrt{16+4}=\sqrt{5}+\sqrt{20}$
$=\sqrt{5}+\sqrt{4(5)}$
$=\sqrt{5}+2\sqrt{5}$
$=3\sqrt{5}$

13. $\sqrt{18y}-3\sqrt{8y}=\sqrt{9(2y)}-3\sqrt{4(2y)}$
$=3\sqrt{2y}-6\sqrt{2y}$
$=-3\sqrt{2y}$

17. $3\sqrt{200}-\sqrt{162}-\sqrt{288}$
$=3\sqrt{100(2)}-\sqrt{81(2)}-\sqrt{144(2)}$
$=30\sqrt{2}-9\sqrt{2}-12\sqrt{2}$
$=9\sqrt{2}$

21. $\sqrt{40}+\sqrt{\frac{5}{2}}=\sqrt{(4)10}+\frac{\sqrt{5}}{\sqrt{2}}\times\frac{\sqrt{2}}{\sqrt{2}}$
$=2\sqrt{10}+\frac{\sqrt{10}}{2}$
$=\frac{4\sqrt{10}+\sqrt{10}}{2}$
$=\frac{5\sqrt{10}}{2}$

25. $\sqrt[3]{81}+\sqrt[3]{3000}=\sqrt[3]{(27)3}+\sqrt[3]{(1000)3}$
$=3\sqrt[3]{3}+10\sqrt[3]{3}$
$=13\sqrt[3]{3}$

29. $$5\sqrt{a^3b}-\sqrt{4ab^5}=5\sqrt{a^2\cdot ab}-\sqrt{4b^4ab}$$
$$=5a\sqrt{ab}-2b^2\sqrt{ab}$$
$$=\left(5a-2b^2\right)\sqrt{ab}$$

33. $$\sqrt[3]{24a^2b^4}-\sqrt[3]{3a^5b}=\sqrt[3]{8b^3\cdot 3a^2b}-\sqrt[3]{a^3\cdot 3a^2b}$$
$$=2b\sqrt[3]{3a^2b}-a\sqrt[3]{3a^2b}$$
$$=(2b-a)\sqrt[3]{3a^2b}$$

37. $$\sqrt[3]{ab^{-1}}-\sqrt[3]{8a^{-2}b^2}=\sqrt[3]{\frac{a}{b}}-\sqrt[3]{\frac{8b^2}{a^2}}$$
$$=\sqrt[3]{\frac{a}{b}\cdot\frac{b^2}{b^2}}-\sqrt[3]{\frac{8b^2}{a^2}\cdot\frac{a}{a}}$$
$$=\sqrt[3]{\frac{ab^2}{b^3}}-\sqrt[3]{\frac{8ab^2}{a^3}}$$
$$=\frac{\sqrt[3]{ab^2}}{b}-\frac{2\sqrt[3]{ab^2}}{a}$$
$$=\frac{a\sqrt[3]{ab^2}-2b\sqrt[3]{ab^2}}{ab}$$
$$=\frac{(a-2b)\sqrt[3]{ab^2}}{ab}$$

41. $$\sqrt{4x+8}+2\sqrt{9x+18}=\sqrt{4(x+2)}+2\sqrt{9(x+2)}$$
$$=2\sqrt{x+2}+6\sqrt{x+2}$$
$$=8\sqrt{x+2}$$

45. $$2\sqrt{\frac{2}{3}}+\sqrt{24}-5\sqrt{\frac{3}{2}}$$
$$=2\frac{\sqrt{2}}{\sqrt{3}}\times\frac{\sqrt{3}}{\sqrt{3}}+\sqrt{4(6)}-5\frac{\sqrt{3}}{\sqrt{2}}\times\frac{\sqrt{2}}{\sqrt{2}}$$
$$=\frac{2\sqrt{6}}{3}+2\sqrt{6}-\frac{5\sqrt{6}}{2}$$
$$=\ \frac{2}{3}+2-\frac{5}{2}\ \sqrt{6}$$
$$\frac{4+12-15}{6}\ \sqrt{6}$$
$$=\frac{\sqrt{6}}{6}$$
$=0.40\,824\,829\ldots$ on a calculator

$$2\sqrt{\frac{2}{3}}+\sqrt{24}-5\sqrt{\frac{3}{2}}$$
$=0.40\,824\,829\ldots$ on a calculator

49. $x^2-2x-2=0$ has roots
$$x=\frac{-b\pm\sqrt{b^2-4ac}}{2a}$$
$$x=\frac{-(-2)\pm\sqrt{(-2)^2-4(1)(-2)}}{2(1)}$$
$$x=\frac{2\pm\sqrt{12}}{2}$$
$$x=\frac{2\pm\sqrt{(4)3}}{2}$$
$$x=\frac{2\pm 2\sqrt{3}}{2}$$
$$x=1\pm\sqrt{3}$$
so $x=1+\sqrt{3}$ is the positive root

$x^2+2x-11=0$ has roots
$$x=\frac{-b\pm\sqrt{b^2-4ac}}{2a}$$
$$x=\frac{-2\pm\sqrt{2^2-4(1)(-11)}}{2(1)}$$
$$x=\frac{-2\pm\sqrt{48}}{2}$$
$$x=\frac{-2\pm\sqrt{(16)3}}{2}$$
$$x=\frac{-2\pm 4\sqrt{3}}{2}$$
$$x=-1\pm 2\sqrt{3}$$
so $x=-1+2\sqrt{3}$ is the positive root

sum of positive roots $=1+\sqrt{3}+(-1+2\sqrt{3})$

sum of positive roots $=3\sqrt{3}$

53.

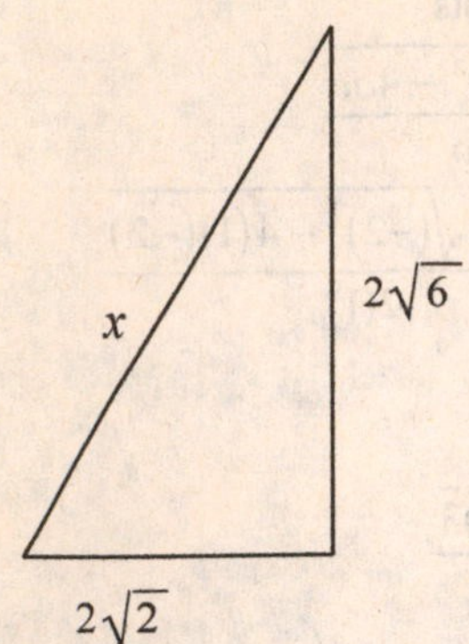

Using Pythagoras' theorem,

$$x^2 = \left(2\sqrt{2}\right)^2 + \left(2\sqrt{6}\right)^2$$
$$x^2 = 8 + 24$$
$$x^2 = 32$$
$$x = \sqrt{32}$$

$$\text{Perimeter} = \sqrt{32} + 2\sqrt{2} + 2\sqrt{6}$$
$$= \sqrt{16(2)} + 2\sqrt{2} + 2\sqrt{6}$$
$$= 4\sqrt{2} + 2\sqrt{2} + 2\sqrt{6}$$
$$= 6\sqrt{2} + 2\sqrt{6} \text{ units}$$

11.5 Multiplication and Division of Radicals

1. $\sqrt{2}\left(3\sqrt{5} - 4\sqrt{8}\right) = 3\sqrt{10} - 4\sqrt{16}$
$= 3\sqrt{10} - 4(4)$
$= 3\sqrt{10} - 16$

5. $\sqrt{5}\sqrt{7} = \sqrt{5(7)} = \sqrt{35}$

9. $\sqrt[3]{4} \cdot \sqrt[3]{2} = \sqrt[3]{4(2)} = \sqrt[3]{8} = 2$

13. $\sqrt{72} \cdot \sqrt{\frac{5}{2}} = \sqrt{\frac{72(5)}{2}} = \sqrt{36(5)} = 6\sqrt{5}$

17. $\left(2 - \sqrt{5}\right)\left(2 + \sqrt{5}\right) = 2^2 - \sqrt{5}^2 = 4 - 5 = -1$

21. $\left(3\sqrt{11} - \sqrt{x}\right)\left(2\sqrt{11} + 5\sqrt{x}\right) = 6(11) + 15\sqrt{11x} - 2\sqrt{11x} - 5(x)$
$= 66 + 13\sqrt{11x} - 5x$

25. $$\frac{\sqrt{6}-9}{\sqrt{6}}=\frac{\sqrt{6}-9}{\sqrt{6}}\cdot\frac{\sqrt{6}}{\sqrt{6}}$$
$$=\frac{6-9\sqrt{6}}{6}$$
$$=\frac{2-3\sqrt{6}}{2}$$

29. $$\left(\sqrt{2a}-\sqrt{b}\right)\left(\sqrt{2a}+3\sqrt{b}\right)=2a+3\sqrt{2ab}-\sqrt{2ab}-3b$$
$$=2a+2\sqrt{2ab}-3b$$

33. $$\frac{1}{\sqrt{7}+\sqrt{3}}=\frac{1}{\sqrt{7}+\sqrt{3}}\cdot\frac{\sqrt{7}-\sqrt{3}}{\sqrt{7}-\sqrt{3}}$$
$$=\frac{\sqrt{7}-\sqrt{3}}{7-\sqrt{21}+\sqrt{21}-3}$$
$$=\frac{\sqrt{7}-\sqrt{3}}{4}$$

37. $$\frac{2\sqrt{3}-5\sqrt{5}}{\sqrt{3}+2\sqrt{5}}=\frac{2\sqrt{3}-5\sqrt{5}}{\sqrt{3}+2\sqrt{5}}\times\frac{\sqrt{3}-2\sqrt{5}}{\sqrt{3}-2\sqrt{5}}$$
$$=\frac{2(3)-4\sqrt{15}-5\sqrt{15}+10(5)}{3-4(5)}$$
$$=\frac{56-9\sqrt{15}}{-17}$$
$$=\frac{9\sqrt{15}-56}{17}$$

41. $$\left(\sqrt[5]{\sqrt{6}-\sqrt{5}}\right)\left(\sqrt[5]{\sqrt{6}+\sqrt{5}}\right)=\sqrt[5]{\left(\sqrt{6}\right)^2+\sqrt{30}-\sqrt{30}-\left(\sqrt{5}\right)^2}$$
$$=\sqrt[5]{6-5}$$
$$=\sqrt[5]{1}$$
$$=1$$

45. $$\frac{\sqrt{x+y}}{\sqrt{x-y}-\sqrt{x}}=\frac{\sqrt{x+y}}{\sqrt{x-y}-\sqrt{x}}\times\frac{\sqrt{x-y}+\sqrt{x}}{\sqrt{x-y}+\sqrt{x}}$$
$$=\frac{\sqrt{x+y}\sqrt{x-y}+\sqrt{x+y}\sqrt{x}}{(x-y)-(x)}$$
$$=\frac{\sqrt{x^2-y^2}+\sqrt{x^2+xy}}{-y}$$
$$=-\frac{\sqrt{x^2-y^2}+\sqrt{x^2+xy}}{y}$$

49. $\left(\sqrt{11}+\sqrt{6}\right)\left(\sqrt{11}-2\sqrt{6}\right)=11-2\sqrt{66}+\sqrt{66}-2(6)$

$$\begin{aligned}&=11-\sqrt{66}-12\\&=-1-\sqrt{66}\\&=-1\left(1+\sqrt{66}\right)\\&=-9.124\,038\,405\cdots \text{ on calculator}\end{aligned}$$

$\left(\sqrt{11}+\sqrt{6}\right)\left(\sqrt{11}-2\sqrt{6}\right)=-9.124\,038\,405\cdots$ on calculator

53. $$2\sqrt{x}+\frac{1}{\sqrt{x}}=\frac{2\sqrt{x}\cdot\sqrt{x}+1}{\sqrt{x}}=\frac{2x+1}{\sqrt{x}}$$

57. $$\frac{5+2\sqrt{2}}{3\sqrt{10}}=\frac{5+2\sqrt{2}}{3\sqrt{10}}\times\frac{5-2\sqrt{2}}{5-2\sqrt{2}}=\frac{25-4(2)}{15\sqrt{10}-6\sqrt{20}}=\frac{17}{15\sqrt{10}-6\sqrt{4(5)}}=\frac{17}{15\sqrt{10}-12\sqrt{5}}$$

61. If $\sqrt{x}+\sqrt{y}=\sqrt{x+y}$ then squaring both sides yields $x+2\sqrt{xy}+y=x+y$ or $\sqrt{xy}=0$. Thus, it is necessary for one of x or y to be 0 for $\sqrt{x}+\sqrt{y}=\sqrt{x+y}$ to hold.

65. The quadratic equation

$ax^2+bx+c=0$

has two solutions

$$x=\frac{-b\pm\sqrt{b^2-4ac}}{2a}$$

If the two solutions are reciprocals of each other, then

$$\frac{-b+\sqrt{b^2-4ac}}{2a}=\frac{1}{\dfrac{-b-\sqrt{b^2-4ac}}{2a}}$$

$$\frac{-b+\sqrt{b^2-4ac}}{2a}=\frac{2a}{-b-\sqrt{b^2-4ac}} \quad \text{cross multiply}$$
$$4a^2=\left(-b+\sqrt{b^2-4ac}\right)\left(-b-\sqrt{b^2-4ac}\right)$$
$$4a^2=b^2+b\sqrt{b^2-4ac}-b\sqrt{b^2-4ac}-\left(b^2-4ac\right)$$
$$4a^2=b^2-b^2+4ac$$
$$4a^2=4ac$$
$$\frac{4a^2}{4a}=c$$
$$a=c$$

69. $m^2+bm+k^2=0$

If we substitute $m=\frac{1}{2}\left(\sqrt{b^2-4k^2}-b\right)$ into the equation

$$\frac{1}{2}\left(\sqrt{b^2-4k^2}-b\right)^2+b\cdot\frac{1}{2}\left(\sqrt{b^2-4k^2}-b\right)+k^2=0$$
$$\frac{1}{4}\left(b^2-4k^2\right)-2b\sqrt{b^2-4k^2}+b^2+\frac{b}{2}\sqrt{b^2-4k^2}-\frac{1}{2}b^2+k^2=0$$
$$\frac{1}{4}\left(b^2-4k^2\right)-\frac{b}{2}\sqrt{b^2-4k^2}+\frac{b^2}{4}+\frac{b}{2}\sqrt{b^2-4k^2}-\frac{1}{2}b^2+k^2=0$$
$$\frac{1}{4}b^2-k^2+\frac{1}{4}b^2-\frac{1}{2}b^2+k^2=0$$
$$\frac{1}{2}b^2-\frac{1}{2}b^2=0$$
$$0=0$$

Therefore, it is a solution

73. $$\frac{2Q}{\sqrt{\sqrt{2}-1}}=\frac{2Q}{\sqrt{\sqrt{2}-1}}\times\frac{\sqrt{\sqrt{2}+1}}{\sqrt{\sqrt{2}+1}}$$
$$=\frac{2Q\sqrt{\sqrt{2}+1}}{\sqrt{2-1}}$$
$$=2Q\sqrt{\sqrt{2}+1}$$

Review Exercises

1. This is false. If $x \neq 0$, then $2x^0 = 1$. If $x = 0$, then $2x^0$ is undefined.

5. This is true: $\frac{1}{\sqrt{3}-\sqrt{2}} = \frac{1}{\sqrt{3}-\sqrt{2}} \times \frac{\sqrt{3}+\sqrt{2}}{\sqrt{3}+\sqrt{2}} = \frac{\sqrt{3}+\sqrt{2}}{3-2} = \sqrt{3}+\sqrt{2}.$

9. $\frac{9c^{-1}}{d^{-3}} = \frac{9d^3}{c}$

13.
$$\begin{aligned} 8(400^{-3/2}) &= \frac{8}{400^{3/2}} \\ &= \frac{8}{\left(20^2\right)^{3/2}} \\ &= \frac{8}{20^3} \\ &= \frac{8}{8000} \\ &= \frac{1}{1000} \end{aligned}$$

17.
$$\begin{aligned} \frac{-8^{2/3}}{49^{-1/2}} &= -\left(2^3\right)^{2/3}(49)^{1/2} \\ &= -(2)^2\left(7^2\right)^{1/2} \\ &= -(2)^2(7) \\ &= -4(7) \\ &= -28 \end{aligned}$$

21.
$$\begin{aligned} \left(-32m^{15}n^{10}\right)^{3/5} &= (-32)^{3/5}\left(m^{15}\right)^{3/5}\left(n^{10}\right)^{3/5} \\ &= ((-2)^5)^{3/5}m^{15(3/5)}n^{10(3/5)} \\ &= (-2)^3m^9n^6 \\ &= -8m^9n^6 \end{aligned}$$

25.
$$\begin{aligned} \frac{3x^{-1}}{3x^{-1}+y^{-1}} &= \frac{\frac{3}{x}}{\frac{3}{x}+\frac{1}{y}} \\ &= \frac{\frac{3}{x}}{\frac{3y+x}{xy}} \\ &= \frac{3}{x}\cdot\frac{xy}{(3y+x)} \\ &= \frac{3y}{3y+x} \quad \text{where } x \neq 0, y \neq 0, 3y+x \neq 0 \end{aligned}$$

29. $\left(x^3 - y^{-3}\right)^{1/3} = \left(x^3 - \dfrac{1}{y^3}\right)^{1/3}$

$= \left(\dfrac{x^3y^3 - 1}{y^3}\right)^{1/3}$

$= \dfrac{\left(x^3y^3 - 1\right)^{1/3}}{\left(y^3\right)^{1/3}}$

$= \dfrac{\left(x^3y^3 - 1\right)^{1/3}}{y}$

$= \dfrac{(xy-1)^{1/3}\,(x^2y^2 + xy + 1)^{1/3}}{y}$

33. $2x(x-1)^{-2} - 2\left(x^2+1\right)(x-1)^{-3} = \dfrac{2x}{(x-1)^2} - \dfrac{2\left(x^2+1\right)}{(x-1)^3}$

$= \dfrac{2x(x-1) - 2\left(x^2+1\right)}{(x-1)^3}$

$= \dfrac{2x^2 - 2x - 2x^2 - 2}{(x-1)^3}$

$= \dfrac{-2x-2}{(x-1)^3}$

$= -\dfrac{2(x+1)}{(x-1)^3}$

37. $\sqrt{ab^5c^2} = \sqrt{b^4c^2ab} = \sqrt{b^4} \cdot \sqrt{c^2} \cdot \sqrt{ab} = b^2c\sqrt{ab}$

41. $\sqrt{84st^3u^{-2}} = \sqrt{4t^2 \frac{1}{u^2}(21st)} = \dfrac{2t}{u}\sqrt{21st}$

45. $\sqrt{\dfrac{8}{27}} = \sqrt{\dfrac{4(2)}{9(3)}}$

$= \dfrac{2}{3}\dfrac{\sqrt{2}}{\sqrt{3}} \times \dfrac{\sqrt{3}}{\sqrt{3}}$

$= \dfrac{2\sqrt{6}}{3(3)}$

$= \dfrac{2\sqrt{6}}{9}$

49. $\sqrt[4]{\sqrt[3]{64}} = \sqrt[12]{2^6} = 2^{6/12} = 2^{1/2} = \sqrt{2}$

53. $\sqrt{63}-2\sqrt{112}-\sqrt{28}=\sqrt{9(7)}-2\sqrt{16(7)}-\sqrt{4(7)}$
$$=3\sqrt{7}-8\sqrt{7}-2\sqrt{7}$$
$$=-7\sqrt{7}$$

57. $\sqrt[3]{8a^4}+b\sqrt[3]{a}=\sqrt[3]{2^3a^3a}+b\sqrt[3]{a}$
$$=2a\sqrt[3]{a}+b\sqrt[3]{a}$$
$$=(2a+b)\sqrt[3]{a}$$

61. $2\sqrt{2}\left(\sqrt{6}-\sqrt{10}\right)=2\sqrt{12}-2\sqrt{20}$
$$=2\sqrt{4(3)}-2\sqrt{4(5)}$$
$$=4\sqrt{3}-4\sqrt{5}$$
$$=4\left(\sqrt{3}-\sqrt{5}\right)$$

65. $\left(2\sqrt{7}-3\sqrt{a}\right)\left(3\sqrt{7}+\sqrt{a}\right)=6(7)+2\sqrt{7a}-9\sqrt{7a}-3a$
$$=42-3a-7\sqrt{7a}$$

69. $\dfrac{\sqrt{3x}}{2\sqrt{3x}-\sqrt{y}}=\dfrac{\sqrt{3x}}{\left(2\sqrt{3x}-\sqrt{y}\right)}\times\dfrac{\left(2\sqrt{3x}+\sqrt{y}\right)}{\left(2\sqrt{3x}+\sqrt{y}\right)}$
$$=\frac{2(3x)+\sqrt{3x}\cdot\sqrt{y}}{4(3x)-(y)}$$
$$=\frac{6x+\sqrt{3xy}}{12x-y}$$

73. $\dfrac{\sqrt{7}-\sqrt{5}}{\sqrt{5}+3\sqrt{7}}=\dfrac{\sqrt{7}-\sqrt{5}}{\sqrt{5}+3\sqrt{7}}\times\dfrac{\sqrt{5}-3\sqrt{7}}{\sqrt{5}-3\sqrt{7}}\dfrac{\pi}{2}$
$$=\frac{\sqrt{35}-3(7)-5+3\sqrt{35}}{5-9(7)}$$
$$=\frac{4\sqrt{35}-21-5}{5-63}$$
$$=\frac{4\sqrt{35}-26}{-58}$$
$$=\frac{2(2\sqrt{35}-13)}{-58}$$
$$=\frac{13-2\sqrt{35}}{29}$$

77. $\sqrt{4b^2+1}$ is already in its simplest form

81. $\left(1+6^{1/2}\right)\left(3^{1/2}+2^{1/2}\right)\left(3^{1/2}-2^{1/2}\right)=\left(1+6^{1/2}\right)\left(\left(3^{1/2}\right)^2-\left(2^{1/2}\right)^2\right)$

$$=\left(1+6^{1/2}\right)(3-2)$$

$$=1+\sqrt{6}$$

85. $\left(\sqrt{7}-2\sqrt{15}\right)\left(3\sqrt{7}-\sqrt{15}\right)=3(7)-\sqrt{105}-6\sqrt{105}+2(15)$

$$=21-7\sqrt{105}+30$$

$$=51-7\sqrt{105}$$

89.

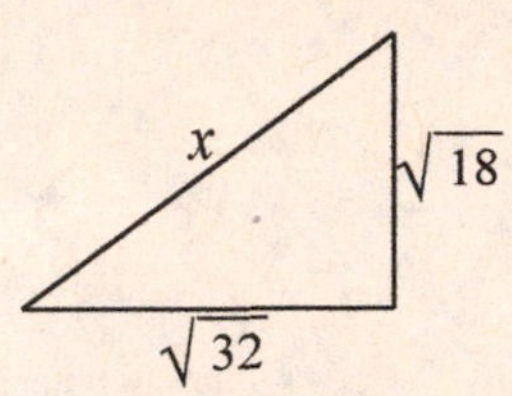

$x^2=\sqrt{18}^2+\sqrt{32}^2$

$x^2=18+32$

$x^2=50$

$x=\sqrt{50}$

perimeter $p=\sqrt{18}+\sqrt{32}+\sqrt{50}$

$p=\sqrt{(9)2}+\sqrt{(16)2}+\sqrt{(25)2}$

$p=3\sqrt{2}+4\sqrt{2}+5\sqrt{2}$

$p=12\sqrt{2}$ units

93. 2010 to 2015 inclusive is a 6-year period

$i=100\quad \frac{C_2}{C_1}^{1/n}\quad -1$

$i=100\quad \frac{237.0}{218.8}^{1/6}\quad -1$

$i=1.3406078\ \%$

$i=1.341\ \%$ (rounded)

97. Let $x =$ the length of the original square side

Let $x+\sqrt{x} =$ the length of the lenghtened square side

Let $A =$ the area of the larger square

Let $A_O = x^2$ be the original area

$$A=\left(x+\sqrt{x}\right)^2 = x^2+2x\sqrt{x}+x$$

$$A_{\text{increase}} = A - A_O$$

$$A_{\text{increase}} = x^2+2x\sqrt{x}+x-x^2$$

$$A_{\text{increase}} = 2x\sqrt{x}+x$$

$$A_{\text{increase}} = x\left(2\sqrt{x}+1\right)$$

$$\%\text{ increase} = \frac{A_{\text{increase}}}{A_O}\times 100\ \%$$

$$\%\text{ increase} = \frac{x\left(2\sqrt{x}+1\right)}{x^2}\times 100\ \%$$

$$\%\text{ increase} = \frac{\left(2\sqrt{x}+1\right)}{x}\times 100\ \%$$

101. $$\frac{\sqrt{A+h}-\sqrt{A}}{h} = \frac{\sqrt{A+h}-\sqrt{A}}{h}\times\frac{\sqrt{A+h}+\sqrt{A}}{\sqrt{A+h}+\sqrt{A}}$$

$$=\frac{(A+h)-(A)}{h\left(\sqrt{A+h}+\sqrt{A}\right)}$$

$$=\frac{h}{h\left(\sqrt{A+h}+\sqrt{A}\right)}$$

$$=\frac{1}{\sqrt{A+h}+\sqrt{A}}$$

105. $$f=\frac{1}{2\pi\sqrt{\frac{LC_1C_2}{C_1+C_2}}}$$

$$f=\frac{1}{2\pi}\times\sqrt{\frac{C_1+C_2}{LC_1C_2}}\cdot\frac{\sqrt{LC_1C_2}}{\sqrt{LC_1C_2}}$$

$$f=\frac{\sqrt{LC_1C_2(C_1+C_2)}}{2\pi LC_1C_2}$$

Chapter 12

Complex Numbers

12.1 Basic Definitions

1. $$-\sqrt{-3}\sqrt{-12} = -\sqrt{-1(3)}\sqrt{-1(12)} = -\left(j\sqrt{3}\right)\left(j\sqrt{12}\right) = -j^2\sqrt{36} = -(-1)6 = 6$$

5. $$\sqrt{-81} = \sqrt{81(-1)} = \sqrt{81}\sqrt{-1} = 9j$$

9. $$\sqrt{-0.36} = \sqrt{0.36(-1)} = \sqrt{0.36}\sqrt{-1} = 0.6j$$

13. $$\sqrt{-\frac{7}{4}} = \frac{\sqrt{-7}}{\sqrt{4}} = \frac{\sqrt{7(-1)}}{2} = \frac{\sqrt{7}\sqrt{-1}}{2} = j\frac{\sqrt{7}}{2}$$

17. **(a)** $$\left(\sqrt{-7}\right)^2 = \left(\sqrt{7(-1)}\right)^2 = \left(\sqrt{7}\cdot\sqrt{-1}\right)^2 = \left(\sqrt{7}\cdot j\right)^2 = \sqrt{7}^2\cdot j^2 = 7(-1) = -7$$

(b) $$\sqrt{(-7)^2} = \sqrt{49} = 7$$

21. $$\sqrt{-\frac{1}{15}}\sqrt{-\frac{27}{5}} = j\sqrt{\frac{1}{15}}j\sqrt{\frac{27}{5}} = j^2\sqrt{\frac{27}{75}} = -\sqrt{\frac{9(3)}{25(3)}} = -\frac{3}{5}$$

25. **(a)** $$-j^6 = -\left(j^2\right)^3 = -(-1)^3 = -(-1) = 1$$

(b) $$(-j)^6 = \left((-j)^2\right)^3 = (-1)^3 = -1$$

29. $$j^{15} - j^{13} = j^{12}\cdot j^3 - j^{12}\cdot j = (1)(-j) - (1)(j) = -j - j = -2j$$

33. $$2 + \sqrt{-9} = 2 + \sqrt{9(-1)} = 2 + 3j$$

37. $$\sqrt{-4j^2} + \sqrt{-4} = \sqrt{(-4)(-1)} + 2j = 2 + 2j$$

41. $$\sqrt{18} - \sqrt{-8} = \sqrt{9\cdot 2} - \sqrt{4\cdot 2}j = 3\sqrt{2} - 2j\sqrt{2}$$

45. $$5j(-3j)(j^2) = -15j^4 = -15(1) = -15$$

49. $$\frac{\sqrt{-9} - 6}{3} = \frac{3j - 6}{3} = -2 + j$$

53. **(a)** $2j$ has $-2j$ as conjugate.

(b) -4 has -4 as conjugate.

57. Rewriting each side in the form $a + bj$ and equating real and imaginary parts:

$$6j - 7 = 3 - x - yj$$
$$6j - 7 - 3 = -x - yj$$
$$6j - 10 = -x - yj$$
$$-x = -10 \quad \text{and} \quad -y = 6$$
$$x = 10 \qquad\qquad y = -6$$

61. $$x^2 + 32 = 0$$
$$x^2 = -32$$
$$x = \pm\sqrt{-32}$$
$$x = \pm\sqrt{(16)(-2)}$$
$$x = \pm 4j\sqrt{2}$$

65. **(a)** Since $j^4 = 1$, any number multiplied by j^4 is unchanged.

(b) Since $j^2 = -1$, any number multiplied by j^2 will change sign.

69. $$j + j^2 + j^3 + j^4 + j^5 + j^6 + j^7 + j^8 = j + (-1) + (-j) + 1 + j + (-1) + (-j) + 1 = j - 1 - j + 1 + j - 1 - j + 1 = 0$$

73. $j(x+yj)=jx+yj^2=-y+xj$, thus the imaginary part of $j(x+yj)$ is x which is the real part of $x+yj$.

12.2 Basic Operations with Complex Numbers

1. $$(7-9j)-(6-4j)=7-9j-6+4j$$
$$=1-5j$$

5. $$(3-7j)+(2-j)=(3+2)+(-7-1)j=5-8j$$

9. $$0.23-(0.46-0.19j)+0.67j$$
$$=0.23-0.46+0.19j+0.67j$$
$$=-0.23+0.86j$$

13. $$(7-j)(7j)=49j-7j^2$$
$$=49j-7(-1)$$
$$=7+49j$$

17. $$\left(\sqrt{-18}\sqrt{-4}\right)(3j)=\left(j\sqrt{(9)(2)}\right)\left(j\sqrt{4}\right)(3j)$$
$$=\left(3j\sqrt{2}\right)(2j)(3j)$$
$$=18j^3\sqrt{2}$$
$$=18(-j)\sqrt{2}$$
$$=-18j\sqrt{2}$$

21. $$j\sqrt{-7}-j^6\sqrt{112}+3j=j\sqrt{7(-1)}-j^6\sqrt{16(7)}+3j$$
$$=j^2\sqrt{7}-4j^6\sqrt{7}+3j$$
$$=(-1)\sqrt{7}-4j^4\left(j^2\right)\sqrt{7}+3j$$
$$=-\sqrt{7}-4(1)(-1)\sqrt{7}+3j$$
$$=-\sqrt{7}+4\sqrt{7}+3j$$
$$=3\sqrt{7}+3j$$

25. $$(8+3j)(8-3j)=8^2-(3j)^2$$
$$=64-(-9)$$
$$=73$$

29. $$\frac{1-j}{3j}=\frac{1-j}{3j}\cdot\frac{-j}{-j}$$
$$=\frac{-j+j^2}{-3j^2}$$
$$=\frac{-j-1}{3}$$
$$=-\frac{1+j}{3}$$

33. $$\frac{j^2-j}{2j-j^8}=\frac{-1-j}{2j-1}\cdot\frac{-2j-1}{-2j-1}$$
$$=\frac{1+3j+2j^2}{1^2+2^2}$$
$$=\frac{1+3j-2}{5}$$
$$=\frac{-1+3j}{5}$$

37. We simplify before squaring:

$$\left(4j^5-5j^4+2j^3-3j^2\right)^2=(4j-5-2j+3)^2$$
$$=(-2+2j)^2$$
$$=4-8j-4$$
$$=-8j$$

41. $$\frac{\left(2-j^3\right)^4}{\left(j^8-j^6\right)}+j=\frac{(2+j)^4}{(1+1)^3}+j$$
$$=\frac{(2+j)^4}{8}+j$$
$$=\frac{(2+j)^2(2+j)^2}{8}+j$$
$$=\frac{(3+4j)(3+4j)}{8}+j$$
$$=\frac{-7+24j}{8}+j$$
$$=-\frac{7}{8}+4j$$

45. Using the quadratic formula, the roots of $x^2-4x+13$ are

$$x=\frac{-4\pm\sqrt{16-52}}{2}=\frac{-4\pm j\sqrt{36}}{2}=-2\pm3j$$

The sum of these roots is

$(-2+3j)+(-2-3j)=-4.$

49. $\frac{1}{3-j}=\frac{1}{3-j}\cdot\frac{3+j}{3+j}$

$=\frac{3+j}{9+1}$

$=\frac{1}{10}(3+j)$

$=\frac{3}{10}+\frac{1}{10}j$

53. $(x+3j)^2=7-24j$

$x^2+6xj-9=7-24j$

$x^2+6xj=16-24j$ requires

$x^2=16$ and $6x=-24$

$x=\pm 4$ $\quad x=-4$

The solution is $x=-4$.

57. $E=I\cdot Z$

$=(0.835-0.427j)(250+170j)$

$=208.75+141.95j-106.75j-72.59j^2$

$=208.75+35.2j+72.59$

$=281+35.2j$ V

61. **(a)** $(a+bj)+(a-bj)=a+bj+a-bj=2a$, a real number.

(b) $(a+bj)-(a-bj)=2bj$, pure imaginary

12.3 Graphical Representation of Complex Numbers

1. Add $5-2j$ and $-2+j$ graphically.

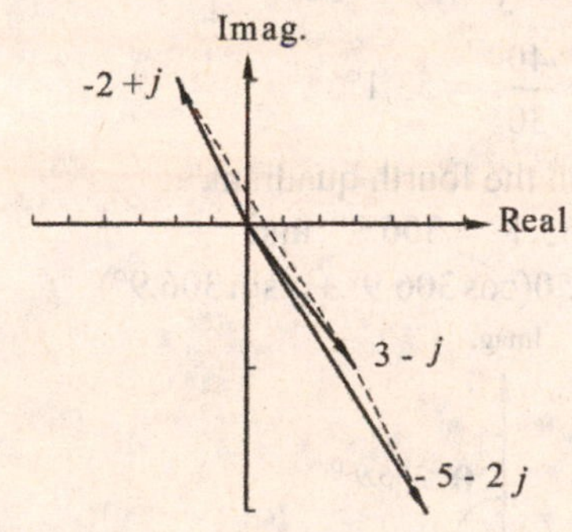

$5-2j+(-2+j)=3-j$

5. $-4-3j$

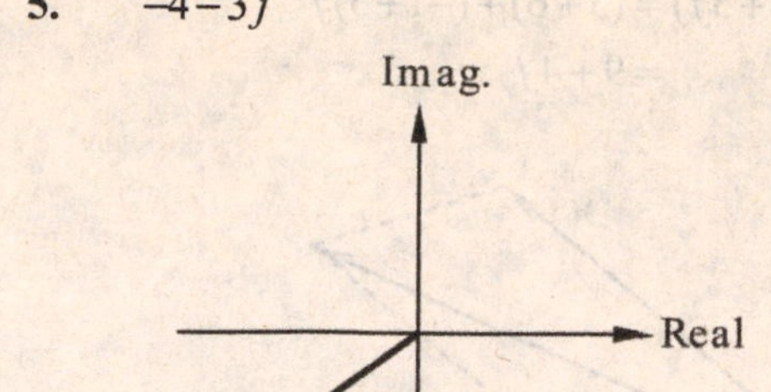

9. $2+3+4j=5+4j$

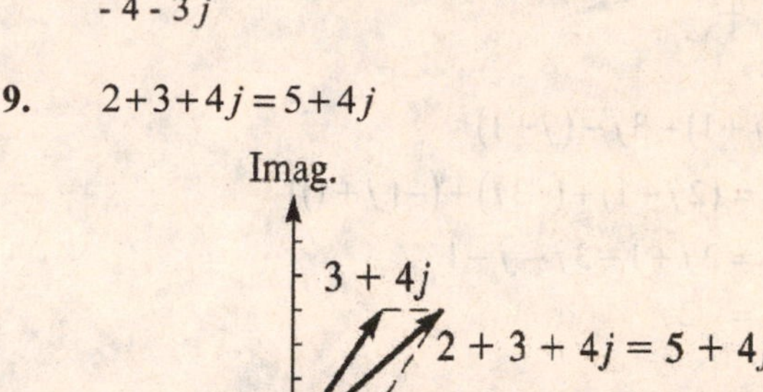

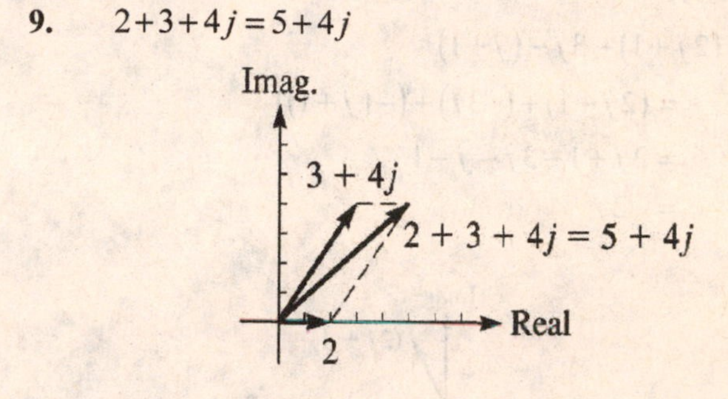

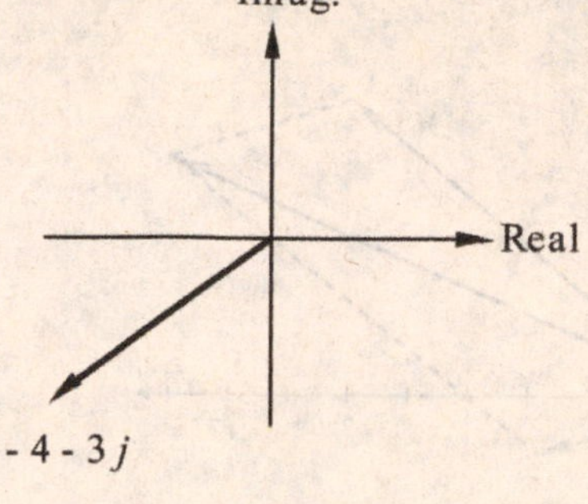

13. $5j-1(1-4j)=5j-1+4j=-1+9j$

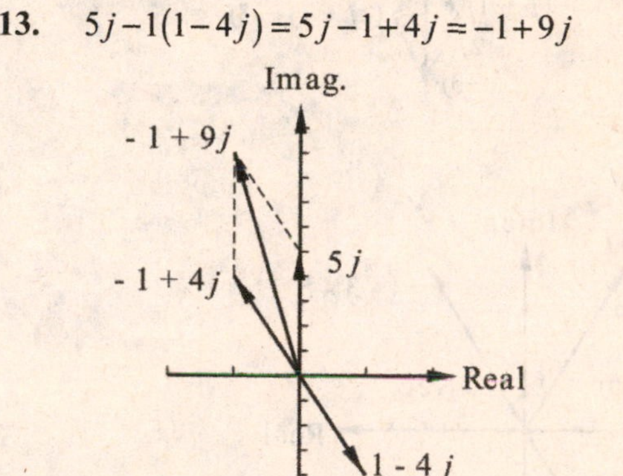

17. $(3-2j)-(4-6j)=3-2j-4+6j=-1+4j$

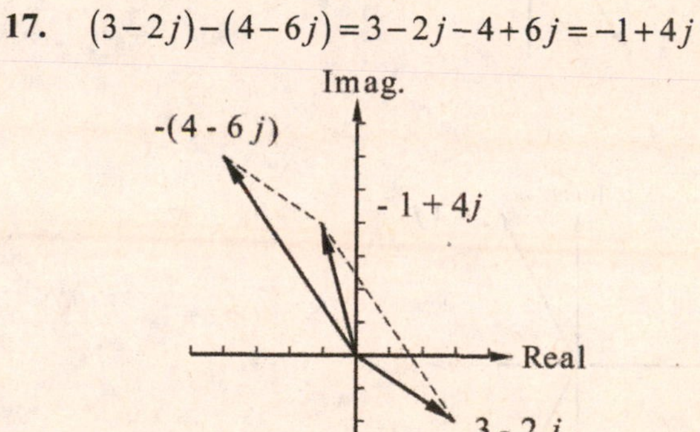

21. 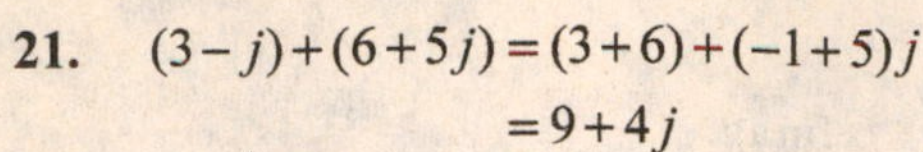

$$(3-j)+(6+5j)=(3+6)+(-1+5)j$$
$$=9+4j$$

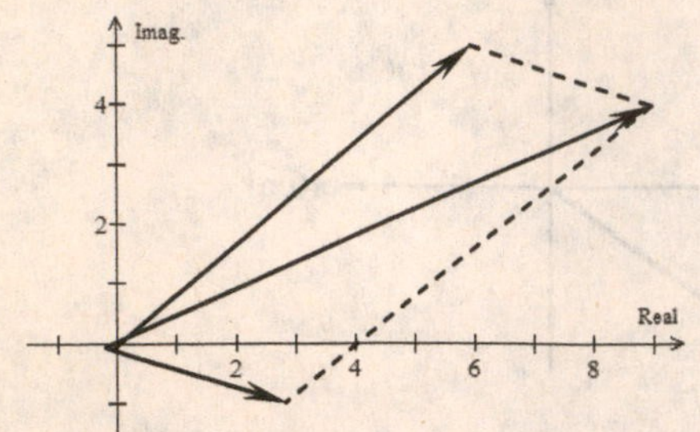

25. $(2j+1)-3j-(j+1)$

$$=(2j+1)+(-3j)+(-(j+1))$$
$$=2j+1-3j-j-1$$
$$=-2j$$

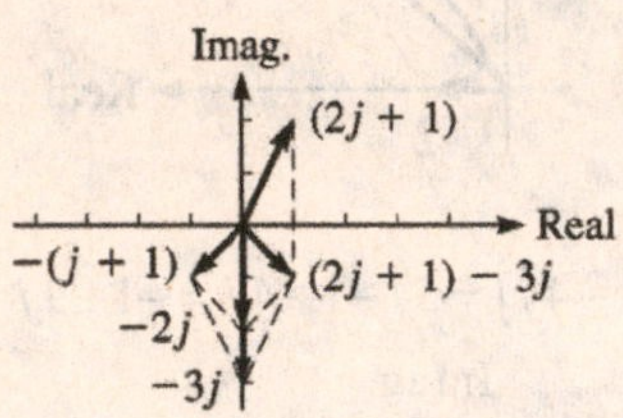

29. $-3-5j$

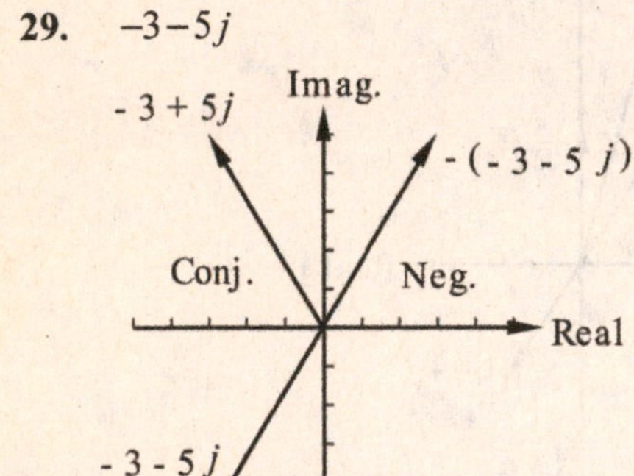

33.

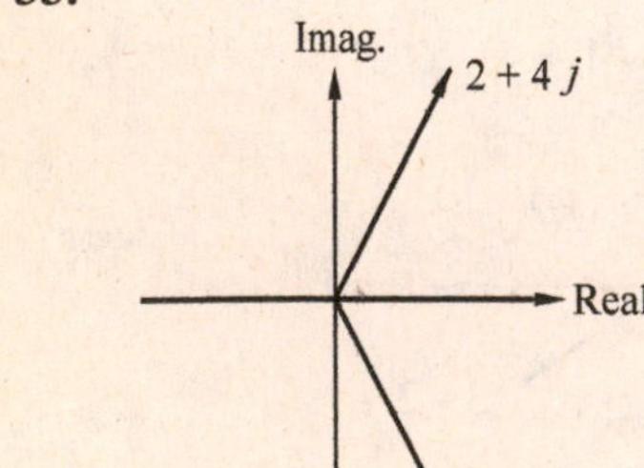

The graph of a complex number and its conjugate are symmetrical with respect to the real axis.

37. $40+10j+50-25j=90-15j$ N

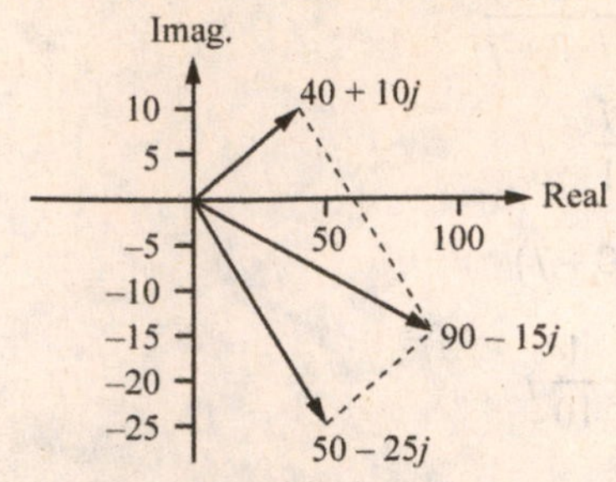

12.4 Polar Form of a Complex Number (GRAPHS)

1. From the rectangular form we have $x=-3$ and $y=4$, so

$$r=\sqrt{(-3)^2+4^2}=5$$

$$\theta_{\text{ref}}=\tan^{-1}\frac{4}{3}=53.1°.$$

Since θ is in the second quadrant, $\theta=180°-53.1°=126.9°$. Therefore, $-3+4j=5(\cos 126.9°+j\sin 126.9°)$

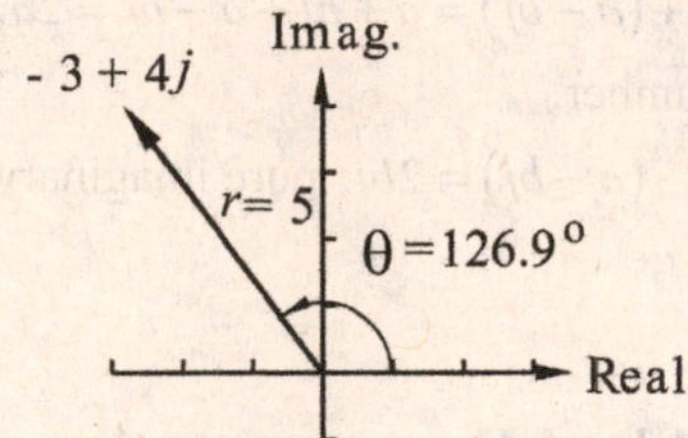

5. From the rectangular form we have $x=30$ and $y=-40$, so

$$r=\sqrt{(30)^2+(-40)^2}=50$$

$$\theta_{\text{ref}}=\tan^{-1}\frac{40}{30}=53.1°.$$

Since θ is in the fourth quadrant, $\theta=360°-53.1°=306.9°$ and $30-40j=50(\cos 306.9°+j\sin 306.9°)$

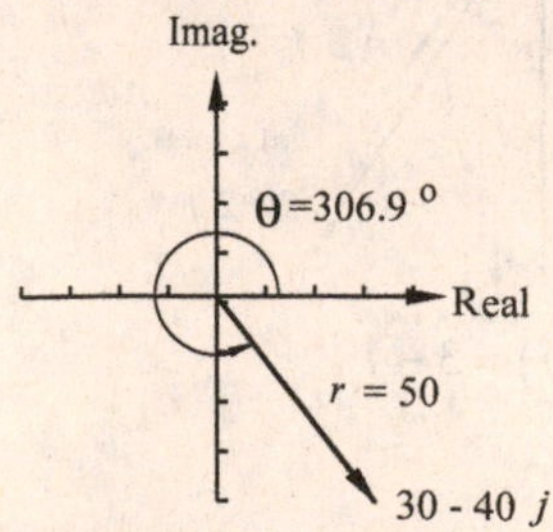

9. From the rectangular form we have $x = -0.55$ and $y = -0.24$, so

$$r = \sqrt{(-0.55)^2 + (-0.24)^2} = 0.60$$

$$\theta_{\text{ref}} = \tan^{-1} \frac{0.24}{0.55} = 24°.$$

Since θ is in the third quadrant,
$\theta = 180° + 24° = 204°$ and
$-0.55 - 0.24j = 0.60(\cos 204° + j \sin 204°)$

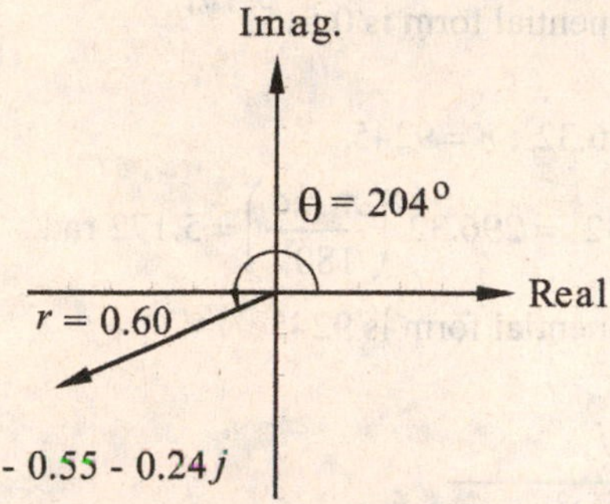

13. From the rectangular form we have $x = 3.514$ and $y = -7.256$, so

$$r = \sqrt{(3.514)^2 + (-7.256)^2} = 8.062$$

$$\theta_{\text{ref}} = \tan^{-1}\left(\frac{7.256}{3.514}\right) = 64.16°.$$

Since θ is in the fourth quadrant,
$\theta = 360° - 64.16° = 295.84°$ and
$3.514 - 7.256j = 8.062(\cos 295.84° + j \sin 295.84°)$

Imag.

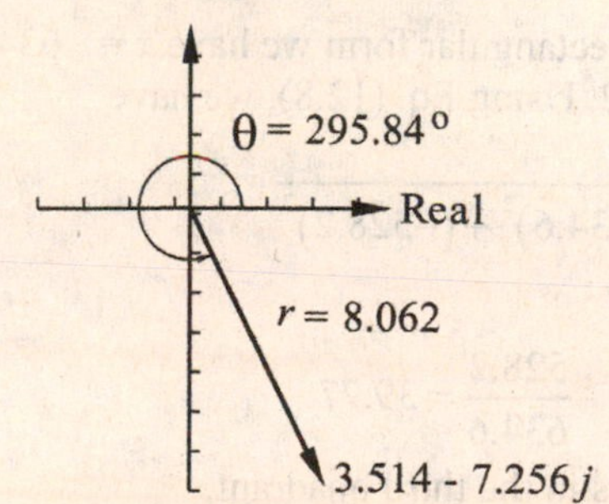

17. From the rectangular form we have $x = 0$ and $y = 9$, so

$$r = \sqrt{0^2 + 9^2} = 9$$

$\theta = 90°$

since y is positive, so $9j = 9(\cos 90° + j \sin 90°)$

Imag.

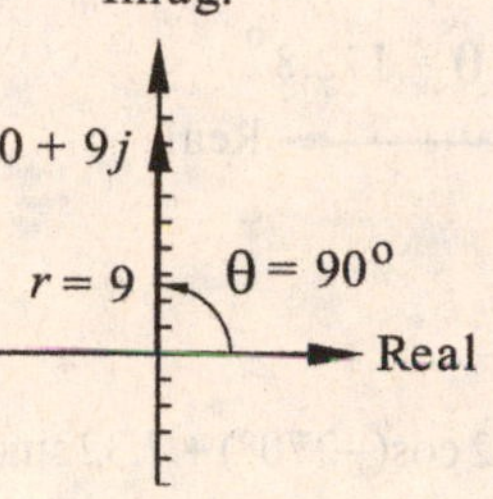

21. $160(\cos 150.0° + j \sin 150.0°)$

$$= 160 \cos 150.0° + 160 \sin 150.0° \cdot j$$
$$= -140 + 80j$$

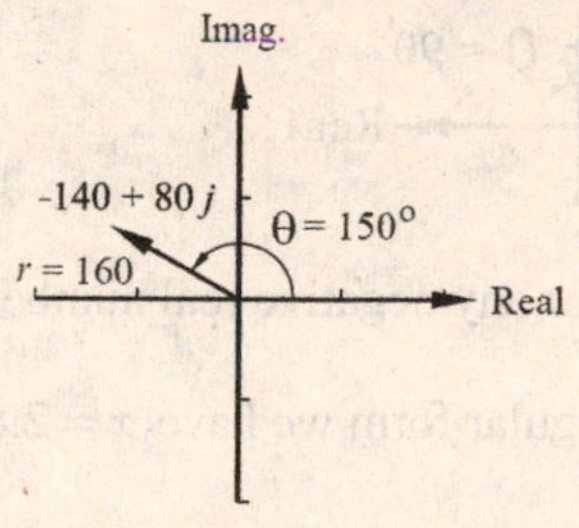

25. $0.08(\cos 360° + j \sin 360°)$

$$= 0.08 \cos 360° + 0.08 \sin 360° \cdot j$$
$$= 0.08(1) + 0.08(0)j$$
$$= 0.08 + 0j$$
$$= 0.08$$

Imag.

θ = 360° 0.08 + 0j

Real

r = 0.08

29. $4.75\underline{/172.8°} = 4.75\cos 172.8° + 4.75\sin 172.8° \cdot j$

$= -4.71 + 0.595j$

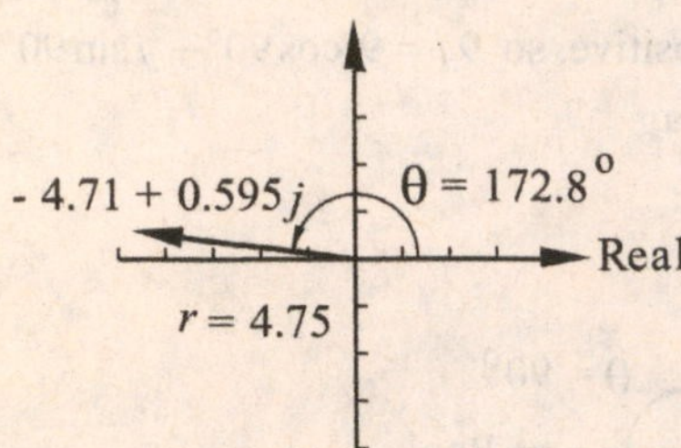

33. $7.32\underline{/-270°} = 7.32\cos(-270°) + 7.32\sin(-270°) \cdot j$

$= 7.32(0) + 7.32(1)j$

$= 7.32j$

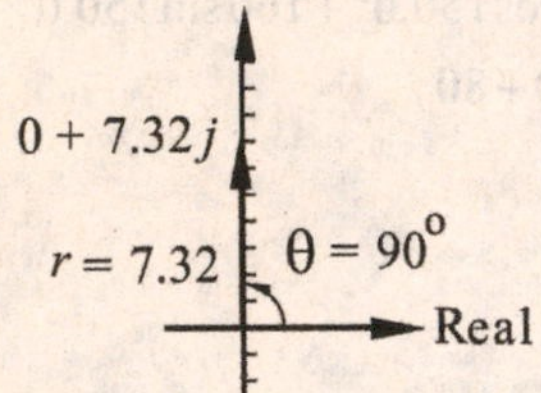

37. The argument for any negative real number is 180°.

41. From the rectangular form we have $x = 2.84$ and $y = -1.04$, so

$$r = \sqrt{2.84^2 + (-1.06)^2} = 3.03$$

$$\theta_{ref} = \tan^{-1}\frac{1.06}{2.84} = 20.5°.$$

Since θ is in the fourth quadrant, $\theta = 360° - 20.5° = 339.5°$. Therefore,

$$2.84 - 1.06j \text{ kV} = 3.03(\cos 339.5° + j\sin 339.5°) \text{ kV}$$

$$= 3.03\underline{/339.5°} \text{ kV}$$

12.5 Exponential Form of a Complex Number

1. $8.50\angle 226.3^\circ$, $r = 8.50$,

$$\theta = 226.3^\circ\left(\frac{\pi \text{ rad}}{180^\circ}\right) = 3.950$$

The exponential form is

$8.50\angle 226.3^\circ = 8.50e^{3.95j}$

5. $0.450\left(\cos 282.3^\circ + j \sin 282.3^\circ\right)$; $r = 0.450$

$$\theta = 282.3^\circ = 282.3^\circ\left(\frac{\pi \text{ rad}}{180^\circ}\right) = 4.93 \text{ rad}$$

The exponential form is $0.450e^{4.93j}$

9. $0.515\angle 198.3^\circ$; $r = 0.515$;

$$\theta = 198.3^\circ = 198.3^\circ\left(\frac{\pi \text{ rad}}{180^\circ}\right) = 3.461 \text{ rad}$$

The exponential form is $0.515e^{3.461j}$

13. $9245\angle 296.32^\circ$; $r = 9245$;

$$\theta = 296.32^\circ = 296.32^\circ\left(\frac{\pi \text{ rad}}{180^\circ}\right) = 5.172 \text{ rad}$$

The exponential form is $9245e^{5.172j}$

17. $-30 + 20j$

$$r = \sqrt{(-30)^2 + 20^2} = 36.1$$

$$\theta_{ref} = \tan^{-1}\frac{20}{30} = 33.7^\circ$$

Since θ is in the second quadrant,

$$\theta = 180^\circ - 33.7^\circ$$

$$= 146.3^\circ \cdot \frac{\pi}{180^\circ}$$

$$= 2.55 \text{ rad}$$

$$-30 + 20j = 36.1e^{2.55j}$$

21. From the rectangular form we have $x = -634.6$ and $y = -528.2$. Using Eq. (12.8), we have

$$r = \sqrt{(-634.6)^2 + (-528.2)^2}$$

$$= 825.7$$

$$\theta_{ref} = \tan^{-1}\frac{528.2}{634.6} = 39.77^\circ$$

Since θ is in the third quadrant,

$$\theta = 180^\circ + 39.77^\circ$$

$$= 219.77^\circ\left(\frac{\pi \text{ rad}}{180^\circ}\right) = 3.836 \text{ rad}$$

The exponential form is $825.7e^{3.836j}$

25. $464e^{1.85j}$; $r = 464$

$$\theta = 1.85 \text{ rad} = \frac{1.85(180^\circ)}{\pi} = 106.0^\circ$$

The polar form is

$464(\cos 106.0^\circ + j \sin 106.0^\circ)$

$x = 464 \cos 106.0^\circ = 464(-0.276) = -128$

$y = 464 \sin 106.0^\circ = 464(0.961) = 446$

The rectangular form is

$-128 + 446j$

29. $1724e^{2.391j}$; $r = 1724$

$$\theta = 2.391 \text{ rad} = \frac{2.391(180^\circ)}{\pi} = 137.0^\circ$$

The polar form is

$1724(\cos 137.0^\circ + j \sin 137.0^\circ)$

$x = 1724 \cos 137.0^\circ = 1724(-0.7312)$

$= -1261$

$y = 1724 \sin 137.0^\circ = 1724(0.6821) = 1176$

The rectangular form is

$-1261 + 1176j$

33. $(6.25e^{3.46j})(4.40e^{1.22j})$

$= 27.50e^{4.68j}$ exponential

$= 27.50\left(\cos\left(4.68 \cdot \frac{180^\circ}{\pi}\right) + j \sin\left(4.68 \cdot \frac{180^\circ}{\pi}\right)\right)$

$= 27.50 \ (\cos 268.1^\circ + j \sin 268.1^\circ)$ polar

$= 0.912 - 27.5j$

37. From the rectangular form we have $x = 3.75$ and $y = 1.10$, so

$r = \sqrt{3.75^2 + 1.10^2} = 3.91$

$$\theta = \tan^{-1} \frac{1.10}{3.75}$$

$$= 16.3^\circ \ \frac{\pi \text{ rad}}{180^\circ}$$

$= 0.285$

$3.75 + 1.10j = 3.91e^{0.285j}$

The magnitude of the impedance is 3.91 ohms.

12.6 Products, Quotients, Powers, and Roots of Complex Numbers

1. $(3.61\angle -56.3°)(1.41\angle 315.0°) = (3.61)(1.41)\angle(-56.3° + 315.0°)$
$= 5.09\angle 258.7°$

5. $[4(\cos 60° + j\sin 60°)][2(\cos 20° + j\sin 20°)] = 4\cdot 2[(\cos(60° + 20°) + j\sin(60° + 20°)]$
$= 8(\cos 80° + j\sin 80°)$

9. $\dfrac{8(\cos 100° + j\sin 100°)}{4(\cos 65° + j\sin 65°)} = \dfrac{8}{4}(\cos(100° - 65°) + j\sin(100° - 65°))$
$= 2(\cos 35° + j\sin 35°)$

13. $\left[0.2\left(\cos 35° + j\sin 35°\right)\right]^3 = 0.2^3\left(\cos 3\cdot 35° + j\sin 3\cdot 35°\right)$
$= 0.008\left(\cos 105° + j\sin 105°\right)$

17. $\dfrac{\left(50\angle 236°\right)\left(2\angle 84°\right)}{125\angle 47°} = \dfrac{100\angle 320°}{125\angle 47°}$
$= \dfrac{4}{5}\angle 273°$
$= \dfrac{4}{5}\left(\cos 273° + j\sin 273°\right)$

21. $2.78\underline{/56.8°} + 1.37\underline{/207.3°}$
$= 2.78(\cos 56.8° + j\sin 56.8°) + 1.37(\cos 207.3° + j\sin 207.3°)$
$= 1.52 + 2.33j - 1.22 - 0.628j$
$= 0.30 + 1.702j$

In polar form:

$r = \sqrt{0.30^2 + 1.703^2} = 1.73$

$\theta = \tan^{-1}\left(\dfrac{1.703}{0.30}\right)$
$= 80.0°$

$2.78\underline{/56.8°} + 1.37\underline{/207.3°} = 1.72\underline{/80.0°}$

25. $3 + 4j$:
$r_1 = \sqrt{3^2 + 4^2}$
$= 5$
$\theta_{ref} = \tan^{-1}\dfrac{4}{3}$
$= 53.1°$
$\theta_1 = 53.1°$

$5 - 12j$:
$r_2 = \sqrt{5^2 + (-12)^2}$
$= 13$
$\theta_{ref} = \tan^{-1}\dfrac{12}{5}$
$= 67.4°$
$\theta_2 = 360° - 67.4°$
$= 292.6$

In polar form:

$(3 + 4j)(5 - 12j) = (5\underline{/53.1°})(13\underline{/292.6°})$
$= 65\underline{/345.7°}$
$= 63.0 - 16.1j$

In rectangular form:

$$(3+4j)(5-12j) = 15-36j+20j-48j^2$$
$$= 15-16j+48$$
$$= 63-16j$$

29. 21: $\quad$ $3-9j$:

$r_1 = \sqrt{21^2+0^2}$ $\quad$ $r_2 = \sqrt{3^2+(-9)^2}$

$= 21$ $\quad$ $= 9.48$

$\theta_1 = 0°$ $\quad$ $\theta_{\text{ref}} = \tan^{-1}\left(\frac{9}{3}\right)$

$= 71.6°$

$\theta_2 = 360° - 71.6°$

$288.4°$

In polar form:

$$\frac{21}{3-9j} = \frac{21\underline{/0°}}{9.48\underline{/288.4°}}$$
$$= 2.22\underline{/-288.4°}$$
$$= 0.70+2.1j$$

In rectangular form:

$$\frac{21}{3-9j}\cdot\frac{3+9j}{3+9j} = \frac{63+189j}{9+81}$$
$$= 0.7+2.1j$$

33. $r = \sqrt{3^2+4^2} = 5$

$$\theta = \tan^{-1}\ \frac{4}{3}$$
$$= 53.1°$$

$$(3+4j)^4 = 5^4(\cos(4(53.1)) + j\sin(4(53.1))$$
$$= 625(\cos 212.5 + j\sin 212.5)$$
$$= -527-336j$$

In rectangular form:

$$(3+4j)^4 = [(3+4j)^2]^2$$
$$= (9+24j+16j^2)^2$$
$$= (-7+24j)^2$$
$$= 49-336j+576j^2$$
$$= 49-336j-336j$$
$$= -527-336j$$

37. $r = \sqrt{3^2 + (-4)^2} = 5$

$\theta_{\text{ref}} = \tan^{-1} \frac{4}{3}$

$= 53.1°.$

$\theta = 360 - 53.1$

$= 306.9°$

So the three cube roots of $3 - 4j$ are

$r_1 = 5^{1/3} \cos \frac{306.9}{3} + j \sin \frac{306.9}{3}$

$= 1.71(\cos 102.3° + j \sin 102.3°)$

$= -0.364 + 1.67j$

$r_2 = 5^{1/3} \cos \frac{306.9 \quad 360}{3} + j \sin \frac{306.9 \quad 360}{3}$

$= 1.71(\cos 222.3° + j \sin 222.3°)$

$= -1.26 - 1.15j$

$r_3 = 5^{1/3} \cos \frac{306.9° + 2(360)}{3} + j \sin \frac{306.9 \quad 2(360)}{3}$

$= 1.71(\cos 342.3° + j \sin 342.3°)$

$= 1.63 - 0.52j$

41. $x^4 - 1 = 0$

$x^4 = 1$

$x = \sqrt[4]{1}$

To find the four fourth roots of 1:

$r = \sqrt{1^2 + 0^2} = 1$

$\theta = 0°$

So the four fourth roots of 1 are

$r_1 = 1^{1/4} \cos \frac{0°}{4} + j \sin \frac{0°}{4}$

$= 1(\cos 0° + j \sin 0°)$

$= 1 + 0j$

$r_2 = 1^{1/4} \cos \frac{0° + 360°}{4} + j \sin \frac{0° + 360°}{4}$

$= 1(\cos 90° + j \sin 90°)$

$= j$

$$r_3 = 1^{1/4} \cos \frac{0° + 2(360)°}{4} + j\sin \frac{0° + 2(360)°}{4}$$
$$= 1(\cos 180° + j\sin 180°)$$
$$= -1 + 0j$$

$$r_4 = 1^{1/4} \cos \frac{0° + 3(360)°}{4} + j\sin \frac{0° + 3(360)°}{4}$$
$$= 1(\cos 270° + j\sin 270°)$$
$$= -j$$

45. $x^5 + 32 = 0$

$$x^5 = -32$$
$$x = \sqrt[5]{-32}$$

To find the five fifth roots of -32:

$r = \sqrt{(-32)^2 + 0^2} = 32$

$\theta = 180°$

So the five fifth roots of -32 are

$$r_1 = 32^{1/5} \cos \frac{180}{5} + j\sin \frac{180}{5}$$
$$= 2(\cos 36° + j\sin 36°)$$
$$= 1.62 + 1.18j$$

$$r_2 = 32^{1/5} \cos \frac{180 \quad 360}{5} + j\sin \frac{180 \quad 360}{5}$$
$$= 2(\cos 108° + j\sin 108°)$$
$$= -0.618 + 1.90j$$

$$r_3 = 32^{1/5} \cos \frac{180° + 2(360)°}{5} + j\sin \frac{180° + 2(360)}{5}$$
$$= 2(\cos 180° + j\sin 180°)$$
$$= -2 + 0j$$

$$r_4 = 32^{1/5} \cos \frac{180 \quad 3(360)}{5} + j\sin \frac{180 \quad 3(360)}{5}$$
$$= 2(\cos 252° + j\sin 252°)$$
$$= -0.618 - 1.90j$$

$$r_5 = 32^{1/5} \cos \frac{180 \quad 4(360)}{5} + j\sin \frac{180 \quad 4(360)}{5}$$
$$= 2(\cos 324° + j\sin 324°)$$
$$= 1.62 - 1.18j$$

49. $$\left(\frac{1}{2}-\frac{\sqrt{3}}{2}j\right)^3 = \left(\frac{1}{2}(1-j\sqrt{3})\right)^3$$
$$=\frac{1}{8}(1-j\sqrt{3})^2(1-j\sqrt{3})$$
$$=\frac{1}{8}(1-2j\sqrt{3}-3)(1-j\sqrt{3})$$
$$=\frac{1}{8}(-2-2j\sqrt{3})(1-j\sqrt{3})$$
$$=\frac{1}{8}(-2-2j\sqrt{3}+2j\sqrt{3}+2\sqrt{3}^2j^2)$$
$$=\frac{1}{8}(-2-6)$$
$$=-1$$

53. $$p=e\cdot i$$
$$=(6.80\underline{/56.3°})(0.0705\underline{/-15.8°})$$
$$=(6.80)(0.0705)\underline{/(56.3-15.8)°}$$
$$=0.479\underline{/40.5°}\text{ W}$$

12.7 An Application to Alternating-current (ac) Circuits

1. $$V_R=IR=2.00(12.0)=24.0\text{ V}$$
$$Z=R+j(X_L-X_C)=12.0+j(16.0)$$
$$|Z|=\sqrt{12.0^2+16.0^2}=20.0\Omega$$
$$V_L=IX_L=2.00(16.0)=32.0\text{ V}$$
$$V_{RL}=IZ=2.00(20.0)=40.0\text{ V}$$
$\theta=\tan^{-1}\frac{X_L}{R}=\tan^{-1}\frac{16.0}{12.0}=53.1°$, voltage leads current.

5. **(a)** $|Z|=\sqrt{R^2+X_L^2}=\sqrt{2250^2+1750^2}=2850\Omega$

(b) $\tan\theta=\frac{1750}{2250};\ \theta=37.9°$

(c) $V_{RLC}=IZ=(0.00575)(2850)=16.4\text{ V}$

9. (a) $X_R=45.0\Omega$
$$X_L=2\pi fL=2\pi(60)(0.0429)=16.2\Omega$$
$$Z=45.0+16.2j$$
$$|Z|=\sqrt{45.0^2+16.2^2}=47.8\Omega$$
(b) $\tan\theta=\frac{16.2}{45.0};\ \theta=19.8°$

13. $$R=25.3\Omega$$
$$X_C=1/(2\pi fC)$$
$$=1/\left(2\pi\left(1.2\times10^6\right)\left(2.75\times10^{-9}\right)\right)=48.2\Omega$$
$$=f=1200\text{ kHz}=1.2\times10^6\text{ Hz}$$
$$Z=R-X_{Cj}=25.3-48.2j$$
$$|Z|=\sqrt{25.3^2+(-48.2)^2}=54.4\Omega$$
$$\tan\theta=\frac{-48.2}{25.3};\ \theta=-62.3°$$

17. $$L=12.5\times10^{-6}\text{ H}$$
$$C=47.0\times10^{-9}\text{ F}$$
$$X_L=X_C$$
$$2\pi fL=\frac{1}{2\pi fC}$$
$$f=\sqrt{\frac{1}{4\pi^2LC}}$$
$$=\sqrt{\frac{1}{4\pi^2\left(12.5\times10^{-6}\right)\left(4.70\times10^{-9}\right)}}$$
$$=208\text{ kHz}$$

21. $$P=VI\cos\theta$$
$$V=225\text{ mV}$$
$$\theta=-18.0°=342°$$
$$Z=47.3\Omega$$
$$V=IZ$$
$$I=\frac{225\times10^{-3}}{47.3}=0.00476\text{ A}$$
$$P=\left(225\times10^{-3}\right)(0.00476)\cos 342°$$
$$=0.00102\text{ W}=1.02\text{ mW}$$

Review Exercises

1. This is false: $\left(\sqrt{-9}\right)^2 = (3j)^2 = 9j^2 = -9.$

This is true:

$(2\angle 120°)^3$

5.
$$= 2^3\angle(3\cdot 120°)$$
$$= 8\angle 360°$$
$$= 8.$$

9. $(18-3j)-(12-5j) = 18-3j-12+5j = 6+2j$

13. $(6-3j)(4+3j) = \left(24+18j-12j-9j^2\right)$
$$= 33+6j$$

17. $\dfrac{6-\sqrt{-16}}{\sqrt{-4}} = \dfrac{6-4j}{2j}\cdot\dfrac{-2j}{-2j}$
$$= \frac{-12j+8j^2}{-4j^2}$$
$$= \frac{-8-12j}{4}$$
$$= -2-3j$$

21. $3x-2j = yj-9$

$3x=-9,\ -2=y$

$x=-\dfrac{9}{3}\quad y=-2$

$x=-3$

$x=-3,\ y=-2$

25.

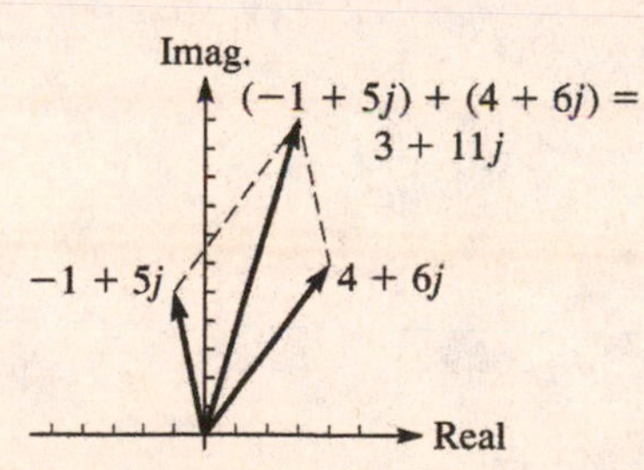

algebraically:

$$(-1+5j)+(4+6j) = -1+5j+4+6j$$
$$= -1+4+5j+6j$$
$$= 3+11j$$

29. From the rectangular form we have $x=1$ and $y=-1$, so

$$r = \sqrt{1^2+(-1)^2}$$
$$= \sqrt{2}$$
$$\theta_{ref} = \tan^{-1}\frac{1}{1}$$
$$= 45°$$

Since θ is in the fourth quadrant,

$$\theta = 360° - 45°$$
$$= 315°\ \frac{\pi \text{ rad}}{180°}$$
$$= 5.50 \text{ rad}$$
$$1-j = \sqrt{2}(\cos 45° + j\sin 45°)$$
$$= \sqrt{2}\underline{/45°}$$
$$= \sqrt{2}e^{5.50j}$$

33. From the rectangular form we have $x=1.07$ and $y=4.55$, so

$$r = \sqrt{1.07^2+4.55^2}$$
$$= 4.67$$
$$\theta = \tan^{-1}\frac{4.55}{1.07}$$
$$= 76.8°\ \frac{\pi \text{ rad}}{180°}$$
$$= 1.34 \text{ rad}$$

$$1.07+4.55j = 4.67(\cos 76.8° + j\sin 76.8°)$$
$$= 4.67\underline{/76.8°}$$
$$= 4.67\,e^{1.34j}$$

37. $2\left(\cos 225^\circ + j\sin 225^\circ\right) = -\sqrt{2}-\sqrt{2}j$

41. $0.62\angle -72^\circ = 0.62\left(\cos\left(-72^\circ\right) + j\sin\left(-72^\circ\right)\right)$
$$= 0.19-0.59j$$

45. Using radian mode:

$2.00e^{0.25j} = 2.00(\cos 0.25 + j\sin 0.25)$
$$= 1.94+0.495j$$

49. $\left[3\left(\cos 32^\circ + j\sin 32^\circ\right)\right]\cdot\left[5\left(\cos 52^\circ + j\sin 52^\circ\right)\right]$
$$= 3\cdot 5\left(\cos\left(32^\circ+52^\circ\right) + j\sin\left(32^\circ+52^\circ\right)\right)$$
$$= 15\left(\cos 84^\circ + j\sin 84^\circ\right)$$

53. $\dfrac{24(\cos 165^\circ + j\sin 165^\circ)}{3(\cos 55^\circ + j\sin 55^\circ)}$

$$= \frac{24}{3}\cos(165^\circ - 55^\circ) + j\sin(165^\circ - 55^\circ)$$
$$= 8(\cos 110^\circ + j\sin 110^\circ)$$

57. $0.983\underline{/47.2^\circ} + 0.366\underline{/95.1^\circ}$

$$= 0.983(\cos 47.2^\circ + j\sin 47.2^\circ) + 0.366(\cos 95.1^\circ + j\sin 95.1^\circ)$$
$$= 0.6679 + 0.7213j - 0.03254 + 0.3646j$$
$$= 0.6354 + 1.0859j$$

In polar form:

$r = \sqrt{0.6354^2 + 1.0859^2} = 1.26$

$\theta = \tan^{-1}\dfrac{1.0859}{0.6354}$

$= 59.7^\circ.$

$0.983\underline{/47.2^\circ} + 0.366\underline{/95.1^\circ} = 1.26\underline{/59.7^\circ}$

61. $\left[2(\cos 16^\circ + j\sin 16^\circ)\right]^{10}$

$$= 2^{10}\left[\cos(10\cdot 16^\circ) + j\sin(10\cdot 16^\circ)\right]$$
$$= 1024(\cos 160^\circ + j\sin 160^\circ)$$

65. $1 - j = \sqrt{2}(\cos 315^\circ + j\sin 315^\circ)$ from Problem 25

$$(1-j)^{10} = \left[\sqrt{2}(\cos 315^\circ + j\sin 315^\circ)\right]^{10}$$
$$= \sqrt{2}^{10}\left(\cos(10\cdot 315^\circ) + j\sin(10\cdot 315^\circ)\right)$$
$$= 32(\cos 3150^\circ + j\sin 3150^\circ)$$
$$= 32(\cos 270^\circ + j\sin 270^\circ), \text{ polar form}$$
$$= 0 - 32j, \text{ rectangular form}$$

$$(1-j)^{10} = \left((1-j)^2\right)^5$$
$$= \left(1 - 2j + j^2\right)^5$$
$$= (-2j)^5$$
$$= -32j^5$$
$$= -32j$$

69. $x^3 + 8 = 0$

$x^3 = -8$

$x = \sqrt[3]{-8}$

To find the three cube roots of –8:

$r = \sqrt{8^2 + 0^2} = 8$

$\theta = 180°$

So the three cube roots of –8 are

$$r_1 = 8^{1/3}\left(\cos\frac{180°}{3} + j\sin\frac{180°}{3}\right)$$
$$= 2(\cos 60° + j\sin 60°)$$
$$= 1 + j\sqrt{3}$$

$$r_2 = 8^{1/3}\left(\cos\frac{180° + 360°}{3} + j\sin\frac{180° + 360°}{3}\right)$$
$$= 2(\cos 180° + j\sin 180°)$$
$$= -2 + 0j$$

$$r_3 = 8^{1/3}\left(\cos\frac{180° + 2(360)°}{3} + j\sin\frac{180° + 2(360)°}{3}\right)$$
$$= 2(\cos 300° + j\sin 300°)$$
$$= 1 - j\sqrt{3}$$

73. From the graph, $x = 40$, $y = 9$. Therefore,

$$r = \sqrt{40^2 + 9^2}$$
$$= 41$$
$$\theta = \tan^{-1}\frac{9}{40}$$
$$= 12.7°$$
$$40 + 9j = 41(\cos 12.7° + j\sin 12.7°)$$

77. $$x^2 - 2x + 4\Big|_{x=5-2j} = (5-2j)^2 - 2(5-2j) + 4$$
$$= 25 - 20j + 4j^2 - 10 + 4j + 4$$
$$= 19 - 16j - 4$$
$$= 15 - 16j$$

81. $$x^2 - 2x + 2 = (1-j)^2 - 2(1-j) + 2$$
$$= (1 - 2j + j^2) - 2 + 2j + 2$$
$$= 1 - 2j - 1 - 2 + 2j + 2$$
$$= 0 \Rightarrow 1 - j \text{ is a solution}$$
$$x^2 - 2x + 2 = (-1-j)^2 - 2(-1-j) + 2$$
$$= (1 + 2j + j^2) + 2 + 2j + 2$$
$$= 4j + 4 \Rightarrow -1 - j \text{ is not a solution}$$

85. $f(x) = 2x - (x-1)^{-1}$

$$f(1+2j) = 2(1+2j) - (1+2j-1)^{-1}$$
$$= 2 + 4j - \frac{1}{2j} \cdot \frac{j}{j}$$
$$f(1+2j) = 2 + 4.5j$$

89. $V_L = 60j,\ V_C = -60j$

$$V = V_R + V_L + V_C$$
$$60 = V_R + 60j - 60j$$
$$V_R = 60 \text{ V}$$

93. $2\pi fL = \frac{1}{2\pi fC} \Rightarrow f = \sqrt{\frac{1}{4\pi^2 LC}}$

$$= \sqrt{\frac{1}{4\pi^2(2.65)(18.3\times10^{-6})}}$$
$$f = 22.9 \text{ Hz}$$

97. $\frac{1}{\mu + j\omega n} = \frac{1}{(\mu + j\omega n)} \cdot \frac{(\mu - j\omega n)}{(\mu - j\omega n)} = \frac{\mu - j\omega n}{\mu^2 + \omega^2 n^2}$

$$= \frac{\mu}{\mu^2 + \omega^2 n^2} - \frac{\omega n}{\mu^2 + \omega^2 n^2} j$$

101. For a positive real number, the argument is always 0°, so the root will be a real number if $(k \cdot 360)/n$ is a multiple of 180, and it will be a pure imaginary number if $(k \cdot 360)/n$ is a multiple of 90, for $0 \le k \le n-1$.
For a negative real number, the argument is always 180°, so the root will be a real number if $(180 + k360)/n$ is a multiple of 180, and it will be a pure imaginary number if $(180 + k \cdot 360)/n$ is a multiple of 90, for $0 \le k \le n-1$.

Chapter 13

Exponential and Logarithmic Functions

13.1 Exponential Functions

1. For $x = -\frac{3}{2}$:

$$y = -2\left(4^x\right)$$
$$= -2\left(4^{-3/2}\right)$$
$$= -2\left(\frac{1}{8}\right)$$
$$= -\frac{1}{4}$$

5. $(2\pi)^{-e} = 0.0067658469...$
$= 0.00677$ (to three significant digits)

9. (a) $y = -7(-5)^{-x}$, $-5 < 0$; not an exponential function.

(b) $y = -7\left(5^{-x}\right)$ is a real number multiple of an exponential function and therefore an exponential function.

13. $y = 4^x$, $x = -2$:

$$y = 4^{-2}$$
$$= \frac{1}{4^2}$$
$$= \frac{1}{16}$$

17. $y = 4^x$

x	−3	−2	−1	0	1	2	3
y	$\frac{1}{64}$	$\frac{1}{16}$	$\frac{1}{4}$	1	4	16	64

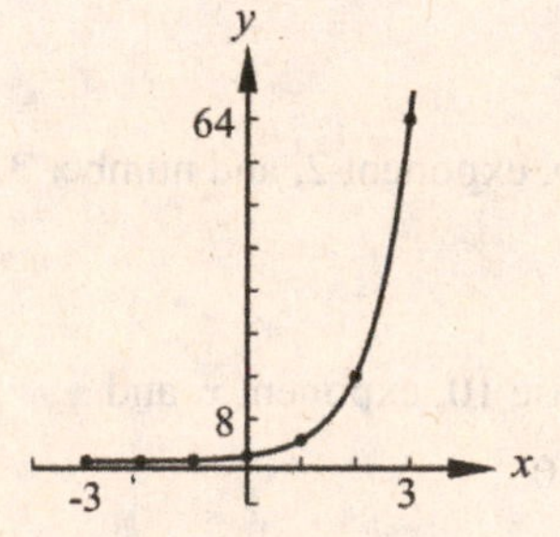

21. $y = 0.5\pi^x$

x	y
-3	0.016
-2	0.051
-1	0.16
0	0.5
1	1.57
2	4.94
3	15.50

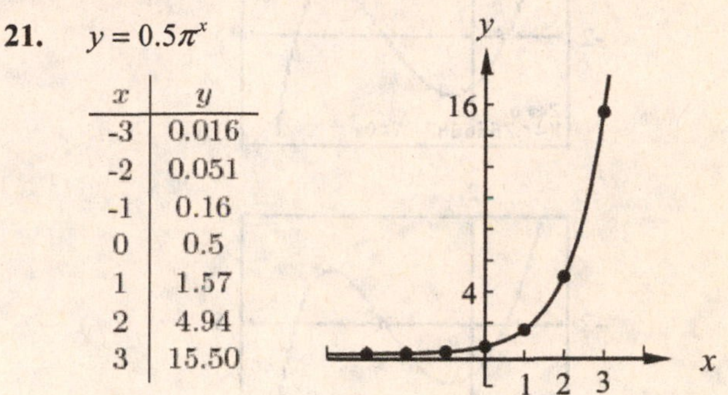

25. $y_1 = 0.1(0.25)^{2x}$

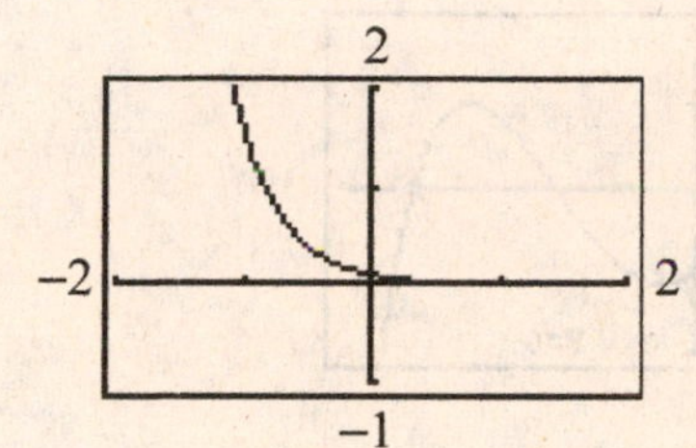

29.

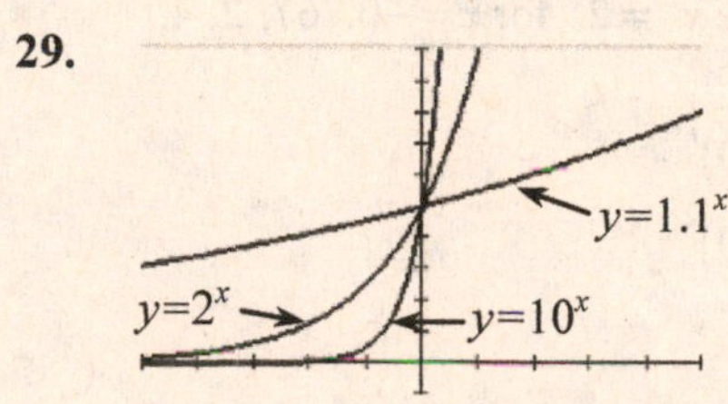

33. $f(x) = b^x$
Use Eq. (11.1) to get

$$f(c+d) = b^{c+d}$$
$$= b^c \cdot b^d$$
$$= f(c) \cdot f(d)$$

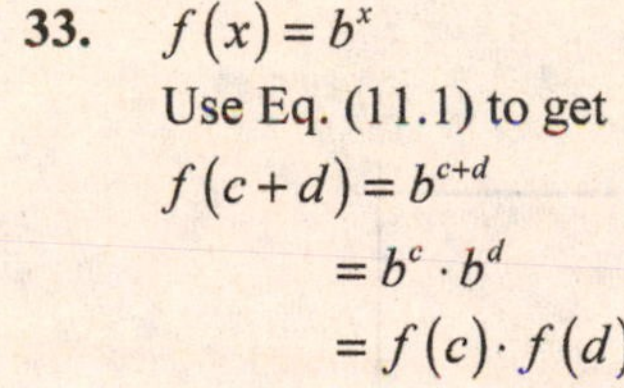

37. Since $x^2 = 2^x \Leftrightarrow x^2 - 2^x = 0$, graph $y_1 = x^2 - 2^x$.

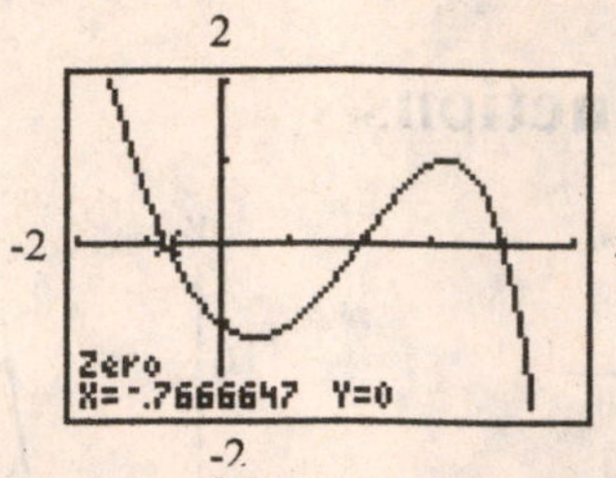

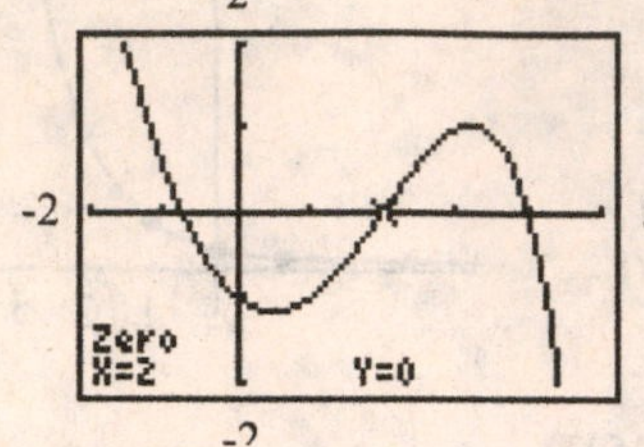

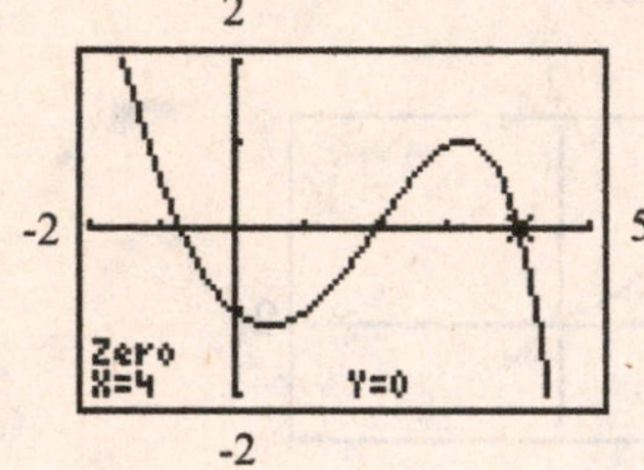

From the graph, $x^2 = 2^x$ for $x = -0.767, 2, 4$.

41. $V = 250(1.0500)^t$

$= 250(1.0500)^4$

$= \$304$

45. $q = 100e^{-10t}$

Graph $y_1 = 100e^{-10x}$.

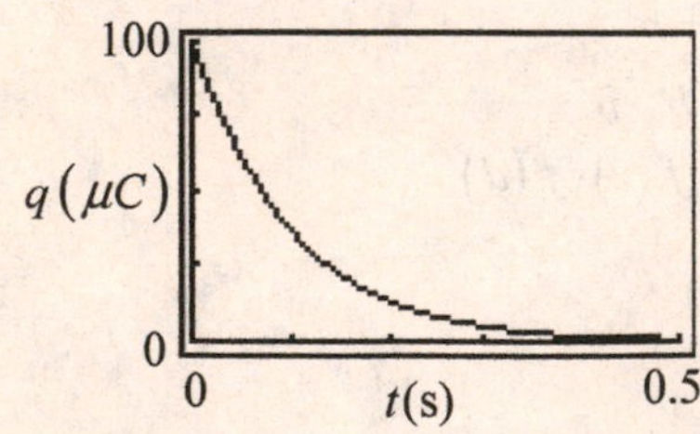

13.2 Logarithmic Functions

1. $32^{4/5} = 16$ in logarithmic form is

$\frac{4}{5} = \log_{32} 16.$

5. $3^4 = 81$ has base 3, exponent 4, and number 81.

$\log_3 81 = 4.$

9. $4^{-2} = \frac{1}{16}$ has base 4, exponent -2, and number $\frac{1}{16}$.

$\log_4 \frac{1}{16} = -2$

13. $8^{1/3} = 2$ has base 8, exponent $\frac{1}{3}$, and number 2.

$\log_8 2 = \frac{1}{3}$

17. $\log_2 32 = 5$ has base 2, exponent 5, and number 32.

$2^5 = 32$

21. $\log_{25} 5 = \frac{1}{2}$ has base 25, exponent $\frac{1}{2}$, and number 5.

$25^{1/2} = 5$

25. $\log_{10} 0.01 = -2$ has base 10, exponent -2, and number 0.1.

$0.01 = 10^{-2}$

29. $\log_4 16 = x$ has base 4, exponent x, and number 16.

$4^x = 16$

$4^x = 4^2$

$x = 2$

33. $\log_7 y = 3$ has base 7, exponent 3, and number y.

$7^3 = y$

$y = 343$

37. $\log_b 3 = 2$ has base b, exponent 2, and number 3.

$b^2 = 3,\ b = \sqrt{3}$

41. $\log_{10} 10^{0.2} = x$ has base 10, exponent x, and number $10^{0.2}$

$10^x = 10^{0.2}$

$x = 0.2$

45. Write $y = \log_3 x$ as $3^y = x$ to find values in the table.

x	y
$\frac{1}{27}$	-3
$\frac{1}{9}$	-2
$\frac{1}{3}$	-1
1	0
3	1
9	2
27	3

49.

$$N = 0.2\log_4 v$$

$$\frac{N}{0.2} = \log_4 v$$

$$4^{N/0.2} = v$$

Use $4^{N/0.2} = v$ to find values in the table.

v	N
$\frac{1}{16}$	-.4
$\frac{1}{4}$	-.2
$\frac{1}{2}$	-.1
1	0
2	.1
4	.2
16	.4

53. Graph $y_1 = -\log(-x)$.

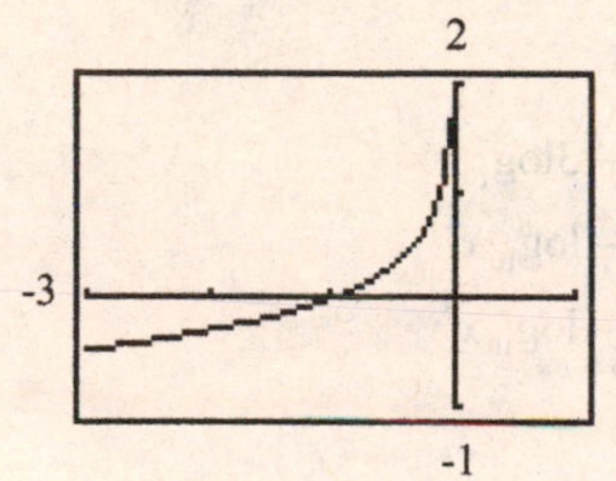

57. **(a)**

$$f(x) = \log_5 x$$

$$f(\sqrt{5}) = \log_5 \sqrt{5}$$

$$= \log_5 5^{1/2}$$

$$N = \log_5 5^{1/2}$$

$$5^N = 5^{1/2}$$

$$N = \frac{1}{2}$$

$$f(\sqrt{5}) = \frac{1}{2}$$

(b) $f(0)$ does not exist.

Logarithms are defined for only positive values.

61.

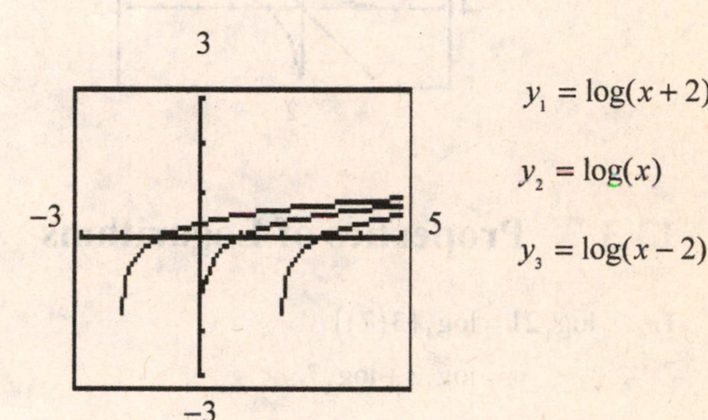

65. The logarithm is defined if and only if $2 - x > 0$, and so the domain is the set of all $x < 2$.

69.

$$\log_e\left(\frac{N}{N_0}\right) = -kt$$

$$e^{-kt} = \frac{N}{N_0}$$

$$N = N_0 e^{-kt}$$

73. $t = N + \log_2 N$ where, $N > 0$ and $t > 0$.

N	t
1	1
2	3
4	6
8	11

77. Solve $y = 10^{x/2}$ for x by changing into logarithmic form. We get $x = 2\log_{10} y$. Interchange x and y to get $y = 2\log_{10} x$, which is the inverse function. Graph $y_1 = 10^{x/2}$, $y_2 = 2\log_{10} x$ and $y_3 = x$. For each graph to look like the mirror image of the other across $y = x$, the calculator window must be square. Do this by using the key sequence ZOOM Zsquare.

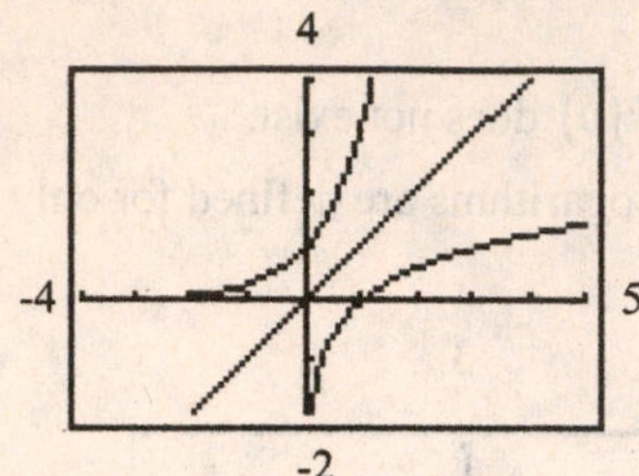

13.3 Properties of Logarithms

1. $\log_4 21 = \log_4 (3(7))$
$= \log_4 3 + \log_4 7$

5. $\log_3 27 = \log_3 3^3$
$= 3\log_3 3$
$= 3(1)$
$= 3$

9. $\log_5 33 = \log_5 (3 \cdot 11)$
$= \log_5 3 + \log_5 11$

13. $\log_2 (a^3) = 3\log_2 a$

17. $10\log_5 \sqrt{t} = 10\log_5 t^{1/2}$
$= 5\log_5 t$

21. $\log_b a + \log_b c = \log_b (ac)$

25. $\log_b \sqrt{x} + \log_b x^2 = \log_b x^2 x^{1/2}$
$= \log_b x^{5/2}$

29. $\log_2 \left(\frac{1}{32}\right) = \log_2 \left(\frac{1}{2^5}\right)$
$= \log_2 2^{-5}$
$= -5\log_2 2$
$= -5$

33. $6\log_7 \sqrt{7} = 6\log_7 7^{1/2}$
$= 3\log_7 7$
$= 3$

37. $\log_3 18 = \log_3 (9 \cdot 2)$
$= \log_3 9 + \log_3 2$
$= \log_3 3^2 + \log_3 2$
$= 2\log_3 3 + \log_3 2$
$= 2 + \log_3 2$

41. $\log_3 \sqrt{6} = \log_3 (3 \cdot 2)^{1/2}$
$= \frac{1}{2}\log_3 (3 \cdot 2)$
$= \frac{1}{2} \cdot [\log_3 3 + \log_3 2]$
$= \frac{1}{2} \cdot [1 + \log_3 2]$

45. $\log_b y = \log_b 2 + \log_b x$
$\log_b y = \log_b (2x)$
$y = 2x$

49. $\log_{10} y = 2\log_{10} 7 - 3\log_{10} x$
$= \log_{10} 7^2 - \log_{10} x^3$
$= \log_{10} 49 - \log_{10} x^3$
$\log_{10} y = \log_{10} \frac{49}{x^3}$
$y = \frac{49}{x^3}$

53. $\log_2 x + \log_2 y = 1$
$\log_2 (xy) = 1$
$2^1 = xy$
$y = \frac{2}{x}$

57. $\log_{10}(x+3) = \log_{10}x + \log_{10}3$
$= \log_{10}(3x)$

Then

$$x+3=3x$$
$$2x=3$$
$$x=\frac{3}{2}$$

This means that $x=\frac{3}{2}$ is the only value for which $\log_{10}(x+3) = \log_{10}x + \log_{10}3$ is true.
For any other x-value,
$\log_{10}(x+3) \neq \log_{10}x + \log_{10}3$
and thus the statement is not true in general.
This can be also be seen from the following graph.

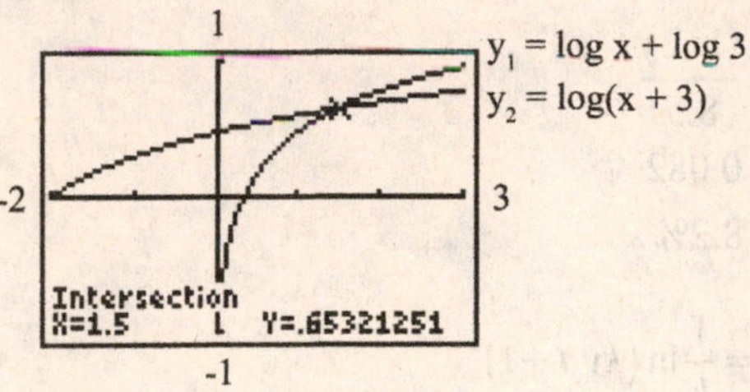

61. $\log_b\sqrt{x^2y^4} = \log_b\left(x^2y^4\right)^{1/2}$
$= \log_b\left(xy^2\right)$
$= \log_b x + 2\log_b y$
$= 2+2(3)=8$

65. Graph $y_1 = \log x - \log\left(x^2+1\right)$, $y_2 = \log\frac{x}{x^2+1}$.

The graphs are the same.

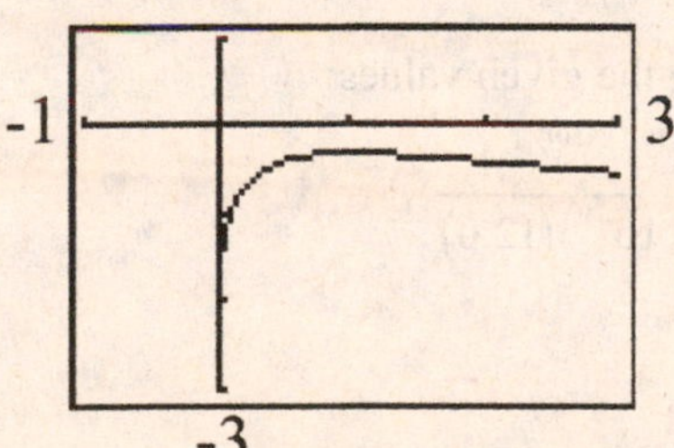

13.4 Logarithms to the Base 10

1. $\log 0.3654 = -0.4372$

5. $\log 9.24\times10^6 = 6.966$

9. $\log\left(\sin 45.2°\right) = -0.149$

13. $10^{1.257} = 18.1$

17. $10^{3.30112} = 2000.4$

21. $\log\left(185^{100}\right) = 100\log(185)$
$= 226.7171728403$
and so $185^{100} = 10^{226.7171728403}$
$= (10^{0.7171728403})\times10^{226}$
$= 5.214\times10^{226}$

25. $\log 14 + \log 0.5 = \log 7$
$\log 14 + \log 0.5 = 0.84509804$
$\log 7 = 0.84509804$

29. $\log 9.00\times10^8 = 8.954$

We could also evaluate as
$\log\left(9.00\times10^8\right) = \log 9.00 + \log 10^8$
$= 0.954+8$
$= 8.954$

33. $\log T = 8$
$T = 10^8$ K

37. $\frac{\log_b x^2}{\log 100} = \frac{2\log_b x}{\log 10^2}$
$= \frac{2\log_b x}{2\log 10}$
$= \log_b x$

41. $R = \log\left(\frac{I}{I_0}\right)$; $I = 1.6\times10^9 I_0$
$R = \log\frac{1.6\times10^9 I_0}{I_0}$
$= \log 1.6\times10^9$
$= 9.2$
The 1964 Alaska earthquake had magnitude 9.2 on the Richter scale.

13.5 Natural Logarithms

1. $\ln 200 = \dfrac{\log 200}{\log e}$

$= 5.298$

5. $\ln 1.562 = \dfrac{\log 1.562}{\log e}$

$= \dfrac{0.1937}{0.4343}$

$= 0.4460$

9. $\log_7 52 = \dfrac{\log 52}{\log 7}$

$= \dfrac{1.716}{0.8451}$

$= 2.03$

13. $\log_{40} 750 = \dfrac{\log 750}{\log 40}$

$= \dfrac{2.875}{1.6021}$

$= 1.795$

17. $\ln 1.394 = 0.3322$

21. $\ln\left(0.012937^4\right) = -17.39066$

25. $e^{0.0084210} = 1.0085$

29. $e^{-23.504} = 6.20\times 10^{-11}$

33. We had shown algebraically in Example 9 of Section 13.2 that the functions are inverses of each other. To verify this using a graphing calculator, we graph $y_1 = 2^x$, $y_2 = \ln x/\ln 2$, and $y_3 = x$. We use the key sequence ZOOM Zsquare and note that $y_1 = 2^x$ and $y_2 = \ln x/\ln 2$ are indeed mirror images of each other—and therefore inverses.

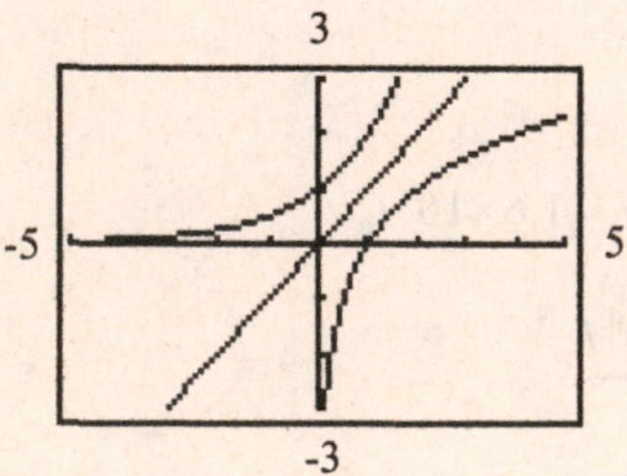

37. $4\ln 3 = \ln 81$

$4\ln 3 = 4.394449155$

$\ln 81 = 4.394449155$

41. $\ln(\log x) = 0$

$e^{\ln(\log x)} = e^0$

$= 1$

$\log x = 1$

$10^{\log x} = 10^1$

$x = 10$

45. $\ln\left(e^2\sqrt{1-x}\right) = \ln\left(e^2\right) + \ln\ (1-x)^{1/2}$

$= 2\ln e + \frac{1}{2}\ln(1-x)$

$= 2 + \frac{1}{2}\ln(1-x)$

49. $i = \dfrac{(\ln 2)}{8.5}$

$= 0.082$

$= 8.2\%$

53. $x = \frac{1}{k}\ln(kv_0t + 1)$

$x = 150$ m,

$k = 6.80\times 10^{-3}$ m

$v_0 = 12.0$ m/s

We first solve for t as a function of x, k, and v_0. We get

$kx = \ln(kv_0t + 1)$

$e^{kx} = kv_0t + 1$

$kv_0t = e^{kx} - 1$

$t = \dfrac{e^{kx} - 1}{kv_0}$

Substituting the given values:

$t = \dfrac{e^{\left(6.80\times 10^{-3}\right)(150)} - 1}{\left(6.80\times 10^{-3}\right)(12.0)}$

$= \dfrac{e^{1.02} - 1}{0.0816}$

$= \dfrac{1.773}{0.0816}$

$= 21.7$ s

13.6 Exponential and Logarithmic Equations

1. $3^{x+2}=5$

$\log 3^{x+2}=\log 5$

$(x+2)\log 3=\log 5$

$x+2=\dfrac{\log 5}{\log 3}$

$x=\dfrac{\log 5}{\log 3}-2$

$x=-0.535$

5. $3.50^x=82.9$

$\log 3.50^x=\log 82.9$

$x\log 3.50=\log 82.9$

$x=\dfrac{\log 82.9}{\log 3.50}$

$=3.526$

9. $6^{x+2}=78$

$\ln 6^{x+2}=\ln 78$

$(x+2)\cdot \ln 6=\ln 78$

$x+2=\dfrac{\ln 78}{\ln 6}$

$x=\dfrac{\ln 78}{\ln 6}-2$

$x=0.432$

13. $0.6^x=2^{x^2}$

$\ln\left(0.6^x\right)=\ln 2^{x^2}$

$x\cdot \ln 0.6=x^2\cdot \ln 2$

$x^2\cdot \ln 2-x\ln 0.6=0$

$x(x\cdot \ln 2-\ln 0.6)=0$

$x=0$ or $x\cdot \ln 2-\ln 0.6=0$

$x=\dfrac{\ln 0.6}{\ln 2}$

$=-0.7$

17. $3\log_8 x=-2$

$\log_8 x=\dfrac{-2}{3}$

$8^{-2/3}=x$

$x=\dfrac{1}{4}$

21. $\log_2 x+\log_2 7=\log_2 21$

$\log_2 7x=\log_2 21$

$7x=21$

$x=3$

25. $\log 12x^2-\log 3x=3$

$\log\dfrac{12x^2}{3x}=3$

$\log 4x=3$

$4x=10^3$

$=1000$

$x=250$

29. $\dfrac{1}{2}\log(x+2)+\log 5=1$

$\log(x+2)^{1/2}+\log 5=1$

$5\left[(x+2)^{1/2}\right]=10$

$\sqrt{(x+2)}=2$

$x+2=4$

$x=2$

33. $15^{-x}=1.326$. Graph $y_1=15^{-x}-1.326$ and use the zero feature to solve.

$x=-0.104$

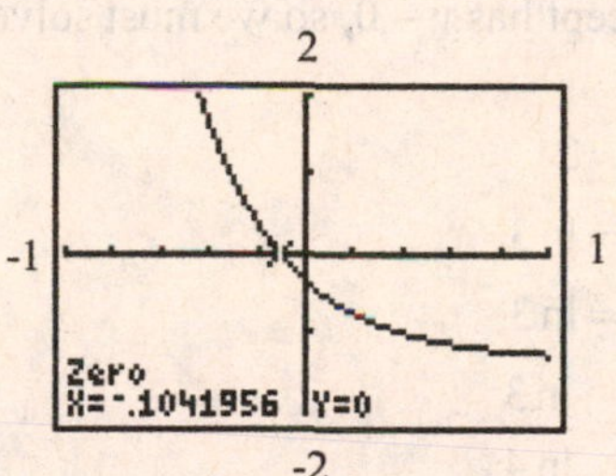

37. $3\ln 2x=2$. Graph $y_1=3\ln 2x-2$ and use the zero feature to solve.

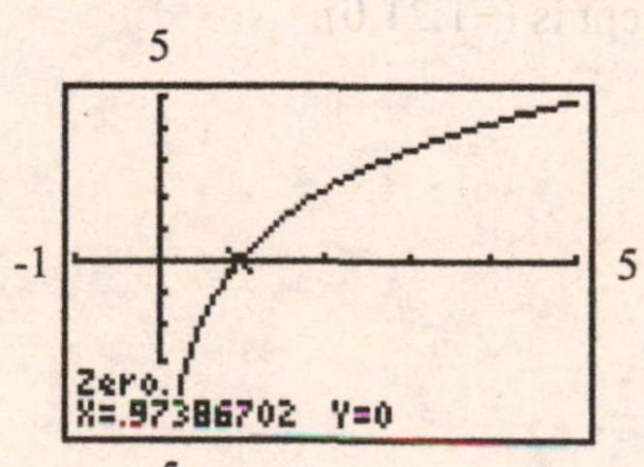

$x=0.974$

5

-1

5

Zero
X=.97386702 Y=0

-5

41. $2^{2x}-2^{x}-6=0$. Graph $y_1=2^{2x}-2^{x}-6$ and use the zero feature to solve.

$x=1.585$

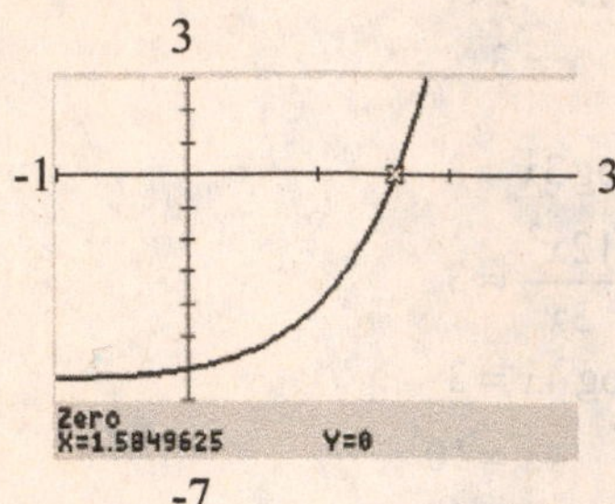

45. Multiplying both sides by e^x:

$$e^x+e^{-x}=3$$

$$e^{2x}+e^0-3e^x=0$$

$$\left(e^x\right)^2-3e^x+1=0, \text{ quadratic in } e^x$$

$$e^x=\frac{-(-3)\pm\sqrt{(-3)^2-4(1)(1)}}{2(1)}$$

$$=\frac{3\pm\sqrt{5}}{2}$$

$$x=\ln\frac{3\pm\sqrt{5}}{2}$$

$$x=\pm 0.9624$$

49. The x-intercept has $y=0$, so we must solve

$$3-4^{x+2}=0$$

$$4^{x+2}=3$$

$$\ln 4^{x+2}=\ln 3$$

$$(x+2)\ln 4=\ln 3$$

$$x+2=\frac{\ln 3}{\ln 4}$$

$$x=\frac{\ln 3}{\ln 4}-2$$

$$=-1.21$$

The x-intercept is $(-1.21,0)$.

53. $T=T_0+(37-T_0)0.97^t$

Substituting $T=27$ and $T_0=22$ and then solving for t we get

$$27=22+(37-22)0.97^t$$

$$5=15\cdot 0.97^t$$

$$0.97^t=\frac{1}{3}$$

$$\ln 0.97^t=\ln\frac{1}{3}$$

$$t\ln 0.97=\ln\frac{1}{3}$$

$$t=\frac{\ln\frac{1}{3}}{\ln 0.97}$$

$$=36$$

Death occurred 36 hours earlier, at noon of the previous day

57. For the 1906 San Francisco earthquake,

$$8.3=\log(I_{SF}/I_0)$$

$$I_{SF}/I_0=10^{8.3}$$

$$I_{SF}=2.00\times 10^8 I_0$$

61.

$$\ln P=t\ln 0.999+\ln P_0$$

$$\ln P-\ln P_0=t\ln 0.999$$

$$\ln\frac{P}{P_0}=\ln 0.999^t;$$

$$\frac{P}{P_0}=0.999^t$$

$$P=P_0(0.999)^t$$

65. We substitute y=5.8 to obtain the equation

$2\left(e^{x/4}+e^{-x/4}\right)=5.8$

Graph $y_1=2\left(e^{x/4}+e^{-x/4}\right)-5.8$ and use the zero feature to solve.

$x=\pm3.66$ m

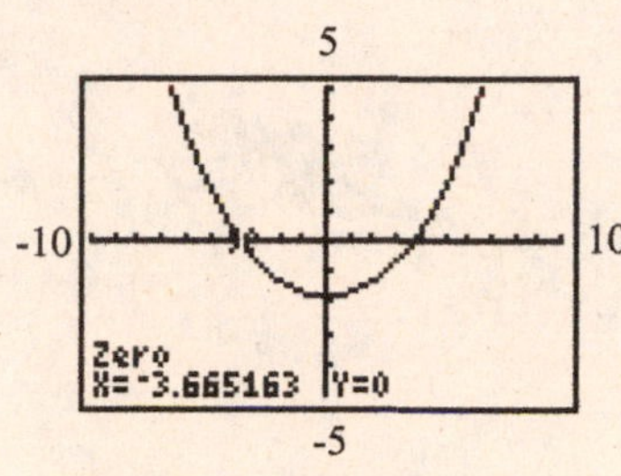

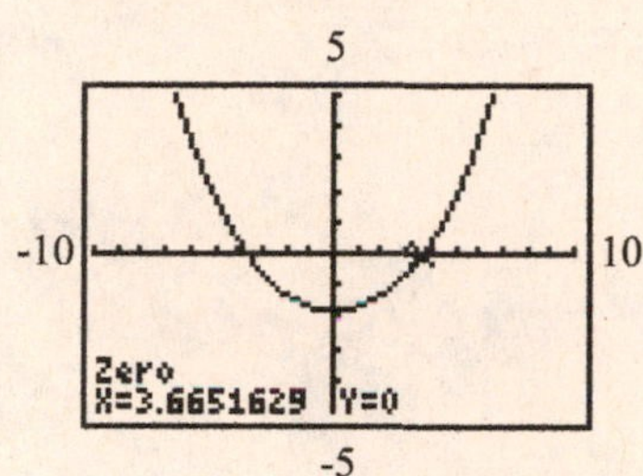

13.7 Graphs on Logarithmic and Semilogarithmic Paper

1. $y=2\left(3^x\right)$

x	−1	0	2	3	4	5
y	0.67	2	18	54	162	486

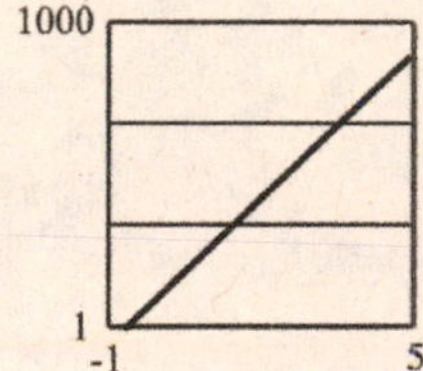

5. $y=5\left(4^{-x}\right)$

x	0	1	2	3	4	5
y	5	5/4	5/16	5/64	5/256	5/1024

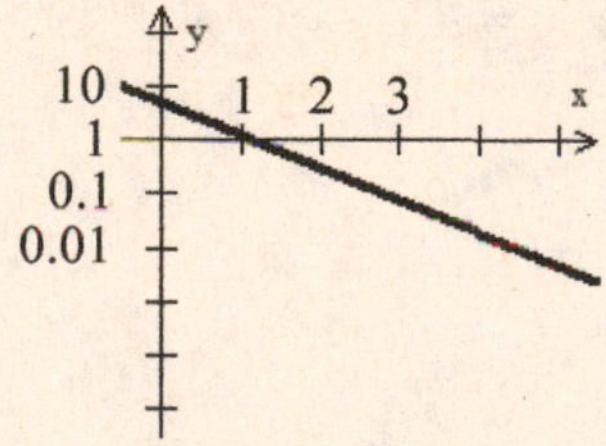

9. $y=2x^3+6x$

x	0	1	2	4	6	8
y	0	8	28	152	468	1072

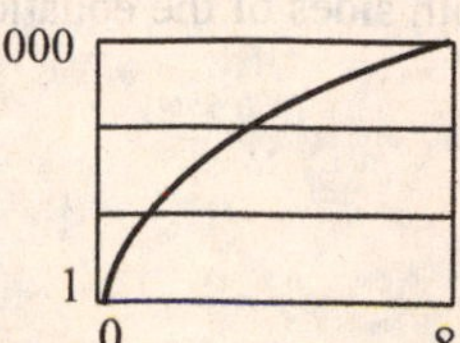

13. $y=x^{2/3}$

x	1	5	10	50	100	500	1000
y	1	2.9	4.6	13.6	21.5	63.0	100

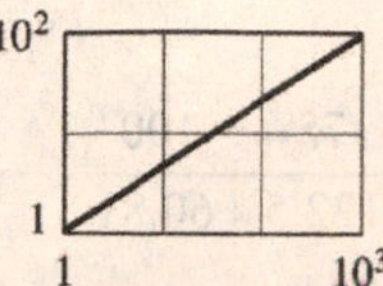

17. $x^2y^2=25$

$y=\sqrt{\frac{25}{x^2}}$

$=\frac{5}{x}$

x	0.1	0.5	1	10	50
y	50	10	5	0.5	0.1

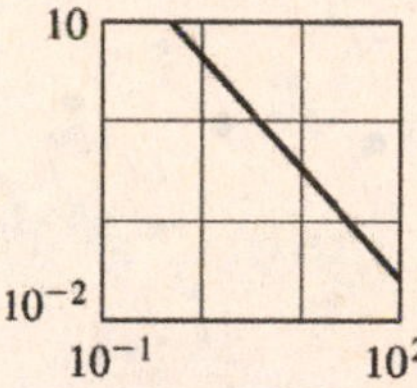

21. $y=3x^6$, log–log paper

x	1	2	3	4
y	3	192	2187	12288

Taking logarithms on both sides of the equation, we have $\log y=\log 3+6\log x$, so we need logarithmic scales along both axes.

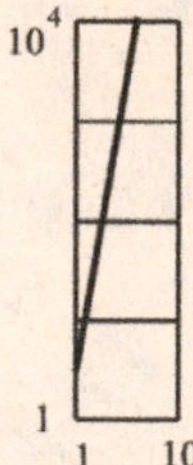

25. $x\sqrt{y}=4,\ y=\dfrac{16}{x^2}$, log− log paper

x	1	25	50	75	100
y	16	0.0256	0.0064	$0.0028\overline{4}$	0.0016

Taking logarithms on both sides of the equation, we have $\log y = \log 16 - 2\log x$, so we need logarithmic scales along both axes.

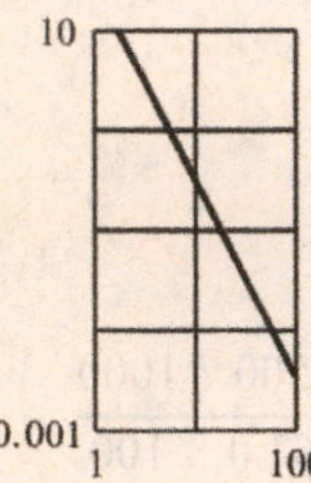

29. $N = N_0 e^{-0.028t}$, $N_0 = 1000$

t	0	25	50	75	100
N	1000	496.6	246.6	122.5	60.81

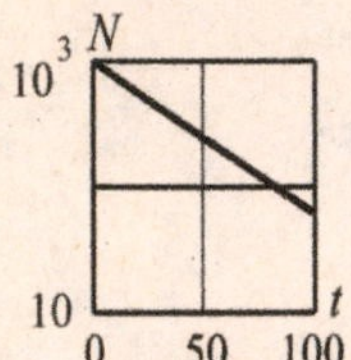

33.

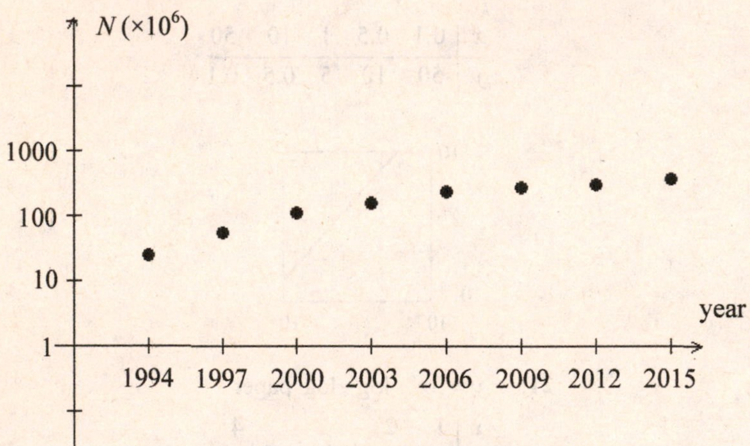

34.

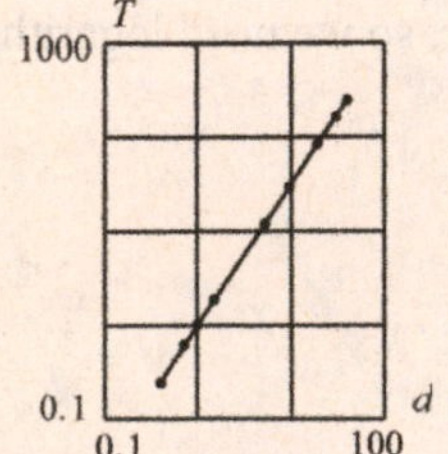

35.

$f(\text{Hz})$	100	200	500	1000
$B(\text{dB})$	40	30	22	20

f	2000	5000	10000
B	18	24	30

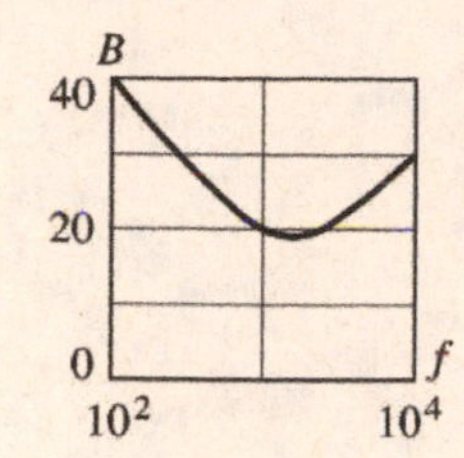

36.

$h(\text{km})$	0	10	20	30	40
$p(\text{kPa})$	101	25	6.3	2.0	0.53

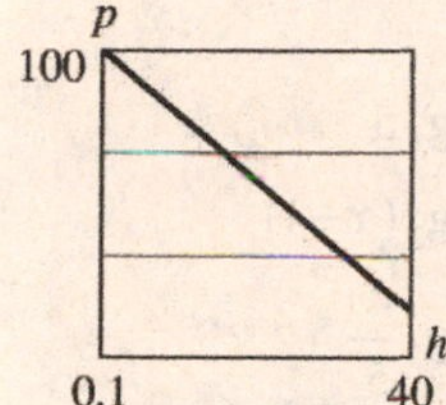

37.

d	0.063	0.13	0.19	0.25	0.38
R	600	190	100	72	46

d	0.50	0.75	1.0	1.5
R	29	17	10	6.0

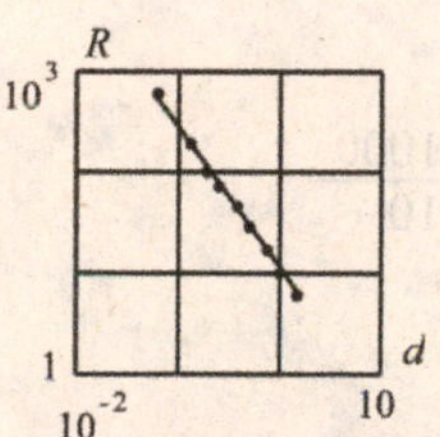

Review Exercises

1. This is true. Any function of the form

$$f(x) = c \cdot b^x,\ b > 0 \text{ and } c \text{ are constants}$$

is considered to be an exponential function.

5. This is false.

$$\frac{\ln x^2}{-\ln e} = \frac{2\ln x}{-1} = -2\ln x.$$

9. Base is 5, exponent is -1, and number is x:

$$\log_5 x = -1$$
$$x = 5^{-1}$$
$$= \frac{1}{5}$$

13. Base is 8, exponent is $x+1$, and number is 32:

$$\log_8 32 = x+1$$
$$8^{x+1} = 32$$
$$= 2^5$$
$$2^{3(x+1)} = 2^5$$
$$3x+3 = 5$$
$$x = \frac{2}{3}$$

17. Base is x, exponent is 1/3, and number is 10:

$$\log_x 10 = \frac{1}{3}$$
$$x^{1/3} = 10$$
$$x = 1000$$

21. $\log_3\left(t^2\right) = 2\log_3 t$

25.

$$\log_4 \sqrt{48} = \log_4 48^{1/2}$$
$$= \frac{1}{2}\log_4\left(16(3)\right)$$
$$= \frac{1}{2}\left(\log_4 16 + \log_4 3\right)$$
$$= \frac{1}{2}\left(\log_4 4^2 + \log_4 3\right)$$
$$= \frac{1}{2}\left(2\log_4 4 + \log_4 3\right)$$
$$= \frac{1}{2}\left(2 + \log_4 3\right)$$
$$= 1 + \frac{1}{2}\log_4 3$$

29.

$$\log_{10}\left(1000x^4\right) = \log_{10} 10^3 + \log_{10} x^4$$
$$= 3\log_{10} 10 + 4\log_{10} x$$
$$= 3 + 4\log_{10} x$$

33.
$$\begin{aligned} 3\ln y &= 2+3\ln x \\ 3\ln y-3\ln x &= 2 \\ \ln y^3-\ln x^3 &= 2 \\ \ln\frac{y^3}{x^3} &= 2 \\ \frac{y^3}{x^3} &= e^2 \\ y^3 &= x^3e^2 \\ y &= xe^{2/3} \end{aligned}$$

37.
$$\begin{aligned} \log_5 x+\log_5 y &= \log_5 3+1 \\ &= \log_5 3+\log_5 5 \\ &= \log_5 15 \\ \log_5(xy) &= \log_5(15) \\ xy &= 15 \\ y &= \frac{15}{x} \end{aligned}$$

41.
$$\begin{aligned} \log_x y &= \ln e^3 \\ \frac{\ln y}{\ln x} &= 3 \\ \ln y &= 3\ln x \\ y &= x^3 \end{aligned}$$

45. $R = 0.2\log_4 r$

Graph $y_1 = 0.2\frac{\ln x}{\ln 4}$.

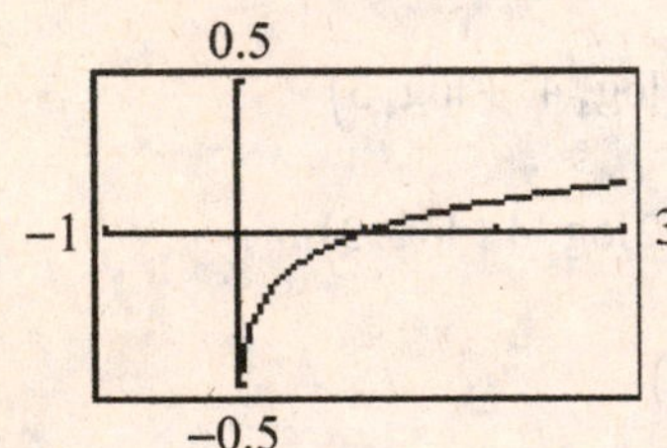

49. $y = 1-e^{-|x|}$

Graph $y_1 = 1-e^{\text{-abs}(x)}$

1

−5 5

−0.5

53. $\ln 87.9 = 4.476$

57.
$$\begin{aligned} \log_{4.25}(0.0067) &= \frac{\ln 0.0067}{\ln 4.25} \\ &= -3.46 \end{aligned}$$

61.
$$\begin{aligned} 3^{x+2} &= 5^x \\ 3^x\cdot 3^2 &= 5^x \\ \left(\frac{3}{5}\right)^x &= \frac{1}{9} \\ x &= \log_{3/5}\frac{1}{9} \\ &= \frac{\ln\frac{1}{9}}{\ln\frac{3}{5}} \\ &= 4.30 \end{aligned}$$

65.
$$\begin{aligned} 2\log_3 2-\log_3(x+1) &= \log_3 5 \\ \log_3 2^2-\log_3 5 &= \log_3(x+1) \\ \log_3(x+1) &= \log_3\frac{4}{5} \\ x+1 &= \frac{4}{5} \\ x &= -\frac{1}{5} \\ &= -0.2 \end{aligned}$$

69. $y = \sqrt[3]{x}$

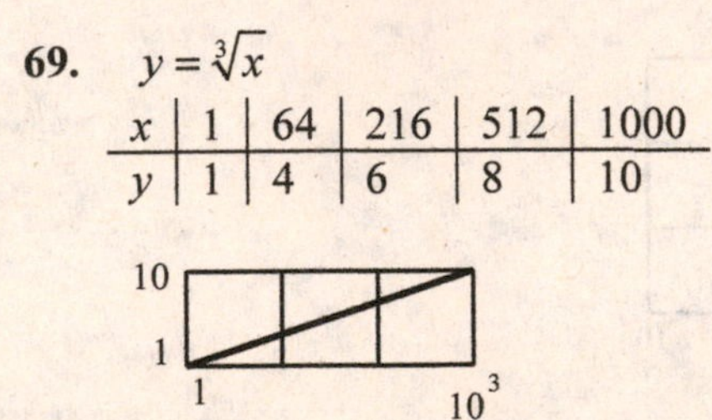

x	1	64	216	512	1000
y	1	4	6	8	10

73.
$$\begin{aligned} 3e^{2\ln 2} &= 3e^{\ln 2^2} \\ &= 3e^{\ln 4} \\ &= 3(4) \\ &= 12 \end{aligned}$$

77.
$$\begin{aligned} \sqrt[3]{\ln e^8}-\sqrt{\log 10^4} &= \sqrt[3]{8\ln e}-\sqrt{4\log 10} \\ &= \sqrt[3]{8}-\sqrt{4} \\ &= 2-2 \\ &= 0 \end{aligned}$$

81. $$7\left(10^{\log 0.1}\right)+6000\left(100^{\log 0.001}\right)$$
$$=7(0.1)+6000\left(10^{2\log(0.001)}\right)$$
$$=0.7+6000\left(10^{\log 0.001^2}\right)$$
$$=0.7+6000\left(0.001^2\right)$$
$$=0.706$$

85. $$\ln M=15.34$$
$$M=e^{15.34}=4{,}593{,}000$$

89. $$\ln\frac{I}{I_0}=-\beta h$$
$$\frac{I}{I_0}=e^{-\beta h}$$
$$I=I_0e^{-\beta h}$$

93. $$400=283e^{0.0085t}$$
$$e^{0.0085t}=\frac{400}{283}$$
$$0.0085t=\ln\left(\frac{400}{283}\right)$$
$$t=\frac{1}{0.0085}\cdot\ln\left(\frac{400}{283}\right)=40.7\text{ years}$$
This model predicts that the U.S. population reach 400 million by 2041.

97. $$2\ln\omega=\ln 3g+\ln\sin\theta-\ln l$$
$$\ln\omega^2=\ln\frac{3g\sin\theta}{l}$$
$$\omega^2=\frac{3g\sin\theta}{l}$$
$$\sin\theta=\frac{\omega^2 l}{3g}$$

101. $$m_1-m_2=2.5\log\frac{b_2}{b_1}$$
$$-1.4-6.0=2.5\log\frac{b_2}{b_1}$$
$$-2.96=\log\frac{b_2}{b_1}$$
$$\frac{b_2}{b_1}=10^{-2.96}$$
$$\frac{b_1}{b_2}=10^{2.96}$$
$$=912$$
$$b_1=912b_2$$
Hence Sirius is 912 times brighter than the faintest stars.

105. $$x=k\left(\ln I_0-\ln I\right)$$
We substitute $k=5.00$ and $I=0.850I_0$ to get
$$x=5.00\left(\ln I_0-\ln 0.850I_0\right)$$
$$x=5.00\ln\frac{I_0}{0.850I_0}$$
$$x=0.813\text{ cm}$$

109. $$\ln n=-0.04t+\ln 20$$
$$\ln n-\ln 20=-0.04t$$
$$\ln\frac{n}{20}=-0.04t$$
$$\frac{n}{20}=e^{-0.04t}$$
$$n=20e^{-0.04t}$$

113. We manipulate the second equation algebraically to obtain the first one.
$$y=(2\ln x)/3+\ln 4-\ln\left(\ln e^2\right)\text{ requires }x>0$$
$$y=\frac{2}{3}\ln x+\ln 4-\ln(2\ln e)$$
$$y=\ln x^{2/3}+\ln 4-\ln 2$$
$$y=\ln x^{2/3}+\ln\frac{4}{2}$$
$$y=\ln x^{2/3}+\ln 2$$
$$y=\ln\left(2x^{2/3}\right)\text{ which only requires }x\neq 0$$
since $2x^{2/3}>0$ for all $x\neq 0$.

(a) The two equations are equivalent for $x > 0$.

(b) The graph of $y = (2\ln x)/3 + \ln 4 - \ln(\ln e^2)$ contains only the right-hand branch of the graph of $y = \ln(2x^{2/3})$.

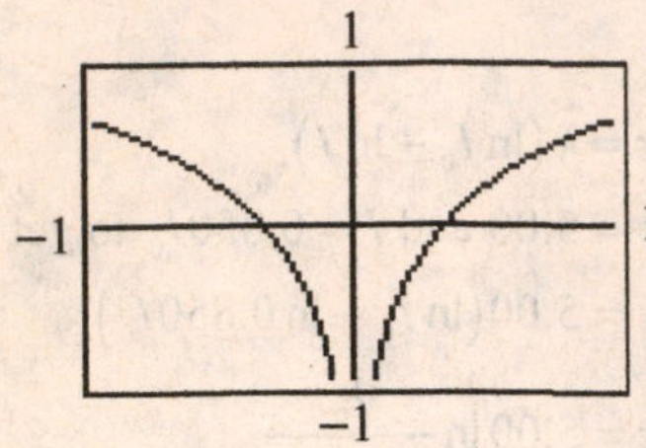

$y = \ln(2x^{2/3})$

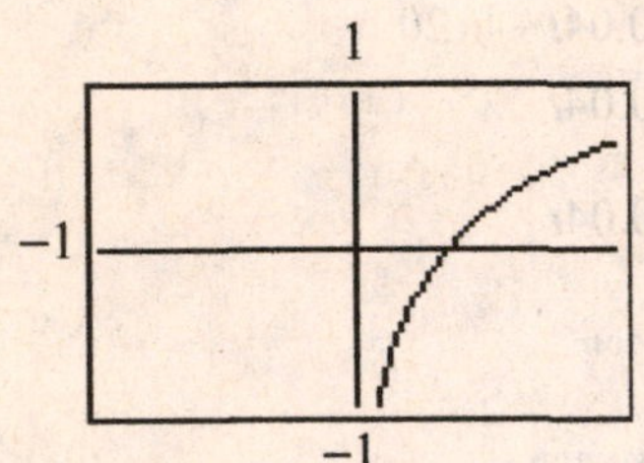

$y = (2\ln x)/3 + \ln 4 - \ln(\ln e^2)$

Chapter 14

Additional Types of Equations and Systems of Equations

14.1 Graphical Solution of Systems of Equations

1. Graph $y_1 = 3x^2 + 6x$.

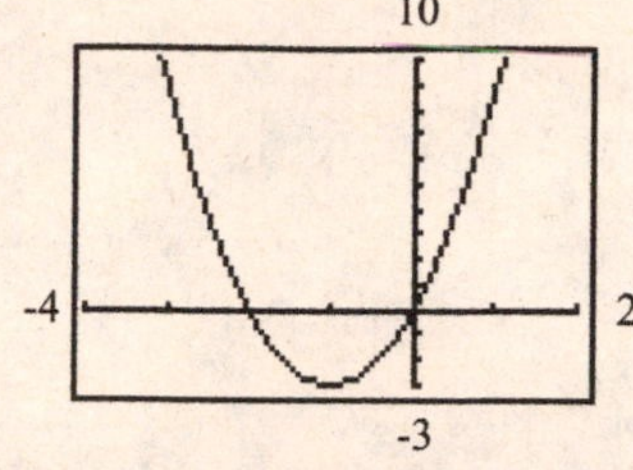

5. $y = \dfrac{x}{2}$

$x^2 + y^2 = 16 \Rightarrow y = \pm\sqrt{16 - x^2}$.

Graph $y_1 = x/2$, $y_2 = \sqrt{16 - x^2}$, and $y_3 = -\sqrt{16 - x^2}$.

Use the intersect feature to solve.

Solutions:

$x = 3.58,\ y = 1.79$

$x = -3.58,\ y = -1.79$

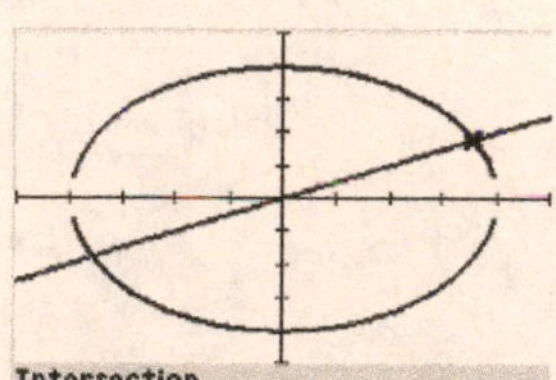

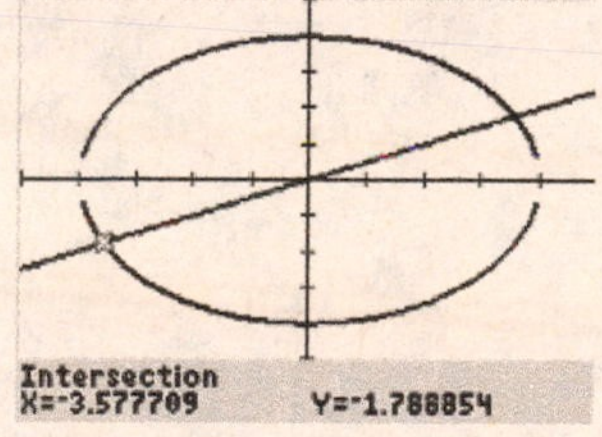

9. $y = x^2 - 2$

$4y = 12x - 17 \quad y \quad \frac{12x - 17}{4}$

Graph $y_1 = x^2 - 2$ and $y_2 = (12x - 17)/4$.

Use the intersect feature to solve.

Solution:

$x = 1.50,\ y = 0.25$

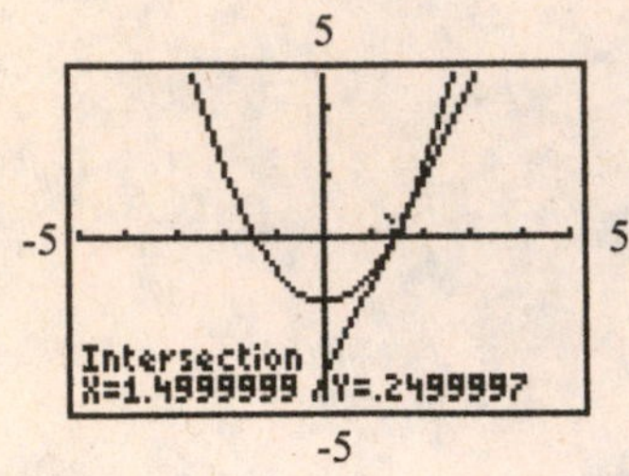

13. $y = x^2 - 3$

$x^2 + y^2 = 25 \quad y \quad \sqrt{25 \quad x^2}$

Graph $y_1 = \sqrt{25 - x^2}$, $y_2 = -\sqrt{25 - x^2}$, $y_3 = x^2 - 3$.

Use the intersect feature to solve.

Solutions:

$x = -2.69,\ y = 4.22$

$x = 2.69,\ y = 4.22$

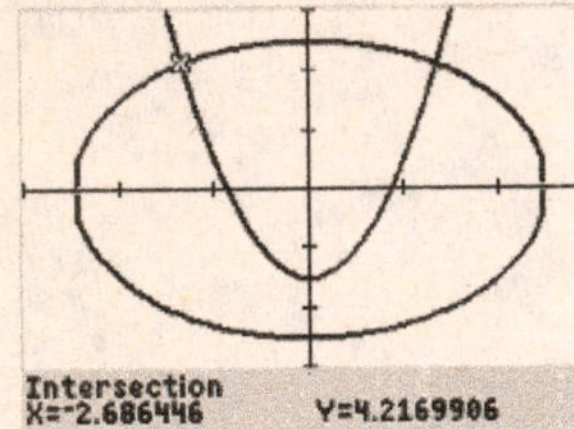

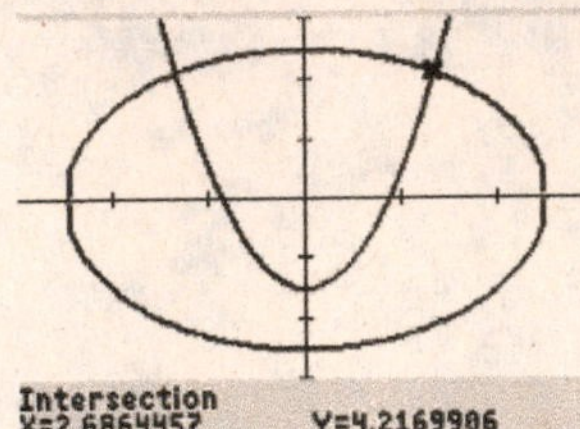

17. $2x^2 + 3y^2 = 19 \quad y \quad \sqrt{(19 \quad 2x^2)/3}$

$x^2 + y^2 = 9 \quad y \quad \sqrt{9 \quad x^2}.$

Graph $y_1 = \sqrt{(19 - 2x^2)/3}$, $y_2 = -\sqrt{(19 - 2x^2)/3}$,

$y_3 = \sqrt{9 - x^2}$; $y_4 = -\sqrt{9 - x^2}$ and use the intersect feature to solve.

Solutions:

$x = -2.83,\ y = 1.00$

$x = 2.83,\ y = 1.00$

$x = -2.83,\ y = -1.00$

$x = 2.83,\ y = -1.00$

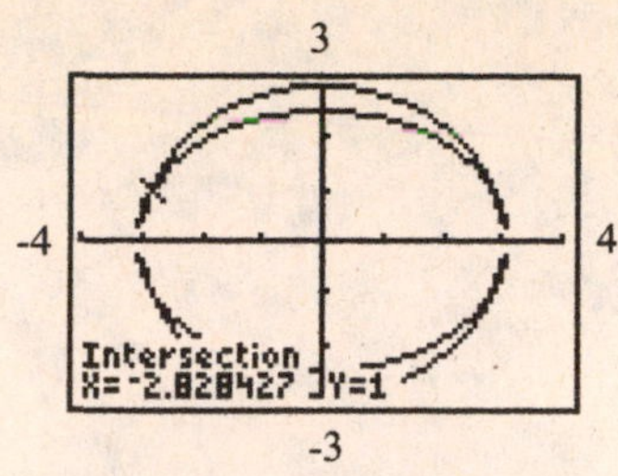

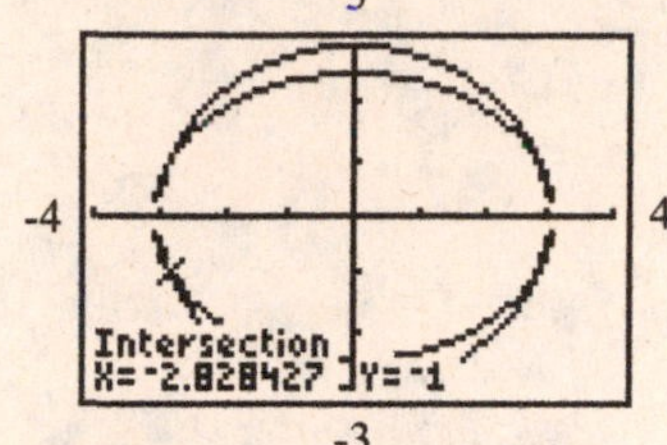

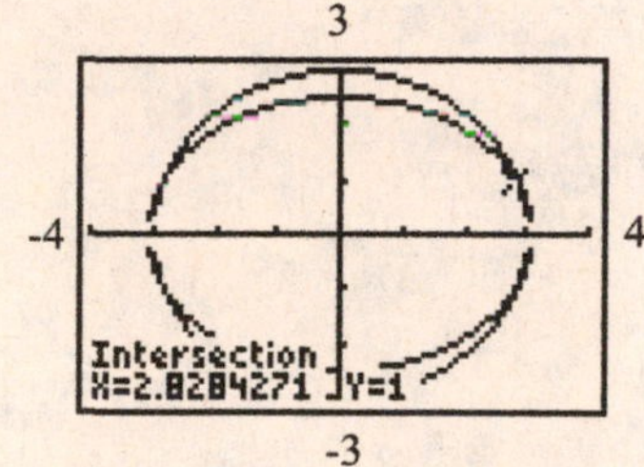

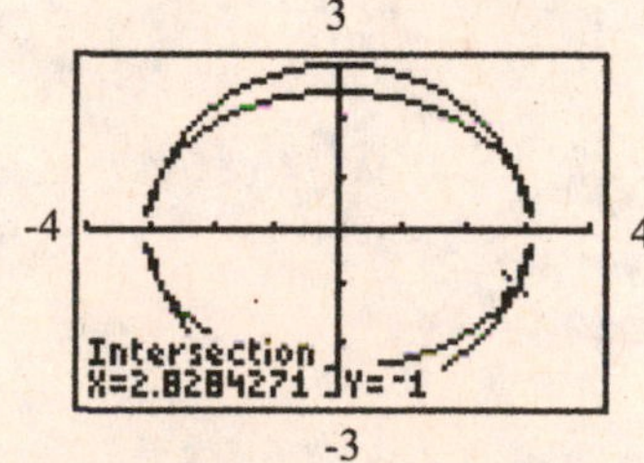

21. $y = x^2/4$

$y = \sin x$

Graph $y_1 = x^2/4$ and $y_2 = \sin x$ and use the intersect feature to solve.

Solutions:

$x = 0.00,\ y = 0.00$

$x = 1.93,\ y = 0.93$

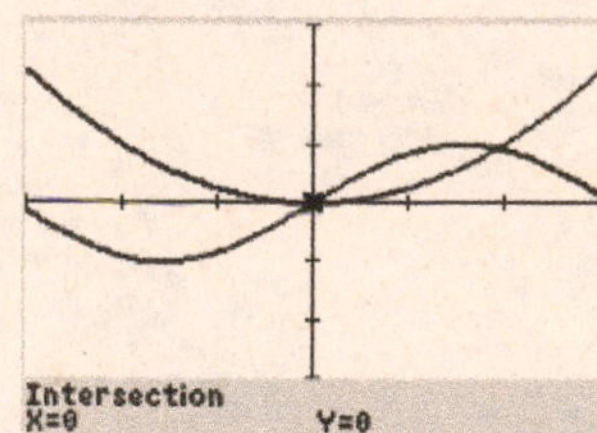

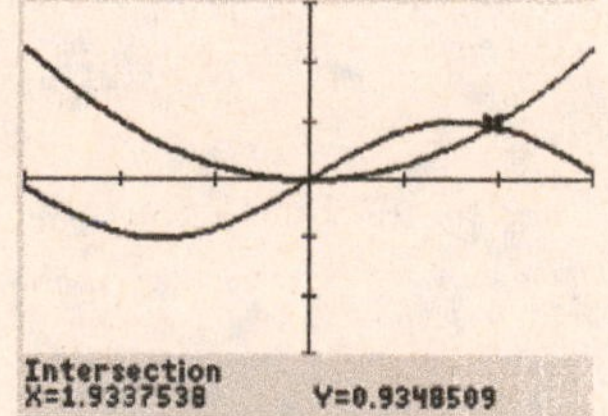

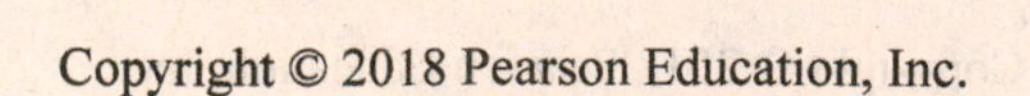

25. $x^2 - y^2 = 7 \Rightarrow y = \pm\sqrt{x^2 - 7},$

$y = 5\log x$

Graph $y_1 = \sqrt{x^2 - 7}$, $y_2 = -\sqrt{x^2 - 7}$, and $y_3 = 5\log x$.

Use the intersect feature to solve.

Solution:

$x = 4.01,\ y = 3.02$

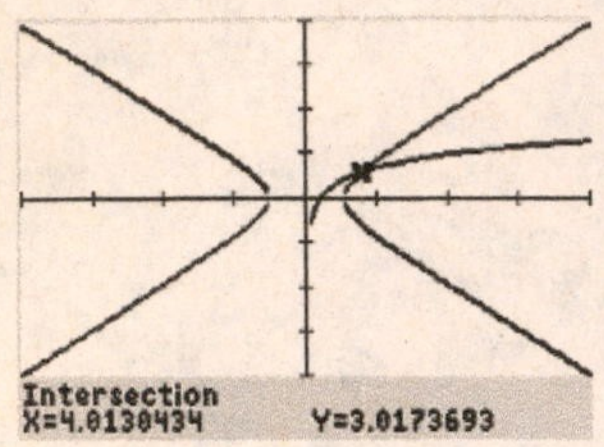

29.

$$10^{x+y} = 150$$

$$(x+y)\log 10 = \log 150$$

$$y = \log 150 - x$$

$$y = x^2$$

Graph $y_1 = \log 150 - x$, $y_2 = x^2$ and use the intersect feature to solve.

Solutions:

$x = -2.06,\ y = 4.23$

$x = 1.06,\ y = 1.12$

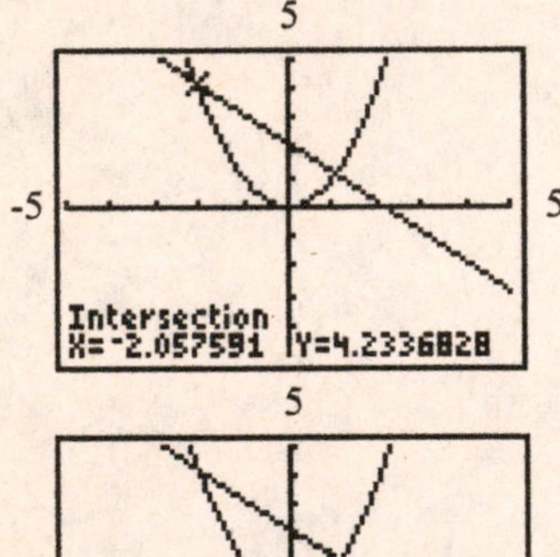

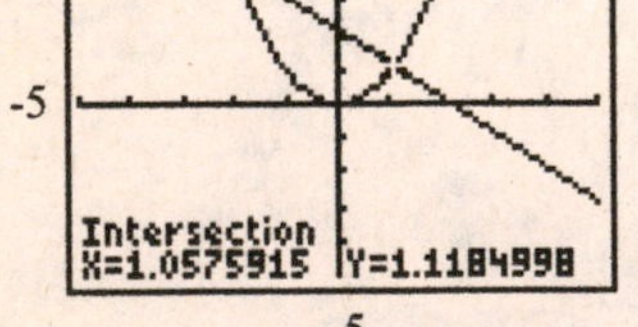

33. Let x = distance east

y = distance north

t = time (in hours) traveling east

Then $x = 45t$,

$$y = 40(5-t)$$
$$= 200 - 40t$$
$$= 200 - \frac{40}{45}x$$
$$= 200 - \frac{8}{9}x$$

$$x^2 + y^2 = 150^2$$
$$y = \sqrt{150^2 - x^2}, y > 0$$

Graph $y_1 = 200 - \frac{8}{9}x,\ x > 0$ and

$y_2 = \sqrt{150^2 - x^2}$, and use the intersect feature to solve.

One solution is $x = 90$ mi, $y = 120$ mi, taking 2 hours east and 3 hours north.

A second solution is $x = 108.6$ mi, $y = 103.4$ mi, taking 2.41 hours east and 2.59 hours north.

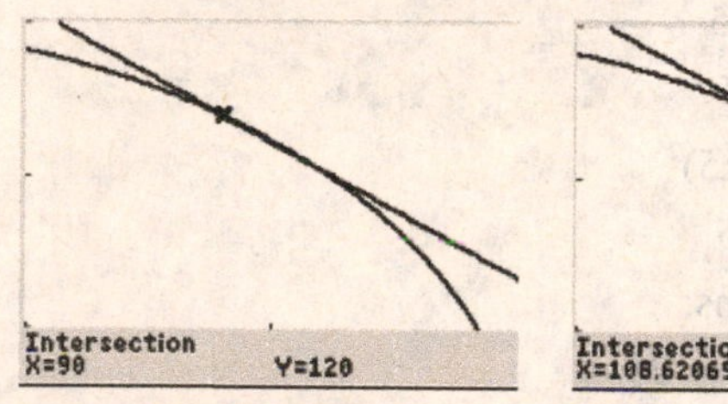

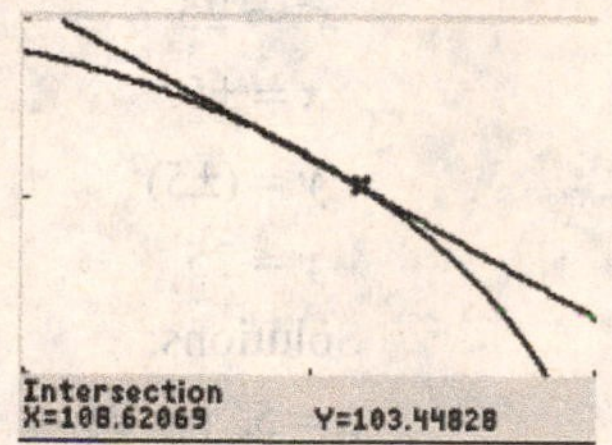

37. $x^2 + y^2 = 41 \quad y \quad \sqrt{41 \quad x^2}$

$y^2 = 20x + 140 \quad y \quad \sqrt{20x \quad 140}$

Graph $y_1 = \sqrt{41 - x^2}$, $y_2 = -\sqrt{41 - x^2}$,

$y_3 = \sqrt{20x + 140}, y_4 = \sqrt{20x + 140}$

From the graph, there is no intersection. No, the meteorite will not strike the earth.

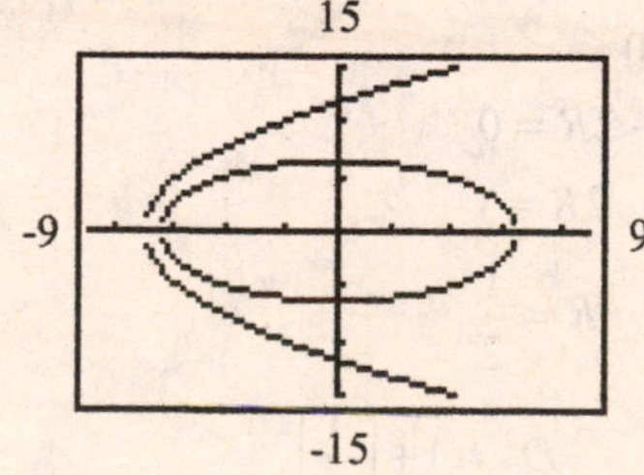

14.2 Algebraic Solution of Systems of Equations

1. $2x+y=4 \Rightarrow y=4-2x.$

Substitute into second equation: $x^2-y^2=4$

$$x^2-(4-2x)^2=4$$
$$x^2-(16-16x+4x^2)=4$$
$$x^2-16+16x-4x^2=4$$
$$3x^2-16x+20=0$$
$$(x-2)(3x-10)=0$$
$$x-2=0 \quad \text{or} \quad 3x-10=0$$
$$x=2 \qquad x=\frac{10}{3}$$
$$y=4-2(2) \qquad y=4-2\left(\frac{10}{3}\right)=-\frac{8}{3}$$

Solutions:

$x=2,\ y=0$

$x=\frac{10}{3},\ y=-\frac{8}{3}$

5. $y=2x+9$

Substitute into: $y=x^2+1$

$$2x+9=x^2+1$$
$$x^2-2x-8=0$$
$$(x-4)(x+2)=0$$
$$x=4 \quad \text{or} \quad x=-2$$
$$y=2(4)+9 \qquad y=2(-2)+9$$
$$y=17 \qquad y=5$$

Solutions:

$x=4,\ y=17$

$x=-2,\ y=5$

9. $x+y=1 \quad y \quad 1 \quad x$

Substitute into: $x^2-y^2=1$

$$x^2-(1-x)^2=1$$
$$x^2-1+2x-x^2=1$$
$$-1+2x=1$$
$$2x=2$$
$$x=1$$
$$y=1-(1)=0$$

Solution:

$x=1,\ y=0$

13. $w+h=6 \Rightarrow h=6-w$

Substitute into: $wh=9$

$$w(6-w)=9$$
$$6w-w^2=9$$
$$w^2-6w+9=0$$
$$(w-3)^2=0$$
$$w-3=0$$
$$w=3$$
$$3+h=6$$
$$h=3$$

Solution:

$w=3,\ h=3$

17. $y=x^2$

Substitute into: $y=3x^2-50$

$$x^2=3x^2-50$$
$$2x^2=50$$
$$x^2=25$$
$$x=\pm 5$$
$$y=(\pm 5)^2$$
$$y=25$$

Solutions:

$x=5,\ y=25$

$x=-5,\ y=25$

Alternatively, subtracting the second equation from the first also results in $2x^2=50$.

21. $D^2-1=R \Rightarrow D^2=1+R$

Substitute into: $D^2-2R^2=1$

$$1+R-2R^2=1$$
$$R-2R^2=0$$
$$R(1-2R)=0$$
$$R=0 \quad \text{or} \quad 1-2R=0$$
$$2R=1$$
$$R=\frac{1}{2}$$
$$D^2=1+(0) \qquad D^2=1+\left(\frac{1}{2}\right)$$
$$=\frac{6}{4}$$
$$D=\pm 1 \qquad D=\frac{\pm\sqrt{6}}{2}$$

Solutions:

$R = 0, D = 1$

$R = 0, D = -1$

$R = \frac{1}{2}, D = \frac{\sqrt{6}}{2}$

$R = \frac{1}{2}, D = \frac{-\sqrt{6}}{2}$

Alternatively, subtracting the first equation from the second also results in $R - 2R^2 = 0$.

25. $x^2 + 3y^2 = 37$

$2x^2 - 9y^2 = 14$

Multiplying the first equation by 3:

$$3x^2 + 9y^2 = 111$$
$$2x^2 - 9y^2 = 14$$
$$5x^2 = 125$$
$$x^2 = 25$$
$$x = \pm 5$$
$$(\pm 5)^2 + 3y^2 = 37$$
$$25 + 3y^2 = 37$$
$$3y^2 = 12$$
$$y^2 = 4$$
$$y = \pm 2$$

Solutions:

$x = 5, y = 2$

$x = 5, y = -2$

$x = -5, y = 2$

$x = -5, y = -2$

29. $x - y = a - b \Rightarrow y = x - (a - b)$

Substituting into: $x^2 - y^2 = a^2 - b^2$

$$x^2 - (x - (a - b))^2 = a^2 - b^2$$
$$x^2 - \left(x^2 - 2(a - b)x + (a - b)^2\right) = a^2 - b^2$$
$$x^2 - x^2 + 2(a - b)x - a^2 + 2ab - b^2 = a^2 - b^2$$
$$2(a - b)x + 2ab - 2a^2 = 0$$
$$(a - b)x - a(a - b) = 0$$
$$x = a$$
$$y = x - (a - b)$$
$$= a - a + b$$
$$y = b$$

Solution:

$x = a, y = b$

33.

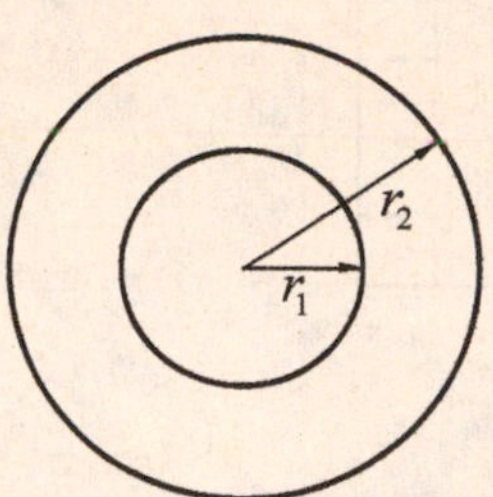

Let r_1 = inner radius, r_2 = outer radius.

$$r_1 = r_2 - 2.00$$
$$\pi r_2^2 - \pi r_1^2 = 37.7$$

Substituting $r_1 = r_2 - 2.00$ into $\pi r_2^2 - \pi r_1^2 = 37.7$

$$\pi r_2^2 - \pi(r_2 - 2.00)^2 = 37.7$$
$$\pi r_2^2 - \pi r_2^2 + 4.00\pi r_2 - 4.00\pi = 37.7$$
$$r_2 = \frac{37.7 + 4.00\pi}{4.00\pi}$$
$$r_2 = 4.00$$

Then $r_1 = r_2 - 2.00$

$= 2.00$

The radii are 2.00 cm and 4.00 cm.

37.

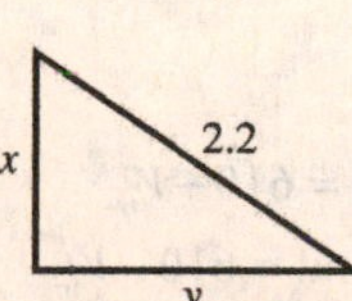

$x + y + 2.2 = 4.6 \Rightarrow y = 2.4 - x$

Substitute into: $x^2 + y^2 = 2.2^2$

$$x^2 + (2.4 - x)^2 = 4.84$$
$$x^2 + 5.76 - 4.8x + x^2 = 4.84$$
$$2x^2 - 4.8x + 0.92 = 0$$
$$x^2 - 2.4x + 0.46 = 0$$

Using the quadratic formula:

$$x = \frac{-(-2.4) \pm \sqrt{(-2.4)^2 - 4(1)(0.46)}}{2(1)}$$
$$= \frac{2.4 \pm \sqrt{3.92}}{2}$$

$x = 2.19$ or $x = 0.21$

$(2.19) + y = 2.40$ $(0.21) + y = 2.40$

$y = 0.21$ $y = 2.19$

The lengths of the sides of the truss are 2.19 m and 0.21 m.

41.

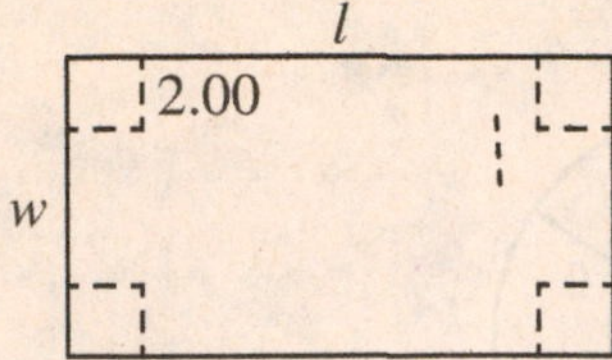

$$wl = 1600 \quad w \quad \frac{216}{l}$$

Substitute into: $2(w - 4.00)(l - 4.00) = 224$

$$\frac{216}{l} - 4.00 \quad (l - 4.00) = 112$$

$$216 - \frac{864}{l} - 4l + 16 = 112$$

$$120l - 4l^2 - 864 = 0$$

$$l^2 - 30l + 216 = 0$$

$$(l - 18)(l - 12) = 0$$

$$l - 18 = 0 \quad \text{or} \quad l - 12 = 0$$

$$l = 18 \qquad\qquad l = 12$$

$$w = \frac{216}{(18)} \qquad w = \frac{216}{(12)}$$

$$= 12 \qquad\qquad = 18$$

The dimensions of the sheet are
18.0 cm. by 12.0 cm.

45. Let V_w = wind velocity

V_{ww} = velocity with wind = $610 + V_w$

V_{aw} = velocity against wind = $610 - V_w$

Use distance = velocity · time, so

$$\text{time} = \frac{\text{distance}}{\text{velocity}}, \text{ and}$$

$$\frac{3660}{V_{ww}} + 1.6 = \frac{3660}{V_{aw}}$$

$$\frac{3660}{610 + V_w} + 1.6 = \frac{3660}{610 - V_w}$$

$$3660(610 - V_w) + 1.6(610 - V_w)(610 + V_w) = 3660(610 + V_w)$$

$$2232600 - 3660V_w + 1.6(610^2 - V_w^2) = 2232600 + 3660V_w$$

$$-1.6V_w^2 - 7320V_w + 1.6(610)^2 = 0$$

$$V_w^2 + 4575V_w - 372100 = 0$$

Using the quadratic formula

$$V_w = \frac{-4575 \pm \sqrt{4575^2 - 4(1)(-372100)}}{2}$$

$$V_w = 80, -4655$$

Solution: the wind velocity is 80 mi/h.

14.3 Equations in Quadratic Form

1. $2x^4 - 7x^2 = 4$

$2x^4 - 7x^2 - 4 = 0$, let $y = x^2$

$2y^2 - 7y - 4 = 0$

$(y-4)(2y+1) = 0$

$y - 4 = 0$ or $2y + 1 = 0$

$y = 4$ $\quad y = -\frac{1}{2}$

$x^2 = 4$ $\quad x^2 = -\frac{1}{2}$

$x = \pm 2$ $\quad x = \pm\sqrt{-\frac{1}{2}}$

$= \pm j\sqrt{\frac{1}{2}\cdot\frac{2}{2}}$

$= \pm j\frac{\sqrt{2}}{2}$

Check:

$2(\pm 2)^4 - 7(\pm 2)^2 = 32 - 28 = 4$

$2\left(\pm j\frac{\sqrt{2}}{2}\right)^4 - 7\left(\pm j\frac{\sqrt{2}}{2}\right)^2 = \frac{1}{2} + \frac{7}{2} = 4$

5. $3x^{-2} - 7x^{-1} - 6 = 0$

Let $y = x^{-1}$, $y^2 = x^{-2}$

$3y^2 - 7y - 6 = 0$

$(3y+2)(y-3) = 0$

$3y + 2 = 0$ or $y - 3 = 0$

$y = -\frac{2}{3}$ $\quad y = 3$

$x^{-1} = -\frac{2}{3}$ $\quad x^{-1} = 3$

$x = -\frac{3}{2}$ $\quad x = \frac{1}{3}$

Check:

$3\left(-\frac{3}{2}\right)^{-2} - 7\left(-\frac{3}{2}\right)^{-1} - 6 = \frac{4}{3} + \frac{14}{3} - 6 = 0$

$3\left(\frac{1}{3}\right)^{-2} - 7\left(\frac{1}{3}\right)^{-1} - 6 = 27 - 21 - 6 = 0$

9. $2x - 5\sqrt{x} + 3 = 0$

Let $y = \sqrt{x}$, $y^2 = x$

$2y^2 - 5y + 3 = 0$

$(2y-3)(y-1) = 0$

$2y - 3 = 0$ or $y - 1 = 0$

$2y = 3$ $\quad y = 1$

$y = \frac{3}{2}$

$\sqrt{x} = \frac{3}{2}$ $\quad \sqrt{x} = 1$

$x = \frac{9}{4}$ $\quad x = 1$

Check:

$2\left(\frac{9}{4}\right) - 5\sqrt{\frac{9}{4}} + 3 = \frac{9}{2} - \frac{15}{2} + \frac{6}{2} = 0$

$2(1) - 5\sqrt{1} + 3 = 2 - 5 + 3 = 0$

13. $x^{2/3} - 2x^{1/3} - 15 = 0$

Let $y = x^{1/3}$, $y^2 = x^{2/3}$

$y^2 - 2y - 15 = 0$

$(y-5)(y+3) = 0$

$y - 5 = 0$ or $y + 3 = 0$

$y = 5$ $\quad y = -3$

$x^{1/3} = 5$ $\quad x^{1/3} = -3$

$x = 125$ $\quad x = -27$

Check:

$125^{2/3} - 2(125)^{1/3} - 15 = 25 - 10 - 15 = 0$

$(-27)^{2/3} - 2(-27)^{1/3} - 15 = 9 + 6 - 15 = 0$

17. $(x-1) - \sqrt{x-1} = 20$

Let $y = \sqrt{x-1}$, $y^2 = x - 1$

$y^2 - y - 20 = 0$

$(y-5)(y+4) = 0$

$y - 5 = 0$ or $y + 4 = 0$

$y = 5$ $\quad y = -4$

$\sqrt{x-1} = 5$ $\quad \sqrt{x-1} = -4$

$x - 1 = 25$ $\quad$ not possible

$x = 26$

Check:

$(26-1) - \sqrt{26-1} = 25 - \sqrt{25}$

$= 25 - 5$

$= 20$

21. $x-3\sqrt{x-2}=6$

Let $y=\sqrt{x-2},\ y^2=x-2\Rightarrow y^2+2=x$

$y^2+2-3y=6$

$y^2-3y-4=0$

$(y-4)(y+1)=0$

$y-4=0$ or $y+1=0$

$y=4$ $\qquad y=-1$

$\sqrt{x-2}=4$ $\qquad \sqrt{x-2}=-1$

$x-2=16$ $\qquad$ (not possible)

$x=18$

Check:

$18-3\sqrt{18-2}=18-3\sqrt{16}=6$

25. $\dfrac{1}{s^2+1}+\dfrac{2}{s^2+3}=1$

Clear denominators:

$s^2+3+2(s^2+1)=(s^2+3)(s^2+1)$

$3s^2+5=s^4+4s^2+3$

$s^4+s^2-2=0$

$y=s^2$

$y^2+y-2=0$

$(y-1)(y+2)=0$

$y=1$ or $y=-2$

$s^2=1$ $\qquad s^2=-2$

$s=\pm 1$ $\qquad s=\pm j\sqrt{2}$

Check:

$\dfrac{1}{(\pm1)^2+1}+\dfrac{2}{(\pm1)^2+3}=\dfrac{1}{2}+\dfrac{2}{4}=1$

$\dfrac{1}{\left(\pm j\sqrt{2}\right)^2+1}+\dfrac{2}{\left(\pm j\sqrt{2}\right)^2+3}=\dfrac{1}{-2+1}+\dfrac{2}{-2+3}$

$=-1+2=1$

29. $x^4-20x^2+64=0$

Let $y=x^2, y^2=x^4$

$y^2-20y+64=0$

$(y-16)(y-4)=0$

$y-16=0$ or $y-4=0$

$y=16$ $\qquad y=4$

$x^2=16$ $\qquad x^2=4$

$x=\pm4$ $\qquad x=\pm2$

Check:

$(\pm4)^4-20(\pm4)^2+64=256-20(16)+64$

$=-64+64=0$

$(\pm2)^4-20(\pm2)^2+64=16-80+64$

$=-64+64=0$

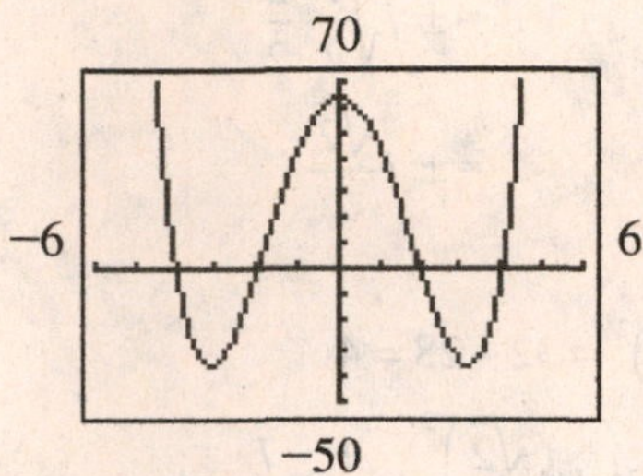

33. $(\log x)^2-3\log x+2=0$

Let $y=\log x$

$y^2-3y+2=0$

$(y-2)(y-1)=0$

$y-2=0$ or $y-1=0$

$y=2$ $\qquad y=1$

$\log x=2$ $\qquad \log x=1$

$x=100$ $\qquad x=10$

Check:

$(\log 100)^2-3\log 100+2=4-6+2=0$

$(\log 10)^2-3\log 10+2=1-3+2=0$

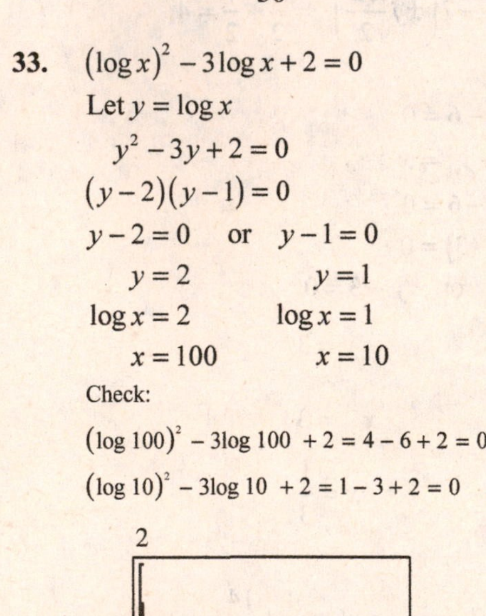

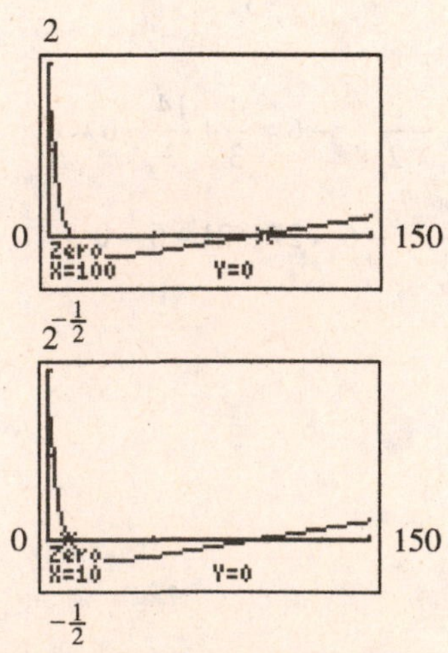

37. $R_T^{-1} = R_1^{-1} + R_2^{-1}$

$1.00^{-1} = R_1^{-1} + \sqrt{R_1}^{-1}$

$1 = R_1^{-1} + R_1^{-1/2}.$

Let $y = R_1^{-1/2}$, $y^2 = R_1^{-1}$, $y > 0$

$y^2 + y - 1 = 0.$

Using the quadratic formula:

$$y = \frac{-1 \pm \sqrt{1^2 - 4(1)(-1)}}{2(1)}$$

$$= \frac{-1 \pm \sqrt{5}}{2}.$$

The negative is rejected since $y > 0$

$$y = \frac{-1+\sqrt{5}}{2}$$

$$R_1^{-1/2} = \frac{-1+\sqrt{5}}{2}$$

$$\sqrt{R_1} = \frac{2}{-1+\sqrt{5}}$$

$$R_1 = \left(\frac{2}{-1+\sqrt{5}}\right)^2$$

$= 2.62\ \Omega$

$R_2 = R_1^{1/2}$

$$= \frac{2}{-1+\sqrt{5}}$$

$R_2 = 1.62\ \Omega$

41. $lw = 1540 \quad w \quad \dfrac{1540}{l}$

Substitute into: $l^2 + w^2 = 60.0^2$

$$l^2 + \frac{1540^2}{l^2} = 60.0^2$$

$l^4 - 3600l^2 + 1540^2 = 0$

Let $y = l^2, y^2 = l^4, y > 0$

$y^2 - 3600y + 1540^2 = 0$

Using the quadratic formula:

$$y = \frac{-(-3600) \pm \sqrt{(-3600)^2 - 4(1)(1540^2)}}{2(1)}$$

$y = 2731.88$ or $y = 868.12$

$l^2 = 2731.88$ $\quad l^2 = 868.12$

$l = 52.27$ $\quad l = 29.46$

$w = \dfrac{1540}{52.27} = 29.46$ $\quad w = \dfrac{1540}{29.46} = 52.27$

The dimensions are 52.3 in by 29.5 in

14.4 Equations with Radicals

1. $2\sqrt{3x-1} = 3$; square both sides

$4(3x-1) = 9$

$12x - 4 = 9$

$12x = 13$

$x = \dfrac{13}{12}$

Check: $2\sqrt{3\left(\dfrac{13}{12}\right) - 1} \stackrel{?}{=} 3$

$3 = 3$

$x = \dfrac{13}{12}$ is the solution.

5. $\sqrt{x-8} = 2$; square both sides

$x - 8 = 4$

$x = 12$

Check:

$\sqrt{12-8} = \sqrt{4} = 2$

$x = 12$ is the solution.

9. $2\sqrt{3x+2} = 6x$

$\sqrt{3x+2} = 3x$; square both sides

$3x + 2 = 9x^2$

$9x^2 - 3x - 2 = 0$

$(3x+1)(3x-2) = 0$

$3x + 1 = 0$ or $3x - 2 = 0$

$3x = -1$ $\quad 3x = 2$

$x = -\dfrac{1}{3}$ $\quad x = \dfrac{2}{3}$

Check:

$\sqrt{3 \cdot \dfrac{1}{3} + 2} \stackrel{?}{=} 3 \cdot \dfrac{-1}{3}$

$\sqrt{-1+2} \stackrel{?}{=} -1$

$1 \neq -1$

$x = -\dfrac{1}{3}$ is not a solution.

Check:

$\sqrt{3 \cdot \dfrac{2}{3} + 2} \stackrel{?}{=} 3 \cdot \dfrac{2}{3}$

$\sqrt{2+2} \stackrel{?}{=} 2$

$2 = 2$

$x = \dfrac{2}{3}$ is the only solution.

13. $\sqrt[3]{y-7}=2$; cube both sides

$y-7=2^3$

$=8$

$y=15$

Check:

$\sqrt[3]{15-7}=\sqrt[3]{8}=2$

$y=15$ is the solution.

17. $\sqrt{x^2-11}=5$; square both sides

$x^2-11=25$

$x^2=36$

$x=\pm 6$

Check:

$\sqrt{(\pm 6)^2-11}=\sqrt{36-11}$

$=\sqrt{25}$

$=5$

The solutions are $x=\pm 6$.

21. $\sqrt{5+\sqrt{x}}=\sqrt{x}-1$; square both sides

$5+\sqrt{x}=x-2\sqrt{x}+1$

$x-3\sqrt{x}-4=0$

Let $y=\sqrt{x}, y^2=x, y\ge 0$

$y^2-3y-4=0$

$(y-4)(y+1)=0$

$y-4=0$ or $y+1=0$

$y=4$ $\quad$ $y=-1$

$x=4^2$ $\quad$ (not possible)

$=16$

Check:

$\sqrt{5+\sqrt{16}}\overset{?}{=}\sqrt{16}-1$

$\sqrt{9}\overset{?}{=}4-1$

$3=3$

The only solution is $x=16$.

25. $2\sqrt{x+2}-\sqrt{3x+4}=1$

$2\sqrt{x+2}=1+\sqrt{3x+4}$; square both sides

$4(x+2)=\left(1+\sqrt{3x+4}\right)^2$

$4x+8=1+2\sqrt{3x+4}+3x+4$

$x+3=2\sqrt{3x+4}$; square both sides

$(x+3)^2=4(3x+4)$

$x^2+6x+9=12x+16$

$x^2-6x-7=0$

$(x-7)(x+1)=0$

$x=7$ or $x=-1$

Check:

$2\sqrt{-1+2}-\sqrt{3(-1)+4}=2\sqrt{1}-\sqrt{1}=1$

$2\sqrt{7+2}-\sqrt{3(7)+4}=2\sqrt{9}-\sqrt{25}=1$

The solutions are $x=-1$ and $x=7$.

29. $\sqrt{6x-5}-\sqrt{x+4}=2$

$\sqrt{6x-5}=\sqrt{x+4}+2$; square both sides

$6x-5=x+4+4\sqrt{x+4}+4$

$5x-13=4\sqrt{x+4}$; square both sides

$25x^2-130x+169=16x+64$

$25x^2-146x+105=0$

Using the quadratic formula:

$$x=\frac{-(-146)\pm\sqrt{(-146)^2-4(25)(105)}}{2(25)}$$

$$=\frac{146\pm 104}{2(25)}$$

$x=5$ or $x=\frac{21}{25}$

Check:

$\sqrt{6(5)-5}-\sqrt{5+4}=\sqrt{25}-\sqrt{9}=2$

$\sqrt{6\left(\frac{21}{25}\right)-5}-\sqrt{\frac{21}{25}+4}=\sqrt{\frac{1}{25}}-\sqrt{\frac{121}{25}}$

$=\frac{1}{5}-\frac{11}{5}=-2\ne 2$

The only solution is $x=5$.

33. $\sqrt{x-\sqrt{2x}}=2$; square both sides

$x-\sqrt{2x}=4$

$x-4=\sqrt{2x}$; square both sides

$x^2-8x+16=2x$

$x^2-10x+16=0$

$(x-2)(x-8)=0$

$x=2 \quad \text{or} \quad x=8$

Check:

$x=2: \quad \sqrt{2-\sqrt{2(2)}}=j\sqrt{2}\neq 2$

$x=8: \quad \sqrt{8-\sqrt{2(8)}}=\sqrt{4}=2$

$x=8$ is the only solution.

37. $\sqrt{2x+1}+3\sqrt{x}=9$

$\sqrt{2x+1}=9-3\sqrt{x}$; square both sides

$2x+1=81-54\sqrt{x}+9x$

$54\sqrt{x}=7x+80$; square both sides

$2916x=49x^2+1120x+6400$

$49x^2-1796x+6400=0$

$(x-4)(49x-1600)=0$

$x-4=0 \quad \text{or} \quad 49x=1600$

$x=4 \qquad x=\frac{1600}{49}$

Check:

$\sqrt{2(4)+1}+3\sqrt{4}\overset{?}{=}9$

$3+6\overset{?}{=}9$

$9=9$

$\sqrt{2\left(\frac{1600}{49}\right)+1}\overset{?}{=}9$

$\frac{57}{7}+\frac{120}{7}\overset{?}{=}9$

$\frac{177}{9}\neq 9$

$x=4$ is the only solution.
We see from the graph that there is only one point of intersection at $x=4$.

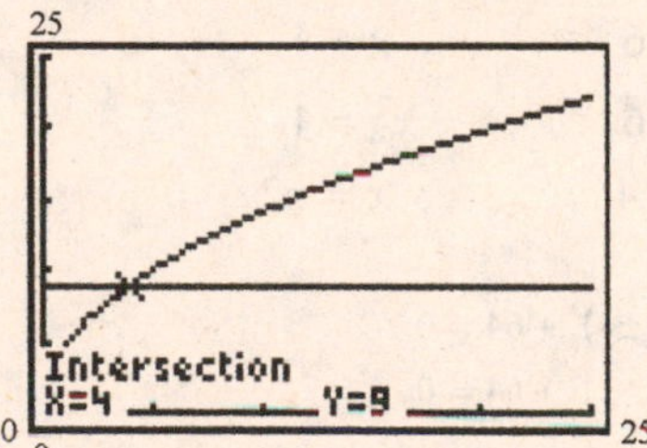

41. $\sqrt{x-1}+x=3$

$\sqrt{x-1}=3-x$; square both sides

$x-1=9-6x+x^2$

$x^2-7x+10=0$

$(x-5)(x-2)=0$

$x-5=0 \quad \text{or} \quad x-2=0$

$x=5 \qquad x=2$

Check:

$\sqrt{5-1}+5=\sqrt{4}+5=7\neq 3$

$\sqrt{2-1}+2=\sqrt{1}+2=3$

The only solution is $x=2$.
To compare this solution with that of Example 4, we write $\sqrt{x-1}=x-3$ vs. $\sqrt{x-1}=3-x$ as $\sqrt{x-1}=x-3$ vs. $-\sqrt{x-1}=x-3$. We see that one represents the positive square root and the other one the negative root. Squaring both sides gives the same quadratic equation for both problems, but the extraneous root for one is the solution for the other and viceversa.

45. $kC=\sqrt{R_1^2-R_2^2}+\sqrt{r_1^2-r_2^2}-A$

$\sqrt{r_1^2-r_2^2}=kC+A-\sqrt{R_1^2-R_2^2}$; square both sides

$r_1^2-r_2^2=\left(kC+A-\sqrt{R_1^2-R_2^2}\right)^2$

$r_1^2=\left(kC+A-\sqrt{R_1^2-R_2^2}\right)^2+r_2^2$

49.

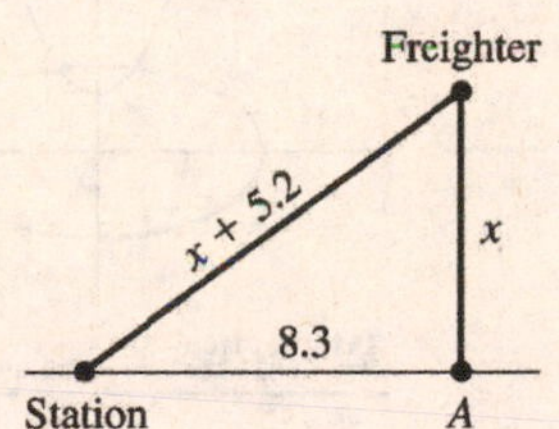

$y=x+5.2$

Substitute into: $y=\sqrt{x^2+8.3^2}$

$x+5.2=\sqrt{x^2+8.3^2}$; square both sides

$x^2+10.4x+27.04=x^2+68.89$

$10.4x=41.85$

$x=4.0$

$y=4.0+5.2$

$=9.2$

Check:

$\sqrt{4.0^2+83.3^2}=9.2$

The station is 9.2 km from the freighter.

Review Exercises

1. This is false. To obtain the entire curve, it is necessary to include $y_2 = -\sqrt{4-2x^2}$ as well.

5. Graph $y = \dfrac{9-x}{2}$, $y_2 = 3x^2$, and use the intersect feature to solve.

Solutions:

$x = -1.31,\ y = 5.16$

$x = 1.14,\ y = 3.93$

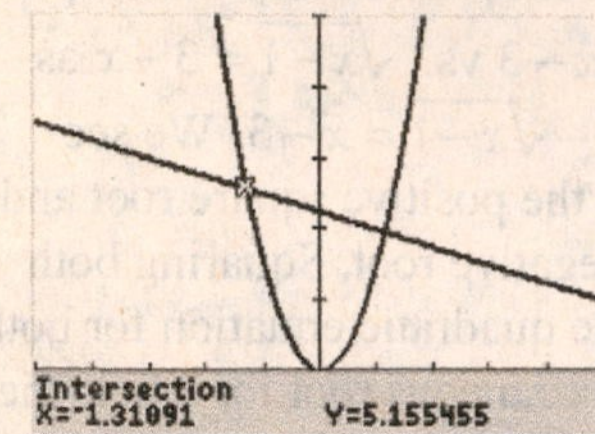

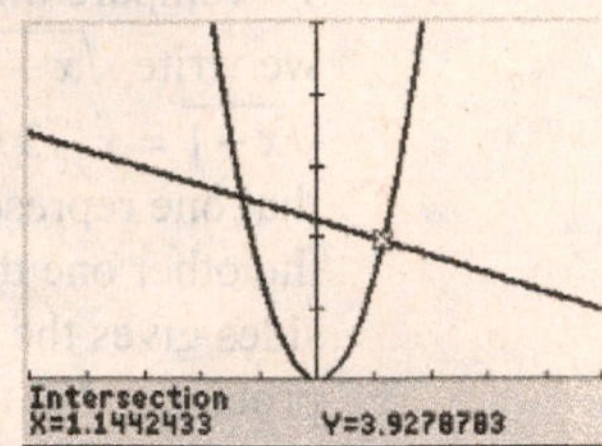

9. Graph $y_1 = x^2+1, y_2 = \sqrt{\dfrac{35-4x^2}{16}}, y_3 = -\sqrt{\dfrac{35-4x^2}{16}}$ and use the intersect feature to solve.

Solutions:

$x = -0.66,\ y = 1.44$

$x = 0.66,\ y = 1.44$

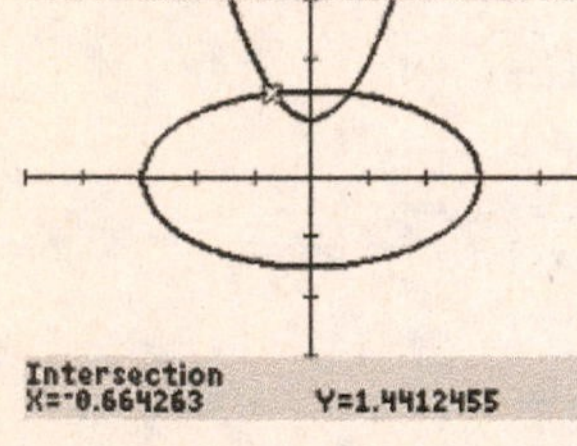

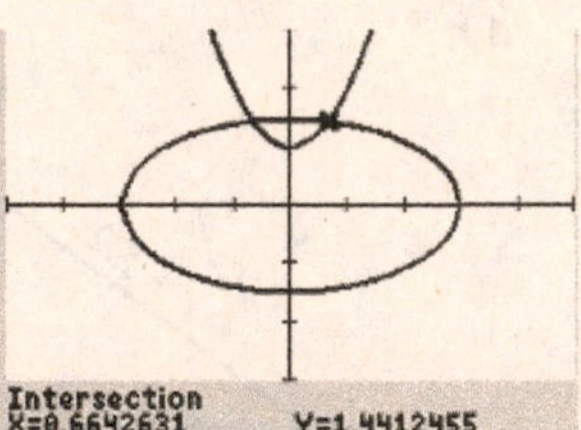

13. Graph $y_1 = x^2 - 2x,\ y_2 = 1 - e^{-x} + x$, and use the intersect feature to solve.

Solutions:

$x = 0,\ y = 0$

$x = 3.29,\ y = 4.26$

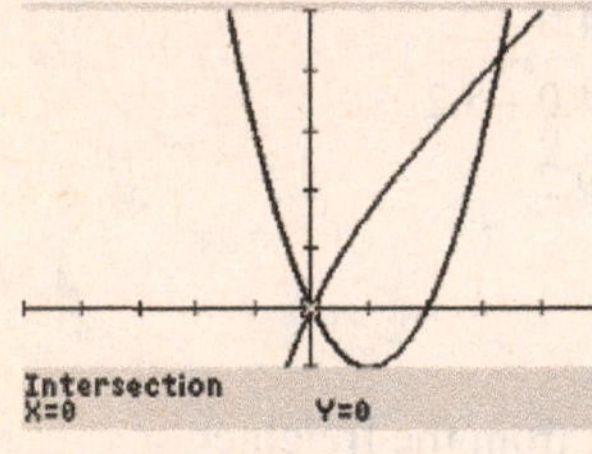

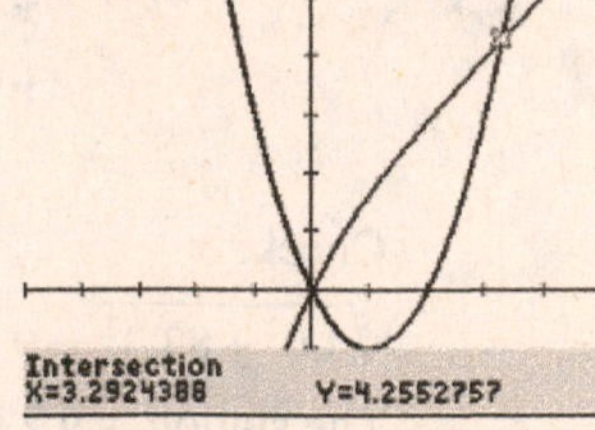

17. Substitute $L^2 = 3R$ into $R^2 + L^2 = 4$.

$$R^2 + 3R = 4$$
$$R^2 + 3R - 4 = 0$$
$$(R+4)(R-1) = 0$$
$$R+4=0 \quad \text{or} \quad R-1=0$$
$$R=-4 \qquad R=1$$
$$L^2 = 3(-4) \qquad L^2 = 2(1)$$
$$L = \pm j\sqrt{12} \qquad L = \pm\sqrt{2}$$

Solutions:

$L = \pm j\sqrt{12}, R = -4$

$L = \pm\sqrt{2},\ R = 1$

21. (1) $4x^2 - 7y^2 = 21$

(2) $x^2 + 2y^2 = 99$

Multiply (1) by 4, (2) by 7 and add:

(1) $8x^2 - 14y^2 = 42$

(2) $7x^2 + 14y^2 = 693$

$$15x^2 = 735$$
$$x^2 = 49$$
$$x = \pm 7$$

(2) $(\pm 7)^2 + 2y^2 = 99$

$$49 + 2y^2 = 99$$
$$2y^2 = 50$$
$$y^2 = 25$$
$$y = \pm 5$$

Solutions:

$x = 7,\ y = \pm 5$

$x = -7,\ y = \pm 5$

25. $x^4 - 20x^2 + 64 = 0$

Let $y = x^2,\ y^2 = x^4$

$$y^2 - 20y + 64 = 0$$
$$(y-16)(y-4) = 0$$
$$y-16=0 \quad \text{or} \quad y-4=0$$
$$y=16 \qquad y=4$$
$$x^2=16 \qquad x^2=4$$
$$x=\pm 4 \qquad x=\pm 2$$

Check:

$(\pm 4)^4 - 20(\pm 4)^2 + 64 = 0$

$(\pm 2)^4 - 20(\pm 2)^2 + 64 = 0$

29. $D^{-2}+4D^{-1}-21=0$

Let $x=D^{-1}$, $x^2=D^{-2}$

$$x^2+4x-21=0$$
$$(x+7)(x-3)=0$$
$$x+7=0 \quad \text{or} \quad x-3=0$$
$$x=-7 \qquad\qquad x=3$$
$$D^{-1}=-7 \qquad\qquad D^{-1}=3$$
$$D=-\frac{1}{7} \qquad\qquad D=\frac{1}{3}$$

Check:

$$\left(-\frac{1}{7}\right)^{-2}+4\left(-\frac{1}{7}\right)^{-1}-21=49-28-21=0$$
$$\left(\frac{1}{3}\right)^{-2}+4\left(\frac{1}{3}\right)^{-1}-21=9-12-21\neq 0$$

33. $2(x+1)^2-5(x+1)-3=0$

$$y=x+1$$
$$2y^2-5y-3=0$$
$$(2y+1)(y-3)=0$$
$$y=-\frac{1}{2} \quad \text{or} \quad y=3$$
$$x=y-1:$$

Solutions: $x=-\dfrac{3}{2}$ or $x=2$

37. $\dfrac{4}{r^2+1}+\dfrac{7}{2r^2+1}=2$

$$4\left(2r^2+1\right)+7\left(r^2+1\right)=2\left(r^2+1\right)\left(2r^2+1\right)$$
$$8r^2+4+7r^2+7=2\left(2r^4+3r^2+1\right)$$
$$15r^2+11=4r^4+6r^2+2$$
$$4r^4-9r^2-9=0$$

Let $y=r^2$, $y^2=r^4$

$$4y^2-9y-9=0$$
$$(4y+3)(y-3)=0$$
$$4y+3=0 \quad \text{or} \quad y-3=0$$
$$y=-\frac{3}{4} \qquad\qquad y=3$$
$$r^2=-\frac{3}{4} \qquad\qquad r^2=3$$
$$r=\pm\frac{j\sqrt{3}}{2} \qquad\qquad r=\pm\sqrt{3}$$

Check:

$$\frac{4}{\left(\pm\frac{j\sqrt{3}}{2}\right)^2+1}+\frac{7}{2\left(\pm\frac{j\sqrt{3}}{2}\right)^2+1}$$
$$=\frac{4}{-\frac{3}{4}+1}+\frac{7}{2\left(-\frac{3}{4}\right)+1}=16-14=2$$
$$\frac{4}{(\pm\sqrt{3})^2+1}+\frac{7}{2(\sqrt{3})^2+1}$$
$$=\frac{4}{3+1}+\frac{7}{6+1}=1+1=2$$

41. $\sqrt{5x+9}+1=x$

$\sqrt{5x+9}=x-1$; square both sides

$$5x+9=x^2-2x+1$$
$$x^2-7x-8=0$$
$$(x-8)(x+1)=0$$
$$x-8=0 \quad \text{or} \quad x+1=0$$
$$x=8 \qquad\qquad x=-1$$

Check:

$$\sqrt{5(8)+9}+1\overset{?}{=}8$$
$$\sqrt{49}+1\overset{?}{=}8$$
$$8=8$$
$$\sqrt{5(-1)+9}+1\overset{?}{=}-1$$
$$\sqrt{4}+1\overset{?}{=}-1$$
$$3\neq -1$$

The only solution is $x=8$.

45. $\sqrt{n+4}+2\sqrt{n+2}=3$

$\sqrt{n+4}=3-2\sqrt{n+2}$; square both sides

$$n+4=9-12\sqrt{n+2}+4n+8$$

$12\sqrt{n+2}=13+3n$; square both sides

$$144n+288=169+78n+9n^2$$
$$9n^2-66n-119=0$$

Use the quadratic formula

$$n=\frac{-(-66)\pm\sqrt{(-66)^2-4(9)(-119)}}{2(9)}$$
$$=\frac{66\pm\sqrt{8640}}{18}$$
$$=\frac{11\pm 5\sqrt{15}}{3}$$

Check:

$$\sqrt{\frac{11+4\sqrt{15}}{3}+4}+2\sqrt{\frac{11+4\sqrt{15}}{3}+2}=10.2\neq 3$$

$$\sqrt{\frac{11-4\sqrt{15}}{3}+4}+2\sqrt{\frac{11-4\sqrt{15}}{3}+2}=3$$

$n=\dfrac{11-4\sqrt{15}}{3}$ is the only solution.

49. $x^3-2x^{3/2}-48=0$

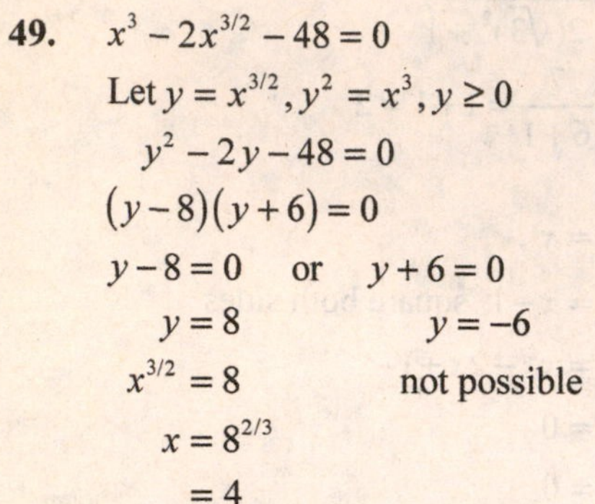

Let $y=x^{3/2}, y^2=x^3, y\geq 0$

$$y^2-2y-48=0$$
$$(y-8)(y+6)=0$$

$y-8=0$ or $y+6=0$

$y=8$ $\quad y=-6$

$x^{3/2}=8$ $\quad$ not possible

$$x=8^{2/3}$$
$$=4$$

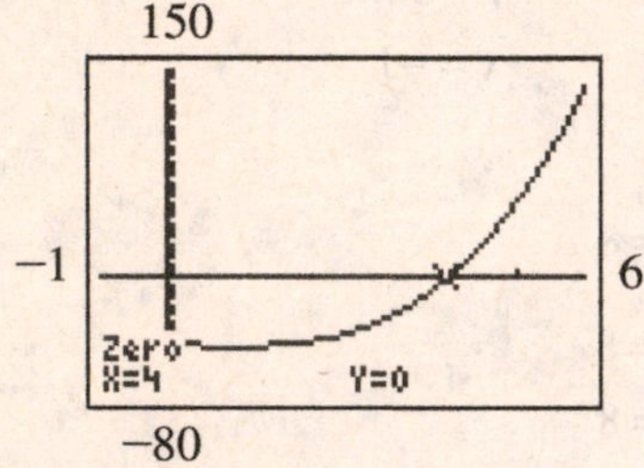

53. $\sqrt[3]{x^3-7}=x-1$; cube both sides

$$x^3-7=x^3-3x^2+3x-1$$
$$3x^2-3x-6=0$$
$$x^2-x-2=0$$

$x-2=0$ or $x+1=0$

$x=2$ $\quad x=-1$

There are two intersections, at $x=2$ and $x=-1$.

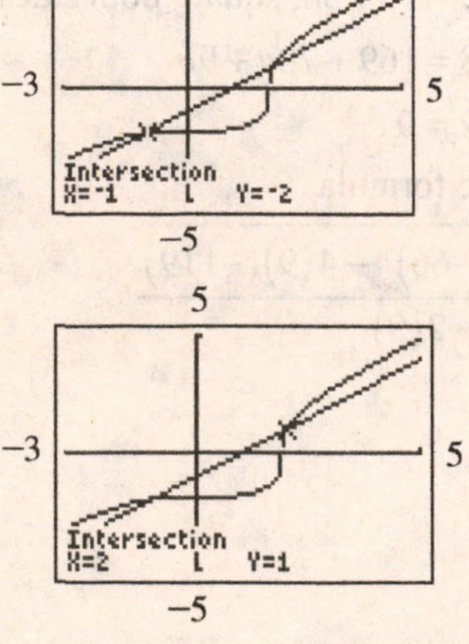

57. $\sqrt{\sqrt{x}-1}=2$; square both sides

$\sqrt{x}-1=4$

$\sqrt{x}=5$; square both sides

$x=25$

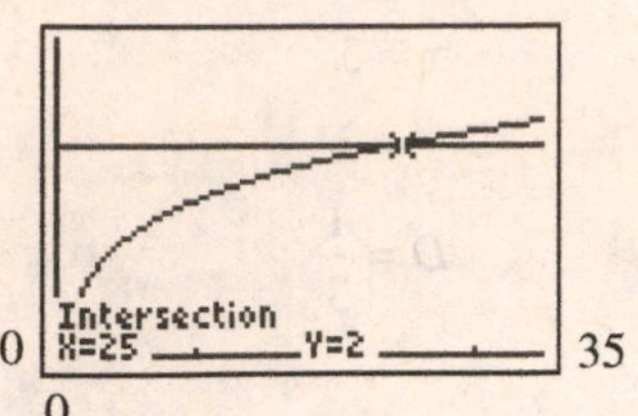

61. Solve $A=\sqrt{\dfrac{HW}{3600}}$ for H

$$A^2=\frac{HW}{3600}$$
$$H=\frac{3600A^2}{W}$$

$A=1.25, W=81.2$:

$$H=\frac{3600(1.25)^2}{81.2}=69.3\text{ cm}$$

65. Substitute $t_2=2t_1$ into $16t_1^2+16t_2^2=45$, where $t_1, t_2>0$

$$16t_1^2+16(2t_1)^2=45$$
$$16t_1^{\,2}+64t_1^{\,2}=45$$
$$80t_1^{\,2}=45$$
$$t_1^{\,2}=\frac{9}{16}; t_1=\frac{3}{4}$$

$t_1=0.75$ s and $t_2=2t_1=1.50$ s

We discard the negative answer since times before the experiment begin are not considered.

69.

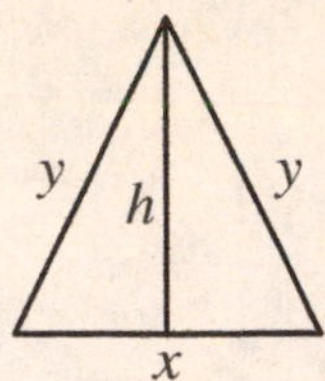

perimeter: $x + 2y = 72 \Rightarrow y = \frac{72 - x}{2}$

area: $\frac{1}{2} x \cdot h = 240 \Rightarrow h = \frac{480}{x}$

$$y^2 = \frac{x^2}{4} + h^2 \Rightarrow y^2 = \frac{x^2}{4} + \frac{480^2}{x^2}$$

Graph $y_1 = 72 - 2x$,

$$y_2 = \sqrt{\frac{x^2}{4} + \frac{480^2}{x^2}}$$

(only the positive root matters here).
Use the intersect feature to solve.

35

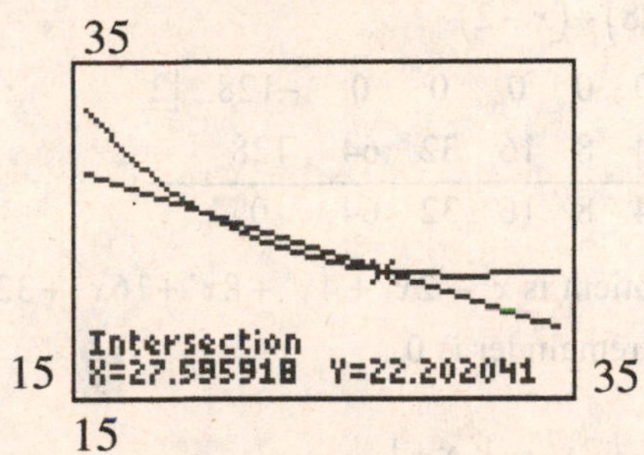

15   35
15

35

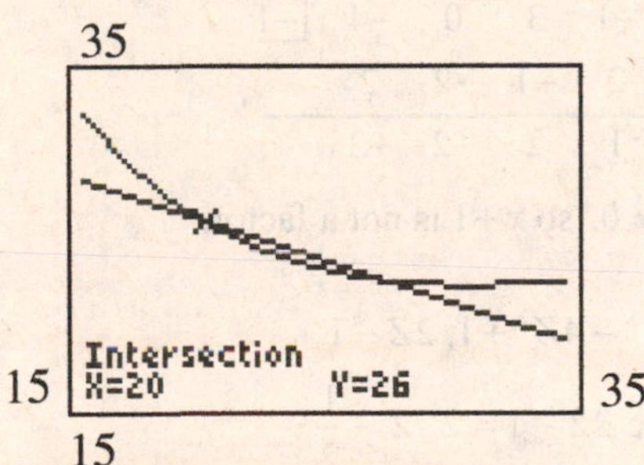

15   35
15

The lengths of the sides the banner can be 27.6 dm, 22.2 dm and 22.2 dm, with height 17.4 dm; or 20 dm, 26 dm and 26 dm, with height 24 dm. Since the height must be longer than the base, the lengths of the sides are 20 dm, 26 dm, and 26 dm, with height 24 dm.

73. Area: $lw = 1770 \Rightarrow w = \frac{1770}{l}$

Substitute into: $l^2 + w^2 = 62^2$

$$l^2 + \frac{1770^2}{l^2} = 62^2$$

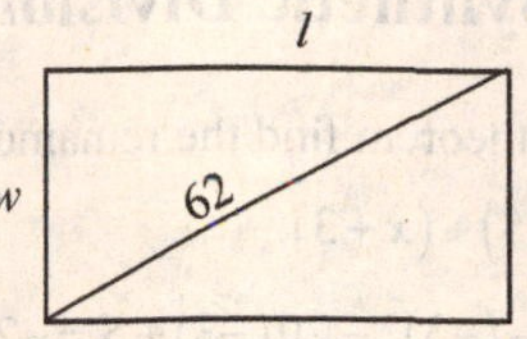

$$l^4 - 62^2 l^2 + 1770^2 = 0$$

Let $y = l^2$

$$y^2 - 62^2 y + 1770^2 = 0$$

$$y = \frac{62^2 \pm \sqrt{\left(-62^2\right)^2 - 4\left(1770^2\right)}}{2}$$

$y = 2269$ or $y = 1175$
$l^2 = 2269$ $l^2 = 1175$
$l = 52$ $l = 34$
$w = 34$ $w = 52$

The dimensions of the rectangle are $l = 52$ mm and $w = 34$ mm.

77. Houston to Mobile: $510 = v_1 \cdot t_1 \quad t_1 \quad \frac{510}{v_1}$

Mobile to Houston: $510 = \left(v_1 + 6.0\right) \cdot t_2$

$$t_2 = \frac{510}{v_1 + 6.0}$$

Substitute into: $t_1 + t_2 = 35$

$$\frac{510}{v_1} + \frac{510}{v_1 + 6.0} = 35$$

$$510\left(v_1 + 6.0\right) + 510 v_1 = 35 v_1 \left(v_1 + 6.0\right)$$

$$35 v_1^{\,2} - 810 v_1 - 3060 = 0$$

Using the quadratic formula
(keeping only the positive root),

$$v_1 = \frac{-(-810) + \sqrt{(-810)^2 - 4(35)(-3060)}}{2(35)}$$

$= 26.45$ mi/hr

$v_2 = v_1 + 6.0 = 32.45$ mi/hr.

Chapter 15

Equations of Higher Degree

15.1 The Remainder and Factor Theorems; Synthetic Division

1. Using the remainder theorem find the remainder, for $(3x^3 - x^2 - 20x + 5) \div (x+3)$.

$R = f(-3) = 3(-3)^3 - (-3)^2 - 20(-3) + 5 = -25$

5.

$$\begin{array}{r} x^2 + 2x + 6 \\ x-2\overline{)x^3 + \qquad 2x - 8} \\ \underline{x^3 - 2x^2} \qquad\quad \\ 2x^2 + 2x \qquad \\ \underline{2x^2 - 4x} \qquad \\ 6x - 8 \\ \underline{6x - 12} \\ 4 \end{array}$$

The remainder is $R = 4$.

9.

$$\begin{array}{r} x^3 - x - 9 \\ 2x-3\overline{)2x^4 - 3x^3 - 2x^2 - 15x - 16} \\ \underline{2x^4 - 3x^3} \qquad\qquad\qquad\quad \\ -2x^2 - 15x \qquad \\ \underline{-2x^2 + 3x} \qquad \\ -18x - 16 \\ \underline{-18x + 27} \\ -43 \end{array}$$

The remainder is $R = -43$.

13. $(2x^4 - 7x^3 - x^2 + 8) \div (x+1);\ r = -1$

$f(-1) = 2 \cdot (-1)^4 - 7 \cdot (-1)^3 - (-1)^2 + 8$

$= 2 + 7 - 1 + 8$

$R = 16$

17. $8x^3 + 2x^2 - 32x - 8,\ x - 2;\ r = 2$

$f(2) = 8(2)^3 + 2(2)^2 - 32(2) - 8$

$= 64 + 8 - 64 - 8 = 0$

$x - 2$ is a factor since $f(2) = R = 0$.

21. $x^{51} - 2x - 1,\ x + 1;\ r = -1$

$f(-1) = (-1)^{51} - 2(-1) - 1$

$= -1 + 2 - 1$

$= 0$

$x + 1$ is a factor since $f(-1) = 0$.

25. $(x^3 + 2x^2 - 3x + 4) \div (x+1)$

$$\begin{array}{rrrr|l} 1 & 2 & -3 & 4 & \underline{-1} \\ & -1 & -1 & 4 & \\ \hline 1 & 1 & -4 & 8 & \end{array}$$

The quotient is $x^2 + x - 4$ and the remainder is 8.

29. $(x^7 - 128) \div (x - 2)$

$$\begin{array}{rrrrrrrr|l} 1 & 0 & 0 & 0 & 0 & 0 & 0 & -128 & \underline{2} \\ & 2 & 4 & 8 & 16 & 32 & 64 & 128 & \\ \hline 1 & 2 & 4 & 8 & 16 & 32 & 64 & 0 & \end{array}$$

The quotient is $x^6 + 2x^5 + 4x^4 + 8x^3 + 16x^2 + 32x + 64$ and the remainder is 0.

33. $2x^5 - x^3 + 3x^2 - 4;\ x + 1$

$$\begin{array}{rrrrrr|l} 2 & 0 & -1 & 3 & 0 & -4 & \underline{-1} \\ & -2 & 2 & -1 & -2 & 2 & \\ \hline 2 & -2 & 1 & 2 & -2 & -2 & \end{array}$$

$R = -2 \neq 0$, so $x + 1$ is not a factor.

37. $2Z^4 - Z^3 - 4Z^2 + 1;\ 2Z - 1$

We write $2Z - 1 = 2\ Z - \dfrac{1}{2}$

$$\begin{array}{rrrrr|l} 2 & -1 & -4 & 0 & 1 & \underline{\tfrac{1}{2}} \\ & 1 & 0 & -2 & -1 & \\ \hline 2 & 0 & -4 & -2 & 0 & \end{array}$$

Since $R = 0$, $Z - \dfrac{1}{2}$ is a factor.

Therefore, $2Z - 1$ is also a factor.

41. $x^4-5x^3-15x^2+5x+14;\ 7$

$$\begin{array}{rrrrr|l} 1 & -5 & -15 & 5 & 14 & \underline{7} \\ & 7 & 14 & -7 & -14 & \\ \hline 1 & 2 & -1 & -2 & 0 & \end{array}$$

$R=0$, 7 is a zero.

45. $f(x)=2x^3+3x^2-19x-4$

$f(x)=(x+4)g(x)$

$g(x)=(2x^3+3x^2-19x-4)\div(x+4)$

$r=-4$

$$\begin{array}{rrrr|l} 2 & 3 & -19 & -4 & \underline{-4} \\ & -8 & 20 & -4 & \\ \hline 2 & -5 & 1 & -8 & \end{array}$$

$g(x)=2x^2-5x+1-\dfrac{8}{x+4}$

49. $f(x)=2x^3+kx^2-x+14;\ x-2$

We want $f(2)=R=0$

$f(2)=2(2)^3+k(2)^2-2+14$

$=16+4k-2+14$

$=28+4k$

$=0$

$4k=-28$

$k=-7$

If $k=-7$ then $x-2$ will be a factor.

53. Suppose r is a zero of $f(x)$, then $f(r)=0$. But $f(r)=-g(r)=0\Rightarrow g(r)=0$, so r is also a zero of $g(x)$. Therefore, if $f(x)=-g(x)$ then $f(x)$ and $g(x)$ have the same zeros.

57. **(a)** $(s^3+5s^2+4s+20)\div(s-2)$

$$\begin{array}{rrrr|l} 1 & 5 & 4 & 20 & \underline{2} \\ & 2 & 14 & 36 & \\ \hline 1 & 7 & 18 & 56 & \end{array}$$

$R=56\neq 0$, so $s-2$ is not a factor.

(b) $(s^3+5s^2+4s+20)\div(s+5)$

$$\begin{array}{rrrr|l} 1 & 5 & 4 & 20 & \underline{-5} \\ & -5 & 0 & -20 & \\ \hline 1 & 0 & 4 & 0 & \end{array}$$

$R=0$, so $s+5$ is a factor.

15.2 The Roots of an Equation

1. $f(x)=(x-1)^3(x^2+2x+1)=0$

$(x-1)^3=0$ or $x^2+2x+1=0$

$x=1$ $\quad (x+1)^2=0$

A triple root $\quad x+1=0$

$x=-1$, a double root

the five roots are 1, 1, 1, −1, −1

5. $(x-5)(x^2+9)=0$

by inspection $x=5,\ x=\pm 3j$

9. $2x^3+11x^2+20x+12=0 \quad r_1=-\dfrac{3}{2}$

$$\begin{array}{rrrr|l} 2 & 11 & 20 & 12 & -\frac{3}{2} \\ & -3 & -12 & -12 & \\ \hline 2 & 8 & 8 & 0 & \end{array}$$

$2x^3+11x^2+20x+12= x+\dfrac{3}{2}\ (2x^2+8x+8)$

$=2\ x+\dfrac{3}{2}\ (x^2+4x+4)$

$=2\ x+\dfrac{3}{2}\ (x+2)(x+2)$

The three roots are $r_1=-\dfrac{3}{2},\ r_2=-2,\ r_3=-2$

13. $t^3-7t^2+17t-15=0\ (r_1=2+j)$

Another root must be the conjugate, $r_2=2-j$

$$\begin{array}{rrrr|l} 1 & -7 & 17 & -15 & \underline{2+j} \\ & 2+j & -11-3j & 15 & \\ \hline 1 & -5+j & 6-3j & 0 & \underline{2-j} \\ & 2-j & -6+3j & & \\ \hline 1 & -3 & 0 & & \end{array}$$

$t^3-7t^2+17t-15=(t-2-j)(t-2+j)(t-3)$

The three roots are $r_1=2+j,\ r_2=2-j,\ r_3=3$

17. $6x^4+5x^3-15x^2+4=0\quad r_1=-\frac{1}{2}, r_2=\frac{2}{3}$

$$\begin{array}{rrrrr|l} 6 & 5 & -15 & 0 & 4 & -\frac{1}{2} \\ & -3 & -1 & 8 & -4 & \\ \hline 6 & 2 & -16 & 8 & 0 & \frac{2}{3} \\ & 4 & 4 & -8 & 0 & \\ \hline 6 & 6 & -12 & 0 & 0 & \end{array}$$

$6x^4+5x^3-15x^2+4$

$=6\ x+\frac{1}{2}\ \ x-\frac{2}{3}\ (x^2+x-2)$

$=6\ x+\frac{1}{2}\ \ x-\frac{2}{3}\ (x+2)(x-1)$

The four roots are $r_1=-\frac{1}{2}, r_2=\frac{2}{3}, r_3=-2, r_4=1$

21. $x^5-3x^4+4x^3-4x^2+3x-1=0$ (1 is a triple root)

$$\begin{array}{rrrrrr|l} 1 & -3 & 4 & -4 & 3 & -1 & 1 \\ & 1 & -2 & 2 & -2 & 1 & \\ \hline 1 & -2 & 2 & -2 & 1 & 0 & 1 \\ & 1 & -1 & 1 & -1 & & \\ \hline 1 & -1 & 1 & -1 & 0 & & 1 \\ & 1 & 0 & 1 & & & \\ \hline 1 & 0 & 1 & 0 & & & \end{array}$$

$x^5-3x^4+4x^3-4x^2+3x-1=(x-1)^3(x^2+1)$

The five roots are 1, 1, 1, $-j$, j.

25. $x^6+2x^5-4x^4-10x^3-41x^2-72x-36=0$

(-1 is a double root; $2j$ is a root)

The complex conjugate $r_4=-2j$ must also be a root.

$$\begin{array}{rrrrrrr|l} 1 & 2 & -4 & -10 & -41 & -72 & -36 & -1 \\ & -1 & -1 & 5 & 5 & 36 & 36 & \\ \hline 1 & 1 & -5 & -5 & -36 & -36 & 0 & -1 \\ & -1 & 0 & 5 & 0 & 36 & 0 & \\ \hline 1 & 0 & -5 & 0 & -36 & 0 & 0 & 2j \\ & 2j & -4 & -18j & 36 & 0 & 0 & \\ \hline 1 & 2j & -9 & -18j & 0 & 0 & 0 & -2j \\ & -2j & 0 & 18j & 0 & 0 & 0 & \\ \hline 1 & 0 & -9 & 0 & 0 & 0 & 0 & \end{array}$$

$x^6+2x^5-4x^4-10x^3-41x^2-72x-36$

$=(x+1)^2(x-2j)(x+2j)(x^2-9)$

The six roots are $-1, -1, 2j, -2j, -3, 3$.

29. Synthetic division:

$$\begin{array}{rrrr|l} 2 & k & -k & -2 & 2 \\ & 4 & 2k+8 & 2k+16 & \\ \hline 2 & k+4 & k+8 & 2k+14 & \end{array}$$

It is necessary that $2k+14=0$ or $k=-7$.

15.3 Rational and Irrational Roots

1. $f(x)=4x^5+x^4+4x^3-x^2+5x+6=0$ has two sign changes and thus no more than two positive roots.

$f(-x)=-4x^5+x^4-4x^3-x^2-5x+6=0$ has three sign changes and thus no more than three negative roots.

5. $x^3+2x^2-5x-6=0$; there are 3 roots.

$f(x)=x^3+x^2-5x-6$; there is exactly one positive root.

$f(-x)=-x^3+x^2+5x-6$; there are at most two negative roots.

Possible rational roots are $\pm1, \pm2, \pm3, \pm6$

Trying -1, we have

$$\begin{array}{rrrr|l} 1 & 2 & -5 & -6 & -1 \\ & -1 & -1 & 6 & \\ \hline 1 & 1 & -6 & 0 & \end{array}$$

Hence, -1 is a root and the remaining factor is $x^2+x-6=(x+3)(x-2)$.

The three roots are:

$r_1=-1, r_2=-3, r_3=2$.

9. $2x^3-3x^2-3x+2=0$; there are three roots.

$f(x)=2x^3-3x^2-3x+2$; there are at most two positive roots.

$f(-x)=-2x^3-3x^2+3x+3$; there is at most one negative root.

Possible rational roots are $\pm\frac{1}{2}, \pm1, \pm2$.

Trying -1 we have:

2	−3	−3	2	$\lfloor -1$
	−2	5	−2	
2	−5	2	0	

Remainder is 0 so -1 is a root.

$$2x^3-3x^2-3x+2=(x+1)(2x^2-5x+2)$$
$$=(x+1)(2x-1)(x-2).$$

The three roots are:

$r_1=-1, r_2=\frac{1}{2}, r_3=2.$

13. $5n^4-2n^3+40n-16=0$; there are four roots. $f(n)$ has 3 sign changes, at most 3 positive roots. $f(-n)=5n^4+2n^3-40n-16$ has one sign change, so exactly one negative root.

Possible rational roots: $\pm 16, \pm\frac{16}{5}, \pm 2, \pm\frac{2}{5}, \pm 4, \pm\frac{4}{5}, \pm 8, \pm\frac{8}{5}$

Trying -2:

5	−2	0	40	−16	$\lfloor -2$
	−10	24	−48	16	
5	−12	24	−8	0	

Remainder is 0, so -2 is a root

$$5n^4-2n^3+40n-16=(n+2)(5n^3-12n^2+24n-8)$$

No other roots are negative. Trying $\frac{2}{5}$:

5	−12	24	−8	$\lfloor \frac{2}{5}$
	2	−4	8	
5	−10	20	0	

Remainder is 0, so $\frac{2}{5}$ is a root

$$5n^4-2n^3+40n-16$$
$$=(n+2)\left(n-\frac{2}{5}\right)(5n^2-10n+20)$$
$$=(n+2)\left(n-\frac{2}{5}\right)5(n^2-2n+4)$$

Using the quadratic formula on the last factor:

$$n=\frac{-(-2)\pm\sqrt{(-2)^2-4(1)(4)}}{2}=1\pm j\sqrt{3}$$

The four roots are: $-2, \frac{2}{5}, 1\pm j\sqrt{3}$

17. $D^5+D^4-9D^3-5D^2+16D+12=0$ has five roots. $f(D)$ has two sign changes and therefore at most two positive roots.

$f(-D)=-D^5+D^4+9D^3-5D^2-16D+12$ has three sign changes and therefore at most three negative roots.

Possible rational roots: $\pm 1, \pm 2, \pm 3, \pm 4, \pm 6, \pm 12$

1	1	−9	−5	16	12	$\lfloor 2$
	2	6	−6	−22	−12	
1	3	−3	−11	−6	0	

Remainder is 0 so 2 is a root

$$D^5+D^4-9D^3-5D^2+16D+12$$
$$=(D-2)(D^4+3D^3-3D^2-11D-6)$$

1	3	−3	−11	−6	$\lfloor 2$
	2	10	14	6	
1	5	7	3	0	

Remainder is 0 so 2 is a root.

$$D^5+D^4-9D^3-5D^2+16D+12$$
$$=(D-2)(D-2)(D^3+5D^2+7D+3)$$

1	5	7	3	$\lfloor -1$
	−1	−4	−3	
1	4	3	0	

Remainder is 0 so -1 is a root.

$$D^5+D^4-9D^3-5D^2+16D+12$$
$$=(D-2)(D-2)(D+1)(D^2+4D+3)$$
$$=(D-2)(D-2)(D+1)(D+1)(D+3)$$

The five roots are: 2, 2, −1, −1, −3.

2	7	7	2	0	$\lfloor -\frac{1}{2}$
	−1	−3	−2	0	
2	6	4	0	0	

The remaining factor is

$$2x^5+5x^4-4x^3-19x^2-16x-4$$
$$=\left(x+\frac{1}{2}\right)(x+1)(x-2)(2x^2+6x+4)$$
$$=\left(x+\frac{1}{2}\right)(x+1)(x-2)2(x+2)(x+1).$$

The five roots are $-1, 2, -\frac{1}{2}, -2, -1.$

21. $2x^3-8x+3=0$

Graph $y_1=2x^3-8x+3$ and use the Zero feature to solve. The Zeros are $-2.17, 0.39, 1.78$.

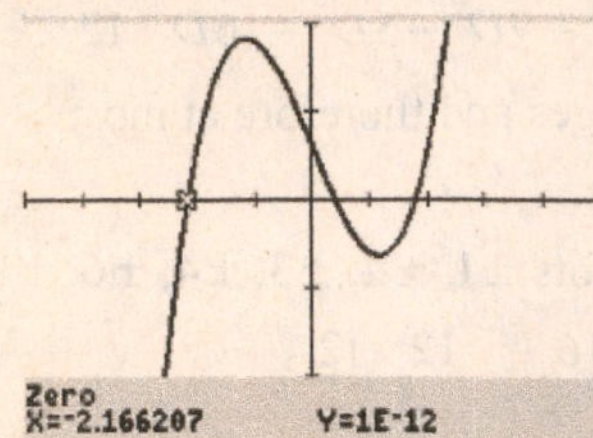

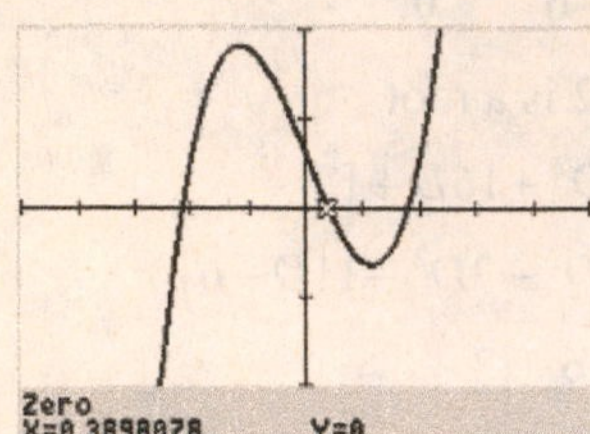

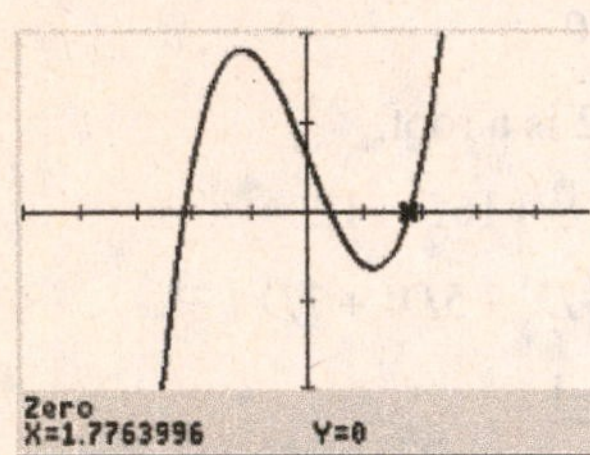

25. Substitute $y=x^4-11x^2$ into $y=12x-4$

$x^4-11x^2=12x-4$

$x^4-11x^2-12x+4=0$; there are 4 roots, with at most 2 positive roots.

$f(-x)=x^4-11x^2+12x+4$, so at most 2 negative roots.

Possible rational roots: $\pm1, \pm2$

Try -2:

$$\begin{array}{rrrrr|r} 1 & 0 & -11 & -12 & 4 & \underline{-2} \\ & -2 & 4 & 14 & -4 & \\ \hline 1 & -2 & -7 & 2 & 0 & \end{array}$$

$$\begin{array}{rrrr|r} 1 & -2 & -7 & 2 & \underline{-2} \\ & -2 & 8 & -2 & \\ \hline 1 & -4 & 1 & 0 & \end{array}$$

$x^4-11x^2-12x+4=(x+2)(x+2)(x^2-4x+1)$

Use quadratic formula:

$$x=\frac{-(-4)\pm\sqrt{(-4)^2-4(1)(1)}}{2(1)}$$

$$=2\pm\sqrt{3}$$

The solutions are:

$x=-2,\ y=-28;$

$x=2+\sqrt{3}, y=20+12\sqrt{3};$

$x=2-\sqrt{3},\ y=20-12\sqrt{3}.$

29. The polynomial $f(x)=2x^4+x^2-22x+8$ has two sign changes and so at most two positive roots.

Furthermore,

$f(-x)=2x^4+x^2+22x+8$

has no sign changes and so $f(x)$ has no negative roots.

The smallest possible rational root is $\frac{1}{2}$ and the largest is $\frac{8}{1}=8$.

33.

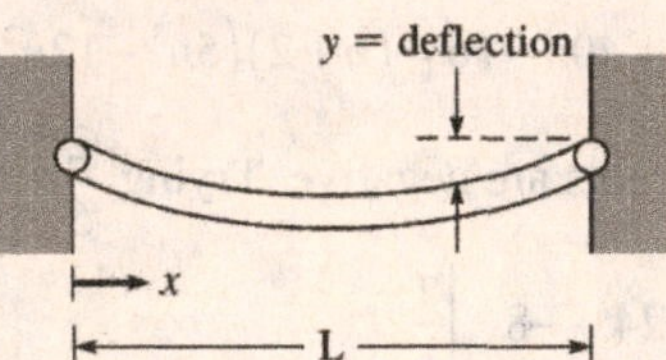

Deflection is zero if $y=0$, so we must solve

$k\left(x^4-2Lx^3+L^3x\right)=0$

$x^4-2Lx^3+L^3x=0$

$x\left(x^3-2Lx^2+L^3\right)=0$, $x=0$ is a root

$x^3-2Lx^2+L^3=0$

We try $x=L$, since at the other attached end of the beam ($x=L$) we suspect that there is no deflection.

$$\begin{array}{rrrr|l} 1 & -2L & 0 & L^3 & \underline{L} \\ & L & -L^2 & -L^3 & \\ \hline 1 & -L & -L^2 & 0, & \end{array}$$

Remainder is 0, so L is a root.

$x^4-2Lx^2+L^3x=x(x-L)\left(x^2-Lx-L^2\right)$

Using the quadratic formula for the remaining factor:

$$x = \frac{-(-L) \pm \sqrt{(-L)^2 - 4(1)(-L^2)}}{2}$$

$$= \frac{L \pm L\sqrt{5}}{2}$$

$$= \frac{L\left(1 \pm \sqrt{5}\right)}{2}$$

$\frac{L}{2}\left(1-\sqrt{5}\right) < 0$ and $\frac{L}{2}\left(1+\sqrt{5}\right) > L$; reject both.

Beam has a deflection of 0 for $x = 0$ and $x = L$.

37.

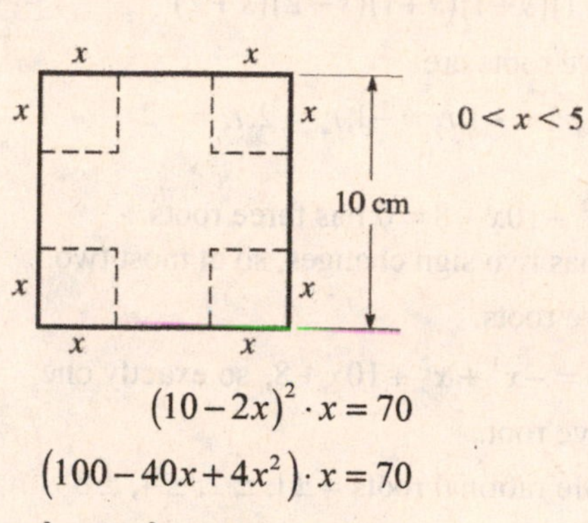

$$(10-2x)^2 \cdot x = 70$$

$$\left(100 - 40x + 4x^2\right) \cdot x = 70$$

$$4x^3 - 40x^2 + 100x - 70 = 0$$

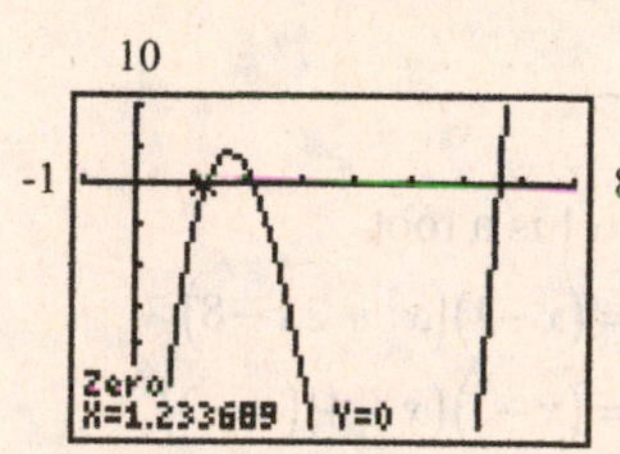

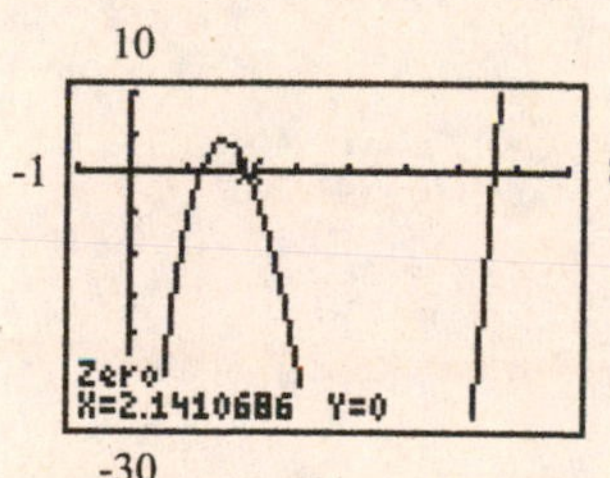

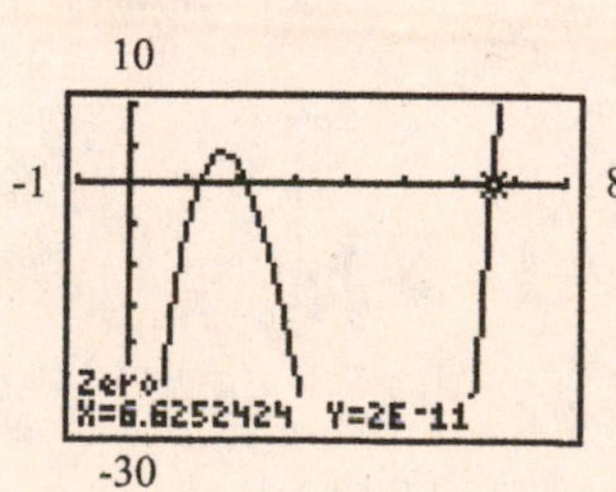

$x = 1.23$ cm, or $x = 2.14$ cm.

$6.63 > 5$ and must be rejected.

41. We must determine where the line $y = x + 3$ intersects the curve

$$y^2 = x^3 - 6x + 9.$$

Substituting $y = x + 3$ into $y^2 = x^3 - 6x + 9$,

$$(x+3)^2 = x^3 - 6x + 9$$

$$x^2 + 6x + 9 = x^3 - 6x + 9$$

$$x^3 - x^2 - 12x = 0$$

$$x(x+3)(x-4) = 0$$

The roots are $x = 0, x = -3, x = 4$ and so the x-coordinate of R is 4. The y-coordinate is $4 + 3 = 7$.

The coordinates of R are (4,7).

Review Exercises

1. This is false. Letting $f(x) = 3x^2 + 5x - 8$, we can use the remainder theorem to find the remainder after dividing $f(x)$ by $x - 2$ by evaluating $f(2) = 14$. The remainder is 14, not 12.

5. $f(x) = 2x^3 - 4x^2 - x + 4$; we evaluate $f(1)$.

$$2(1)^3 - 4(1)^2 - (1) + 4 = 1$$

$\left(2x^3 - 4x^2 - x + 4\right) \div (x-1)$ has remainder 1

9. $f(x) = x^4 + x^3 + x^2 - 2x - 3$; we evaluate $f(-1)$.

$$(-1)^4 + (-1)^3 + (-1)^2 - 2(-1) - 3 = 0$$

$f(-1) = 0$, so $x + 1$ is a factor of $x^4 + x^3 + x^2 - 2x - 3$

13.

1	4	5	1	$\lfloor 1$
	1	5	10	
1	5	10	11	

$$\left(x^3 + 4x^2 + 5x + 1\right) \div (x-1)$$

$$= \left(x^2 + 5x + 10\right) + \frac{11}{x-1}$$

17.

$$\begin{array}{rrrrr|l} 1 & 3 & -20 & -2 & 56 & \underline{-6} \\ & -6 & 18 & 12 & -60 & \\ \hline 1 & -3 & -2 & 10 & -4 & \end{array}$$

$$x^4+3x^3-20x^2-2x+56 \div x+6$$
$$=x^3-3x^2-2x+10+\frac{-4}{x+6}$$

21.

$$\begin{array}{rrrr|l} 1 & 4 & 0 & -9 & \underline{-3} \\ & -3 & -3 & 9 & \\ \hline 1 & 1 & -3 & 0 & \end{array}$$

remainder = 0; therefore, -3 is a root of y^3+4y^2-9

25.

$$\begin{array}{rrrr|l} 1 & -4 & -7 & 10 & \underline{5} \\ & 5 & 5 & -10 & \\ \hline 1 & 1 & -2 & 0 & \end{array}$$

$$x^3-4x^2-7x+10=(x-5)(x^2+x-2)$$
$$=(x-5)(x+2)(x-1)$$

The three roots are $r_1=5$, $r_2=-2$, $r_3=1$.

29.

$$\begin{array}{rrrrr|l} 4 & 0 & -1 & -18 & 9 & \underline{1/2} \\ & 2 & 1 & 0 & -9 & \\ \hline 4 & 2 & 0 & -18 & 0 & \underline{3/2} \\ & 6 & 12 & 18 & & \\ \hline 4 & 8 & 12 & 0 & & \end{array}$$

$$4p^4-p^2-18p+9= \quad p-\frac{1}{2} \quad p-\frac{3}{2} \quad 4(p^2+2p+3)$$

Use the quadratic formula for the remaining factor:

$$p=\frac{-2\pm\sqrt{2^2-4(1)(3)}}{2}$$
$$=-1\pm j\sqrt{2}$$

The four roots are

$$r_1=\frac{1}{2}, r_2=\frac{3}{2}, r_3=-1+j\sqrt{2}, r_4=-1-j\sqrt{2}$$

33.

$$\begin{array}{rrrrrr|l} 1 & 3 & -1 & -11 & -12 & -4 & \underline{-1} \\ & -1 & -2 & 3 & 8 & 4 & \\ \hline 1 & 2 & -3 & -8 & -4 & 0 & \underline{-1} \\ & -1 & -1 & 4 & 4 & & \\ \hline 1 & 1 & -4 & -4 & 0 & & \underline{-1} \\ & -1 & 0 & 4 & & & \\ \hline 1 & 0 & -4 & 0 & & & \end{array}$$

$$s^5+3s^4-s^3-11s^2-12s-4$$
$$=(s+1)(s+1)(s+1)(s^2-4)$$
$$=(s+1)(s+1)(s+1)(s-2)(s+2)$$

The five roots are

$$r_1=-1, r_2=-1, r_3=-1, r_4=2, r_5=-2$$

37. $x^3+x^2-10x+8=0$ has three roots.

$f(x)$ has two sign changes, so at most two positive roots.

$f(-x)=-x^3+x^2+10x+8$, so exactly one negative root.

Possible rational roots $=\pm1, \pm2, \pm4, \pm8$

$$\begin{array}{rrrr|l} 1 & 1 & -10 & 8 & \underline{1} \\ & 1 & 2 & -8 & \\ \hline 1 & 2 & -8 & 0 & \end{array}$$

Remainder is 0, so 1 is a root

$$x^3+x^2-10x+8=(x-1)(x^2+2x-8)$$
$$=(x-1)(x+4)(x-2)$$

The three roots are $1, -4, 2$

41. $6x^3 - x^2 - 12x - 5 = 0$ has three roots. $f(x)$ has one sign change, so exactly one positive root. $f(-x) = -6x^3 - x^2 + 12x - 5$ has two sign changes, so at most two negative roots.

Possible rational roots $= \dfrac{\pm 1, \pm 5}{\pm 1, \pm 2, \pm 3, \pm 6}$

6	−1	−12	−5	|5/3
	10	15	5	
6	9	3	0	

Remainder is 0, so $\frac{5}{3}$ is a root.

$$6x^3 - x^2 - 12x - 5 = x - \frac{5}{3}\left(6x^2 + 9x + 3\right)$$
$$= 3\ x - \frac{5}{3}\left(2x^2 + 3x + 1\right)$$
$$= 3\ x - \frac{5}{3}\left(2x + 1\right)\left(x + 1\right)$$

The three roots are $\frac{5}{3}, -\frac{1}{2}, -1$

45. The graph appears to have x-intercepts at $-2, -1,$ and 4.

Using synthetic division,

3	−7	−26	16	32	|4
	12	20	−24	−32	
3	5	−6	−8	0	|−1
	−3	−2	8		
3	2	−8	0		|−2
	−6	8			
3	−4	0			

so

$$3x^4 - 7x^3 - 26x^2 + 16x + 32$$
$$= (x-4)(x+1)(x+2)(3x-4)$$

corresponding to the four roots $4, -1, -2, \frac{4}{3}$.

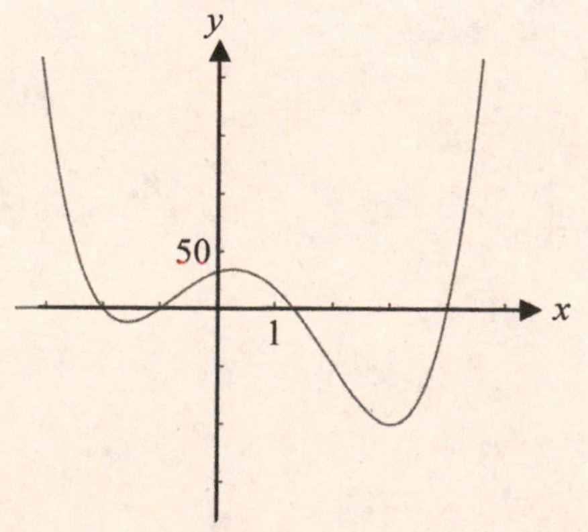

49. Since there are six roots, one of which is real, and the complex roots occur in conjugate pairs, the number of complex roots is zero, two or four.

53. A polynomial of degree 3 has 3 roots. Since j is a root, so is $-j$. A degree 3 polynomial equation that has 5 and j as roots is

$$(x-j)(x+j)(x-5) = 0$$
$$(x^2+1)(x-5) = 0$$
$$x^3 - 5x^2 + x - 5 = 0$$

57.

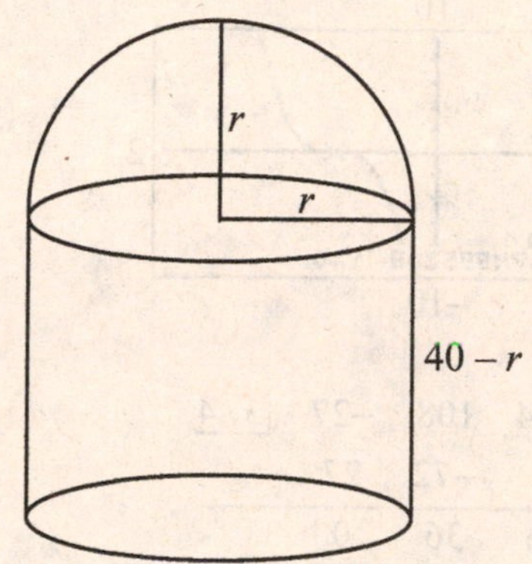

We solve:

$$(40-r)\pi r^2 + \frac{2}{3}\pi r^3 = 15500$$
$$\frac{1}{3}\pi r^3 - 40\pi r^2 + 15500 = 0$$

There are three roots, at most two of which are positive.

Graph $y_1 = \frac{1}{3}\pi x^3 - 40\pi x^2 + 15500$ for $0 < x < 40$ and use the zero feature to solve, finding $r = 11.7$ ft.

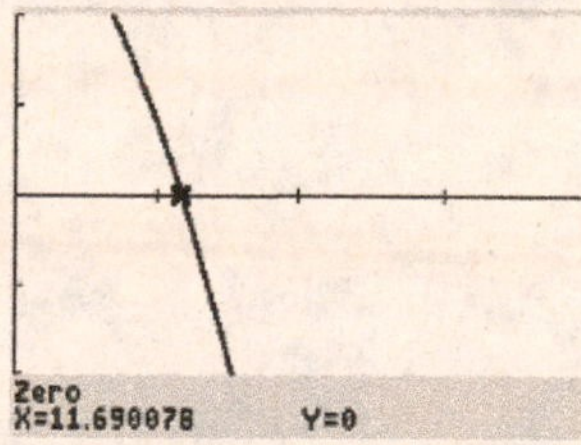

61. $f(d) = 64d^3 - 144d^2 + 108d - 27$ has $n = 3$ and therefore three roots. $f(d)$ has three sign changes and therefore at most three positive roots. $f(-d) = -64d^3 - 144d^2 + 108d - 27$ has two sign changes and therefore at most two negative roots.

Possible rational roots

$$= \frac{\pm 1, \pm 3, \pm 9, \pm 27}{\pm 1, \pm 2, \pm 4, \pm 8, \pm 16, \pm 32, \pm 64}$$

From the graph, the root: is at $d = 0.75$.

We verify that 3/4 is a root.

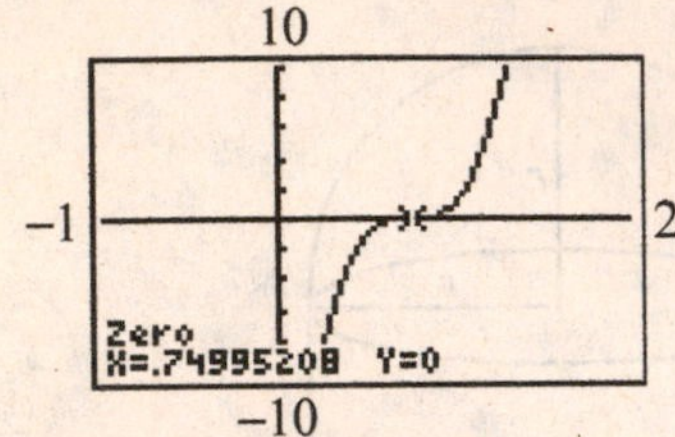

64	−144	108	−27	3/4
	48	−72	27	
64	−96	36	0	

The remainder is 0, so 3/4 is indeed a root.

Using the quadratic formula for the remaining factor:

$$x = \frac{96 \pm \sqrt{(-96)^2 - 4(64)(36)}}{2(64)}$$

$$= \frac{3}{4}$$

Therefore, $\frac{3}{4}$ is a repeated (multiplicity = 3) root, and the diameter is $d = 0.75$ cm.

65. $h = r + 3.2$, h and $r > 0$.

$$680 = \pi r^2 h$$

$$= \pi r^2 (r + 3.2)$$

$$= \pi r^3 + 3.2\pi r^2$$

$\pi r^3 + 3.2\pi r^2 - 680 = 0$ has one sign change and therefore one positive root. From the graph $r = 5.1$ m and $h = 5.1 + 3.2 = 8.3$ m.

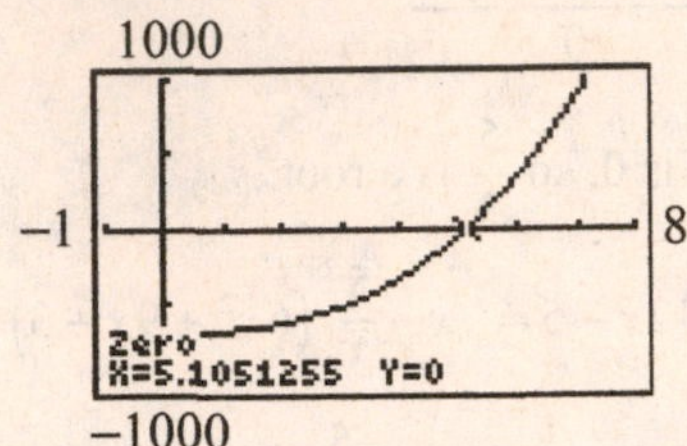

69. The area is

$A(x) = 2x(4 - x^2) = 8x - 2x^3$

Plotting $y_1 = 8x - 2x^3$, and using the maximum feature, we find the area is maximized at $x = 1.155$ m.

The maximum area is 6.158 m^2.

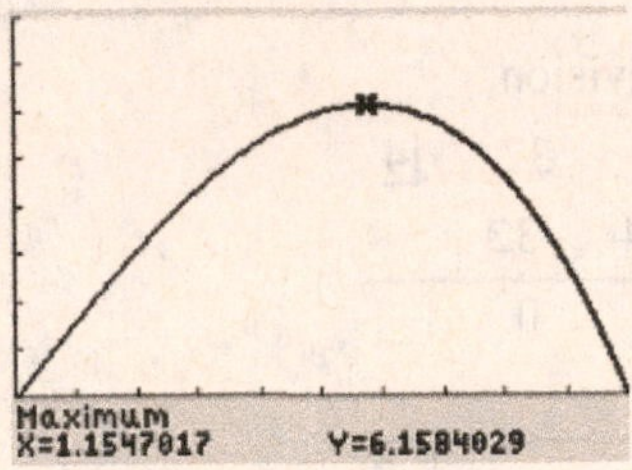

Chapter 16

Matrices; Systems of Linear Equations

16.1 Matrices: Definitions and Basic Operations

1. $\begin{bmatrix} 8 & 1 & -5 & 9 \\ 0 & -2 & 3 & 7 \end{bmatrix} + \begin{bmatrix} -3 & 6 & 4 & 0 \\ 6 & 6 & -2 & 5 \end{bmatrix}$

$= \begin{bmatrix} 8+(-3) & 1+6 & -5+4 & 9+0 \\ 0+6 & -2+6 & 3+(-2) & 7+5 \end{bmatrix}$

$= \begin{bmatrix} 5 & 7 & -1 & 9 \\ 6 & 4 & 1 & 12 \end{bmatrix}$

5. $\begin{bmatrix} x & 2y & z \\ r/4 & -s & -5t \end{bmatrix} = \begin{bmatrix} -2 & 10 & -9 \\ 12 & -4 & 5 \end{bmatrix}$

$x = -2,\quad 2y = 10,\quad z = -9$

$y = 5,$

$\frac{r}{4} = 12,\quad -s = -4,\quad -5t = 5$

$r = 48\quad s = 4\quad t = -1$

9. $\begin{bmatrix} x-3 & x+y \\ x-z & y+z \\ x+t & y-t \end{bmatrix} = \begin{bmatrix} 5 & 3 \\ 4 & -1 \end{bmatrix}$

This equality is not valid since the matrices have different dimensions.

13. $\begin{bmatrix} 50+(-55) & -82+82 \\ -34+30 & 57+14 \\ -15+26 & 62+(-70) \end{bmatrix} = \begin{bmatrix} -5 & 0 \\ -4 & 71 \\ 11 & -8 \end{bmatrix}$

17. Since A and C do not have the same number of columns, they cannot be added.

21. $-2C + D = -2\begin{bmatrix} -1 & 4 & -7 \\ 2 & -6 & 11 \end{bmatrix} + \begin{bmatrix} 7 & 9 & -6 \\ -4 & 0 & 8 \end{bmatrix}$

$= \begin{bmatrix} 9 & 1 & 8 \\ -8 & 12 & -14 \end{bmatrix}$

25. $B - 3A = \begin{bmatrix} 3 & 12 \\ -9 & -6 \end{bmatrix} - 3\begin{bmatrix} 6 & -3 \\ 4 & -5 \end{bmatrix} + = \begin{bmatrix} -15 & 21 \\ -21 & 6 \end{bmatrix}$

29. $C + 3D = \begin{bmatrix} -1 & 4 & -7 \\ 2 & -6 & 11 \end{bmatrix} + 3\begin{bmatrix} 7 & 9 & -6 \\ -4 & 0 & 8 \end{bmatrix}$

$= \begin{bmatrix} -1 & 4 & -7 \\ 2 & -6 & 11 \end{bmatrix} + \begin{bmatrix} 21 & 27 & -18 \\ -12 & 0 & 24 \end{bmatrix}$

$= \begin{bmatrix} 20 & 31 & -25 \\ -10 & -6 & 35 \end{bmatrix}$

33. $-6B - 4A = -6\begin{matrix} 3 & 12 \\ -9 & -6 \end{matrix} - 4\begin{matrix} 6 & -3 \\ 4 & -5 \end{matrix} +$

$= \begin{matrix} -42 & -60 \\ 38 & 56 \end{matrix}$

37. $-(A-B) = -\begin{bmatrix} -1-4 & 2+1 & 3+3 & 7-0 \\ 0-5 & -3-0 & -1+1 & 4-1 \\ 9-1 & -1-11 & 0-8 & -2-2 \end{bmatrix}$

$= \begin{bmatrix} 5 & -3 & -6 & -7 \\ 5 & 3 & 0 & -3 \\ -8 & 12 & 8 & 4 \end{bmatrix}$

$B - A = \begin{bmatrix} 4+1 & -1-2 & -3-3 & 0-7 \\ 5-0 & 0+3 & -1+1 & 1-4 \\ 1-9 & 11+1 & 8-0 & 2+2 \end{bmatrix}$

$= \begin{bmatrix} 5 & -3 & -6 & -7 \\ 5 & 3 & 0 & -3 \\ -8 & 12 & 8 & 4 \end{bmatrix}$

41. $\begin{bmatrix} 96 & 75 & 0 & 0 \\ 62 & 44 & 24 & 0 \\ 0 & 35 & 68 & 78 \end{bmatrix} + 2\begin{bmatrix} 96 & 75 & 0 & 0 \\ 62 & 44 & 24 & 0 \\ 0 & 35 & 68 & 78 \end{bmatrix}$

$= \begin{bmatrix} 288 & 225 & 0 & 0 \\ 186 & 132 & 72 & 0 \\ 0 & 105 & 204 & 234 \end{bmatrix}$

16.2 Multiplication of Matrices

1. $A=\begin{bmatrix}1 & 2\\ 0 & -3\\ 2 & 1\end{bmatrix}, B=\begin{bmatrix}-1 & 6 & 5 & -2\\ 3 & 0 & 1 & -4\end{bmatrix}$

$$AB=\begin{bmatrix}1 & 2\\ 0 & -3\\ 2 & 1\end{bmatrix}\begin{bmatrix}-1 & 6 & 5 & -2\\ 3 & 0 & 1 & -4\end{bmatrix}$$

$$AB=\begin{bmatrix}-1+6 & 6+0 & 5+2 & -2-8\\ 0-9 & 0+0 & 0-3 & 0+12\\ -2+3 & 12+0 & 10+1 & -4-4\end{bmatrix}$$

$$AB=\begin{bmatrix}5 & 6 & 7 & -10\\ -9 & 0 & -3 & 12\\ 1 & 12 & 11 & -8\end{bmatrix}$$

5. $\begin{bmatrix}9 & -1 & 3\\ 7 & 0 & 2\end{bmatrix}\begin{bmatrix}40\\ -15\\ 20\end{bmatrix}=\begin{bmatrix}9(40)+(-1)(-15)+3(20)\\ 7(40)+\quad 0(-15)+2(20)\end{bmatrix}$

$$=\begin{bmatrix}435\\ 320\end{bmatrix}$$

9. $\begin{matrix}-1 & 7\\ 3 & 5\\ 10 & -1\\ -5 & 12\end{matrix}\quad\begin{matrix}2 & 1\\ 5 & -3\end{matrix}$

$$=\begin{matrix}-1(2)+7(5) & -1(1)+7(-3)\\ 3(2)+5(5) & 3(1)+5(-3)\\ 10(2)+(-1)(5) & 10(1)+(-1)(-3)\\ -5(2)+(12)(5) & -5(1)+12(-3)\end{matrix}$$

$$=\begin{matrix}33 & -22\\ 31 & -12\\ 15 & 13\\ 50 & -41\end{matrix}$$

13. $\begin{bmatrix}-7.1 & 2.3 & 0.5\\ -3.8 & -2.4 & 4.9\end{bmatrix}\begin{bmatrix}6.5 & -5.2\\ 4.9 & 1.7\\ -1.8 & 6.9\end{bmatrix}=\begin{bmatrix}-35.78 & 44.28\\ -45.28 & 49.49\end{bmatrix}$

17. $AB = \begin{bmatrix} -10 & 25 & 40 \\ 42 & -5 & 0 \end{bmatrix} \begin{bmatrix} 6 \\ -15 \\ 12 \end{bmatrix}$

$= \begin{bmatrix} (-10)(6)+(25)(-15)+(40)(12) \\ (42)(6)+(-5)(-15)+(0)(12) \end{bmatrix}$

$= \begin{bmatrix} 45 \\ 327 \end{bmatrix}$

BA is not possible because the number of columns in *B* is not equal to the number of rows in *A*.

21. $AI = \begin{matrix} 3 & 9 & -15 \\ 8 & 0 & 4 \\ 6 & -12 & 24 \end{matrix} \quad \begin{matrix} 1 & 0 & 0 \\ 0 & 1 & 0 \\ 0 & 0 & 1 \end{matrix}$

$= \begin{matrix} 3(1)+9(0)+(-15)(0) & 3(0)+9(1)+(-15)(0) & 3(0)+9(0)+(-15)(1) \\ 8(1)+0(0)+4(0) & 8(0)+0(1)+4(0) & 8(0)+0(0)+4(1) \\ 6(1)+(-12)(0)+24(0) & 6(0)+(-12)(1)+24(0) & 6(0)+(-12)(0)+24(1) \end{matrix}$

$= \begin{matrix} 3 & 9 & -15 \\ 8 & 0 & 4 \\ 6 & -12 & 24 \end{matrix}$

$= A$

$IA = \begin{bmatrix} 1 & 0 & 0 \\ 0 & 1 & 0 \\ 0 & 0 & 1 \end{bmatrix} \begin{bmatrix} 3 & 9 & -15 \\ 8 & 0 & 4 \\ 6 & -12 & 24 \end{bmatrix}$

$= \begin{bmatrix} 1(3)+0(8)+0(6) & 1(9)+0(0)+0(-12) & 1(-15)+0(4)+0(24) \\ 0(3)+1(8)+0(6) & 0(9)+1(0)+0(-12) & 0(-15)+1(4)+0(24) \\ 0(3)+0(8)+1(6) & 0(9)+0(0)+1(-12) & 0(-15)+0(4)+1(24) \end{bmatrix}$

$= \begin{bmatrix} 3 & 9 & -15 \\ 8 & 0 & 4 \\ 6 & -12 & 24 \end{bmatrix}$

$= A$

Therefore, $AI = IA = A$

25. $AB = \begin{bmatrix} 1 & -2 & 3 \\ 2 & -5 & 7 \\ -1 & 3 & -5 \end{bmatrix}\begin{bmatrix} 4 & -1 & 1 \\ 3 & -2 & -1 \\ 1 & -1 & -1 \end{bmatrix}$

$$= \begin{bmatrix} 1(4)+(-2)(3)+3(1) & 1(-1)+(-2)(-2)+3(-1) & 1(1)+(-2)(-1)+3(-1) \\ 2(4)+(-5)(3)+7(1) & 2(-1)+(-5)(-2)+7(-1) & 2(1)+(-5)(-1)+7(-1) \\ -1(4)+3(3)+(-5)(1) & -1(-1)+3(-2)+(-5)(-1) & -1(1)+3(-1)+(-5)(-1) \end{bmatrix}$$

$$= \begin{bmatrix} 1 & 0 & 0 \\ 0 & 1 & 0 \\ 0 & 0 & 1 \end{bmatrix}$$

$$BA = \begin{bmatrix} 4 & -1 & 1 \\ 3 & -2 & -1 \\ 1 & -1 & -1 \end{bmatrix}\begin{bmatrix} 1 & -2 & 3 \\ 2 & -5 & 7 \\ -1 & 3 & -5 \end{bmatrix}$$

$$= \begin{bmatrix} (4)(1)+(-1)(2)+(1)(-1) & (4)(-2)+(-1)(-5)+(1)(3) & (4)(3)+(-1)(7)+(1)(-5) \\ (3)(1)+(-2)(2)+(-1)(-1) & (3)(-2)+(-2)(-5)+(-1)(3) & (3)(3)+(-2)(7)+(-1)(-5) \\ (1)(1)+(-1)(2)+(-1)(-1) & (1)(-2)+(-1)(-5)+(-1)(3) & (1)(3)+(-1)(7)+(-1)(-5) \end{bmatrix}$$

$$= \begin{bmatrix} 1 & 0 & 0 \\ 0 & 1 & 0 \\ 0 & 0 & 1 \end{bmatrix}$$

Therefore, $B = A^{-1}$ since $AB = BA = I$.

29. $$\begin{matrix} 3 & 1 & 2 \\ 1 & -3 & 4 \\ 2 & 2 & 1 \end{matrix} \quad \begin{matrix} -1 \\ 2 \\ 1 \end{matrix} = \begin{matrix} 3(-1)+1(2)+2(1) \\ 1(-1)+(-3)(2)+4(1) \\ 2(-1)+2(2)+1(1) \end{matrix}$$

$$= \begin{matrix} 1 \\ -3 \\ 3 \end{matrix}$$

$$\neq \begin{matrix} 1 \\ -3 \\ 1 \end{matrix};$$

A is not the proper matrix of solution values.

33. $$B^2 = \begin{bmatrix} 1 & -2 & -6 \\ -3 & 2 & 9 \\ 2 & 0 & -3 \end{bmatrix}\begin{bmatrix} 1 & -2 & -6 \\ -3 & 2 & 9 \\ 2 & 0 & -3 \end{bmatrix} = \begin{bmatrix} -5 & -6 & -6 \\ 9 & 10 & 9 \\ -4 & -4 & -3 \end{bmatrix}$$

$$B^3 = B^2B = \begin{bmatrix} -5 & -6 & -6 \\ 9 & 10 & 9 \\ -4 & -4 & -3 \end{bmatrix}\begin{bmatrix} 1 & -2 & -6 \\ -3 & 2 & 9 \\ 2 & 0 & -3 \end{bmatrix} = \begin{bmatrix} 1 & -2 & -6 \\ -3 & 2 & 9 \\ 2 & 0 & -3 \end{bmatrix} = B$$

37. $$I = \begin{bmatrix} 1 & 0 \\ 0 & 1 \end{bmatrix}, -I = \begin{bmatrix} -1 & 0 \\ 0 & -1 \end{bmatrix}$$

$$(-I)^2 = \begin{bmatrix} -1 & 0 \\ 0 & -1 \end{bmatrix}\begin{bmatrix} -1 & 0 \\ 0 & -1 \end{bmatrix} = \begin{bmatrix} (-1)(-1)+(0)(0) & (-1)(0)+0(-1) \\ 0(-1)+(-1)(0) & 0(0)+(-1)(-1) \end{bmatrix} = \begin{bmatrix} 1 & 0 \\ 0 & 1 \end{bmatrix} = I$$

41. We have

$$\begin{bmatrix} 0.40 & 0.60 \end{bmatrix}\begin{bmatrix} 0.92 & 0.08 \\ 0.14 & 0.86 \end{bmatrix} = \begin{bmatrix} 0.452 & 0.548 \end{bmatrix}$$

Company A has 45.2% of the market share after one year.

45. $$\begin{bmatrix} V_2 \\ i_2 \end{bmatrix} = \begin{bmatrix} 1 & 0 \\ -\frac{1}{R} & 1 \end{bmatrix}\begin{bmatrix} V_1 \\ i_1 \end{bmatrix}$$

$$\begin{bmatrix} V_2 \\ i_2 \end{bmatrix} = \begin{bmatrix} 1(V_1)+0(i_1) \\ -\frac{1}{R}(V_1)+1(i_1) \end{bmatrix}$$

$$V_2 = V_1$$

$$i_2 = -\frac{V_1}{R} + i_1$$

16.3 Finding the Inverse of a Matrix

1. $A=\begin{bmatrix}2 & -3\\ 4 & -5\end{bmatrix}$

$\det A = 2(-5)-(-3)(4) = -10+12 = 2$

$A^{-1}=\frac{1}{2}\begin{bmatrix}-5 & 3\\ -4 & 2\end{bmatrix}=\begin{bmatrix}-\frac{5}{2} & \frac{3}{2}\\ -2 & 1\end{bmatrix}$

Check: $AA^{-1}=\begin{bmatrix}2 & -3\\ 4 & -5\end{bmatrix}\begin{bmatrix}-\frac{5}{2} & \frac{3}{2}\\ -2 & 1\end{bmatrix}=\begin{bmatrix}1 & 0\\ 0 & 1\end{bmatrix}$

5. $\begin{bmatrix}-1 & 5\\ 4 & 10\end{bmatrix}$

Find the determinant of the original matrix.

$\begin{vmatrix}-1 & 5\\ 4 & 10\end{vmatrix} = -30 \neq 0$

Interchange the elements of the principal diagonal and change the signs of the off-diagonal elements.

$\begin{bmatrix}10 & -5\\ -4 & -1\end{bmatrix}$

Divide each element of the second matrix by -30.

$-\frac{1}{30}\begin{bmatrix}10 & -5\\ -4 & -1\end{bmatrix}=\begin{bmatrix}-\frac{1}{3} & \frac{1}{6}\\ \frac{2}{15} & \frac{1}{30}\end{bmatrix}$

9. $\begin{bmatrix}-30 & -45\\ 26 & 50\end{bmatrix}$

Find the determinant of the original matrix.

$\begin{vmatrix}-30 & -45\\ 26 & 50\end{vmatrix} = -2670 \neq 0$

Interchange the elements of the principal diagonal and change the signs of the off-diagonal elements.

$\begin{bmatrix}50 & 45\\ -26 & -30\end{bmatrix}$

Divide each element of the second matrix by -2670.

$-\frac{1}{2670}\begin{bmatrix}50 & 45\\ -26 & -30\end{bmatrix}=\begin{bmatrix}-\frac{5}{267} & -\frac{9}{534}\\ \frac{13}{1335} & \frac{1}{89}\end{bmatrix}$

13. $\left[\begin{array}{cc|cc}2 & 4 & 1 & 0\\ -1 & -1 & 0 & 1\end{array}\right] R1 \to 1/2\,R1 \left[\begin{array}{cc|cc}1 & 2 & \frac{1}{2} & 0\\ -1 & -1 & 0 & 1\end{array}\right]$

$\left[\begin{array}{cc|cc}1 & 2 & \frac{1}{2} & 0\\ -1 & -1 & 0 & 1\end{array}\right] R2 \to R2+R1 \left[\begin{array}{cc|cc}1 & 2 & \frac{1}{2} & 0\\ 0 & 1 & \frac{1}{2} & 1\end{array}\right]$

$\left[\begin{array}{cc|cc}1 & 2 & \frac{1}{2} & 0\\ 0 & 1 & \frac{1}{2} & 1\end{array}\right] R1 \to R1-2R2 \left[\begin{array}{cc|cc}1 & 0 & -\frac{1}{2} & -2\\ 0 & 1 & \frac{1}{2} & 1\end{array}\right];$

$A^{-1}=\begin{bmatrix}-\frac{1}{2} & -2\\ \frac{1}{2} & 1\end{bmatrix}$

17. $\left[\begin{array}{ccc|ccc}1 & -3 & -2 & 1 & 0 & 0\\ -2 & 7 & 3 & 0 & 1 & 0\\ 1 & -1 & -3 & 0 & 0 & 1\end{array}\right] \begin{matrix}R2 \to 2R1+R2\\ R3 \to -R1+R3\end{matrix}$

$\left[\begin{array}{ccc|ccc}1 & -3 & -2 & 1 & 0 & 0\\ 0 & 1 & -1 & 2 & 1 & 0\\ 0 & 2 & -1 & -1 & 0 & 1\end{array}\right] \begin{matrix}R1 \to 3R2+R1\\ R3 \to -2R2+R3\end{matrix}$

$\left[\begin{array}{ccc|ccc}1 & 0 & -5 & 7 & 3 & 0\\ 0 & 1 & -1 & 2 & 1 & 0\\ 0 & 0 & 1 & -5 & -2 & 1\end{array}\right] \begin{matrix}R2 \to R3+R2\\ R1 \to 5R3+R1\end{matrix}$

$\left[\begin{array}{ccc|ccc}1 & 0 & 0 & -18 & -7 & 5\\ 0 & 1 & 0 & -3 & -1 & 1\\ 0 & 0 & 1 & -5 & -2 & 1\end{array}\right];$

$A^{-1}=\begin{bmatrix}-18 & -7 & 5\\ -3 & -1 & 1\\ -5 & -2 & 1\end{bmatrix}$

21.

```
[A]
         [[2  8]
          [-1 6]]
[A]-1▸Frac
     [[3/10 -2/5]
      [1/20 1/10]]
█
```

25.

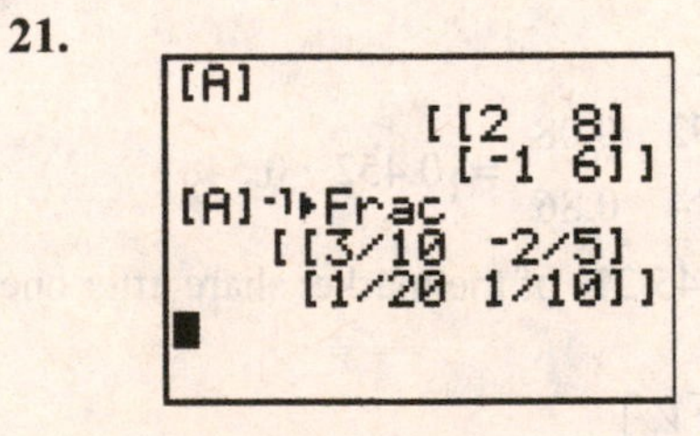

```
       [[2  4 0 ]
        [3  4 -2]
        [-1 1 2 ]]
[A]-1
[[2.5   -2   -2]
 [-1    1    1 ]
 [1.75 -1.5 -1]]
█
```

29. $A^{-1}=\begin{matrix}\frac{3}{2} & 2\\ \frac{1}{2} & 1\end{matrix}$

$BA^{-1}=\begin{matrix}8 & -2\\ 3 & 4\end{matrix}\;\begin{matrix}\frac{3}{2} & 2\\ \frac{1}{2} & 1\end{matrix} = \begin{matrix}11 & 14\\ \frac{13}{2} & 10\end{matrix}$

33. Using Gauss-Jordan or a calculator, $A^{-1} = \begin{matrix} 3 & -3 & 1 \\ -2 & 2 & -1 \\ -4 & 5 & -2 \end{matrix}$

$$CA^{-1} = \begin{matrix} 5 & -1 & 0 \\ 2 & -2 & 1 \\ -3 & 0 & 4 \end{matrix} \begin{matrix} 3 & -3 & 1 \\ -2 & 2 & -1 \\ -4 & 5 & -2 \end{matrix} = \begin{matrix} 17 & -17 & 6 \\ 6 & -5 & 2 \\ -25 & 29 & -11 \end{matrix}$$

37. $\dfrac{1}{ad-bc} \cdot \begin{matrix} a & b \\ c & d \end{matrix} \cdot \begin{matrix} d & -b \\ -c & a \end{matrix}$

$$= \frac{1}{ad-bc} \cdot \begin{matrix} ad-bc & -ab+ab \\ cd-cd & -bc+ad \end{matrix}$$

$$= \frac{1}{ad-bc} \cdot \begin{matrix} ad-bc & 0 \\ 0 & ad-bc \end{matrix}$$

$$= \begin{matrix} \frac{ad-bc}{ad-bc} & 0 \\ 0 & \frac{ad-bc}{ad-bc} \end{matrix}$$

$$= \begin{matrix} 1 & 0 \\ 0 & 1 \end{matrix}$$

41. $\begin{matrix} a_{11} & a_{12} \\ a_{21} & a_{22} \end{matrix}$

Find the determinant of the original matrix.

$$\begin{vmatrix} a_{11} & a_{12} \\ a_{21} & a_{22} \end{vmatrix} = a_{11}a_{22} - a_{12}a_{21}$$

Assume $a_{11}a_{22} - a_{12}a_{21} \neq 0$.

Interchange the elements of the principal diagonal and change the signs of the off-diagonal elements.

$$\begin{matrix} a_{22} & -a_{12} \\ -a_{21} & a_{11} \end{matrix}$$

$$A^{-1} = \frac{1}{a_{11}a_{22} - a_{12}a_{21}} \begin{bmatrix} a_{22} & -a_{12} \\ -a_{21} & a_{11} \end{bmatrix}$$

$$V = A^{-1}I = \frac{1}{a_{11}a_{22} - a_{12}a_{21}} \begin{bmatrix} a_{22} & -a_{12} \\ -a_{21} & a_{11} \end{bmatrix} \begin{bmatrix} i_1 \\ i_2 \end{bmatrix}$$

$$= \frac{1}{a_{11}a_{22} - a_{12}a_{21}} \begin{bmatrix} a_{22}i_1 & -a_{12}i_2 \\ -a_{21}i_1 & a_{11}i_2 \end{bmatrix}$$

$$v_1 = \frac{a_{22}i_1 - a_{12}i_2}{a_{11}a_{22} - a_{12}a_{21}}, \; v_2 = \frac{-a_{21}i_1 + a_{11}i_2}{a_{11}a_{22} - a_{12}a_{21}}$$

16.4 Matrices and Linear Equations

1. $2x - y = 7$

$5x - 3y = 19$

$$A = \begin{matrix} 2 & -1 \\ 5 & -3 \end{matrix}, \ C = \begin{matrix} 7 \\ 19 \end{matrix}, \ A^{-1} = \begin{matrix} 3 & -1 \\ 5 & -2 \end{matrix}$$

$$A^{-1}C = \begin{matrix} 3 & -1 & 7 \\ 5 & -2 & 19 \end{matrix}$$

$$= \begin{matrix} 2 \\ -3 \end{matrix}$$

The solution is $x = 2$, $y = -3$.

5. $x + 2y = 7$

$2x + 5y = 11$

$$A = \begin{bmatrix} 1 & 2 \\ 2 & 5 \end{bmatrix}; \ C = \begin{bmatrix} 7 \\ 11 \end{bmatrix}; \ A^{-1} = \begin{bmatrix} 5 & -2 \\ -2 & 1 \end{bmatrix}$$

$$A^{-1}C = \begin{bmatrix} 5 & -2 \\ -2 & 1 \end{bmatrix}\begin{bmatrix} 7 \\ 11 \end{bmatrix}$$

$$= \begin{bmatrix} 13 \\ -3 \end{bmatrix}$$

The solution is $x = 13$, $y = -3$.

9. $$A = \begin{matrix} 2 & -3 \\ 4 & -5 \end{matrix}; \ C = \begin{matrix} 3 \\ 4 \end{matrix}; \ \begin{vmatrix} 2 & -3 \\ 4 & 5 \end{vmatrix} = -2$$

$$A^{-1} = \begin{matrix} -\frac{5}{2} & \frac{3}{2} \\ -2 & 1 \end{matrix}$$

$$A^{-1}C = \begin{matrix} -\frac{5}{2} & \frac{3}{2} & 3 \\ -2 & 1 & 4 \end{matrix}$$

$$= \begin{matrix} -\frac{3}{2} \\ -2 \end{matrix}$$

The solution is $x = -\dfrac{3}{2}$, $y = -2$.

13. $$\left[\begin{array}{ccc|ccc} 1 & 2 & 2 & 1 & 0 & 0 \\ 4 & 9 & 10 & 0 & 1 & 0 \\ -1 & 3 & 7 & 0 & 0 & 1 \end{array}\right] \begin{matrix} \\ R2 \to -4R1 + R2 \\ R3 \to R1 + R3 \end{matrix}$$

$$\left[\begin{array}{ccc|ccc} 1 & 2 & 2 & 1 & 0 & 0 \\ 0 & 1 & 2 & -4 & 1 & 0 \\ 0 & 5 & 9 & 1 & 0 & 1 \end{array}\right] \begin{matrix} R1 \to R1 - 2R2 \\ \\ R3 \to R3 - 5R2 \end{matrix}$$

$$\begin{array}{ccc|ccc} 1 & 0 & -2 & 9 & -2 & 0 \\ 0 & 1 & 2 & -4 & 1 & 0 \\ 0 & 0 & -1 & 21 & -5 & 1 \end{array} \begin{matrix} \\ \\ R3 \to -R3 \end{matrix}$$

$$\begin{array}{ccc|ccc} 1 & 0 & -2 & 9 & -2 & 0 \\ 0 & 1 & 2 & -4 & 1 & 0 \\ 0 & 0 & 1 & -21 & 5 & -1 \end{array} \begin{matrix} R1 \to R1 + 2R3 \\ R2 \to R2 - 2R3 \\ \\ \end{matrix}$$

$$\left[\begin{array}{ccc|ccc} 1 & 0 & 0 & -33 & 8 & -2 \\ 0 & 1 & 0 & 38 & -9 & 0 \\ 0 & 0 & 1 & -21 & 5 & -1 \end{array}\right]$$

$$A^{-1} = \begin{bmatrix} -33 & 8 & -2 \\ 38 & -9 & 2 \\ -21 & 5 & -1 \end{bmatrix}$$

$$A^{-1}C = \begin{bmatrix} -33 & -8 & -2 \\ 38 & -9 & 2 \\ -21 & 5 & -1 \end{bmatrix}\begin{bmatrix} -4 \\ -18 \\ -7 \end{bmatrix}$$

$$= \begin{bmatrix} 2 \\ -4 \\ 1 \end{bmatrix}$$

The solution is $x = 2$, $y = -4$, $z = 1$.

17. $$A = \begin{bmatrix} 3 & -1 \\ 7 & 2 \end{bmatrix}; A^{-1} = \begin{bmatrix} \frac{2}{13} & \frac{1}{13} \\ \frac{-7}{13} & \frac{3}{13} \end{bmatrix}; C = \begin{bmatrix} 4 \\ 18 \end{bmatrix}.$$

$$A^{-1}C = \begin{bmatrix} \frac{2}{13} & \frac{1}{13} \\ \frac{-7}{13} & \frac{3}{13} \end{bmatrix}\begin{bmatrix} 4 \\ 18 \end{bmatrix} = \begin{bmatrix} 2 \\ 2 \end{bmatrix}$$

The solution is $x = 2, y = 2$

21. $$A = \begin{bmatrix} 2 & -1 & -1 \\ 4 & -3 & 2 \\ 3 & 5 & 1 \end{bmatrix}; A^{-1} = \begin{bmatrix} \frac{13}{57} & \frac{4}{57} & \frac{5}{57} \\ \frac{-2}{57} & \frac{-5}{57} & \frac{8}{57} \\ \frac{-29}{57} & \frac{13}{57} & \frac{2}{57} \end{bmatrix}.$$

$$A^{-1}C = \begin{bmatrix} \frac{13}{57} & \frac{4}{57} & \frac{5}{57} \\ \frac{-2}{57} & \frac{-5}{57} & \frac{8}{57} \\ \frac{-29}{57} & \frac{13}{57} & \frac{2}{57} \end{bmatrix}\begin{bmatrix} 7 \\ 4 \\ -10 \end{bmatrix}$$

$$= \begin{bmatrix} 1 \\ -2 \\ -3 \end{bmatrix}$$

The solution is $x = 1, y = -2, z = -3$.

25. $A=\begin{bmatrix}1 & -5 & 2 & -1\\ 3 & 1 & -3 & 2\\ 4 & -2 & 1 & -1\\ -2 & 3 & -1 & 4\end{bmatrix}; C=\begin{bmatrix}-18\\ 17\\ -1\\ 11\end{bmatrix};$

$$A^{-1}=\begin{bmatrix}\frac{-7}{85} & \frac{2}{85} & \frac{23}{85} & \frac{3}{85}\\ \frac{-47}{170} & \frac{-23}{170} & \frac{33}{170} & \frac{4}{85}\\ \frac{-13}{170} & \frac{-57}{170} & \frac{67}{170} & \frac{21}{85}\\ \frac{5}{34} & \frac{1}{34} & \frac{3}{34} & \frac{5}{17}\end{bmatrix}$$

$$A^{-1}C=\begin{bmatrix}\frac{-7}{85} & \frac{2}{85} & \frac{23}{85} & \frac{3}{85}\\ \frac{-47}{170} & \frac{-23}{170} & \frac{33}{170} & \frac{4}{85}\\ \frac{-13}{170} & \frac{-57}{170} & \frac{67}{170} & \frac{21}{85}\\ \frac{5}{34} & \frac{1}{34} & \frac{3}{34} & \frac{5}{17}\end{bmatrix}\begin{bmatrix}-18\\ 17\\ -1\\ 11\end{bmatrix}$$

$$=\begin{bmatrix}2\\ 3\\ -2\\ 1\end{bmatrix}$$

The solution is $x=2, y=3, z=-2, t=1$.

29. $x^2+y=2$

$2x^2-y=10$

$$A=\begin{bmatrix}1 & 1\\ 2 & -1\end{bmatrix}, C=\begin{bmatrix}2\\ 10\end{bmatrix}, \begin{vmatrix}1 & 1\\ 2 & -1\end{vmatrix}=-3$$

$$A^{-1}=\begin{bmatrix}\frac{1}{3} & \frac{1}{3}\\ \frac{2}{3} & -\frac{1}{3}\end{bmatrix}$$

$$A^{-1}C=\begin{bmatrix}\frac{1}{3} & \frac{1}{3}\\ \frac{2}{3} & -\frac{1}{3}\end{bmatrix}\begin{bmatrix}2\\ 10\end{bmatrix}$$

$$=\begin{bmatrix}4\\ -2\end{bmatrix}$$

The solution for the linear system is $x^2=4$, $y=-2$.

Therefore $x=\pm 2$

The solutions are $x=2$, $y=-2$ and $x=-2$, $y=-2$.

33. $A=\begin{bmatrix}\sin 47.2^\circ & \sin 64.4^\circ\\ \cos 47.2^\circ & -\cos 64.4^\circ\end{bmatrix}; C=\begin{bmatrix}2540\\ 0\end{bmatrix};$

$$A^{-1}=\frac{1}{-0.930}\begin{bmatrix}-\cos 64.4^\circ & -\sin 64.4^\circ\\ -\cos 47.2^\circ & \sin 47.2^\circ\end{bmatrix}$$

$$A^{-1}C=\frac{1}{-0.930}\begin{bmatrix}-\cos 64.4^\circ & -\sin 64.4^\circ\\ -\cos 47.2^\circ & \sin 47.2^\circ\end{bmatrix}\begin{bmatrix}2540\\ 0\end{bmatrix}$$

$$=\begin{bmatrix}1180\\ 1860\end{bmatrix} \text{ (rounded to three significant digits)}$$

$F_1=1180$ N, $F_2=1860$ N

37. $x+y=48$

$0.2x+0.5y=0.25(48)$

$=12$

$$A=\begin{matrix}1 & 1\\ 0.2 & 0.5\end{matrix}, C=\begin{matrix}48\\ 12\end{matrix}, \begin{vmatrix}1 & 1\\ 0.2 & 0.5\end{vmatrix}=0.3$$

$$A^{-1}=\begin{matrix}\frac{5}{3} & -\frac{10}{3}\\ -\frac{2}{3} & \frac{10}{3}\end{matrix}$$

$$A^{-1}C=\begin{matrix}\frac{5}{3} & -\frac{10}{3} & 48\\ -\frac{2}{3} & \frac{10}{3} & 12\end{matrix}$$

$$=\begin{matrix}40\\ 8\end{matrix}$$

40 mL of 20% acid and 8 mL of 50% acid will produce 48 mL of 25% acid.

16.5 Gaussian Elimination

1. $3x-2y=3 \quad R1\rightarrow\frac{1}{3}R1$

$2x+y=4$

$x-\frac{2}{3}y=1$

$2x+\ y=4 \quad R2\rightarrow R2-2R1$

$x-\frac{2}{3}y=1$

$\frac{7}{3}y=2 \quad R2\rightarrow\frac{3}{7}R2$

$x-\frac{2}{3}y=1$

$y=\frac{6}{7}$

$x-\frac{2}{3}\left(\frac{6}{7}\right)=1$

$x \quad =\frac{11}{7}$

The solution is $x=\frac{11}{7}$, $y=\frac{6}{7}$.

5.

$$\begin{array}{l} 4x-3y=2 \quad R1\to \frac{1}{4}R1 \\ \underline{-2x+4y=3} \\ x-\frac{3}{4}y=\frac{1}{2} \\ \underline{-2x+4y=3 \quad R2\to 2R1+R2} \\ x-\frac{3}{4}y=\frac{1}{2} \\ \underline{\frac{5}{2}y=4 \quad R2\to \frac{2}{5}R2} \\ x-\frac{3}{4}y=\frac{1}{2} \\ \underline{y=\frac{8}{5}} \\ x-\frac{3}{4}\left(\frac{8}{5}\right)=\frac{1}{2} \\ x \qquad =\frac{17}{10} \end{array}$$

The solution is $x=\frac{17}{10},\ y=\frac{8}{5}$.

9.

$$\begin{array}{l} x+3y+3z=-3 \quad R2\to -2R1+R2 \\ 2x+2y+z=-5 \quad R3\to 2R1+R3 \\ \underline{-2x-y+4z=6} \\ x+3y+3z=-3 \\ -4y-5z=1 \quad R2\to -\frac{1}{4}R2 \\ \underline{5y+10z=0} \\ x+3y+3z=-3 \\ y+\frac{5}{4}z=-\frac{1}{4} \quad R3\to -5R2+R3 \\ \underline{5y+10z=0} \\ x+3y+3z=-3 \\ y+\frac{5}{4}z=-\frac{1}{4} \\ \underline{\frac{15}{4}z=\frac{5}{4}} \\ z=\frac{1}{3} \\ \underline{y+\frac{5}{4}\left(\frac{1}{3}\right)=-\frac{1}{4}} \\ y=-\frac{2}{3} \\ x+3\left(-\frac{2}{3}\right)+3\left(\frac{1}{3}\right)=-3 \\ x=-2 \end{array}$$

The solution is $x=-2,\ y=-\frac{2}{3},\ z=\frac{1}{3}$

13.

$$\begin{array}{l} x-4y+z=2 \\ \underline{2x-y+4z=-4 \quad R2\to -2R1+R2} \\ x-4y+z=2 \\ \underline{7y+2z=-8} \\ y=-\frac{8}{7}-\frac{2}{7}z \end{array}$$

The value of z can be chosen arbitrarily, so there is an unlimited number of solutions. For example, if $z=3$, then $y=-2$ and $x=-9$. If $z=-4$, then $y=0$ and $x=6$.

17.

$$\begin{array}{l} x+3y+z=4 \\ 2x-6y-3z=10 \quad R2\to -2R1+R2 \\ \underline{4x-9y+3z=4 \quad R3\to -4R1+R3} \\ x+3y+z=4 \\ -12y-5z=2 \quad R2\to -\frac{1}{12}R2 \\ \underline{-21y-z=-12} \\ x+3y+z=4 \\ y+\frac{5}{12}z=-\frac{1}{6} \quad R3\to R3+21R2 \\ \underline{-21y-z=-12} \\ x+3y+z=4 \\ y+\frac{5}{12}z=-\frac{1}{6} \\ \underline{\frac{31}{4}z=-\frac{31}{2}} \\ z=-2 \\ -12y-5(-2)=2 \\ y=\frac{2}{3} \\ x+3\left(\frac{2}{3}\right)+(-2)=4 \\ x=4 \end{array}$$

The solution is $x=4,\ y=\frac{2}{3},\ z=-2$.

21.

$$\begin{aligned} 6x+10y&=-4 \quad R1\to \tfrac{1}{2}R1\\ 24x-18y&=13 \quad R2\to -4R1+R2\\ 15x-33y&=19 \quad R3\to -\tfrac{5}{2}R1+R3\\ 6x+68y&=-33 \quad R4\to -R1+R4 \end{aligned}$$

$$\begin{aligned} 3x+5y&=-2\\ -58y&=29 \quad R2\to -\tfrac{1}{58}R2\\ -58y&=29 \quad R3\to R3-R2\\ 58y&=-29 \quad R4\to R4+R2 \end{aligned}$$

$$\begin{aligned} 3x+5y&=-2\\ y&=-\tfrac{1}{2}\\ 0&=0\\ 0&=0 \end{aligned}$$

$$3x+5\left(-\tfrac{1}{2}\right)=-2$$
$$x=\tfrac{1}{6}$$

The solution is $x=\frac{1}{6}$, $y=-\frac{1}{2}$.

25.

$$\begin{aligned} 7x+5y-3z&=16\\ 3x-5y+2z&=-8\\ 7s+14t-21u&=0 \end{aligned}$$

The matrix

$$\begin{matrix} 7 & 5 & -3 & 16\\ 3 & -5 & 2 & -8\\ 7 & 14 & -21 & 0 \end{matrix}$$

reduces to

$$\begin{matrix} 1 & 0 & 0 & 1\\ 0 & 1 & 0 & 3\\ 0 & 0 & 1 & 2 \end{matrix}$$

and so the solution is $x=1, y=3, z=2$.

29.

$$\begin{aligned} a_1x+b_1y&=c_1\\ a_2x+b_2y&=c_2 \quad R2\to -\frac{a_2}{a_1}R1+R2 \end{aligned}$$

$$a_1x+b_1y=c_1$$
$$\left(\frac{-a_2b_1}{a_1}+b_2\right)y=\frac{-a_2c_1}{a_1}+c_2$$

$$y=\frac{-\frac{a_2c_1}{a_1}+c_2}{-\frac{a_2b_1}{a_1}+b_2}=\frac{a_1c_2-a_2c_1}{a_1b_2-a_2b_1}=\frac{\begin{vmatrix} a_1 & c_1\\ a_2 & c_2\end{vmatrix}}{\begin{vmatrix} a_1 & b_1\\ a_2 & b_2\end{vmatrix}}$$

$$a_1x+b_1\cdot\frac{a_1c_2-a_2c_1}{a_1b_2-a_2b_1}=c_1$$

$$x=\frac{-b_1\left(a_1c_2-a_2c_1\right)}{a_1b_2-a_2b_1}+\frac{c_1\left(a_1b_2-a_2b_1\right)}{a_1\left(a_1b_2-a_2b_1\right)}$$

$$x=\frac{-a_1b_1c_2+a_2b_1c_1+a_1b_2c_1-a_2b_1c_1}{a_1\left(a_1b_2-a_2b_1\right)}$$

$$x=\frac{a_1\left(b_2c_1-b_1c_2\right)}{a_1\left(a_1b_2-a_2b_1\right)}=\frac{\begin{vmatrix} c_1 & b_1\\ c_2 & b_2\end{vmatrix}}{\begin{vmatrix} a_1 & b_1\\ a_2 & b_2\end{vmatrix}}$$

33.

$$\begin{aligned} x+y+z&=650\\ -x+2y-z&=10 \quad R2\to R1+R2\\ 3x+2y+2z&=1550 \quad R3\to -3R1+R3 \end{aligned}$$

$$\begin{aligned} x+y+z&=650\\ 3y&=660 \quad R2\to \tfrac{1}{3}R2\\ -y-z&=-400 \quad R3\to R3+\tfrac{1}{3}R2 \end{aligned}$$

$$\begin{aligned} x+y+z&=650\\ y&=220\\ -z&=-180 \end{aligned}$$

$$x+220+180=650$$
$$x=250$$

The production rates are 250 parts/h, 220 parts/h, and 180 parts/h.

16.6 Higher-order Determinants

1. $\begin{vmatrix} 3 & 0 & 0 & 2 \\ 1 & 0 & -1 & 4 \\ -3 & 1 & 2 & -2 \\ 2 & -1 & 0 & -1 \end{vmatrix} = 3\begin{vmatrix} 0 & -1 & 4 \\ 1 & 2 & -2 \\ -1 & 0 & -1 \end{vmatrix} - 2\begin{vmatrix} 1 & 0 & -1 \\ -3 & 1 & 2 \\ 2 & -1 & 0 \end{vmatrix}$

$= 3\ -1\begin{vmatrix} -1 & 4 \\ 0 & -1 \end{vmatrix} + (-1)\begin{vmatrix} -1 & 4 \\ 2 & -2 \end{vmatrix}\ - 2\ 1\begin{vmatrix} 1 & 2 \\ -1 & 0 \end{vmatrix} + (-1)\begin{vmatrix} -3 & 1 \\ 2 & -1 \end{vmatrix}$

$= 3(-1+6) - 2(2-1)$

$= 13$

5. $\begin{vmatrix} 3 & -2 & 4 & 2 \\ 5 & -1 & 2 & -1 \\ 3 & -2 & 4 & 2 \\ 0 & 3 & -6 & 0 \end{vmatrix} = 0$

Row 1 and Row 3 are identical, so the determinant is zero.

9. $\begin{vmatrix} 2 & -3 & 3 \\ -4 & 1 & 9 \\ 1 & -3 & -6 \end{vmatrix} = 120,$

Column 3 has been multiplied by 3, so the value of the determinant was multiplied by 3.

13. Expand by first row,

$\begin{vmatrix} 3 & 1 & 0 \\ -2 & 3 & -1 \\ 4 & 2 & 5 \end{vmatrix} = 3\begin{vmatrix} 3 & -1 \\ 2 & 5 \end{vmatrix} - (1)\begin{vmatrix} -2 & -1 \\ 4 & 5 \end{vmatrix} + 0\begin{vmatrix} -2 & 3 \\ 4 & 2 \end{vmatrix}$

$= 3(3(5) - 2(-1)) - (-2(5) - 4(-1))$

$= 57$

17. Expand by first column,

$\begin{vmatrix} 5 & 3 & 0 & 5 \\ 4 & 2 & 1 & 2 \\ 3 & 2 & -2 & 2 \\ 0 & 1 & 2 & -1 \end{vmatrix} = 5\begin{vmatrix} 2 & 1 & 2 \\ 2 & -2 & 2 \\ 1 & 2 & -1 \end{vmatrix} - 4\begin{vmatrix} 3 & 0 & 5 \\ 2 & -2 & 2 \\ 1 & 2 & -1 \end{vmatrix} + 3\begin{vmatrix} 3 & 0 & 5 \\ 2 & 1 & 2 \\ 1 & 2 & -1 \end{vmatrix}$

$= 5\ 2\begin{vmatrix} -2 & 2 \\ 2 & -1 \end{vmatrix} - 1\begin{vmatrix} 2 & 2 \\ 1 & -1 \end{vmatrix} + 2\begin{vmatrix} 2 & -2 \\ 1 & 2 \end{vmatrix}\ - 4\ 3\begin{vmatrix} -2 & 2 \\ 2 & -1 \end{vmatrix} + 5\begin{vmatrix} 2 & -2 \\ 1 & 2 \end{vmatrix}\ + 3\ 3\begin{vmatrix} 1 & 2 \\ 2 & -1 \end{vmatrix} + 5\begin{vmatrix} 2 & 1 \\ 1 & 2 \end{vmatrix}$

$= 5(-4+4+12) - 4(-6+30) + 3(-15+15)$

$= -36$

$+3\begin{vmatrix} 3 & -3 & 5 \\ 2 & 1 & 2 \\ 1 & 2 & -1 \end{vmatrix} - 0\begin{vmatrix} 3 & -3 & 5 \\ 2 & 1 & 2 \\ 2 & -2 & 2 \end{vmatrix}$

$= 1(12) - 4(12) + 3(-12)$

$= -72$

21. $3x+y+z \quad = -1$
$2x \quad +2t = 0$
$2y-z+3t = 1$
$2z-3t = 1$

$$\begin{vmatrix}3&1&1&0\\2&0&0&2\\0&2&-1&3\\0&0&2&-3\end{vmatrix} = \begin{vmatrix}3&1&1&0\\0&-\frac{1}{3}&-\frac{1}{3}&1\\0&2&-1&3\\0&0&2&-3\end{vmatrix}$$

$$= 3\begin{vmatrix}-\frac{1}{3}&-\frac{1}{3}&1\\2&-1&3\\0&2&-3\end{vmatrix}$$

$$= 3\begin{vmatrix}-\frac{1}{3}&-\frac{1}{3}&1\\0&-3&9\\0&2&-3\end{vmatrix}$$

$$= 3(-\tfrac{1}{3})(9-18)$$

$$= 9$$

$$x = \frac{\begin{vmatrix}-1&1&1&0\\0&0&0&1\\1&2&-1&3\\1&0&2&-3\end{vmatrix}}{9}$$

$$= (1)\frac{\begin{vmatrix}-1&1&1\\1&2&-1\\1&0&2\end{vmatrix}}{9}$$

$$= \frac{1(-3)+2(-3)}{9}$$

$$= -1$$

$$y = \frac{\begin{vmatrix}3&-1&1&0\\1&0&0&1\\0&1&-1&3\\0&1&2&-3\end{vmatrix}}{9}$$

$$= \frac{\begin{vmatrix}3&-1&1&0\\0&\frac{1}{3}&-\frac{1}{3}&1\\0&1&-1&3\\0&1&2&-3\end{vmatrix}}{9}$$

$$= \frac{\begin{vmatrix}3&-1&1&0\\0&1&-1&3\\0&1&-1&3\\0&1&2&-3\end{vmatrix}}{9}$$

$$= 0$$

$$z = \frac{\begin{vmatrix}3&1&-1&0\\1&0&0&1\\0&2&1&3\\0&0&1&-3\end{vmatrix}}{9}$$

$$= \frac{\begin{vmatrix}3&1&-1&0\\0&-\frac{1}{3}&\frac{1}{3}&1\\0&2&1&3\\0&0&1&-3\end{vmatrix}}{9}$$

$$= \frac{3}{9}\begin{vmatrix}-\frac{1}{3}&\frac{1}{3}&1\\2&1&3\\0&1&-3\end{vmatrix}$$

$$= \frac{3}{9}\begin{vmatrix}-\frac{1}{3}&\frac{1}{3}&1\\0&3&9\\0&1&-3\end{vmatrix}$$

$$= \frac{1}{3}\left(-\frac{1}{3}\right)(-18)$$

$$= 2$$

$$t = \frac{\begin{vmatrix}3&1&1&-1\\1&0&0&0\\0&2&-1&1\\0&0&2&1\end{vmatrix}}{9}$$

$$= -\frac{\begin{vmatrix}1&1&-1\\2&-1&1\\0&2&1\end{vmatrix}}{9}$$

$$= -\frac{\begin{vmatrix}1&1&-1\\0&-3&3\\0&2&1\end{vmatrix}}{9}$$

$$= \frac{-1(-3-6)}{9}$$

$$= 1$$

The solution is $x=-1,\ y=0,\ z=2,\ t=1$

25.
$$\begin{aligned} 2x+y+z \quad &= 2 \\ 3y-z+2t &= 4 \\ y+2z+t &= 0 \\ 3x+ \quad 2z \quad &= 4 \end{aligned}$$

$$\begin{vmatrix} 2 & 1 & 1 & 0 \\ 0 & 3 & -1 & 2 \\ 0 & 1 & 2 & 1 \\ 3 & 0 & 2 & 0 \end{vmatrix} = \begin{vmatrix} 2 & 1 & 1 & 0 \\ 0 & 3 & -1 & 2 \\ 0 & 1 & 2 & 1 \\ 0 & -\frac{3}{2} & \frac{1}{2} & 0 \end{vmatrix}$$

$$= \begin{vmatrix} 2 & 1 & 1 & 0 \\ 0 & 3 & -1 & 2 \\ 0 & 0 & \frac{7}{3} & \frac{1}{3} \\ 0 & 0 & 0 & 1 \end{vmatrix}$$

$$= (2)(3)(\tfrac{7}{3})(1)$$
$$= 14$$

$$x = \frac{\begin{vmatrix} 2 & 1 & 1 & 0 \\ 4 & 3 & -1 & 2 \\ 0 & 1 & 2 & 1 \\ 4 & 0 & 2 & 0 \end{vmatrix}}{14} = \frac{\begin{vmatrix} 2 & 1 & 1 & 0 \\ 0 & 1 & -3 & 2 \\ 0 & 1 & 2 & 1 \\ 0 & -2 & 0 & 0 \end{vmatrix}}{14} = \frac{\begin{vmatrix} 2 & 1 & 1 & 0 \\ 0 & 1 & -3 & 2 \\ 0 & 0 & 5 & -1 \\ 0 & 0 & -6 & 4 \end{vmatrix}}{14} = \frac{\begin{vmatrix} 2 & 1 & 1 & 0 \\ 0 & 1 & -3 & 2 \\ 0 & 0 & 5 & -1 \\ 0 & 0 & 0 & \frac{14}{5} \end{vmatrix}}{14} = \frac{28}{14} = 2$$

$$y = \frac{\begin{vmatrix} 2 & 2 & 1 & 0 \\ 0 & 4 & -1 & 2 \\ 0 & 0 & 2 & 1 \\ 3 & 4 & 2 & 0 \end{vmatrix}}{14} = \frac{\begin{vmatrix} 2 & 2 & 1 & 0 \\ 0 & 4 & -1 & 2 \\ 0 & 0 & 2 & 1 \\ 0 & 1 & \frac{1}{2} & 0 \end{vmatrix}}{14} = \frac{\begin{vmatrix} 2 & 2 & 1 & 0 \\ 0 & 4 & -1 & 2 \\ 0 & 0 & 2 & 1 \\ 0 & 0 & \frac{3}{4} & -\frac{1}{2} \end{vmatrix}}{14} = \frac{\begin{vmatrix} 2 & 2 & 1 & 0 \\ 0 & 4 & -1 & 2 \\ 0 & 0 & 2 & 1 \\ 0 & 0 & 0 & -\frac{7}{8} \end{vmatrix}}{14} = \frac{(2)(4)(2)(-\frac{7}{8})}{14} = -1$$

$$z = \frac{\begin{vmatrix} 2 & 1 & 2 & 0 \\ 0 & 3 & 4 & 2 \\ 0 & 1 & 0 & 1 \\ 3 & 0 & 4 & 0 \end{vmatrix}}{14} = \frac{\begin{vmatrix} 2 & 1 & 2 & 0 \\ 0 & 3 & 4 & 2 \\ 0 & 1 & 0 & 1 \\ 0 & -\frac{3}{2} & 1 & 0 \end{vmatrix}}{14} = \frac{\begin{vmatrix} 2 & 1 & 2 & 0 \\ 0 & 3 & 4 & 2 \\ 0 & 0 & -\frac{4}{3} & \frac{1}{3} \\ 0 & 0 & 3 & 1 \end{vmatrix}}{14} = \frac{\begin{vmatrix} 2 & 1 & 2 & 0 \\ 0 & 3 & 4 & 2 \\ 0 & 0 & -\frac{4}{3} & \frac{1}{3} \\ 0 & 0 & 0 & \frac{7}{4} \end{vmatrix}}{14} = \frac{(2)(3)(-\frac{4}{3})(\frac{7}{4})}{14} = -1$$

$$t = \frac{\begin{vmatrix} 2 & 1 & 1 & 2 \\ 0 & 3 & -1 & 4 \\ 0 & 1 & 2 & 0 \\ 3 & 0 & 2 & 4 \end{vmatrix}}{14} = \frac{\begin{vmatrix} 2 & 1 & 1 & 2 \\ 0 & 3 & -1 & 4 \\ 0 & 1 & 2 & 0 \\ 0 & -\frac{3}{2} & \frac{1}{2} & 1 \end{vmatrix}}{14} = \frac{\begin{vmatrix} 2 & 1 & 1 & 2 \\ 0 & 3 & -1 & 4 \\ 0 & 0 & \frac{7}{3} & -\frac{4}{3} \\ 0 & 0 & 0 & 3 \end{vmatrix}}{14} = \frac{(2)(3)(\frac{7}{3})(3)}{14} = 3$$

The solution is $x = 2$, $y = -1$, $z = -1$, $t = 3$.

29.
$$\begin{vmatrix} c & b & c \\ f & e & f \\ i & h & i \end{vmatrix} = 0$$

Two columns are identical, so the determinant is zero.

33.
$$\begin{aligned} I_A+I_B+I_C+I_D+I_E&=0\\ -2I_A+3I_B&=0\\ 3I_B-3I_C&=6\\ -3I_C+I_D&=0\\ -I_D+2I_E&=0 \end{aligned}$$

$$\begin{vmatrix} 1&1&1&1&1\\ -2&3&0&0&0\\ 0&3&-3&0&0\\ 0&0&-3&1&0\\ 0&0&0&-1&2 \end{vmatrix}=-96$$

$$I_A=\frac{\begin{vmatrix} 0&1&1&1&1\\ 0&3&0&0&0\\ 6&3&-3&0&0\\ 0&0&-3&1&0\\ 0&0&0&-1&2 \end{vmatrix}}{-96}=\frac{-198}{-96}=\frac{33}{16}\text{ A}\qquad I_B=\frac{\begin{vmatrix} 1&0&1&1&1\\ -2&0&0&0&0\\ 0&6&-3&0&0\\ 0&0&-3&1&0\\ 0&0&0&-1&2 \end{vmatrix}}{-96}=\frac{-132}{-96}=\frac{11}{8}\text{ A}$$

$$I_C=\frac{\begin{vmatrix} 1&1&0&1&1\\ -2&3&0&0&0\\ 0&3&6&0&0\\ 0&0&0&1&0\\ 0&0&0&-1&2 \end{vmatrix}}{-96}=\frac{60}{-96}=-\frac{5}{8}\text{ A}$$

$$I_D=\frac{\begin{vmatrix} 1&1&1&0&1\\ -2&3&0&0&0\\ 0&3&-3&6&0\\ 0&0&-3&0&0\\ 0&0&0&0&2 \end{vmatrix}}{-96}=\frac{180}{-96}=-\frac{15}{8}\text{ A}$$

$$I_E=\frac{\begin{vmatrix} 1&1&1&1&0\\ -2&3&0&0&0\\ 0&3&-3&0&6\\ 0&0&-3&1&0\\ 0&0&0&-1&0 \end{vmatrix}}{-96}=\frac{90}{-96}=-\frac{15}{16}\text{ A}$$

37.
$$\begin{aligned} x+y+z+w&=6\\ 10x\qquad\quad -w&=0\\ x-y-z\qquad&=0\\ x+y\qquad\quad&=0.8, \end{aligned}$$

$$\begin{vmatrix} 1&1&1&1\\ 10&0&0&-1\\ 1&-1&-1&0\\ 1&1&0&0 \end{vmatrix}=-12$$

$$x=\frac{\begin{vmatrix} 6&1&1&1\\ 0&0&0&-1\\ 0&-1&-1&0\\ 0.8&1&0&0 \end{vmatrix}}{-12}=\frac{-6}{-12}=0.5\qquad y=\frac{\begin{vmatrix} 1&6&1&1\\ 10&0&0&-1\\ 1&0&-1&0\\ 1&0.8&0&0 \end{vmatrix}}{-12}=\frac{-3.6}{-12}=0.3$$

$$z=\frac{\begin{vmatrix} 1&1&6&1\\ 10&0&0&-1\\ 1&-1&0&0\\ 1&1&0.8&0 \end{vmatrix}}{-12}=\frac{-2.4}{-12}=0.2\qquad w=\frac{\begin{vmatrix} 1&1&1&6\\ 10&0&0&0\\ 1&-1&-1&0\\ 1&1&0&0.8 \end{vmatrix}}{-12}=\frac{-60}{-12}=5$$

The solution is $x = 0.5$ ppm SO_2, $y = 0.3$ ppm NO, $z = 0.2$ ppm NO_2, $w = 5.0$ ppm CO.

Review Exercises

1. This is false. The entry in the lower right corner should be 4.

5. This is false. The given matrix equation demonstrates that $x = 3, y = 7$ solves the system of equations

$$2x - y = -1$$
$$5x - 3y = -6$$

9.
$$2x = 4 \qquad a = 7 - x$$
$$x = 2 \qquad = 7 - 2$$
$$= 5$$
$$3y = -9 \qquad b = -4 - y$$
$$y = -3 \qquad = -4 - (-3)$$
$$= -1$$
$$2z = 8 \qquad c = 2 + z$$
$$z = 4 \qquad = 2 + 4$$
$$= 6$$

13.
$$A + B = \begin{matrix} 2 & -3 \\ 4 & 1 \\ -5 & 0 \\ 2 & -3 \end{matrix} + \begin{matrix} -1 & 0 \\ 4 & -6 \\ -3 & -2 \\ 1 & -7 \end{matrix} = \begin{matrix} 1 & -3 \\ 8 & -5 \\ -8 & -2 \\ 3 & -10 \end{matrix}$$

17.
$$2A - 3B = 2\begin{bmatrix} 2 & -3 \\ 4 & 1 \\ -5 & 0 \\ 2 & -3 \end{bmatrix} - 3\begin{bmatrix} -1 & 0 \\ 4 & -6 \\ -3 & -2 \\ 1 & -7 \end{bmatrix} = \begin{bmatrix} 7 & -6 \\ -4 & 20 \\ -1 & 6 \\ 1 & 15 \end{bmatrix}$$

21.
$$\begin{matrix} 0 & 0.6 \\ 0.2 & 0.0 \\ 0.4 & -0.1 \end{matrix} \quad \begin{matrix} 0.1 & -0.4 & 0.5 \\ 0.5 & 0.1 & 0.0 \end{matrix}$$

$$= \begin{matrix} 0.0 + 0.3 & 0.0 + 0.06 & 0.0 + 0.0 \\ 0.02 + 0.0 & -0.08 + 0.0 & 0.1 + 0.0 \\ 0.04 - 0.05 & -0.16 - 0.01 & 0.2 - 0.0 \end{matrix}$$

$$= \begin{matrix} 0.3 & 0.06 & 0.0 \\ 0.02 & -0.08 & 0.1 \\ -0.01 & -0.17 & 0.2 \end{matrix}$$

$$= \begin{matrix} 0.3 & 0.1 & 0.0 \\ 0.0 & -0.1 & 0.1 \\ 0.0 & -0.2 & 0.2 \end{matrix}$$

(with the correct precision)

25. Find the determinant of the original matrix.

$$\begin{vmatrix} -0.8 & -0.1 \\ 0.4 & -0.7 \end{vmatrix} = 0.6$$

Interchange elements of principal diagonal and change signs of off-diagonal elements.

$$\begin{matrix} -0.7 & 0.1 \\ -0.4 & -0.8 \end{matrix}$$

Divide each element of second matrix by 0.6.

$$\frac{1}{0.6}\begin{matrix} -0.7 & 0.1 \\ -0.4 & -0.8 \end{matrix} = \begin{matrix} -\frac{7}{6} & \frac{1}{6} \\ -\frac{2}{3} & -\frac{4}{3} \end{matrix}$$

29.
$$\left.\begin{matrix} 2 & -4 & 3 \\ 4 & -6 & 5 \\ -2 & 1 & -1 \end{matrix}\right| \begin{matrix} 1 & 0 & 0 \\ 0 & 1 & 0 \\ 0 & 0 & 1 \end{matrix}$$

$$\begin{matrix} \frac{1}{2}R1 \to R1 \\ -4R1 + R2 \to R2 \\ 2R1 + R3 \to R3 \end{matrix} \quad \left.\begin{matrix} 1 & -2 & \frac{3}{2} \\ 0 & 2 & -1 \\ 0 & -3 & 2 \end{matrix}\right| \begin{matrix} \frac{1}{2} & 0 & 0 \\ -2 & 1 & 0 \\ 1 & 0 & 1 \end{matrix}$$

$$\frac{1}{2}R2 \to R2 \quad \left.\begin{matrix} 1 & -2 & \frac{3}{2} \\ 0 & 1 & -\frac{1}{2} \\ 0 & -3 & 2 \end{matrix}\right| \begin{matrix} \frac{1}{2} & 0 & 0 \\ -1 & \frac{1}{2} & 0 \\ 1 & 0 & 1 \end{matrix}$$

$$\begin{matrix} 2R2 + R1 \to R1 \\ 3R2 + R3 \to R3 \end{matrix} \quad \left.\begin{matrix} 1 & 0 & \frac{1}{2} \\ 0 & 1 & -\frac{1}{2} \\ 0 & 0 & \frac{1}{2} \end{matrix}\right| \begin{matrix} -\frac{3}{2} & 1 & 0 \\ -1 & \frac{1}{2} & 0 \\ -2 & \frac{3}{2} & 1 \end{matrix}$$

$$\begin{matrix} -R3 + R1 \to R1 \\ R3 + R2 \to R2 \\ 2R3 \to R3 \end{matrix} \quad \left.\begin{matrix} 1 & 0 & 0 \\ 0 & 1 & -\frac{1}{2} \\ 0 & 0 & 1 \end{matrix}\right| \begin{matrix} \frac{1}{2} & -\frac{1}{2} & -1 \\ -3 & 2 & 1 \\ -4 & 3 & 2 \end{matrix}$$

$$A^{-1} = \begin{bmatrix} \frac{1}{2} & -\frac{1}{2} & -1 \\ -3 & 2 & 1 \\ -4 & 3 & 2 \end{bmatrix}$$

33. $A=\begin{bmatrix}33 & 52\\ 45 & -62\end{bmatrix}; C=\begin{bmatrix}-450\\ 1380\end{bmatrix}; \begin{vmatrix}33 & 52\\ 45 & -62\end{vmatrix}=-4386$

$$A^{-1}=\begin{bmatrix}\frac{62}{4386} & \frac{52}{4386}\\ \frac{45}{4386} & -\frac{33}{4386}\end{bmatrix}$$

$$A^{-1}C=\begin{bmatrix}\frac{62}{4386} & \frac{52}{4386}\\ \frac{45}{4386} & -\frac{33}{4386}\end{bmatrix}\begin{bmatrix}-450\\ 1380\end{bmatrix}$$

$$=\begin{bmatrix}10\\ -15\end{bmatrix}$$

The solution is $x=10, y=-15$.

37. $\begin{bmatrix}x\\ y\\ z\end{bmatrix}=\begin{bmatrix}1 & 2 & 3\\ 3 & -4 & -3\\ 7 & -6 & 6\end{bmatrix}^{-1}\begin{bmatrix}1\\ 2\\ 2\end{bmatrix}$

$$=\begin{bmatrix}\frac{7}{15} & \frac{1}{3} & -\frac{1}{15}\\ \frac{13}{30} & \frac{1}{6} & -\frac{2}{15}\\ -\frac{1}{9} & -\frac{2}{9} & \frac{1}{9}\end{bmatrix}\begin{bmatrix}1\\ 2\\ 2\end{bmatrix}$$

$$=\begin{bmatrix}1\\ \frac{1}{2}\\ -\frac{1}{3}\end{bmatrix}$$

The solution is $x=1, y=\frac{1}{2}, z=-\frac{1}{3}$.

41.
$2u-3v+2w=7$
$3u+v-3w=-6 \quad \frac{1}{2}R1\to R1$
$u+4v+w=-13$

$u-1.5v+w=3.5$
$3u+v-3w=-6 \quad -3R1+R2\to R2$
$u+4v+w=-13 \quad -1R1+R3\to R3$

$u-1.5v+w=3.5$
$5.5v-6w=-16.5 \quad \frac{1}{5.5}R3\to R3$
$5.5v=-16.5$

$v=-3$
$5.5(-3)-6w=-16.5$
$w=0$
$u-1.5(-3)+0=3.5$
$u=-1$

The solution is $u=-1, v=-3, w=0$.

45.
$2x+3y-z=10 \quad R1\leftrightarrow R2$
$x-2y+6z=-6$
$5x+4y+4z=14$

$x-2y+6z=-6 \quad -2R1+R2\to R2$
$2x+3y-z=10 \quad -5R1+R3\to R3$
$5x+4y+4z=14$

$x-2y+6z=-6$
$7y-13z=22$
$14y-26z=44 \quad -2R2+R3\to R3$

$x-2y+6z=-6$
$7y-13z=22$
$0=0$

The value of z can be chosen arbitrarily, so there is an unlimited number of solutions.

49.
$x+2y+3z=1$
$3x-4y-3z=2$
$7x-6y+6z=2$

$$\begin{vmatrix}1 & 2 & 3\\ 3 & -4 & -3\\ 7 & -6 & 6\end{vmatrix}=-90$$

$$x=\frac{\begin{vmatrix}1 & 2 & 3\\ 2 & -4 & -3\\ 2 & -6 & 6\end{vmatrix}}{-90}=\frac{-90}{-90}=1$$

$$y=\frac{\begin{vmatrix}1 & 1 & 3\\ 3 & 2 & -3\\ 7 & 2 & 6\end{vmatrix}}{-90}=\frac{-45}{-90}=\frac{1}{2}$$

$$z=\frac{\begin{vmatrix}1 & 2 & 1\\ 3 & -4 & 2\\ 7 & -6 & 2\end{vmatrix}}{-90}=\frac{30}{-90}=-\frac{1}{3}$$

The solution is $x=1, y=\frac{1}{2}, z=-\frac{1}{3}$.

53.
$$\begin{aligned} 2x-3y+z-\ t&=-8\\ 4x+\quad 3z+2t&=-3\\ 2y-3z-\ t&=12\\ x-y-\ z+\ t&=\ 3 \end{aligned}$$

$$A=\begin{bmatrix} 2 & -3 & 1 & -1\\ 4 & 0 & 3 & 2\\ 0 & 2 & -3 & -1\\ 1 & -1 & -1 & 1 \end{bmatrix}, C=\begin{bmatrix} -8\\ -3\\ 12\\ 3 \end{bmatrix}; A^{-1}=\begin{bmatrix} \frac{1}{7} & \frac{16}{91} & \frac{20}{91} & \frac{1}{91}\\ -\frac{1}{7} & \frac{12}{91} & \frac{15}{91} & -\frac{22}{91}\\ 0 & \frac{1}{13} & -\frac{2}{13} & -\frac{4}{13}\\ -\frac{2}{7} & \frac{3}{91} & -\frac{19}{91} & \frac{40}{91} \end{bmatrix}$$

$$A^{-1}C=\begin{bmatrix} \frac{1}{7} & \frac{16}{91} & \frac{20}{91} & \frac{1}{91}\\ -\frac{1}{7} & \frac{12}{91} & \frac{15}{91} & -\frac{22}{91}\\ 0 & \frac{1}{13} & -\frac{2}{13} & -\frac{4}{13}\\ -\frac{2}{7} & \frac{3}{91} & -\frac{19}{91} & \frac{40}{91} \end{bmatrix}\begin{bmatrix} -8\\ -3\\ 12\\ 3 \end{bmatrix}$$

$$=\begin{bmatrix} 1\\ 2\\ -3\\ 1 \end{bmatrix}$$

The solution is $x=1, y=2, z=-3, t=1$.

57.
$$\begin{aligned} 4r-\ s+8t-2u+4v&=-1\\ 3r+2s-4t+3u-\ v&=\ 4\\ 3r+3s+2t+5u+6v&=13\\ 6r-s+\ 2t-2u+\ v&=\ 0\\ r-2s+4t\ -3u+\ 3v&=\ 1 \end{aligned}$$

$$A=\begin{matrix} 4 & -1 & 4 & -2 & 4\\ 3 & 2 & -4 & 3 & -1\\ 3 & 3 & 2 & 5 & 2\\ 6 & -1 & 2 & -2 & 1\\ 1 & -2 & 4 & -3 & 3 \end{matrix}, C=\begin{matrix} -1\\ 4\\ 13\\ 0\\ 1 \end{matrix}$$

$$A^{-1}=\begin{matrix} -\frac{2}{15} & \frac{1}{10} & \frac{1}{30} & \frac{1}{6} & \frac{2}{15}\\ \frac{43}{15} & -\frac{7}{5} & -\frac{7}{15} & -\frac{1}{3} & -\frac{58}{15}\\ -\frac{1}{30} & -\frac{7}{20} & \frac{2}{15} & \frac{1}{6} & -\frac{13}{60}\\ -\frac{26}{15} & \frac{4}{5} & \frac{13}{30} & \frac{1}{6} & \frac{67}{30}\\ \frac{4}{15} & \frac{3}{10} & -\frac{1}{15} & -\frac{1}{3} & \frac{7}{30} \end{matrix}$$

$$A^{-1}C=\begin{matrix} -\frac{2}{15} & \frac{1}{10} & \frac{1}{30} & \frac{1}{6} & \frac{2}{15}\\ \frac{43}{15} & -\frac{7}{5} & -\frac{7}{15} & -\frac{1}{3} & -\frac{58}{15}\\ -\frac{1}{30} & -\frac{7}{20} & \frac{2}{15} & \frac{1}{6} & -\frac{13}{60}\\ -\frac{26}{15} & \frac{4}{5} & \frac{13}{30} & \frac{1}{6} & \frac{67}{30}\\ \frac{4}{15} & \frac{3}{10} & -\frac{1}{15} & -\frac{1}{3} & \frac{7}{30} \end{matrix}\begin{matrix} -1\\ 4\\ 13\\ 0\\ 1 \end{matrix}$$

$$=\begin{matrix} \frac{11}{10}\\ -\frac{92}{5}\\ \frac{3}{20}\\ \frac{64}{5}\\ \frac{3}{10} \end{matrix}$$

61. $B = \begin{matrix} 0 & 1 & 0 \\ 0 & 0 & 1 \\ 1 & 0 & 0 \end{matrix}$

$B^2 = \begin{matrix} 0 & 1 & 0 & 0 & 1 & 0 \\ 0 & 0 & 1 & 0 & 0 & 1 \\ 1 & 0 & 0 & 1 & 0 & 0 \end{matrix}$

$= \begin{matrix} 0 & 0 & 1 \\ 1 & 0 & 0 \\ 0 & 1 & 0 \end{matrix}$

$B^3 = B^2 B$

$= \begin{matrix} 0 & 0 & 1 & 0 & 1 & 0 \\ 1 & 0 & 0 & 0 & 0 & 1 \\ 0 & 1 & 0 & 1 & 0 & 0 \end{matrix}$

$= \begin{matrix} 1 & 0 & 0 \\ 0 & 1 & 0 \\ 0 & 0 & 1 \end{matrix}$

$= I$

65. Expanding by the first row:

$$\begin{vmatrix} 1 & 4 & 0 & -3 \\ 3 & 1 & 2 & 5 \\ -2 & -2 & -4 & 1 \\ -1 & 6 & 3 & -4 \end{vmatrix}$$

$$= 1\begin{vmatrix} 1 & 2 & 5 \\ -2 & -4 & 1 \\ 6 & 3 & -4 \end{vmatrix} - 4\begin{vmatrix} 3 & 2 & 5 \\ -2 & -4 & 1 \\ -1 & 3 & -4 \end{vmatrix}$$

$$-(-3)\begin{vmatrix} 3 & 1 & 2 \\ -2 & -2 & -4 \\ -1 & 6 & 3 \end{vmatrix}$$

$= 99 - 4(-29) + 3(36)$

$= 323$

69. $\begin{vmatrix} 1 & 4 & 0 & -3 \\ 3 & 1 & 2 & 5 \\ -2 & -2 & -4 & 1 \\ -1 & 6 & 3 & -4 \end{vmatrix} = 323$

73. $\begin{bmatrix} n & 1+n \\ 1-n & -n \end{bmatrix}^2$

$= \begin{bmatrix} n & 1+n \\ 1-n & -n \end{bmatrix}\begin{bmatrix} n & 1+n \\ 1-n & -n \end{bmatrix}$

$= \begin{bmatrix} n^2+(1+n)(1-n) & n(1+n)(1+n) \\ n(1-n)-n(1-n) & (1-n)(1+n)+n^2 \end{bmatrix}$

$= \begin{bmatrix} n^2+1-n^2 & 0 \\ 0 & 1-n^2+n^2 \end{bmatrix}$

$= \begin{bmatrix} 1 & 0 \\ 0 & 1 \end{bmatrix}$

$= I$

77. $A = \begin{bmatrix} 1 & -2 \\ 0 & 3 \end{bmatrix}, B = \begin{bmatrix} -3 & 1 \\ 2 & -1 \end{bmatrix}$

$(A+B)(A-B)$

$= \left(\begin{bmatrix} 1 & -2 \\ 0 & 3 \end{bmatrix} + \begin{bmatrix} -3 & 1 \\ 2 & -1 \end{bmatrix}\right)\left(\begin{bmatrix} 1 & -2 \\ 0 & 3 \end{bmatrix} - \begin{bmatrix} -3 & 1 \\ 2 & -1 \end{bmatrix}\right)$

$= \begin{bmatrix} -2 & -1 \\ 2 & 2 \end{bmatrix}\begin{bmatrix} 4 & -3 \\ -2 & 4 \end{bmatrix}$

$= \begin{bmatrix} -6 & 2 \\ 4 & 2 \end{bmatrix}$

$A^2 - B^2 = \begin{bmatrix} 1 & -2 \\ 0 & 3 \end{bmatrix}^2 - \begin{bmatrix} -3 & 1 \\ 2 & -1 \end{bmatrix}^2$

$= \begin{bmatrix} 1 & -8 \\ 0 & 9 \end{bmatrix} - \begin{bmatrix} 11 & -4 \\ -8 & 3 \end{bmatrix}$

$= \begin{bmatrix} -10 & -4 \\ 8 & 6 \end{bmatrix}$

$(A+B)(A-B) \neq A^2 - B^2$

81. $A = \begin{bmatrix} 2 & 3 \\ 3 & 2 \end{bmatrix}, C = \begin{bmatrix} 26 \\ 24 \end{bmatrix}, \begin{vmatrix} 2 & 3 \\ 3 & 2 \end{vmatrix} = -5;$

$A^{-1} = \begin{bmatrix} -\frac{2}{5} & \frac{3}{5} \\ \frac{3}{5} & -\frac{2}{5} \end{bmatrix}$

$A^{-1}C = \begin{bmatrix} -\frac{2}{5} & \frac{3}{5} \\ \frac{3}{5} & -\frac{2}{5} \end{bmatrix}\begin{bmatrix} 26 \\ 24 \end{bmatrix}$

$= \begin{bmatrix} 4 \\ 6 \end{bmatrix}$

The solution is $R_1 = 4\,\Omega$, $R_2 = 6\,\Omega$.

85. $2R_1 + 3R_2 = 26$

$\underline{3R_1 + 2R_2 = 24} \quad -\frac{3}{2}R1 + R2 \to R2$

$2R_1 + 3R_2 = 26$

$-\frac{5}{2}R_2 = -15$

$R_2 = 6$

$2R_1 + 3(6) = 26$

$R_1 = 4$

The solution is $R_1 = 4\,\Omega$, $R_2 = 6\,\Omega$.

89. $110t - d = 0$, suspect

$135t - d = \dfrac{135(3.0)}{60} = 6.75$, police

$A = \begin{matrix} 110 & -1 \\ 135 & -1 \end{matrix}, C = \begin{matrix} 0 \\ 6.75 \end{matrix}, A^{-1} = \begin{matrix} -0.04 & 0.04 \\ -5.4 & 4.4 \end{matrix}$

$A^{-1}C = \begin{matrix} -0.04 & 0.04 \\ -5.4 & 4.4 \end{matrix} \begin{matrix} 0 \\ 6.75 \end{matrix}$

$= \begin{matrix} 0.27 \\ 29.7 \end{matrix} = \begin{matrix} t \\ d \end{matrix}$

$t = 0.27, \quad t - \frac{3.0}{60} = 0.22$

The police overtake the suspect 0.22 h, or 13.2 minutes, after passing the intersection.

93.

	Standard Transmission	Automatic Transmission
4 cylinders	12 000	15 000
6 cylinders	24 000	8000
8 cylinders	4000	30 000

$A = \begin{bmatrix} 12\,000 & 15\,000 \\ 24\,000 & 8000 \\ 4000 & 30\,000 \end{bmatrix}$

$B = \begin{bmatrix} 15\,000 & 20\,000 \\ 12\,000 & 3000 \\ 2000 & 22\,000 \end{bmatrix}$

$A + B = \begin{bmatrix} 27\,000 & 35\,000 \\ 36\,000 & 11\,000 \\ 6000 & 52\,000 \end{bmatrix}$

97. Answers may vary, but the basic idea is that the matrix entries show the inventory of a particular product in a particular store in a compact way. It is easy to put information together for a single product throughout all stores, and for all products in a single store.

Chapter 17

Inequalities

17.1 Properties of Inequalities

1. The inequality $x+1<0$ is true for all values of x less than -1.Therefore, the values of x that satisfy this inequality are written as $x<-1$, or as the interval $(-\infty,-1)$.

5.
$$4<9$$
$$4+5<9+5$$
$$9<14$$
(property 1)

9.
$$4<9$$
$$\frac{4}{-1}>\frac{9}{-1}$$
$$-4>-9$$
(property 3)

13. $x>-2$

17. $1<x<7$

21. $x<1$ or $3<x\le 5$

25. x is greater than 0 and less than or equal to 9.

29. $x<3$
$(-\infty,3)$

3 x

33. $0\le x<5$
$[0,5)$

0 5 x

37. $x<-1$ or $1\le x<4$
$(-\infty,-1)$ or $[1,4)$

-1 1 4 x

41. $t\le -5$ and $t\ge -5$
Only the value $t=-5$ satisfies both of these inequalities.

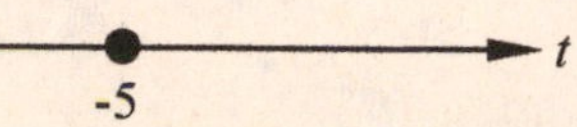

45. Suppose $0<a<b$.
Since a and b are both positive, and the power 2 is a positive integer, then by property 4,
$$a^2<b^2$$
Hence this is an absolute inequality since it is always true for a,b satisfying the original inequality.

49. For $x>0,\ y<0$ we have
$xy<0<|x||y|$
Multiplying both members by 2 does not change the sense of the inequality, so
$2xy<|x||y|$
We add x^2+y^2 (which is the same as $|x|^2+|y|^2$) to both members, and the sense of the inequality does not change
$x^2+2xy+y^2<|x|^2+2|x||y|+|y|^2$
Factoring and taking the square root of two positive numbers does not change the sense of the inequality. Hence we conclude
$(x+y)^2<(|x|+|y|)^2$
$|x+y|<|x|+|y|$

53. The diagonals can range from a minimum of
$$\sqrt{110^2+70^2}=\sqrt{17000}\text{ yd}$$
to a maximum of
$$\sqrt{120^2+80^2}=\sqrt{20800}\text{ yd}$$
and so the length of the diagonal d in yards satisfies the inequality
$$\sqrt{17000}\le d\le\sqrt{20800}.$$

57. $18000\le v\le 25000$ mi/h

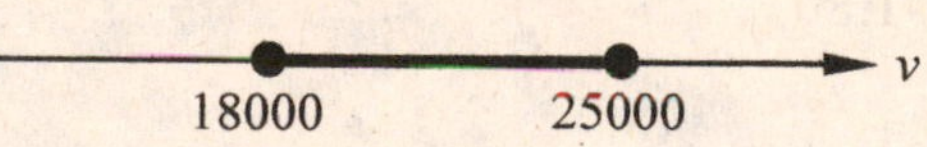

61. $5\text{ h}\le t\le 6\text{ h}$

17.2 Solving Linear Inequalities

1. $25-2x\ge 15$

$-2x\ge -10$

$x\le 5,$

or $(-\infty,5]$

5. $x-3>-4$

$x>-4+3$

$x>-1,$

or $(-1,\infty)$

-1 x

9. $3x-5\le -11$

$3x\le -11+5$

$3x\le -6$

$x\le -2,$

or $(-\infty,-2]$

-2 x

13. $\dfrac{4x-5}{2}\le x$

$4x-5\le 2x$

$2x\le 5$

$x\le \dfrac{5}{2},$

or $\left(-\infty,\frac{5}{2}\right]$

5/2 x

17. $2.50(1.50-3.40x)<3.84-8.45x$

$3.75-8.50x<3.84-8.45x$

$-0.05x<0.09$

$x>-1.80,$

or $(-1.80,\infty)$

-1.80 x

21. $-1<2x+1<3$

$-2<2x<2$

$-1<x<1,$

or $(-1,1)$

-1 1 x

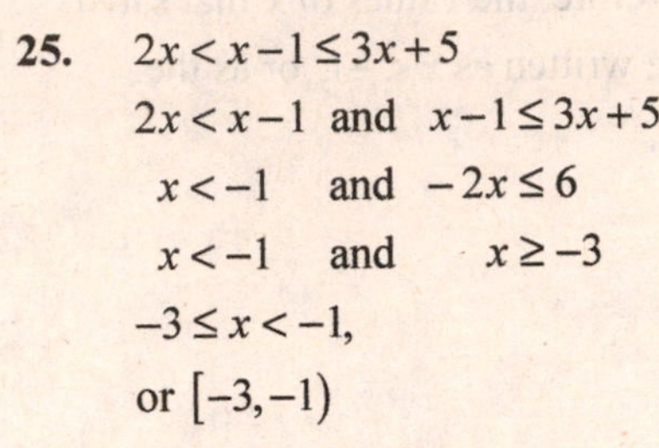

25. $2x<x-1\le 3x+5$

$2x<x-1$ and $x-1\le 3x+5$

$x<-1$ and $-2x\le 6$

$x<-1$ and $x\ge -3$

$-3\le x<-1,$

or $[-3,-1)$

-3 -1 x

29. The solution is $x<\frac{5}{2}$ with graph.

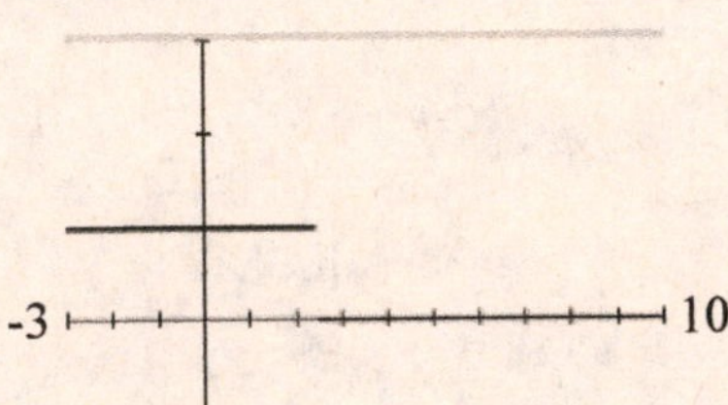

33. The solution is $-2<x<2$ with graph.

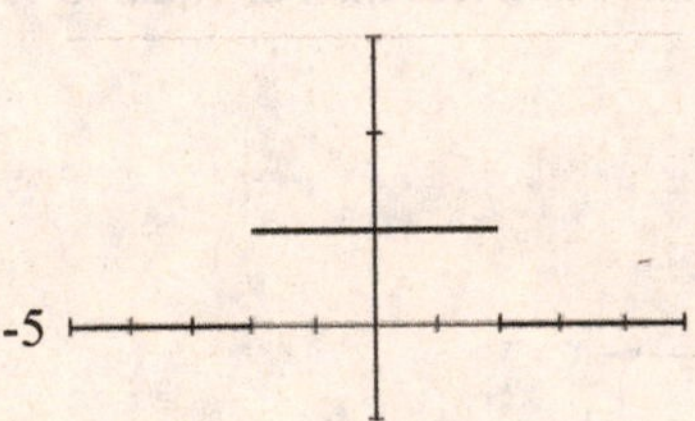

37. The solution is $x\le 0$ with graph.

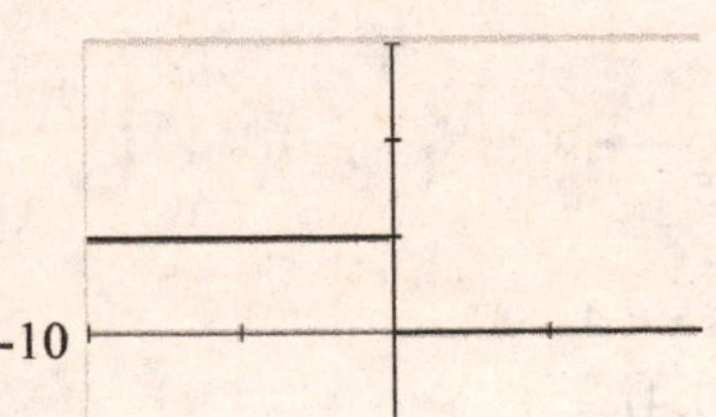

41. $x^2 - kx + 9 = 0$ has roots $\dfrac{k \pm \sqrt{k^2 - 36}}{2}$ which will be imaginary for $k^2 - 36 < 0$, or $(k+6)(k-6) < 0$.

This requires:

$(k+6>0 \text{ and } k-6<0)$ or $(k+6<0 \text{ and } k-6>0)$

$(k>-6 \text{ and } k<6)$ or $(k<-6 \text{ and } k>6)$

$(-6<k<6)$ or (no values in common)

Therefore, the roots are imaginary for $-6 < k < 6$, or $(-6, 6)$.

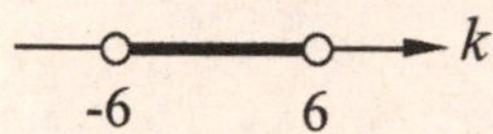

45. $|5-(-2)| = |7| = 7$

$|-5-|-2|| = |-5-2| = |-7| = 7$

Thus,

$|5-(-2)| = |-5-|-2||$

49. $25n > 350 + 15n$

$10n > 350$

for $n > 35$ h, the second position pays more.

n

35

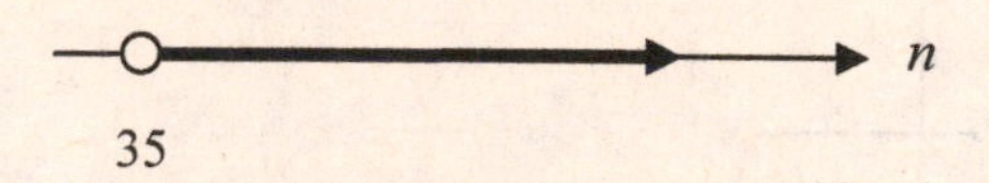

53. $100 < 130(1.42)w < 150$

$0.542 \text{ m} < w < 0.813 \text{ m}$,

or $(0.542 \text{ m}, 0.813 \text{ m})$

w

0.542 0.813

57. $0 \le x \le 800 - 300$

$0 \le x \le 500$

$y = x + 400 - 200$

$= x + 200$

$x = y - 200$

$0 \le y - 200 \le 500$

$200 \le y \le 700$,

or $[200, 700]$

y

200 700

17.3 Solving Nonlinear Inequalities

1. $x^2 + 3 > 4x$

$x^2 - 4x + 3 > 0$

$(x-3)(x-1) > 0$

The critical values are 1, 3.

Interval	$(x-3)(x-1)$		Sign
$x<1$ $(-\infty,1)$	−	−	+
$1<x<3$ $(1,3)$	−	+	−
$x>3$ $(3,\infty)$	+	+	+

$x^2 - 4x + 3 > 0$ when $x < 1$ or $x > 3$, or in $(-\infty, 1)$ or $(3, \infty)$.

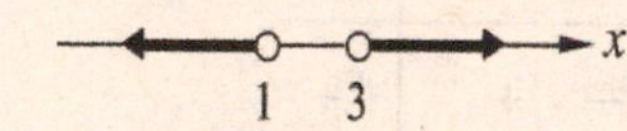

5. $2x^2 \ge 4x$

$2x^2 - 4x \ge 0$

$2x(x-2) \ge 0$

The critical values are $x = 0$, $x = 2$.

Interval	$2x(x-2)$		Sign
$x<0$ $(-\infty,0)$	−	−	+
$0<x<2$ $(0,2)$	+	−	−
$x>2$ $(2,\infty)$	+	+	+

$2x(x-2) \ge 0$ for $x \le 0$ or $2 \le x$, or in $(-\infty, 0]$ or $[2, \infty)$.

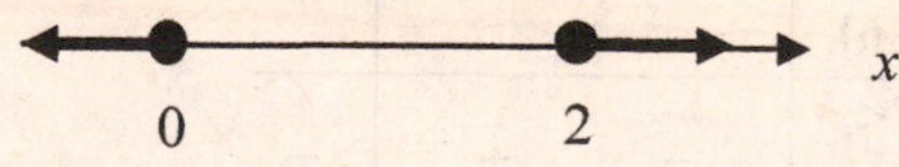

9. $x^2+4x\le -4$

$x^2+4x+4\le 0$

$(x+2)^2\le 0$

$(x+2)^2=0$

for $x=-2$ and $(x+2)^2>0$ for all other values of x. Therefore, $(x+2)(x+2)\le 0$ has only $x=-2$, for a solution.

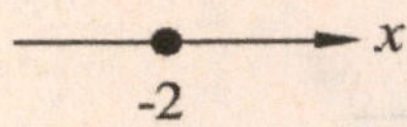

13. $x^3+x^2-2x<0$

$x(x^2+x-2)<0$

$x(x-1)(x+2)<0$

The critical values are $x=0$, $x=1$, and $x=-2$.

Interval	$(x)(x-1)(x+2)$	Sign
$x<-2$ $(-\infty,-2)$	− − −	−
$-2<x<0$ $(-2,0)$	− − +	+
$0<x<1$ $(0,1)$	+ − +	−
$x>1$ $(1,\infty)$	+ + +	+

$x(x-1)(x+2)>0$ for $x<-2$, or $0<x<1$, or in $(-\infty,-2)$ or $(0,1)$.

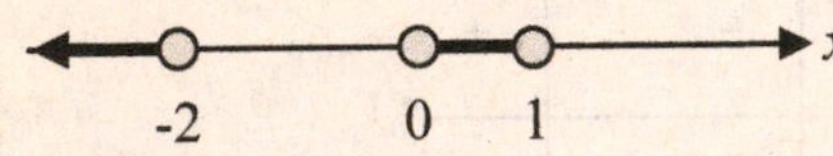

17. $\frac{2x-3}{x+6}\le 0 \quad (x\ne -6)$

The critical values are $x=\frac{3}{2}$, $x=-6$.

Interval	$(2x-3)\div(x+6)$	Sign
$x<-6$ $(-\infty,-6)$	− −	+
$-6<x<\frac{3}{2}$ $(-6,\frac{3}{2})$	− +	−
$x>\frac{3}{2}$ $(\frac{3}{2},\infty)$	+ +	+

$\frac{2x-3}{x+6}\le 0$ for $-6<x\le\frac{3}{2}$, or in $(-6,\frac{3}{2}]$.

$$-6 \qquad 3/2$$

21. $\frac{x}{x+3}>1$

$\frac{x}{x+3}-1>0$

$\frac{x-(x+3)}{x+3}>0$

$\frac{-3}{x+3}>0$

The critical value is -3.

Interval	$x+3$	$\frac{-3}{x+3}$
$x<-3$ $(-\infty,-3)$	−	+
$x>-3$ $(-3,\infty)$	+	−

$\frac{-3}{x+3}>0$ for $x<-3$, or in $(-\infty,-3)$.

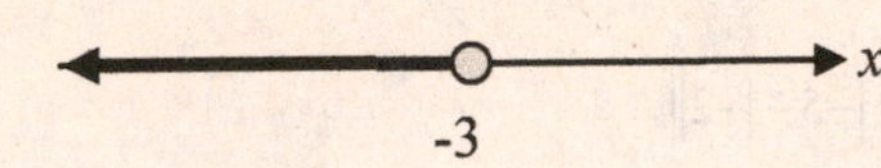

25. $\frac{T-8}{3-T}\le 0$

Solution: $T<3$ or $T\ge 8$

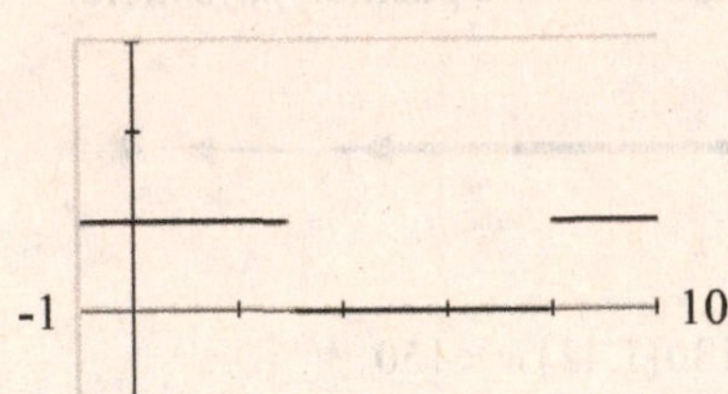

29. $\frac{x^4(9-x)(x-5)(2-x)}{(4-x)^5}\ge 0$

Solution: $2\le x<4$ or $5\le x\le 9$

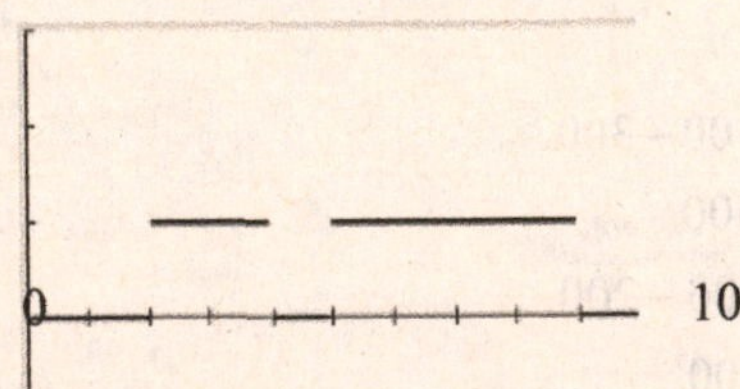

33. $\sqrt{-2x-x^2} = \sqrt{-x(2+x)}$ is real if $-x(2+x) \geq 0$

The critical values are $x = 0$, $x = -2$.

Interval	$-x(2+x)$	Sign
$x < -2$ $(-\infty, -2)$	$+ \quad -$	$-$
$-2 < x < 0$ $(-2, 0)$	$+ \quad +$	$+$
$x > 0$ $(0, \infty)$	$- \quad +$	$-$

$-x(2+x) \geq 0$ for $-2 \leq x \leq 0$, or in $[-2, 0]$.

37. To solve $x^4 < x^2 - 2x - 1$ using a graphing calculator, let $y_1 = x^4 - x^2 + 2x + 1$.

$y < 0$ for $-1.39 < x < -0.43$, or in $(-1.39, -0.43)$.

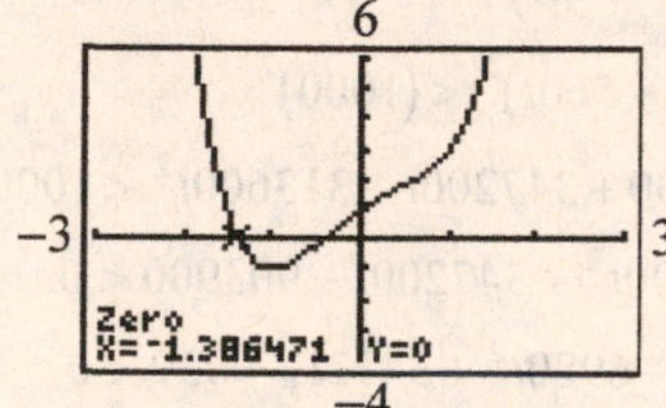

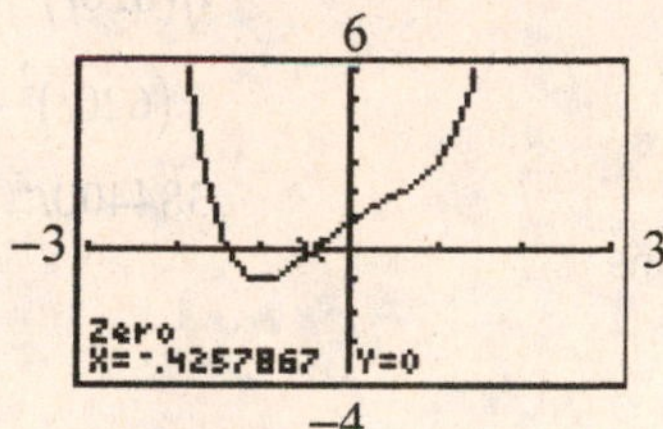

41. To solve $\sin x < 0.1x^2 - 1$ using a graphing calculator, let $y_1 = \sin x - 0.1x^2 + 1$.

$y < 0$ for $x < -4.43$, $-3.11 < x < -1.08$, $x > 3.15$, or in $(-\infty, -4.43)$ or $(-3.11, -1.08)$ or $(3.15, \infty)$.

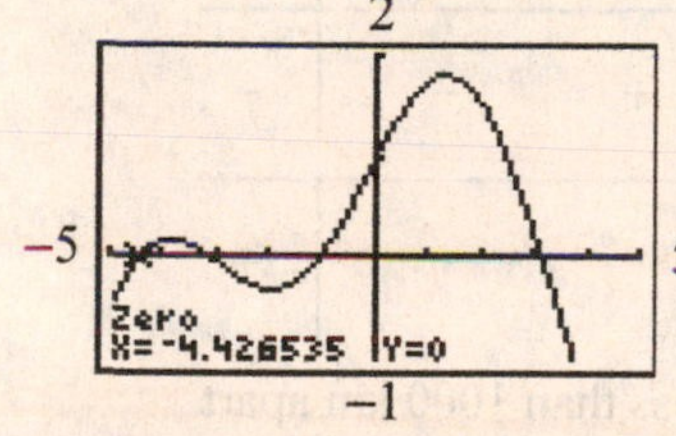

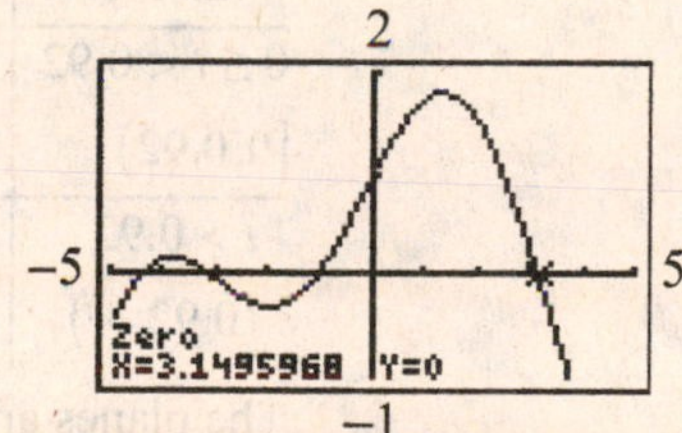

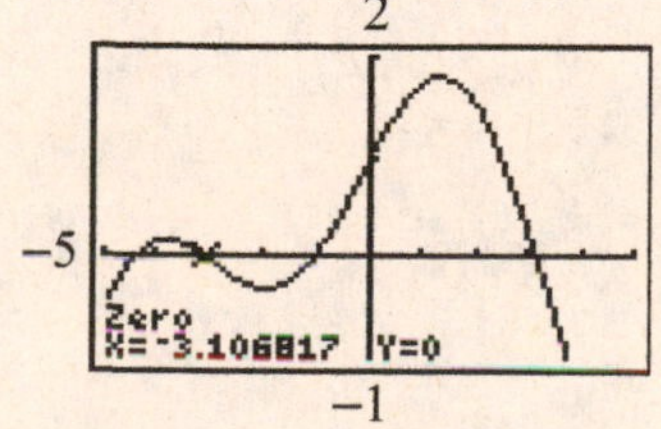

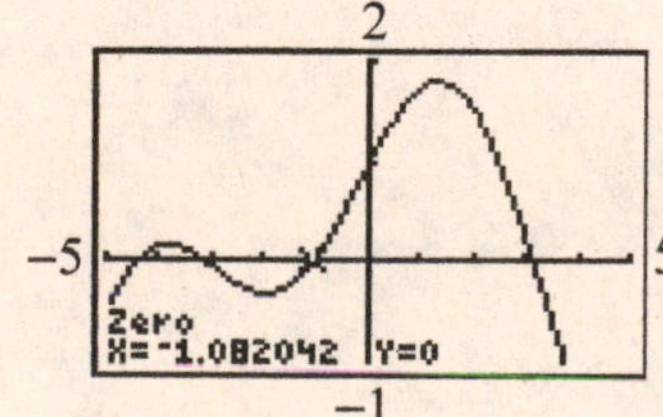

45. An inequality of the form $ax^2+bx+c<0$ with $a>0$ with the solution $-1<x<4$ has

$x>-1$ and $x<4$

$(x+1)>0$ $(x-4)<0$

from which $(x+1)(x-4)<0$ and $x^2-3x-4<0$.

49. For $a=b$, $(x-a)(x-b)<0$ becomes

$$(x-a)(x-a)<0$$

$$(x-a)^2<0 \text{ which is never true.}$$

$(x-a)(x-b)<0$ does not have a solution for $a=b$.

For $a\neq b$ the critical values are a, b. Without loss of generality, it may be assumed that $a<b$.

Interval	$x-a$	$x-b$	$(x-a)(x-b)$
$x<a$ $(-\infty,a)$	−	−	+
$a<x<b$ (a,b)	+	−	−
$x>b$ (b,∞)	+	+	+

$(x-a)(x-b)<0$ for $a<x<b$, or in (a,b).

Therefore, $(x-a)(x-b)<0$ will have solutions for all values of a and b such that $a\neq b$.

53. $S=\dfrac{100t}{t^2+100}$

For sales to be at least 4000 ($S\geq 4$), we solve

$$\frac{100t}{t^2+100}\geq 4$$

$$\frac{100t-4(t^2+100)}{t^2+100}\geq 0$$

$$\frac{-4t^2+100t-400}{t^2+100}\geq 0$$

The denominator is always positive, so the inequality will be true when

$$-4t^2+100t-400\geq 0$$

$$t^2-25t+100\leq 0$$

$$(t-5)(t-20)\leq 0$$

The critical values are 5,20.

Interval	$(t-5)(t-20)$		Sign
$0<t<5$ $(0,5)$	−	−	+
$5<t<20$ $(5,20)$	+	−	−
$t>20$ $(20,\infty)$	+	+	+

$t^2-25t+100\leq 0$ for $5\leq t\leq 20$ weeks, or in $[5 \text{ weeks}, 20 \text{ weeks}]$.

57. $w=\dfrac{r^2w_0}{(r+h)^2}<100$

$$(r+h)^2>\frac{r^2w_0}{100}$$

$$r+h>\sqrt{\frac{r^2w_0}{100}}$$

$$h>\sqrt{\frac{r^2w_0}{100}}-r$$

$$h>\sqrt{\frac{6380^2(200)}{100}}-6380$$

$$h>2640 \text{ m}$$

61. Let t = time from take off. The distance between planes is given by

$\sqrt{(620t)^2+(310+560t)^2}$, so we must solve

$$\sqrt{(620t)^2+(310+560t)^2}<1000$$

$$(620t)^2+(310+560t)^2<(1000)^2$$

$$384400t^2+96100+347200t+313600t^2<1000000$$

$$698000t^2+347200t-903900<0$$

$$6980t^2+3472t-9039<0$$

Use the quadratic formula to solve for t:

$$t=\frac{-3472\pm\sqrt{3472^2-4(6980)(-9039)}}{2(6980)}$$

$$=-1.41, 0.92$$

Time cannot be negative, so we have

Interval	$(t+1.41)(t-0.92)$		Sign
$0\leq t<0.92$ $[0,0.92)$	+	−	−
$t>0.92$ $(0.92,\infty)$	+	+	+

The planes are less than 1000 km apart for $0\leq t<0.92$ h, or in $[0, 0.92 \text{ h})$.

17.4 Inequalities Involving Absolute Values

1. $|2x-1|<5$

$-5<2x-1<5$

$-4<2x<6$

$-2<x<3$

$(-2,3)$

-2 3 x

5. $|5x+4|>6$

$5x+4<-6$ or $5x+4>6$

$5x<-10$ or $5x>2$

$x<-2$ or $x>\frac{2}{5}$

$(-\infty,-2)$ or $\left(\frac{2}{5},\infty\right)$

-2 2/5 x

9. $|3-4x|>3$

$3-4x<-3$ or $3-4x>3$

$-4x<-6$ or $-4x>0$

$x>\frac{3}{2}$ or $x<0$

$(-\infty,0)$ or $\left(\frac{3}{2},\infty\right)$

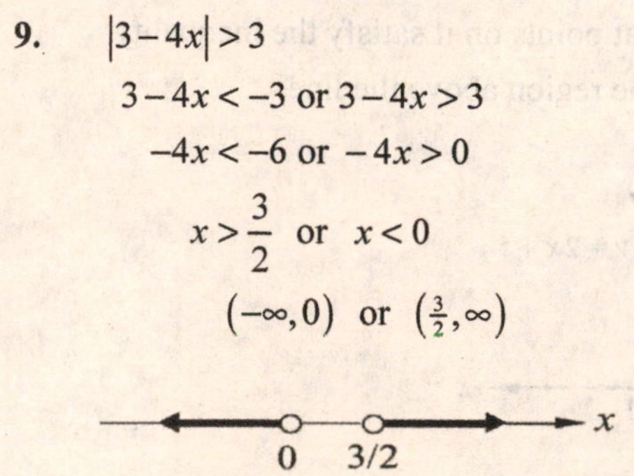

0 3/2 x

13. $|20x+85|\le 46$

$-46\le 20x+85\le 46$

$-131\le 20x\le -39$

$-6.55\le x\le -1.95$

$[-6.55,-1.95]$

-6.55 -1.95 x

17. $8+3|3-2x|<11$

$3|3-2x|<3$

$|3-2x|<1$

$-1<3-2x<1$

$-4<-2x<-2$

$2>x>1$

$1<x<2$

$(1,2)$

1 2 x

21. $\left|\frac{3R}{5}+1\right|<8$

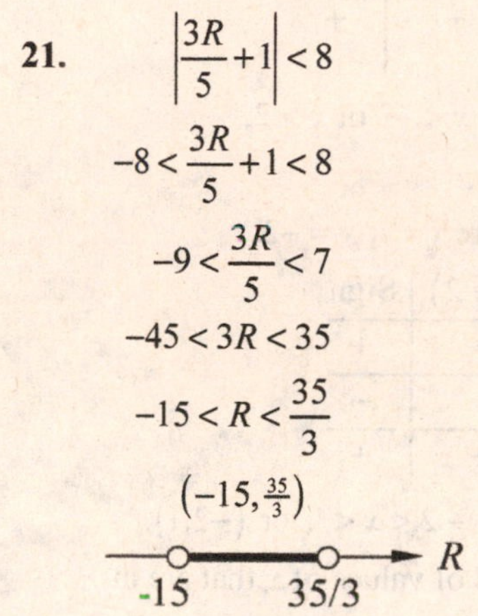

$-8<\frac{3R}{5}+1<8$

$-9<\frac{3R}{5}<7$

$-45<3R<35$

$-15<R<\frac{35}{3}$

$\left(-15,\frac{35}{3}\right)$

-15 35/3 R

25. $|2x-5|<3$

Solution: $1<x<4$

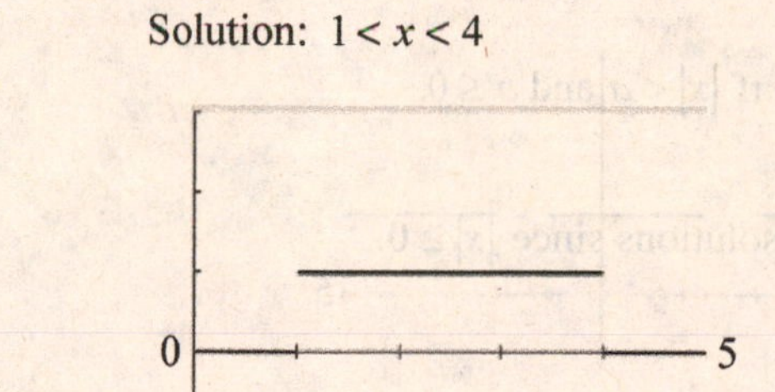

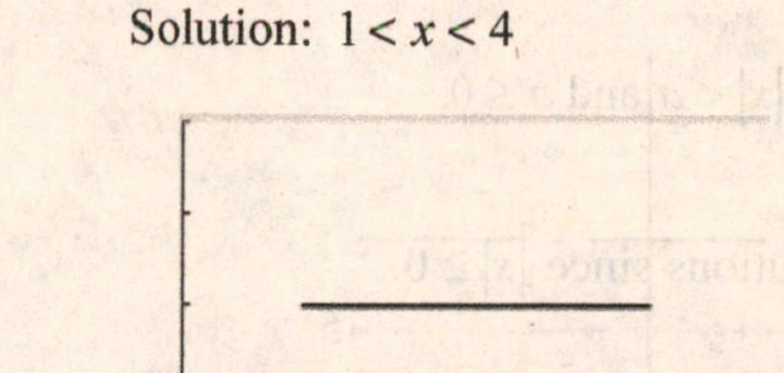

29. $|x^2+x-4|>2$

$x^2+x-4>2$ or $x^2+x-4<-2$

$x^2+x-6>0$ or $x^2+x-2<0$

(A) $(x+3)(x-2)>0$

(B) $(x-1)(x+2)<0$

(A) Critical values are $x=-3$, $x=2$.

Interval	$(x+3)(x-2)$		Sign
$x<-3$ $(-\infty,-3)$	−	−	+
$-3<x<2$ $(-3,2)$	+	−	−
$x>2$ $(2,\infty)$	+	+	+

$(x+3)(x-2)>0$ for $x<-3$ or $x>2$,

or $(-\infty,-3)$ or $(2,\infty)$.

(B) Critical values are $x=1$, $x=-2$.

Interval	$(x-1)(x+2)$		Sign
$(-\infty,-2)$	−	−	+
$(-2,1)$	−	+	−
$(1,\infty)$	+	+	+

$(x-1)(x+2)<0$ for $-2<x<1$, or $(-2,1)$.

The solution consists of values of x that are in (A) or (B): $x<-3$, $-2<x<1$, $x>2$, that is, $(-\infty,-3)$ or $(-2,1)$ or $(2,\infty)$.

33. Solve for x if $|x|<a$ and $a\le 0$.

$|x|<a\le 0$

$|x|<0$, no solutions since $|x|\ge 0$.

-5 5

37. $|x-1|<4$

$-4<x-1<4$

add 5 to all expressions:

$1<x+4<9$ and so $a=1, b=9$

41. $|p-2\,000\,000|\le 200\,000$

$-200\,000\le p-2\,000\,000\le 200\,000$

$1\,800\,000\le p\le 2\,200\,000$ barrels

$[1\,800\,000, 2\,200\,000]$ barrels

The production will be at least 1 800 000 barrels but not greater than 2 200 000 barrels.

45. $3.675-0.002\le d\le 3.675+0.002$

$-0.002\le d-3.675\le 0.002$

$|d-3.675|\le 0.002$ cm

17.5 Graphical Solution of Inequalities with Two Variables

1. $y<3-x$

Graph $y=3-x$. Used a dashed line to indicate that points on the line do not satisfy the inequality. Shade in the region below the line.

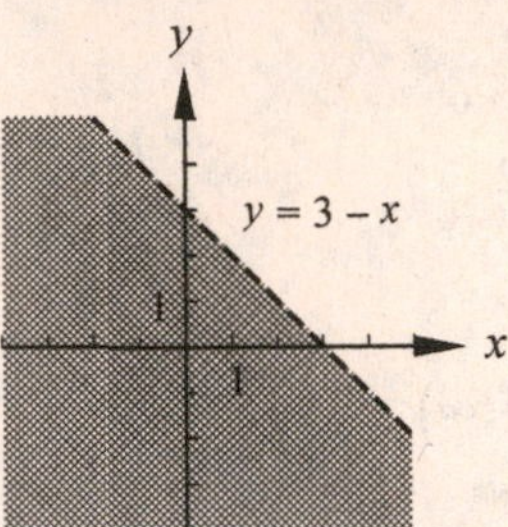

5. $y\ge 2x+5$

Graph $y=2x+5$. Use a solid line to indicate that points on it satisfy the inequality. Shade in the region above the line.

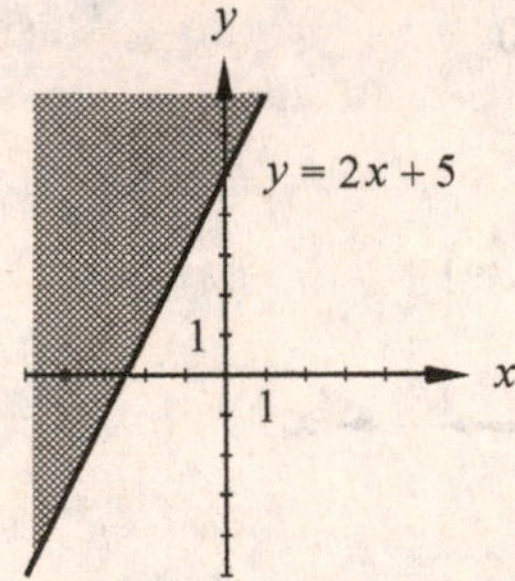

9. $4y<x^2$

Graph $y=\frac{1}{4}x^2$. Use a dashed curve to indicate that points on it do not satisfy the inequality. Shade in the region below the curve.

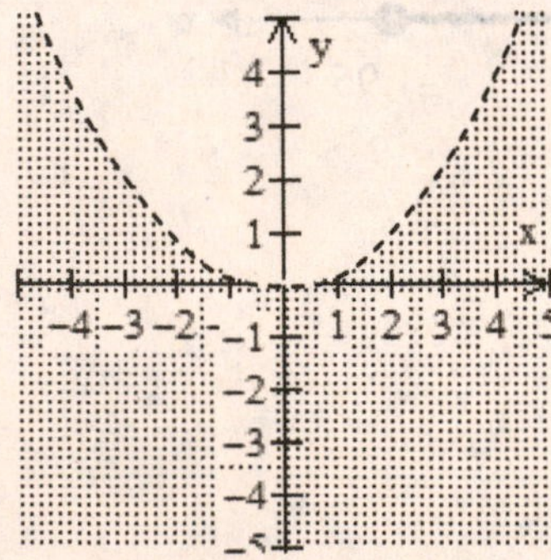

13. $y < 32x - x^4$

Graph $y = 32x - x^4$. Use a dashed curve to indicate that points on it do not satisfy the inequality. Shade in the region below the curve.

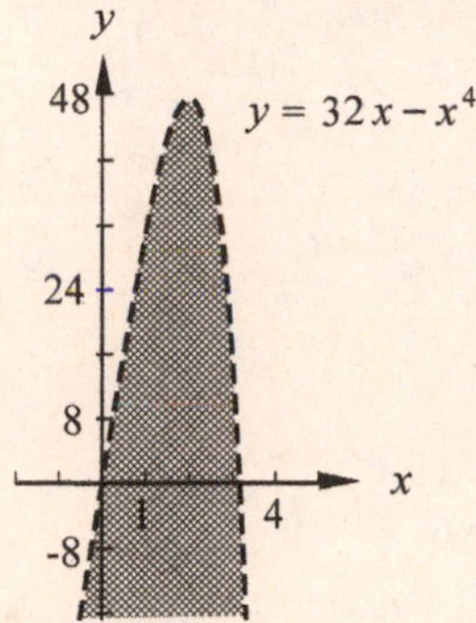

17. $y > 1 + \sin 2x$

Graph $y = 1 + \sin 2x$. Use a dashed curve to indicate that the points on it do not satisfy the inequality. Shade in the region above the curve.

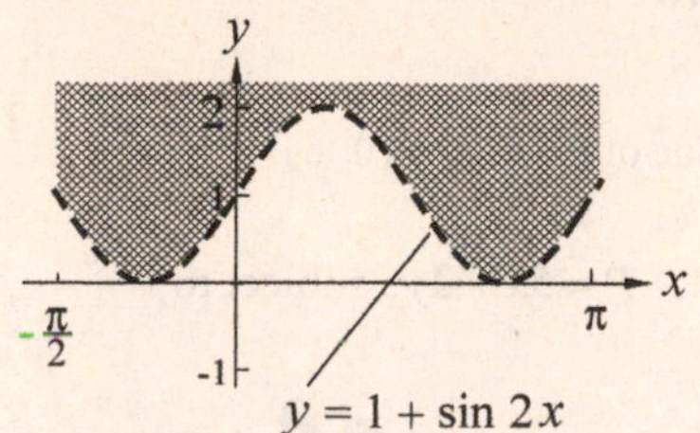

21. $|x| < |y|$.

For $y > 0$, $|x| < |y|$ becomes $y > |x|$.

Graph $y = |x|$ with dashed line and shade in the region above the graph.

For $y < 0$, $|x| < |y|$ becomes $-y > |x|$, so that $y < -|x|$. Graph $y = -|x|$ with dashed line and shade in the region below the graph.

The solution consists of both regions.

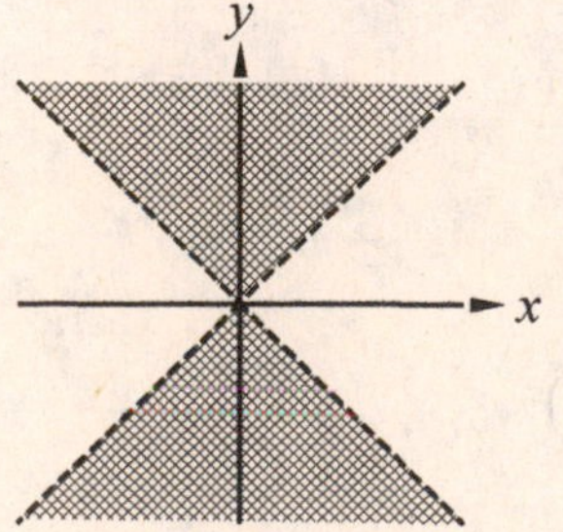

25. $y \le 2x^2$ and $y > x - 2$. Graph $y = 2x^2$ using a solid curve. Shade the region below the curve. Graph $y = x - 2$ using a dashed line. Shade the region above the line. The region where the shadings overlap satisfies both inequalities.

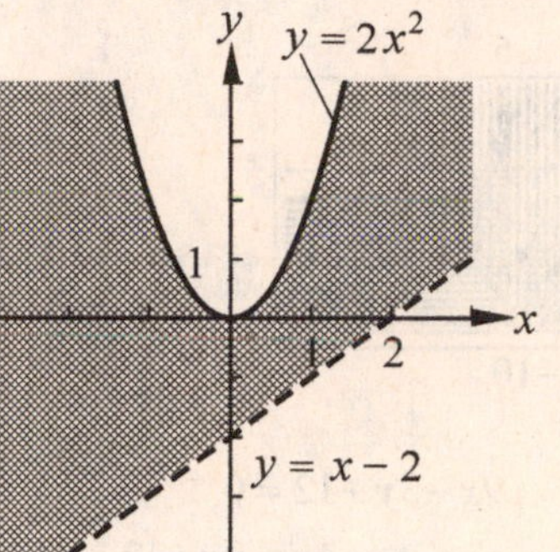

29. $y \ge 0$ and $y \le \sin x$; $0 \le x \le 3\pi$. Graph $y = \sin x$ using a solid curve. Shade the region below the curve and above the x-axis for $0 \le x \le 3\pi$.

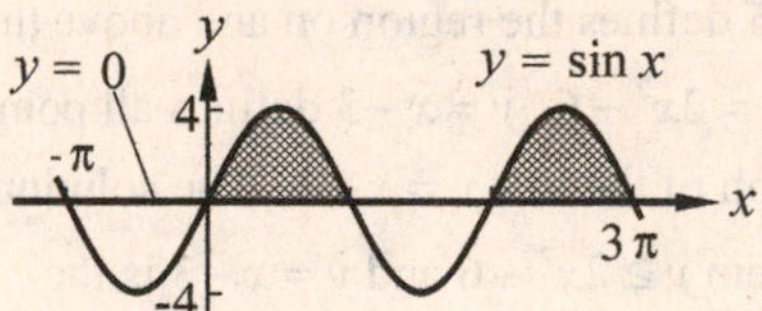

33. $y > 0, x < 0, y \le x$

There are no values of x and y that satisfy all three of these inequalities, so there is nothing to shade.

37. $y \ge 1 - x^2$. Graph $y_1 = 1 - x^2$.

The boundary line is solid and the region above the graph is shaded in.

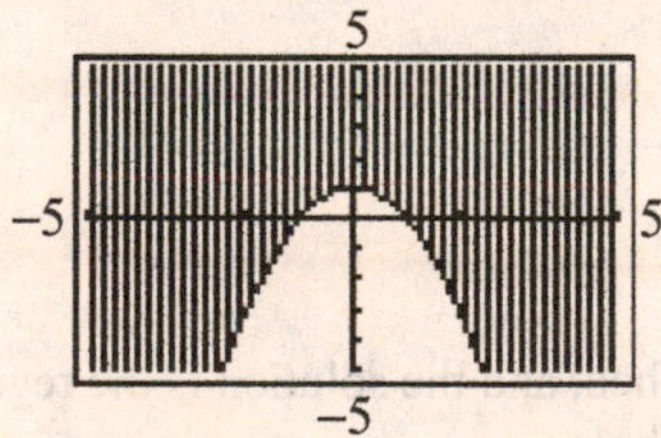

41. $y > x^2 + 2x - 8$. Graph $y_1 = x^2 + 2x - 8$.
$y < \frac{1}{x} - 2$. Graph $y_2 = \frac{1}{x} - 2$.
The boundary lines are dashed. The solution is the intersection of the region above the parabola and the region below the hyperbola.

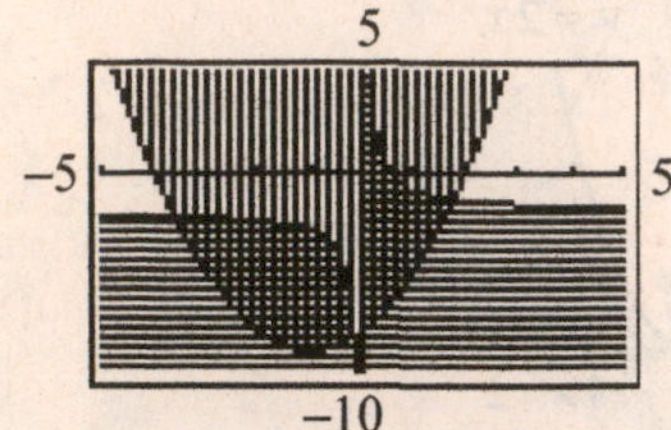

45.
$$9x - 3y + 12 = 0$$
$$3y = 9x + 12$$
$$y = 3x + 4$$
The region below the line $y = 3x + 4$ is defined by the inequality $y < 3x + 4$.

49. $y \ge 2x^2 - 6$ defines the region on and above the parabola $y = 2x^2 - 6$. $y = x - 3$ defines all points on the graph of the line $y = x - 3$. The solution to the system $y \ge 2x^2 - 6$ and $y = x - 3$ is the intersection of the graph of the parabola and the line together with the points on the line above the parabola.

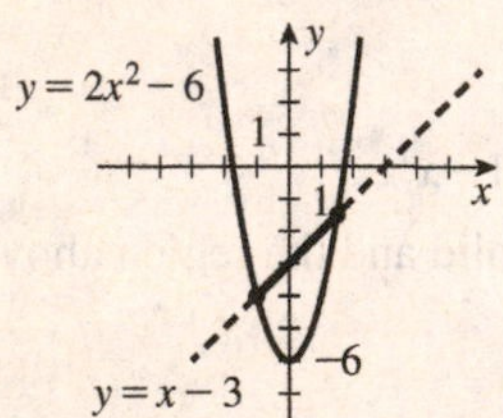

53. $P_R = Ri^2$
$R = 0.5\Omega$
$P > P_R$
$P > 0.5i^2$
The line is dashed, and the solution is the region above the parabola.

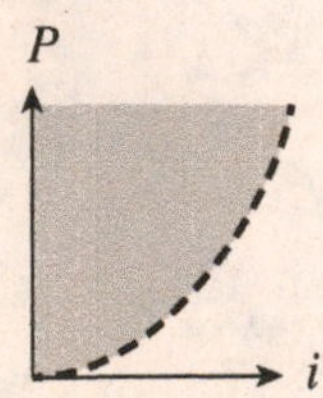

17.6 Linear Programming

1. Maximize $F = 2x + 3y$, subject to $x \ge 0$, $y \ge 0$, $x + y \le 6$, $2x + y \le 8$.

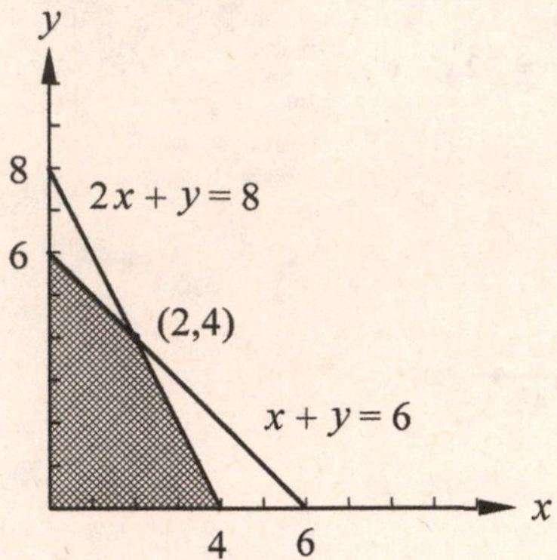

point	value of F
(0, 0)	0
(0, 6)	18
(2, 4)	16
(4, 0)	8

Maximum value of F is 18 at (0, 6)

5. Maximum P: $P = 5x + 2y$, subject to
$x \ge 0$, $y \ge 0$
$2x + y \le 6$

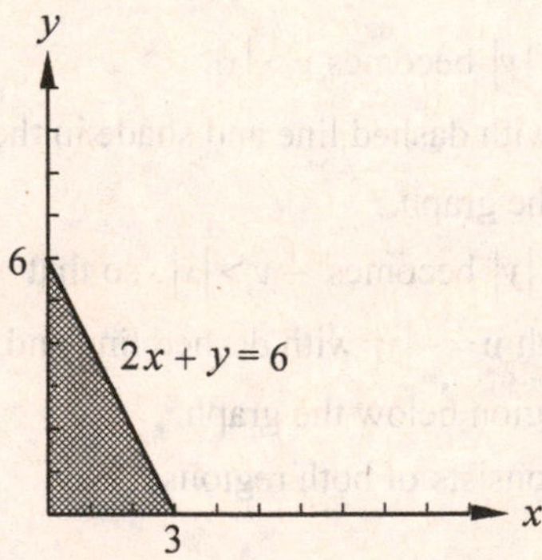

Vertex	$P = 5x + 2y$
(0, 0)	0
(0, 6)	12
(3, 0)	15

max $P = 15$ at (3, 0)

9. Minimum C: $C = 5x + 6y$, subject to

$x \geq 0,\ y \geq 0$

$x + y \geq 5$

$x + 2y \geq 7$

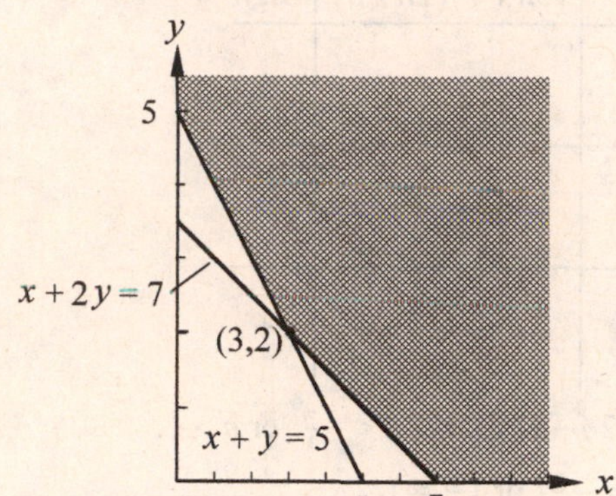

Vertex	$C = 5x + 6y$
(0, 5)	30
(3, 2)	27
(7, 0)	35

min $C = 27$ at $(3, 2)$

13. Maximum P: $P = 8x + 6y$, subject to

$x \geq 0,\ y \geq 0$

$2x + 5y \leq 10$

$4x + 2y \leq 12$

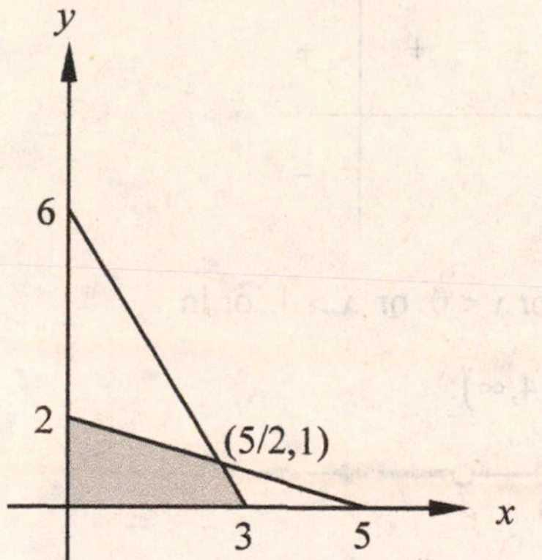

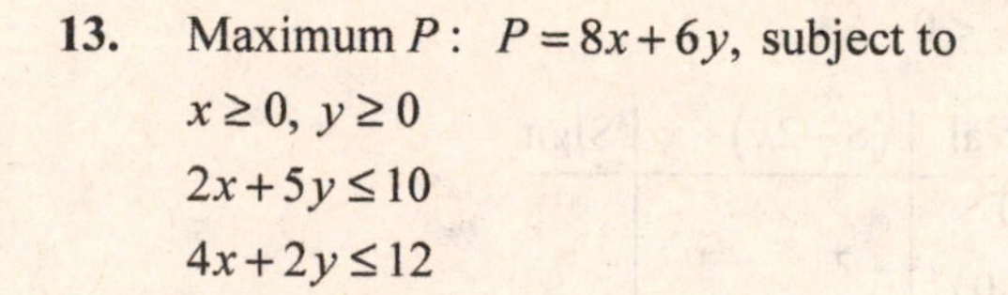

Vertex	$P = 8x + 6y$
(0, 0)	0
(0, 2)	12
$\left(\frac{5}{2}, 1\right)$	26
(3, 0)	24

max $P = 26$ at $\left(\frac{5}{2}, 1\right)$

17. x = number of newspaper ads

y = number of radio ads

Maximum $I = 8000x + 6000y$,

subject to

$x \geq 0,\ y \geq 0$

$50x + 150y \leq 9000$

$100x \leq 150y$

Reduce the last two inequalities by a factor of 50:

$x + 3y \leq 180$

$2x \leq 3y$

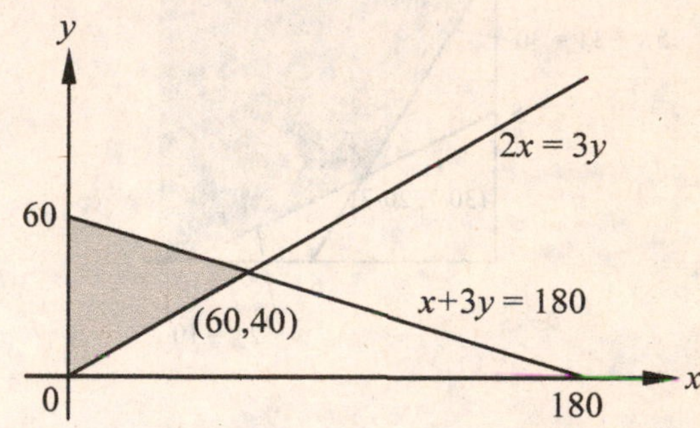

Vertex	$I = 8000x + 6000y$
(0, 0)	0
(0, 60)	360000
(60, 40)	720000

max $I = 720000$ voters by buying 60 newspaper ads and 40 radio ads.

21. $x =$ servings of cereal A,
$y =$ servings of cereal B
Minimum $C = 12x + 18y$, subject to
$x \ge 0,\ y \ge 0$
$x + 2y \ge 10$
$5x + 3y \ge 30$

Vertex	$C = 12x + 18y$
(0,10)	180
(10,0)	120
$\left(\frac{30}{7}, \frac{20}{7}\right)$	$\frac{720}{7}$

Minimum $C = \frac{720}{7}$, when $x = \dfrac{30}{7},\ y = \dfrac{20}{7}$.

Since each serving is 30 g, the optimal solution is $\dfrac{900}{7}$ g of cereal A, and $\dfrac{600}{7}$ g of cereal B.

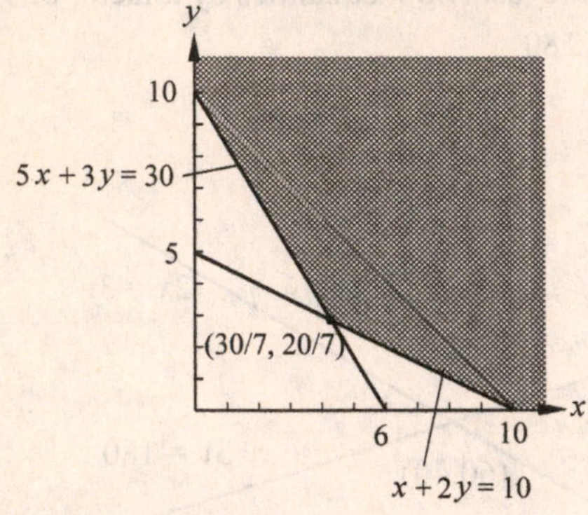

Review Exercises

1. This is false. A set of inequalities written $1 < x < -3$ is interpreted as $1 < x$ and $x < -3$, which is inconsistent.

5. This is false. One should also include the points on the line $y = x + 1$ as well.

9. $4 < 2x - 1 < 11$
$5 < 2x < 12$
$\dfrac{5}{2} < x < 6$

5/2 ——— 6 → x

13. $6n^2 - n > 35$
$6n^2 - n - 35 > 0$
$(3n+7)(2n-5) > 0$

Interval	$(3n+7)$	$(2n-5)$	Sign
$n < -\frac{7}{3}$ $\left(-\infty, -\frac{7}{3}\right)$	−	−	+
$-\frac{7}{3} < n < \frac{5}{2}$ $\left(-\frac{7}{3}, \frac{5}{2}\right)$	+	−	−
$n > \frac{5}{2}$ $\left(\frac{5}{2}, \infty\right)$	+	+	+

$(3n+7)(2n-5) > 0$ for $n < -\frac{7}{3}$ or $n > \frac{5}{2}$, or in $\left(-\infty, -\frac{7}{3}\right)$ or $\left(\frac{5}{2}, \infty\right)$.

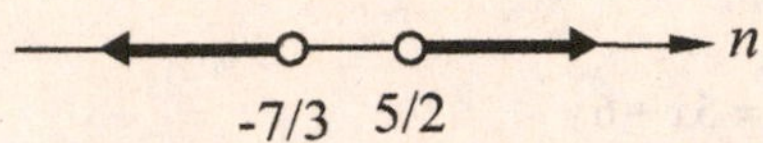

17. $\dfrac{8}{x} < 2$
$\dfrac{8}{x} - 2 < 0$
$\dfrac{8 - 2x}{x} < 0$

Interval	$(8-2x)$	$\div x$	Sign
$x < 0$ $(-\infty, 0)$	+	−	−
$0 < x < 4$ $(0, 4)$	+	+	+
$x > 4$ $(4, \infty)$	−	+	−

$\dfrac{8-2x}{x} < 0$ for $x < 0$ or $x > 4$, or in $(-\infty, 0)$ or $(4, \infty)$.

← 0 4 → x

21. $|3-5x|>7$

$3-5x<-7$ or $3-5x>7$

$-5x<-10$ $\quad$ $-5x>4$

$x>2$ $\quad$ $x<-\frac{4}{5}$

$(2,\infty)$ or $\left(-\infty,-\frac{4}{5}\right)$

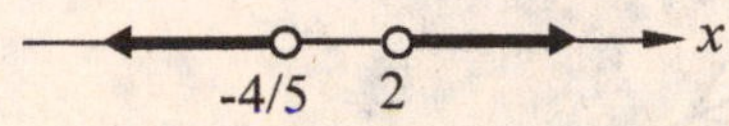

25. $2\le\frac{4n-2}{3}<3$

Solution: $2\le n<\frac{11}{4}$

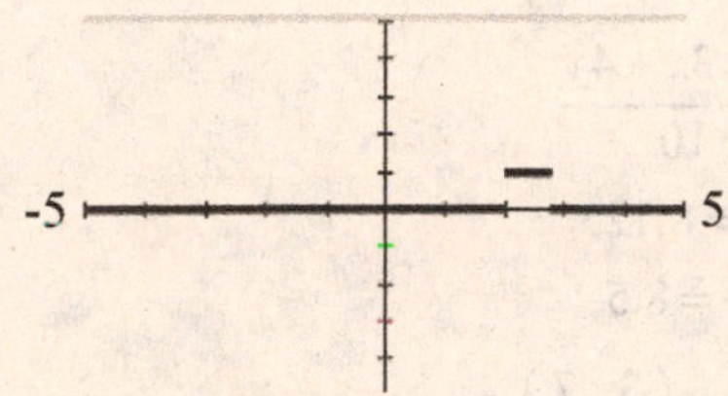

29. $|x-30|>48$

Solution: $x<-18$ or $x>78$

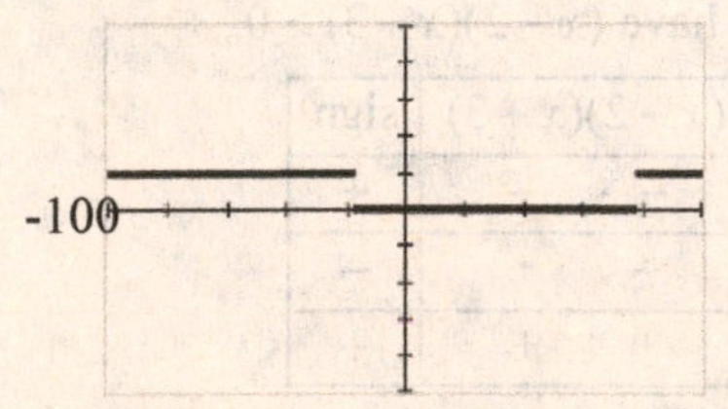

33. $e^{-t}>0.5$

$e^{-t}-0.5>0$

Graph $y_1=e^{-x}-0.5$ and use zero feature to find the root.

$e^{-t}-0.5>0$ for $t<0.69$, or in $(-\infty,0.69)$.

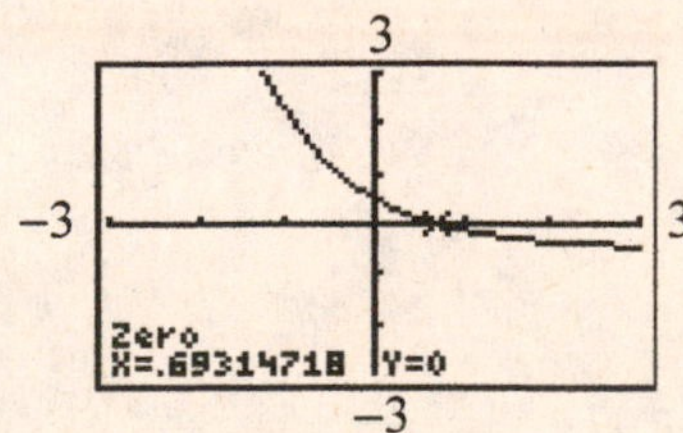

37. $4y-6x-8\le 0$

$y\le\frac{3x+4}{2}$

Graph $y=\frac{3x+4}{2}$. Use a solid line to indicate that the line is part of the solution. Shade in the region below the line.

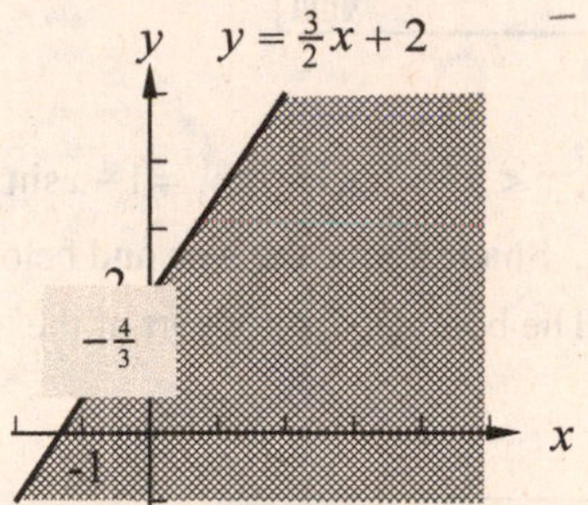

41. $y-|x+1|<0$

$y<|x+1|$

Graph $y=|x+1|$. Use a dashed line to indicate that the boundary line is not part of the solution. Shade in the region below the line.

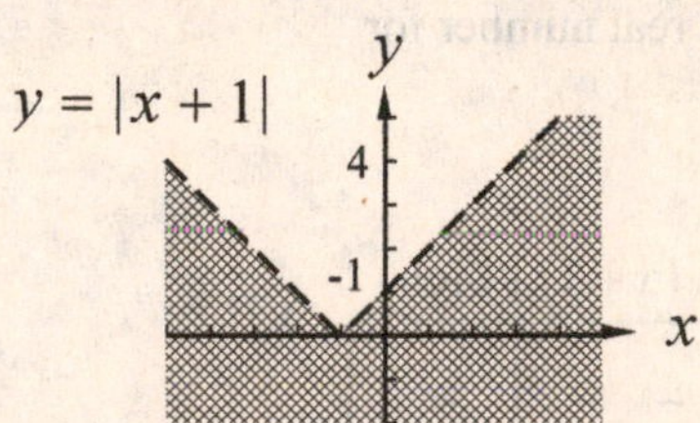

45. $y\le\frac{4}{x^2+1}$, $y<x-3$.

Graph $y=\frac{4}{x^2+1}$. Use a solid line and shade in the region below the curve.

Graph $y=x-3$. Use a dashed line and shade in the region below the line.

The solution is the intersection of the region below the curve and the region below the line.

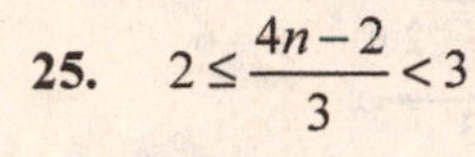

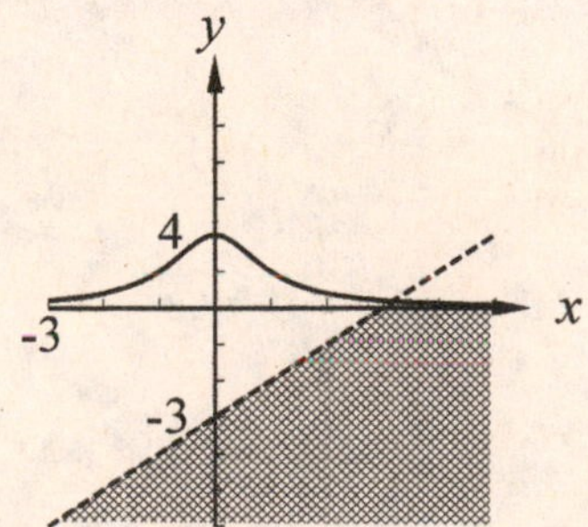

49. $y > 8 + 7x - x^2$. Graph $y_1 = 8 + 7x - x^2$ and shade above curve. Boundary line is not part of solution.

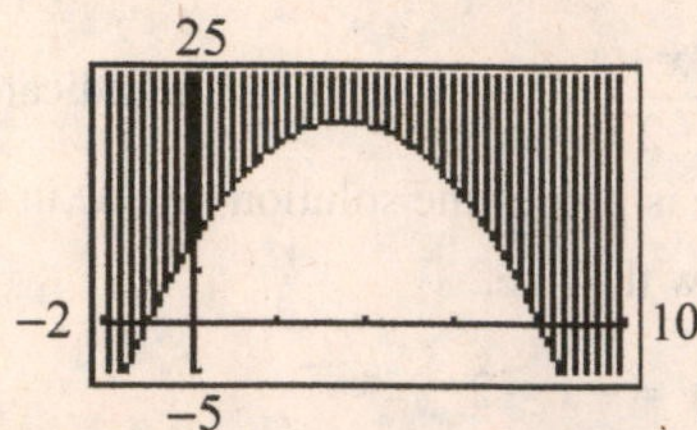

53. $y > 1 - x\sin 2x$, $y < 5 - x^2$. Graph $y_1 = 1 - x\sin 2x$ and $y_2 = 5 - x^2$. Shade above the sine and below the parabola. The boundary is not part of the solution.

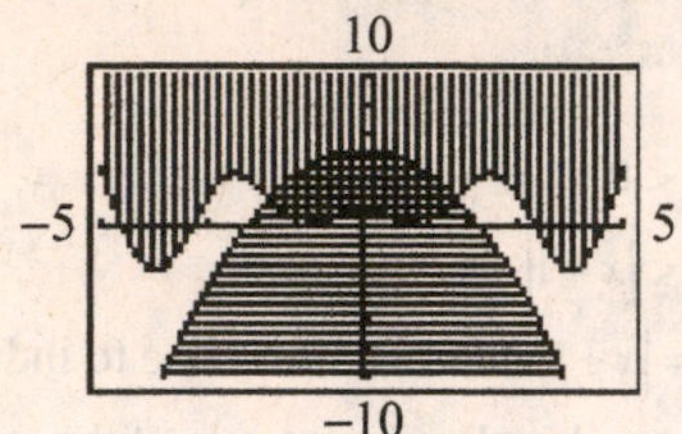

57. $\sqrt{x^2 + 3x}$ is a real number for

$x^2 + 3x \geq 0$

$x(x+3) \geq 0$

Interval	$x(x+3)$	Sign
$x < -3$ $(-\infty, -3)$	$-\ -$	+
$-3 < x < 0$ $(-3, 0)$	$-\ +$	−
$x > 0$ $(0, \infty)$	$+\ +$	+

$\sqrt{x^2 + 3x}$ is a real number for $x \leq -3$ or $x \geq 0$, or in $(-\infty, -3]$ or $[0, \infty)$.

61. Minimize $C: C = 3x + 4y$, subject to

$x \geq 0,\ y \geq 1$

$2x + 3y \geq 6$

$4x + 2y \geq 5$

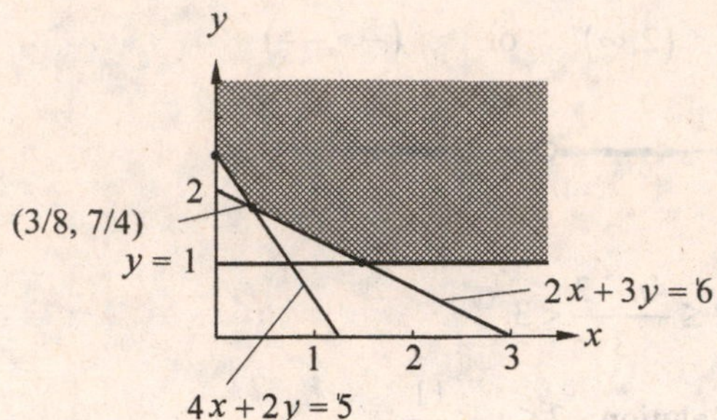

Vertex	$C = 3x + 4y$
$\left(0, \frac{5}{2}\right)$	10
$\left(\frac{3}{8}, \frac{7}{4}\right)$	$\frac{65}{8} = 8.125$
$\left(\frac{3}{2}, 1\right)$	$\frac{17}{2} = 8.5$

min $C = \dfrac{65}{8}$ at $\left(\dfrac{3}{8}, \dfrac{7}{4}\right)$

65. We want $x^2 + 3x > 2x + 6$ or $x^2 + x - 6 > 0$. Factoring, we have $(x-2)(x+3) > 0$.

	$(x-2)$	$(x+3)$	sign
$x < -3$	−	−	+
$-3 < x < 2$	−	+	−
$2 < x$	+	+	+

and so the graph of $y = x^2 + 3x$ is above that of $y = 2x + 6$ when $x < -3$ or when $x > 2$.

69. The solution is the part of the parabola $y = x^2$ that is below the graph of $y = 1 - x^2$, without including the intersection of the two curves.

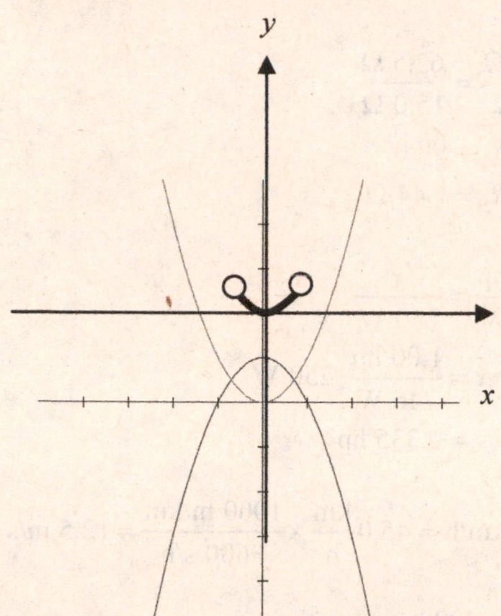

73. There are no values of x and y that can satisfy both inequalities at once. The region is empty.

77. $2x + 5y = 50 \quad y = \dfrac{50 - 2x}{5}$

$$5 < y < 8$$
$$5 < \frac{50 - 2x}{5} < 8$$
$$25 < 50 - 2x < 40$$
$$-25 < -2x < -10$$
$$12.5 > x > 5$$
$$5 < x < 12.5$$
$$(\$5, \$12.50)$$

The cost of production for the first type is between \$5 and \$12.50.

81. $0.8x + 0.9y = 360000, \quad 0 \le x \le 261000$

$$x = \frac{360000 - 0.9y}{0.8}$$
$$0 \le \frac{360000 - 0.9y}{0.8} \le 261000$$
$$0 \le 360000 - 0.9y \le 208800$$
$$-360000 \le -0.9y \le -151200$$
$$400000 \ge y \ge 168000$$
$$168000 \le y \le 400000 \text{ Btu}$$
$$[168000 \text{ Btu}, 400000 \text{ Btu}]$$

85. $E = 100\left(1 - r^{-0.4}\right)$

$$100\left(1 - r^{-0.4}\right) > 50$$
$$100 - 100r^{-0.4} > 50$$
$$r^{-0.4} < \frac{1}{2}$$
$$r^{0.4} > 2$$
$$r > 2^{\frac{1}{0.4}}$$
$$r > 5.7$$

89. x = the number of regular models

y = the number of deluxe models

$x + y \le 450$

$P = 8x + 15y$ is the profit

t = time spent on one regular model

$2t$ = time spent on one deluxe model

xt = time spent on regular models

$2yt$ = time spent on deluxe models

$xt + 2yt$ = total time = $600t \Rightarrow$

$x + 2y \le 600$

Maximize $P = 8x + 15y$, subject to

$x \ge 0,\ y \ge 0$

$x + y \le 450,\ x + 2y \le 600$

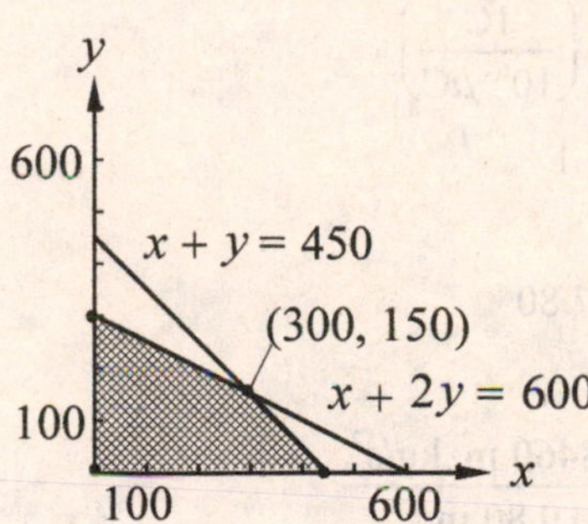

vertex	$P = 8x + 15y$
(0, 0)	0
(0, 300)	4500
(300, 150)	4650
(450, 0)	3600

Produce 300 regular, 150 deluxe for maximum profit

Chapter 18

Variation

18.1 Ratio and Proportion

1. The ratio of the length to the width is $\frac{24 \text{ ft}}{16 \text{ ft}} = \frac{3}{2}$.

5. $\frac{96 \text{ h}}{3 \text{ days}} = \frac{96 \text{ h}}{72 \text{ h}}$
$= \frac{4}{3}$

9. $\frac{0.175 \text{ kg}}{3500 \text{ mg}} = \frac{0.175 \text{ kg}}{3500 \text{ mg}} \cdot \left(\frac{10^6 \text{ mg}}{1 \text{ kg}}\right)$
$= 50$

13. $\frac{2.6 \text{ W}}{9.6 \text{ W}} = 0.27$
$= 27\%$

17. $C = \frac{q}{V}$
$= \frac{5.00 \ \mu\text{C}}{200 \text{ V}} \cdot \left(\frac{1\text{C}}{10^6 \ \mu\text{C}}\right)$
$= 2.5 \times 10^{-8} \text{ F}$

21. $\frac{487 \text{ lb/ft}^3}{62.4 \text{ lb/ft}^3} = 7.80$

25. $\frac{8460 \text{ N}}{9.80 \text{ m/s}^2} = \frac{8460 \text{ m} \cdot \text{kg/s}^2}{9.80 \text{ m/s}^2}$
$= 863 \text{ kg}$

29. $\frac{6.00 \ \Omega}{R_2 \ \Omega} = \frac{62.5 \ \Omega}{15.0 \ \Omega}$
$62.5R_2 = 90.0$
$R_2 = 1.44 \ \Omega$

33. $\frac{1.00 \text{ hp}}{746 \text{ W}} = \frac{x}{250 \text{ W}}$
$x = \frac{1.00 \text{ hp}}{746 \text{ W}} \cdot 250 \text{ W}$
$= 0.335 \text{ hp}$

37. $45.0 \text{ km/h} = 45.0 \frac{\text{km}}{\text{h}} \times \frac{1000 \text{ m/km}}{3600 \text{ s/h}} = 12.5 \text{ m/s}$

41. $\frac{5058 - 4998}{5058} = 0.0118624$
The percent error is 1.2%

45. Let x be the size of the PIP.
$\frac{1}{3} = \frac{x}{1160 - x}$
$1160 - x = 3x$
$1160 = 4x$
$x = 290 \text{ in}^2$
The remaining area is $1160 - 290 = 870 \text{ in}^2$

49. $\frac{17 \text{ defectives}}{500 \text{ total chips}} = \frac{595}{x}$
$17x = 297\ 500$
$x = 17\ 500 \text{ chips}$

18.2 Variation

1. $c = kd$
$c = \pi d$
$k = \pi$

5. $v = kr$

9. $p = \frac{k}{\sqrt{A}}$

13. The area varies directly as the square of the radius.

17. $V = kH^2$
$48 = k \cdot 4^2$
$k = \frac{48}{4^2} = 3;$
$V = 3H^2$

21. $y = kx$
$200 = k(16)$
$k = 12.5$
$y = 12.5x$
$y = 12.5(10)$
$y = 125$

25. $y = \frac{kx}{z}$
$60 = \frac{k(4)}{10}$
$k = 150$
$y = \frac{150x}{z}$
$y = \frac{150(6)}{5}$
$y = 180$

29. $A = k_1x,\ B = k_2x$
$A + B = k_1x + k_2x$
$= (k_1 + k_2)x,$
so that $A + B$ varies directly as x.

33. $H = km$
$2.93 \times 10^5 = k(875)$
$k = 335$ J/g
$H = 335m$
$H = 335(625)$
$= 2.09 \times 10^5$ J

37. $F = ks$
$10.0 = k(4.00)$
$k = \frac{10.0}{4.00}$
$= 2.50$ N/cm
$F = 2.50s$
$6 = 2.50s$
$s = 2.40$ cm

41. $t = \frac{k}{A}$
$2.0 = \frac{k}{48}$
$k = 48(2.0)$
$= 96$
$t = \frac{96}{A}$
$t = \frac{96}{68}$
$= 1.4$ h

45. $P = ks^3$
$5200 = k(12.0)^3$
$k = \frac{5200}{12.0^3}$
$P = \frac{5200}{12.0^3}(15.0)^3$
$= 10156.25$
10200 hp will propel the ship at 15.0 mi/h

49. $F = \frac{k}{d}$

$0.750 \text{ N} = \frac{k}{1.25 \text{ cm}}$

$k = 0.938 \text{ N}\cdot\text{cm}$

$F = \frac{0.938}{d}$

$F = \frac{\frac{15}{16}}{1.75}$

$F = 0.536 \text{ N}$

53. $R = \frac{kl}{A}$

$0.200 = \frac{k(225)}{0.0500}$

$k = 4.44\times10^{-5}$

$R = \frac{4.44\times10^{-5} l}{A}$

57. $P = kRI^2$

$10.0 = k(40.0)(0.500)^2$

$k = 1$

$P = RI^2$

$P = (20.0)(2.00)^2$

$= 80.0 \text{ W}$

61. *Note*: Make sure calculator is in rad mode.

$x = k\omega^2(\cos\omega t)$

$-11.4 = k(0.524^2)\cos[(1.00)(0.524)]$

$k = \frac{-11.4}{0.524^2\cos(0.524)}$

$= -48.0$

$x = -48.0\omega^2(\cos\omega t)$

$x = (-48.0)(0.524^2)\cos(0.524(2.00))$

$= -6.58 \text{ ft/s}^2$

Review Exercises

1. This is true. $\frac{25 \text{ cm}}{50 \text{ mm}} = \frac{25 \text{ cm}}{5 \text{ cm}} = 5.$

5. $\frac{840 \text{ mg}}{3 \text{ g}} = \frac{840 \text{ mg}}{3000 \text{ mg}} = 0.28$

9. $\pi = \frac{c}{d}$

$\frac{4.2736}{1.3603} = 3.1417$

13. $p = \frac{F}{A}$

$= \frac{42.3 \text{ lb}}{(2.25 \text{ in})^2}$

$= 8.36 \text{ lb/in}^2$

17. $\text{Commission rate} = \frac{\text{Commission}}{\text{Selling price}}$

$= \frac{17100}{380000}$

$= 0.045$

$= 4.5\%$

21. $\frac{1.3 \text{ in}}{20 \text{ mi}} = \frac{6.0 \text{ in}}{x}$

$x = 92.3 \text{ mi}$

25. $3.0 \text{ min} = 180 \text{ s}$

$\frac{60 \text{ pages}}{45 \text{ s}} = \frac{x}{180 \text{ s}}$

$x = \frac{(60)(180)}{45}$

$x = 240 \text{ pages}$

29. $\frac{25.0 \text{ ft}}{2.00 \text{ in}} = \frac{x}{5.75 \text{ in}};$

$x = \frac{(25.0)(5.75)}{2.00}$

$x = 71.9 \text{ ft}$

33. $\frac{80.0}{98.0}=\frac{x}{37.0}$

$x=\frac{(80.0)(37.0)}{98.0}$

$x=30.2\text{ kg}$

37. $y=kx^2$

$27=k(3^2)$

$k=3$

$y=3x^2$

41. $\frac{F_1}{F_2}=\frac{L_2}{L_1}$

$\frac{4.50}{6.75}=\frac{L_2}{17.5}$

$(6.75)L_2=(4.50)(17.5)$

$L_2=11.7\text{ in}$

45. $R=kA$

$850=k(900)$

$k=0.94$

$R=0.94\text{ A}$

49. F varies inversely as L, so

$F=\frac{k}{L}$

$250=\frac{k}{22}$

$k=5500$

$F=\frac{5500}{L}$

Check:

$F=\frac{5500}{10}$

$=550\text{ N}$

53. $v=k\sqrt{p}$

$120=k\sqrt{100}$

$k=12$

$v=12\sqrt{p}$

$v=12\sqrt{p}$

$=96\text{ ft/s}$

57. $P=kA$

$30.0=k(8.00)$

$k=3.75$

$P=3.75A$

$P=3.75(6.00)$

$=22.5\text{ hp}$

61. $t=k(\log N)^2$

For 8000 numbers:

$t_1=k(\log 8000)^2$

For 2000 numbers:

$t_2=k(\log 2000)^2$

The ratio of t_1 over t_2 is :

$\frac{k(\log 8000)^2}{k(\log 2000)^2}=1.4$

t_1 is 1.4 times longer than t_2.

65. $f=\frac{v}{\lambda}$

$90.9\times10^6=\frac{v}{3.29}$

$v=90.9\times10^6(3.29)$

$=299\times10^6$

$=2.99\times10^8\text{ m/s}$

69. $r=k\sqrt{\lambda}$

$3.56=k\sqrt{575}$

$k=\frac{3.56}{\sqrt{575}}$

$=0.148$

$r=0.148\sqrt{\lambda}$

$r=0.148\left(\sqrt{483}\right)$

$r=3.25\text{ cm}$

73. $d=kv^2$

$52=k(32)^2$

$k=\frac{52}{32^2}$

$=0.051$

$d=0.051v^2$

$d=0.051\cdot(55)^2$

$=154\text{ ft}$

77.

$$A = k(1+r)^n$$
$$= k(1+0.04)^n$$
$$= k(1.04)^n$$
$$185.03 = k(1.04)^{10}$$
$$k = 125$$
$$A = 125(1.04)^n$$
$$A = 125(1.04)^0$$

= \$125 is the value of the original investment

$$A = 125(1.04)^{40}$$

= \$600.13 is the current value of the investment

81.

$$I = k\cos^2\theta$$
$$0.025 = k\cos^2 12.0^\circ$$
$$k = \frac{0.025}{\cos^2 12.0^\circ}$$
$$= 0.026$$
$$I = 0.026\cos^2\theta$$
$$I = 0.026\cos^2 20.0^\circ$$
$$= 0.023 \text{ W/m}^2$$

Chapter 19

Sequences and the Binomial Theorem

19.1 Arithmetic Sequences

1. $a_1 = 5,\ a_{32} = -88,\ n = 32$

$$-88 = 5 + (32-1)d$$
$$31d = -93$$
$$d = -3$$

5. $a_1 = 2.5,\ a_5 = -1.5$

$$a_1 + 4d = -1.5$$
$$4d = -1.5 - 2.5$$
$$d = -1$$

2.5, 1.5, 0.5, −0.5, −1.5

9. $\frac{8}{3}, \frac{5}{3}, \frac{2}{3}, \cdots n = 9$

$$\frac{8}{3} + d = \frac{5}{3}$$
$$d = -1$$
$$a_9 = \frac{8}{3} + (9-1)(-1) = \frac{8}{3} - 8$$
$$a_9 = -\frac{16}{3}$$

13. $d = \frac{a_3 - a_1}{3-1} = \frac{5b-b}{2} = 2b$

$$a_{25} = b + (25-1)(2b)$$
$$= b + 48b$$
$$= 49b$$

17. $d = -\frac{5}{2} - (-2)$

$$= -\frac{1}{2}$$
$$a_{10} = -2 + (10-1)\ -\frac{1}{2}$$
$$= -6.5$$
$$S_{10} = \frac{10}{2}(-2 - 6.5)$$
$$= 5(-8.5)$$
$$= \frac{-85}{2}$$

21. $\frac{40}{3} = \frac{20}{2}\left(\frac{5}{3} + a_{20}\right)$

$$\frac{80}{60} = \frac{5}{3} + a_{20}$$
$$\frac{4}{3} - \frac{5}{3} = a_{20}$$
$$a_{20} = -\frac{1}{3}$$
$$-\frac{1}{3} = \frac{5}{3} + (20-1)d$$
$$-\frac{1}{3} - \frac{5}{3} = 19d$$
$$-2 = 19d$$
$$d = -\frac{2}{19}$$

25. $-23.1 = 7.4 + (n-1)(-0.5)$

$$-30.5 = -0.5n + 0.5$$
$$-0.5n = -31.0$$
$$n = 62$$
$$S_{62} = \frac{62}{2}(7.4 - 23.1)$$
$$= -486.7$$

29. ln 3, ln 6, ln 12,...is an arithmetic sequence since

$$d = \ln 2x - \ln x$$
$$= \ln \frac{2x}{x}$$
$$= \ln 2$$
$$a_5 = a_1 + (5-1)d$$
$$= \ln 3 + 4\ln 2$$
$$= \ln 3 + \ln 2^4$$
$$= \ln 3 + \ln 16$$
$$= \ln(3 \cdot 16)$$
$$= \ln 48$$

33. a_1, b, c, a_4, a_5

$$b+d=c \Rightarrow d=c-b$$

$$\begin{aligned} a_1 &= b-d \\ &= b-(c-b) \\ &= 2b-c \\ a_4 &= c+d \\ &= c+(c-b) \\ &= 2c-b \\ a_5 &= c+2d \\ &= c+2(c-b) \\ &= 3c-2b \end{aligned}$$

37. $3-x,\ -x,\ \sqrt{9-2x}$

$$\begin{aligned} d &= -x-(3-x) \\ &= -3 \end{aligned}$$

$-x-3=\sqrt{9-2x}$; square both sides

$$x^2+6x+9=9-2x$$

$$x^2+8x+9=0$$

$$x(x+8)=0$$

$$x=-8 \text{ or } x=0$$

Check: $x=-8$

$$-(-8)-3=\sqrt{9-2(-8)}$$

$$5=5$$

Check: $x=0$

$$0-3=\sqrt{9-2(0)}$$

$$-3\neq 3$$

The only solution is $x=-8$, which gives the sequence 11, 8, 5,... with $d=-3$.

41. Area losses for next 8 years are: 500 m^2; 600 m^2, 700 m^2,..., $a_8=500+7(100)=1200$ m^2

$$S_8=\frac{8}{2}(500+1200)=6800 \text{ m}^2$$

Area in 8 years $=9500-6800=2700$ m^2

45. $a_1=12,\ d=4,\ S_n=300$

$$\begin{aligned} a_n &= 12+(n-1)4 \\ &= 8+4n \end{aligned}$$

Substitute into: $300=\frac{n}{2}(12+a_n)$

$$300=\frac{n}{2}(12+8+4n)$$

$$600=12n+8n+4n^2$$

$$4n^2+20n-600=0$$

$$n^2+5n-150=0$$

$$(n-10)(n+15)=0$$

$n=10$ or $n=-15$ (not possible)

There are 10 rows of seats.

49. 16, 48, 80, 112,... is an AS with $a_1=16,\ d=32$

$$a_{10}=16+(10-1)(32)=304$$

After 10 seconds, the total distance is

$$\begin{aligned} S_{10} &= \frac{10}{2}(16+304) \\ &= 1600 \text{ ft} \end{aligned}$$

53.

$$\begin{aligned} S_n &= \frac{n}{2}(a_1+a_n) \\ &= \frac{n}{2}\left[a_1+(a_1+(n-1)d)\right] \\ &= \frac{n}{2}\left[2a_1+(n-1)d\right] \end{aligned}$$

19.2 Geometric Sequences

1. Find a_{10} for $a_2 = 3,\ a_4 = 9$

Use $n = 3$ for a new sequence starting at the second term, so $a_1 = 3, a_3 = 9$ and

$9 = 3r^2$

$r = \sqrt{3}$

Find the first term of the original sequence:

$3 = a_1\left(\sqrt{3}\right)^{2-1}$

$a_1 = \sqrt{3}$

Find the 10th term:

$a_{10} = \sqrt{3}\left(\sqrt{3}\right)^{10-1} = \sqrt{3}\left(\sqrt{3}\right)^9 = 243$

5. $\frac{1}{6}, \frac{1}{6}\cdot 3, \frac{1}{6}\cdot 3^2, \frac{1}{6}\cdot 3^3, \frac{1}{6}\cdot 3^4$

$\frac{1}{6}, \frac{1}{2}, \frac{3}{2}, \frac{9}{2}, \frac{27}{2}$

9. $r = -25 \div 125 = -\frac{1}{5},\ a_1 = 125,\ n = 7$

$a_7 = 125\left(-\frac{1}{5}\right)^{7-1} = \frac{1}{125}$

13. $10^{100}, -10^{98}, 10^{96}, \ldots; n = 51$

$r = \frac{-10^{98}}{10^{100}} = -10^{-2}$

$a_{51} = 10^{100}\left(-10^{-2}\right)^{50}$

$= 10^{100}(-1)^{100}(10^{-100})$

$= 1$

17. 384, 192, 96, ⋯

$r = \frac{192}{384} = \frac{1}{2}$

$S_7 = \frac{384\left(1-\left(\frac{1}{2}\right)^7\right)}{1-\frac{1}{2}} = 762$

21. $r = \frac{1}{4} \div \frac{1}{16} = 4, a_6 = \left(\frac{1}{16}\right)(4)^{6-1} = \left(\frac{1}{16}\right)(4)^5 = 64$

$S_6 = \frac{\frac{1}{16}\left(1-4^6\right)}{1-4} = \frac{\frac{1}{16}(1-4096)}{-3} = \frac{4095}{48} = \frac{1365}{16}$

25. $27 = a_1 r^{4-1}$

$a_1 = \frac{27}{r^3}$

$40 = a_1 \frac{\left(1-r^4\right)}{1-r}$

$= a_1 \frac{\left(1+r^2\right)(1+r)(1-r)}{1-r}$

$= a_1\left(1+r^2\right)(1+r)$

Substitute a from first equation in second equation:

$40 = \frac{27}{r^3}\left(1+r^2\right)(1+r)$

$40r^3 = 27 + 27r + 27r^2 + 27r^3$

$13r^3 - 27r^2 - 27r - 27 = 0$

There is one change of sign so there is exactly one positive root. There are at most two negative roots. We first test the possible integer roots:

$\pm 1, \pm 3, \pm 9, \pm 27$

Using synthetic division, 3 gives a remainder of zero (see Section 15.3):

13	−27	−27	−27	|3
	39	36	27	
13	12	9	0	

The remaining roots are solutions to the equation

$13r^2 + 12r + 9 = 0$

Using the quadratic formula:

$r = \frac{-12 \pm \sqrt{-288}}{26}$

Since the other solutions are complex, the only solution is $r = 3$. Now

$27 = a_1\left(3^{4-1}\right)$

$27a_1 = 27$

$a_1 = 1$

29. $\frac{3^{x+1}}{3} = 3^x$

$\frac{3^{2x+1}}{3^{x+1}} = 3^x$

The sequence 3, 3^{x+1}, 3^{2x+1}, ⋯ is a geometric sequence since every term in the sequence can be obtained from the preceding one by multiplying it by 3^x.

Therefore

$a_1 = 3,\ r = 3^x$

$a_{20} = 3 \cdot (3^x)^{20-1}$

$= 3^{19x+1}$

33. $a_1 = x^2$

$r = -x$

$S_n = a_1 \frac{1-r^n}{1-r}$

$= x^2 \frac{1-(-x)^n}{1-(-x)}$

$= (1-(-x)^n)\frac{x^2}{1+x}$

37. $r = 1 - 0.125 = 0.875,\ a_1 = 3.27$ mA, $n = 9.2$

$a_{9.2} = 3.27(0.875)^{8.2} = 1.09$ mA

41. $a_1 = 9800, r = 1 - 0.1 = 0.9$

$n = \frac{4000}{800} = 5$ periods of 800 years

$a_6 = (9800)0.9^5$

$= 5800^\circ$C

45. $a_1 = 40000$

$r = 1 - 0.10 = 0.90$

$a_{11} = a_1 r^{10} = 40000(0.90^{10}) = 13947.14$

The car has a value of \$13947

49. The first option can be modeled with an arithmetic sequence where $a_1 = 5{,}000{,}000$ and $d = 400{,}000$.

Then $a_6 = a_1 + 5d = 7{,}000{,}000$

$$S_6 = \frac{6}{2}(a_1 + a_6) = 36{,}000{,}000$$

The second option can be modeled with a geometric sequence where $a_1 = 5{,}000{,}000$ and $r = 1.05$.

Then

$$S_6 = a_1 \frac{1-r^6}{1-r} = 34{,}000{,}000$$

The first option pays \$2,000,000 more.

53. Let a, ar, ar^2, ..., ar^{n-1} be the given G.S. Squaring each term gives a^2, a^2r^2, a^2r^4,..., a^2r^{2n-2} which is also a G.S. with common ratio r^2.

19.3 Infinite Geometric Series

1. Given the G.S. $4 + \frac{1}{2} + \frac{1}{16} + \frac{1}{128} + \cdots$ find the sum.

$a_1 = 4,\ r = \frac{1}{8}$

$S = \frac{a}{1-r} = \frac{4}{1-\frac{1}{8}}$

$S = \frac{32}{7}$

5. $a_1 = 0.5, S = 0.625$

$0.625 = \frac{0.5}{1-r}$

$1 - r = 0.8$

$r = \frac{1}{5}$

9. $a_1 = 4,\ r = \frac{7}{8}$

$S = \frac{4}{1-\frac{7}{8}}$

$= 32$

13. $a_1 = 2+\sqrt{3},\ r = \dfrac{1}{2+\sqrt{3}}$

$$S = \frac{2+\sqrt{3}}{1-\frac{1}{2+\sqrt{3}}}$$

$$= \frac{2+\sqrt{3}}{\frac{2+\sqrt{3}-1}{2+\sqrt{3}}}$$

$$= \frac{\left(2+\sqrt{3}\right)\left(2+\sqrt{3}\right)}{1+\sqrt{3}}$$

$$= \frac{7+4\sqrt{3}}{1+\sqrt{3}} \times \frac{1-\sqrt{3}}{1-\sqrt{3}}$$

$$= \frac{7-7\sqrt{3}+4\sqrt{3}-12}{1-3}$$

$$= \frac{-5-3\sqrt{3}}{-2}$$

$$= \frac{1}{2}\left(5+3\sqrt{3}\right)$$

17. $0.499\,99\ldots = 0.4+0.09+0.009+0.0009+\cdots$

$a_1 = 0.09, r = 0.01$

$$S = 0.4+\frac{0.09}{1-\frac{1}{10}}$$

$$= 0.5$$

$$= \frac{1}{2}$$

21. $0.0273273273\ldots = 0.0273+0.0000273+\ldots$

$a_1 = 0.0273,\ r = 0.001$

$$S = \frac{0.0273}{1-0.001}$$

$$= \frac{273}{9990}$$

$$= \frac{91}{3330}$$

25. $a_1 = \dfrac{5}{36}$

$r = \dfrac{25}{36}$

$$S = \frac{5/36}{1-(25/36)} = \frac{5/36}{11/36} = \frac{5}{11}$$

29. $a_1 = 5.882$ g

The sequence is approximately geometric, since the ratios of successive terms are constant (to four significant digits).

$$r = \frac{5.782}{5.882} = \frac{5.684}{5.782} = \frac{5.587}{5.684} = 0.9830$$

$$S = \frac{5.882}{1-0.9830}$$

$$= \frac{5.882}{0.0170}$$

$$= 346 \text{ g}$$

33.

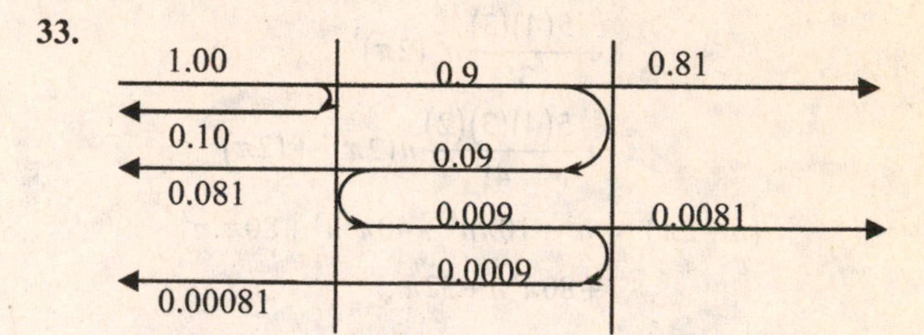

0.81 is transmitted during the first pass, 0.0081 is transmitted after one reflection in the first pane, 0.000081 is transmitted after reflections in panes 1, 2, and 1, and so on. We have $a_1 = 0.81$ and $r = 0.01$, implying

$$S = \frac{0.81}{1-0.01} = \frac{0.81}{0.99} = \frac{9}{11} = 0.81818181\ldots$$

and so 81.8% of the light is transmitted through the second pane.

19.4 The Binomial Theorem

1. $$(2x+3)^5=(2x)^5+5(2x)^4(3)+\frac{5(4)}{2!}(2x)^3(3)^2+\frac{5(4)(3)}{3!}(2x)^2(3)^3+\frac{5(4)(3)(2)}{4!}(2x)^1(3)^4+(3)^5$$
$$(2x+3)^5=32x^5+240x^4+720x^3+1080x^2+810x+243$$

5. $$(2x-3)^4=(2x)^4+4(2x)^3(-3)+\frac{4(3)}{2}(2x)^2(-3)^2+\frac{4(3)(2)}{6}(2x)(-3)^3+\frac{4(3)(2)(1)}{24}(-3)^4$$
$$=16x^4-96x^3+216x^2-216x+81$$

9. $$(n+2\pi)^5=n^5+5n^4(2\pi)+\frac{5(4)}{2!}n^3(2\pi)^2+\frac{5(4)(3)}{3!}n^2(2\pi)^3+\frac{5(4)(3)(2)}{4!}n(2\pi)^4+(2\pi)^5$$
$$(n+2\pi)^5=n^5+10\pi n^4+40\pi^2n^3+80\pi^3n^2+80\pi^4n+32\pi^5$$

13. From Pascal's triangle, the coefficients for $n=4$ are 1, 4, 6, 4, 1.
$$(5x-3)^4=\left[5x+(-3)\right]^4$$
$$=1(5x)^4+4(5x)^3(-3)+6(5x)^2(-3)^2+4(5x)(-3)^3+(-3)^4$$
$$=625x^4-1500x^3+1350x^2-540x+81$$

17. $$(x+2)^{10}=x^{10}+10x^9(2)+\frac{(10)(9)}{2}x^8(2)^2+\frac{(10)(9)(8)}{6}x^7(2)^3+\cdots$$
$$=x^{10}+20x^9+180x^8+960x^7+\cdots$$

21. $$\left(x^{1/2}-4y\right)^{12}=\left(x^{1/2}\right)^{12}+12\left(x^{1/2}\right)^{11}(-4y)+\frac{12\cdot11}{2!}\left(x^{1/2}\right)^{10}(-4y)^2+\frac{12\cdot11\cdot10}{3!}\left(x^{1/2}\right)^9(-4y)^3$$
$$=x^6-48x^{11/2}y+1056x^5y^2-14\ 080x^{9/2}y^3+\cdots$$

25. $$(1.05)^6=(1+0.05)^6$$
$$=1^6+6(1)^5(0.05)+\frac{6(5)}{2!}(1)^4(0.05)^2$$
$$=1.3375$$
$=1.338$ to 3 decimal places using three terms

From a calculator, $(1.05)^6=1.340$ to 3 decimal places.

29. $$(1+x)^8=1+8x+\frac{8(7)}{2}x^2+\frac{8(7)(6)}{6}x^3+\cdots$$
$$=1+8x+28x^2+56x^3+\cdots$$

33. $$\sqrt{1+x}=(1+x)^{1/2}=1+\frac{1}{2}x+\frac{\frac{1}{2}\left(-\frac{1}{2}\right)}{2}x^2+\frac{\frac{1}{2}\left(-\frac{1}{2}\right)\left(-\frac{3}{2}\right)}{6}x^3+\cdots$$
$$=1+\frac{1}{2}x-\frac{1}{8}x^2+\frac{1}{16}x^3\cdots$$

37. (a) $17!+4!=3.557\times10^{14}$

(b) $21!=5.109\times10^{19}$

(c) $17!\times4!=8.536\times10^{15}$

(d) $68!=2.480\times10^{96}$

41. The term involving b^5 will be the sixth term.

$r=5$, $n=8$

The sixth term is $\dfrac{8(7)(6)(5)(4)}{5(4)(3)(2)}a^3b^5=56a^3b^5$.

45. The given expression is the expansion of
$$((2x-1)+(3-2x))^3$$
using the binomial theorem. We have
$$((2x-1)+(3-2x))^3=2^3=8.$$

49. $1 = (0.95 + 0.05)^4$
$= 0.95^4 + 4(0.95^3)(0.05) + 6(0.95^2)(0.05^2) + 4(0.95)(0.05^3) + 0.05^4$
The probability that exactly three engines function properly is
$4(0.95^3)(0.05) = 0.171$

53. $V = A(1-r)^5$; expand $(1-r)^5$

$$1^5 + 5(1)^4(-r) + \frac{5(4)}{2}(1)^3(-r)^2$$
$$+\frac{5(4)(3)}{6}(1)^2(-r)^3 + \frac{5(4)(3)(2)}{24}(1)(-r)^4 + (-r)^5$$
$$= 1 - 5r + 10r^2 - 10r^3 + 5r^4 - r^5$$
$$A(1-r)^5 = A(1 - 5r + 10r^2 - 10r^3 + 5r^4 - r^5)$$

57. $(1-t)^4 P_0 + 4(1-t)^3 tP_1 + 6(1-t)^2 t^2 P_2 + 4(1-t)t^3 P_3 + t^4 P_4$

Review Exercises

1. This is false. If $a_1 = 5$ and $d = 3$, then
$a_{10} = a_1 + (10-1)d = 5 + 9(3) = 32.$

5. This is an A.S. with $d = 6 - 1 = 5, a_1 = 1$, so
$$a_{17} = 1 + (17-1)5$$
$$= 1 + 80$$
$$= 81$$

9. This is an A.S. with $d = 8 - 3.5 = 4.5, a_1 = -1$, so
$$a_{16} = -1 + (16-1)(4.5)$$
$$= 66.5$$

13. $$S_{15} = \frac{15}{2}(-4+17)$$
$$= \frac{15}{2}(13)$$
$$= \frac{195}{2}$$

17. $$a_9 = 17 + (9-1)(-2)$$
$$= 17 - 16$$
$$= 1$$
$$S_9 = \frac{9}{2}(1+17) = 81$$

21. $a_1 = 80,\ a_n = -25,\ S_n = 220$

$$220 = \frac{n}{2}(80-25)$$
$$440 = 55n$$
$$n = 8$$
$$-25 = 80 + (8-1)d$$
$$= 80 + 7d$$
$$d = \frac{-105}{7}$$
$$= -15$$

25. $S_{12} = \frac{12}{2}(-1+32)$

$$= 186$$

29. $r = \frac{1.02^{-1}}{1} = 1.02^{-1}$

$$S = \frac{1}{1-1.02^{-1}}$$
$$= 51$$

33. $0.07272727\cdots = 0.072 + 0.00072 + \ldots$

$$a_1 = 0.072, r = 0.01$$
$$S = \frac{0.072}{1-0.01}$$
$$= \frac{4}{55}$$

37. $(x^2+4)^5 = (x^2)^5 + \frac{5}{1}(x^2)^4(4) + \frac{5\cdot 4}{1\cdot 2}(x^2)^3(4)^2$

$$+\frac{5\cdot 4\cdot 3}{1\cdot 2\cdot 3}(x^2)^2(4)^3 + \frac{5\cdot 4\cdot 3\cdot 2}{1\cdot 2\cdot 3\cdot 4}(x^2)(4)^4$$
$$+4^5$$
$$= x^{10} + 20x^8 + 160x^6 + 640x^4 + 1280x^2$$
$$+1024$$

41. $p^2 - \frac{q}{6} \quad ^{9}$

$$= (p^2)^9 + \frac{9}{1}(p^2)^8 \ -\frac{q}{6} \ + \frac{9\cdot 8}{1\cdot 2}(p^2)^7 \ -\frac{q}{6} \ ^{2}$$
$$+\frac{9\cdot 8\cdot 7}{1\cdot 2\cdot 3}(p^2)^6 \ -\frac{q}{6} \ ^{3} + \cdots$$
$$= p^{18} - \frac{3}{2}p^{16}q + p^{14}q^2 - \frac{7}{18}p^{12}q^3 + \cdots$$

45. $(1+x^2)^{1/2} = 1^{1/2} + \frac{\frac{1}{2}}{1}1^{-1/2}(x^2)^1 + \frac{\frac{1}{2}\left(\frac{1}{2}-1\right)^{-3/2}}{1\cdot 2}1^{-3/2}(x^2)^2$

$$+\frac{\frac{1}{2}\left(\frac{1}{2}-1\right)\left(\frac{1}{2}-2\right)}{1\cdot 2\cdot 3}1^{-5/2}(x^2)^3 + \cdots$$
$$= 1 + \frac{1}{2}x^2 - \frac{1}{8}x^4 + \frac{1}{16}x^6 - \cdots$$

49. $(2-4x)^{-3} = 2^{-3} + \frac{-3}{1}2^{-4}(-4x)$

$$+\frac{(-3)(-4)}{1\cdot 2}2^{-5}(-4x)^2$$
$$+\frac{(-3)(-4)(-5)}{1\cdot 2\cdot 3}2^{-6}(-4x)^3 + \cdots$$
$$= \frac{1}{8} + \frac{3}{4}x + 3x^2 + 10x^3 + \cdots$$

53. $b = a + d$

$$d = b - a$$
$$a_5 = a + (5-1)(b-a)$$
$$= 4b - 3a$$

57. $a_1 = -5,\ a_{n+1} = a_n - 3,\ n = 1, 2, 3$

$$d = a_{n+1} - a_n$$
$$= -3$$
$$a_n = a_1 + (n-1)d$$
$$= -5 + (n-1)(-3)$$
$$= -5 - 3n + 3$$
$$= -2 - 3n$$

61. $\sqrt{30} = 5(1+0.2)^{1/2}$

$$\sqrt{30} = 5\left[1 + \frac{1}{2}(0.2) + \frac{\frac{1}{2}\left(\frac{1}{2}-1\right)}{2!}(0.2)^2\right]$$
$$\sqrt{30} = 5.475$$
$$= 5.5 \text{ using three terms}$$

65. $a_1 = x+1, a_2 = x^2 - 2, a_3 = 4x - 2$

$$d = a_2 - a_1 = x^2 - 2 - (x+1) = x^2 - x - 3$$
$$d = a_3 - a_2 = 4x - 2 - (x^2 - 2) = -x^2 + 4x$$
$$x^2 - x - 3 = -x^2 + 4x$$
$$2x^2 - 5x - 3 = 0$$
$$(2x+1)(x-3)$$
$$x = -\frac{1}{2} \text{ or } x = 3$$

69. n = lens thickness in mm;

This is a G.S. with $a_1 = 100$,

$r = 1 - 0.12$

$= 0.88$

Find n so that

$20 = 100(0.88)^n$

$\log 0.20 = n \log 0.88$

$n = \dfrac{\log 0.20}{\log 0.88}$

$= 12.6$ mm

73. This is a G.S. with $a_1 = 8600$ and

$r = 1 - .01$

$= 0.99$

We are interested in $n = 12 \times 6 + 1 = 61$.

$V_{61} = 8600(0.99)^{60}$

$= \$4700$

77. This is an A.S. with

$d = 48 - 16$

$= 32$

The distance travelled during the 20th second is

$a_{20} = 16 + (20-1)32$

$= 624$ ft

81. The value of each investment is

$a_n = 1000\left(1 + \dfrac{0.075}{2}\right)^n$

$= 1000(1.0375)^n$,

where n = twice the number of invested years.

The total value is the sum of all the investments' values, which are as follows:

$1000(1.0375)^{40}$, $1000(1.0375)^{38}$, $1000(1.0375)^{36}$,... $1000(1.0375)^2$

This is a G.S. with $a_1 = 1000(1.0375)^{40}$ and

$r = 1.0375^{-2}$

$$S_{20} = \frac{\left[1000(1.0375)^{40}\right]\left[1-\left(1.0375^{-2}\right)^{20}\right]}{1-(1.0375)^{-2}}$$

$$= \frac{\left[1000(1.0375)^{40}\right]\left[1-1.0375^{-40}\right]}{\left[1-1.0375^{-2}\right]}$$

$= \$47\ 340.80$

85. $50, 50\left(\dfrac{2}{3}\right), 50\left(\dfrac{2}{5}\right), 50\left(\dfrac{2}{7}\right), \cdots, 50\left(\dfrac{2}{2n+1}\right), \cdots$

temperature after 12 h $= 50\left(\dfrac{2}{2(12)+1}\right)$

$= 4^\circ\text{C}$

89. The level of erosion is an A.S. with $a_1 = 1.2$ and $d = 0.1$. To reach the wall, the sum of the first n terms of the sequence will have to be 48.3 m.

$48.3 = \dfrac{n}{2}(1.2 + (1.2 + (n-1)(0.1)))$

$96.6 = 2.3n + 0.1n^2$

$0 = n^2 + 23n - 966$

$n = \dfrac{-23 \pm \sqrt{23^2 - 4(-966)}}{2}$

$= \dfrac{-23 \pm \sqrt{4393}}{2}$

$n = -44.6,\ 21.6$

The water will reach the wall during the 22nd year.

93. This is a G.S. with $a_1 = A$ and

$r = 1 - 0.10$

$= 0.9$

After n years,

$0.1A = A(0.9)^{n-1}$

$0.9^n = 0.09$

$n = \dfrac{\ln 0.09}{\ln 0.9}$

$= 22.85$

After 22 years the production will be 10% of first year's production.

97. Reciprocals of the terms of $a, ar, ar^2, \cdots, ar^{n-1}, \cdots$

are $\dfrac{1}{a}, \dfrac{1}{ar}, \dfrac{1}{ar^2}, \cdots, \dfrac{1}{ar^{n-1}}, \cdots$ or

$\dfrac{1}{a}, \dfrac{1}{a}\cdot\dfrac{1}{r}, \ldots, \dfrac{1}{a}\cdot\dfrac{1}{r^{n-1}}, \cdots$

which is a geometric sequence with first term $\dfrac{1}{a}$

and common ratio $\dfrac{1}{r}$.

Chapter 20

Additional Topics in Trigonometry

20.1 Fundamental Trigonometric Identities

1. We simplify the right side:

$$\frac{\tan x}{\sec x} = \frac{\frac{\sin x}{\cos x}}{\frac{1}{\cos x}}$$
$$= \frac{\sin x}{\cos x} \cdot \frac{\cos x}{1}$$
$$= \sin x$$

5. Verify $\sin^2\theta + \cos^2\theta = 1$ for $\theta = \frac{4\pi}{3}$

$$\left(\sin\frac{4\pi}{3}\right)^2 = \left(-\frac{1}{2}\sqrt{3}\right)^2 = \frac{3}{4}$$
$$\left(\cos\frac{4\pi}{3}\right)^2 = \left(-\frac{1}{2}\right)^2 = \frac{1}{4}$$
$$\frac{3}{4} + \frac{1}{4} = 1$$

9. $\cos\theta\cot\theta(\sec\theta - 2\tan\theta)$

$$= \cos\theta\cot\theta\left(\frac{1}{\cos\theta} - 2\frac{1}{\cot\theta}\right)$$
$$= \cot\theta - 2\cos\theta$$

13.
$$\sin x + \sin x\tan^2 x = \sin x\left(1 + \tan^2 x\right)$$
$$= \sin x\sec^2 x$$
$$= \sin x \cdot \frac{1}{\cos x} \cdot \sec x$$
$$= \tan x\sec x$$

17.
$$\csc^4 y - 1 = \left(\csc^2 y + 1\right)\left(\csc^2 y - 1\right)$$
$$= \left(\csc^2 y + 1\right)\left(\cot^2 y\right)$$

21.
$$\sin x\sec x = \sin x\frac{1}{\cos x}$$
$$= \frac{\sin x}{\cos x}$$
$$= \tan x$$

25.
$$\sin x\left(1 + \cot^2 x\right) = \sin x\left(\csc^2 x\right)$$
$$= \sin x\left(\frac{1}{\sin^2 x}\right)$$
$$= \frac{1}{\sin x}$$
$$= \csc x$$

29.
$$\cot\theta\sec^2\theta - \frac{1}{\tan\theta} = \frac{\cos\theta}{\sin\theta}\frac{1}{\cos^2\theta} - \frac{\cos\theta}{\sin\theta}$$
$$= \frac{1}{\sin\theta\cos\theta} - \frac{\cos\theta}{\sin\theta}$$
$$= \frac{1 - \cos^2\theta}{\sin\theta\cos\theta}$$
$$= \frac{\sin^2\theta}{\sin\theta\cos\theta}$$
$$= \frac{\sin\theta}{\cos\theta}$$
$$= \tan\theta$$

33.
$$\cos^2 x - \sin^2 x = 1 - \sin^2 x - \sin^2 x$$
$$= 1 - 2\sin^2 x$$

37.
$$2\sin^4 x - 3\sin^2 x + 1 = \left(2\sin^2 x - 1\right)\left(\sin^2 x - 1\right)$$
$$= \left(2\sin^2 x - 1\right)\left(-\cos^2 x\right)$$
$$= \cos^2 x\left(1 - 2\sin^2 x\right)$$

41.
$$\cot x(\sec x - \cos x) = \frac{\cos x}{\sin x} \cdot \frac{1}{\cos x} - \frac{\cos x}{\sin x} \cdot \cos x$$
$$= \frac{1}{\sin x} - \frac{\cos^2 x}{\sin x}$$
$$= \frac{1 - \cos^2 x}{\sin x}$$
$$= \frac{\sin^2 x}{\sin x}$$
$$= \sin x$$

45.
$$\frac{\cos x + \sin x}{1 + \tan x} = \frac{\cos x + \sin x}{1 + \frac{\sin x}{\cos x}}$$
$$= \frac{(\cos x + \sin x)}{\frac{\cos x + \sin x}{\cos x}}$$
$$= \cos x$$

49. We use $1+\cot^2 x = \csc^2 x$

$$1+\frac{1}{\tan^2 x} = \csc^2 x$$

$$\frac{1}{\tan^2 x} = \csc^2 x - 1$$

$$\tan^2 x = \frac{1}{\csc^2 x - 1}$$

$$\tan x = \sqrt{\frac{1}{\csc^2 x - 1}}$$

We use the positive square root because the angle is in the first quadrant.

53.

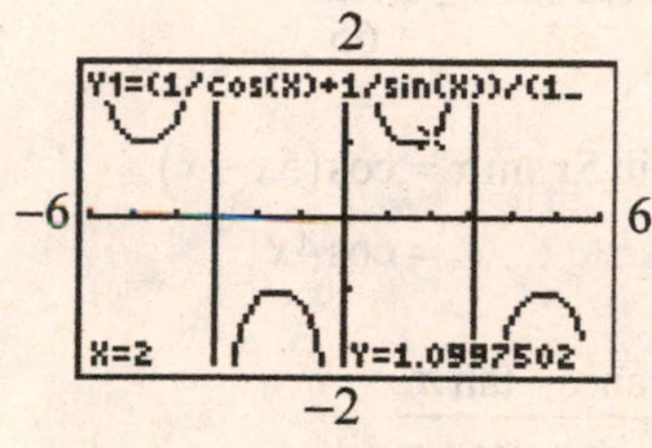

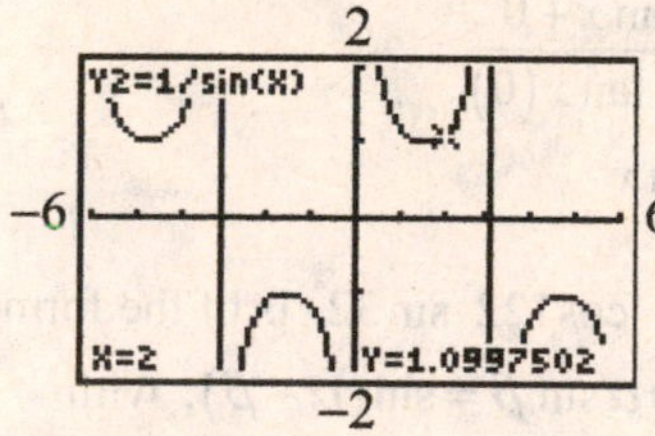

57. No. $\frac{2\cos^2 x - 1}{\sin x \cos x} \neq \tan x - \cot x$

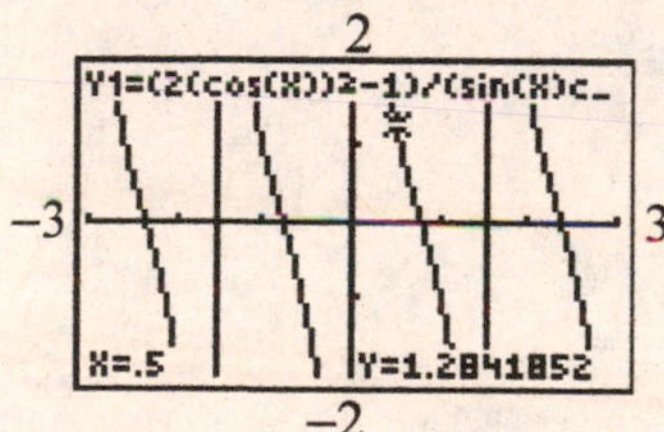

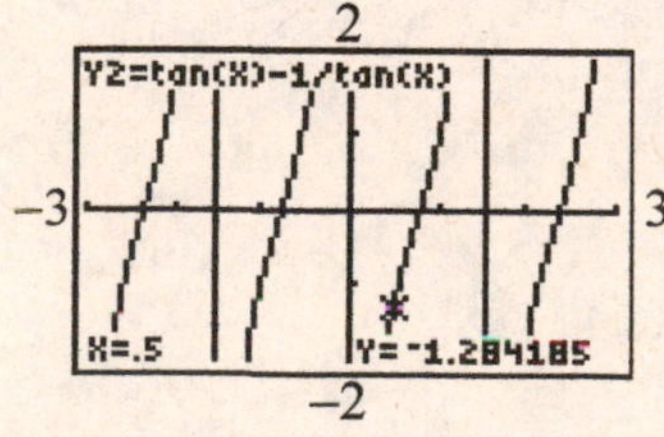

61.

$$\begin{aligned} l &= a\csc\theta + a\sec\theta \\ &= a(\csc\theta + \sec\theta) \\ &= a\ \frac{1}{\sin\theta} + \frac{1}{\cos\theta} \\ &= a\ \frac{1}{\sin\theta} + \frac{\sin\theta}{\cos\theta}\cdot\frac{1}{\sin\theta} \\ &= a\ \frac{1}{\sin\theta} + \frac{\tan\theta}{\sin\theta} \\ &= a\frac{(1+\tan\theta)}{\sin\theta} \end{aligned}$$

65.

$$\begin{aligned} &\sin^2 x\left(1-\sec^2 x\right) + \cos^2 x\left(1+\sec^4 x\right) \\ &= \sin^2 x - \sin^2 x\sec^2 x + \cos^2 x + \cos^2 x\sec^4 x \\ &= \sin^2 x - \frac{\sin^2 x}{\cos^2 x} + \cos^2 x + \frac{\cos^2 x}{\cos^4 x} \\ &= \sin^2 x - \tan^2 x + \cos^2 x + \sec^2 x \\ &= 1 - \tan^2 x + \sec^2 x \\ &= 1 - \left(\sec^2 x - 1\right) + \sec^2 x \\ &= 1 - \sec^2 x + 1 + \sec^2 x \\ &= 2 \end{aligned}$$

69.

$$\begin{aligned} \sec^2\theta + \csc^2\theta &= \left(\frac{r}{x}\right)^2 + \left(\frac{r}{y}\right)^2 \\ &= \frac{r^2}{x^2} + \frac{r^2}{y^2} \\ &= \frac{r^2y^2 + r^2x^2}{x^2y^2} \\ &= \frac{r^2\left(y^2+x^2\right)}{x^2y^2} \\ &= \frac{r^4}{x^2y^2} \\ &= \sec^2\theta\csc^2\theta \end{aligned}$$

73. $x = 2\tan\theta;$

$$\sqrt{4+x^2} = \sqrt{4+(2\tan\theta)^2}$$
$$= \sqrt{4+4\tan^2\theta}$$
$$= \sqrt{4(1+\tan^2\theta)}$$
$$= \sqrt{4\sec^2\theta}$$
$$= 2\sec\theta$$

20.2 The Sum and Difference Formulas

1. $\sin\alpha = \frac{12}{13}(\alpha \text{ in first quadrant})$ and $\sin\beta = -\frac{3}{5}$ for β in third quadrant.

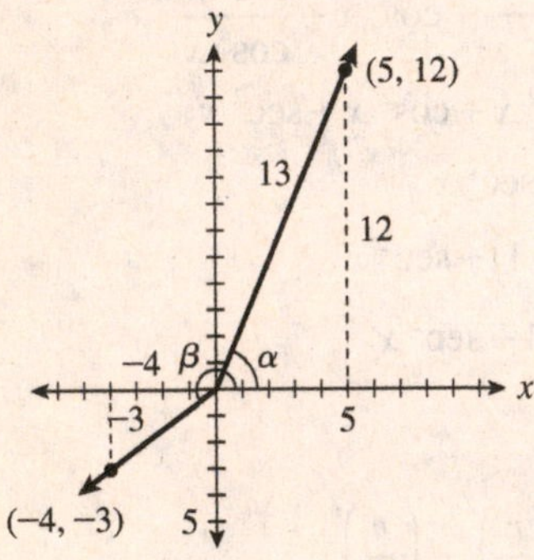

$$\cos(\alpha+\beta) = \cos\alpha\cos\beta - \sin\alpha\sin\beta$$
$$= \frac{5}{13}\cdot\frac{-4}{5} - \frac{12}{13}\cdot\frac{-3}{5} = \frac{16}{65}$$

5. Given $15^\circ = 60^\circ - 45^\circ$

$$\cos(\alpha-\beta) = \cos\alpha\cos\beta + \sin\alpha\sin\beta$$
$$\cos 15^\circ = \cos(60^\circ - 45^\circ)$$
$$= \cos 60^\circ \cos 45^\circ + \sin 60^\circ \sin 45^\circ$$
$$= \frac{1}{2}\times\frac{\sqrt{2}}{2} + \frac{\sqrt{3}}{2}\times\frac{\sqrt{2}}{2}$$
$$= \frac{\sqrt{2}}{4} + \frac{\sqrt{6}}{4}$$
$$= 0.966$$

9. Using the results of Exercise 7:

$$\cos\alpha = \frac{3}{5}$$
$$\cos\beta = -\frac{12}{13}$$
$$\sin\alpha = \frac{4}{5}$$
$$\sin\beta = \frac{5}{13}$$
$$\cos(\alpha+\beta) = \cos\alpha\cos\beta - \sin\alpha\sin\beta$$
$$= \frac{3}{5}\left(-\frac{12}{13}\right) - \frac{4}{5}\left(\frac{5}{13}\right)$$
$$= \frac{-36-20}{65} = -\frac{56}{65}$$

13. $\cos 5x\cos x + \sin 5x\sin x = \cos(5x - x)$

$$= \cos 4x$$

17. $\tan(x-\pi) = \frac{\tan x + \tan\pi}{1-\tan x\tan\pi}$

$$= \frac{\tan x + 0}{1 - \tan x(0)}$$
$$= \tan x$$

21. $\sin 122^\circ \cos 32^\circ - \cos 122^\circ \sin 32^\circ$ is of the form $\sin\alpha\cos\beta - \cos\alpha\sin\beta = \sin(\alpha-\beta)$, with $\alpha = 122^\circ$ and $\beta = 32^\circ$ so

$$\sin 122^\circ \cos 32^\circ - \cos 122^\circ \sin 32^\circ$$
$$= \sin(122^\circ - 32^\circ)$$
$$= \sin 90^\circ$$
$$= 1$$

25. $\sin(x+y)\sin(x-y)$

$= (\sin x\cos y + \cos x\sin y)(\sin x\cos y - \cos x\sin y)$

$= \sin^2 x\cos^2 y - \cos^2 x\sin^2 y$

$= \sin^2 x(1-\sin^2 y) - (1-\sin^2 x)(\sin^2 y)$

$= \sin^2 x - \sin^2 x\sin^2 y - \sin^2 y + \sin^2 x\sin^2 y$

$= \sin^2 x - \sin^2 y$

29.

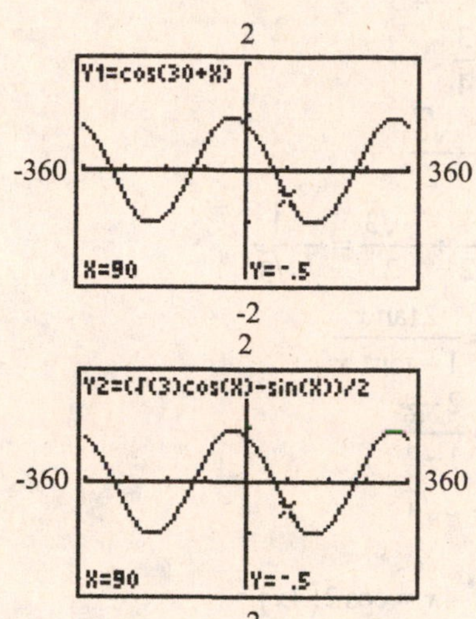

33. $\tan(\alpha \pm \beta)$

$$= \frac{\sin(\alpha\pm\beta)}{\cos(\alpha\pm\beta)} = \frac{\sin\alpha\cos\beta \pm \cos\alpha\sin\beta}{\cos\alpha\cos\beta \mp \sin\alpha\sin\beta}$$

(divide numerator and denominator by $\cos\alpha\cos\beta$)

$$= \frac{\dfrac{\sin\alpha\cos\beta}{\cos\alpha\cos\beta} \pm \dfrac{\cos\alpha\sin\beta}{\cos\alpha\cos\beta}}{\dfrac{\cos\alpha\cos\beta}{\cos\alpha\cos\beta} \mp \dfrac{\sin\alpha\sin\beta}{\cos\alpha\cos\beta}}$$

$$= \frac{\dfrac{\sin\alpha}{\cos\alpha} \pm \dfrac{\sin\beta}{\cos\beta}}{1 \mp \dfrac{\sin\alpha}{\cos\alpha} \times \dfrac{\sin\beta}{\cos\beta}}$$

$$= \frac{\tan\alpha \pm \tan\beta}{1 \mp \tan\alpha\tan\beta}$$

37. $\alpha+\beta = x;\ \alpha-\beta = y;\ \alpha = \frac{1}{2}(x+y);\ \beta = \frac{1}{2}(x-y)$

We will use Eq. (20.14) in its equivalent form

$\sin(\alpha+\beta) + \sin(\alpha-\beta) = 2\sin\alpha\cos\beta.$

$\sin x + \sin y = \sin(\alpha+\beta) + \sin(\alpha-\beta)$

$= 2\sin\alpha\cos\beta$

$= 2\sin\frac{1}{2}(x+y)\cos\frac{1}{2}(x-y)$

41. $\sin(x+30°)\cos x - \cos(x+30°)\sin x$

$= \sin((x+30°)-x)$

$= \sin 30°$

$= \frac{1}{2}$

45. To use Eq. (20.9), we must express 75° as a sum of angles whose sines and cosines we already know. Therefore, we write 75°=30° + 45°.

$\sin 75° = \sin(30° + 45°)$

$= \sin 30°\cos 45° + \sin 45°\cos 30°$

$= \frac{1}{2}\cdot\frac{\sqrt{2}}{2} + \frac{\sqrt{2}}{2}\cdot\frac{\sqrt{3}}{2}$

$= \frac{\sqrt{2}+\sqrt{6}}{4}$

To use Eq. (20.11), we must express 75° as the difference of two angles whose sines and cosines we already know. In this case we write 75°=135° − 60°. (The sine and cosine of 135° are found by using 45° as reference angle in the second quadrant.)

$\sin 75° = \sin(135° - 60°)$

$= \sin 135°\cos 60° - \sin 60°\cos 135°$

$= \frac{\sqrt{2}}{2}\cdot\frac{1}{2} - \frac{\sqrt{3}}{2}\cdot\left(-\frac{\sqrt{2}}{2}\right)$

$= \frac{\sqrt{2}+\sqrt{6}}{4}$

49. $V = 20\sqrt{2}\sin(120\pi t + \pi/4)$
$= 20\sqrt{2}(\sin 120\pi t \cos \pi/4 + \sin \pi/4 \cos 120\pi t)$
$= 20\sqrt{2}\left(\frac{1}{\sqrt{2}}\right)(\sin 120\pi t + \cos 120\pi t)$
$= 20\sin 120\pi t + 20\cos 120\pi t$
$= V_1 + V_2$

53. $\tan\alpha(R + \cos\beta) = \sin\beta$
$R\tan\alpha + \tan\alpha\cos\beta = \sin\beta$
$R\tan\alpha = \sin\beta - \tan\alpha\cos\beta$
$R = \frac{\sin\beta - \tan\alpha\cos\beta}{\tan\alpha}$
$= \frac{\sin\beta - \frac{\sin\alpha}{\cos\alpha}\cos\beta}{\frac{\sin\alpha}{\cos\alpha}}$
$= \frac{\sin\beta\cos\alpha - \sin\alpha\cos\beta}{\sin\alpha}$
$= \frac{\sin(\beta - \alpha)}{\sin\alpha}$

20.3 Double-Angle Formulas

1. If $\alpha = \frac{\pi}{3}$,
$\tan\frac{2\pi}{3} = \tan\left(2\cdot\frac{\pi}{3}\right) = \frac{2\tan\frac{\pi}{3}}{1-\tan^2\frac{\pi}{3}}$
$= \frac{2(\sqrt{3})}{1-(\sqrt{3})^2}$
$= -\sqrt{3}$

5. $60^\circ = 2(30^\circ)$; $\sin 2\alpha = 2\sin\alpha\cos\alpha$
$\sin 2(30^\circ) = 2\sin 30^\circ \cos 30^\circ$
$= 2\left(\frac{1}{2}\right)\left(\frac{\sqrt{3}}{2}\right)$
$= \frac{\sqrt{3}}{2}$

9. $\sin 100^\circ = 0.9848$
$\sin 100^\circ = \sin 2(50^\circ)$
$= 2\sin 50^\circ \cos 50^\circ$
$= 0.9848$

13. $\tan\frac{2\pi}{7} = 1.254$ (calculator in radian mode)
$\tan\frac{2\pi}{7} = \frac{2\tan\frac{\pi}{7}}{1-\tan^2\frac{\pi}{7}} = 1.254$

17. $\sin x = 0.5$ (QII)
$\cos^2 x = 1 - \sin^2 x$
$= 1 - \frac{1}{4}$
$= \frac{3}{4}$
$\cos x = -\frac{\sqrt{3}}{2}$
$\tan x = \frac{1}{2} \div \frac{-\sqrt{3}}{2} = -\frac{1}{\sqrt{3}}$
$\tan 2x = \frac{2\tan x}{1-\tan^2 x}$
$= \frac{2\cdot\frac{-1}{\sqrt{3}}}{1-\frac{1}{3}}$
$= -\sqrt{3}$

21. $1 - 2\sin^2 4x = \cos 2(4x)$
$= \cos 8x$

25. $8\sin^2 2x - 4 = -4(1 - 2\sin^2 2x)$
$= -4\cos 2(2x)$
$= -4\cos 4x$

29. $\frac{\sin 3x}{\sin x} - \frac{\cos 3x}{\cos x} = \frac{\sin 3x\cos x - \sin x\cos 3x}{\sin x\cos x}$
$= \frac{\sin(3x - x)}{\sin x\cos x}$
$= \frac{\sin 2x}{\sin x\cos x}$
$= \frac{2\sin x\cos x}{\sin x\cos x}$
$= 2$

33. $\frac{\cos x - \tan x\sin x}{\sec x} = \frac{\cos x - \frac{\sin x\sin x}{\cos x}}{\frac{1}{\cos x}}$
$= \frac{\cos^2 x - \sin^2 x}{\cos x} \times \frac{\cos x}{1}$
$= \cos^2 x - \sin^2 x$
$= \cos 2x$

37. $$1-\cos 2\theta = 1-\left(1-2\sin^2\theta\right)$$
$$= 2\sin^2\theta$$
$$= \frac{2}{\csc^2\theta}$$
$$= \frac{2}{1+\cot^2\theta}$$

41. Both graphs are the same.

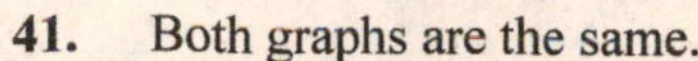

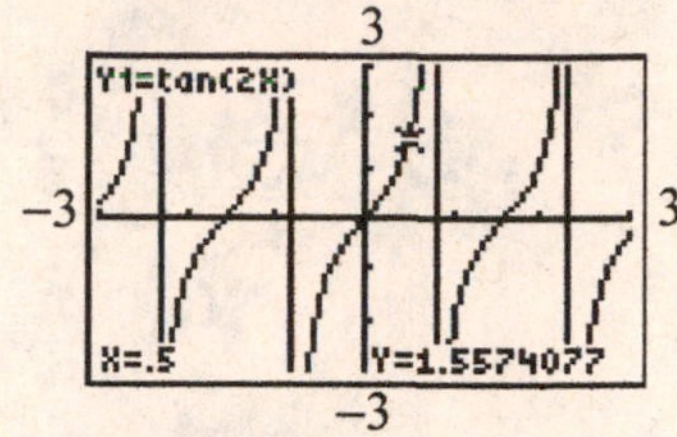

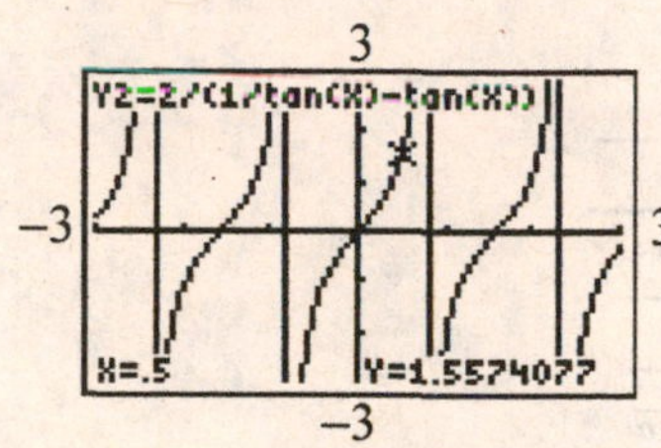

45. $$\sin 3x$$
$$= \sin(2x+x)$$
$$= \sin 2x\cos x + \cos 2x\sin x$$
$$= (2\sin x\cos x)(\cos x) + \left(\cos^2 x - \sin^2 x\right)(\sin x)$$
$$= 2\sin x\cos^2 x + \sin x\cos^2 x - \sin^3 x$$
$$= 3\sin x\cos^2 x - \sin^3 x$$
$$= 3\sin x\left(1-\sin^2 x\right) - \sin^3 x$$
$$= 3\sin x - 4\sin^3 x$$

49. $$\cos 2x + \sin 2x\tan x = \cos^2 x - \sin^2 x + 2\sin x\cos x \cdot \frac{\sin x}{\cos x}$$
$$= \cos^2 x - \sin^2 x + 2\sin^2 x$$
$$= \cos^2 x + \sin^2 x$$
$$= 1$$

53. $$y = 4\sin x\cos x = 2(2\sin x\cos x)$$
$$y = 2\sin 2x$$
$$A = 2,\ \text{period } = \frac{\pi}{2} = \pi$$

57.

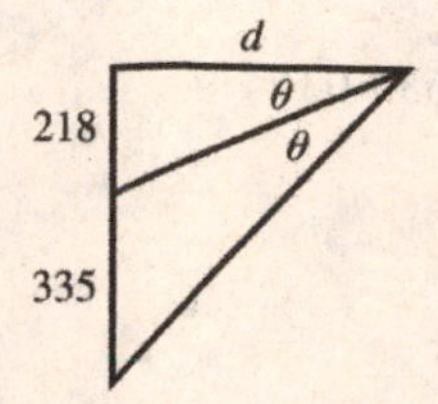

$$\tan\theta = \frac{218}{d}$$
$$\tan 2\theta = \frac{553}{d}$$
$$= \frac{2\tan\theta}{1-\tan^2\theta}$$
$$\frac{553}{d} = \frac{2\cdot\frac{218}{d}}{1-\left(\frac{218}{d}\right)^2}$$
$$1-\left(\frac{218}{d}\right)^2 = \frac{436}{553}$$
$$\left(\frac{218}{d}\right)^2 = 1-\frac{436}{553}$$
$$\frac{218}{d} = \sqrt{1-\frac{436}{553}}$$
$$d = \frac{218}{\sqrt{1-\frac{436}{553}}}$$
$$= 474 \text{ m}$$

61. $$P = vi\sin\omega t\sin\ \ \omega t - \frac{\pi}{2}$$
$$= vi\sin\omega t\ \ \sin\omega t\cos\frac{\pi}{2} - \cos\omega t\sin\frac{\pi}{2}$$
$$= vi\sin\omega t\ \ \sin\omega t(0) - \cos\omega t(1)$$
$$= -vi\sin\omega t\cos\omega t$$
$$= -\frac{1}{2}vi(2\sin\omega t\cos\omega t)$$
$$= -\frac{1}{2}vi\sin 2\omega t$$

20.4 Half-angle Formulas

1. $\sqrt{\frac{1+\cos 114^\circ}{2}} = \cos\frac{1}{2}(114^\circ) = \cos 57^\circ$

$\sqrt{\frac{1+\cos 114^\circ}{2}} = 0.544\ 639\ 035$

$\cos 57^\circ = 0.544\ 639\ 035$

5. $\sin 105^\circ = \sin\frac{210^\circ}{2}$

$= \sqrt{\frac{1-\cos 210^\circ}{2}}$

$= \sqrt{\frac{1-\left(-\frac{\sqrt{3}}{2}\right)}{2}}$

$= \frac{\sqrt{2+\sqrt{3}}}{2}$

$= 0.966$

9. $\sqrt{\frac{1-\cos 236^\circ}{2}} = \sin\frac{1}{2}(236^\circ)$

$= \sin 118^\circ$

$= 0.883$

$\sqrt{\frac{1-\cos 236^\circ}{2}} = 0.883$

13. $\sin\frac{\alpha}{2} = \sqrt{\frac{1-\cos\alpha}{2}}$

$\sqrt{\frac{1-\cos 8x}{2}} = \sin\frac{8x}{2}$

$= \sin 4x$

17. $\sqrt{4-4\cos 10\theta} = \sqrt{\frac{(2)(4)(1-\cos 10\theta)}{2}}$

$= 2\sqrt{2}\sin\frac{10\theta}{2}$

$= 2\sqrt{2}\sin 5\theta$

21. $0 < \frac{\alpha}{2} < 45^\circ$, so $\sin\frac{\alpha}{2} > 0$

$\sin\frac{\alpha}{2} = \sqrt{\frac{1-\cos\alpha}{2}}$

$= \sqrt{\frac{1-\frac{12}{13}}{2}}$

$= \sqrt{\frac{1}{26}}$

$= \sqrt{\frac{1}{26}\frac{26}{26}}$

$= \frac{1}{26}\sqrt{26}$

25. $\csc\frac{\alpha}{2} = \frac{1}{\sin\frac{\alpha}{2}}$

$= \frac{1}{\pm\sqrt{\frac{1-\cos\alpha}{2}}}$

$= \pm\sqrt{\frac{2}{1-\cos\alpha}}$

$= \pm\sqrt{\frac{2}{1-\frac{1}{\sec\alpha}}}$

$= \pm\sqrt{\frac{2\sec\alpha}{\sec\alpha-1}}$

29. $\frac{1-\cos\alpha}{2\sin\frac{\alpha}{2}} = \frac{1-\cos\alpha}{2\sqrt{\frac{1-\cos\alpha}{2}}} \times \frac{\sqrt{\frac{1-\cos\alpha}{2}}}{\sqrt{\frac{1-\cos\alpha}{2}}}$

$= \frac{(1-\cos\alpha)\sqrt{\frac{1-\cos\alpha}{2}}}{2\left(\frac{1-\cos\alpha}{2}\right)}$

$= \sqrt{\frac{1-\cos\alpha}{2}}$

$= \sin\frac{\alpha}{2}$

33. 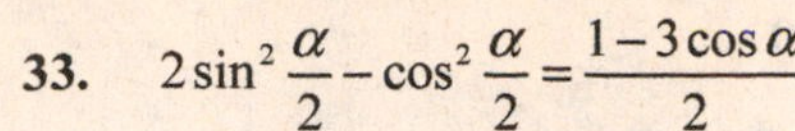

$$2\sin^2\frac{\alpha}{2}-\cos^2\frac{\alpha}{2}=\frac{1-3\cos\alpha}{2}$$

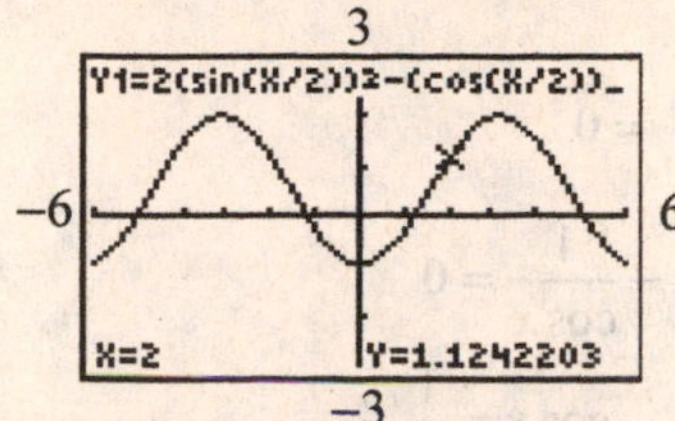

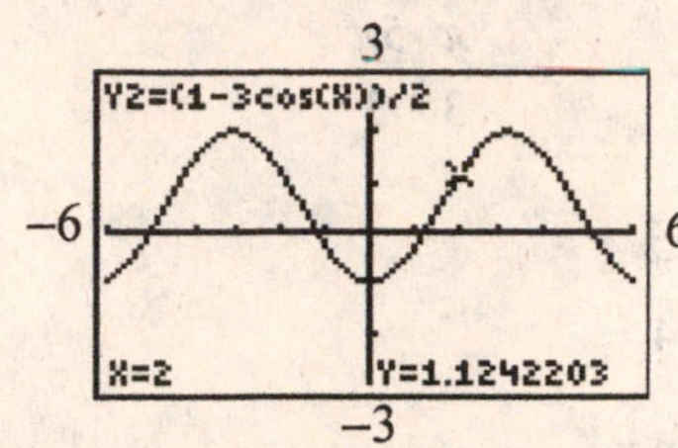

37.

$$\sin\frac{\theta}{2}=\sqrt{\frac{1-\cos\theta}{2}}$$
$$=\frac{3}{5}$$
$$1-\cos\theta=2\cdot\frac{9}{25}$$
$$\cos\theta=1-\frac{18}{25}$$
$$=\frac{7}{25}$$
$$\sin\theta=\pm\sqrt{1-\frac{49}{625}}$$
$$=\pm\frac{24}{25}$$
$$\tan\theta=\pm\frac{24}{7}$$

41. $\sin\theta=\frac{4}{5}$

$$\cos\theta=-\sqrt{1-\sin^2\theta}\ \text{(QII)}$$
$$=-\sqrt{1-\frac{16}{25}}=-\frac{3}{5}$$
$$\cos\frac{\theta}{2}=\sqrt{\frac{1+\cos\theta}{2}}\qquad 45°<\frac{\theta}{2}<90°$$
$$=\sqrt{\frac{1-3/5}{2}}$$
$$=\sqrt{\frac{1}{5}}$$

45.

$$\sin^2\omega t=\left(\sqrt{\frac{1-\cos 2\omega t}{2}}\right)^2$$
$$=\frac{1-\cos 2\omega t}{2}$$

20.5 Solving Trigonometric Equations

1. $\tan\theta-1=0,\ 0\le\theta<2\pi$

$$\tan\theta=1$$
$$\theta=\tan^{-1}1$$
$$\theta=\frac{\pi}{4},\ \frac{5\pi}{4}$$

5. $\sin x-1=0,\ 0\le x<2\pi;$

$$\sin x=1$$
$$x=\frac{\pi}{2}$$

9. $4-\sec^2 x=0;\ \ 0\le x<2\pi$

$$4\cos^2 x-1=0$$
$$4\cos^2 x=1$$
$$\cos^2 x=\frac{1}{4}$$
$$\cos x=\pm\frac{1}{2}$$
$$x=\frac{\pi}{3},\ \frac{2\pi}{3},\ \frac{4\pi}{3},\ \frac{5\pi}{3}$$

13. $\sin 2x\sin x+\cos x=0,\ \ 0\le x<2\pi$

$$(2\sin x\cos x)(\sin x)+\cos x=0$$
$$2\sin^2 x\cos x+\cos x=0$$
$$\cos x\left(2\sin^2 x+1\right)=0$$
$$\cos x=0\quad\text{or}\quad 2\sin^2 x+1=0$$
$$x=\frac{\pi}{2},\ \frac{3\pi}{2}\qquad 2\sin^2 x=-1$$

no real solution

17. $4\tan x - \sec^2 x = 0$

$4\tan x - (1 + \tan^2 x) = 0$

$4\tan x - 1 - \tan^2 x = 0$

$\tan^2 x - 4\tan x + 1 = 0$

Let $y = \tan x$

$y^2 - 4y + 1 = 0$

$y = \dfrac{4 \pm \sqrt{16 - 4(1)(1)}}{2}$

$= 2 \pm \sqrt{3}$

$\tan x = 2 + \sqrt{3}$ or $\tan x = 2 - \sqrt{3}$

$x = 1.31, 4.45$ $\quad x = 0.260, 3.40$

21. $\tan x + 1 = 0$

$\tan x = -1$

$x_{\text{ref}} = \dfrac{\pi}{4}$

(tan negative QII, QIV)

$x = \pi - \dfrac{\pi}{4} = \dfrac{3\pi}{4}$

$x = 2\pi - \dfrac{\pi}{4} = \dfrac{7\pi}{4}$

25. $4 - 3\csc^2 x = 0$

$4\sin^2 x - 3 = 0$

$4\sin^2 x = 3$

$\sin^2 x = \dfrac{3}{4};$

$\sin x = \pm\dfrac{\sqrt{3}}{2}$

$x_{\text{ref}} = \dfrac{\pi}{3}$

(sin positive or negative—all quadrants)

$x = \dfrac{\pi}{3}, \dfrac{2\pi}{3}, \dfrac{4\pi}{3}, \dfrac{5\pi}{3}$

29. $2\sin x - \tan x = 0$

$2\sin x - \dfrac{\sin x}{\cos x} = 0$

$\sin x \quad 2 - \dfrac{1}{\cos x} \quad = 0$

$\sin x = 0$ or $2 - \dfrac{1}{\cos x} = 0$

$x = 0, \pi$ $\quad \cos x = \dfrac{1}{2}$

$x = \dfrac{\pi}{3}, \dfrac{5\pi}{3}$

33. $\tan x + 3\cot x = 4$

$\tan x + \dfrac{3}{\tan x} = 4$

$\tan^2 x + 3 = 4\tan x$

$\tan^2 x - 4\tan x + 3 = 0;$

$(\tan x - 1)(\tan x - 3) = 0$

$\tan x = 1$ $\quad \tan x = 3$

$x = \dfrac{\pi}{4}, \dfrac{5\pi}{4}$ $\quad x = 1.25, \pi + 1.25$

37. $2\sin 2x - \cos x \sin^3 x = 0$

$2(2\sin x\cos x) - \cos x\sin^3 x = 0$

$\sin x\cos x(4 - \sin^2 x) = 0$

$\sin x\cos x = 0$ $\quad 4 - \sin^2 x = 0$

$\sin x = 0 \quad \cos x = 0$ $\quad \sin x = \pm 2$

$x = 0, \pi$ $\quad x = \dfrac{\pi}{2}, \dfrac{3\pi}{2}$ $\quad$ no solution

41. $\sin\theta + \cos\theta + \tan\theta + \cot\theta + \sec\theta + \csc\theta = f(\theta) = 1$

θ is a positive acute angle $\Rightarrow 0 < \theta < \dfrac{\pi}{2}$ for which

$0 < \sin\theta < 1$, $0 < \cos\theta < 1$, $\tan\theta > 0$, $\cot\theta > 0$,

$\sec\theta > 1$, $\csc\theta > 1$ which implies $f(\theta) = \sin\theta$

$+ \cos\theta + \tan\theta + \cot\theta + \sec\theta + \csc\theta > 1 \neq 1$.

$f(\theta) = 1$ has no solution.

45. Let the three sides be $a, 2a, b$ and the angles opposite to the sides be $\theta, \theta+60°, 120° - 2\theta,$ respectively.

By the law of sines,

$$\frac{\sin\theta}{a} = \frac{\sin(\theta+60°)}{2a}$$

$$2\sin\theta = \sin\theta\cos 60° + \cos\theta\sin 60°$$

$$\frac{3}{2}\sin\theta = \frac{\sqrt{3}}{2}\cos\theta$$

$$\tan\theta = \frac{\sqrt{3}}{3}$$

$$\theta = 30°$$

The other two angles are $\theta+60° = 90°$ and $120° - 2\theta = 60°$.

49. $y = 2.30\cos 0.1t - 1.35\sin 0.2t$

$$2.30\cos 0.1t - 1.35\sin 0.2t = 0$$

$$2.30\cos 0.1t - 1.35(2\sin 0.1t\cos 0.1t) = 0$$

$$\cos 0.1t(2.30 - 2.70\sin 0.1t) = 0$$

$\cos 0.1t = 0$ or $\sin 0.1t = 2.30/2.70$

For $\cos 0.1t = 0$,

$0.1t = \frac{\pi}{2}$ or $0.1t = \frac{3\pi}{2}$

$t = 15.7$ s $\quad t = 47.1$ s

For $\sin 0.1t = 2.30/2.70$,

$0.1t = 1.0195$ or $0.1t = \pi - 1.0195$

$t = 10.2$ s $\quad t = 21.2$ s

53.

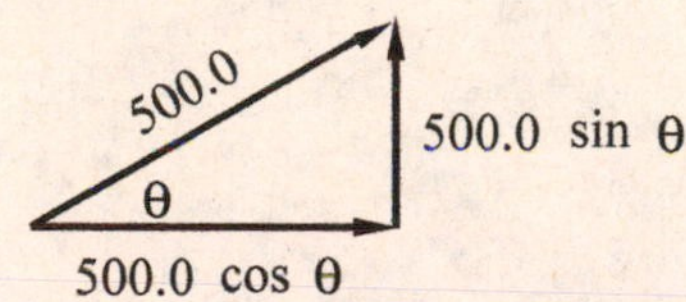

$$500.0\cos\theta + 500.0\sin\theta = 700.0$$

$$\sqrt{1-\sin^2\theta} + \sin\theta = 1.400$$

$$\sqrt{1-\sin^2\theta} = 1.400 - \sin\theta \text{ (square both sides)}$$

$$1 - \sin^2\theta = 1.400^2 - 2.800\sin\theta + \sin^2\theta$$

$$2\sin^2\theta - 2.800\sin\theta + 0.96 = 0$$

$$\sin^2\theta - 1.4\sin\theta + 0.48 = 0$$

Let $y = \sin\theta$

$$y^2 - 1.4y + 4.8 = 0$$

$$(y-0.6)(y-0.8) = 0$$

$y = 0.6$ or $y = 0.8$

$\sin\theta = 0.6$ $\quad$ $\sin\theta = 0.8$

$\cos\theta = \sqrt{1-0.6^2}$ $\quad$ $\cos\theta = \sqrt{1-0.8^2}$

$= 0.8$ $\quad$ $= 0.6$

The components are

$500.0(0.6) = 300$ N

and

$500(0.8) = 400$ N.

57. $2\sin 2x = x^2 + 1;\ -x^2 + 2\sin 2x - 1 = 0$

Graph $y_1 = -x^2 + 2\sin 2x - 1$. Use the Zero feature to solve. $x = 0.29,\ 0.95$.

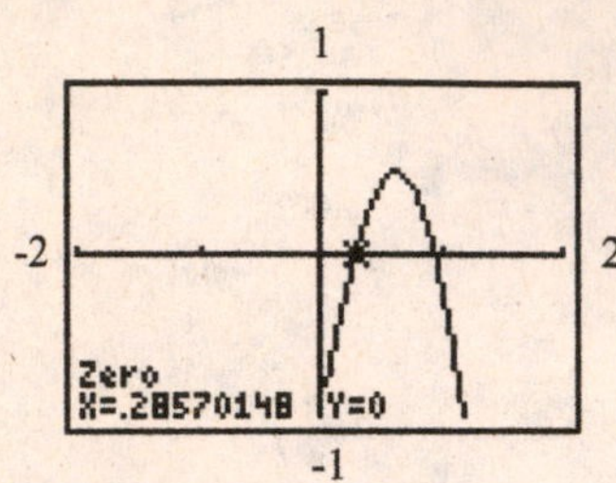

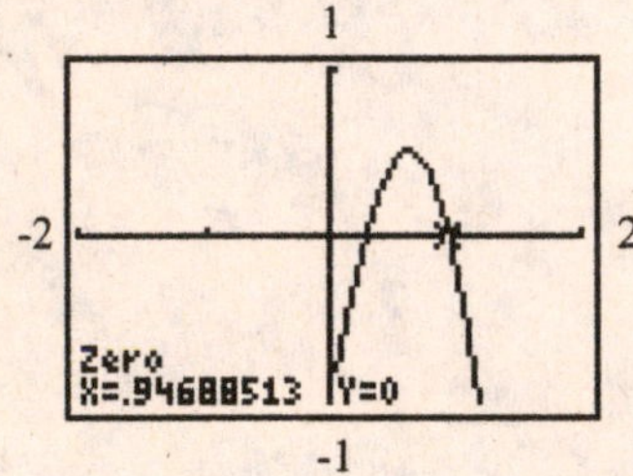

61. $x\tan x = 2.00,\ 0 < x < \dfrac{\pi}{2};\ x\tan x - 2.00 = 0$

Graph $y_1 = x\tan x - 2.00$. Use the Zero feature to solve.

$x = 1.08$.

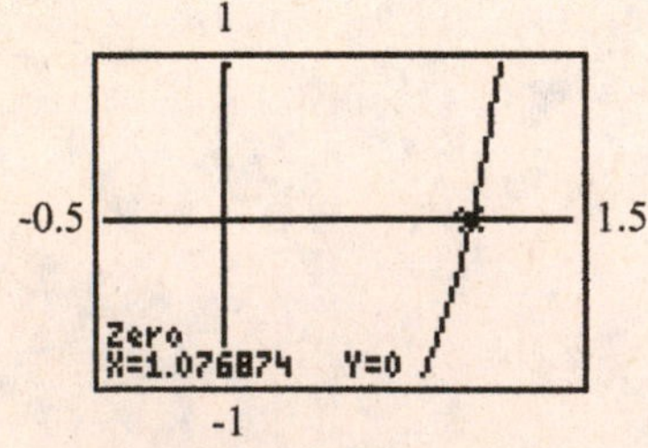

20.6 The Inverse Trigonometric Functions

1. $y = \tan^{-1} 3A$ is read as "y is the angle whose tangent is $3A$." In this case, $3A = \tan y$.

5. y is the angle whose tangent is $5x$.

9. y is five times the angle whose cosine is $2x-1$.

13. $\tan^{-1} 1 = \frac{\pi}{4}$ since $\tan \frac{\pi}{4} = 1$ and $-\frac{\pi}{2} < \frac{\pi}{4} < \frac{\pi}{2}$

17. Let $\sec^{-1} 0.5 = x$, then

$$\sec x = 0.5$$
$$\frac{1}{\cos x} = 0.5$$
$$\cos x = 2$$

and since $-1 \le \cos x \le 1$ there is no value for x.

21. $\sin\left(\tan^{-1}\sqrt{3}\right) = \sin\frac{\pi}{3} = \frac{1}{2}\sqrt{3}$

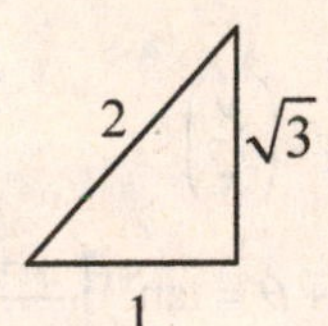

25. $\cos\left[\tan^{-1}(-5)\right] = \frac{1}{\sqrt{26}}$

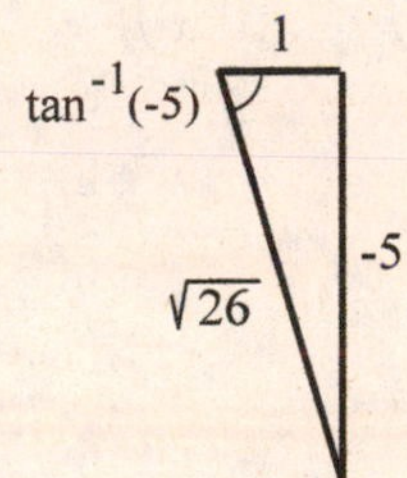

29. $\tan^{-1}(-2.8229) = -1.2303$

Note $-\frac{\pi}{2} < -1.2303 < \frac{\pi}{2}$

33. $\tan\left[\cos^{-1}(-0.6281)\right] = \tan 2.250 = -1.239$

37.
$$y = \sin 3x$$
$$3x = \sin^{-1} y$$
$$x = \frac{1}{3}\sin^{-1} y$$

41.
$$1 - y = \cos^{-1}(1-x)$$
$$\cos(1-y) = 1-x$$
$$x = 1 - \cos(1-y)$$

45. $\sin\left(\sec^{-1}\frac{x}{4}\right) = \sin\theta = \frac{\sqrt{x^2-16}}{x}$.

In a triangle, θ is set up such that its sec is $\frac{x}{4}$. This gives an adjacent side 4, hypotenuse x, and opposite side $\sqrt{x^2-16}$. Therefore, the sine is the opposite $\sqrt{x^2-16}$ over the hypotenuse.

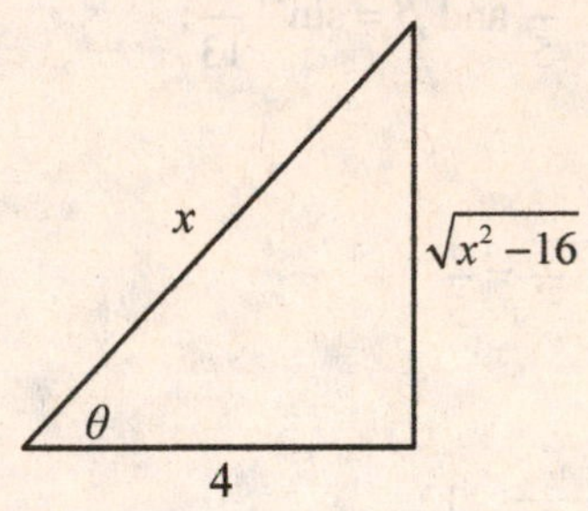

49.
$$\sin\left(2\sin^{-1} x\right) = \sin 2\theta = 2\sin\theta\cos\theta$$
$$= 2\left(\frac{x}{1}\right)\left(\frac{\sqrt{1-x^2}}{1}\right)$$
$$= 2x\sqrt{1-x^2}$$

In a triangle, θ is set up such that its sine is x. This gives an opposite side of x, hypotenuse 1, and adjacent side $\sqrt{1-x^2}$.

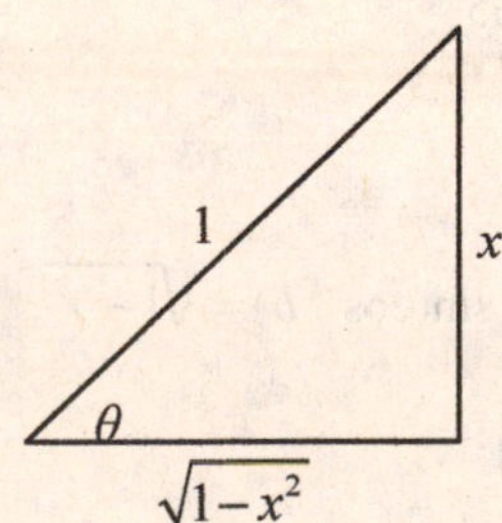

53.

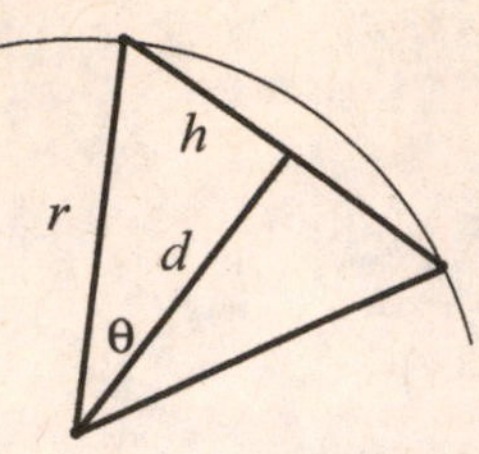

$$r^2 = h^2 + d^2 \Rightarrow h = \sqrt{r^2 - d^2}$$

$$\cos\theta = \frac{d}{r} \Rightarrow \theta = \cos^{-1}\frac{d}{r}$$

$$A_{\text{segment}} = A_{\text{sector}} - A_{\text{triangle}}$$

$$A_{\text{segment}} = \frac{1}{2}r^2(2\theta) - 2\left(\frac{1}{2}dh\right)$$

$$A_{\text{segment}} = r^2\theta - d\sqrt{r^2 - d^2} = r^2\cos^{-1}\frac{d}{r} - d\sqrt{r^2 - d^2}$$

57. Let $\alpha = \sin^{-1}\frac{3}{5}$ and $\beta = \sin^{-1}\frac{5}{13}$;

$$\sin\alpha = \frac{3}{5}$$

$$\cos\alpha = \sqrt{1 - \tfrac{9}{25}} = \frac{4}{5}$$

$$\sin\beta = \frac{5}{13}$$

$$\cos\beta = \sqrt{1 - \tfrac{25}{169}} = \frac{12}{13}$$

$$\sin^{-1}\frac{3}{5} + \sin^{-1}\frac{5}{13} = \alpha + \beta$$

$$\sin(\alpha+\beta) = \sin\alpha\cos\beta + \cos\alpha\sin\beta$$

$$= \frac{3}{5}\left(\frac{12}{13}\right) + \frac{4}{5}\left(\frac{5}{13}\right)$$

$$= \frac{36}{65} + \frac{20}{65} = \frac{56}{65}$$

Since $\sin(\alpha+\beta) = \frac{56}{65}$,

$$\alpha + \beta = \sin^{-1}\frac{56}{65}.$$

61. Using $\cos(\sin^{-1} b) = \sin(\cos^{-1} b) = \sqrt{1 - b^2}$,

$$\sin(\sin^{-1} x + \cos^{-1} y)$$

$$= \sin(\sin^{-1} x)\cos(\cos^{-1} y) + \cos(\sin^{-1} x)\sin(\cos^{-1} y)$$

$$= xy + \sqrt{1 - x^2}\sqrt{1 - y^2}$$

65. $y = \sin^{-1} x + \sin^{-1}(-x)$

$$\sin y = \sin(\sin^{-1} x + \sin^{-1}(-x))$$

$$\sin y = \sin(\sin^{-1} x)\cos(\sin^{-1}(-x)) + \cos(\sin^{-1} x)\sin(\sin^{-1}(-x))$$

$$= x\cos(\sin^{-1} x) - x\cos(\sin^{-1} x)$$

$$= 0$$

Here we have used the fact that if $-\frac{\pi}{2} \le x \le \frac{\pi}{2}$, $\cos(\sin^{-1}(-x)) = \cos(\sin^{-1} x)$.

69. $\tan B = \frac{y}{b} \Rightarrow y = b\tan B$

$$\tan A = \frac{y}{a} = \frac{b\tan B}{a}$$

$$A = \tan^{-1}\frac{b\tan B}{a}$$

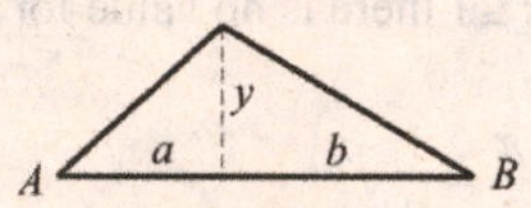

73. $\tan\alpha = \frac{y}{x} \Rightarrow a = \tan^{-1}\left(\frac{y}{x}\right)$

$$\tan(\alpha+\theta) = \frac{y+50}{x} \Rightarrow \alpha + \theta = \tan^{-1}\left(\frac{y+50}{x}\right)$$

$$\tan^{-1}\left(\frac{y}{x}\right) + \theta = \tan^{-1}\left(\frac{y+50}{x}\right)$$

$$\theta = \tan^{-1}\left(\frac{y+50}{x}\right) - \tan^{-1}\left(\frac{y}{x}\right)$$

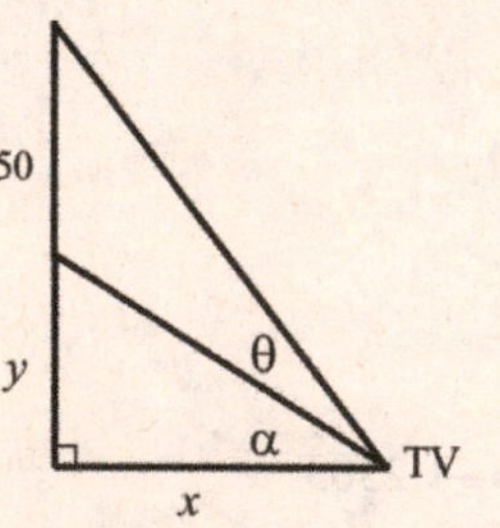

Review Exercises

1. This is false. The definition of the tangent function yields $\tan 2\theta = \dfrac{\sin 2\theta}{\cos 2\theta}$.

5. This is false because the range of the inverse sine function is $-\dfrac{\pi}{2} \le \sin^{-1} x \le \dfrac{\pi}{2}$.

9.
$$\begin{aligned}\sin 315^\circ &= \sin(360^\circ - 45^\circ)\\ &= \sin 360^\circ \cos 45^\circ - \sin 45^\circ \cos 360^\circ\\ &= 0\left(\frac{\sqrt{2}}{2}\right) - \frac{\sqrt{2}}{2}(1)\\ &= -\frac{\sqrt{2}}{2}\end{aligned}$$

13.
$$\begin{aligned}\tan 60^\circ &= \tan(2(30^\circ))\\ &= \frac{2\tan 30^\circ}{1-\tan^2 30^\circ}\\ &= \frac{2\left(\frac{1}{\sqrt{3}}\right)}{1-\left(\frac{1}{\sqrt{3}}\right)^2}\\ &= \frac{3}{\sqrt{3}}\\ &= \sqrt{3}\end{aligned}$$

17.
$$\begin{aligned}2\sin\frac{\pi}{14}\cos\frac{\pi}{14} &= \sin\left(2\left(\frac{\pi}{14}\right)\right)\\ &= \sin\left(\frac{\pi}{7}\right)\\ 0.434 &= 0.434\end{aligned}$$

21.
$$\begin{aligned}\frac{2\tan 18^\circ}{1-\tan^2 18^\circ} &= \tan(2(18^\circ))\\ &= \tan 36^\circ\\ 0.7265 &= 0.7265\end{aligned}$$

25.
$$\begin{aligned}4\sin 7x\cos 7x &= 2(2\sin 7x\cos 7x)\\ &= 2\sin 14x\end{aligned}$$

29.
$$\begin{aligned}\sqrt{2+2\cos 2x} &= \sqrt{\frac{4(1+\cos 2x)}{2}}\\ &= 2\sqrt{\frac{1+\cos 2x}{2}}\\ &= 2\cos\frac{2x}{2}\\ &= 2\cos x\end{aligned}$$

33. $\cos^{-1} 0.8629 = 0.5298$ since $\cos 0.5298 = 0.8629$

37.
$$\begin{aligned}\sin^{-1}\left[\sin\left(\frac{7\pi}{6}\right)\right] &= \sin^{-1}\left(-\frac{1}{2}\right)\\ &= -\frac{\pi}{6}\end{aligned}$$

41.
$$\begin{aligned}\sin x(\csc x - \sin x) &= \sin x\left(\frac{1}{\sin x} - \sin x\right)\\ &= 1-\sin^2 x\\ &= \cos^2 x,\ x \ne 0 + n\pi\end{aligned}$$

45.
$$\begin{aligned}2\csc 2x\cot x &= \frac{2}{\sin 2x}\frac{\cos x}{\sin x}\\ &= \frac{2\cos x}{2\sin x\cos x\sin x}\\ &= \frac{1}{\sin^2 x}\\ &= \csc^2 x\\ &= 1+\cot^2 x\end{aligned}$$

49.
$$\begin{aligned}\sin\frac{\theta}{2}\cos\frac{\theta}{2} &= \frac{1}{2}\left(2\sin\frac{\theta}{2}\cos\frac{\theta}{2}\right)\\ &= \frac{1}{2}\sin\theta\end{aligned}$$

53.
$$\begin{aligned}\sin x\tan x + \cos x &= \sin x\frac{\sin x}{\cos x} + \cos x\\ &= \frac{\sin^2 x + \cos^2 x}{\cos x}\\ &= \frac{1}{\cos x}\\ &= \sec x\end{aligned}$$

57.
$$\begin{aligned}\frac{\sin 2x\sec x}{2} &= \frac{2\sin x\cos x\frac{1}{\cos x}}{2}\\ &= \sin x\end{aligned}$$

61.

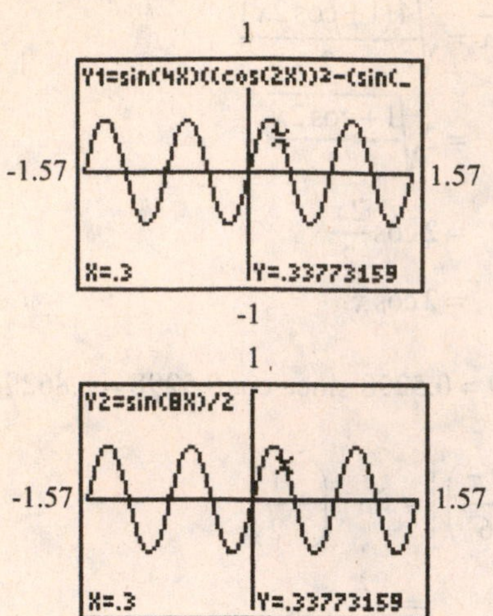

65.

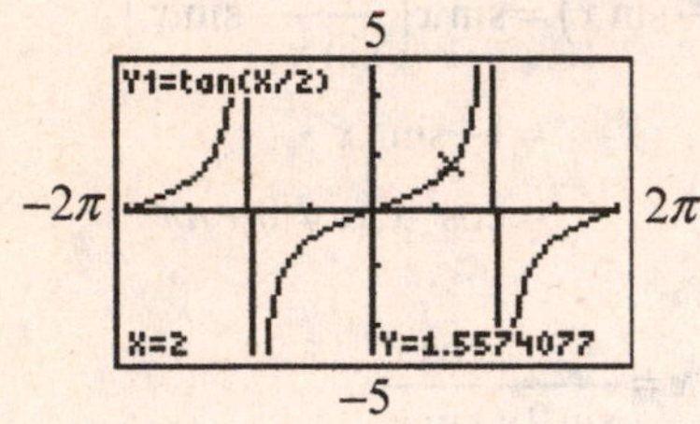

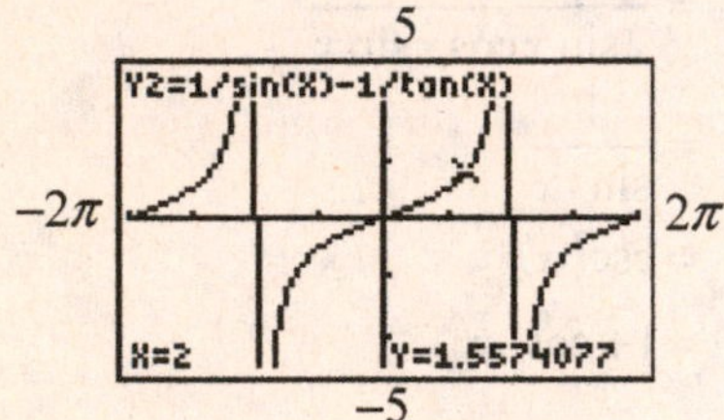

69.

$$y = \frac{\pi}{4} - 3\sin^{-1} 5x$$

$$\sin^{-1} 5x = \frac{\frac{\pi}{4} - y}{3}$$

$$5x = \sin\frac{\frac{\pi}{4} - y}{3}$$

$$x = \frac{1}{5}\sin\frac{\frac{\pi}{4} - y}{3}$$

73. $2\left(1 - 2\sin^2 x\right) = 1$

$$2 - 4\sin^2 x = 1$$

$$\sin^2 x = \frac{1}{4}$$

$$\sin x = \frac{1}{2} \quad \text{or} \quad \sin x = -\frac{1}{2}$$

$$x = \frac{\pi}{6}, \frac{5\pi}{6} \qquad x = \frac{7\pi}{6}, \frac{11\pi}{6}$$

77.

$$\sin x = \sin\frac{x}{2}$$

$$2\sin\frac{x}{2}\cos\frac{x}{2} = \sin\frac{x}{2}$$

$$\sin\frac{x}{2} \quad 2\cos\frac{x}{2} - 1 \quad = 0$$

$$0 \le x < 2\pi \qquad 0 \quad \frac{x}{2} \quad \pi$$

$$\sin\frac{x}{2} = 0 \text{ or } 2\cos\frac{x}{2} - 1 = 0$$

$$\frac{x}{2} = 0 \qquad \cos\frac{x}{2} = \frac{1}{2}$$

$$x = 0 \qquad \frac{x}{2} = \frac{\pi}{3}$$

$$x = \frac{2\pi}{3}$$

81.

$$\sin^2\frac{x}{2} - \cos x + 1 = 0$$

$$\frac{1 - \cos x}{2} - \cos x + 1 = 0$$

$$1 - \cos x - 2\cos x + 2 = 0$$

$$3\cos x = 3$$

$$\cos x = 1$$

$$x = 0$$

85. $\sin x \cos x - 1 = \cos x - \sin x$

$$\sin x \cos x - \cos x - 1 + \sin x = 0$$

$$\cos x\left(\sin x - 1\right) + \left(\sin x + 1\right) = 0$$

$$\left(\sin x - 1\right)\left(\cos x + 1\right) = 0$$

$$\sin x = 1 \quad \text{or} \quad \cos x = -1$$

$$x = \frac{\pi}{2} \qquad x = \pi$$

Conditional equation.

89. $2\tan^{-1}x + x^2 = 3$

$2\tan^{-1}x + x^2 - 3 = 0$

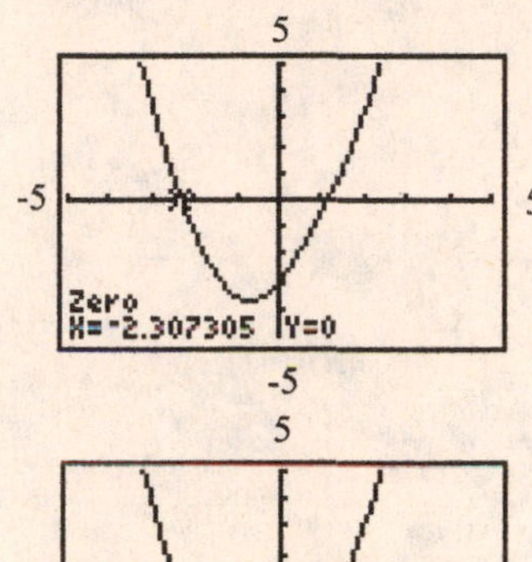

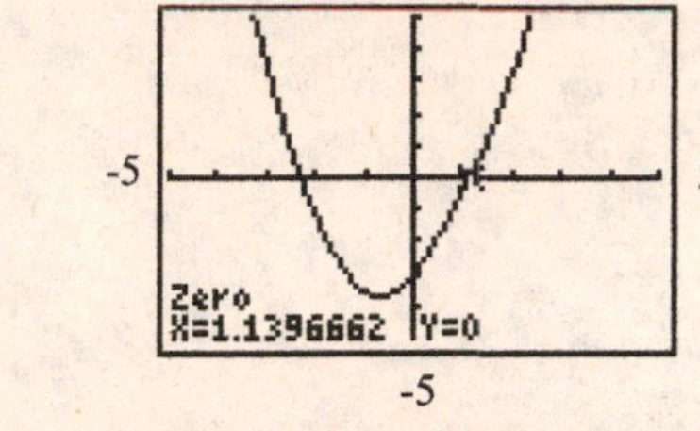

93. $\cos\left(\sin^{-1}\frac{x}{5}\right) = \frac{\sqrt{25-x^2}}{5}$

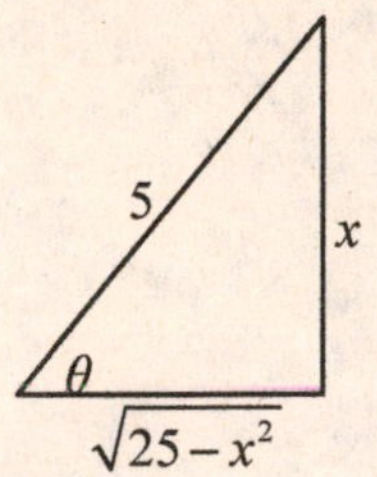

97. $\sqrt{4-x^2} = \sqrt{4-(2\cos\theta)^2}$

$= 2\sqrt{1-\cos^2\theta}$

$= 2\sqrt{\sin^2\theta}$

$= 2\sin\theta$

101. The hypotenuse is 5 and so

$\sin 2\theta = 2\sin\theta\cos\theta$

$= 2\cdot\frac{3}{5}\cdot\frac{4}{5}$

$= \frac{24}{25}$

105. $(\cos\theta + j\sin\theta)^2$

$= \cos^2\theta + 2j\sin\theta\cos\theta + j^2\sin^2\theta$

$= \cos^2\theta - \sin^2\theta + 2j\sin\theta\cos\theta$

$= \cos 2\theta + j\sin 2\theta$

109. $\sin 2x > 2\cos x, \quad 0 \le x < 2\pi$

$2\sin x\cos x - 2\cos x > 0$

$\cos x(\sin x - 1) > 0$, there are two cases

I $\quad \cos x > 0$ and $\sin x - 1 > 0$

$0 < x < \frac{\pi}{2}$ $\qquad$ $\sin x > 1$, no solution

$\frac{3\pi}{2} < x < 2\pi$

II $\quad \cos x < 0$ and $\sin x - 1 < 0$

$\frac{\pi}{2} < x < \frac{3\pi}{2}$ $\qquad$ $\sin x < 1$

$0 < x < \frac{\pi}{2}$ or $\frac{\pi}{2} < x < 2\pi$

The solution is $\frac{\pi}{2} < x < \frac{3\pi}{2}$

113. $\sin 2A = 2\sin A\cos A$

$= 2\cdot\frac{a}{c}\cdot\frac{b}{c}$

$= \frac{2ab}{c^2}$

117. $r = \frac{k}{2}\csc^2\frac{\theta}{2}$

$= \frac{k}{2\sin^2\frac{\theta}{2}}$

$r = \frac{k}{2\,\frac{1-\cos\theta}{2}}$

$r = \frac{k}{1-\cos\theta}$

121. $(2\cos\alpha\cos\beta)^{-1} - \tan\alpha\tan\beta$

$= (2\cos^2\alpha)^{-1} - \tan^2\alpha$ for $\alpha = \beta$

$= \frac{1}{2\cos^2\alpha} - \frac{2\sin^2\alpha}{2\cos^2\alpha}$

$= \frac{1-2\sin^2\alpha}{2\cos^2\alpha}$

$= \frac{\cos 2\alpha}{2\cos^2\alpha}$

125.

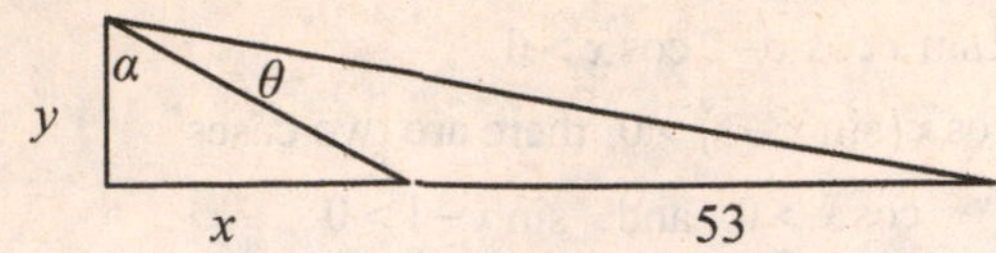

(a) $\tan\alpha = \dfrac{x}{y}$

$$\tan(\alpha+\theta) = \frac{x+53}{y}$$

$$\tan(\alpha+\theta) = \frac{\tan\alpha + \tan\theta}{1-\tan\alpha\tan\theta}$$

$$\frac{x+53}{y} = \frac{\frac{x}{y}+\tan\theta}{1-\frac{x}{y}\tan\theta} = \frac{x+y\tan\theta}{y-x\tan\theta}$$

$$(x+53)(y-x\tan\theta) = y(x+y\tan\theta)$$

$$xy - x^2\tan\theta + 53y - 53x\tan\theta = xy + y^2\tan\theta$$

$$(x^2+y^2+53x)\tan\theta = 53y$$

$$\tan\theta = \frac{53y}{x^2+y^2+53x}$$

$$\theta = \tan^{-1}\ \frac{53y}{x^2+y^2+53x}$$

(b) For $x = 28, y = 12$

$$\theta = \tan^{-1}\left(\frac{53\cdot 12}{28^2+12^2+53\cdot 28}\right)$$
$$= 14.8°$$

129. $2\sin\theta\cos^2\theta - \sin^3\theta = 0, \quad 0 < \theta < 90^\circ$

$$\sin\theta\left(2\cos^2\theta - \sin^2\theta\right) = 0$$

$$\sin\theta\left(2\cos^2\theta - \left(1-\cos^2\theta\right)\right) = 0$$

$$\sin\theta\left(3\cos^2\theta - 1\right) = 0$$

$\sin\theta = 0$ or $\cos^2\theta = \dfrac{1}{3}$

no solution $\quad \cos\theta = \sqrt{\dfrac{1}{3}}$

$$\theta = 54.7^\circ$$

Chapter 21

Plane Analytic Geometry

21.1 Basic Definitions

1. The distance between $(3,-1)$ and $(-2,5)$ is

$$d=\sqrt{(3-(-2))^2+(-1-5)^2}$$
$$=\sqrt{61}$$

5. Given $(x_1, y_1)=(3,8); (x_2, y_2)=(-1,-2)$

$$d=\sqrt{(x_2-x_1)^2+(y_2-y_1)^2}$$
$$=\sqrt{(-1-3)^2+(-2-8)^2}$$
$$=\sqrt{(-4)^2+(-10)^2}$$
$$=\sqrt{16+100}$$
$$=\sqrt{116}$$
$$=\sqrt{4\times 29}$$
$$=2\sqrt{29}$$

9. Given $(x_1, y_1)=(-12,20); (x_2, y_2)=(32,-13)$

$$d=\sqrt{(x_2-x_1)^2+(y_2-y_1)^2}$$
$$=\sqrt{(32+12)^2+(-13-20)^2}$$
$$=\sqrt{(44)^2+(-33)^2}$$
$$=\sqrt{1936+1089}$$
$$=\sqrt{3025}$$
$$=55$$

13. Given $(x_1, y_1)=(1.22,-3.45)$;
$(x_2, y_2)=(-1.07,-5.16)$

$$d=\sqrt{(x_2-x_1)^2+(y_2-y_1)^2}$$
$$=\sqrt{(-1.07-1.22)^2+(-5.16-(-3.45))^2}$$
$$=\sqrt{(-2.29)^2+(-5.16+3.45)^2}$$
$$=\sqrt{(-2.29)^2+(-1.71)^2}$$
$$=\sqrt{8.1682}$$
$$=2.86$$

17. Given $(x_1, y_1)=(4,-5); (x_2, y_2)=(4,-8)$

$$m=\frac{y_2-y_1}{x_2-x_1}=\frac{-8-(-5)}{4-4}$$

Since $x_2-x_1=4-4=0$, the slope is undefined.

21. Given $(x_1, y_1)=\left(\sqrt{32}, \sqrt{18}\right)$;
$(x_2, y_2)=\left(-\sqrt{50}, \sqrt{8}\right)$

$$m=\frac{y_2-y_1}{x_2-x_1}=\frac{\sqrt{8}-\sqrt{18}}{-\sqrt{50}-\sqrt{32}}$$
$$=\frac{2\sqrt{2}-3\sqrt{2}}{-5\sqrt{2}-4\sqrt{2}}$$
$$=\frac{-\sqrt{2}}{-9\sqrt{2}}$$
$$=\frac{1}{9}$$

25. Given $\alpha=30^\circ$; $m=\tan\alpha$, $0^\circ<\alpha<180^\circ$

$$\tan 30^\circ=\frac{\sqrt{3}}{3}$$

29. Given $m=0.364$; $m=\tan\alpha$; $0.364=\tan\alpha$;
$\alpha=20.0^\circ$

33. Given $(x_1, y_1)=(6,-1); (x_2, y_2)=(4,3)$
$(x_3, y_3)=(-5,2); (x_4, y_4)=(-7,6)$

$$m_1=\frac{y_2-y_1}{x_2-x_1}=\frac{3-(-1)}{4-6}=\frac{4}{-2}=-2$$
$$m_2=\frac{y_4-y_3}{x_4-x_3}=\frac{6-2}{-7-(-5)}=\frac{4}{-2}=-2$$

$m_1=m_2$ so the lines are parallel.

37. Given distance between $(-1, 3)$ and $(11, k)$ is 13.

$$d = \sqrt{(x_1 - x_2)^2 + (y_1 - y_2)^2}$$
$$13 = \sqrt{(-1-11)^2 + (3-k)^2}$$
$$= \sqrt{(-12)^2 + (3-k)^2}$$
$$169 = 144 + (3-k)^2$$
$$(3-k)^2 = 25$$
$$3 - k = \pm 5$$
$$-k = -3 \pm 5$$
$$k = -2, 8$$

41. $$d_1 = \sqrt{(9-7)^2 + [4-(-2)]^2}$$
$$= \sqrt{2^2 + 6^2} = \sqrt{40} = 2\sqrt{10}$$
$$d_2 = \sqrt{(9-3)^2 + (4-2)^2} = \sqrt{6^2 + 2^2}$$
$$= \sqrt{40} = 2\sqrt{10}$$

$d_1 = d_2$ so the triangle is isosceles.

45. $$d_1 = \sqrt{(3-5)^2 + (-1-3)^2} = \sqrt{(-2)^2 + (-4)^2}$$
$$= \sqrt{4+16} = \sqrt{20}$$
$$m_1 = \frac{y - y_1}{x - x_1} = \frac{5-3}{3-(-1)} = \frac{5-3}{3+1} = \frac{2}{4} = \frac{1}{2}$$
$$d_2 = \sqrt{(5-1)^2 + (3-5)^2} = \sqrt{(4)^2 + (-2)^2}$$
$$= \sqrt{16+4} = \sqrt{20}$$
$$m_2 = \frac{y - y_1}{x - x_1} = \frac{5-1}{3-5} = \frac{4}{-2} = -2$$
$$m_1 = \frac{-1}{m_2},\ m_1 \perp m_2$$
$$A = \frac{1}{2}d_1 d_2 = \frac{1}{2}\sqrt{20}\sqrt{20} = \frac{1}{2}(20) = 10$$

49. $$\left(\frac{-4+6}{2}, \frac{9+1}{2}\right) = \left(\frac{2}{2}, \frac{10}{2}\right) = (1, 5)$$

53. The distance between (x, y) and $(0, 0) = 3$.

$$\sqrt{(x-0)^2 + (y-0)^2} = 3$$
$$x^2 + y^2 = 9$$

57. $$m = \frac{y_2 - y_1}{x_2 - x_1} = \frac{5-0}{-2-x} = 3$$
$$x = -\frac{11}{3}$$

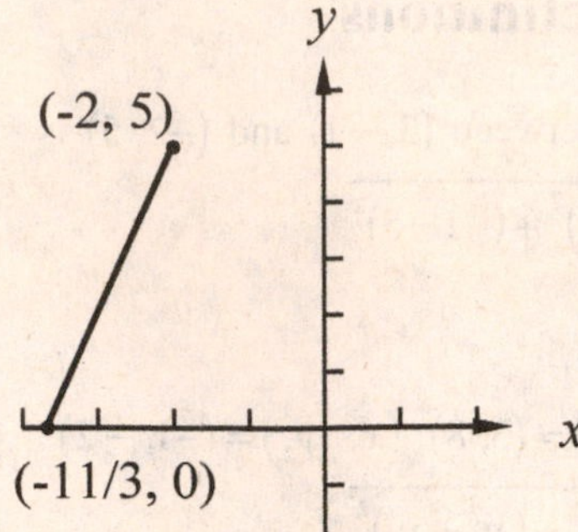

61. We compute the distances:

$$d_1 = \sqrt{(-1-2)^2 + (6-1)^2}$$
$$= \sqrt{9+25}$$
$$= \sqrt{34}$$
$$d_2 = \sqrt{(5-2)^2 + (5-1)^2}$$
$$= \sqrt{9+16}$$
$$= 5$$

The two distances are different and so the points are not equidistant from (2,1).

65. Treating Seattle as the origin (0,0), Denver is located at (1350, −900) and Edmonton is located at (620, 640).

The distance between them is

$$d = \sqrt{(1350-620)^2 + (-900-640)^2}$$
$$= 1704.25937 \text{ km}$$

Denver and Edmonton are 1700 km apart.

21.2 The Straight Line

1. $m = -1/2, (x_1, y_1) = (4, -1)$

$$y - y_1 = m(x - x_1)$$
$$y - (-1) = -\frac{1}{2}(x - 4)$$
$$y + 1 = -\frac{1}{2}x + 2$$
$$2y + 2 = -x + 4$$
$$2y - x - 2 = 0$$

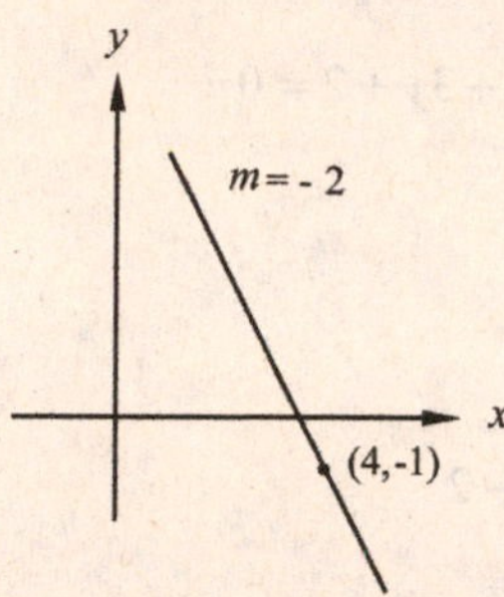

5. Given $m = 4; (x_1, y_1) = (-3, 8)$

$$y - y_1 = m(x - x_1)$$
$$y - 8 = 4[x - (-3)]$$
$$= 4(x + 3)$$
$$= 4x + 12$$
$$y = 4x + 20 \text{ or } 4x - y + 20 = 0$$

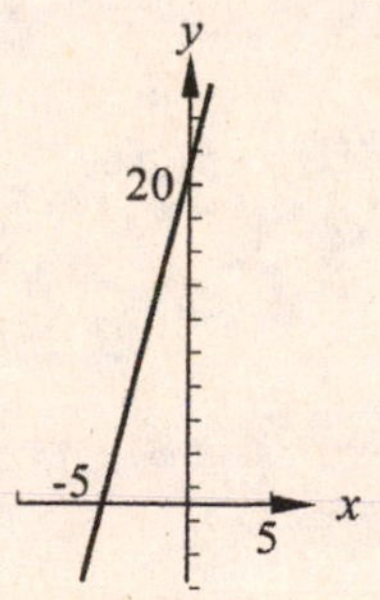

9. Given $(x_1, y_1) = (-7, 12)$ $\alpha = 45^\circ$

$$m = \tan\alpha = \tan 45^\circ = 1$$
$$y - y_1 = m(x - x_1)$$
$$y - 12 = 1(x + 7)$$
$$= x + 7$$
$$y = x + 19 \text{ or } x - y + 19 = 0$$

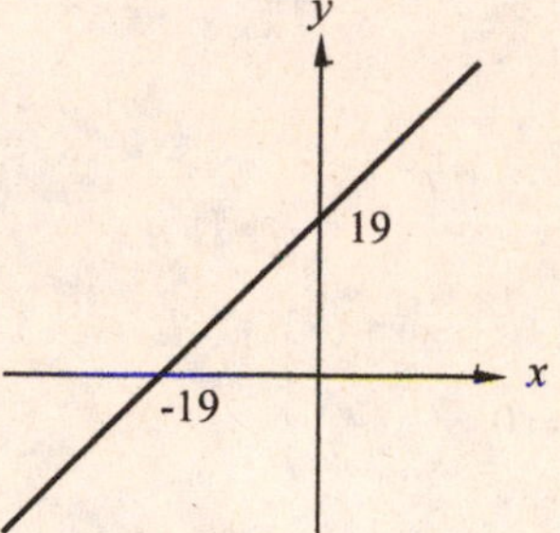

13. Parallel to y-axis and 3 units left of y-axis.

$$x = -3$$

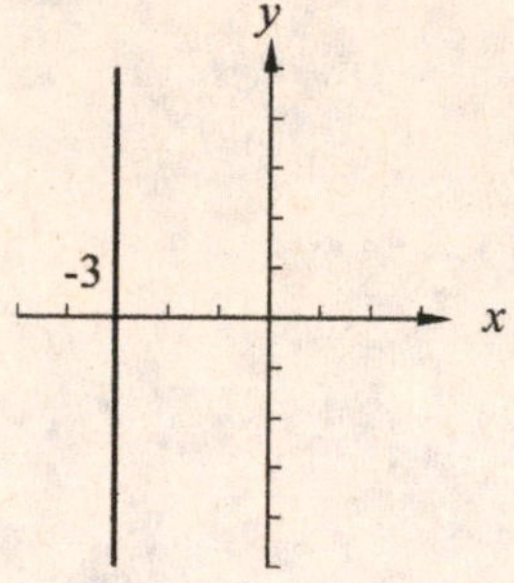

17. $\frac{\alpha - 2}{0 - 5} = \frac{2 - 0}{5 - \alpha}$

$$\alpha^2 - 7\alpha = 0$$
$$\alpha(\alpha - 7) = 0$$
$$\alpha = 7$$
$$y - y_1 = m(x - x_1)$$
$$y - 2 = \frac{7 - 2}{-5}(x - 5)$$
$$y = -x + 7 \text{ or } x + y - 7 = 0$$

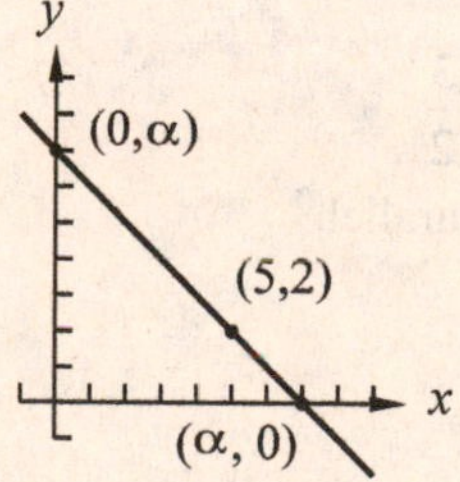

21. Given $4x - y = 8$,

$y = 4x - 8$, so $m = 4$, $b = -8$

When $x = 0$, $y = -8$

$y = 0$, $x = 2$

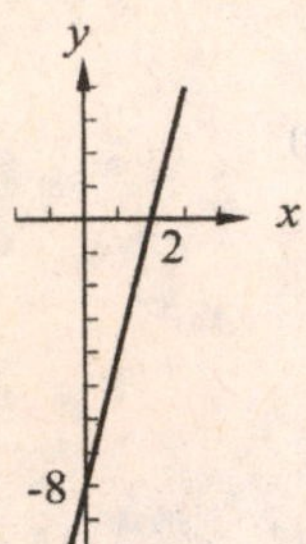

25. Given $3x - 2y - 1 = 0$

$$y = \frac{-3}{-2}x + \frac{1}{-2};$$

$$y = \frac{3}{2}x - \frac{1}{2}, \text{ so } m = \frac{3}{2}, b = -\frac{1}{2}$$

When $x = 0, y = -\frac{1}{2}$

$$y = 0, x = \frac{1}{3}$$

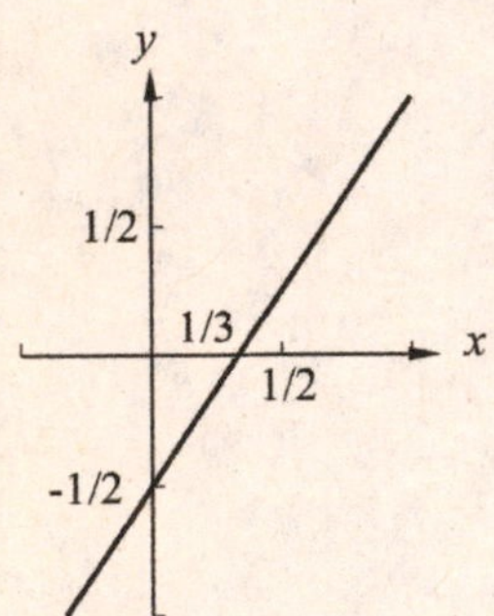

29. $3x - 2y + 5 = 0;$

$-2y = -3x - 5$

$$y = \frac{-3}{-2}x + \frac{-5}{-2}$$

$$y = \frac{3}{2}x + \frac{5}{2}; \text{slope } m_1 = \frac{3}{2}$$

$4y = 6x - 1$

$$y = \frac{6}{4}x - \frac{1}{4}; \text{ slope } m_2 = \frac{3}{2}$$

$m_1 = m_2$ so the lines are parallel.

33. $5x + 2y = 3$

$$y = \frac{-5}{2}x + \frac{3}{2}, m_1 = -\frac{5}{2}$$

$10y = 7 - 4x$

$$y = \frac{-4}{10}x + \frac{7}{10}, m_2 = -\frac{2}{5}$$

$$m_1 \cdot m_2 = \frac{-5}{2} \cdot \frac{-2}{5} = 1 \neq -1$$

$m_1 \neq m_2$

The lines are neither perpendicular nor parallel.

37. Given: $4x - ky = 6 \| 6x + 3y + 2 = 0$

$6x + 3y + 2 = 0$

$3y = -6x - 2$

$$y = \frac{-6}{3}x - \frac{2}{3}$$

$$y = -2x - \frac{2}{3}; \text{ slope is } -2$$

$4x - ky = 6$

$-ky = -4x + 6$

$$y = \frac{-4}{-k}x + \frac{6}{-k}$$

$$y = \frac{4}{k}x - \frac{6}{k}; \text{ slope is } \frac{4}{k}$$

Since the lines are parallel, the slopes are equal.

$$\frac{4}{k} = -2$$

$k = -2$

41. $4x - 3y + 12 = 0$

$3y = 4x + 12$

(1) $\quad y = \frac{4}{3}x + 4$

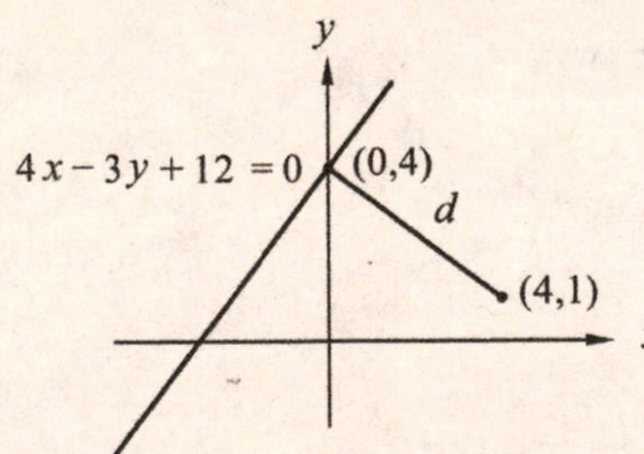

The perpendicular line from $(4, 1)$ to the line $4x - 3y + 12 = 0$ has $m = -\frac{3}{4}$ and equation

$$y - 1 = -\frac{3}{4}(x - 4)$$

$4y - 4 = -3x + 12$

$3x + 4y = 16$, substituting for y from (1)

$$3x+4\left(\frac{4}{3}x+4\right)=16$$
$$3x+\frac{16}{3}x+16=16$$
$$x=0$$
$$y=\frac{4}{3}(0)+4=4$$

The lines intersect at $(0,\ 4)$

$$d=\sqrt{(4-0)^2+(1-4)^2}=\sqrt{16+9}$$
$$d=5$$

45. **(a)** $ax+by+c=0 \quad y \quad \frac{a}{b}x \quad \frac{c}{b}$

$ax+by+d=0 \quad y \quad \frac{a}{b}x \quad \frac{d}{b}$

$m_1=m_2=-\frac{a}{b}$ lines are parallel

(b) $ax+by+c=0 \quad y \quad \frac{a}{b}x \quad \frac{c}{a}$

$bx-ay+d=0 \quad y \quad \frac{b}{a}x \quad \frac{d}{a}$

$m_1\cdot m_2=-\frac{a}{b} \quad \frac{b}{a} \quad =-1$

lines are perpendicular.

49. By expanding along the top row,

$$\begin{vmatrix} y & x & b \\ m & 1 & 0 \\ 1 & 0 & 1 \end{vmatrix}=y(1-0)-x(m-0)+b(0-1)$$
$$=y-mx-b$$

and the given equation reduces to

$$y-mx-b=0$$

or

$$y=mx+b,$$

the equation of a line in slope-intercept form.

53. $v=mT+b,$

An increase of 0.607 m/s for every increase of 1°C means that the slope is $m=0.607$. The line goes through the point (20.0,343). Therefore,

$$343=0.607(20)+b$$
$$b=331$$
$$v=0.607T+331$$

57. $t=0$ months: $w=30$ mg

Rate of growth is -2 mg/month

Therefore,

$$w=-2t+30$$

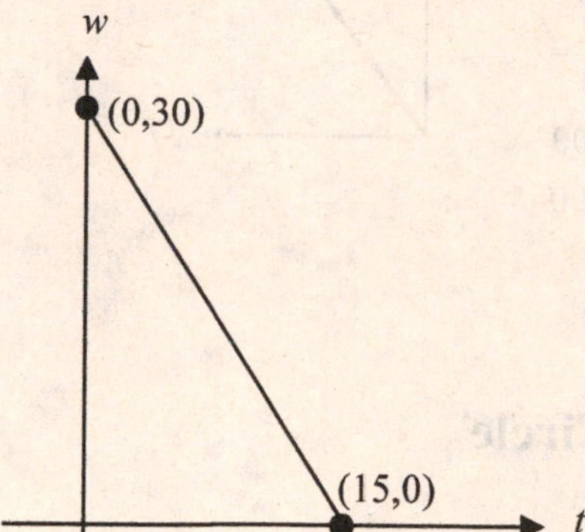

61. Start is 6:30 = 0 min

Therefore, $(t_1,\ n_1)=(30,\ 45)$; $(t_2,\ n_2)=(90,\ 115)$

$$m=\frac{n_2-n_1}{t_2-t_1}=\frac{115-45}{90-30}=\frac{7}{6}$$
$$n-n_1=m(t-t_1)$$
$$n-45=\frac{7}{6}(t-30)$$
$$n-45=\frac{7}{6}t-35$$
$$n=\frac{7}{6}t+10$$

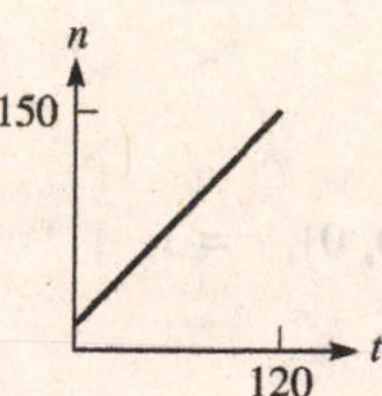

Therefore, n at 6:30 $(t=0)=10$;

n at 8:30$(t=120)=150$

The slope represents the increase in number of cars per minute for every minute increase between 6:30 and 8:30.

65. $n = 1200\sqrt{t} + 0$

$m = 1200$

$b = 0$

t	$\sqrt{t}$	h
0	0	0
1	1	1200
4	2	2400

n ; (4, 2400) ; $\sqrt{t}$

21.3 The Circle

1. $(x-1)^2 + (y+1)^2 = 16$

$(x-1)^2 + (y-(-1))^2 = 16$

has center at $(1, -1)$ and $r = 4$

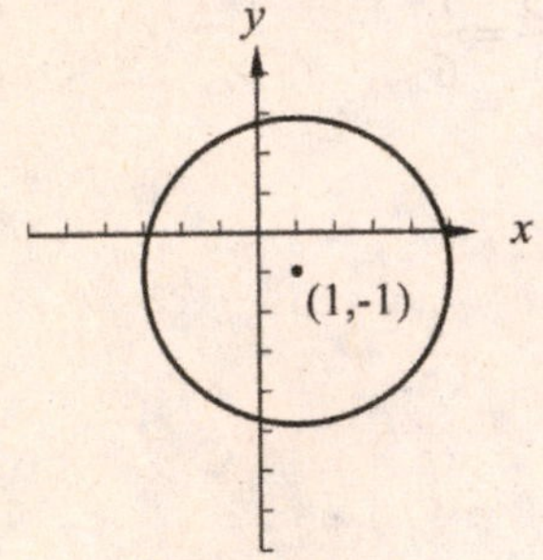

5. $(x-2)^2 + (y-1)^2 = 25$

$C(2, 1)$, radius is 5.

9. $(x-h)^2 + (y-k)^2 = r^2$; $C(0, 0)$, $r = 3$

$(x-0)^2 + (y-0)^2 = 3^2$

$x^2 + y^2 = 9$

$x^2 + y^2 - 9 = 0$

13. $(x-h)^2 + (y-k)^2 = r^2$; $C(12, -15)$, $r = 18$

$(x-12)^2 + (y-(-15))^2 = 18^2$

$(x-12)^2 + (y+15)^2 = 324$

$x^2 + y^2 - 24x + 30y + 45 = 0$

17. Concentric with $(x-2)^2 + (y-1)^2 = 4$ gives center at $(2, 1)$. The standard equation is

$(x-2)^2 + (y-1)^2 = r^2$

Since it is satisfied by $(4, -1)$ we get

$r = \sqrt{(4-2)^2 + (-1-1)^2}$

$= 2\sqrt{2}$

The equations are

$(x-2)^2 + (y-1)^2 = 8$

$x^2 + y^2 - 4x - 2y - 3 = 0$

21. The center has coordinates

$\dfrac{-4+0}{2}, \dfrac{4+0}{2} = (-2, 2)$

The radius is 2. The equations are

$(x+2)^2 + (y-2)^2 = 2^2$

$x^2 + y^2 + 4x - 4y + 4 = 0$

25. $x^2 + (y-3)^2 = 4$ is the same as

$(x-0)^2 + (y-3)^2 = 2^2$, so

$C(0, 3), r = 2$

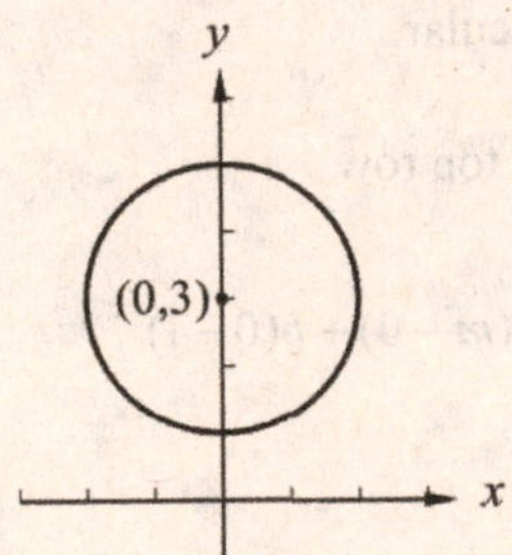

29. $2x^2 + 2y^2 - 16 = 4x$

$x^2 + y^2 - 2x - 8 = 0$

$x^2 - 2x + 1 + y^2 = 8 + 1$

$(x-1)^2 + (y-0)^2 = 3^2$

$C(1, 0)$, $r = 3$

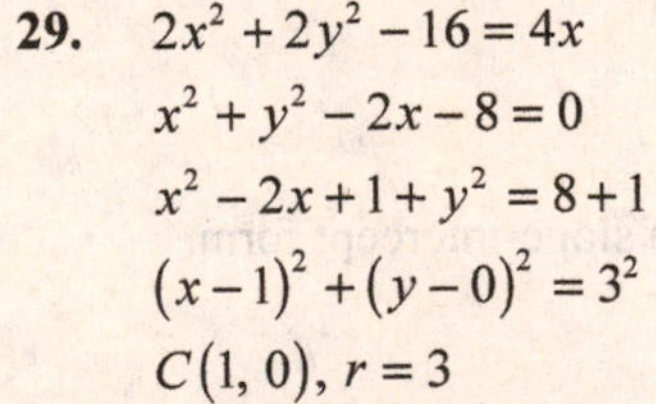

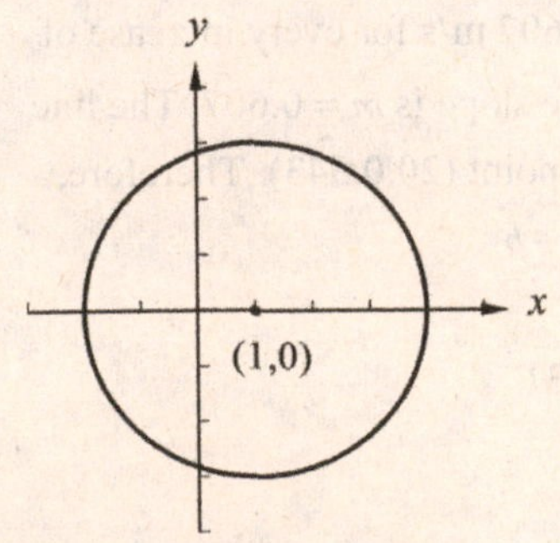

33. $4x^2+4y^2-9=16y$

$4x^2+4y^2-16y=9$

$x^2+y^2-4y=\frac{9}{4}$

$x^2+y^2-4y+4=\frac{9}{4}+\frac{16}{4}$

$(x-0)^2+(y-2)=\left(\frac{5}{2}\right)^2$

$C(0,2),\ r=\frac{5}{2}$

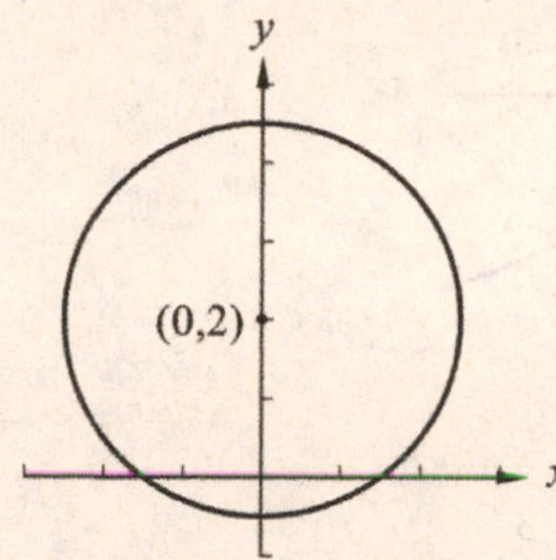

37. Replace x by $-x$:

$(-x)^2+y^2=100;\ x^2+y^2=100$

Symmetric to y-axis.

Replace x by $-x$ and y by $-y$:

$(-x)^2+(-y)^2=100;\ x^2+y^2=100$

Symmetric to origin.

Replace y by $-y$:

$x^2+(-y)^2=100;\ x^2+y^2=100$

Symmetric to x-axis.

41. This circle is centered at the origin. We find a set of four corners of a square by intersecting the circle with the lines $y=x$ and $y=-x$. In each case, we arrive at the equation $2x^2=32$, implying $x=\pm4$; in turn, $y=\pm4$ as well. The square has side length 8 (the distance between (–4, 4) and (4,4)) and so it has area 64.

45. The larger circle has area 25π while the smaller circle has area 9π. Since the smaller circle is entirely enclosed within the larger circle, the area between them is 16π.

49. The viewing window may not have the same horizontal and vertical scales.

53. **(a)** $y=\sqrt{9-(x-2)^2}$

$y^2=9-(x-2)^2$

$(x-2)^2+(y-0)^2=3^2$, which is a circle with centre at (2, 0) and radius = 3.

$y=\sqrt{9-(x-2)^2}$ represents the top half of this circle, a semicircle.

(b) $y=-\sqrt{9-(x-2)^2}$ represents the bottom half of the same circle, also a semicircle.

(c) Yes, for each x in the domain there is only one y.

57. $(x-h)^2+(y-k)^2=p$

If $p>0$, the equation is a circle with radius $\sqrt{p}$

If $p=0$, the equation is the point (h,k).

If $p<0$, the graph does not exist since $(x-h)^2+(y-k)^2\geq 0\neq p<0$.

61. $x^2+y^2=14.5$ is a circle with center (0, 0), $r=\sqrt{14.5}$

$x^2+y^2-19.6y+86=0$

$x^2+y^2-19.6y+96.04=-86+96.04=10.04$

$(x-0)^2+(y-9.8)^2=10.04$

This is a circle with center (0, 9.8), and radius $\sqrt{10.04}$

distance between circles $=9.8-\sqrt{10.04}-\sqrt{14.5}$

$=2.82$ in

65. The center of the porthole is on the y-axis. Since the top is 6 ft below the surface, and the radius is $r=2/2=1$ ft, the center has coordinates $(0,-7)$.

The equation is

$x^2+(y+7)^2=1^2$,

$x^2+y^2+14y+49=1$

or

$x^2+y^2+14y+48=0$.

21.4 The Parabola

1.

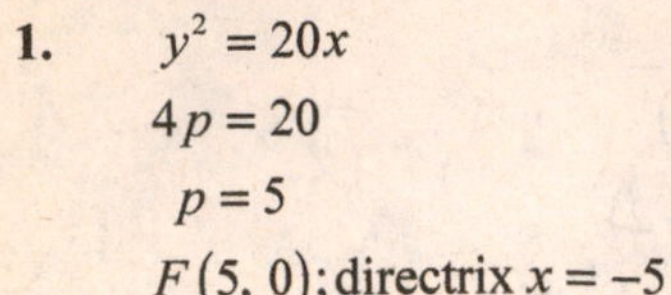

$y^2 = 20x$

$4p = 20$

$p = 5$

$F(5, 0)$; directrix $x = -5$

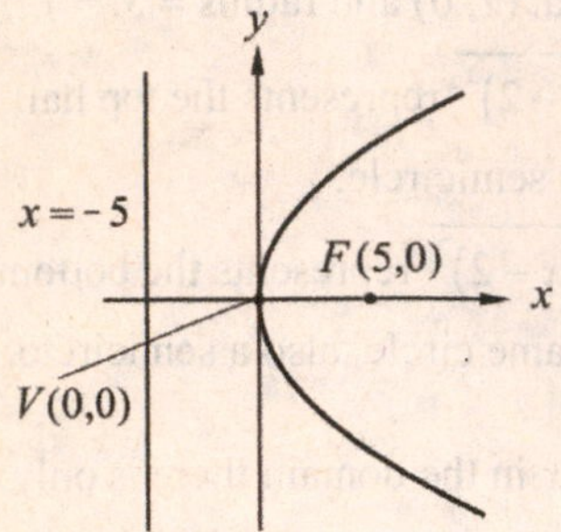

5.

$y^2 = 4x$

$4p = 4$

$p = 1$

$F(1, 0)$; directrix $x = -1$

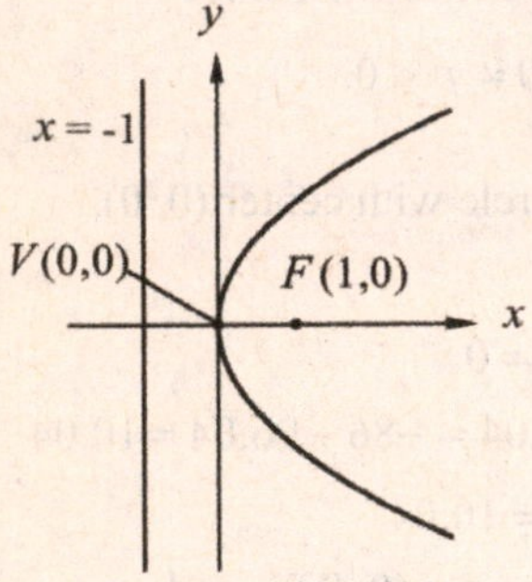

9.

$x^2 = 72y$

$4p = 72$

$p = 18$

$F(0, 18)$; directrix $y = -18$

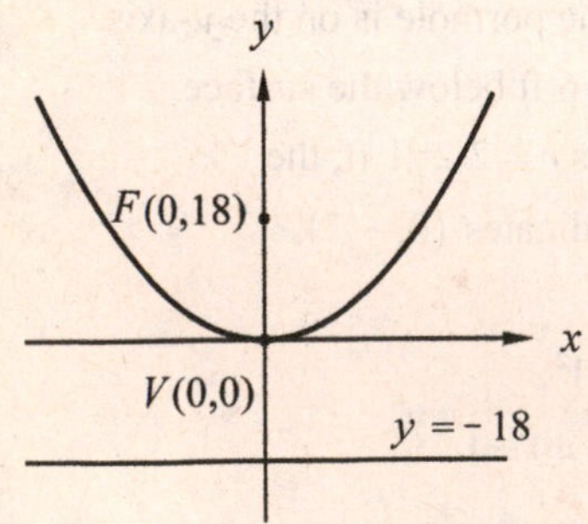

13.

$2y^2 - 5x = 0$

$2y^2 = 5x$

$4p = 5$

$p = \frac{5}{8}$

$F\left(\frac{5}{8}, 0\right)$; directrix $x = -\frac{5}{8}$

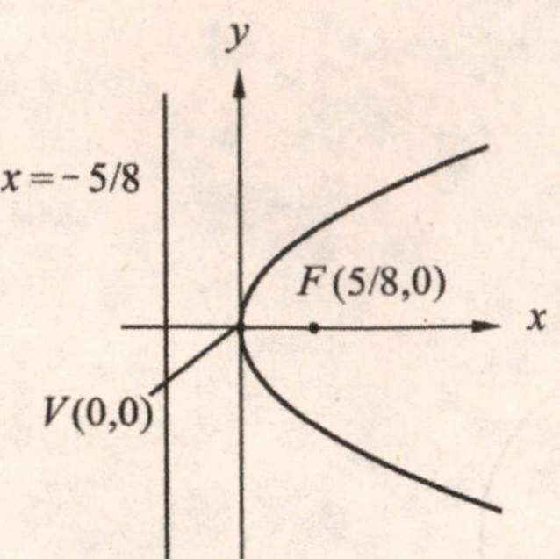

17.

$F(3, 0); V(0,0)$

directrix $x = -3$, $p = 3$

$y^2 = 4px$

$y^2 = 4(3)x$

$y^2 = 12x$

21.

$V(0, 0)$, directrix $y = -0.16$

$F(0, 0.16)$, $p = 0.16$

$x^2 = 4(0.16)y$

$x^2 = 0.64y$

25.

$V(0, 0)$, axis $x = 0$

Substitute $(-1, 8)$ into $x^2 = 4py$:

$(-1)^2 = 4p(8)$

$1 = 32p$

$p = \frac{1}{32}$

Therefore,

$x^2 = 4 \cdot \frac{1}{32} y$

$x^2 = \frac{1}{8} y.$

29. Passes through $(3,3),(12,6)$

Axis is the x-axis.

Substitute $(3,3)$ into $y^2 = 4px$:

$3^2 = 4p(3)$

$4p = 3$

$y^2 = 3x$

33. $F(6, 1)$; directrix $x = 0$;

For graphing, note that the vertex is $V(3, 1)$.

$d_1 = d_2$

$d_1 = |x - 0| = |x|$

$d_2 = \sqrt{(x-6)^2 + (y-1)^2}$

$|x| = \sqrt{(x-6)^2 + (y-1)^2}$; square both sides

$x^2 = (x-6)^2 + (y-1)^2$

$x^2 = x^2 - 12x + 36 + y^2 - 2y + 1$

$0 = -12x + 36 + y^2 - 2y + 1$

$y^2 - 2y - 12x + 37 = 0$

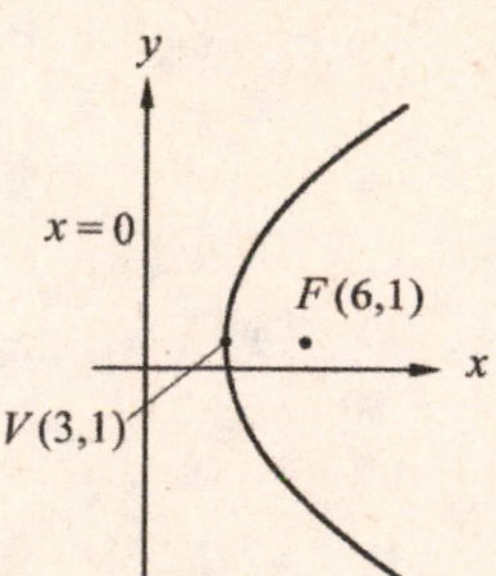

37. Vertex at $(2, -3)$

Focus is 2 units from vertex at $(4, -3)$.

$(y+3)^2 = 8(x-2)$

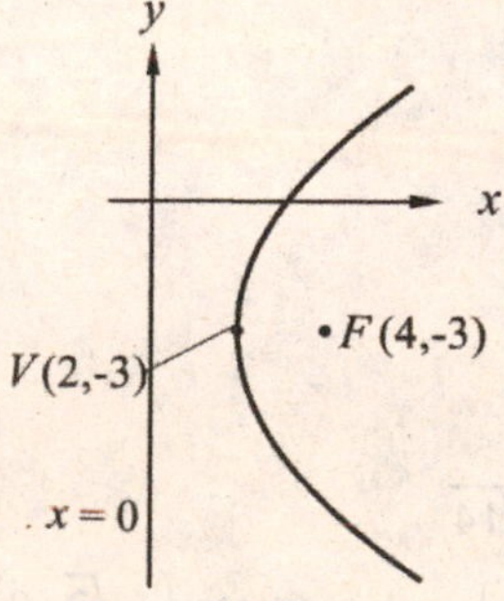

41. Passes through $(2,2),(8,4)$

Axis is the x-axis.

Substitute $(2,2)$ into $y^2 = 4px$:

$2^2 = 4p(2)$

$4p = 2$

$y^2 = 2x$

45. Let the vertex of the parabola be at the origin. $y^2 = 4px$. A point on the parabola will be $(65.0, 17.9)$. Substitute this into the equation and solve for p.

$17.9^2 = 4p(65.0)$

$p = 1.23$

The focal length is 1.23 ft.

49. Place the vertex at the origin. Then the equation of the parabola is

$y^2 = 4px$

Substitute $(0.00625, 1.20)$ into the equation:

$(1.20)^2 = 4p(0.00625)$

$p = 57.6$

The focal length is 57.6 m.

53. $y^2 = 4px$

Substitute $(6.5, 7.5)$:

$7.5^2 = 4p(6.5)$

$p = 2.16$ cm

The filament should be placed 2.16 cm from the vertex.

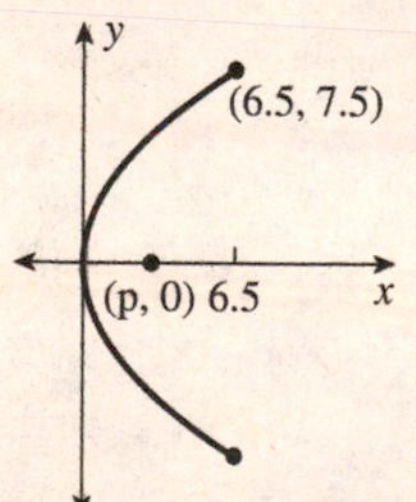

57. The path of the ship channel is a parabola with focus at $(0, 2)$ and vertex $(0, 0)$ and directrix $y = -2$. Therefore, $p = 2$.

The parabola is of the type

$x^2 = 4py$;

therefore, $x^2 = 8y$.

21.5 The Ellipse

1. $\frac{x^2}{25}+\frac{y^2}{36}=1,$

$a^2=36,\ a=6,$

$b^2=25,\ b=5$

$V(0,\pm 6)$, minor axis: $(\pm 5,\ 0)$

$a^2=b^2+c^2$

$36=25+c^2$

$c=\sqrt{11}$

$F\left(0,\pm\sqrt{11}\right)$

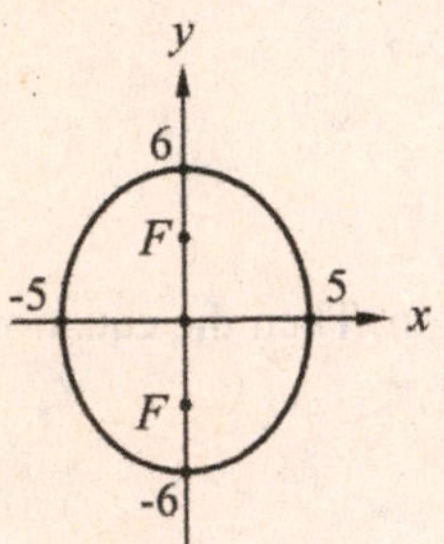

5. $\frac{x^2}{25}+\frac{y^2}{144}=1$

$a^2=144,\ a=12,$

$b^2=25,\ b=5$

$V(0,\pm 12)$, minor axis: $(\pm 5,\ 0)$

$a^2=b^2+c^2$

$144=25+c^2$

$c=\sqrt{119}=10.9$

$F\left(0,\pm\sqrt{119}\right)$

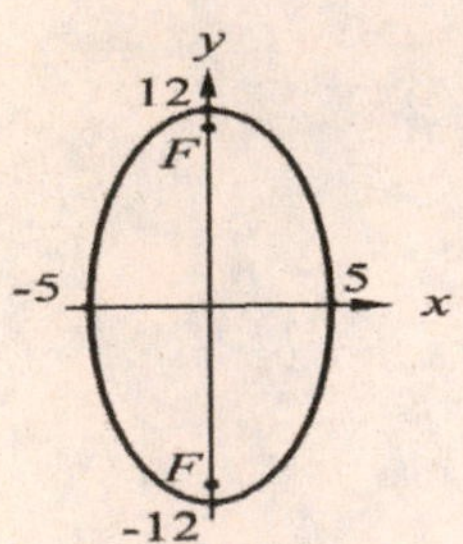

$V(0,\pm 12)$, $F\left(0,\pm\sqrt{119}\right)$, x-intercepts $(\pm 5,\ 0)$

9. $4x^2+9y^2=324$

$\frac{x^2}{81}+\frac{y^2}{36}=1$

$a^2=81,\ a=9$

$b^2=36, b=6$

$c^2=81-36$

$=45,\ c=3\sqrt{5}$

$V(\pm 9,\ 0)$, $F\left(\pm 3\sqrt{5},\ 0\right)$,

y-intercepts $(0,\pm 6)$

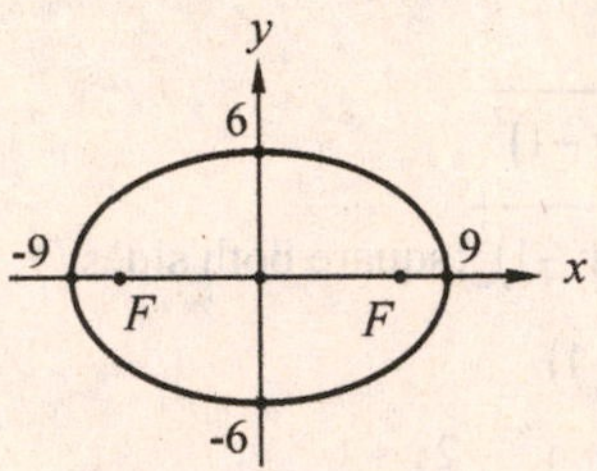

13. $y^2=8\left(2-x^2\right)$

$8x^2+y^2=16$

$\frac{8x^2}{16}+\frac{y^2}{16}=1$

$\frac{x^2}{2}+\frac{y^2}{16}=1$

$\frac{y^2}{16}+\frac{x^2}{2}=1$

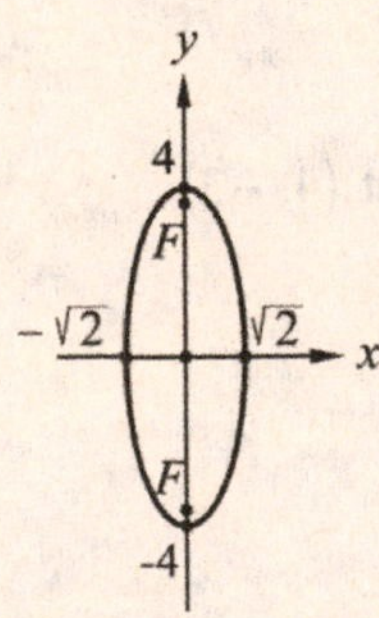

$a^2=16, a=4$

$b^2=2, b=\sqrt{2}$

$c^2=16-2=14, c=\sqrt{14}$

$V(0,\pm 4)$, $F\left(0,\pm\sqrt{14}\right)$, x-intercepts $\left(\pm\sqrt{2},\ 0\right)$

17. $V(15, 0)$; $F(9, 0)$

$a = 15, a^2 = 225;$

$c = 9, c^2 = 81;$

$a^2 - c^2 = b^2$

$b^2 = 144;$

$\frac{x^2}{a^2} + \frac{y^2}{b^2} = 1;$

$\frac{x^2}{225} + \frac{y^2}{144} = 1$

$144x^2 + 225y^2 = 32\ 400$

21. $F(8, 0) \Rightarrow c = 8$ and the major axis is the x – axis.

end of minor axis: $(0, 12) \Rightarrow b = 12$

$a^2 = b^2 + c^2$

$a^2 = 12^2 + 8^2$

$a = \sqrt{208}$

$\frac{x^2}{\sqrt{208}^2} + \frac{y^2}{12^2} = 1$

$\frac{x^2}{208} + \frac{y^2}{144} = 1$

25. $(x_1, y_1) = (2, 2), (x_2, y_2) = (1, 4)$

$\frac{x^2}{b^2} + \frac{y^2}{a^2} = 1$

Substitute: $\frac{4}{b^2} + \frac{4}{a^2} = 1$

Therefore, $4a^2 + 4b^2 = a^2b^2$

$\frac{1}{b^2} + \frac{16}{a^2} = 1$

Therefore, $a^2 + 16b^2 = a^2b^2$

$16b^2 = a^2b^2 - a^2 = a^2(b^2 - 1)$

$a^2 = \frac{16b^2}{b^2 - 1}$

Substitute:

$4\frac{16b^2}{b^2 - 1} + 4b^2 = \frac{16b^2}{b^2 - 1}b^2$

$64b^2 + 4b^4 - 4b^2 = 16b^4$

$-12b^4 + 60b^2 = 0$

$12b^2(-b^2 + 5) = 0$

$b^2 = 5$

$a^2 = \frac{16(5)}{4} = 20$

The equation is $\frac{y^2}{20} + \frac{x^2}{5} = 1$

or: $5y^2 + 20x^2 = 100$

$4x^2 + y^2 = 20$

29. $4x^2 + 9y^2 = 40, \quad y^2 = 4x$

$4x^2 + 9(4x) = 40$

$x^2 + 9x - 10 = 0$

$(x + 10)(x - 1) = 0$

$x = -10$ or $x = 1$

$y^2 = 4(-10)$ $y^2 = 4(1)$

$y^2 = -40$, no solution $y = \pm 2$

Graphs intersect at $(1, 2), (1, -2)$

33. $4x^2 + 3y^2 + 16x - 18y + 31 = 0$; solve for y

$3y^2 - 18y + (4x^2 + 16x + 31) = 0$

$y = \frac{18 \pm \sqrt{(-18)^2 - 4(3)(4x^2 + 16x + 31)}}{2(3)}$

$= \frac{18 + \sqrt{-48x^2 - 192x - 48}}{2}$

$y_1 = 3 + \frac{\sqrt{-12x^2 - 48x - 12}}{3}$

$y_{2(3)} = 3 - \frac{\sqrt{-12x^2 - 48x - 12}}{3}$

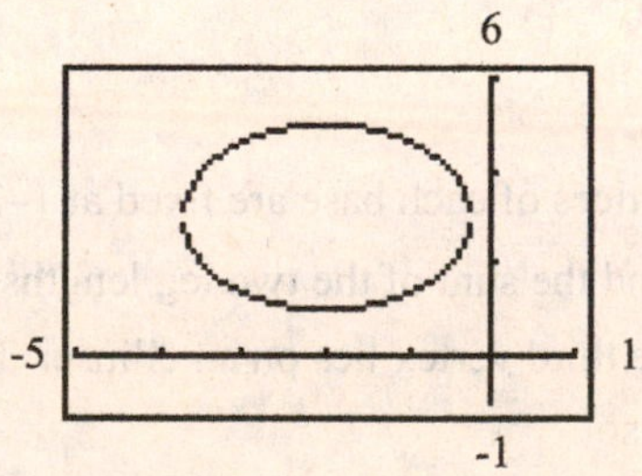

6

-5   1

-1

37. $x^2 + k^2y^2 = 1$

Therefore, $\frac{x^2}{1} + \frac{y^2}{\frac{1}{k^2}} = 1$

The vertices will be on the y-axis if the denominator of y^2 is greater than that of x^2, so

$\frac{1}{k^2} > 1$

$k^2 < 1$

Therefore, $|k| < 1$

41. $x = 2\sin t, \qquad y = 3\cos t$

$x^2 = 4\sin^2 t \qquad y^2 = 9\cos^2 t$

$\frac{x^2}{4} + \frac{y^2}{9} = \sin^2 t + \cos^2 t$

$\frac{x^2}{4} + \frac{y^2}{9} = 1$

This is an ellipse with centre at $(0,0)$, vertices at $(0,3)$ and $(0,-3)$ and minor axis ending at $(-2,0)$ and $(2,0)$.

45. $P = Ri^2$

$P_T = R_1i_1^2 + R_2i_2^2$

$64 = 2i_1^2 + 8i_2^2$

$\frac{2}{64}i_1^2 + \frac{8}{64}i_2^2 = 1$

$\frac{i_1^2}{32} + \frac{i_2^2}{8} = 1$

$a^2 = 32;\ a = \sqrt{32} = 5.7$

$b^2 = 8;\ b = \sqrt{8} = 2.8$

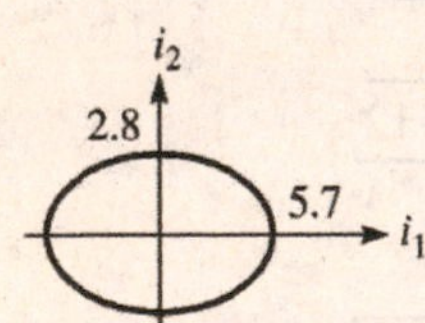

49. If the two vertices of each base are fixed at $(-3, 0)$ and $(3, 0)$, and the sum of the two leg lengths is also fixed, the third vertex lies on an ellipse. The base is 6 cm, so

$d_1 + d_2 = 14 \text{ cm} - 6 \text{ cm} = 8 \text{ cm}$

$(-3, 0)$ and $(3, 0)$ are foci $(-c, 0)$ and $(c, 0)$

$d_1 + d_2 = 2a = 8;\ a = 4$

$a^2 - c^2 = b^2$

$4^2 - 3^2 = b^2$

$b^2 = 7,\ a^2 = 16$

The equation is $\frac{x^2}{16} + \frac{y^2}{7} = 1,\ 7x^2 + 16y^2 = 112$

53. $2a = 64, a = 32,$

$b = 18$

Let the centre be at the origin. The equation of the ellipse is

$\frac{x^2}{32^2} + \frac{y^2}{18^2} = 1$

If $x = 22$,

$\frac{22^2}{32^2} + \frac{y^2}{18^2} = 1$

$y = 13$ ft

21.6 The Hyperbola

1. $\frac{y^2}{16} - \frac{x^2}{4} = 1$

$a^2 = 16,\ a = 4$

$b^2 = 4,\ b = 2$

$c^2 = a^2 + b^2 = 20$

$c = 2\sqrt{5}$

$V(0, \pm 4)$

conjugate axis: $(\pm 2, 0)$

$F\left(0, \pm 2\sqrt{5}\right)$

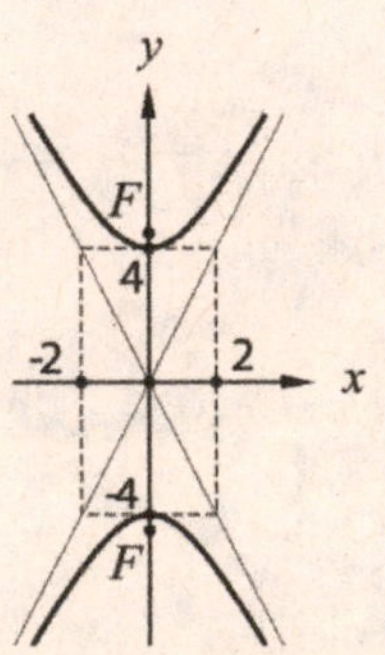

5. $\frac{y^2}{9}-\frac{x^2}{1}=1$

$a^2=9,\ a=3$

$b^2=1,\ b=1$

$c^2=a^2+b^2$

$c^2=10$

$c=\sqrt{10}$

$V(0,\pm 3),\ F\left(0,\pm\sqrt{10}\right)$

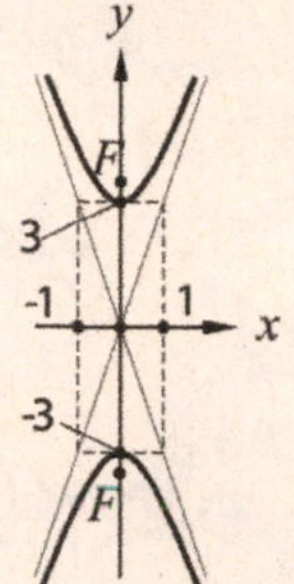

9. $9x^2-y^2=4$

$\frac{9x^2}{4}-\frac{y^2}{4}=1;$

$\frac{x^2}{4/9}-\frac{y^2}{4}=1$

$a^2=\frac{4}{9},a=\frac{2}{3}$

$b^2=4,b=2$

$c^2=4+\frac{4}{9}=\frac{40}{9}$

$c=\frac{\sqrt{40}}{3}$

$V\left(\pm\frac{2}{3},0\right),\ F\left(\pm\frac{\sqrt{40}}{3},0\right)$

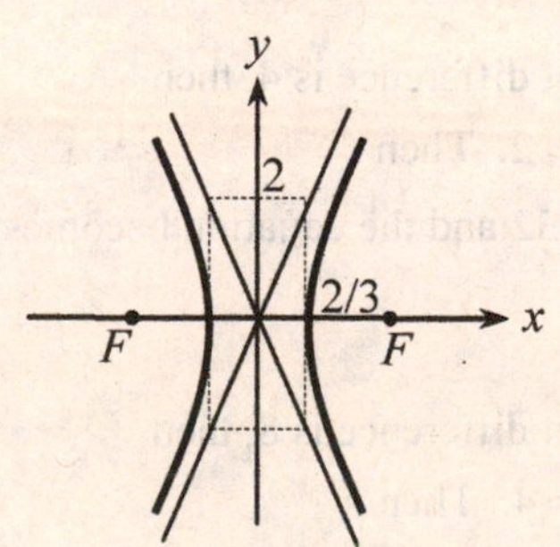

13. $y^2=4\left(x^2+1\right)$

$y^2-4x^2=4$

$\frac{y^2}{4}-\frac{x^2}{1}=1$

$a^2=4,a=2$

$b^2=1,b=1$

$c^2=5,c=\sqrt{5}$

$V(0,\pm 2),\ F\left(0,\pm\sqrt{5}\right)$

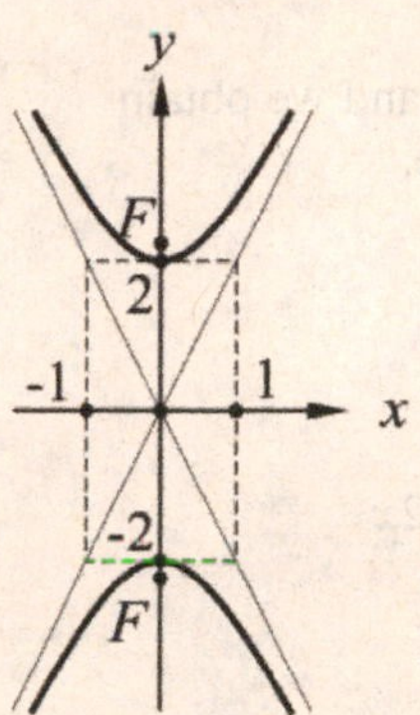

17. $V(3,0);\ F(5,0)$

$a=3;\ c=5;\ a^2=9;\ c^2=25$

$b^2=c^2-a^2=25-9=16$

$\frac{x^2}{a^2}-\frac{y^2}{b^2}=1$

$\frac{x^2}{9}-\frac{y^2}{16}=1;$

$16x^2-9y^2=144$

21. (x,y) is $(2,3)$; $F(2,0),(-2,0)$; $c=\pm 2,\ c^2=4$

$d_1=\sqrt{(2-(-2))^2+(3-0)^2}$

$=\sqrt{4^2+3^2}=\sqrt{16+9}$

$=\sqrt{25}=5$

$d_2=\sqrt{(2-2)^2+(3-0)^2}$

$=\sqrt{0+9}=\sqrt{9}=3$

$d_1-d_2=2a;\ 5-3=2a;\ 2=2a;\ 1=a;\ a^2=1$

$c^2=4;\ b^2=c^2-a^2=3$

$\frac{x^2}{a^2}-\frac{y^2}{b^2}=1;\ \frac{x^2}{1}-\frac{y^2}{3}=1$

$3x^2-y^2=3$

Alternatively, solve the system

$a^2 + b^2 = 4$

$\dfrac{4}{a^2} - \dfrac{9}{b^2} = 1$

Substitute $a^2 = 4 - b^2$:

$\dfrac{4}{4-b^2} - \dfrac{9}{b^2} = 1$

$4b^2 - 36 + 9b^2 = 4b^2 - b^4$

$b^4 + 9b^2 - 36 = 0$

$\left(b^2 + 12\right)\left(b^2 - 3\right) = 0$

$b^2 = 3$ is the only solution, and we obtain the same equation as before.

25. $V(1, 0)$

$a = 1, a^2 = 1$

Asymptote $y = \dfrac{b}{a}x = \dfrac{b}{1}x = 2x$

$b = 2, b^2 = 4$

$\dfrac{x^2}{1} - \dfrac{y^2}{4} = 1$

29. $xy = 2; \; y = \dfrac{2}{x}$

x	y
$\pm\frac{1}{2}$	± 4
± 1	± 2
± 2	± 1
± 4	$\pm\frac{1}{2}$
± 8	$\pm\frac{1}{4}$

33. $y^2 - x^2 = 5$

Substitute $y^2 = x^2 + 5$ into $2x^2 + y^2 = 17$:

$2x^2 + x^2 + 5 = 17$

$3x^2 = 12$

$x^2 = 4$

$x = \pm 2$

$y^2 = x^2 + 5$

$= 4 + 5$

$= 9$

$y = \pm 3$

points of intersection $(2, \pm 3), (-2, \pm 3)$

37. $x^2 - 4y^2 + 4x + 32y - 64 = 0$; solve for y

$4y^2 - 34y + \left(-x^2 - 4x + 64\right) = 0$

$$y = \frac{32 \pm \sqrt{(-32)^2 - 4(4)\left(-x^2 - 4x + 64\right)}}{2(4)}$$

$$= \frac{32 \pm \sqrt{16x^2 + 64x}}{8}$$

$y_1 = 4 + 0.5\sqrt{x^2 + 4x}, \; y_2 = 4 - 0.5\sqrt{x^2 + 4x}$

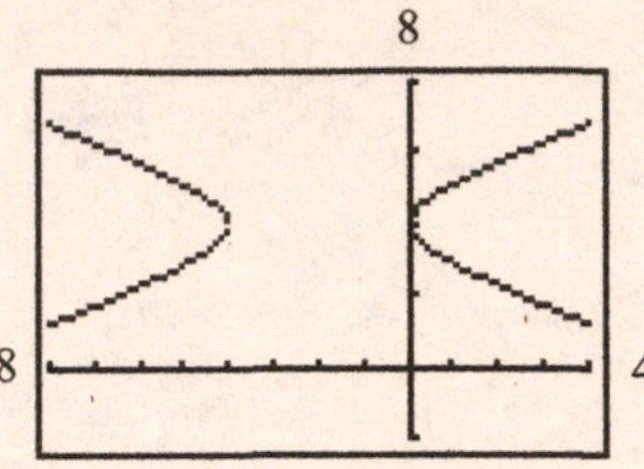

41. $\dfrac{x^2}{9} - \dfrac{y^2}{16} = 1$

implies $a = 3, b = 4$

and so this hyperbola's transverse axis is length 6 along the $x-$axis and its conjugate axis is length 8 along the $y-$axis.

We want the hyperbola whose transverse axis is length 8 along the $y-$axis and whose conjugate axis is length 6 along the $x-$axis.

The desired hyperbola has equation $\dfrac{y^2}{16} - \dfrac{x^2}{9} = 1$

45. We have

$\dfrac{x^2}{a^2} - \dfrac{y^2}{b^2} = 1$

where $b^2 = c^2 - a^2$.

Here, $c = 6$.

(a) If the constant difference is 4, then $4 = 2a$, or $a = 2$. Then $b^2 = 36 - 4 = 32$ and the equation becomes

$\dfrac{x^2}{4} - \dfrac{y^2}{32} = 1.$

(b) If the constant difference is 8, then $8 = 2a$, or $a = 4$. Then $b^2 = 36 - 16 = 20$ and the equation becomes

$\dfrac{x^2}{16} - \dfrac{y^2}{20} = 1.$

49. $600 = vt$

$$v = \frac{600}{t}$$

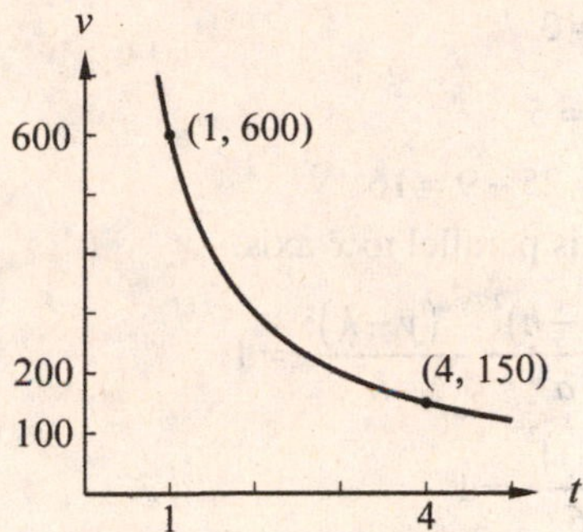

53.

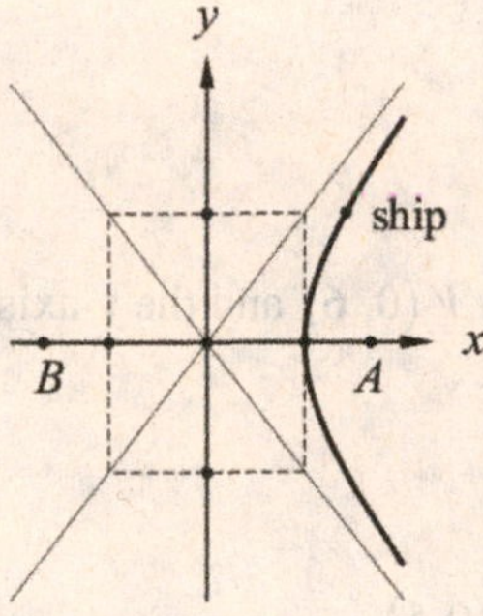

$d_1 - d_2 = \text{constant}$

Let t_1 be time for signal to go from B to ship.

Let t_2 be time for signal to go from A to ship.

Then $t_2 = t_1 - 1.20$ ms

$v = \frac{s}{t}$, therefore, $s = vt$

Therefore, $d_1 = 300t_1$ and $d_2 = 300(t_1 - 1.20)$

$d_1 - d_2 = 2a$

$300t_1 - 300(t_1 - 1.20) = \text{constant} = 360 \text{ km} = 2a$

Therefore, the ship could lie anywhere on the hyperbolic arc sketched.

Foci at $(\pm 300, 0)$; therefore, $c = 300$

Vertices at $(\pm 180, 0)$; therefore, $a = 180$; therefore, $b = 240$

21.7 Translation of Axes

1. $\frac{(x-3)^2}{25} - \frac{(y-2)^2}{9} = 1$, hyperbola: $a = 5$, $b = 3$

Center: $(3, 2)$. Transverse axis parallel to x-axis.

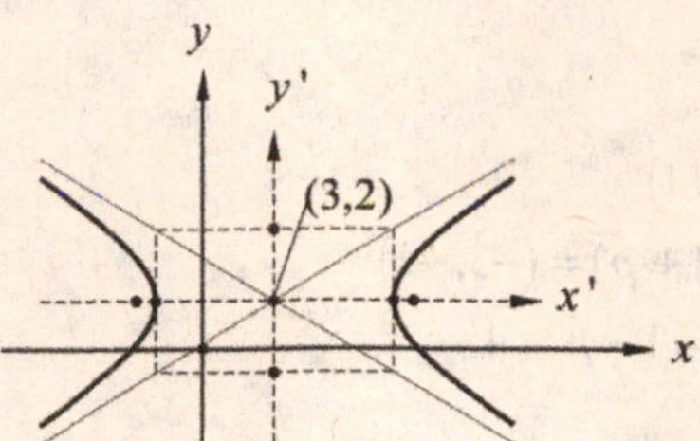

5. $\frac{(x-1)^2}{4} - \frac{(y-2)^2}{25} = 1$; hyperbola with transverse axis parallel to x-axis

Center: $(1, 2)$;

$a^2 = 4, a = 2$;

$b^2 = 25, b = 5$

Vertices are at $(1 \pm 2, 2)$ or at $(-1, 2)$ and at $(3, 2)$.

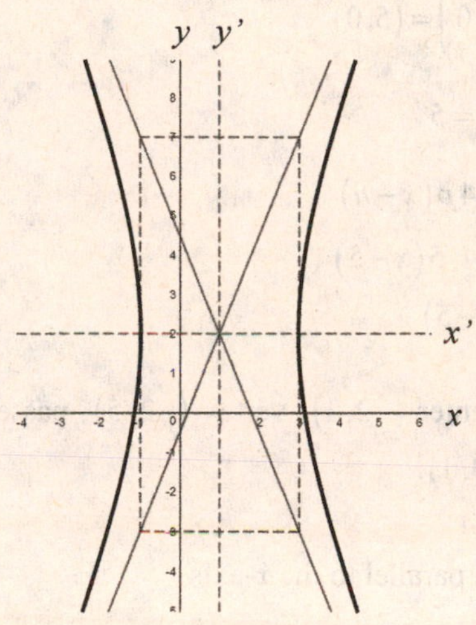

9. $(x+3)^2 = -12(y-1)$; parabola with axis parallel to the y-axis

$x' = x+3;\ y' = y-1$

$(x')^2 = -12y'$

Origin O' at $(h, k) = (-3, 1)$

$4p = -12$

$p = -3$

Vertex $(-3, 1)$

Focus $(-3, 1+p) = (-3, -2)$

Directrix $y = 1 - p = 4$

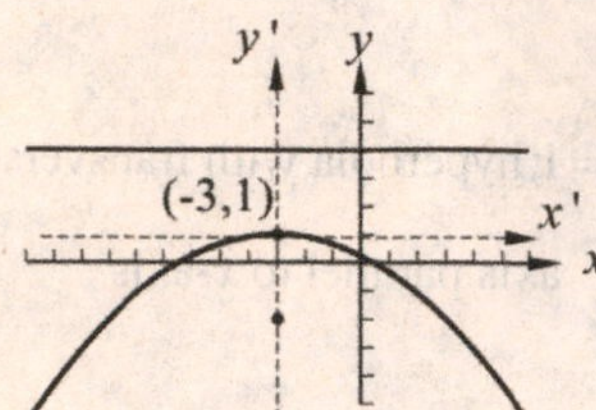

13. $F(10, 0)$, axis, directrix are coordinate axis

If the directrix is a coordinate axis, it can only be the y-axis, so the axis of the parabola is the x-axis

$V\left(\frac{10+0}{2}, 0\right) = (5, 0)$

$p = \frac{10-0}{2} = 5$

$(y-k)^2 = 4p(x-h)$

$(y-0)^2 = 4\cdot 5(x-5)$

$y^2 = 20(x-5)$

17. Ellipse: center $(-2, 1)$, vertex $(-2, 5)$, passes through $(0, 1)$.

$a = 5 - 1 = 4$

Major axis parallel to the x-axis

$$\frac{(y-1)^2}{4^2} + \frac{(x+2)^2}{b^2} = 1$$

Substitute $(0,1)$:

$$\frac{(1-1)^2}{4^2} + \frac{(0+2)^2}{b^2} = 1$$

$$b^2 = 4$$

$$\frac{(y-1)^2}{16} + \frac{(x+2)^2}{4} = 1$$

21. Hyperbola: $V(2,1)$, $V(-4, 1)$, $F(-6, 1)$

Center: $\frac{-4+2}{2}, 1 = (-1, 1)$

$a = \frac{2-(-4)}{2} = 3$

$c = -1 - (-6) = 5$

$b^2 = c^2 - a^2 = 25 - 9 = 16$

Transverse axis parallel to x-axis.

Therefore, $\frac{(x-h)^2}{a^2} - \frac{(y-k)^2}{b^2} = 1$

$$\frac{(x+1)^2}{9} - \frac{(y-1)^2}{16} = 1$$

or $16x^2 - 9y^2 + 32x + 18y - 137 = 0$

25. $x^2 + 4y = 24$

$x^2 = -4y + 24$

$x^2 = -4(y-6)$

Parabola with vertex at $V(0, 6)$ and the y-axis as its axis

$4p = -4$

$p = -1$

Focus is at $(0, 6-1) = (0, 5)$

Directrix is $y = 6 + 1 = 7$

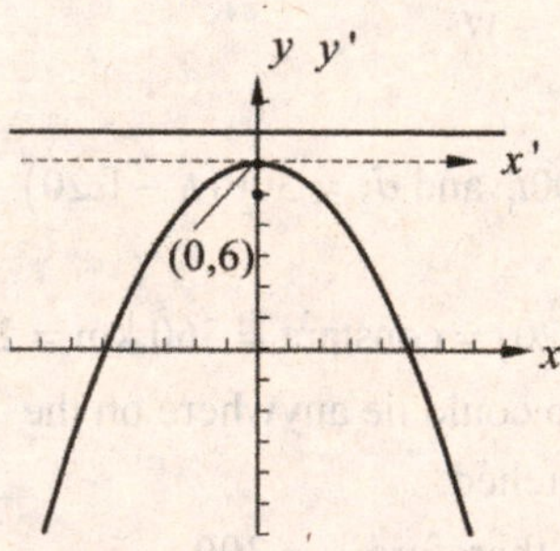

29. $9x^2 - y^2 + 8y = 7$

$9x^2 - (y^2 - 8y + 16) = 7 - 16$

$9x^2 - (y-4)^2 = -9$

$\frac{9x^2}{-9} - \frac{(y-4)^2}{-9} = 1$

$-x^2 + \frac{(y-4)^2}{9} = 1$

$\frac{(y-4)^2}{9} - \frac{x^2}{1} = 1$

Hyperbola whose transverse axis is the y-axis, with center at $(0, 4)$

Vertices are at $(0, 4 \pm 3)$ or at $(0, 7)$ and $(0, 1)$.

The ends of the conjugate axis are at $(0 \pm 1, 4)$ or at $(1, 4)$ and $(-1, 4)$.

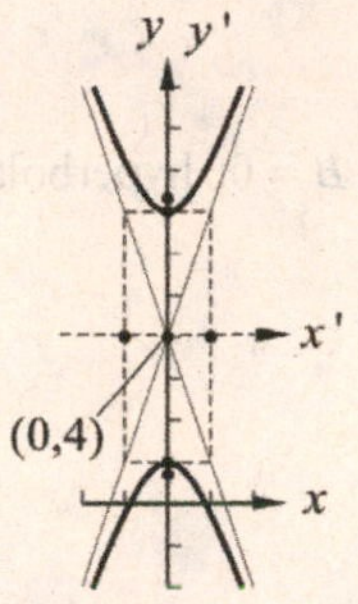

33. $4x^2 - y^2 + 32x + 10y + 35 = 0$

$4(x^2 + 8x) - (y^2 - 10y) = -35$

$4(x^2 + 8x + 16) - (y^2 - 10y + 25) = -35 + 64 - 25$

$\frac{(x+4)^2}{1^2} - \frac{(y-5)^2}{2^2} = 1$

Hyperbola with center at $(-4, 5)$ and transverse axis parallel to the x-axis.

Vertices are at $(-4 \pm 1, 5)$ or at $(-3, 5)$ and at $(-5, 5)$.

The conjugate axis ends at $(-4, 5 \pm 2)$ or at $(-4, 7)$ and $(-4, 3)$.

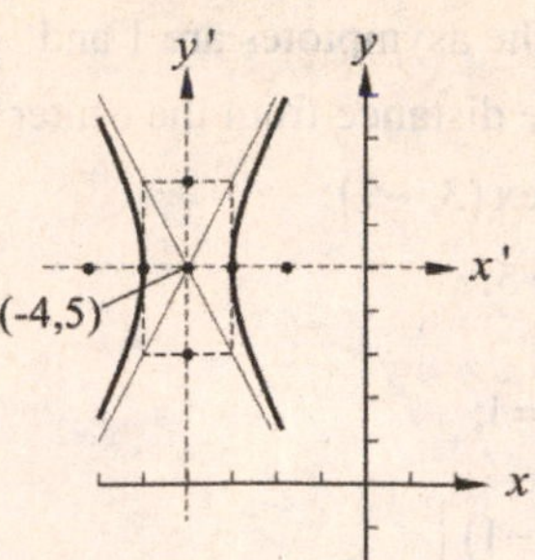

37. $5x^2 - 3y^2 + 95 = 40x$

$5(x^2 - 8x + 16) - 3y^2 = -95 + 80 = -15$

$\frac{y^2}{5} - \frac{(x-4)^2}{3} = 1$

Hyperbola with center at $(4, 0)$ and transverse axis parallel to the y-axis.

Vertices are at $(4, 0 \pm \sqrt{5}) = (4, \pm 2.24)$

Conjugate axis ends at $(4 \pm \sqrt{3}, 0)$ or at $(5.73, 0)$ and $(2.27, 0)$.

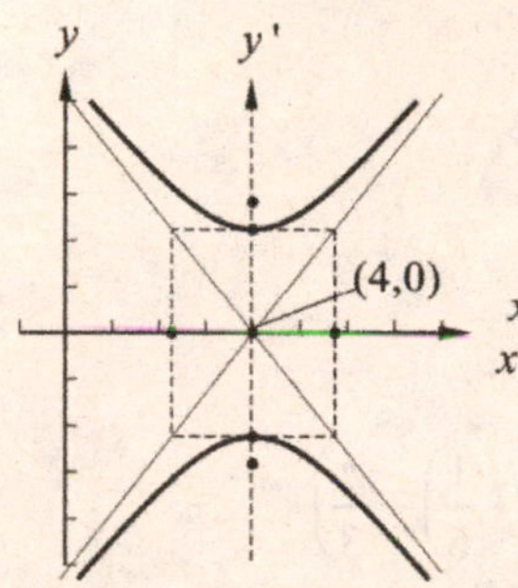

41. Hyperbola: asymptotes: $x - y = -1$ or $y = x + 1$, and $x + y = -3$ or $y = -x - 3$; vertex $(3, -1)$. The center is at the point of intersection of the asymptotes. The equations for the asymptotes are solved simultaneously:

$y = x + 1$

$\underline{y = -x - 3}$

$2y = -2$

$y = -1$

$x = -1 - 1$

$x = -2.$

Therefore, the coordinates of the center are $(-2, -1)$.

Since the slopes of the asymptotes are 1 and -1, $a = b$, where a is the distance from the center $(-2, -1)$ to the vertex $(3, -1)$;

$a = 3-(-2) = 5$, $b = 5$.

$$\frac{(x-h)^2}{a^2}-\frac{(y-k)^2}{b^2}=1;$$

$$\frac{[x-(-2)]^2}{25}-\frac{[y-(-1)]^2}{25}=1$$

$$\frac{(x+2)^2}{25}-\frac{(y+1)^2}{25}=1$$

$$x^2+4x+4-(y^2+2y+1)=25;$$

$$x^2+4x+4-2y-1=25$$

$$x^2-y^2+4x-2y-22=0$$

45. Parabola: vertex and focus on x-axis.
Vertex will be $(h,0)$ and axis will be the x-axis.

$$y^2=4p(x-h)$$

49. $i = 2+\sin\left(2\pi t-\frac{\pi}{3}\right)$

$$i = i'+2$$

$$t = t'+\frac{1}{6}$$

$$i'+2 = 2+\sin\left(2\pi\left(t'+\frac{1}{6}\right)-\frac{\pi}{3}\right)$$

$$i' = \sin\left(2\pi t'+\frac{\pi}{3}-\frac{\pi}{3}\right)$$

$$i' = \sin(2\pi t')$$

53. First ellipse:

$a = 4$, $b = 3$, therefore $c = \sqrt{7}$

Center $(0, 0)$

$$\frac{y^2}{a^2}+\frac{x^2}{b^2}=1$$

$$\frac{y^2}{16}+\frac{x^2}{9.0}=1$$

Second ellipse:

$a = 4$, $b = 3$, therefore $c = \sqrt{7}$

Center $(7.0, 0.0)$

$$\frac{(x-h)^2}{a^2}+\frac{(y-k)^2}{b^2}=1$$

$$\frac{(x-7)^2}{16}+\frac{y^2}{9.0}=1$$

21.8 The Second-degree Equation

1. $2x^2 = 3+2y^2$

$2x^2-2y^2-3=0$, A, C have different signs, $B = 0$, hyperbola

5. $2x^2-y^2-1=0$

A and C have different signs, $B = 0$; hyperbola

9. $2.2x^2-x-y=1.6$

$A \neq 0$; $C = 0$; $B = 0$; parabola

13. $36x^2 = 12y(1-3y)+1$

$$36x^2 = 12y-36y^2+1$$

$$36x^2+36y^2-12x-1=0,$$

$A = C$, $B = 0$, circle

17. $2xy+x-3y=6$

$$2xy+x-3y-6=0$$

$A = 0$; $B \neq 0$; $C = 0$; hyperbola

21. $(x+1)^2+(y+1)^2 = 2(x+y+1)$

$$x^2+2x+1+y^2+2y+1=2x+2y+2$$

$x^2+y^2=0$, point $(0, 0)$ is the only solution

25. $x^2 = 8(y - x - 2)$

$x^2 = 8y - 8x - 16$

$x^2 + 8x - 8y + 16 = 0$

$A \neq 0;\ B = 0; C = 0;$ parabola

$x^2 + 8x - 8y + 16 = 0$

$x^2 + 8x + 16 = 8y$

$(x+4)^2 = 4(2)y$

$p = 2$

Vertex $(-4, 0)$, focus $(-4, 2)$, directrix $y = -2$.

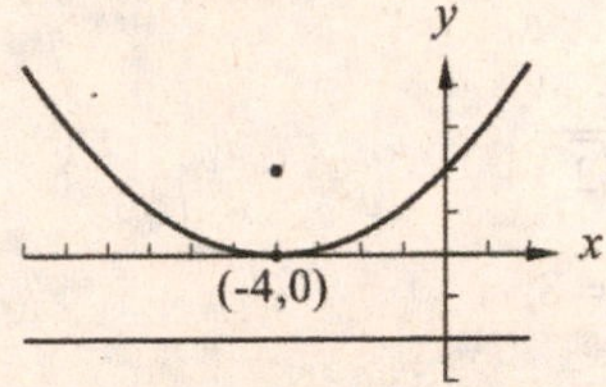

29. $y^2 + 42 = 2x(10 - x)$

$y^2 + 42 = 20x - 2x^2$

$y^2 + 2x^2 - 20x + 42 = 0;$

$A \neq C$, same sign, $B = 0$, ellipse

$\frac{y^2}{2} + x^2 - 10x = -21$

$\frac{y^2}{2} + x^2 - 10x + 25 = -21 + 25$

$\frac{y^2}{2} + (x-5)^2 = 4$

$\frac{y^2}{8} + \frac{(x-5)^2}{4} = 1$

(h, k) at $(5, 0)$, $V\left(5, \pm 2\sqrt{2}\right)$

$a = \sqrt{8} = 2\sqrt{2};\ b = 2$

Minor axis ends at $(5 \pm 2, 0)$ or at $(7,0)$ and $(3,0)$.

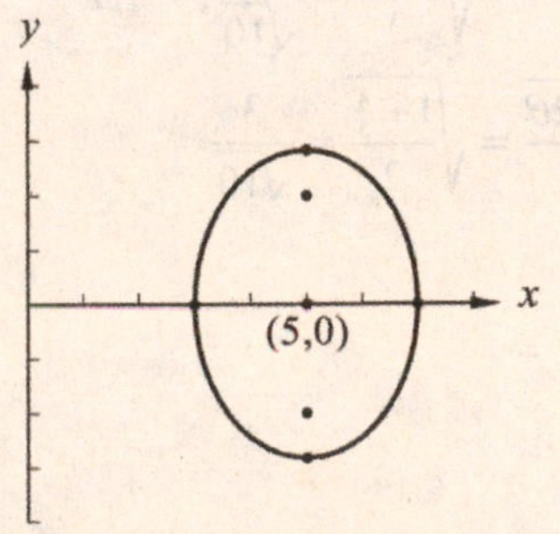

33. $4(y^2 - 4x - 2) = 5(4y - 5)$

$4y^2 - 16x - 8 = 20y - 25$

$4y^2 - 20y - 16x + 17 = 0$

$A = 0;\ C = 4;\ B = 0;$ parabola

$y^2 - 5y - 4x + \frac{17}{4} = 0$

To graph, solve for y.

$y_1 = \frac{5 + \sqrt{25 - 4\left(-4x + \frac{17}{4}\right)}}{2} = \frac{5 + \sqrt{16x + 8}}{2}$

$y_2 = \frac{5 - \sqrt{16x + 8}}{2}$

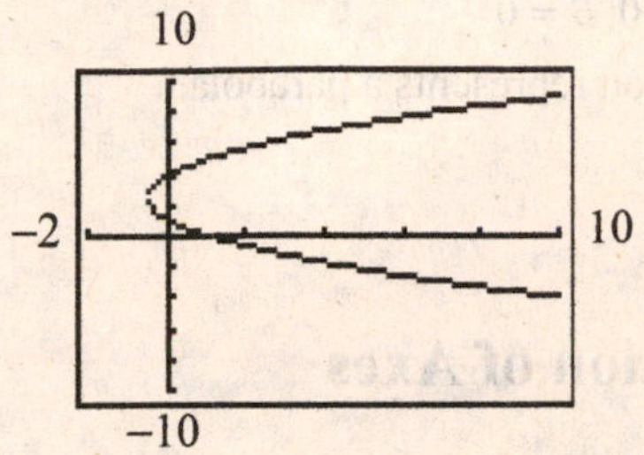

37. $x^2 + ky^2 = a^2$

(a) If $k = 1$,

$x^2 + (1)y^2 = a^2$

$x^2 + y^2 = a^2$ ($A = C, B = 0$, circle)

(b) If $k < 0$,

$x^2 - |k|y^2 = a^2$

$\frac{x^2}{a^2} - \frac{y^2}{a^2/|k|} = 1$ ($A \neq C$, different sign, $B = 0$, hyperbola)

(c) If $k > 0 (k \neq 1)$

$\frac{x^2}{a^2} + \frac{y^2}{a^2/k} = 1$ ($A \neq C$, same sign, $B = 0$, ellipse)

41. Beam is perpendicular to floor. We have a circle.
*See conic section diagrams, Fig. 21.89 in the text.

45.

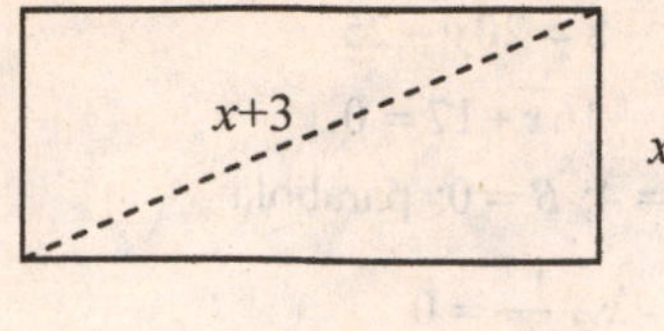

By Pythagoras,

$x^2 + y^2 = (x+3)^2$

$x^2 + y^2 = x^2 + 6x + 9$

$y^2 - 6x - 9 = 0$

$A = 0, C \neq 0, B = 0$

The equation represents a parabola.

21.9 Rotation of Axes

1.

$x^2 - y^2 = 25, \theta = 45^\circ;$

$x = x'\cos 45^\circ - y'\sin 45^\circ$

$= \frac{x'}{\sqrt{2}} - \frac{y'}{\sqrt{2}}$

$y = x'\sin 45^\circ + y'\cos 45^\circ$

$= \frac{x'}{\sqrt{2}} + \frac{y'}{\sqrt{2}}$

$x^2 - y^2 = \left(\frac{x'}{\sqrt{2}} - \frac{y'}{\sqrt{2}}\right)^2 - \left(\frac{x'}{\sqrt{2}} + \frac{y'}{\sqrt{2}}\right)^2 = 25;$

$\frac{x'^2}{2} - \frac{2x'y'}{2} + \frac{y'^2}{2} - \frac{x'^2}{2} - \frac{2x'y'}{2} - \frac{y'^2}{2} = 25$

$2x'y' + 25 = 0$, hyperbola

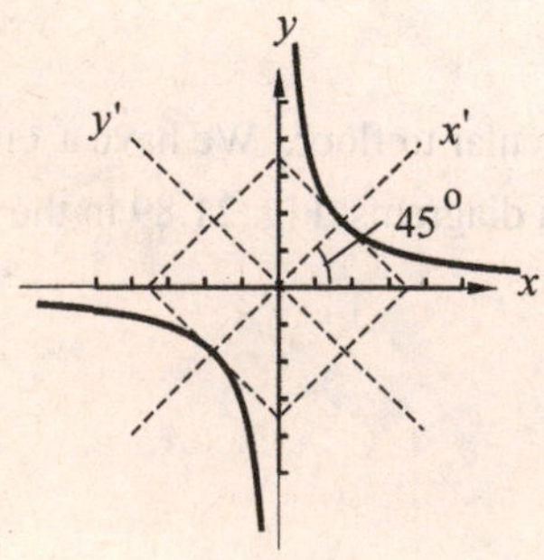

5. $x^2 + 2xy + x - y - 3 = 0$

$B^2 - 4AC = 2^2 - 4(1)(0) = 4 > 0$, hyperbola

9. $13x^2 + 10xy + 13y^2 + 6x - 42y - 27 = 0$

$B^2 - 4AC = 10^2 - 4(13)(13) = -576 < 0$, ellipse

13. $xy = 8$

$\cos 2\theta = 0$ and so $\theta = 45°$.

$\sin\theta = \frac{1}{\sqrt{2}};$

$\cos\theta = \frac{1}{\sqrt{2}};$

$x = \frac{x' - y'}{\sqrt{2}}, y = \frac{x' + y'}{\sqrt{2}}$

$\left(\frac{x' - y'}{\sqrt{2}}\right)\left(\frac{x' + y'}{\sqrt{2}}\right) = 8;$

$x'^2 - y'^2 = 16$, hyperbola

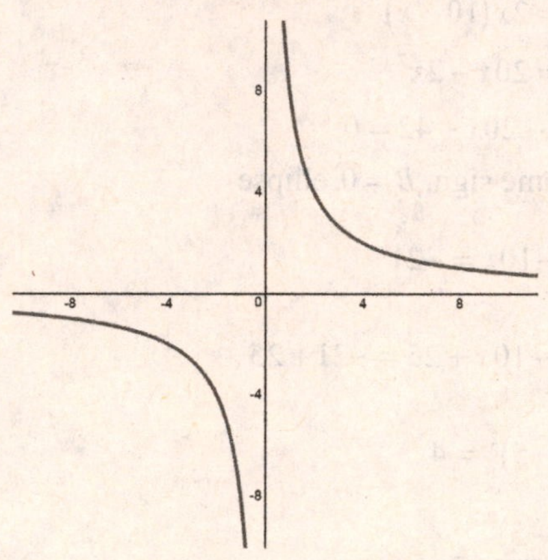

17. $11x^2 - 6xy + 19y^2 = 20$

$\tan 2\theta = \frac{B}{A-C} = \frac{-6}{11-19} = \frac{6}{8} = \frac{3}{4};$

$\cos 2\theta = \frac{4}{5}$

$\sin\theta = \sqrt{\frac{1 - \cos 2\theta}{2}} = \sqrt{\frac{1 - \frac{4}{5}}{2}} = \frac{1}{\sqrt{10}};$

$\cos\theta = \sqrt{\frac{1 + \cos 2\theta}{2}} = \sqrt{\frac{1 + \frac{4}{5}}{2}} = \frac{3}{\sqrt{10}}$

5
3
2θ
4

$$x = \frac{3x' - y'}{\sqrt{10}},\ y = \frac{x' + 3y'}{\sqrt{10}}$$

$$11\left(\frac{3x' - y'}{\sqrt{10}}\right)^2 - 6\left(\frac{3x' - y'}{\sqrt{10}}\right)\left(\frac{x' + 3y'}{\sqrt{10}}\right)$$

$$+19\left(\frac{x' + 3y'}{\sqrt{10}}\right)^2 = 20$$

$$11\left(9x'^2 - 6x'y' + y'^2\right) - 6\left(3x'^2 + 9x'y' - x'y' - 3y'^2\right)$$

$$+19\left(x'^2 + 6x'y' + 9y'^2\right) = 200$$

$$100x'^2 + 200y'^2 = 200$$

$$x'^2 + 2y'^2 = 2,\text{ ellipse}$$

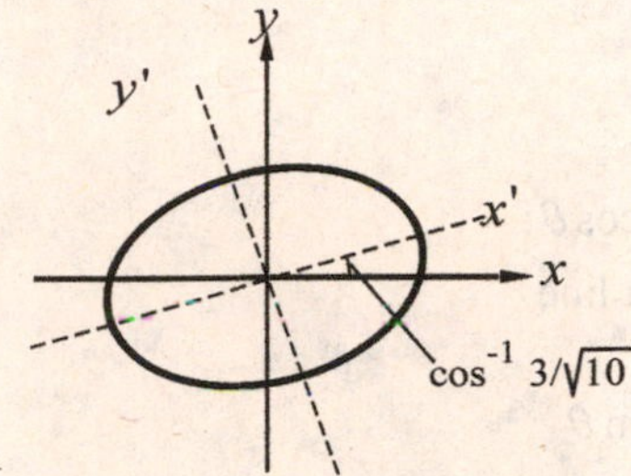

21. $B^2 - 4AC = 1 - 8 = -7 < 0$ indicates the curve should be an ellipse. We observe that

$$2x^2 + xy + y^2 = \frac{7}{4}x^2 + \frac{1}{4}x^2 + xy + y^2$$

$$= \frac{7}{4}x^2 + \left(\frac{1}{2}x + y\right)^2$$

and in order for this to be 0, both x and y must be 0.

The locus is the origin.

25. $x = x'\cos 90° - y'\sin 90° = -y'$

$y = x'\sin 90° + y'\cos 90° = x'$

$y = 2^x$ becomes

$x' = 2^{-y'}$

or

$y' = -\log_2 x'$

21.10 Polar Coordinates

1. $\left(3, \frac{\pi}{3}\right)$ and $\left(3, -\frac{5\pi}{3}\right)$ represent the same point.

$\left(-3, \frac{\pi}{3}\right)$ and $\left(3, -\frac{2\pi}{3}\right)$ represent the same point on the opposite side of the pole.

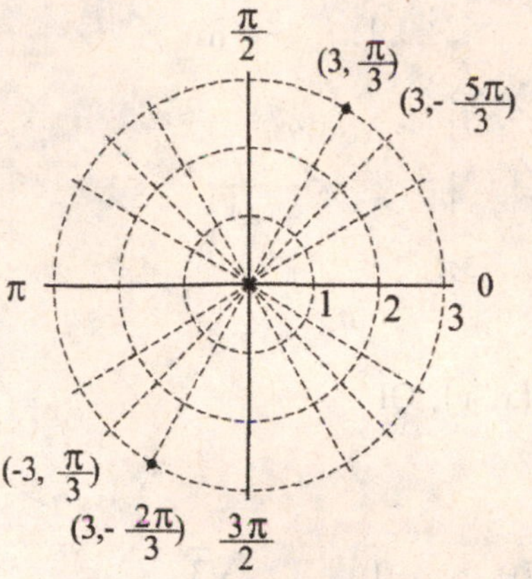

5. $\left(3, \frac{\pi}{6}\right)$; $r = 3,\ \theta = \frac{\pi}{6}$

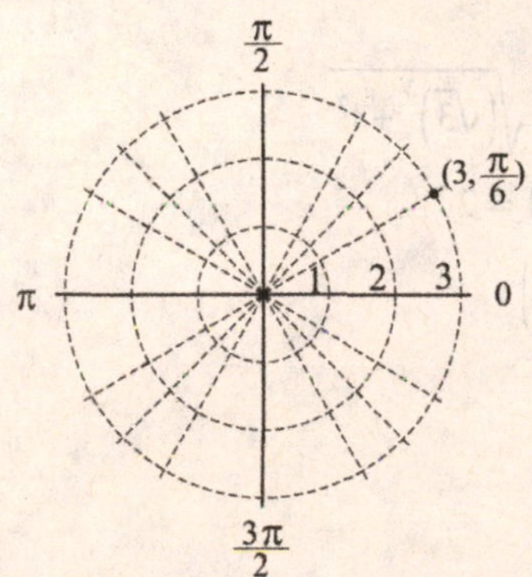

9. $\left(-8, \frac{7\pi}{6}\right)$; negative r is reversed in direction from positive r.

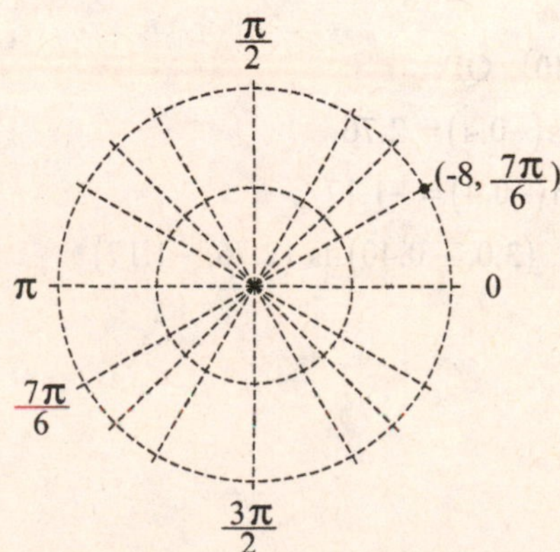

13. $(2, 2)$

$\frac{2}{\pi} = 0.64$, so $2 = 0.64\pi$

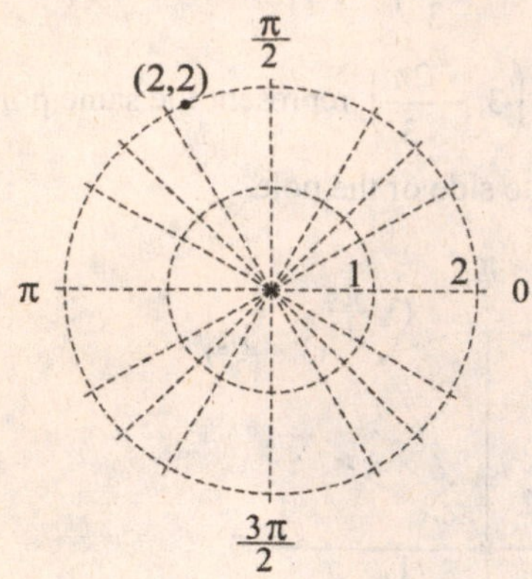

17. $(\sqrt{3}, 1)$ is (x, y), QI

$\tan\theta = \frac{y}{x}$

$\theta = \tan^{-1}\frac{y}{x} = \tan^{-1}\frac{1}{\sqrt{3}} = \tan^{-1}\frac{\sqrt{3}}{3};$

$\theta = 30^\circ = \frac{\pi}{6}$

$r = \sqrt{x^2 + y^2} = \sqrt{(\sqrt{3})^2 + 1^2}$

$= \sqrt{3+1} = \sqrt{4} = 2$

(r, θ) is $\left(2, \frac{\pi}{6}\right)$

21. $(0, 4)$

$\theta = \frac{\pi}{2}$

$r = \sqrt{0^2 + 4^2} = 4$

Therefore, $(0, 4)$ is $\left(4, \frac{\pi}{2}\right)$

25. $(3.0, -0.40)$, QIV

$x = 3.0\cos(-0.4) = 2.76$

$y = 3.0\sin(-0.4) = -1.17$

Therefore, $(3.0, -0.40)$ is $(2.76, -1.17)$

29. $x = 3$

$r\cos\theta = x = 3$

$r = \frac{3}{\cos\theta} = 3\sec\theta$

33. $x^2 + (y-2)^2 = 4$

$x^2 + y^2 - 4y + 4 = 4$

$r^2 - 4\cdot r\sin\theta = 0$

$r = 4\sin\theta$

37. $x^2 + y^2 = 6y$

$r^2 = 6\cdot r\sin\theta$

$r = 6\sin\theta$

41. $r\cos\theta = 4$

Subsitute $x = r\cos\theta$:

$x = 4$, a straight line

45. $r = 4\cos\theta + 2\sin\theta$

$r^2 = 4r\cos\theta + 2r\sin\theta$

$x^2 + y^2 = 4x + 2y$

$x^2 + y^2 - 4x - 2y = 0$, circle

49. As the graph shows, the point $(2, 3\pi/4)$ is on the curve $r = 2\sin 2\theta$ even though $(2, 3\pi/4)$ is not a solution to $r = 2\sin 2\theta$. $(2, 3\pi/4)$ and $(-2, 7\pi/4)$ are the same point and $(-2, 7\pi/4)$ is a solution to $r = 2\sin 2\theta$.

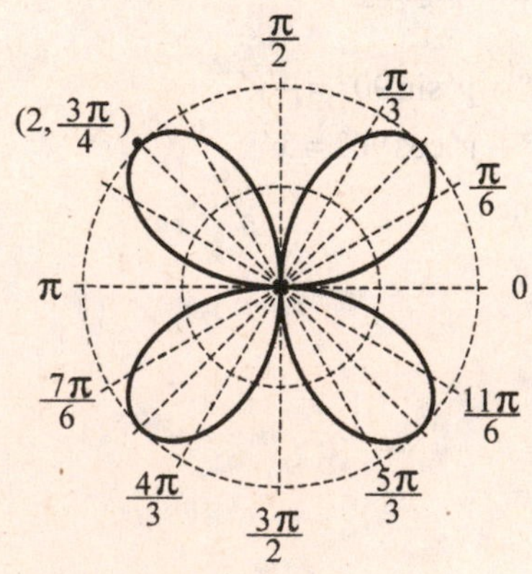

53. Each pair of vertices subtends a central angle of $\frac{\pi}{3}$ at the pole. The coordinates of the other vertices are $(2, 0), \left(2, \frac{\pi}{3}\right), \left(2, \frac{2\pi}{3}\right), \left(2, \frac{4\pi}{3}\right), \left(, \frac{5\pi}{3}\right)$

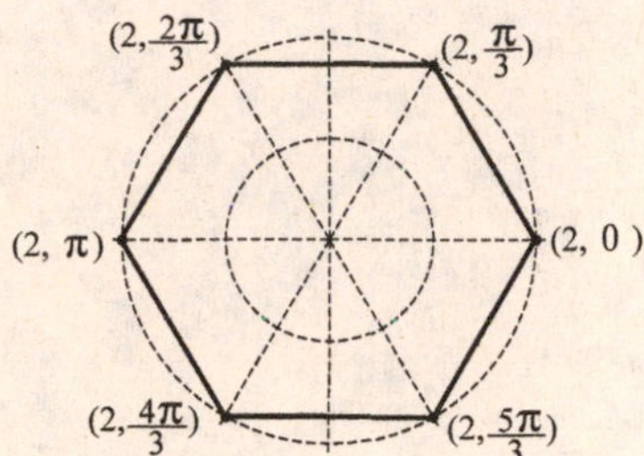

57. $r = 5\cos\theta$

$r^2 = 5r\cos\theta$

$x^2 + y^2 = 5x$

$x^2 - 5x + y^2 = 0$

21.11 Curves in Polar Coordinates

1. The graph of $\theta = \frac{5\pi}{6}$ is a straight line through the pole. $\theta = \frac{5\pi}{6}$ for all possible values of r.

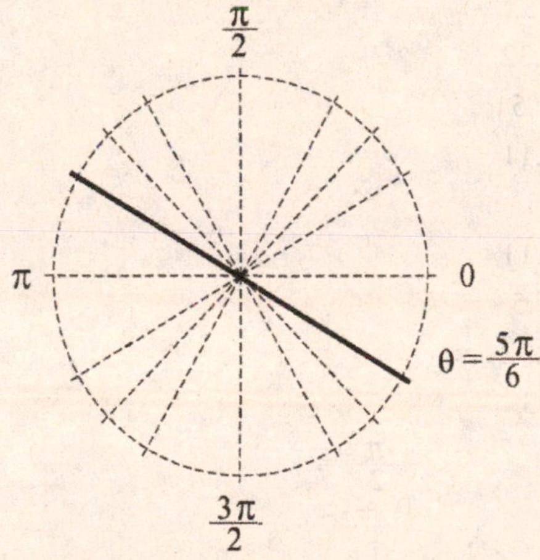

5. $r = 5$ for all θ. Graph is a circle with radius 5.

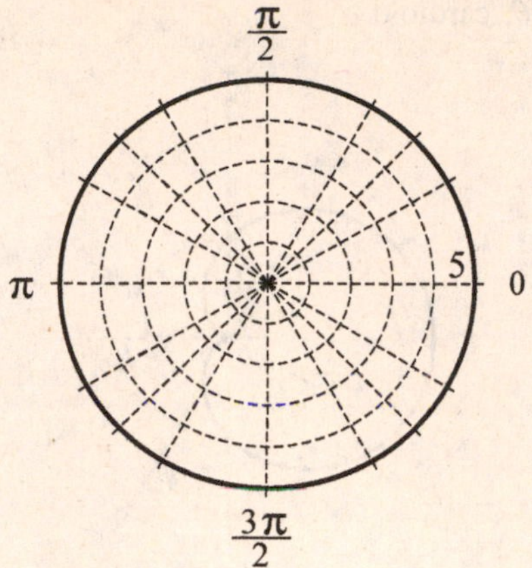

9. $r = 4\sec\theta = \frac{4}{\cos\theta}$

$r\cos\theta = 4$

$x = 4$

This is a vertical line at $x = 4$

θ	r
0	4
$\frac{\pi}{6}$	4.6
$\frac{\pi}{4}$	5.7
$\frac{\pi}{3}$	8
$\frac{\pi}{2}$	*
$\frac{2\pi}{3}$	−8
$\frac{3\pi}{4}$	−5.7
$\frac{5\pi}{6}$	4.6
π	−4
$\frac{5\pi}{4}$	−5.7
$\frac{3\pi}{2}$	*
$\frac{7\pi}{4}$	5.7
2π	4

* denotes undefined

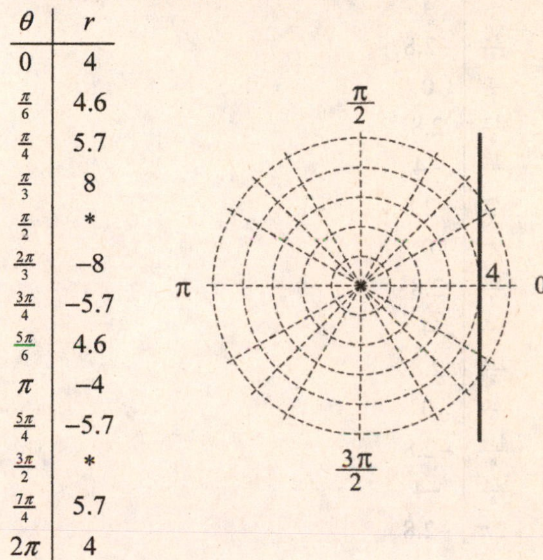

13. $1 - r = \cos\theta$

$r = 1 - \cos\theta$; cardioid

θ	r
0	0
$\frac{\pi}{4}$	0.3
$\frac{\pi}{2}$	1
$\frac{3\pi}{4}$	1.7
π	2
$\frac{5\pi}{4}$	1.7
$\frac{3\pi}{2}$	1
$\frac{7\pi}{4}$	0.3

17. $r = 4\sin 2\theta$; rose (4 petals)

θ	r
0	0
$\frac{\pi}{8}$	2.8
$\frac{\pi}{4}$	4
$\frac{3\pi}{8}$	−2.8
$\frac{\pi}{2}$	0
$\frac{5\pi}{8}$	2.8
$\frac{3\pi}{4}$	−4
$\frac{7\pi}{8}$	−2.8
π	0
$\frac{9\pi}{8}$	2.8
$\frac{5\pi}{4}$	4
$\frac{11\pi}{8}$	2.8
$\frac{3\pi}{2}$	0
$\frac{13\pi}{8}$	−2.8
$\frac{7\pi}{4}$	−4
2π	−2.8

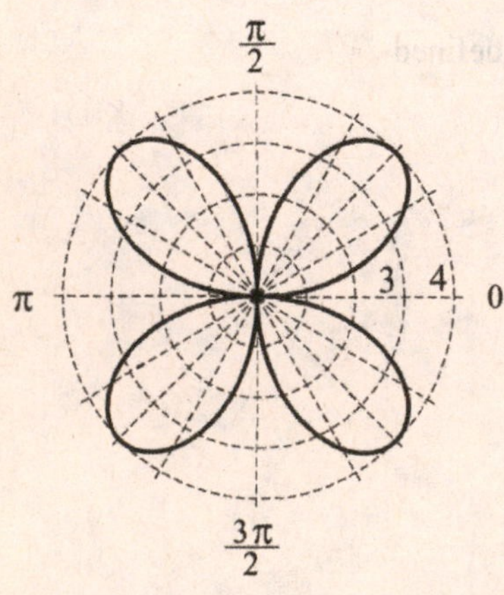

21. $r = 2^{\theta}$; spiral

θ	r
0	1
$\frac{\pi}{4}$	1.7
$\frac{\pi}{2}$	3.0
$\frac{3\pi}{4}$	5.1
π	8.8
$\frac{5\pi}{4}$	15.2
$\frac{3\pi}{2}$	26.2
$\frac{7\pi}{4}$	45.2
2π	77.9

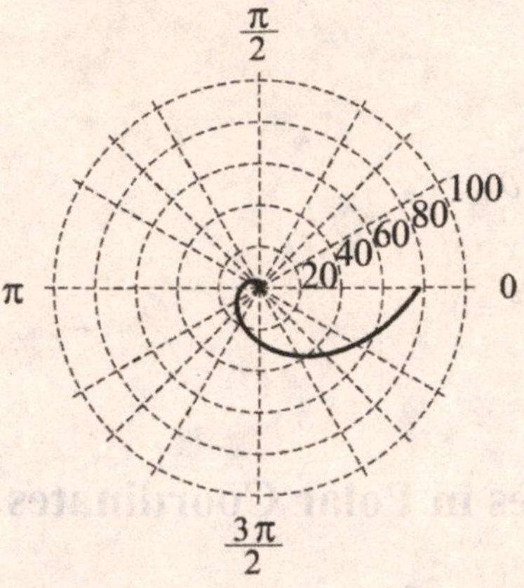

25. $r = \dfrac{3}{2 - \cos\theta}$; ellipse

θ	r
0	3
$\frac{\pi}{4}$	2.32
$\frac{\pi}{2}$	1.5
$\frac{3\pi}{4}$	1.11
π	1
$\frac{5\pi}{4}$	1.11
$\frac{3\pi}{2}$	1.5
$\frac{7\pi}{4}$	2.32
2π	3

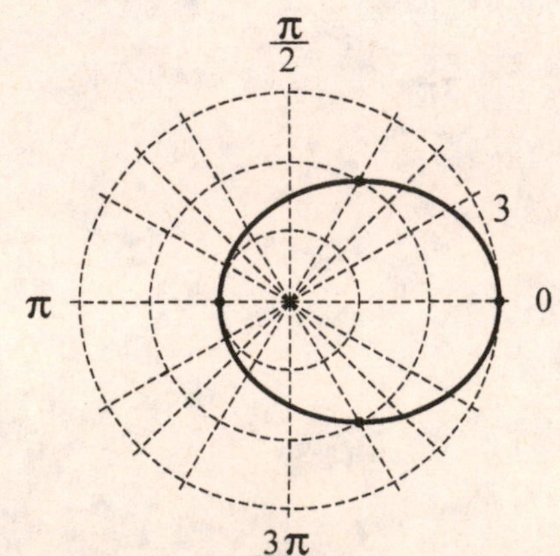

29. $r = 4\cos\frac{1}{2}\theta$

θ	r	θ	r
0	4.0	$\frac{13\pi}{6}$	−3.9
$\frac{\pi}{6}$	3.9	$\frac{9\pi}{4}$	−3.7
$\frac{\pi}{4}$	3.7	$\frac{7\pi}{3}$	−3.5
$\frac{\pi}{3}$	3.5	$\frac{5\pi}{2}$	−2.8
$\frac{\pi}{2}$	2.8	$\frac{8\pi}{3}$	−2.0
$\frac{2\pi}{3}$	2.0	$\frac{11\pi}{4}$	−1.5
$\frac{3\pi}{4}$	1.5	$\frac{17\pi}{6}$	−1.0
$\frac{5\pi}{6}$	1.0	3π	0
π	0	$\frac{19\pi}{6}$	1.0
$\frac{7\pi}{6}$	−1.0	$\frac{13\pi}{4}$	1.5
$\frac{5\pi}{4}$	−1.5	$\frac{10\pi}{3}$	2.0
$\frac{4\pi}{3}$	−2.0	$\frac{7\pi}{2}$	2.8
$\frac{3\pi}{2}$	−2.8	$\frac{11\pi}{3}$	3.5
$\frac{5\pi}{3}$	−3.5	$\frac{15\pi}{4}$	3.7
$\frac{7\pi}{4}$	−3.7	$\frac{23\pi}{6}$	3.9
$\frac{11\pi}{6}$	−3.9	4π	4.0
2π	−4.0		

33. $r = \theta \quad (-20 \le \theta \le 20)$

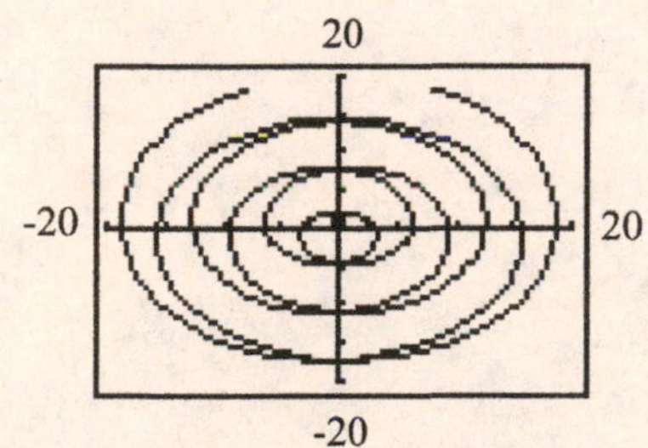

37. $r = 3\cos 4\theta$

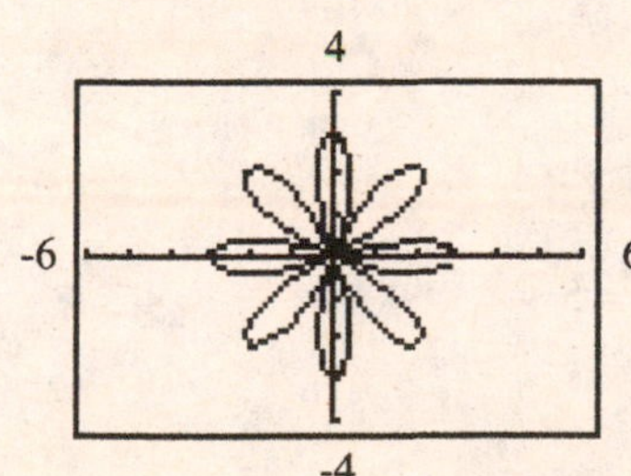

41. $r = \cos\theta + \sin 2\theta$

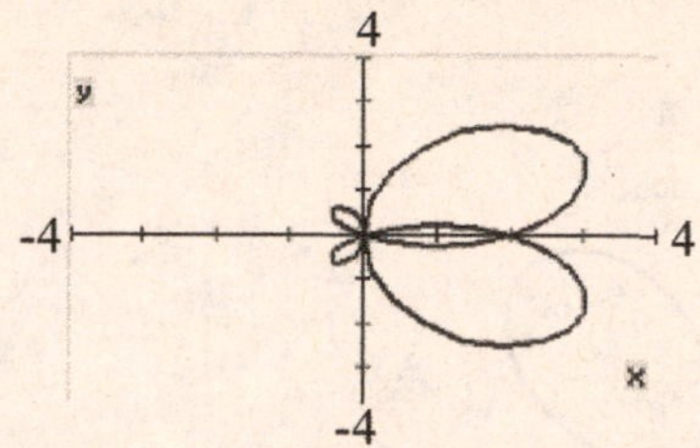

45. From the calculator screen the curves intersect at $(0, 0)$ and $(1, 1)$, where the tangent lines are horizontal and vertical, showing the curves intersect at right angles.

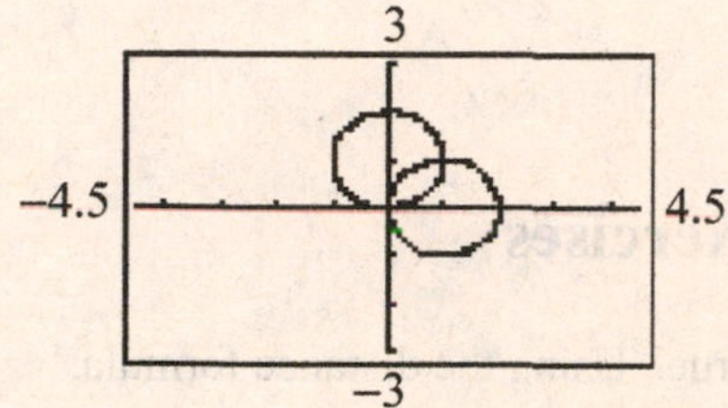

49. $r = 4.0 - \sin\theta$

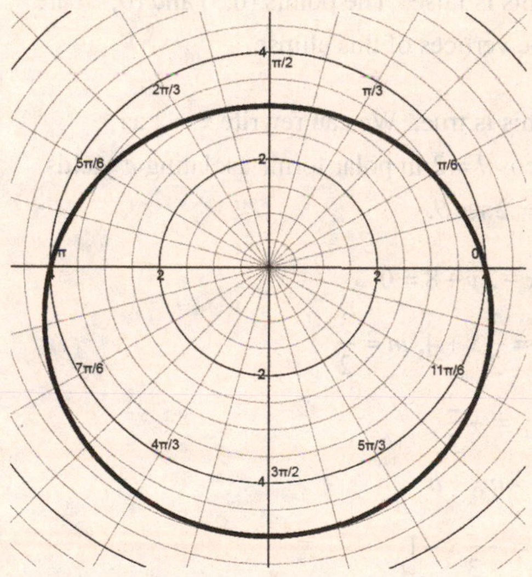

53. $R = \dfrac{\sin^2\theta}{(1-0.5\cos\theta)^2}$

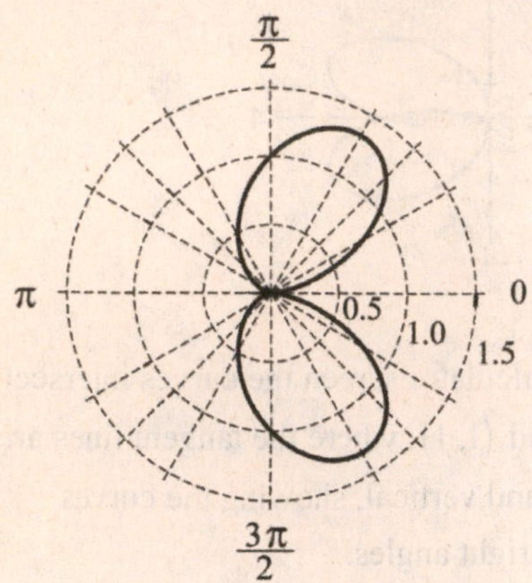

Review Exercises

1. This is true. Using the distance formula, $d = \sqrt{(4-3)^2 + (-3-(-4))^2} = \sqrt{2}.$

5. This is false. The points (0,3) and (0,-3) are the vertices of this ellipse.

9. This is true. We can rewrite $x = 2$ as $r\cos\theta = 2$ in polar form. Isolating r yields $r = 2\sec\theta.$

13. $3x - 2y + 8 = 0$

$y = \frac{3}{2}x + 4,\ m = \frac{3}{2}$

$m_\perp = -\frac{2}{3}$

$y = mx + b$

$y = -\frac{2}{3}x - 1$

$2x + 3y + 3 = 0$

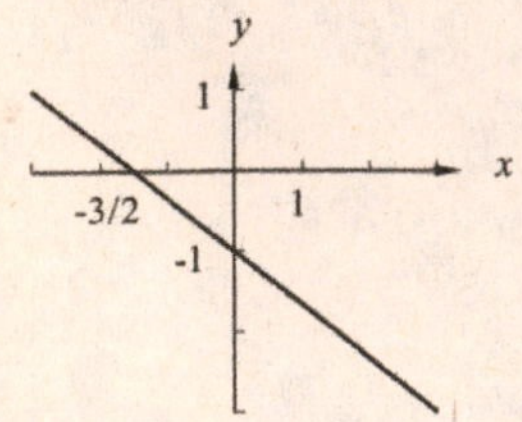

17. If the focus is $(-3,0)$ and the vertex is $(0,0)$, the axis of the parabola is the x-axis. From the definition,

$\sqrt{(x+3)^2 + (y-0)^2} = \sqrt{(x-3)^2}$

$x^2 + 6x + 9 + y^2 = x^2 - 6x + 9$

$y^2 = -12x$

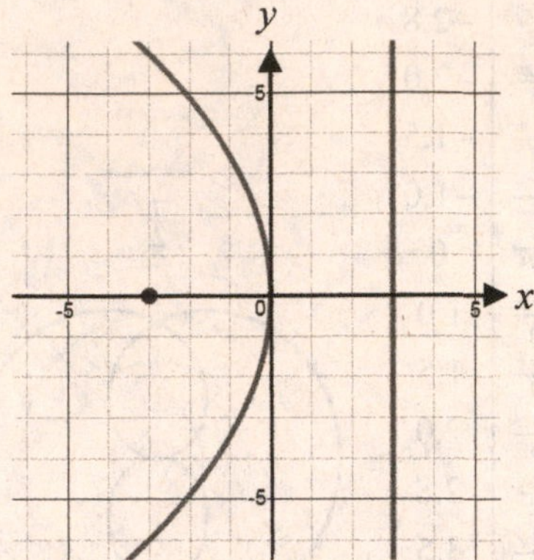

21. $V(0,13), C(0,0), 2b = 24$

The transverse axis is the x-axis, with $a = 13$.
The conjugate axis is the y-axis, with $b = 12$.

$\frac{y^2}{a^2} - \frac{x^2}{b^2} = 1$

$\frac{y^2}{13^2} - \frac{x^2}{12^2} = 1$

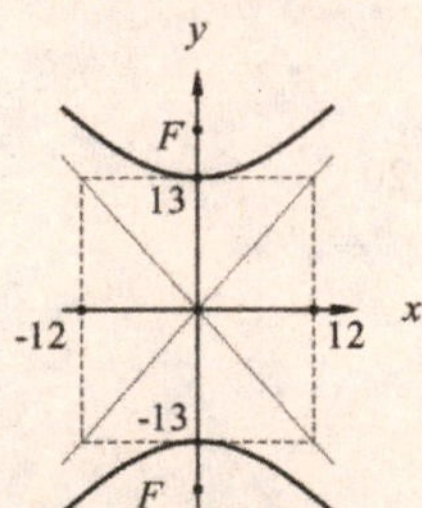

25. $x^2 = -20y$

$4p = -20$

$p = -5$

$F(0, -5),\ D: y = 5$

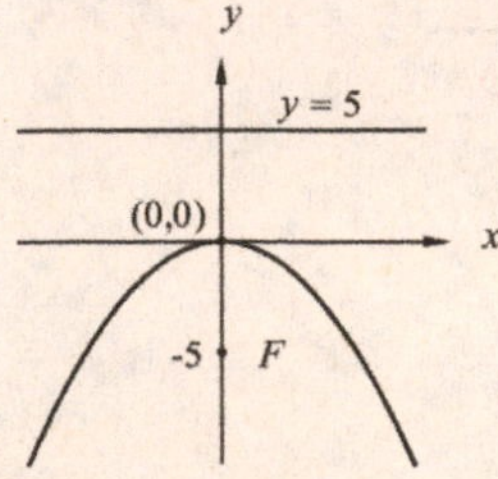

29. $4x^2 - 25y^2 = 0.25$

$16x^2 - 100y^2 = 1$

$\dfrac{x^2}{\frac{1}{16}} - \dfrac{y^2}{\frac{1}{100}} = 1$

$a^2 = \dfrac{1}{16},\ a = \dfrac{1}{4}$

$b^2 = \dfrac{1}{100}, b = \dfrac{1}{10}$

$V\left(\pm\dfrac{1}{4},\ 0\right)$

$c^2 = a^2 + b^2 = \dfrac{1}{100} + \dfrac{1}{16} = \dfrac{29}{400}$

$c = \dfrac{\sqrt{29}}{20}$

$F\left(\pm\dfrac{\sqrt{29}}{20},\ 0\right)$

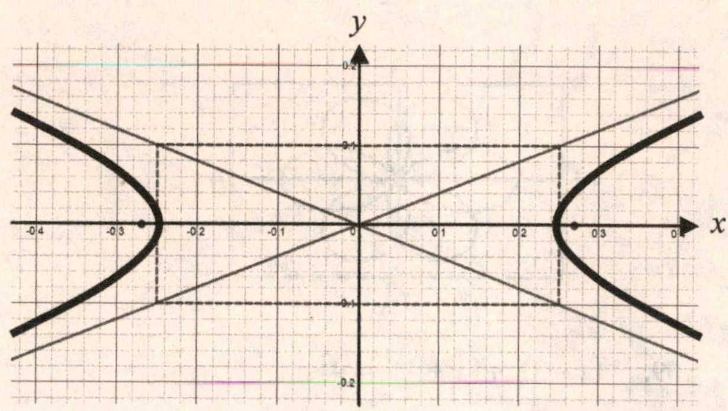

33. $4x^2 + y^2 - 16x + 2y + 13 = 0$

$4\left(x^2 - 4x + 4\right) + y^2 + 2y + 1 = -13 + 16 + 1$

$4(x-2)^2 + (y+1)^2 = 4$

$C(2, -1)$

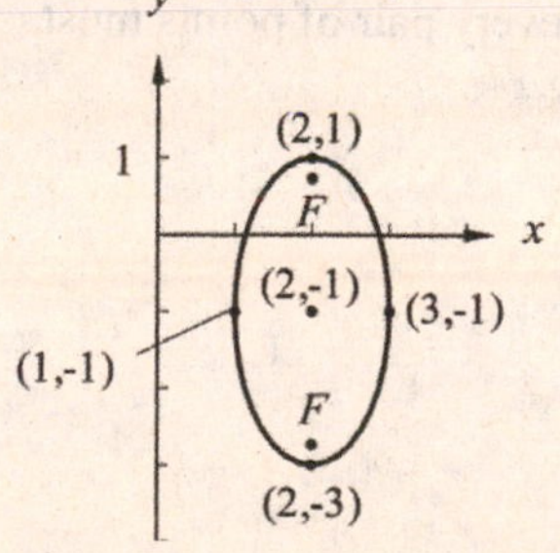

37. $r = 4(1 + \sin\theta)$

Let $\theta = 0$ to 2π in steps of $\dfrac{\pi}{12}$.

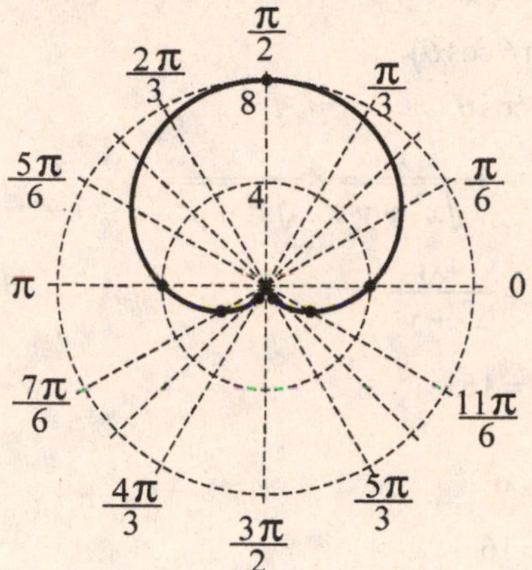

41. $r = \dfrac{3}{\sin\theta + 2\cos\theta}$

Let $\theta = 0$ to 2π in steps of $\dfrac{\pi}{12}$.

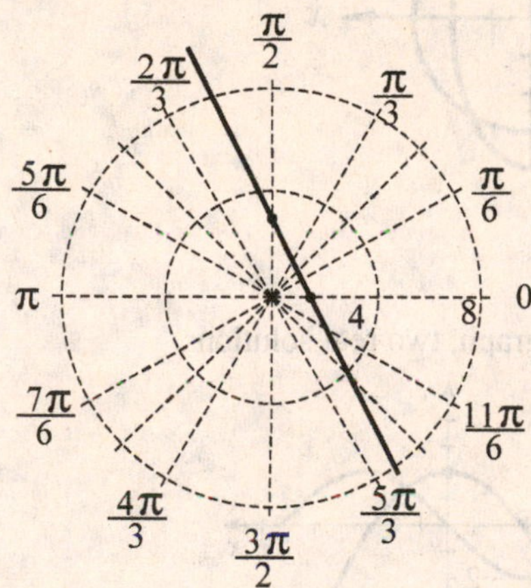

45. Given $y = 2x$

$\dfrac{y}{x} = 2$

$\tan\theta = 2$

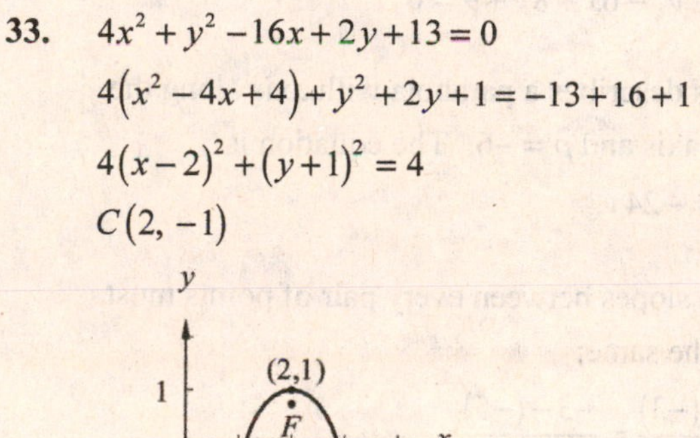

$\theta = \tan^{-1} 2 = 1.11$

49. Given $r = 2\sin 2\theta$; $r = \sqrt{x^2 + y^2}$

$$\sin\theta = \frac{y}{r} = \frac{y}{\sqrt{x^2 + y^2}};\ \cos\theta = \frac{x}{r} = \frac{x}{\sqrt{x^2 + y^2}}$$

$$r = 2(2\sin\theta\cos\theta)$$

$$r = 4\sin\theta\cos\theta$$

$$\sqrt{x^2 + y^2} = 4\frac{y}{\sqrt{x^2 + y^2}} \times \frac{x}{\sqrt{x^2 + y^2}}$$

$$\sqrt{x^2 + y^2} = \frac{4xy}{x^2 + y^2}$$

$$(x^2 + y^2)^3 = 16x^2y^2$$

53. $x^2 + y^2 = 9$

$4x^2 + y^2 = 16$

From the graph, there are four real solutions.

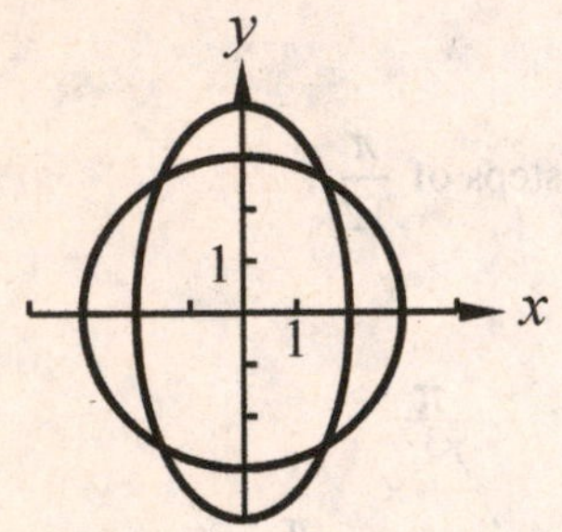

57. $y = 2\sin x$

$y = 2 - x^2$

From the graph, two real solutions.

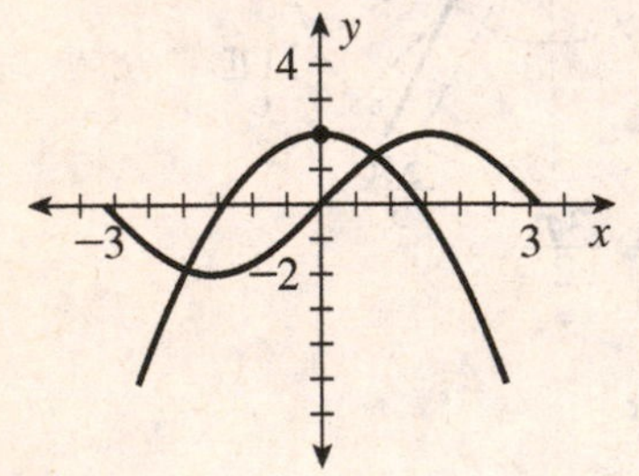

61. $2x^2 + 2y^2 + 4y - 3 = 0$, use quadratic formula

$$y = \frac{-4 \pm \sqrt{4^2 - 4(2)(2x^2 - 3)}}{2(2)}$$

$$= -1 \pm \frac{1}{2}\sqrt{10 - 4x^2}$$

Graph $y_1 = -1 + \frac{1}{2}\sqrt{10 - 4x^2}$, $y_2 = -1 - \frac{1}{2}\sqrt{10 - 4x^2}$

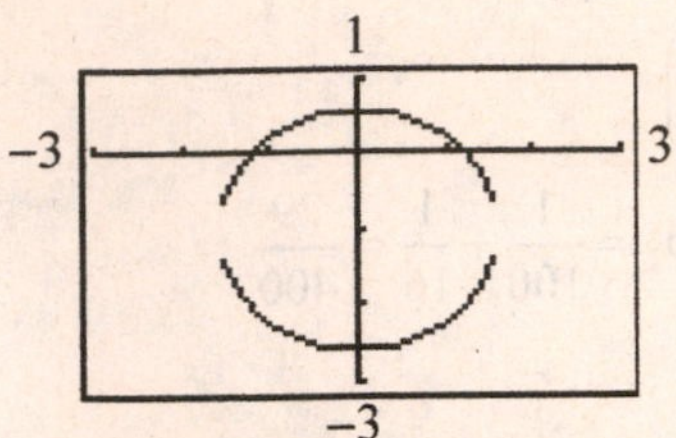

65. Graph $r_1 = 3\cos\frac{3\theta}{2}$

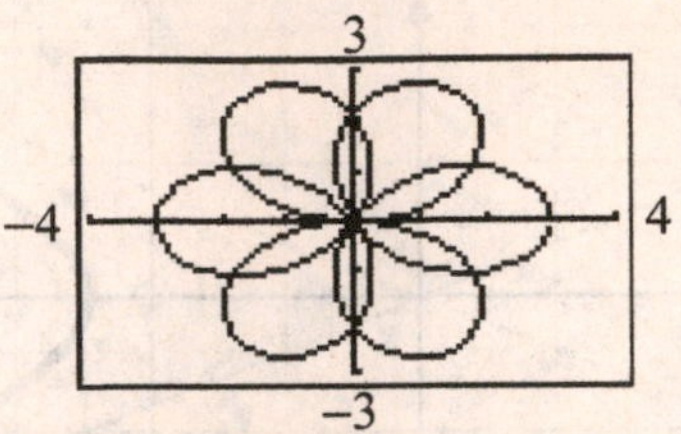

69. $(x-3)^2 + (y+4)^2 = 4^2$

$x^2 - 6x + 9 + y^2 + 8y + 16 = 16$

$x^2 + y^2 - 6x + 8y + 9 = 0$

73. This describes a parabola with axis along the y – axis and $p = -6$. The equation is $x^2 = -24y$.

77. The slopes between every pair of points must be the same:

$$\frac{x - (-3)}{13 - 3} = \frac{-3 - (-5)}{3 - (-2)}$$

$$x = 1$$

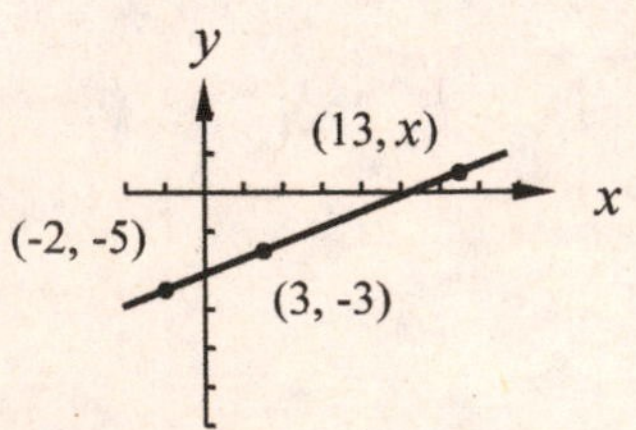

81. $a^2+b^2=(-3-2)^2+(11+1)^2+(14-2)^2+(4+1)^2$

$a^2+b^2=338$

$c^2=(14+3)^2+(4-11)^2=338,$

points form a right triangle

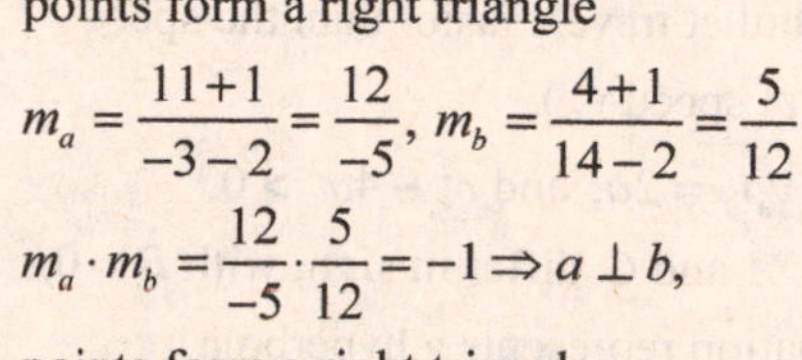

$m_a=\dfrac{11+1}{-3-2}=\dfrac{12}{-5},\ m_b=\dfrac{4+1}{14-2}=\dfrac{5}{12}$

$m_a\cdot m_b=\dfrac{12}{-5}\cdot\dfrac{5}{12}=-1\Rightarrow a\perp b,$

points form a right triangle

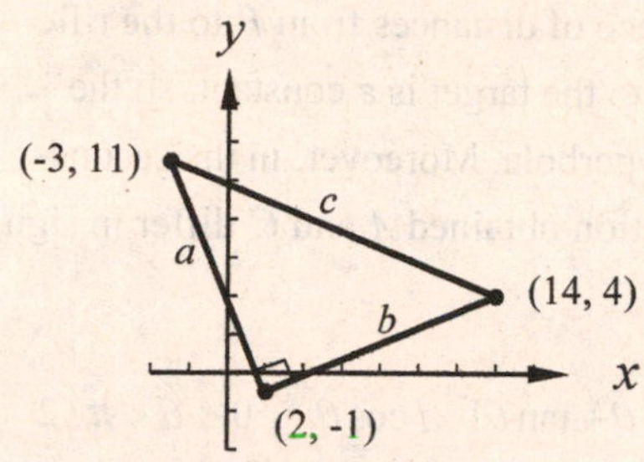

85.

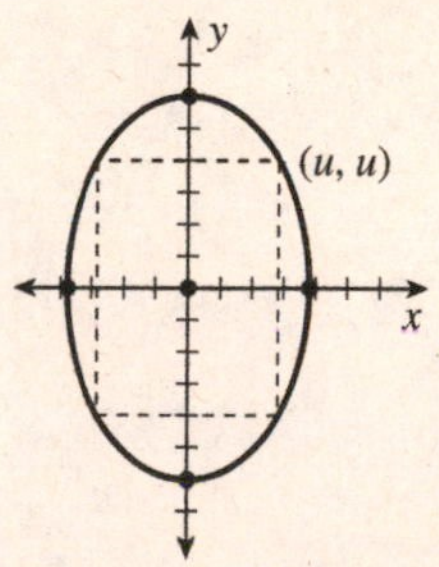

Let (u,u) be the corner of the square in the first quadrant (the corner is on the line $y=x$). Then (u,u) satisfies the equation of the ellipse:

$7u^2+2u^2=18$

$9u^2=18$

$u=\sqrt{2}$

Square has side $=2\sqrt{2}$

Area of square $=\left(2\sqrt{2}\right)^2=8$

89. 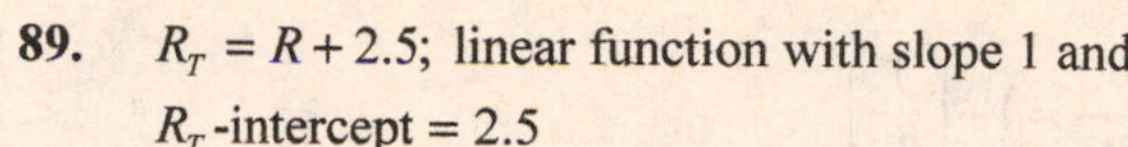
$R_T=R+2.5$; linear function with slope 1 and R_T-intercept $=2.5$

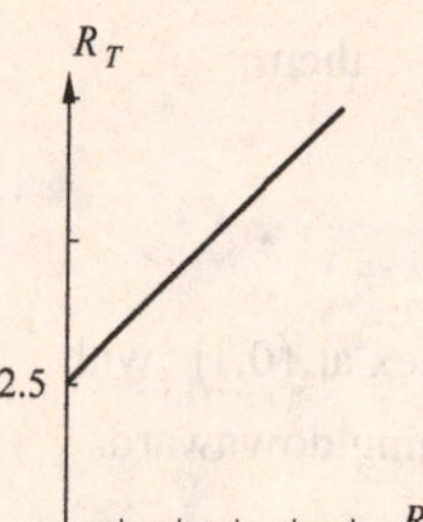

93. 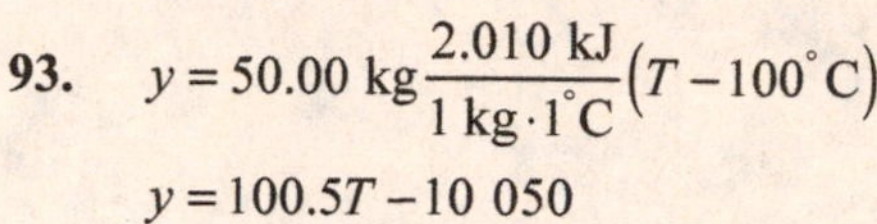
$y=50.00\text{ kg}\dfrac{2.010\text{ kJ}}{1\text{ kg}\cdot 1^\circ\text{C}}\left(T-100^\circ\text{C}\right)$

$y=100.5T-10\ 050$

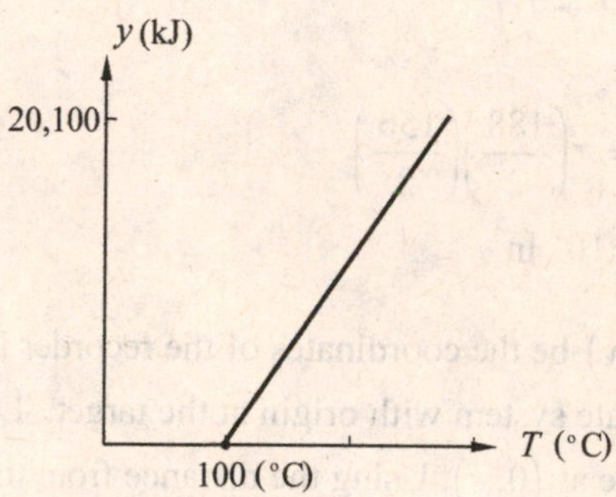

97. The intersection of the cone and the highway is a circle. Its radius is $r=490\tan 7^\circ$

$A=\pi r^2=\pi\left(490\tan 7^\circ\right)^2$

$A=11{,}000\text{ ft}^2$

101. $V=\begin{cases}12000-1250t, & 0\le t\le 4\\ 7000-1000(t-4), & 4\le t\le 11\end{cases}$

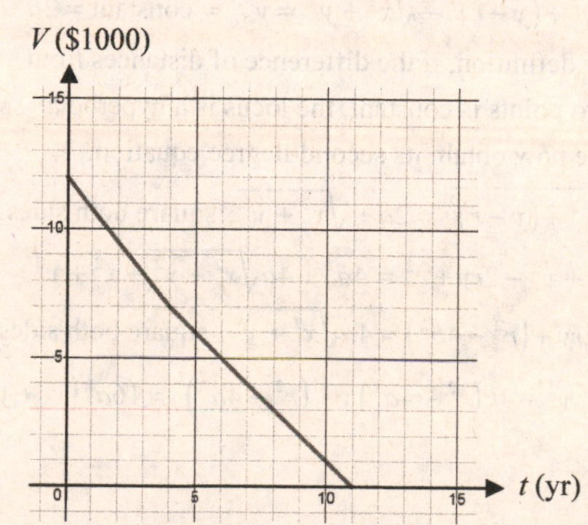

105. $\frac{H_T}{H_0} = 1 - \left(\frac{T}{T_0}\right)^2$

If we let $y = \frac{H_T}{H_0}$ and $x = \frac{T}{T_0}$, then

$y = 1 - x^2$

$x^2 = -(y-1)$

This is a parabola with vertex at $(0,1)$, with axis on the y-axis and opening downward.

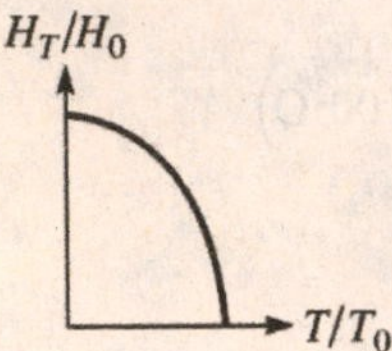

108. $A = \pi ab = \pi\left(\frac{188}{2}\right)\left(\frac{156}{2}\right)$

$A = 2.30 \times 10^4\ \text{m}^2$

115. Let $P(x, y)$ be the coordinates of the recorder in a coordinate system with origin at the target. Let the rifle be at $(0, r)$. Using the distance from the recorder to the rifle, we have

$\sqrt{x^2 + (y-r)^2} = v_s(t_0 + t_1)$ where t_0 is the time for the bullet to reach the target and t_1 is the time for sound to reach the detector from the target.

Using the distance from the recorder to the target,

$\sqrt{x^2 + y^2} = v_s t_1$

Subtracting

$\sqrt{x^2 + (y-r)^2} - \sqrt{x^2 + y^2} = v_s t_0 = \text{constant} = 2a$

By definition, if the difference of distances from two points is constant, the locus is a hyperbola.

We now obtain its second-degree equation.

$\sqrt{x^2 + (y-r)^2} = 2a + \sqrt{x^2 + y^2}$; square both sides

$x^2 + y^2 - 2ry + r^2 = 4a^2 + 4a\sqrt{x^2 + y^2} + x^2 + y^2$

$-2ry + (r^2 - 4a^2) = 4a\sqrt{x^2 + y^2}$; square both sides

$4r^2y^2 - 4r(r^2 - 4a^2)y + (r^2 - 4a^2)^2 = 16a^2(x^2 + y^2)$

$-16a^2x^2 + 4(r^2 - 4a^2)y^2 - 4r(r^2 - 4a^2)y$

$+ (r^2 - 4a^2)^2 = 0$

which has the form $Ax^2 + Cy^2 + Ey + F = 0$

Since the bullet travels faster than the speed of sound (at speed v_B),

$r = v_B t_0 > v_s t_0 = 2a$, and $r^2 - 4a^2 > 0$.

Therefore, A and C differ in sign, with $B = 0$, so the equation represents a hyperbola.

Summary:

The difference of distances from P to the rifle and from P to the target is a constant, so the locus is a hyperbola. Moreover, in the second-degree equation obtained A and C differ in sign with $B = 0$.

119. $r = 200(\sec\theta + \tan\theta)^{-5} / \cos\theta, \quad 0 < \theta < \pi/2$

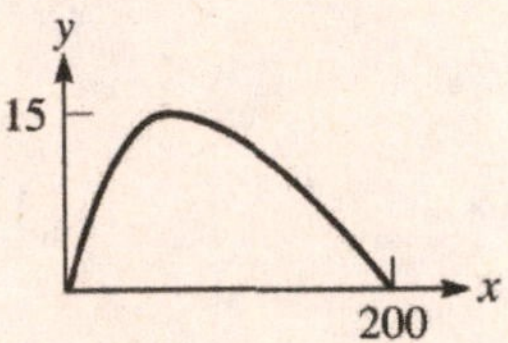

Chapter 22

Introduction to Statistics

22.1 Graphical Displays of Data

1.

Birth Weight. (kg)	1.0 − 2.0	2.0 − 3.0	3.0 − 4.0	4.0 − 5.0
Freq.	1	8	20	3
Rel. Freq. (%)	3.1	25.0	62.5	9.4

5. Qualitative

9. A pie chart is not suitable since the relative frequencies do not sum to 100%.

13.

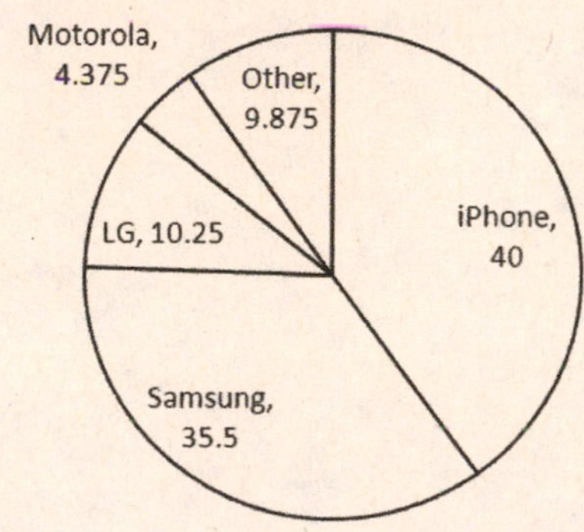

17.

Frequency
10
9
8
7
6
5
4
3
2
1
0
141-144 144-147 147-150 150-153 153-156

21.

Number of apps	Frequency
50 – 60	1
60 – 70	1
70 – 80	2
80 – 90	5
90 – 100	7
100 – 110	9
110 – 120	4
120 – 130	1

25.

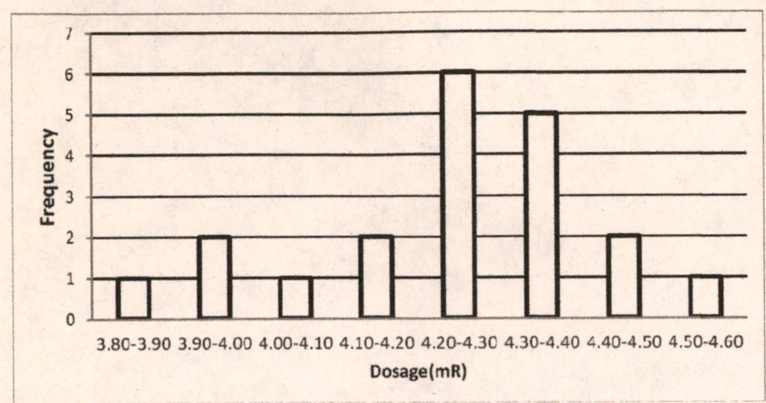

29.

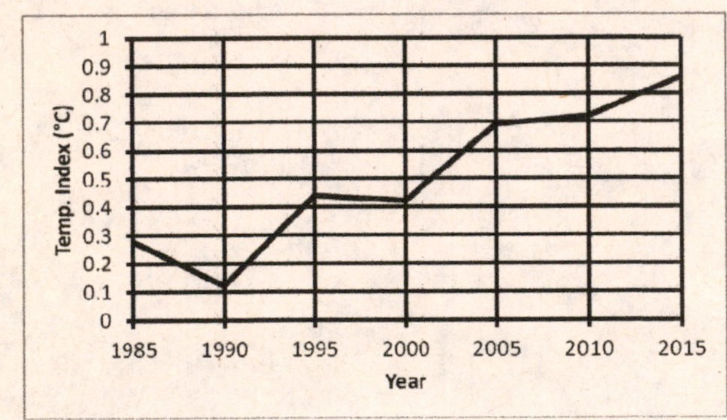

22.2 Measures of Central Tendency

1. 1, 2, 2, 3, 4, 4, 4, 6, 7, 7, 8, 9, 9, 11
There are 14 numbers. The median is halfway between the seventh and eighth numbers.

$$\text{median} = \frac{4+6}{2} = 5$$

5. Arrange the numbers in numerical order:
2, 3, 3, 3, 4, 4, 4, 4, 4, 5, 5, 6, 6, 6, 7, 7
There are 15 numbers. The middle number is the eighth observation. Since the eighth number is 4, the median is 4.

9. Arrange the numbers in numerical order:
2, 3, 3, 3, 4, 4, 4, 4, 4, 5, 5, 6, 6, 6, 7, 7
The arithmetic mean is:

$$\bar{x}$$

$$= \frac{2+3+3+3+4+4+4+4+4+5+5+6+6+6+7+7}{15}$$

$$= \frac{69}{15} = 4.6$$

13. Arrange the numbers in numerical order:
2, 3, 3, 3, 4, 4, 4, 4, 4, 5, 5, 6, 6, 6, 7, 7
The observation that occurs most frequently is 4.
The mode is 4.

17. Arrange in ascending order:
143,144,144,144,145,146,146,146,147,147,
148,148,148,148,149,149,149,150,151,153
$n = 20$;
The median is between observations 10 and 11, or 147.5.

21. Arrange in ascending order:
58,67,76,78,82,85,86,86,89,91,92,93,93,95,97,
99,101,101,103,103,105,105,107,107,109,110,111,112,115,126
$n = 30$;
The mean is $x = \frac{\sum x}{30} = 96.1$.

25. Arrange in ascending order:
3.83,3.90,3.96,4.09,4.15,4.18,4.21,4.23,4.25,4.26,
4.27,4.29,4.33,4.34,4.36,4.36,4.37,4.41,4.44,4.51
$n = 20$;
The mode is the most common observation, or 4.36.

29. Arrange in ascending order:
525,550,575,575,600,600,625,
625,700,700,700,750,750,800
$n = 14$;
The median is 625 and the mode is 700.

33. We use the formula

$$\bar{x} = \frac{xf}{f}$$

$$= \frac{0.0057(18) + 0.0058(36) + 0.0059(50) + 0.0060(65) + 0.0061(31)}{18 + 36 + 50 + 65 + 31}$$

$$= \frac{1.1855}{200}$$

$$= 0.00593$$

37. The midrange is $\frac{525 + 800}{2} = 662.50$

41. We replace the final salary ($700) with $4000.

Arrange in ascending order:

525, 550, 575, 575, 600, 600, 625,

625, 700, 700, 750, 750, 800, 4000

$n = 14$;

The new mean is 2262.50.

22.3 Standard Deviation

1.

x	$x-\bar{x}$	$(x-\bar{x})^2$
6	2	4
5	1	1
4	0	0
7	3	9
6	2	4
2	−2	4
1	−3	9
1	−3	9
5	1	1
3	−1	1
40		42

$\bar{x} = 40/10 = 4$

$$\frac{\sum(x-\bar{x})^2}{n-1} = \frac{42}{10-1} = \frac{42}{9}$$

$$s = \sqrt{\frac{42}{9}} = 2.2$$

5.

x	$x-\bar{x}$	$(x-\bar{x})^2$	x^2
0.45	−0.055	0.003 03	0.2025
0.46	−0.045	0.002 03	0.2116
0.47	−0.035	0.001 23	0.2209
0.48	−0.025	6.3×10^{-4}	0.2304
0.48	−0.025	6.3×10^{-4}	0.2304
0.49	−0.015	2.3×10^{-4}	0.2401
0.49	−0.015	2.3×10^{-4}	0.2401
0.49	−0.015	2.3×10^{-4}	0.2401
0.50	−0.005	2.5×10^{-5}	0.25
0.51	0.005	2.5×10^{-5}	0.2601
0.53	0.025	6.3×10^{-4}	0.2809
0.53	0.025	6.3×10^{-4}	0.2809
0.53	0.025	6.3×10^{-4}	0.2809
0.55	0.045	0.00203	0.3025
0.55	0.045	0.00203	0.3025
0.57	0.045	0.00203	0.3249

$$\sum x = 8.08,\ \bar{x} = \frac{\sum x}{n} = \frac{8.08}{16} = 0.505$$

$$\sum\left(x-\bar{x}\right)^2 = 0.0184, \sum x^2 = 4.0988$$

Using Eq. (22.2):

$$s = \sqrt{\frac{\sum\left(x-\bar{x}\right)^2}{n-1}}$$

$$= \sqrt{\frac{0.0184}{15}} = 0.0340$$

9. Using Eq. (22.3):

$$s = \sqrt{\frac{n\sum x^2 - \left(\sum x\right)^2}{n(n-1)}}$$

$$= \sqrt{\frac{16(4.0988) - (8.08)^2}{16(15)}}$$

$$= 0.0350$$

13.

```
1-Var Stats
x̄=0.505
Σx=8.08
Σx²=4.0988
Sx=0.0350238014
σx=0.0339116499
n=16
minX=0.45
↓Q1=0.48
```

17.

```
1-Var Stats
x̄=96.06666667
Σx=2882
Σx²=283168
Sx=14.74363295
σx=14.49582315
n=30
minX=58
↓Q1=86
```

Using Eq. (22.23):

$$s = \sqrt{\frac{n\sum x^2 - \left(\sum x\right)^2}{n(n-1)}}$$

$$= \sqrt{\frac{30(283168) - (2882)^2}{30(30-1)}}$$

$$= 14.7$$

21.

```
1-Var Stats
x̄=862.2
Σx=862200
Σx²=764940000
Sx=146.8765901
σx=146.8031335
n=1000
minX=500
↓Q1=800
```

Using Eq. (22.23):

$$s = \sqrt{\frac{n\sum x^2 - \left(\sum x\right)^2}{n(n-1)}}$$

$$= \sqrt{\frac{1000(764940000) - (862200)^2}{1000(1000-1)}}$$

$$= 147$$

22.4 Normal Distributions

1. $\mu = 10$, $\sigma = 5$ and $\mu = 20$, $\sigma = 5$ result in the same curve, with the first centred at $x = 10$ and the second centered at $x = 20$.

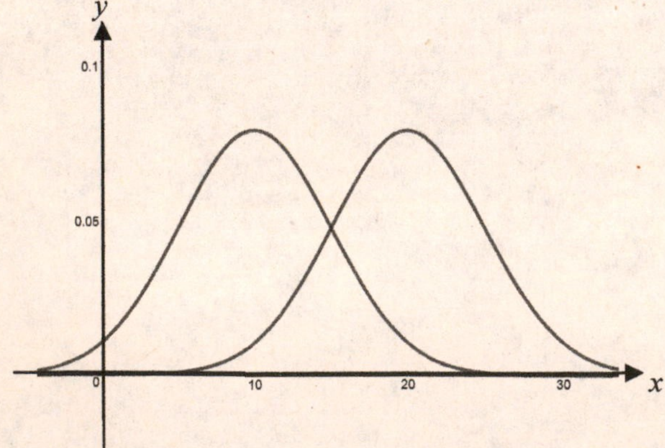

5. 95% of bags have a weight between 58 lb and 62 lb; such a weight is within two standard deviations from the mean.

9. $\mu = 200, \sigma = 15.$

$$z = \frac{218-200}{15} = 1.2$$

The area under the curve to the right of 1.2 is $0.5000 - 0.3849 = 0.1151$, so 11.5% of test scores are expected to be above 218.

13. $\mu = 1.50, \sigma = 0.05.$ x between 1.45 and 1.55

$$z = \frac{1.45-1.50}{0.05} = -1,\ z = \frac{1.55-1.50}{0.05} = 1$$

The area under the curve between -1 and 1 is $0.3413 + 0.3413 = 0.6826$, or 68% by the empirical rule.

17. $\mu = 100\ 000, \sigma = 10\ 000.$ x between 85 000 and 100 000

$$z = \frac{85\ 000 - 100\ 000}{10\ 000} = -1.5$$

$$z = \frac{100\ 000 - 100\ 000}{10\ 000} = 0$$

The area under the curve between -1.5 and 0 is 0.4332, so 43% of tires that are expected to last between 85 000 km and 100 000 km.

21. $\sigma_{\bar{x}} = \frac{\sigma}{\sqrt{n}} = \frac{10\ 000}{\sqrt{100}} = 1000.$

The z-score for the sample is

$$\frac{98200-100000}{1000} = -1.8.$$

The area under the curve to the left of -1.8 is 0.5000-0.4641=0.0359 and so the mean for a sample of 100 tires is expected to be below 98200 about 3.6% of the time.

25. A z-value of 0.7 corresponds to 0.2580 and so .5000+0.2580=0.7580 (or 75.8%) is the area under the curve to the left of 0.7. By symmetry, 75.8% of the area under the normal curve is to the right of $z = -0.7$.

$\mu = 16.0$, $\sigma = 4.0$.

29. $z_1 = \dfrac{12.0-16.0}{4.0} = -1.0, z_2 = \dfrac{18.0-16.0}{4.0} = 0.5.$

The area under the curve to the left of $z_1 = -1.0$ is the same as the area to the right of $z = 1.0$, or 0.3413.

The area to the left of $z_2 = 0.5$ is $0.5000 + 0.1915 = 0.6915$.

The area between z_1 and z_2 is therefore $0.6915 - 0.3413 = 0.3502$, implying 35% of residents live between 12 and 18 km from the center of the city.

22.5 Statistical Process Control

1. With the new observations, the mean of Subgroup 1 changes to 497.8, and the range R of the subgroup stays the same ($R = 9$). Knowing the sum of the means from before, we can recalculate the new sum without constructing a complete table by subtracting 502.2 (the previous mean for Subgroup 1) and adding 497.8 (the new mean for Subgroup 1). Therefore,

$$\overline{\overline{x}} = \frac{9998 - 502.2 + 497.8}{20}$$

$$= 499.7$$

Hence, $\text{UCL}(\overline{x}) = \overline{\overline{x}} + A_2\overline{R} = 499.7 + 0.577(8.7) = 504.7 \text{ mg}$

$\text{LCL}(\overline{x}) = \overline{\overline{x}} - A_2\overline{R} = 499.7 - 0.577(8.7) = 494.7 \text{ mg}$

5.

Hour	Torques	(N · m)	of Five	Engines	
1	366	352	354	360	362
2	370	374	362	366	356
3	358	357	365	372	361
4	360	368	367	359	363
5	352	356	354	348	350
6	366	361	372	370	363
7	365	366	361	370	362
8	354	363	360	361	364
9	361	358	356	364	364
10	368	366	368	358	360
11	355	360	359	362	353
12	365	364	357	367	370
13	360	364	372	358	365
14	348	360	352	360	354
15	358	364	362	372	361
16	360	361	371	366	346
17	354	359	358	366	366
18	362	366	367	361	357
19	363	373	364	360	358
20	372	362	360	365	367

Subgroup	Mean $\bar{x}$	Range R
1	358.8	14
2	365.6	18
3	362.6	15
4	363.4	9
5	352.0	8
6	366.4	11
7	364.8	9
8	360.4	10
9	360.6	8
10	364.0	10
11	357.8	9
12	364.6	13
13	363.8	14
14	354.8	12
15	363.4	14
16	360.8	25
17	360.6	12
18	362.6	10
19	363.6	15
20	365.2	12
Sum	7235.8	248
Mean	361.79	12.4

$$\text{CL}: \bar{\bar{x}} = 361.79 \text{ N}\cdot\text{m}$$

$$\text{UCL}(\bar{x}) = \bar{\bar{x}} + A_2\bar{R} = 361.79 + 0.577(12.4)$$
$$= 368.9 \text{ N}\cdot\text{m}$$

$$\text{LCL}(\bar{x}) = \bar{\bar{x}} - A_2\bar{R} = 361.79 - 0.577(12.4)$$
$$= 354.6 \text{ N}\cdot\text{m}$$

9.

Subgroup	Output	Voltages	Five	Adaptors	
1	9.03	9.08	8.85	8.92	8.90
2	9.05	8.98	9.20	9.04	9.12
3	8.93	8.96	9.14	9.06	9.00
4	9.16	9.08	9.04	9.07	8.97
5	9.03	9.08	8.93	8.88	8.95
6	8.92	9.07	8.86	8.96	9.04
7	9.00	9.05	8.90	8.94	8.93
8	8.87	8.99	8.96	9.02	9.03
9	8.89	8.92	9.05	9.10	8.93
10	9.01	9.00	9.09	8.96	8.98
11	8.90	8.97	8.92	8.98	9.03
12	9.04	9.06	8.94	8.93	8.92
13	8.94	8.99	8.93	9.05	9.10
14	9.07	9.01	9.05	8.96	9.02
15	9.01	8.82	8.95	8.99	9.04
16	8.93	8.91	9.04	9.05	8.90
17	9.08	9.03	8.91	8.92	8.96
18	8.94	8.90	9.05	8.93	9.01
19	8.88	8.82	8.89	8.94	8.88
20	9.04	9.00	8.98	8.93	9.05
21	9.00	9.03	8.94	8.92	9.05
22	8.95	8.95	8.91	8.90	9.03
23	9.12	9.04	9.01	8.94	9.02
24	8.94	8.99	8.93	9.05	9.07

Subgroup	Mean $\overline{\overline{x}}$	Range R
1	8.956	0.23
2	9.078	0.22
3	9.018	0.21
4	9.064	0.19
5	8.974	0.20
6	8.970	0.21
7	8.964	0.15
8	8.974	0.16
9	8.978	0.21
10	9.008	0.13
11	8.960	0.13
12	8.978	0.14
13	9.002	0.17
14	9.022	0.11
15	8.962	0.22
16	8.966	0.15
17	8.980	0.17
18	8.966	0.15
19	8.882	0.12
20	9.000	0.12
21	8.988	0.13
22	8.948	0.13
23	9.026	0.18
24	8.996	0.14
Sum	215.66	3.97
Mean	8.986	0.1654

CL: $\overline{\overline{x}} = 8.986$ V

$$\text{UCL}(\overline{x}) = \overline{\overline{x}} + A_2\overline{R} = 8.986 + 0.577(0.1654) = 9.081 \text{ V}$$

$$\text{LCL}(\overline{x}) = \overline{\overline{x}} - A_2\overline{R} = 8.986 - 0.577(0.1654) = 8.891 \text{ V}$$

13. CL: $\mu = 2.725$ in

$$\text{UCL}(\overline{x}) = \mu + A\sigma = 2.725 + 1.342(0.0032) = 2.729 \text{ in}$$

$$\text{LCL}(\overline{x}) = \mu - A\sigma = 2.725 - 1.342(0.0032) = 2.721 \text{ in}$$

17.

Week	Accounts with Errors	Proportion with Errors
1	52	0.052
2	36	0.036
3	27	0.027
4	58	0.058
5	44	0.044
6	21	0.021
7	48	0.048
8	63	0.063
9	32	0.032
10	38	0.038
11	27	0.027
12	43	0.043
13	22	0.022
14	35	0.035
15	41	0.041
16	20	0.020
17	28	0.028
18	37	0.037
19	24	0.024
20	42	0.042
Total	738	

$$\text{CL: } \overline{p} = \frac{738}{1000(20)} = 0.0369$$

$$\sigma_{\overline{p}} = \sqrt{\frac{\overline{p}(1-\overline{p})}{n}} = \sqrt{\frac{0.0369(1-0.0369)}{1000}} = 0.005\,96$$

$$\text{UCL}(p) = 0.0369 + 3(0.005\,96) = 0.0548$$

$$\text{LCL}(p) = 0.0369 - 3(0.005\,96) = 0.0190$$

22.6 Linear Regression

1.

x	y	xy	x^2
1	3	3	1
2	7	14	4
3	9	27	9
4	9	36	16
5	12	60	25
15	40	140	55

$n = 5$

$$m = \frac{n\sum xy - \sum x \sum y}{n\sum x^2 - \left(\sum x\right)^2} = \frac{5(140) - 15(40)}{5(55) - 15^2} = 2$$

$$b = \frac{\sum x^2 \sum y - \sum xy \sum x}{n\sum x^2 - \left(\sum x\right)^2} = \frac{55(40) - 140(15)}{5(55) - 15^2} = 2$$

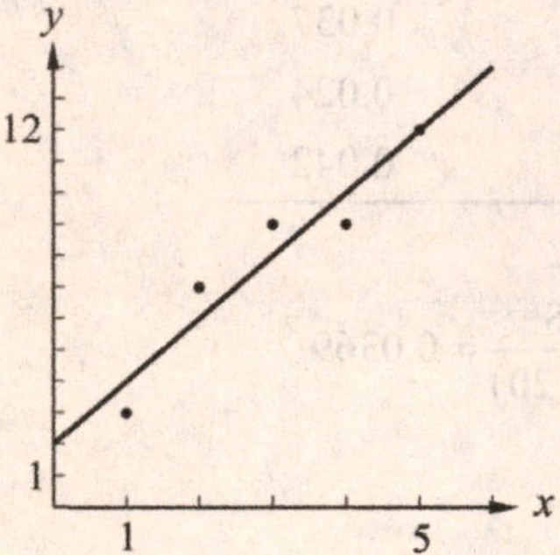

The equation of the least-squares line is $y = 2x + 2$.

5.

t(h)	1.0	2.0	4.0	8.0	10.0	12.0
y(mg/dL)	8.7	8.4	7.7	7.3	5.7	5.2

t	y	ty	t^2
1.0	8.7	8.7	1.0
2.0	8.4	16.8	4.0
4.0	7.7	30.8	16.0
8.0	7.3	58.4	64.0
10.0	5.7	57	100
12.0	5.2	62.4	144
37.0	43	234.1	329

$n = 6$

$$m = \frac{n\sum ty - \sum t \sum y}{n\sum t^2 - \left(\sum t\right)^2} = \frac{6(234.1) - 37.0(43)}{6(329) - 37^2} = -0.308$$

$$b = \frac{\sum t^2 \sum y - \sum ty \sum t}{n\sum t^2 - \left(\sum t\right)^2} = \frac{329(43) - 234.1(37)}{6(329) - 37^2} = 9.07$$

The least-squares line is

$y = -0.308t + 9.07$

To plot the line we use two points:

t	y
0	9.07
5	7.53

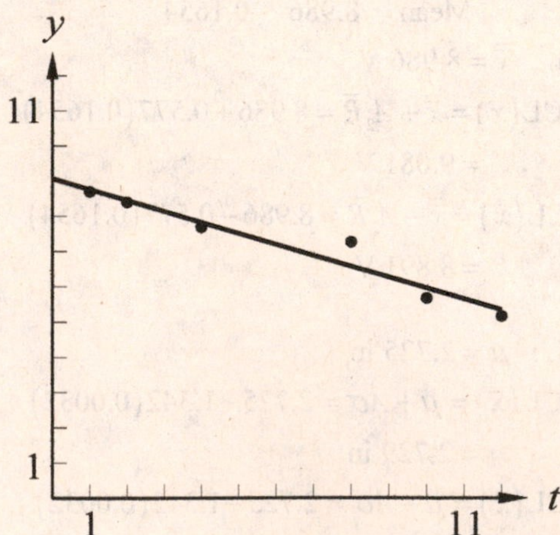

9.

x	h	$xh/(10)^3$	$x^2/(10)^3$
0	0	0	0
500	1130	565	250
1000	2250	2250	1000
1500	3360	5040	2250
2000	4500	9000	4000
2500	5600	14 000	6250
7500	16 840	30 855	13 750

$n = 6$

$$m = \frac{n\sum xh - \sum x \sum h}{n\sum x^2 - \left(\sum x\right)^2}$$

$$= \frac{6(30\,855\times10^3) - (7500)(16\,840)}{6(13{,}750\times10^3) - (7500)^2} = 2.24$$

$$b = \frac{\sum x^2 \sum y - \sum xy \sum x}{n\sum x^2 - \left(\sum x\right)^2}$$

$$= \frac{(13\,750\times10^3)(16\,840) - (30\,855\times10^3)(7500)}{6(13\,750\times10^3) - (7500)^2}$$

$= 5.24$

The least-squares line is

$h = 2.24x + 5.24$

To plot the line we use two points:

x	h
750	1690
2250	5045

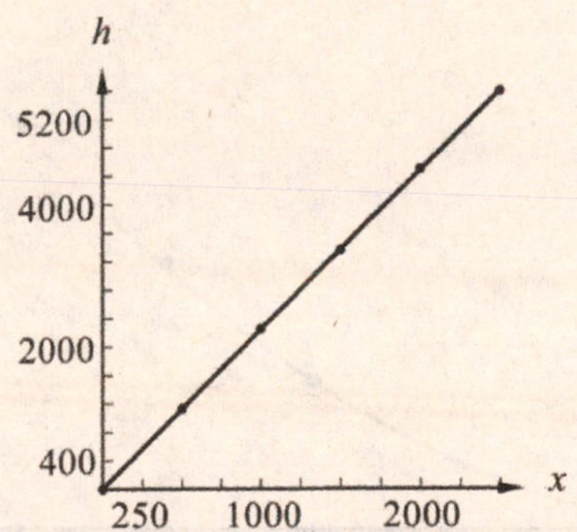

13.

f	V	fV	f^2
0.550	0.350	0.192 50	0.302 500
0.605	0.600	0.363 00	0.366 025
0.660	0.850	0.561 00	0.435 600
0.735	1.10	0.808 50	0.540 225
0.805	1.45	1.167 25	0.648 025
0.880	1.80	1.584 00	0.774 400
4.235	6.15	4.676 25	3.066 775

$n = 6$

$$m = \frac{n\sum fV - \sum f \sum V}{n\sum f^2 - \left(\sum f\right)^2}$$

$$= \frac{6(4.676\,25) - (4.235)(6.15)}{6(3.066\,775) - (4.235)^2} = 4.32$$

$$b = \frac{\sum f^2 \sum V - \sum fV \sum f}{n\sum f^2 - \left(\sum f\right)^2}$$

$$= \frac{(3.066\,775)(6.15) - (4.676\,25)(4.235)}{6(3.066\,775) - (4.235)^2} = -2.03$$

The least-squares line is

$V = 4.32f - 2.03$

To find the threshold frequency, we set V to 0 and solve for f,

$$f_0 = \frac{2.03}{4.32} = 0.470 \text{ PHz}$$

To plot the line we use two points:

f (PHz)	V
0.600	0.562
0.800	1.426

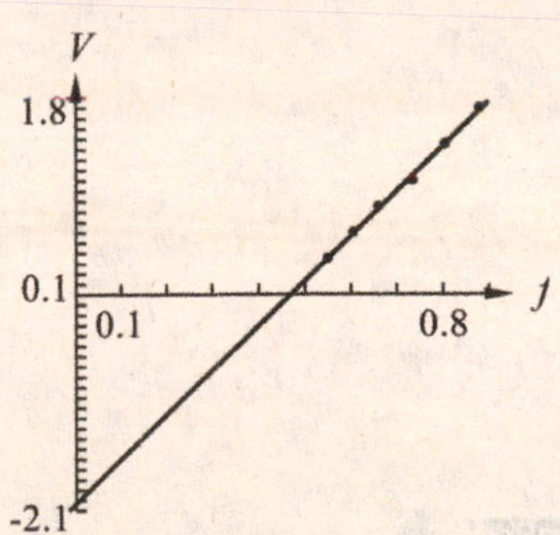

The threshold frequency is 0.469.

17.

t	T	t^2	T^2
0.0	20.5	0	420.25
1.0	20.6	1.0	424.36
2.0	20.9	4.0	436.81
3.0	21.3	9.0	453.69
4.0	21.7	16.0	470.89
5.0	22.0	25.0	484.00
15.0	127.0	55.0	2690.00

$$s_t = \sqrt{\frac{n\left(\sum t^2\right) - \left(\sum t\right)^2}{n(n-1)}}$$

$$= \sqrt{\frac{6(55.0) - (15.0)^2}{6(5)}}$$

$$= 1.8708$$

$$s_T = \sqrt{\frac{n\left(\sum T^2\right) - \left(\sum T\right)^2}{n(n-1)}}$$

$$= \sqrt{\frac{6(2690) - (127)^2}{6(5)}}$$

$$= 0.6055$$

$m = 0.32$

$$r = m\frac{s_t}{s_T} = 0.32 \cdot \frac{1.8708}{0.6055} = 0.9886$$

$r^2 = 0.9774$

The linear relationship is very strong.

22.7 Nonlinear Regression

1. $y = ab^x$

x	y
0	350
10	570
20	929
30	1513
40	2464

```
ExpReg
y=a*b^x
a=349.9981679
b=1.050004339
```

$y = 350(1.05^x)$

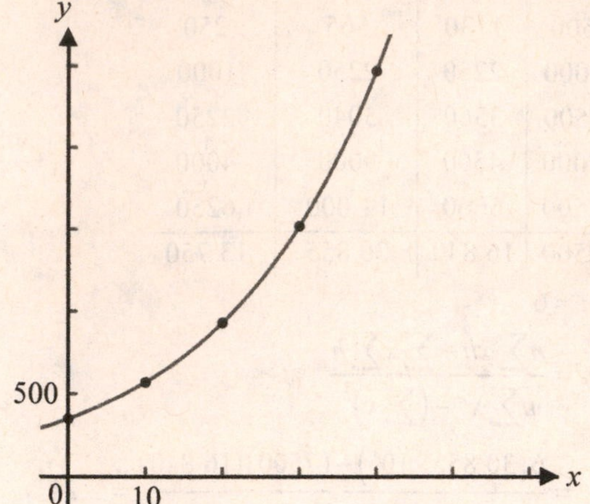

5. $y = at^2 + bt + c$

t	y
1.0	6.0
2.0	23
3.0	55
4.0	98
5.0	148

```
QuadReg
y=ax²+bx+c
a=5.5
b=2.9
c=-3.2
```

$y = 5.5t^2 + 2.9t - 3.2$

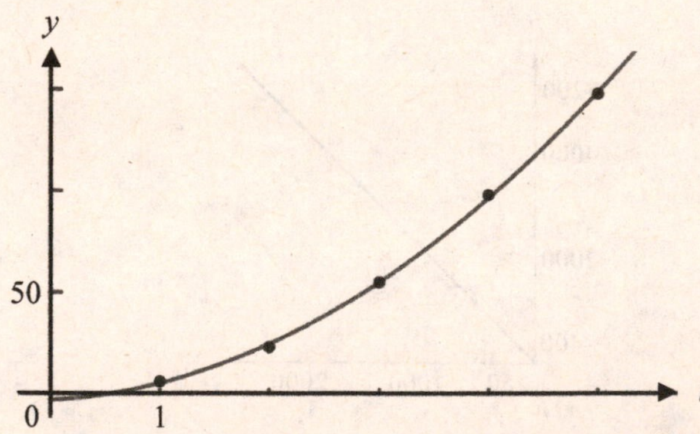

At $t = 2.5$, $y = 5.5(2.5)^2 + 2.9(2.5) - 3.2 = 38.425$

We predict a distance of 38 cm at $t = 2.5$ s.

9. $y = ax^b$

f	T
500	220
1000	120
1500	77
2000	50
2500	43
3000	30

```
PwrReg
y=a*x^b
a=217524.7227
b=-1.09743478
```

$T = 217500f^{-1.097}$

When $f = 3500, T = 217500(3500)^{-1.097} = 28$ J. This is an extrapolation since 3500 is out of the range of the given measurements.

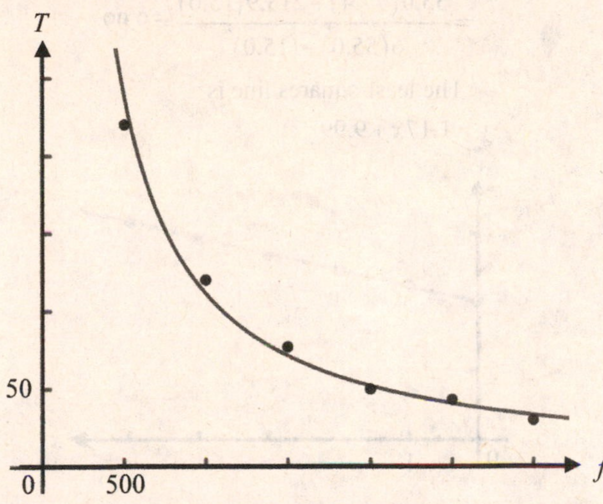

13.

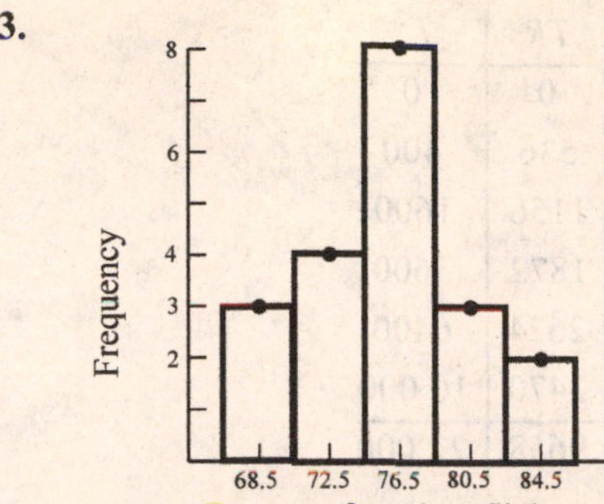

17. $\overline{x} = \frac{\sum x}{n} = \frac{1.61}{12} = 0.134$ mL/L

21. There are 121 observations, so the median is observation 61, which falls in the 700 W class. Therefore, the median is 700 W.

25. $\overline{x} = \frac{\sum xf}{\sum f} = \frac{843}{200} = 4.2$ particles

29. Given $\mu = 100, \sigma = 15$ scores between 70 and 130 fall within 2 standard deviations of the mean. The empirical rule states 95% of scores should be in this range.

33. CL: $\overline{p} = \frac{540}{500 \cdot 20} = 0.0540$

$$\sigma_{\overline{p}} = \sqrt{\frac{\overline{p}(1-\overline{p})}{n}} = \sqrt{\frac{0.054(1-0.054)}{500}} = 0.010\ 11$$

$\text{UCL}(p) = 0.054 + 3(0.010\ 11) = 0.0843$

$LCL(p) = 0.054 - 3(0.010\ 11) = 0.0237$

37. The area under the curve between -1.6 and 2.1 is $0.4452 + 0.4821 = 0.9273$ from Table 22.1.

41. $\mu = 2.20,\ \sigma = 0.50.$ x between 1.50 and 2.50

$$z = \frac{1.50 - 2.20}{0.50} = -1.4,\ z = \frac{2.50 - 2.20}{0.50} = 0.6$$

The area under the curve between -1.4 and 0.6 is $0.4192 + 0.2257 = 0.6449$, so 64.5% of the readings are expected to be between 1.50 μg/m^3 and 2.50 μg/m^3.

Review Exercises

1. This is true because $\frac{3}{25} = 0.12$, or 12%.

5. This is false. Substituting $x = 2$ into $y = 2x + 4$ yields $y = 8$. The deviation is $8 - 6 = 2$.

9. $\overline{x} = \frac{\sum x}{n} = \frac{1523}{20} = 76.2$ %

45.

T	R	TR	T^2
0.0	25.0	0	0
20.0	26.8	536	400
40.0	28.9	1156	1600
60.0	31.2	1872	3600
80.0	32.8	2624	6400
100	34.7	3470	10 000
300	179.4	9658	22 000

$n = 6$

$$m = \frac{n \quad TR - \quad T \quad R}{n \quad T^2 - \left(\quad T\right)^2}$$

$$= \frac{6(9658) - (300)(179.4)}{6(22\,000) - (300)^2} = 0.0983$$

$$b = \frac{T^2 \quad R - \quad TR \quad T}{n \quad T^2 - \left(\quad T\right)^2}$$

$$= \frac{(22\,000)(179.4) - (9658)(300)}{6(22\,000) - (300)^2} = 25.0$$

The least-squares line is

$R = 0.0983T + 25.0$

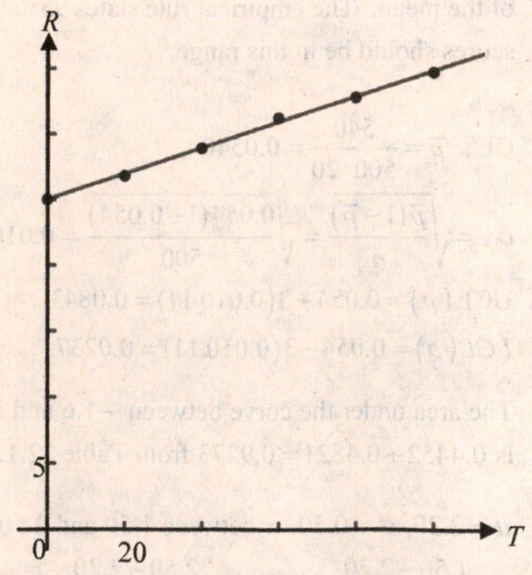

49.

x-Load (kg)	y-length (cm)	xy	x^2
0.0	10.0	0.0	0.0
1.0	11.2	11.2	1.0
2.0	12.3	24.6	4.0
3.0	13.4	40.2	9.0
4.0	14.6	58.4	16.0
5.0	15.9	79.5	25.0
15.0	77.4	213.9	55.0

$n = 6$

$$m = \frac{n\sum xy - \left(\sum x\right)\left(\sum y\right)}{n\sum x^2 - \left(\sum x\right)^2}$$

$$= \frac{6(213.9) - 15(77.4)}{6(55.0) - (15.0)^2} = 1.17$$

$$b = \frac{\left(\sum x^2\right)\left(\sum y\right) - \left(\sum xy\right)\left(\sum x\right)}{n\sum x^2 - \left(\sum x\right)^2}$$

$$= \frac{55.0(77.4) - 213.9(15.0)}{6(55.0) - (15.0)^2} = 9.99$$

The least-squares line is

$y = 1.17x + 9.99$

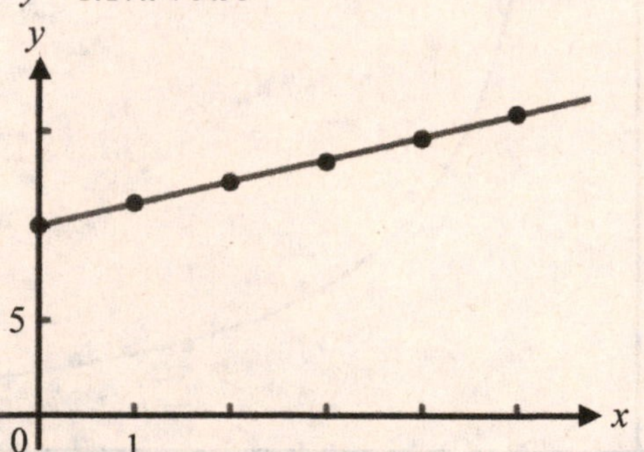

53.

x(m)	y(m)
0	15
100	17
200	23
300	33
400	47
500	65

```
QuadReg
y=ax²+bx+c
a=2E-4
b=0
c=15
```

$y = 0.0002x^2 + 15$

57. The power equation is found to be
$y = 1.39x^{0.878}$

61. Collect data from a random sample of workers regarding income and education.
A summary of descriptive measures should be done, including both measures of centre and of spread.
The equation of the least-squares curve can be obtained and then used to predict income based on education.

Chapter 23

The Derivative

23.1 Limits

1. $f(x)=\frac{1}{x+2}$

The function is not continuous at $x=-2$, due to a division by zero error at that point, making the function undefined at $x=-2$. The condition for continuity that the function must exist at that point is not satisfied.

5. $f(x)=3x^2-98x$

The function is continuous over all values of x. The graph is linear, no points are undefined, and there are no jumps of gaps in the value of the function, as any small change in x produces only a small change in the value of the function. Continuous on $x\in(-\infty,\infty)$ or $x\in\mathbb{R}$.

9. $f(x)=\frac{x}{\sqrt{x-2}}$

The function is not continuous at $x=2$ because the function is undefined due to a division by zero error. The function will also be discontinuous anywhere the argument of the square root is negative (if the sign of the numerator differs from the denominator). These are values of $x<2$. {Continuous on $x\in(2,\infty)$ }

13. Function is continuous for all values of x except $x=2$, since there is a "jump" in the graph, i.e. there is a large change in the value of the function for a small change in the value of x near $x=2$. So the function is continuous on intervals $x<2$ and $x>2$. {Continuous on $x\in(-\infty,2)$, $x\in(2,\infty)$ }

17. **(a)** $f(2)=-1$

(b) $\lim_{x\to2}f(x)$ does not exist since $\lim_{x\to2^-}f(x)=-1$ and $\lim_{x\to2^+}f(x)=2$ do not agree.

21. $f(x)=\begin{cases}x^2 & \text{for } x<2\\ 5 & \text{for } x\geq 2\end{cases}$

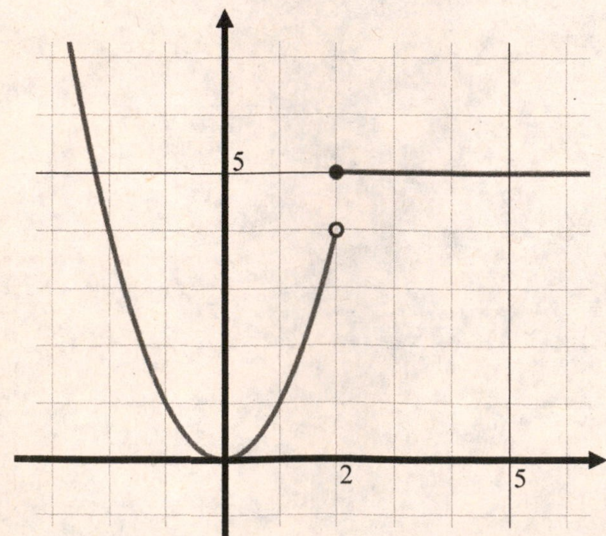

Function is continuous for all values of x except $x = 2$ since there is a "jump" in the graph, i.e. there is a large change in the value of the function for a small change in the value of x near $x = 2$. So the function is continuous on intervals $x < 2$ and $x \geq 2$. {Continuous on $x \in (-\infty, 2)$, $x \in [2, \infty)$ }

25. Find $\lim\limits_{x\to 1}\dfrac{x^3 - x}{x-1}$

x	0.900	0.990	0.999	1.001	1.010	1.100
$f(x)$	1.710	1.970	1.997	2.003	2.030	2.310

$\lim\limits_{x\to 1^-}\dfrac{x^3 - x}{x-1} = 2$ and $\lim\limits_{x\to 1^+}\dfrac{x^3 - x}{x-1} = 2$

Therefore, $\lim\limits_{x\to 1}\dfrac{x^3 - x}{x-1} = 2$

29. Find $\lim\limits_{x\to\infty}\dfrac{2x+1}{5x-3}$

x	10	100	1000	10 000
$f(x)$	0.446 81	0.404 43	0.400 44	0.400 04

Therefore, $\lim\limits_{x\to\infty}\dfrac{2x+1}{5x-3} = 0.4 = \dfrac{2}{5}$

33. $\lim\limits_{x\to 0}\dfrac{6x^2 + x}{x} = \dfrac{0^2 + 0}{0} = \dfrac{0}{0} = \text{indeterminate}$

Try factoring.

$\lim\limits_{x\to 0}\dfrac{6x^2 + x}{x} = \lim\limits_{x\to 0}\dfrac{x(6x+1)}{x}$

$= \lim\limits_{x\to 0}(6x+1)$

$= 0 + 1$

$\lim\limits_{x\to 0}\dfrac{6x^2 + x}{x} = 1$

37. $\lim\limits_{h\to 3}\dfrac{h^3 - 27}{h-3} = \dfrac{(3)^3 - 27}{3-3} = \dfrac{0}{0} = \text{indeterminate}$

Try factoring.

$\lim\limits_{h\to 3}\dfrac{h^3 - 27}{h-3} = \lim\limits_{h\to 3}\dfrac{(h-3)(h^2 + 3h + 9)}{(h-3)}$

$= \lim\limits_{h\to 3}(h^2 + 3h + 9)$

$= (3)^2 + 3(3) + 9$

$\lim\limits_{h\to 3}\dfrac{h^3 - 27}{h-3} = 27$

41. $\lim\limits_{p\to -1}\sqrt{p}(p+1.3) = \sqrt{-1}(-1+1.3)$ has no real solution.

The root is not real for domain $p < 0$, so $\lim\limits_{p\to -1}\sqrt{p}(p+1.3)$ does not exist.

45. $\lim_{h\to 0}\frac{\sqrt{9+h}-3}{h}=\frac{\sqrt{9}-3}{0}=\frac{0}{0}=$ indeterminate

Use conjugates:

$$\lim_{h\to 0}\frac{\sqrt{9+h}-3}{h}=\lim_{h\to 0}\left(\frac{\sqrt{9+h}-3}{h}\cdot\frac{\sqrt{9+h}+3}{\sqrt{9+h}+3}\right)$$

$$=\lim_{h\to 0}\left(\frac{(9+h)-9}{(\sqrt{9+h}+3)h}\right)$$

$$=\lim_{h\to 0}\left(\frac{h}{(\sqrt{9+h}+3)h}\right)$$

$$=\lim_{h\to 0}\left(\frac{1}{\sqrt{9+h}+3}\right)$$

$$=\frac{1}{\sqrt{9+0}+3}$$

$$=\frac{1}{6}$$

49. $\lim_{t\to\infty}\frac{\sqrt{t^2+16}}{t+1}=\frac{\infty}{\infty}=$ indeterminate

Try getting the entire function under the root.

$$\lim_{t\to\infty}\frac{\sqrt{t^2+16}}{t+1}=\lim_{t\to\infty}\frac{\sqrt{t^2+16}}{\sqrt{(t+1)^2}}$$

$$=\lim_{t\to\infty}\sqrt{\frac{t^2+16}{t^2+2t+1}}$$

$$=\sqrt{\lim_{t\to\infty}\frac{t^2+16}{t^2+2t+1}}$$

Now try dividing numerator and denominator by the highest power of t.

$$=\sqrt{\lim_{t\to\infty}\frac{\frac{t^2}{t^2}+\frac{16}{t^2}}{\frac{t^2}{t^2}+\frac{2t}{t^2}+\frac{1}{t^2}}}$$

$$=\sqrt{\lim_{t\to\infty}\frac{1+\frac{16}{t^2}}{1+\frac{2}{t}+\frac{1}{t^2}}}$$

$$=\sqrt{\frac{1+0}{1+0+0}}$$

$$\lim_{t\to\infty}\frac{\sqrt{t^2+16}}{t+1}=1$$

53. **Table Method**

x	10	100	1000
$f(x)$	2.165	2.011	2.001

$$\therefore \lim_{x\to\infty}\frac{2x^2+x}{x^2-3}=2$$

Analytical Method

$$\lim_{x\to\infty}\frac{2x^2+x}{x^2-3}=\frac{\infty}{\infty}=\text{indeterminate}$$

Try dividing numerator and denominator by the highest power of x.

$$\lim_{x\to\infty}\frac{2x^2+x}{x^2-3}=\lim_{x\to\infty}\frac{\frac{2x^2}{x^2}+\frac{x}{x^2}}{\frac{x^2}{x^2}-\frac{3}{x^2}}$$

$$=\lim_{x\to\infty}\frac{2+\frac{1}{x}}{1-\frac{3}{x^2}}=\frac{2+0}{1-0}$$

$$\lim_{x\to\infty}\frac{2x^2+x}{x^2-3}=2$$

57. The object's temperature T (in °C) decreases 10% every minute. This means that 90% of the temperature remains after each minute. This produces a geometric sequence with first term 100°C and ratio 0.900.

$T=100(0.9)^t$ where t is time (in min)

We can find the limits of the temperature as $t\to 10$ and as $t\to\infty$

$\lim_{t\to 10} T=\lim_{t\to 10} 100(0.9)^t$

$=100(0.9)^{10}$

$=100(0.9)^{10}$

$\lim_{t\to 10} T=34.9$ °C

$\lim_{t\to\infty} T=\lim_{t\to\infty} 100(0.9)^t$

$=100\lim_{t\to\infty}(0.9)^t$

$=100(0)$

$\lim_{t\to\infty} T=0$ °C

These limits can be confirmed with a table:

t (min)	0	1	2	3	4	5	6	7	8	9	10	50	100
T (°C)	100.0	90.0	81.0	72.9	65.6	59.0	53.1	47.8	43.0	38.7	34.9	0.515	0.003

61.

x	1.9000	1.9900	1.9990	1.9999	2.0001	2.0010	2.0100	2.1000
$f(x)$	2.6787	2.7630	2.7716	2.7725	2.7727	2.7735	2.7822	2.8709

$$\therefore \lim_{x\to 2}\frac{2^x-4}{x-2}\approx 2.77$$

65. **(a)** $\lim_{x\to 2^-} f(x)=-1$

(b) $\lim_{x\to 2^+} f(x)=2$

(c) $\lim_{x\to 2} f(x)$ does not exist. The limits from above and below must agree for the limit to exist.

69. Remember that

$$|a| = \begin{cases} a & \text{for } a \geq 0 \\ -a & \text{for } a < 0 \end{cases}$$

$$\lim_{x\to 0^-} \frac{x}{|x|} = \lim_{x\to 0^-} \frac{x}{-x} = \lim_{x\to 0^-} -1 = -1$$

$$\lim_{x\to 0^+} \frac{x}{|x|} = \lim_{x\to 0^+} \frac{x}{x} = \lim_{x\to 0^+} 1 = 1$$

Since the limits from above and from below do not agree, and since the function is not defined at $x = 0$, the function is not continuous at $x = 0$.

23.2 The Slope of a Tangent to a Curve

1. Point P has the coordinates $(3,18)$. The coordinates of any point Q can be expressed as $(3+h, f(3+h))$, so in this case, $x = 3$.

$$f(x) = x^2 + 3x$$

$$m_{PQ} = \frac{f(3+h) - f(3)}{(3+h) - 3}$$

$$m_{PQ} = \frac{(3+h)^2 + 3(3+h) - (3^2 + 3(3))}{h}$$

$$= \frac{6h + h^2 + 3h}{h}$$

$$= 9 + h$$

$$m_{\tan} = \lim_{h\to 0}(9+h)$$

$$m_{\tan} = 9$$

5. $y = 2x^2 + 5x$ $\quad P(-2, -2)$

Point	*Q*1	*Q*2	*Q*3	*Q*4	*P*
x_2	−1.5	−1.9	−1.99	−1.999	−2
y_2	−3	−2.28	−2.0298	−2.002 998	−2
$y_2 - y$	−1	−0.28	−0.0298	−0.002 998	
$x_2 - x$	0.5	0.1	0.01	0.001	
$m=\Delta y/\Delta x$	−2	−2.8	−2.98	−2.998	

The slope of PQ approaches −3 as Q approaches P.

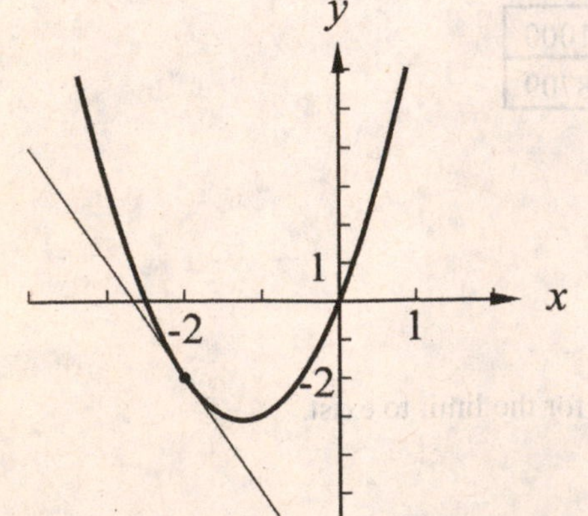

9. $f(x) = 2x^2 + 5x$

$$m_{PQ} = \frac{2(-2+h)^2 + 5(-2+h) - (-2)}{h}$$

$$= \frac{-8h + 2h^2 + 5h}{h}$$

$$= -8 + 2h + 5$$

$$m_{\tan} = \lim_{h \to 0}(-8 + 2h + 5) = -3$$

13. $f(x) = 2x^2 + 5x$

$$m_{PQ} = \frac{\left(2(x_1+h)^2 + 5(x_1+h)\right) - \left(2x_1^2 + 5x_1\right)}{h}$$

$$= \frac{4x_1h + 2h^2 + 5h}{h}$$

$$= 4x_1 + 2h + 5$$

$$m_{tan} = \lim_{h \to 0}\left(4x_1 + 2h + 5\right)$$

$$m_{tan} = 4x_1 + 5$$

$$m_{tan}\Big|_{x=-2} = -3$$

$$m_{tan}\Big|_{x=0.5} = 7$$

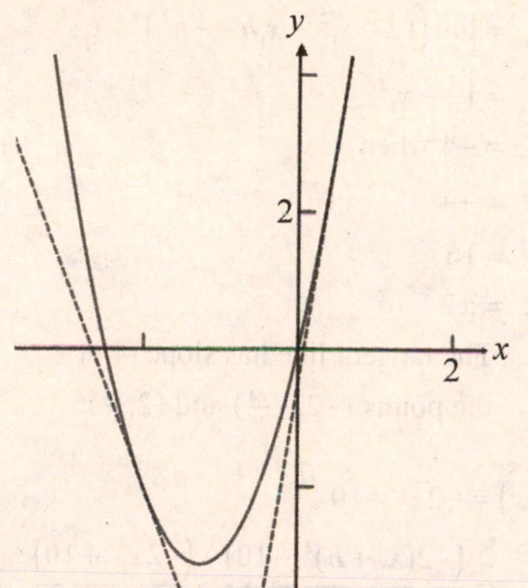

17. $f(x) = 6x - x^2$

$$m_{PQ} = \frac{\left(6(x_1+h) - (x_1+h)^2\right) - \left(6x_1 - x_1^2\right)}{h}$$

$$= \frac{6h - 2x_1h - h^2}{h}$$

$$= 6 - 2x_1 - h$$

$$m_{tan} = \lim_{h \to 0}\left(6 - 2x_1 - h\right)$$

$$m_{tan} = 6 - 2x_1$$

$$m_{tan}\Big|_{x=-1} = 8$$

$$m_{tan}\Big|_{x=1} = 4$$

$$m_{tan}\Big|_{x=3} = 0$$

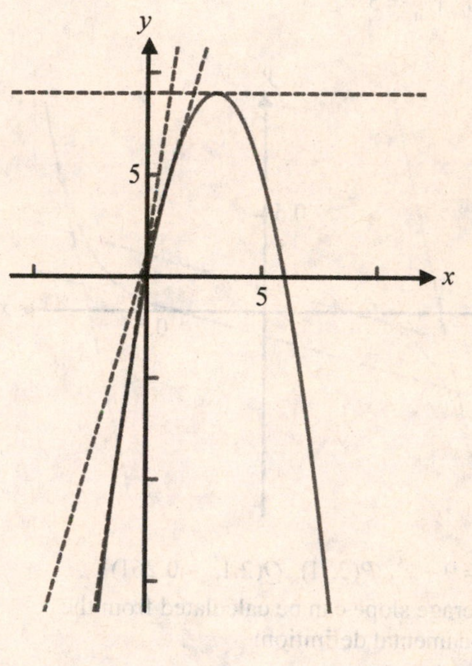

21. $f(x) = x^5$

$$m_{PQ} = \frac{\left((x_1+h)^5\right) - \left(x_1^5\right)}{h}$$
$$= \frac{5x_1^4h + 10x_1^3h^2 + 10x_1^2h^3 + 5x_1h^4 + h^5}{h}$$
$$= 5x_1^4 + 10x_1^3h + 10x_1^2h^2 + 5x_1h^3 + h^4$$
$$m_{tan} = \lim_{h\to 0}\left(5x_1^4 + 10x_1^3h + 10x_1^2h^2 + 5x_1h^3 + h^4\right)$$
$$m_{tan} = 5x_1^4$$
$$m_{tan}\big|_{x=-1} = 5$$
$$m_{tan}\big|_{x=-0.5} = \tfrac{5}{16}$$
$$m_{tan}\big|_{x=0.5} = \tfrac{5}{16}$$
$$m_{tan}\big|_{x=1} = 5$$

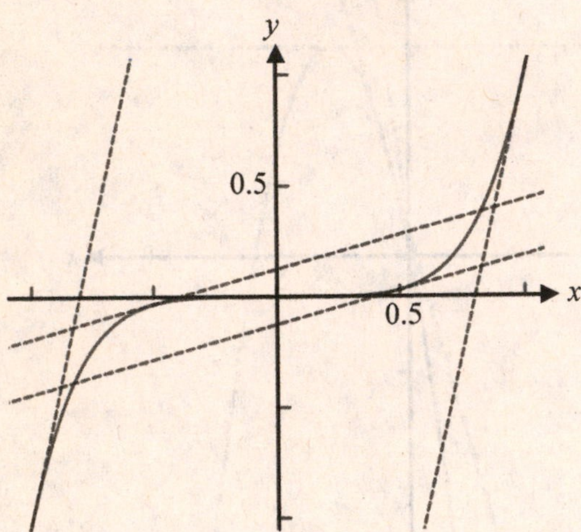

25. $y = 9 - x^3$, $P(2, 1)$, $Q(2.1, -0.261)$

Average slope can be calculated from the fundamental definition:

$$m = \frac{\text{rise}}{\text{run}} = \frac{\Delta y}{\Delta x} = \frac{y_2 - y_1}{x_2 - x_1}$$
$$m = \frac{-0.261 - 1}{2.1 - 2}$$
$$m = \frac{-1.261}{0.1}$$
$$m = -12.6$$

Instantaneous slope can be found from the process in example 3:

$$f(x) = 9 - x^3$$
$$m_{PQ} = \frac{\left(9 - (x_1+h)^3\right) - \left(9 - x_1^3\right)}{h}$$
$$= \frac{-3x_1^2h - 3x_1h^2 - h^3}{h}$$
$$= -3x_1^2 - 3x_1h - h^2$$
$$m_{tan} = \lim_{h\to 0}\left(-3x_1^2 - 3x_1h - h^2\right)$$
$$m_{tan} = -3x_1^2$$
$$m_{tan}\big|_{x=2} = -12$$

29. $f(x) = 12x - \frac{1}{3}x^3$

$$m_{PQ} = \frac{\left(12(x_1+h) - \frac{1}{3}(x_1+h)^3\right) - \left(12x_1 - \frac{1}{3}x_1^3\right)}{h}$$
$$= \frac{12h - x_1^2h - x_1h^2 - \frac{1}{3}h^3}{h}$$
$$= 12 - x_1^2 - x_1h - \tfrac{1}{3}h^2$$
$$m_{tan} = \lim_{h\to 0}\left(12 - x_1^2 - x_1h - \tfrac{1}{3}h^2\right)$$
$$m_{tan} = 12 - x_1^2$$
$m_{tan} = -4$ when
$$12 - x_1^2 = -4$$
$$x_1^2 = 16$$
$$x_1 = \pm 2$$

The tangent line has slope -4 at the points $(-2, -\frac{64}{3})$ and $(2, \frac{64}{3})$.

33. $f(x) = -2x^2 + 10$

$$m_{PQ} = \frac{\left(-2(x_1+h)^2 + 10\right) - \left(-2x_1^2 + 10\right)}{h}$$
$$= \frac{-4x_1h - 2h^2}{h}$$
$$= -4x_1 - 2h$$
$$m_{tan} = \lim_{h\to 0}\left(-4x_1 - 2h\right)$$
$$m_{tan} = -4x_1$$
$$m_{tan}\big|_{x=1.5} = -6$$

$$\tan\theta = \frac{1}{6}$$
$$\theta = \tan^{-1}\frac{1}{6}$$
$$\theta = 9.46^\circ$$

6 θ

1

23.3 The Derivative

1.

$$y = 4x^2 + 3x$$
$$f(x+h) = 4(x+h)^2 + 3(x+h)$$
$$f(x+h) = 4(x^2 + 2xh + h^2) + 3x + 3h$$
$$f(x+h) = 4x^2 + 8xh + 4h^2 + 3x + 3h$$
$$f(x+h) - f(x) = 4x^2 + 8xh + 4h^2 + 3x + 3h - 4x^2 - 3x$$
$$f(x+h) - f(x) = 8xh + 4h^2 + 3h$$
$$\frac{f(x+h) - f(x)}{h} = \frac{8xh + 4h^2 + 3h}{h}$$
$$\frac{f(x+h) - f(x)}{h} = 8x + 4h + 3$$
$$\lim_{h\to 0}\frac{f(x+h) - f(x)}{h} = \lim_{h\to 0}(8x + 4h + 3)$$
$$f'(x) = 8x + 3$$

5.

$$y = 1 - 7x$$
$$f'(x) = \lim_{h\to 0}\frac{f(x+h) - f(x)}{h}$$
$$f'(x) = \lim_{h\to 0}\frac{1 - 7(x+h) - (1 - 7x)}{h}$$
$$f'(x) = \lim_{h\to 0}\frac{1 - 7x - 7h - 1 + 7x}{h}$$
$$f'(x) = \lim_{h\to 0}\frac{-7h}{h}$$
$$f'(x) = \lim_{h\to 0}(-7)$$
$$f'(x) = -7$$

9.

$$y = 3x^2$$
$$f'(x) = \lim_{h\to 0}\frac{f(x+h) - f(x)}{h}$$
$$f'(x) = \lim_{h\to 0}\frac{3(x+h)^2 - 3x^2}{h}$$
$$f'(x) = \lim_{h\to 0}\frac{3x^2 + 6xh + 3h^2 - 3x^2}{h}$$
$$f'(x) = \lim_{h\to 0}\frac{6xh + 3h^2}{h}$$
$$f'(x) = \lim_{h\to 0}(6x + 3h)$$
$$f'(x) = 6x$$

13. $y = 8x - 2x^2$

$$f'(x) = \lim_{h\to 0}\frac{f(x+h) - f(x)}{h}$$

$$f'(x) = \lim_{h\to 0}\frac{8(x+h) - 2(x+h)^2 - (8x - 2x^2)}{h}$$

$$f'(x) = \lim_{h\to 0}\frac{8x + 8h - 2x^2 - 4xh - 2h^2 - 8x + 2x^2}{h}$$

$$f'(x) = \lim_{h\to 0}\frac{8h - 4xh - 2h^2}{h}$$

$$f'(x) = \lim_{h\to 0}(8 - 4x - 2h)$$

$$f'(x) = 8 - 4x$$

$$f'(x) = 4(2 - x)$$

17. $y = 5x^3 + 4$

$$f'(x) = \lim_{h\to 0}\frac{f(x+h) - f(x)}{h}$$

$$f'(x) = \lim_{h\to 0}\frac{5(x+h)^3 + 4 - (5x^3 + 4)}{h}$$

$$f'(x) = \lim_{h\to 0}\frac{5x^3 + 15x^2h + 15xh^2 + 5h^3 + 4 - 5x^3 - 4}{h}$$

$$f'(x) = \lim_{h\to 0}\frac{15x^2h + 15xh^2 + 5h^3}{h}$$

$$f'(x) = \lim_{h\to 0}(15x^2 + 15xh + 5h^2)$$

$$f'(x) = 15x^2$$

21. $y = x + \frac{4}{3x}$

$$f'(x) = \lim_{h\to 0}\frac{f(x+h) - f(x)}{h}$$

$$f'(x) = \lim_{h\to 0}\frac{\left(x + h + \frac{4}{3(x+h)}\right) - \left(x + \frac{4}{3x}\right)}{h}$$

$$f'(x) = \lim_{h\to 0}\frac{x + h + \frac{4}{3(x+h)} - x - \frac{4}{3x}}{h}$$

$$f'(x) = \lim_{h\to 0}\frac{h + \frac{4}{3(x+h)} - \frac{4}{3x}}{h}$$

$$f'(x) = \lim_{h\to 0}\frac{\frac{h(3x)(x+h) + 4x - 4(x+h)}{3x(x+h)}}{h}$$

$$f'(x) = \lim_{h\to 0}\frac{3x^2h + 3xh^2 + 4x - 4x - 4h}{3xh(x+h)}$$

$$f'(x) = \lim_{h\to 0}\frac{3x^2h + 3xh^2 - 4h}{3xh(x+h)}$$

$$f'(x) = \lim_{h\to 0}\frac{3x^2 + 3xh - 4}{3x(x+h)}$$

$$f'(x) = \frac{3x^2 - 4}{3x(x)}$$

$$f'(x) = \frac{3x^2 - 4}{3x^2}$$

$$f'(x) = 1 - \frac{4}{3x^2}$$

25. $y = 3x^2 - 2x \quad \text{Point}(-1, 5)$

$$f(x+h) - f(x) = 3(x+h)^2 - 2(x+h) - (3x^2 - 2x)$$

$$f(x+h) - f(x) = 3x^2 + 6xh + 3h^2 - 2x - 2h - 3x^2 + 2x$$

$$f(x+h) - f(x) = 6xh + 3h^2 - 2h$$

$$\frac{f(x+h) - f(x)}{h} = 6x + 3h - 2$$

$$f'(x) = \lim_{h\to 0}(6x + 3h - 2)$$

$$f'(x) = 6x - 2$$

$$f'(x) = 2(3x - 1)$$

$$f'(x)\big|_{x=-1} = 2(-3 - 1)$$

$$f'(x)\big|_{x=-1} = -8$$

29. $y = 1 + \frac{2}{x}$

$$\frac{f(x+h) - f(x)}{h} = \frac{1 + \frac{2}{x+h} - 1 - \frac{2}{x}}{h}$$

$$\frac{f(x+h) - f(x)}{h} = \frac{\frac{2}{x+h} - \frac{2}{x}}{h}$$

$$\frac{f(x+h) - f(x)}{h} = \frac{2(x) - 2(x+h)}{hx(x+h)}$$

$$\frac{f(x+h) - f(x)}{h} = \frac{-2h}{hx(x+h)}$$

$$\frac{f(x+h) - f(x)}{h} = \frac{-2}{x(x+h)}$$

$$f'(x) = \lim_{h\to 0}\frac{-2}{x(x+h)}$$

$$f'(x) = -\frac{2}{x^2}$$

This function is differentiable for all $x \neq 0$, since the derivative expression is defined at all x except $x = 0$.

33.

$$y = x^2 - 4x$$
$$f(x+h) - f(x) = (x+h)^2 - 4(x+h) - (x^2 - 4x)$$
$$f(x+h) - f(x) = x^2 + 2xh + h^2 - 4x - 4h - x^2 + 4x$$
$$f(x+h) - f(x) = 2xh + h^2 - 4h$$
$$\frac{f(x+h) - f(x)}{h} = 2x + h - 4$$
$$f'(x) = \lim_{h\to 0}(2x + h - 4)$$
$$f'(x) = 2x - 4$$
$$f'(x) = 2(x-2)$$

If the slope of the tangent line is 6, then

$6 = 2(x-2)$

$3 = x - 2$

$x = 5$

$y = 5^2 - 4(5) = 5$

Point (5, 5)

37.

$$y = \sqrt{x+1}$$
$$f(x+h) = \sqrt{x+h+1}$$
$$f(x+h) - f(x) = \sqrt{x+h+1} - \sqrt{x+1}$$
$$\frac{f(x+h) - f(x)}{h} = \frac{\sqrt{x+h+1} - \sqrt{x+1}}{h}$$
$$\lim_{h\to 0}\frac{f(x+h) - f(x)}{h} = \lim_{h\to 0}\frac{\sqrt{x+h+1} - \sqrt{x+1}}{h} = \frac{0}{0}\text{(indeterminate)}$$
$$\lim_{h\to 0}\frac{f(x+h) - f(x)}{h} = \lim_{h\to 0}\frac{\sqrt{x+h+1} - \sqrt{x+1}}{h}\cdot\frac{\sqrt{x+h+1} + \sqrt{x+1}}{\sqrt{x+h+1} + \sqrt{x+1}}$$
$$\frac{dy}{dx} = \lim_{h\to 0}\frac{(\sqrt{x+h+1})^2 - (\sqrt{x+1})^2}{h\,(\sqrt{x+h+1} + \sqrt{x+1})}$$
$$\frac{dy}{dx} = \lim_{h\to 0}\frac{(x+h+1) - (x+1)}{h\,(\sqrt{x+h+1} + \sqrt{x+1})}$$
$$\frac{dy}{dx} = \lim_{h\to 0}\frac{h}{h\,(\sqrt{x+h+1} + \sqrt{x+1})}$$
$$\frac{dy}{dx} = \lim_{h\to 0}\frac{1}{\sqrt{x+h+1} + \sqrt{x+1}}$$
$$\frac{dy}{dx} = \frac{1}{\sqrt{x+1} + \sqrt{x+1}}$$
$$\frac{dy}{dx} = \frac{1}{2\sqrt{x+1}}$$

This function is differentiable for all values of $x + 1 > 0$ or $x > -1$.

23.4 The Derivative as an Instantaneous Rate of Change

1.

$$s(t) = 48t - 16t^2$$
$$s(t+h) - s(t) = 48(t+h) - 16(t+h)^2 - (48t - 16t^2)$$
$$s(t+h) - s(t) = 48t + 48h - 16t^2 - 32ht - 16h^2 - 48t + 16t^2$$
$$s(t+h) - s(t) = 48h - 32ht - 16h^2$$
$$\frac{s(t+h) - s(t)}{h} = 48 - 32t - 16h$$
$$\frac{ds}{dt} = \lim_{h\to 0}(48 - 32t - 16h)$$
$$\frac{ds}{dt} = 48 - 16t$$
$$\left.\frac{ds}{dt}\right|_{t=2} = 48 - 16(2) = 16 \text{ ft/s}$$
$$\left.\frac{ds}{dt}\right|_{t=4} = 48 - 16(4) = -16 \text{ ft/s}$$

5. $y = \dfrac{16}{3x+1}$ Point $(-3, -2)$

$$\frac{dy}{dx} = \lim_{h\to 0}\frac{f(x+h) - f(x)}{h}$$
$$\frac{dy}{dx} = \lim_{h\to 0}\frac{\frac{16}{3(x+h)+1} - \frac{16}{3x+1}}{h}$$
$$\frac{dy}{dx} = \lim_{h\to 0}\frac{\frac{16(3x+1) - 16(3x+3h+1)}{(3x+3h+1)(3x+1)}}{h}$$
$$\frac{dy}{dx} = \lim_{h\to 0}\frac{48x + 16 - 48x - 48h - 16}{h(3x+3h+1)(3x+1)}$$
$$\frac{dy}{dx} = \lim_{h\to 0}\frac{-48h}{h(3x+3h+1)(3x+1)}$$
$$\frac{dy}{dx} = \lim_{h\to 0}\frac{-48}{(3x+3h+1)(3x+1)}$$
$$\frac{dy}{dx} = -\frac{48}{(3x+1)^2}$$
$$\left.\frac{dy}{dx}\right|_{x=-3} = -\frac{48}{(3(-3)+1)^2}$$
$$\left.\frac{dy}{dx}\right|_{x=-3} = -\frac{48}{64} = -\frac{3}{4}$$

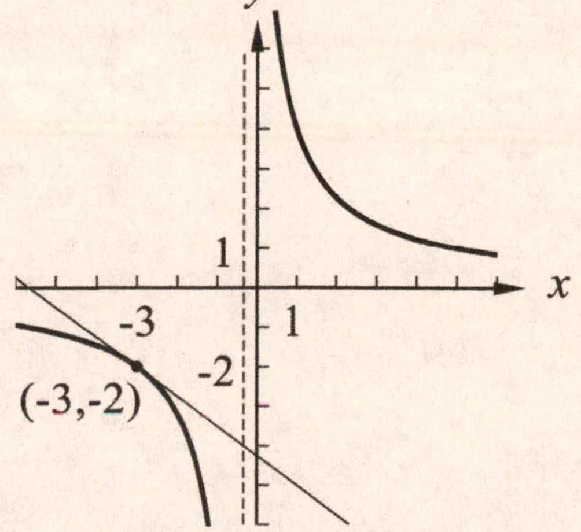

y
x
1
-3
1
-2
(-3,-2)

9. $s = 3t^2 - 4t;\ t = 2$

$s = 3(2)^2 - 4(2) = 4$

$t(\text{s})$	1.0	1.5	1.9	1.99	1.999
$s(\text{ft})$	−1.0	0.75	3.23	3.9203	3.9092003
$4 - s(\text{ft})$	5.0	3.25	0.77	0.0797	0.007997
$h = 2 - t(\text{s})$	1.0	0.5	0.1	0.01	0.001
$v = \frac{4-s}{h}(\text{ft/s})$	5.00	6.50	7.70	7.97	7.997

$v = 8.00$ ft/s when $t = 2$ s

13. $s = 3t^2 - 4t$

$$v = \lim_{h\to 0}\frac{s(t+h) - s(t)}{h}$$
$$v = \lim_{h\to 0}\frac{3(t+h)^2 - 4(t+h) - 3t^2 + 4t}{h}$$
$$v = \lim_{h\to 0}\frac{3t^2 + 6th + 3h^2 - 4t - 4h - 3t^2 + 4t}{h}$$
$$v = \lim_{h\to 0}\frac{6th + 3h^2 - 4h}{h}$$
$$v = \lim_{h\to 0}(6t + 3h - 4)$$
$$v = 6t - 4$$
$$v = 2(3t - 2)$$
$$v\big|_{t=2} = 2(3(2) - 2) = 8.00 \text{ ft/s}$$

17. $s = 12t^2 - t^3$

$$v = \lim_{h\to 0}\frac{s(t+h) - s(t)}{h}$$
$$v = \lim_{h\to 0}\frac{12(t+h)^2 - (t+h)^3 - 12t^2 + t^3}{h}$$
$$v = \lim_{h\to 0}\frac{12t^2 + 24th + 12h^2 - (t^3 + 3t^2h + 3th^2 + h^3) - 12t^2 + t^3}{h}$$
$$v = \lim_{h\to 0}\frac{24th + 12h^2 - 3t^2h - 3th^2 - h^3}{h}$$
$$v = \lim_{h\to 0}(24t + 12h - 3t^2 - 3th - h^2)$$
$$v = 24t - 3t^2$$
$$v = 3t(8 - t)$$

21. $s = 7t^2 - \frac{2}{t+1}$

$$v = \lim_{h\to 0} \frac{s(t+h) - s(t)}{h}$$

$$v = \lim_{h\to 0} \frac{7(t+h)^2 - \frac{2}{t+h+1} - 7t^2 + \frac{2}{t+1}}{h}$$

$$v = \lim_{h\to 0} \frac{7t^2 + 14th + 7h^2 - 7t^2 - \frac{2(t+1)-2(t+h+1)}{(t+h+1)(t+1)}}{h}$$

$$v = \lim_{h\to 0} \quad \frac{14th + 7h^2}{h} + \frac{2h}{h(t+h+1)(t+1)}$$

$$v = \lim_{h\to 0} \quad 14t + 7h + \frac{2}{(t+h+1)(t+1)}$$

$$v = 14t + \frac{2}{(t+1)^2}$$

25. $s = t^3 + 15t$

$$v = \frac{ds}{dt} = \lim_{h\to 0} \frac{s(t+h) - s(t)}{h}$$

$$v = \lim_{h\to 0} \frac{(t+h)^3 + 15(t+h) - t^3 - 15t}{h}$$

$$v = \lim_{h\to 0} \frac{t^3 + 3t^2h + 3th^2 + h^3 + 15t + 15h - t^3 - 15t}{h}$$

$$v = \lim_{h\to 0} \frac{3t^2h + 3th^2 + h^3 + 15h}{h}$$

$$v = \lim_{h\to 0} \left(3t^2 + 3th + h^2 + 15\right)$$

$$v = 3t^2 + 15$$

$$a = \frac{dv}{dt} = \lim_{h\to 0} \frac{v(t+h) - v(t)}{h}$$

$$a = \lim_{h\to 0} \frac{3(t+h)^2 + 15 - 3t^2 - 15}{h}$$

$$a = \lim_{h\to 0} \frac{3t^2 + 6th + 3h^2 + 15 - 3t^2 - 15}{h}$$

$$a = \lim_{h\to 0} \frac{6th + 3h^2}{h}$$

$$a = \lim_{h\to 0} (6t + 3h)$$

$$a = 6t$$

29. $c = 2\pi r$

$$\frac{dc}{dr} = \lim_{h\to 0} \frac{c(r+h) - c(r)}{h}$$

$$\frac{dc}{dr} = \lim_{h\to 0} \frac{2\pi(r+h) - 2\pi r}{h}$$

$$\frac{dc}{dr} = \lim_{h\to 0} \frac{2\pi h}{h}$$

$$\frac{dc}{dr} = \lim_{h\to 0} 2\pi$$

$$\frac{dc}{dr} = 2\pi = 6.28 \text{ cm / cm}$$

33. $q = 30 - 2t$

$$i = \frac{dq}{dt} = \lim_{h\to 0} \frac{q(t+h) - q(t)}{h}$$

$$i = \lim_{h\to 0} \frac{30 - 2(t+h) - 30 + 2t}{h}$$

$$i = \lim_{h\to 0} \frac{-2h}{h}$$

$$i = \lim_{h\to 0} (-2)$$

$$i = -2$$

37. $P = 500 + 250m^2$

$$\frac{dP}{dm} = \lim_{h\to 0} \frac{P(m+h) - P(m)}{h}$$

$$\frac{dP}{dm} = \lim_{h\to 0} \frac{500 + 250(m+h)^2 - 500 - 250m^2}{h}$$

$$\frac{dP}{dm} = \lim_{h\to 0} \frac{250m^2 + 500mh + 250h^2 - 250m^2}{h}$$

$$\frac{dP}{dm} = \lim_{h\to 0} \frac{500mh + 250h^2}{h}$$

$$\frac{dP}{dm} = \lim_{h\to 0} (500m + 250h)$$

$$\frac{dP}{dm} = 500m$$

$$\left.\frac{dP}{dm}\right|_{m=0.920} = 500(0.920)$$

$$\left.\frac{dP}{dm}\right|_{m=0.920} = 460 \text{ W} = 0.460 \text{ kW}$$

41. $V = \dfrac{48}{t+3}$

$$\frac{dV}{dt} = \lim_{h\to 0}\frac{V(t+h)-V(t)}{h}$$

$$\frac{dV}{dt} = \lim_{h\to 0}\frac{\frac{48}{t+h+3}-\frac{48}{t+3}}{h}$$

$$\frac{dV}{dt} = \lim_{h\to 0}\frac{\frac{48(t+3)-48(t+h+3)}{(t+h+3)(t+3)}}{h}$$

$$\frac{dV}{dt} = \lim_{h\to 0}\frac{48t+144-48t-48h-144}{h(t+h+3)(t+3)}$$

$$\frac{dV}{dt} = \lim_{h\to 0}\frac{-48h}{h(t+h+3)(t+3)}$$

$$\frac{dV}{dt} = \lim_{h\to 0}\frac{-48}{(t+h+3)(t+3)}$$

$$\frac{dV}{dt} = -\frac{48}{(t+3)^2}$$

$$\left.\frac{dV}{dt}\right|_{t=3} = -\frac{48}{(3+3)^2} = -1.33 \text{ (thousands of \$/year)}$$

$$\left.\frac{dV}{dt}\right|_{t=3} = -\$1330\text{/year}$$

45. $r = k\sqrt{\lambda}$

If $r = 3.72\times 10^{-2}$ m when $\lambda = 592\times 10^{-9}$ m then

$$k = \frac{r}{\sqrt{\lambda}} = \frac{3.72\times 10^{-2}\text{ m}}{\sqrt{592\times 10^{-9}\text{ m}}} = 48.3484\sqrt{\text{m}}$$

$$\frac{dr}{d\lambda} = \lim_{h\to 0}\frac{r(\lambda+h)-r(\lambda)}{h}$$

$$\frac{dr}{d\lambda} = \lim_{h\to 0}\frac{k\sqrt{\lambda+h}-k\sqrt{\lambda}}{h}\cdot\frac{k\sqrt{\lambda+h}+k\sqrt{\lambda}}{k\sqrt{\lambda+h}+k\sqrt{\lambda}}$$

$$\frac{dr}{d\lambda} = \lim_{h\to 0}\frac{k^2(\lambda+h)-k^2(\lambda)}{h\left(k\sqrt{\lambda+h}+k\sqrt{\lambda}\right)}$$

$$\frac{dr}{d\lambda} = \lim_{h\to 0}\frac{k^2h}{h\left(k\sqrt{\lambda+h}+k\sqrt{\lambda}\right)}$$

$$\frac{dr}{d\lambda} = \lim_{h\to 0}\frac{k^2}{\left(k\sqrt{\lambda+h}+k\sqrt{\lambda}\right)}$$

$$\frac{dr}{d\lambda} = \frac{k^2}{2k\sqrt{\lambda}} = \frac{k}{2\sqrt{\lambda}}$$

$$\frac{dr}{d\lambda} = \frac{48.3484}{2\sqrt{\lambda}}$$

$$\frac{dr}{d\lambda} = \frac{24.2}{\sqrt{\lambda}}\text{ m/m}$$

23.5 Derivatives of Polynomials

1. $v = r^9$

$$\frac{dv}{dr} = \frac{d\left(r^9\right)}{dr}$$

$$\frac{dv}{dr} = 9r^{9-1}$$

$$\frac{dv}{dr} = 9r^8$$

5. $y = x^5$

$$\frac{dy}{dx} = \frac{d\left(x^5\right)}{dx}$$

$$\frac{dy}{dx} = 5x^4$$

9. $y = 5x^4 - 3\pi$

$$\frac{dy}{dx} = 5\frac{d\left(x^4\right)}{dx} - \frac{d(3\pi)}{dx}$$

$$\frac{dy}{dx} = 5\left(4x^3\right) - 0$$

$$\frac{dy}{dx} = 20x^3$$

13. $p = 5r^3 - 2r + 12$

$$\frac{dp}{dr} = 5\frac{d\left(r^3\right)}{dr} - 2\frac{d(r)}{dr} + 12\frac{d(1)}{dr}$$

$$\frac{dp}{dr} = 5\left(3r^2\right) - 2(1) + 0$$

$$\frac{dp}{dr} = 15r^2 - 2$$

17. $f(x) = -6x^7 + 5x^3 + \pi^2$

$$f'(x) = -6\frac{d\left(x^7\right)}{dx} + 5\frac{d\left(x^3\right)}{dx} + \frac{d\left(\pi^2\right)}{dx}$$

$$f'(x) = -6\left(7x^6\right) + 5\left(3x^2\right) + 0$$

$$f'(x) = -42x^6 + 15x^2$$

$$f'(x) = -3x^2(14x^4 - 5)$$

21. $y = 6x^2 - 8x + 1$

$\frac{dy}{dx} = 6(2x^1) - 8(1) + 0$

$\frac{dy}{dx} = 12x - 8$

$\left.\frac{dy}{dx}\right|_{x=2} = 12(2) - 8$

$\left.\frac{dy}{dx}\right|_{x=2} = 16$

25. $y = 2x^6 - 4x^2$

$\frac{dy}{dx} = 2(6x^5) - 4(2x)$

$\frac{dy}{dx} = 12x^5 - 8x$

$\left.\frac{dy}{dx}\right|_{x=-1} = 12(-1)^5 - 8(-1)$

$\left.\frac{dy}{dx}\right|_{x=-1} = -4$

29. $s = 6t^5 - 5t + 2$

$v = \frac{ds}{dt} = 6(5t^4) - 5(1) + 0$

$v = 30t^4 - 5$

$v = 5(6t^4 - 1)$

33. $s = 2t^3 - 4t^2$

$v = \frac{ds}{dt} = 2(3t^2) - 4(2t)$

$v = 6t^2 - 8t$

$v|_{t=4} = 6(4)^2 - 8(4)$

$v|_{t=4} = 64.0 \text{ m/s}$

37. $y = 3x^2 - 6x$

$\frac{dy}{dx} = 3(2x) - 6(1)$

$\frac{dy}{dx} = 6x - 6$

$\frac{dy}{dx} = 6(x-1)$ if slope is parallel to x-axis, slope is 0, factor the derivative

$0 = 6(x-1)$

$x = 1$

41. Parabola

$y = 2x^2 - 7x$

$\frac{dy}{dx} = 2(2x) - 7(1)$

$\frac{dy}{dx} = 4x - 7$

Line

$x - 3y = 16$

$-3y = -x + 16$

$y = \frac{1}{3}x - \frac{16}{3}$

This line has slope 1/3. Perpendicular lines have negative reciprocal slopes, so any curve perpendicular to this one at this location will have instantaneous slope –3.

So, for the parabola to be perpendicular to the given line:

$\frac{dy}{dx} = -3$

$-3 = 4x - 7$

$4 = 4x$

$x = 1$

45. In general, the volume of a cylinder is $V = \pi r^2 h$, where r is the radius and h is the height of the cylinder.

In the particular case where the height is 20 times the radius, $h = 20\,r$, so

$V = \pi r^2(20r) = 20\pi r^3$

Taking the derivative with respect to a change in variable r yields

$\frac{dV}{dr} = \frac{d}{dr}(20\pi r^3)$

$\frac{dV}{dr} = 20\pi(3r^2)$

$\frac{dV}{dr} = 60\pi r^2$

$\left.\frac{dV}{dr}\right|_{r=3.0} = 60\pi(3.0)^2 = 540\pi \ \frac{\text{mm}^3}{\text{mm}}$

49. $R = 16.0 + 0.450T + 0.0125T^2$

$$\frac{dR}{dT} = 0 + 0.450(1) + 0.0125(2T)$$

$$\frac{dR}{dT} = 0.450 + 0.0250T$$

$$\left.\frac{dR}{dT}\right|_{T=115\,^\circ\text{C}} = 0.450 + 0.0250(115)$$

$$\left.\frac{dR}{dT}\right|_{T=115\,^\circ\text{C}} = 3.325\,\frac{\Omega}{^\circ\text{C}}$$

$$\left.\frac{dR}{dT}\right|_{T=115\,^\circ\text{C}} = 3.32\,\frac{\Omega}{^\circ\text{C}}$$

53. $h = 0.000104x^4 - 0.0417x^3 + 4.21x^2 - 8.33x$

$$\frac{dh}{dx} = 0.000104(4x^3) - 0.0417(3x^2) + 4.21(2x) - 8.33(1)$$

$$\frac{dh}{dx} = 0.000416x^3 - 0.1251x^2 + 8.42x - 8.33$$

$$\left.\frac{dh}{dx}\right|_{x=120\text{ km}} = 0.000416(120)^3 - 0.1251(120)^2 + 8.42(120) - 8.33$$

$$\left.\frac{dh}{dx}\right|_{x=120\text{ km}} = -80.5\,\frac{\text{m}}{\text{km}}$$

23.6 Derivatives of Products and Quotients of Functions

1. $p(x) = (5 - 3x^2)(3 - 2x)$

Identify this as a product of two functions u and v, where

$u = 5 - 3x^2$ $\qquad$ $v = 3 - 2x$

$\frac{du}{dx} = -3(2x) = -6x$ $\qquad$ $\frac{dv}{dx} = -2$

The product rule derivative is

$$\frac{d(u \cdot v)}{dx} = u \cdot \frac{dv}{dx} + v \cdot \frac{du}{dx}$$

$$\frac{d(p(x))}{dx} = (5 - 3x^2) \cdot (-2) + (3 - 2x) \cdot (-6x)$$

$$p' = -10 + 6x^2 - 18x + 12x^2$$

$$p' = 18x^2 - 18x - 10$$

$$p' = 2(9x^2 - 9x - 5)$$

5. $s = (3t+2)(2t-5)$

Identify this as a product of two functions u and v, where

$u = 3t+2$ and $v = 2t-5$

$\frac{du}{dt} = 3$ $\frac{dv}{dt} = 2$

$$\frac{d(u \cdot v)}{dt} = u \cdot \frac{dv}{dt} + v \cdot \frac{du}{dt}$$

$$\frac{ds}{dt} = (3t+2)(2) + (2t-5)(3)$$

$$\frac{ds}{dt} = 6t + 4 + 6t - 15$$

$$\frac{ds}{dt} = 12t - 11$$

9. $y = (2x-7)(5-2x)$

$u = (2x-7)$ and $v = (5-2x)$

$\frac{du}{dx} = 2$ $\frac{dv}{dx} = -2$

$$\frac{d(u \cdot v)}{dx} = u \cdot \frac{dv}{dx} + v \cdot \frac{du}{dx}$$

$$\frac{dy}{dx} = (2x-7) \cdot (-2) + (5-2x) \cdot (2)$$

$$\frac{dy}{dx} = -4x + 14 + 10 - 4x$$

$$\frac{dy}{dx} = -8x + 24$$

$$\frac{dy}{dx} = -8(x-3)$$

$$y = (2x-7)(5-2x)$$

$$y = 10x - 4x^2 - 35 + 14x$$

$$y = -4x^2 + 24x - 35$$

$$\frac{dy}{dx} = -8x + 24$$

$$\frac{dy}{dx} = -8(x-3)$$

13. $y = \frac{x}{8x+3}$

$u = x$ and $v = 8x+3$

$\frac{du}{dx} = 1$ $\frac{dv}{dx} = 8$

$$\frac{d\frac{u}{v}}{dx} = \frac{v \cdot \frac{du}{dx} - u \cdot \frac{dv}{dx}}{v^2}$$

$$\frac{dy}{dx} = \frac{(8x+3) \cdot (1) - x \cdot (8)}{(8x+3)^2}$$

$$\frac{dy}{dx} = \frac{8x + 3 - 8x}{(8x+3)^2}$$

$$\frac{dy}{dx} = \frac{3}{(8x+3)^2}$$

17. $y = \frac{6x^2}{3-2x}$

$u = 6x^2$ and $v = 3-2x$

$\frac{du}{dx} = 12x$ $\frac{dv}{dx} = -2$

$$\frac{d\frac{u}{v}}{dx} = \frac{v \cdot \frac{du}{dx} - u \cdot \frac{dv}{dx}}{v^2}$$

$$\frac{dy}{dx} = \frac{(3-2x) \cdot (12x) - (6x^2) \cdot (-2)}{(3-2x)^2}$$

$$\frac{dy}{dx} = \frac{36x - 24x^2 + 12x^2}{(3-2x)^2}$$

$$\frac{dy}{dx} = \frac{36x - 12x^2}{(3-2x)^2}$$

$$\frac{dy}{dx} = \frac{12x(3-x)}{(3-2x)^2}$$

21. $f(x)=\dfrac{3x+8}{x^2+4x+2}$

$u=3x+8$ and $v=x^2+4x+2$

$\dfrac{du}{dx}=3$ $\dfrac{dv}{dx}=2x+4$

$$\frac{d\frac{u}{v}}{dx}=\frac{v\cdot\frac{du}{dx}-u\cdot\frac{dv}{dx}}{v^2}$$

$$\frac{df(x)}{dx}=\frac{(x^2+4x+2)\cdot(3)-(3x+8)\cdot(2x+4)}{(x^2+4x+2)^2}$$

$$\frac{df(x)}{dx}=\frac{3x^2+12x+6-6x^2-12x-16x-32}{(x^2+4x+2)^2}$$

$$\frac{df(x)}{dx}=\frac{-3x^2-16x-26}{(x^2+4x+2)^2}$$

25. $y=(3x-1)(4-7x)$

$u=3x-1$ and $v=4-7x$

$\dfrac{du}{dx}=3$ $\dfrac{dv}{dx}=-7$

$$\frac{d(u\cdot v)}{dx}=u\cdot\frac{dv}{dx}+v\cdot\frac{du}{dx}$$

$$\frac{dy}{dx}=(3x-1)\cdot(-7)+(4-7x)\cdot(3)$$

$$\frac{dy}{dx}=-21x+7+12-21x$$

$$\frac{dy}{dx}=-42x+19$$

$$\left.\frac{dy}{dx}\right|_{x=3}=-42(3)+19$$

$$\left.\frac{dy}{dx}\right|_{x=3}=-107$$

29. $y=\dfrac{3x-5}{2x+3}$

$u=3x-5$ and $v=2x+3$

$\dfrac{du}{dx}=3$ $\dfrac{dv}{dx}=2$

$$\frac{d\frac{u}{v}}{dx}=\frac{v\cdot\frac{du}{dx}-u\cdot\frac{dv}{dx}}{v^2}$$

$$\frac{dy}{dx}=\frac{(2x+3)\cdot(3)-(3x-5)\cdot(2)}{(2x+3)^2}$$

$$\frac{dy}{dx}=\frac{6x+9-6x+10}{(2x+3)^2}$$

$$\frac{dy}{dx}=\frac{19}{(2x+3)^2}$$

$$\left.\frac{dy}{dx}\right|_{x=-2}=\frac{19}{(2(-2)+3)^2}$$

$$\left.\frac{dy}{dx}\right|_{x=-2}=19$$

33. If $v=c$, the product rule

$$\frac{d(u\cdot v)}{dx}=u\cdot\frac{dv}{dx}+v\cdot\frac{du}{dx}$$

becomes

$$\frac{d(u\cdot c)}{dx}=u\cdot\frac{dc}{dx}+c\cdot\frac{du}{dx}$$

$$\frac{d(c\cdot u)}{dx}=u\cdot(0)+c\cdot\frac{du}{dx}$$

$$\frac{d(c\cdot u)}{dx}=c\cdot\frac{du}{dx}$$

This is Eq. (23.10), the derivative of a constant multiplied by a function.

37. $y=x^2\cdot f(x)$

$u=x^2$ and $v=f(x)$

$\dfrac{du}{dx}=2x$ $\dfrac{dv}{dx}=\dfrac{df(x)}{dx}$

$$\frac{d(u\cdot v)}{dx}=u\cdot\frac{dv}{dx}+v\cdot\frac{du}{dx}$$

$$\frac{dy}{dx}=x^2\cdot\frac{df(x)}{dx}+f(x)\cdot 2x$$

$$\frac{dy}{dx}=x^2\cdot f'(x)+2x\cdot f(x)$$

41. Method 1

$$y = \frac{x^2(1-2x)}{3x-7} = \frac{u}{v}$$

The numerator has a product rule in it.

$$\frac{dy}{dx} = \frac{(3x-7)\cdot\ x^2\cdot(-2)+(1-2x)\cdot(2x)\ -x^2(1-2x)\cdot(3)}{(3x-7)^2}$$

$$\frac{dy}{dx} = \frac{(3x-7)(-2x^2+2x-4x^2)-3x^2+6x^3}{(3x-7)^2}$$

$$\frac{dy}{dx} = \frac{(3x-7)(-6x^2+2x)-3x^2+6x^3}{(3x-7)^2}$$

$$\frac{dy}{dx} = \frac{-18x^3+6x^2+42x^2-14x-3x^2+6x^3}{(3x-7)^2}$$

$$\frac{dy}{dx} = \frac{-12x^3+45x^2-14x}{(3x-7)^2}$$

Method 2

$$y = \frac{x^2-2x^3}{3x-7}$$

$$\frac{dy}{dx} = \frac{(3x-7)\cdot(2x-6x^2)-(x^2-2x^3)\cdot(3)}{(3x-7)^2}$$

$$\frac{dy}{dx} = \frac{6x^2-18x^3-14x+42x^2-3x^2+6x^3}{(3x-7)^2}$$

$$\frac{dy}{dx} = \frac{-12x^3+45x^2-14x}{(3x-7)^2}$$

45.

$$y = \frac{x}{x^2+1}$$

$u = x$ and $v = x^2+1$

$\frac{du}{dx} = 1$ and $\frac{dv}{dx} = 2x$

$$\frac{d\ \frac{u}{v}}{dx} = \frac{v\cdot\frac{du}{dx}-u\cdot\frac{dv}{dx}}{v^2}$$

$$\frac{dy}{dx} = \frac{(x^2+1)\cdot(1)-(x)\cdot(2x)}{(x^2+1)^2}$$

$$\frac{dy}{dx} = \frac{x^2+1-2x^2}{(x^2+1)^2}$$

$$\frac{dy}{dx} = \frac{-x^2+1}{(x^2+1)^2}$$

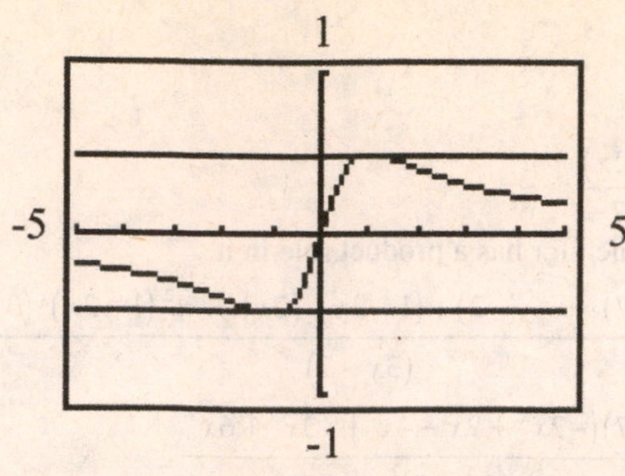

If the slope is supposed to be zero, then

$$0=\frac{-x^2+1}{\left(x^2+1\right)^2}$$
$$0=-x^2+1$$
$$x^2=1$$
$$x=\pm 1$$

49. 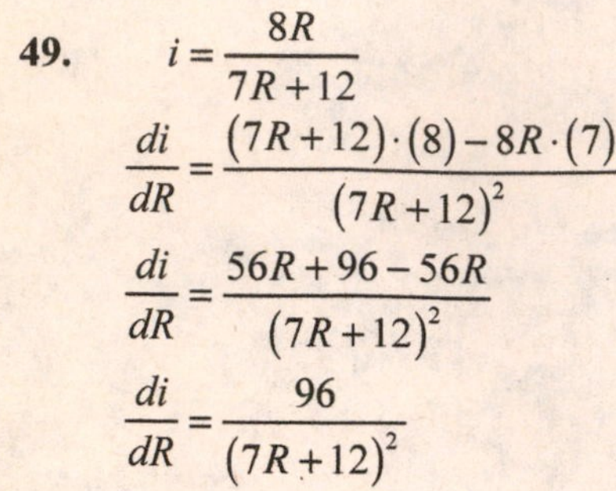

$$i=\frac{8R}{7R+12}$$
$$\frac{di}{dR}=\frac{(7R+12)\cdot(8)-8R\cdot(7)}{(7R+12)^2}$$
$$\frac{di}{dR}=\frac{56R+96-56R}{(7R+12)^2}$$
$$\frac{di}{dR}=\frac{96}{(7R+12)^2}$$

53.

$$T=\frac{2t}{0.05t+1}-20$$
$$\frac{dT}{dt}=\frac{(0.05t+1)\cdot(2)-2t\cdot(0.05)}{(0.05t+1)^2}-0$$
$$\frac{dT}{dt}=\frac{0.1t+2-0.1t}{(0.05t+1)^2}$$
$$\frac{dT}{dt}=\frac{2}{(0.05t+1)^2}$$
$$\left.\frac{dT}{dt}\right|_{t=6.00\text{ h}}=\frac{2}{(1.3)^2}$$
$$\left.\frac{dT}{dt}\right|_{t=6.00\text{ h}}=1.18^\circ\text{C/h}$$

57.

$$P=\frac{E^2r}{R^2+2Rr+r^2};\text{ where } E \text{ and } R \text{ are constants}$$
$$\frac{dP}{dr}=\frac{\left(R^2+2Rr+r^2\right)\cdot E^2-E^2r\cdot(0+2R+2r)}{\left(R^2+2Rr+r^2\right)^2}$$
$$\frac{dP}{dr}=\frac{E^2R^2+2E^2Rr+E^2r^2-2E^2Rr-2E^2r^2}{\left(R^2+2Rr+r^2\right)^2}$$
$$\frac{dP}{dr}=\frac{E^2R^2-E^2r^2}{\left(R^2+2Rr+r^2\right)^2}$$
$$\frac{dP}{dr}=\frac{E^2\left(R^2-r^2\right)}{\left[(R+r)^2\right]^2}$$
$$\frac{dP}{dr}=\frac{E^2(R-r)(R+r)}{(R+r)^4}$$
$$\frac{dP}{dr}=\frac{E^2(R-r)}{(R+r)^3}$$

23.7 The Derivative of a Power of a Function

1. $p(x) = (2 + 3x^3)^4$

Identify this as a power of function u and n, where

$$u = 2 + 3x^3 \qquad \text{and} \qquad n = 4$$

$$\frac{du}{dx} = 9x^2$$

$$\frac{d(u^n)}{dx} = nu^{n-1} \cdot \frac{du}{dx}$$

$$\frac{dp(x)}{dx} = 4(2 + 3x^3)^3 (9x^2)$$

$$\frac{dp(x)}{dx} = 36x^2 (2 + 3x^3)^3$$

5.

$$y = 4\sqrt{x}$$

$$y = 4x^{1/2}$$

Recognize form $y = u^n$ with $n = 1/2$

$$\frac{dy}{dx} = \frac{4}{2}x^{1/2-1}$$

$$\frac{dy}{dx} = 2x^{-1/2}$$

$$\frac{dy}{dx} = 2 \quad \frac{1}{x^{1/2}}$$

$$\frac{dy}{dx} = \frac{2}{\sqrt{x}}$$

9. $y = \dfrac{3}{\sqrt[3]{x}} + 4x^2$

$$y = \frac{3}{x^{1/3}} + 4x^2$$

$$y = 3x^{-1/3} + 4x^2$$

$$\frac{dy}{dx} = 3 \quad -\frac{1}{3}x^{-1/3-1} \quad + 8x$$

$$\frac{dy}{dx} = -1x^{-4/3} + 8x$$

$$\frac{dy}{dx} = -\frac{1}{x^{4/3}} + 8x$$

13. $y = (4x^2 + 3)^5$

$$\frac{dy}{dx} = 5(4x^2 + 3)^4 (8x)$$

$$\frac{dy}{dx} = 40x(4x^2 + 3)^4$$

17. $y = \left(2x^3 - 3\right)^{1/3}$

$$\frac{dy}{dx} = \frac{1}{3}\left(2x^3 - 3\right)^{-2/3}\left(6x^2\right)$$

$$\frac{dy}{dx} = \frac{2x^2}{\left(2x^3 - 3\right)^{2/3}}$$

21. $y = 4\left(2x^4 - 5\right)^{0.75}$

$$\frac{dy}{dx} = 4(0.75)\left(2x^4 - 5\right)^{-0.25}\left(8x^3\right)$$

$$\frac{dy}{dx} = \frac{24x^3}{\left(2x^4 - 5\right)^{0.25}}$$

25. $u = v\sqrt{8v+5}$

$$u = v\left(8v+5\right)^{1/2}$$

$$\frac{du}{dv} = v\ \frac{1}{2}\ \left(8v+5\right)^{-1/2}(8) + \left(8v+5\right)^{1/2}(1)$$

$$\frac{du}{dv} = 4v\left(8v+5\right)^{-1/2} + \left(8v+5\right)^{1/2}$$

$$\frac{du}{dv} = \left(8v+5\right)^{-1/2}\left[4v + (8v+5)\right]$$

$$\frac{du}{dv} = \frac{12v+5}{\left(8v+5\right)^{1/2}}$$

29. $y = \dfrac{6x\sqrt{x+2}}{x+4}$

$$y = \frac{6x\left(x+2\right)^{1/2}}{x+4}$$

The numerator contains a product rule, to a whole function is a quotient rule.

$$\frac{dy}{dx} = \frac{(x+4)\ \ 6x \cdot \frac{1}{2}\left(x+2\right)^{-1/2} + \left(x+2\right)^{1/2}(6)\ \ - 6x\left(x+2\right)^{1/2}(1)}{\left(x+4\right)^2}$$

$$\frac{dy}{dx} = \frac{3x\left(x+4\right)\left(x+2\right)^{-1/2} + 6\left(x+4\right)\left(x+2\right)^{1/2} - 6x\left(x+2\right)^{1/2}}{\left(x+4\right)^2}$$

$$\frac{dy}{dx} = \frac{3\left(x+2\right)^{-1/2}\ \ x\left(x+4\right) + 2\left(x+4\right)\left(x+2\right) - 2x\left(x+2\right)}{\left(x+4\right)^2}$$

$$\frac{dy}{dx} = \frac{3\left(x^2 + 4x + 2x^2 + 4x + 8x + 16 - 2x^2 - 4x\right)}{\left(x+4\right)^2\sqrt{x+2}}$$

$$\frac{dy}{dx} = \frac{3\left(x^2 + 12x + 16\right)}{\left(x+4\right)^2\sqrt{x+2}}$$

33. $y = \sqrt{3x+4}$

$y = (3x+4)^{1/2}$

$u = 3x+4 \quad \text{and} \quad n = \frac{1}{2}$

$$\frac{du}{dx} = 3$$

$$\frac{dy}{dx} = \frac{1}{2}(3x+4)^{-1/2}(3)$$

$$\frac{dy}{dx} = \frac{3}{2}(3x+4)^{-1/2}$$

$$\frac{dy}{dx} = \frac{3}{2\sqrt{3x+4}}$$

$$\left.\frac{dy}{dx}\right|_{x=7} = \frac{3}{2\sqrt{3(7)+4}}$$

$$\left.\frac{dy}{dx}\right|_{x=7} = \frac{3}{2\sqrt{25}} = \frac{3}{10}$$

37. **(a)** $y = \frac{1}{x^3}$

$$\frac{dy}{dx} = \frac{v\dfrac{du}{dx} - u\dfrac{dv}{dx}}{v^2}$$

$$\frac{dy}{dx} = \frac{x^3(0) - 1(3x^2)}{x^6}$$

$$\frac{dy}{dx} = \frac{-3x^2}{x^6}$$

$$\frac{dy}{dx} = -\frac{3}{x^4}$$

(b) $y = x^{-3}$

$$\frac{dy}{dx} = -3x^{-4}$$

$$\frac{dy}{dx} = -\frac{3}{x^4}$$

41. $x+3y-12=0$

$$y=-\frac{1}{3}x+4$$

$$\text{slope } m=-\frac{1}{3}$$

If this line is ever perpendicular to the curve, it must have negative reciprocal slope.

This means the slope of the tangent line must somewhere be 3 if perpendicularity occurs

$$y=\sqrt{2x+3}$$

$$y=(2x+3)^{1/2}$$

$$\frac{dy}{dx}=\frac{1}{2}(2x+3)^{-1/2}(2)$$

$$\frac{dy}{dx}=\frac{1}{\sqrt{2x+3}}$$

$$3=\frac{1}{\sqrt{2x+3}}$$

$$\sqrt{2x+3}=\frac{1}{3}$$

$$2x+3=\frac{1}{9}$$

$$2x=\frac{1}{9}-\frac{27}{9}$$

$$2x=-\frac{26}{9}$$

$$x=-\frac{13}{9}$$

$$y=\sqrt{2\left(-\tfrac{13}{9}\right)+3}=\frac{1}{3}$$

$y=\sqrt{2x+3}$ at $\left(-\frac{13}{9},\frac{1}{3}\right)$ is perpendicular to $x+3y-12=0$.

45. $v=k\sqrt{w}$

If $v=88$ ft/s when $w=16$ lb/ft^2

$$k=\frac{v}{\sqrt{w}}$$

$$k=\frac{88}{\sqrt{16}}=22$$

$$v=22w^{1/2}$$

$$\frac{dv}{dw}=22(1/2)w^{-1/2}$$

$$\frac{dv}{dw}=\frac{11}{\sqrt{w}}$$

49. $P = \frac{k}{V^{3/2}}$; but $P = 300$ kPa when $V = 100$ cm^3

$$k = PV^{3/2}$$
$$k = 300(100)^{3/2} = 300{,}000$$
$$P = \frac{300\,000}{V^{3/2}}$$
$$P = 300\,000V^{-3/2}$$
$$\frac{dP}{dV} = 300\,000\ -\frac{3}{2}\ V^{-5/2}$$
$$\frac{dP}{dV} = -450\,000V^{-5/2} = \frac{-450\,000}{V^{5/2}}$$
$$\left.\frac{dP}{dV}\right|_{V=100} = \frac{-450\,000}{100^{5/2}} = \frac{-450\,000}{100\,000}$$
$$\left.\frac{dP}{dV}\right|_{V=100} = -4.50 \text{ kPa/cm}^3$$

53. $H = \frac{4000}{\sqrt{t^6+100}}$ where $(-6 < t < 6)$

$$H = 4000\left(t^6+100\right)^{-1/2}$$
$$\frac{dH}{dt} = 4000\ -\frac{1}{2}\ \left(t^6+100\right)^{-3/2}\left(6t^5\right)$$
$$\frac{dH}{dt} = \frac{-12{,}000t^5}{\left(t^6+100\right)^{3/2}}$$
$$\left.\frac{dH}{dt}\right|_{t=4} = \frac{-12{,}000(4)^5}{\left(4^6+100\right)^{3/2}}$$
$$\left.\frac{dH}{dt}\right|_{t=4} = -45.2 \text{ W/}\left(\text{m}^2\cdot\text{h}\right)$$

57. $D^2 = w^2 + l^2$

$$D^2 = w^2 + (w+2)^2$$
$$D^2 = w^2 + w^2 + 4w + 4$$
$$D^2 = 2w^2 + 4w + 4$$
$$D = \sqrt{2w^2+4w+4}$$
$$D = \left(2w^2+4w+4\right)^{1/2}$$
$$\frac{dD}{dw} = \frac{1}{2}\left(2w^2+4w+4\right)^{-1/2}(4w+4)$$
$$\frac{dD}{dw} = \frac{2w+2}{\left(2w^2+4w+4\right)^{1/2}}$$
$$\frac{dD}{dw} = \frac{2(w+1)}{2^{1/2}\left(w^2+2w+2\right)^{1/2}}$$
$$\frac{dD}{dw} = \frac{\sqrt{2}(w+1)}{\sqrt{w^2+2w+2}}$$

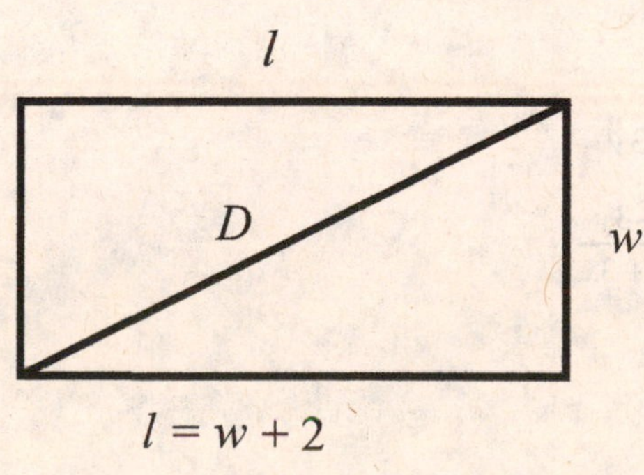

23.8 Differentiation of Implicit Functions

1.
$$y^3 + 2x^2 = 5$$
$$\frac{d}{dx}\left(y^3 + 2x^2\right) = \frac{d}{dx}(5)$$
$$3y^2 \cdot \frac{dy}{dx} + 4x = 0$$
$$3y^2 \cdot \frac{dy}{dx} = -4x$$
$$\frac{dy}{dx} = -\frac{4x}{3y^2}$$

5.
$$4y - 3x^2 = x$$
$$\frac{d}{dx}\left(4y - 3x^2\right) = \frac{d}{dx}(x)$$
$$4\frac{dy}{dx} - 6x = 1$$
$$4\frac{dy}{dx} = 1 + 6x$$
$$\frac{dy}{dx} = \frac{1+6x}{4}$$

9.
$$y^5 = x^2 - 1$$
$$\frac{d}{dx}\left(y^5\right) = \frac{d}{dx}\left(x^2 - 1\right)$$
$$5y^4 \frac{dy}{dx} = 2x + 0$$
$$\frac{dy}{dx} = \frac{2x}{5y^4}$$

13.
$$y + 3xy - 4 = 0$$
$$\frac{d}{dx}(y + 3xy - 4) = \frac{d}{dx}(0)$$
$$1\frac{dy}{dx} + 3\left(x\frac{dy}{dx} + y(1)\right) - 0 = 0$$
$$\frac{dy}{dx} + 3x\frac{dy}{dx} + 3y = 0$$
$$\frac{dy}{dx}(1+3x) = -3y$$
$$\frac{dy}{dx} = \frac{-3y}{1+3x}$$

17.

$$\frac{3x^2}{y^2+1}+y=3x+1$$

$$\frac{d}{dx}\ \frac{3x^2}{y^2+1}+y\ =\frac{d}{dx}(3x+1)$$

$$\frac{\left(y^2+1\right)6x-3x^2\left(2y\frac{dy}{dx}\right)}{\left(y^2+1\right)^2}+1\cdot\frac{dy}{dx}=3+0$$

$$\frac{6x}{y^2+1}-\frac{6x^2y}{\left(y^2+1\right)^2}\frac{dy}{dx}+1\cdot\frac{dy}{dx}=3$$

$$\frac{dy}{dx}\ -\frac{6x^2y}{\left(y^2+1\right)^2}+1\ =3-\frac{6x}{y^2+1}$$

$$\frac{dy}{dx}=\frac{3-\frac{6x}{y^2+1}}{-\frac{6x^2y}{\left(y^2+1\right)^2}+1}$$

$$\frac{dy}{dx}=\frac{\frac{3(y^2+1)-6x}{y^2+1}}{\frac{-6x^2y+\left(y^2+1\right)^2}{\left(y^2+1\right)^2}}$$

$$\frac{dy}{dx}=\frac{3(y^2+1)-6x}{y^2+1}\cdot\frac{\left(y^2+1\right)^2}{-6x^2y+\left(y^2+1\right)^2}$$

$$\frac{dy}{dx}=\frac{3\left(y^2+1-2x\right)\left(y^2+1\right)}{\left(y^2+1\right)^2-6x^2y}$$

21.

$$2\left(x^2+1\right)^3+\left(y^2+1\right)^{1/2}=17$$

$$\frac{d}{dx}\ 2\left(x^2+1\right)^3+\left(y^2+1\right)^{1/2}\ =\frac{d}{dx}(17)$$

$$6\left(x^2+1\right)^2(2x)+\frac{1}{2}\left(y^2+1\right)^{-1/2}2y\frac{dy}{dx}=0$$

$$y\left(y^2+1\right)^{-1/2}\frac{dy}{dx}=-12x\left(x^2+1\right)^2$$

$$\frac{dy}{dx}=\frac{-12\left(x^2+1\right)^2}{y\left(y^2+1\right)^{-1/2}}$$

25.
$$5y^4+7=x^4-3y$$
$$\frac{d}{dx}\left(5y^4+7\right)=\frac{d}{dx}\left(x^4-3y\right)$$
$$20y^3\frac{dy}{dx}+0=4x^3-3\frac{dy}{dx}$$
$$\frac{dy}{dx}\left(20y^3+3\right)=4x^3$$
$$\frac{dy}{dx}=\frac{4x^3}{20y^3+3}$$
$$\left.\frac{dy}{dx}\right|_{(3,-2)}=\frac{4(3)^3}{20(-2)^3+3}=-\frac{108}{157}$$

29.
$$x^2+y^2=4x$$
$$\frac{d}{dx}\left(x^2+y^2\right)=\frac{d}{dx}(4x)$$
$$2x+2y\frac{dy}{dx}=4$$
$$\frac{dy}{dx}=\frac{4-2x}{2y}$$
$$\frac{dy}{dx}=\frac{2-x}{y}$$

If tangent is horizontal, slope is zero
$$0=\frac{2-x}{y}$$
$$0=2-x$$
$$x=2$$
$$2^2+y^2=4(2)$$
$$y^2=4$$
$$y=\pm 2$$

The function will have a horizontal tangent at (2, 2) and (2, -2).

33.
$$\omega^2=(LC)^{-1}-R^2L^{-2}$$
$$\frac{d}{dL}\left(\omega^2\right)=\frac{d}{dL}\left((LC)^{-1}-R^2L^{-2}\right)$$
$$2\omega\frac{d\omega}{dL}=(-1)(LC)^{-2}(C)+2R^2L^{-3}$$
$$=\frac{2R^2}{L^3}-\frac{C}{L^2C^2}$$
$$=\frac{2R^2C-L}{L^3C}$$
$$\frac{d\omega}{dL}=\frac{2R^2C-L}{2\omega L^3C}$$

37. $PV = n\ RT + aP - \frac{bP}{T}$ where a, b, n, R, V are constant

$$PV = nRT + naP - nbPT^{-1}$$
$$\frac{d}{dT}(PV) = \frac{d}{dT}\left(nRT + naP - nbPT^{-1}\right)$$
$$V\frac{dP}{dT} = nR + na\frac{dP}{dT} - nbP\left(-T^{-2}\right) + T^{-1}\ \ -nb\frac{dP}{dT}$$
$$\frac{dP}{dT}\ V - na + \frac{nb}{T}\ = nR + \frac{nbP}{T^2}$$
$$\frac{dP}{dT}\ \frac{VT - naT + nb}{T}\ = \frac{nRT^2 + nbP}{T^2}$$
$$\frac{dP}{dT} = \frac{nRT^2 + nbP}{T^2} \cdot \frac{T}{VT - naT + nb}$$
$$\frac{dP}{dT} = \frac{nRT^2 + nbP}{T\left(VT - naT + nb\right)}$$

41. $r^2 = 2rR + 2R - 2r$

$$\frac{d}{dr}\left(r^2\right) = \frac{d}{dr}(2rR + 2R - 2r)$$
$$2r = (2r)\frac{dR}{dr} + R(2) + 2\frac{dR}{dr} - 2$$
$$2r - 2R + 2 = \frac{dR}{dr}(2r + 2)$$
$$\frac{dR}{dr} = \frac{2(r - R + 1)}{2(r + 1)}$$
$$\frac{dR}{dr} = \frac{r - R + 1}{r + 1}$$

23.9 Higher Derivatives

1. $y = 5x^3 - 2x^2$

$$y' = 15x^2 - 4x$$
$$y'' = 30x - 4$$
$$y''' = 30$$
$$y^{(4)} = 0$$
$$y^{(n)} = 0 \text{ for } n \geq 4$$

5. $f(x)=x^3-6x^4$

$f'(x)=3x^2-24x^3$

$f''(x)=6x-72x^2$

$f'''(x)=6-144x$

9. $f(r)=r(4r+9)^3$

$f'(r)=r\cdot 3(4r+9)^2(4)+(4r+9)^3$

$f'(r)=12r(4r+9)^2+(4r+9)^3$

$f'(r)=(4r+9)^2(12r+(4r+9))$

$f'(r)=(4r+9)^2(16r+9)$

$f''(r)=(4r+9)^2(16)+(16r+9)(2)(4r+9)(4)$

$f''(r)=16(4r+9)^2+8(16r+9)(4r+9)$

$f''(r)=8(4r+9)(2(4r+9)+16r+9)$

$f''(r)=8(4r+9)(24r+27)$

$f''(r)=24(4r+9)(8r+9)$

$f'''(r)=24(4r+9)(8)+(8r+9)(24)(4)$

$f'''(r)=768r+1728+768r+864$

$f'''(r)=1536r+2592$

$f'''(r)=96(16r+27)$

13. $y=5x+8\sqrt{x}$

$y=5x+8x^{1/2}$

$y'=5+4x^{-1/2}$

$y''=-2x^{-3/2}$

17. $f(p)=\dfrac{4.8\pi}{\sqrt{1+2p}}$

$f(p)=4.8\pi(1+2p)^{-1/2}$

$f'(p)=-2.4\pi(1+2p)^{-3/2}(2)$

$f'(p)=-4.8\pi(1+2p)^{-3/2}$

$f''(p)=-4.8\pi\ \dfrac{-3}{2}\ (1+2p)^{-5/2}(2)$

$f''(p)=\dfrac{14.4\pi}{(1+2p)^{5/2}}$

21. $y=\left(3x^2-1\right)^5$

$y'=5\left(3x^2-1\right)^4(6x)$

$y'=30x\left(3x^2-1\right)^4$

$y''=30x(4)\left(3x^2-1\right)^3(6x)+\left(3x^2-1\right)^4(30)$

$y''=720x^2\left(3x^2-1\right)^3+30\left(3x^2-1\right)^4$

$y''=30\left(3x^2-1\right)^3\left(24x^2+(3x^2-1)\right)$

$y''=30\left(3x^2-1\right)^3\left(27x^2-1\right)$

25. $u=\dfrac{v^2}{4v+15}$

$u'=\dfrac{(4v+15)2v-v^2(4)}{(4v+15)^2}$

$u'=\dfrac{8v^2+30v-4v^2}{(4v+15)^2}$

$u'=\dfrac{4v^2+30v}{(4v+15)^2}$

$u'=\left(4v^2+30v\right)\left(4v+15\right)^{-2}$

$u''=\left(4v^2+30v\right)(-2)\left(4v+15\right)^{-3}(4)+\left(4v+15\right)^{-2}(8v+30)$

$u''=(4v+15)^{-3}\ \ (-8)\left(4v^2+30v\right)+(4v+15)(8v+30)$

$u''=\dfrac{-32v^2-240v+32v^2+120v+120v+450}{(4v+15)^3}$

$u''=\dfrac{450}{(4v+15)^3}$

29. $f(x)=\sqrt{x^2+9}$

$f(x)=\left(x^2+9\right)^{1/2}$

$f'(x)=\dfrac{1}{2}\left(x^2+9\right)^{-1/2}(2x)$

$f'(x)=x\left(x^2+9\right)^{-1/2}$

$f''(x)=x\ \ -\dfrac{1}{2}\left(x^2+9\right)^{-3/2}(2x)\ \ +\left(x^2+9\right)^{-1/2}(1)$

$f''(x)=-x^2\left(x^2+9\right)^{-3/2}+\left(x^2+9\right)^{-1/2}$

$f''(x)=\left(x^2+9\right)^{-3/2}\ \ -x^2+(x^2+9)$

$f''(x)=\dfrac{9}{\left(x^2+9\right)^{3/2}}$

$f''(4)=\dfrac{9}{(25)^{3/2}}$

$f''(4)=\dfrac{9}{125}$

33. $v = t(8-t)^5$

$$\frac{dv}{dt} = t(5)(8-t)^4(-1) + (8-t)^5(1)$$

$$\frac{dv}{dt} = -5t(8-t)^4 + (8-t)^5$$

$$\frac{dv}{dt} = (8-t)^4[-5t + (8-t)]$$

$$\frac{dv}{dt} = (8-t)^4[-6t+8]$$

$$\frac{d^2v}{dt^2} = (8-t)^4(-6) + [-6t+8](4)(8-t)^3(-1)$$

$$\frac{d^2v}{dt^2} = -6(8-t)^4 - 4[-6t+8](8-t)^3$$

$$\frac{d^2v}{dt^2} = 2(8-t)^3[-3(8-t) - 2(-6t+8)]$$

$$\frac{d^2v}{dt^2} = 2(8-t)^3[-24+3t+12t-16]$$

$$\frac{d^2v}{dt^2} = 2(8-t)^3(15t-40)$$

$$\left.\frac{d^2v}{dt^2}\right|_{t=2} = 2(6)^3(30-40)$$

$$\left.\frac{d^2v}{dt^2}\right|_{t=2} = -4320$$

37. $s = \dfrac{16}{0.5t^2+1}$

$$s = 16(0.5t^2+1)^{-1}$$

$$v = \frac{ds}{dt} = 16(-1)(0.5t^2+1)^{-2}(t)$$

$$v = -16t(0.5t^2+1)^{-2}$$

$$a = \frac{dv}{dt} = -16(0.5t^2+1)^{-2} + (-16t)(-2)(0.5t^2+1)^{-3}(t)$$

$$a = \frac{32t^2 - 16(0.5t^2+1)}{(0.5t^2+1)^3}$$

$$a|_{t=2.00\text{ s}} = \frac{128-16(3)}{(2+1)^3} = \frac{80}{27} = 2.96\text{ m/s}^2$$

41. $y = (1-2x)^4$

$$y' = 4(1-2x)^3(-2) = -8(1-2x)^3$$

$$y'' = -8(3)(1-2x)^2(-2) = 48(1-2x)^2$$

$$y''|_{x=1} = 48(1-2)^2 = 48$$

45. Let $f(x) = ax^3 + bx^2 + cx + d$.
Then $f'(x) = 3ax^2 + 2bx + c$, $f''(x) = 6ax + 2b$,
and $f'''(x) = 6a$.
If $f'''(-1) = 12 = 6a$ then $a = 2$.
If $f''(-1) = -14 = 6(2)(-1) + 2b = -12 + 2b$ then $b = -1$.
If $f'(-1) = 8 = 3(2)(-1)^2 + 2(-1)(-1) + c = 6 + 2 + c$ then $c = 0$.
If $f(-1) = 9 = 2(-1)^3 - (-1)^2 + 0(-1) + d = -3 + d$ then $d = 12$.
The desired polynomial is $f(x) = 2x^3 - x^2 + 12$.

49.
$$V = L\,\frac{d^2q}{dt^2}$$
$$q = \sqrt{2t+1} - 1$$
$$q = (2t+1)^{1/2} - 1$$
$$\frac{dq}{dt} = \frac{1}{2}(2t+1)^{-1/2}(2)$$
$$\frac{dq}{dt} = (2t+1)^{-1/2}$$
$$\frac{d^2q}{dt^2} = -\frac{1}{2}(2t+1)^{-3/2}(2)$$
$$\frac{d^2q}{dt^2} = \frac{-1}{(2t+1)^{3/2}}$$
$$V = 1.60\,\frac{-1}{(2t+1)^{3/2}}$$
$$V = -\frac{1.60}{(2t+1)^{3/2}}$$

Review Exercises

1. This is false.
$$\lim_{x\to 3}\frac{x^2-3x}{x-3} = \lim_{x\to 3}\frac{x(x-3)}{x-3} = \lim_{x\to 3} x = 3$$

5. This is false.
$$\frac{d}{dx}(3-x)^3 = 3(3-x)^2(-1)$$
from the power rule.

9. $\lim_{x\to 4}(8-3x) = 8 - 3(4) = -4$

13. $\lim_{x\to 2}\frac{4x-8}{x^2-4}=\lim_{x\to 2}\frac{4(x-2)}{(x-2)(x+2)}$

$\lim_{x\to 2}\frac{4x-8}{x^2-4}=\lim_{x\to 2}\frac{4}{x+2}$

$\lim_{x\to 2}\frac{4x-8}{x^2-4}=\frac{4}{2+2}=1$

17. $\lim_{x\to\infty}\frac{2+\frac{2}{x}}{3-\frac{1}{x^2}}=\frac{2+0}{3-0}=\frac{2}{3}$

21. $y=7+5x$

$\frac{dy}{dx}=\lim_{h\to 0}\frac{f(x+h)-f(x)}{h}$

$\frac{dy}{dx}=\lim_{h\to 0}\frac{7+5(x+h)-7-5x}{h}$

$\frac{dy}{dx}=\lim_{h\to 0}\frac{5h}{h}$

$\frac{dy}{dx}=\lim_{h\to 0}5$

$\frac{dy}{dx}=5$

25. $y=\frac{2}{x^2}$

$\frac{dy}{dx}=\lim_{h\to 0}\frac{f(x+h)-f(x)}{h}$

$\frac{dy}{dx}=\lim_{h\to 0}\frac{\frac{2}{(x+h)^2}-\frac{2}{x^2}}{h}$

$\frac{dy}{dx}=\lim_{h\to 0}\frac{2x^2-2(x+h)^2}{hx^2(x+h)^2}$

$\frac{dy}{dx}=\lim_{h\to 0}\frac{2x^2-2x^2-4xh-2h^2}{hx^2(x+h)^2}$

$\frac{dy}{dx}=\lim_{h\to 0}\frac{-4xh-2h^2}{hx^2(x+h)^2}$

$\frac{dy}{dx}=\lim_{h\to 0}\frac{-4x-2h}{x^2(x+h)^2}$

$\frac{dy}{dx}=\frac{-4x}{x^4}=-\frac{4}{x^3}$

29. $y=2x^7-3x^2+5$

$\frac{dy}{dx}=2(7x^6)-3(2x)+0$

$\frac{dy}{dx}=14x^6-6x$

$\frac{dy}{dx}=2x(7x^5-3)$

33. $f(y)=\frac{12y}{1-5y}$

$\frac{df(y)}{dy}=\frac{(1-5y)(12)-12y(-5)}{(1-5y)^2}$

$\frac{df(y)}{dy}=\frac{12-60y+60y}{(1-5y)^2}$

$\frac{df(y)}{dy}=\frac{12}{(1-5y)^2}$

37. $y=\frac{3\pi}{(5-2x^2)^{3/4}}$

$y=3\pi(5-2x^2)^{-3/4}$

$\frac{dy}{dx}=3\pi\left(\frac{-3}{4}\right)(5-2x^2)^{-7/4}(-4x)$

$\frac{dy}{dx}=\frac{9\pi x}{(5-2x^2)^{7/4}}$

41. $y=\frac{\sqrt{4x+3}}{2x}=\frac{(4x+3)^{1/2}}{2x}$

$\frac{dy}{dx}=\frac{2x\cdot\frac{1}{2}(4x+3)^{-1/2}(4)-(4x+3)^{1/2}(2)}{(2x)^2}$

$\frac{dy}{dx}=\frac{4x(4x+3)^{-1/2}-2(4x+3)^{1/2}}{4x^2}$

$\frac{dy}{dx}=\frac{2(4x+3)^{-1/2}\ 2x-(4x+3)}{4x^2}$

$\frac{dy}{dx}=\frac{(-2x-3)}{2x^2(4x+3)^{1/2}}$

$\frac{dy}{dx}=-\frac{2x+3}{2x^2\sqrt{4x+3}}$

45.

$$y = \frac{4}{x} + 2\sqrt[3]{x}$$

$$y = 4x^{-1} + 2x^{1/3}$$

$$\frac{dy}{dx} = 4(-1)x^{-2} + 2\ \frac{1}{3}x^{-2/3}$$

$$\frac{dy}{dx} = \frac{-4}{x^2} + \frac{2}{3x^{2/3}}$$

$$\left.\frac{dy}{dx}\right|_{x=8} = \frac{-4}{8^2} + \frac{2}{3(8)^{2/3}}$$

$$\left.\frac{dy}{dx}\right|_{x=8} = \frac{-4}{64} + \frac{2}{3(4)}$$

$$\left.\frac{dy}{dx}\right|_{x=8} = \frac{-1}{16} + \frac{1}{6} = \frac{-3}{48} + \frac{8}{48} = \frac{5}{48}$$

49.

$$y = 3x^4 - \frac{1}{x} = 3x^4 - x^{-1}$$

$$y' = 12x^3 + x^{-2}$$

$$y'' = 36x^2 - 2x^{-3}$$

$$y'' = 2x^{-3}\left(18x^5 - 1\right)$$

$$y'' = \frac{2\left(18x^5 - 1\right)}{x^3}$$

53. The slopes of the three functions are

$$\frac{d}{dx}\ \frac{1}{x} = -\frac{1}{x^2}$$

$$\frac{d}{dx}\ \frac{1}{x^2} = -\frac{2}{x^3}$$

$$\frac{d}{dx}\ \frac{1}{\sqrt{x}} = \frac{d}{dx}\left(x^{-1/2}\right)$$

$$= -\frac{1}{2x^{3/2}}$$

The derivative that is the most negative as x approaches 0 is the derivative of $1/\sqrt{x}$, so it is the function that increases most rapidly.

57.

$$y = \frac{2\left(x^2 - 4\right)}{x-2}$$

$$y = \frac{2(x+2)(x-2)}{(x-2)}$$

$$y = 2(x+2), x \neq 2$$

$$\lim_{x\to 2} 2(x+2) = 2(2+2) = 8.00$$

Just to the left of $x = 2$, the trace feature gives $x = 1.9787234$, $y = 7.9574468$ and just to the right of $x = 2$, the trace feature gives $x = 2.0319149$, $y = 8.0638298$ which would appear to give $y = 8$ for $x = 2$. However, using the value feature shows there is no y-value for $x = 2$.

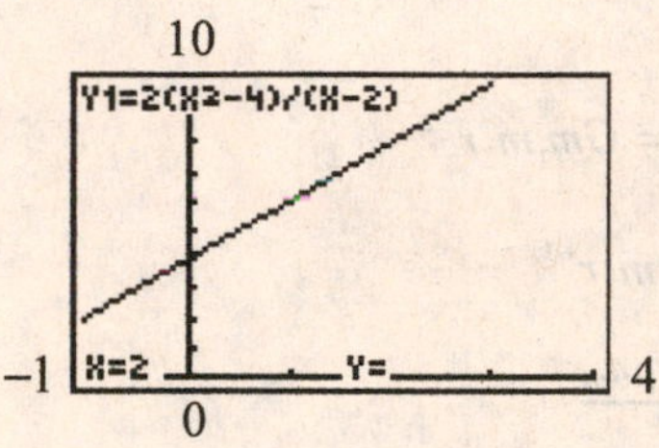

61.

$$y = 7x^4 - x^3$$

$$\frac{dy}{dx} = 28x^3 - 3x^2$$

$$\left.\frac{dy}{dx}\right|_{(-1,8)} = 28(-1)^3 - 3(-1)^2$$

$$\left.\frac{dy}{dx}\right|_{(-1,8)} = -31.0$$

65.

$$A = 5000(1+0.250i)^8$$

$$\frac{dA}{di} = 5000(8)(1+0.250i)^7(0.250)$$

$$\frac{dA}{di} = 10\,000(1+0.250i)^7$$

69.

$$y = 0.0015x^2 + C$$
$$\frac{dy}{dx} = 2(0.0015)x$$
$$\frac{dy}{dx} = 0.003x$$

If this slope is along a line $y = 0.3x - 10$, it must have slope 0.3

$$0.3 = 0.003x$$
$$x = 100$$

To be tangent to $y = 0.3x - 10$, it must intersect that curve

$$y = 0.3(100) - 10$$
$$y = 20$$

The curve must pass through point (100, 20)

$$y = 0.0015x^2 + C$$
$$20 = 0.0015(100)^2 + C$$
$$20 = 15 + C$$
$$C = 5$$

73. $F = \frac{Gm_1m_2}{r^2} = Gm_1m_2r^{-2}$

$$\frac{dF}{dr} = -2Gm_1m_2r^{-3}$$
$$\frac{dF}{dr} = -\frac{2Gm_1m_2}{r^3}$$

77. $r_f = \frac{2(R^3 - r^3)}{3(R^2 - r^2)}$

The numerator and denominator each has a factor of $(R - r)$:

$$r_f = \frac{2(R^2 + Rr + r^2)(R - r)}{3(R + r)(R - r)}$$
$$r_f = \frac{2(R^2 + Rr + r^2)}{3(R + r)}$$
$$\frac{dr_f}{dR} = \frac{3(R + r)(2)(2R + r) - 2(R^2 + Rr + r^2)(3)}{(3(R + r))^2}$$
$$\frac{dr_f}{dR} = \frac{12R^2 + 18Rr + 6r^2 - 6R^2 - 6Rr - 6r^2}{9(R + r)^2}$$
$$\frac{dr_f}{dR} = \frac{6R^2 + 12Rr}{9(R + r)^2}$$
$$\frac{dr_f}{dR} = \frac{2R(R + 2r)}{3(R + r)^2}$$

81. $y = kx(x^4 + 450x^2 - 950)$

$$y = kx^5 + 450kx^3 - 950kx$$
$$\frac{dy}{dx} = 5kx^4 + 1350kx^2 - 950k$$
$$\frac{dy}{dx} = 5k(x^4 + 270x^2 - 190)$$

85. $$T=\frac{10(1-t)}{0.5t+1}$$

$$\frac{dT}{dt}=\frac{(0.5t+1)(-10)-10(1-t)(0.5)}{(0.5t+1)^2}$$

$$\frac{dT}{dt}=\frac{-5t-10-5+5t}{(0.5t+1)^2}$$

$$\frac{dT}{dt}=\frac{-15}{(0.5t+1)^2}$$

89. $$A=lw=75.0$$

$$l=\frac{75.0}{w}$$

$$p=2l+2w$$

$$p=\frac{150}{w}+2w$$

$$\frac{dp}{dw}=\frac{-150}{w^2}+2$$

$$\frac{dp}{dw}=2\ \ 1-\frac{75.0}{w^2}$$

93. $$A=\pi r^2$$

$$\frac{dA}{dr}=2\pi r$$

$$\left.\frac{dA}{dr}\right|_{r=1.8}=3.6\pi=11.3$$

97. Using Pythagoras' Theorem

$$r^2=x^2+y^2$$

Where $y = 0.500$ km constant and $\frac{dx}{dt} = 400$ km/h

$$\frac{d}{dt}\left(r^2\right)=\frac{d}{dt}\left(x^2+y^2\right)$$

$$2r\frac{dr}{dt}=2x\cdot\frac{dx}{dt}+0$$

$$\frac{dr}{dt}=\frac{x}{r}\cdot\frac{dx}{dt}$$

After $t = 0.600$ min the plane has moved

$$x=(400\text{ km/h})(0.600\text{ min})\ \ \frac{1\text{ h}}{60\text{ min}}\ \ =4.00\text{ km}$$

$$r^2=4^2+0.5^2=4.0311\text{ km}$$

$$\frac{dr}{dt}=\frac{4.00\text{ km}}{4.0311\text{ km}}\cdot 400\text{ km/h}=397\text{ km/h}$$

Chapter 24

Applications of the Derivative

24.1 Tangents and Normals

1. $x^2+4y^2=17,\ (1,\ 2)$

$2x+8yy'=0$

$y'=\left.\dfrac{-x}{4y}\right|_{(1,\ 2)}$

$=\dfrac{-1}{4(2)}=-\dfrac{1}{8}$

$y-2=-\dfrac{1}{8}(x-1)$

$8y-16=-x+1$

$x+8y-17=0$

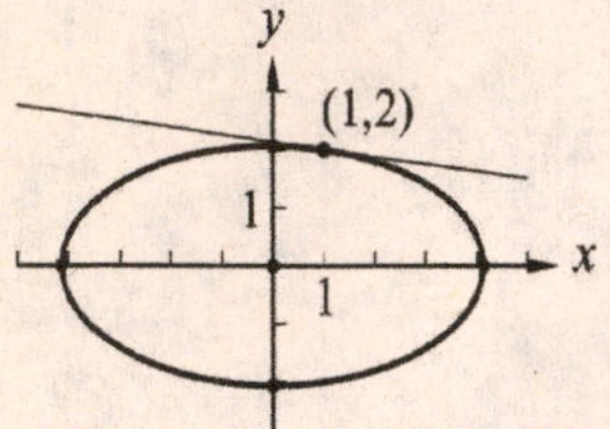

5. $y=\dfrac{1}{x^2+1}$ at $-1,\ \dfrac{1}{2}$

$y=\left(x^2+1\right)^{-1}$

$\dfrac{dy}{dx}=-\left(x^2+1\right)^{-2}(2x)$

$m_{\tan}=\left.\dfrac{-2x}{\left(x^2+1\right)^2}\right|_{x=-1}$

$=\dfrac{1}{2}$

Eq. T.L.:

$y-\dfrac{1}{2}=\dfrac{1}{2}(x+1)$

$2y-1=x+1$

$y=\dfrac{1}{2}x+1$

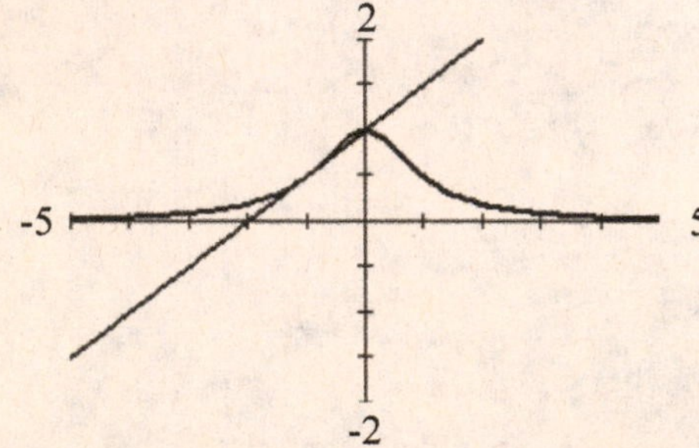

9. $y^2(2-x)=x^3$ at $(1,\ 1)$

$y^2=\dfrac{x^3}{2-x}$

$2y\dfrac{dy}{dx}=\dfrac{(2-x)(3x^2)-x^3(-1)}{(2-x)^2}$

$m_{\tan}=\left.\dfrac{dy}{dx}\right|_{(x,y)=(1,1)}$

$2m_{\tan}=\dfrac{(1)(3)-(1)(-1)}{1}=4$

$m_{\tan}=2$

$m_{\text{normal}}=-\dfrac{1}{2}$

Eq. of normal:

$y-1=-\dfrac{1}{2}(x-1)$

$2y+x-3=0$

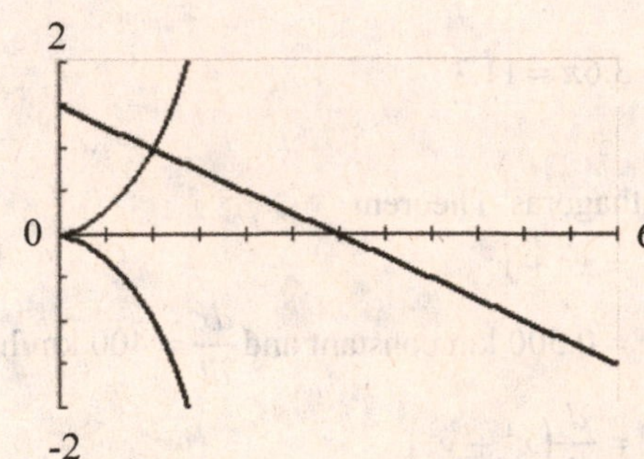

13. $y=(2x-1)^3$; normal line $m=-\dfrac{1}{24},\ x>0$

Therefore, $m_{\tan}=24$

$m_{\tan}=3(2x-1)^2(2)$

$=6(x-1)^2$

$6(2x-1)^2=24$

$(2x-1)^2=4$

$2x-1=\pm2$

$x=\frac{3}{2}$ or $x=-\frac{1}{2}$ (reject $x<0$)

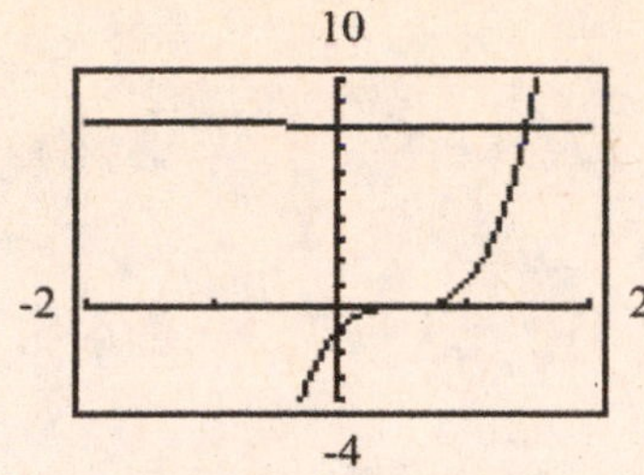

$x=\frac{3}{2},\ y=8$

Eq. of N.L.:

$y-8=-\frac{1}{24}\ x-\frac{3}{2}$

$24y-192=-x+\frac{3}{2}$

$2x+48y-387=0$

17. $y=x+2x^2-x^4$

$y'=1+4x-4x^3$

At $(1,\ 2)$, $y'=1+4(1)-4(1)^3=1$

Eq. of T.L. at $(1,\ 2)$: $y-2=1(x-1)$

$y=x+1$

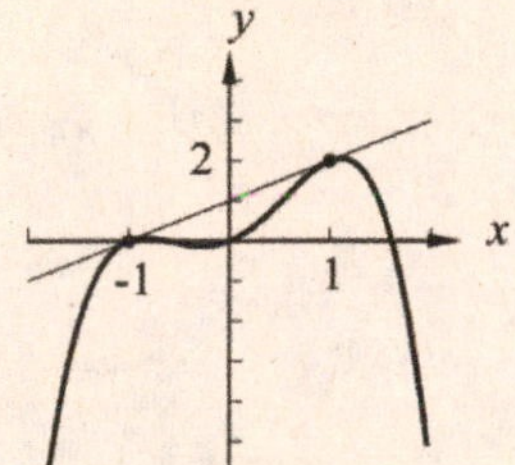

At $(-1,\ 0)$, $y'=1+4(-1)-4(-1)^3=1$

Eq. of TL at $(-1,\ 0)$:

$y-0=1(x-(-1))$

$y=x+1$

The tangent lines are the same: $y=x+1$

21. Given $x^2+y^2=25$, we have

$2x+2y\frac{dy}{dx}=0$

$\frac{dy}{dx}=-\frac{x}{y}.$

The tangent line at $(3,4)$ has slope $m=-\frac{3}{4}$ and equation $y-4=-\frac{3}{4}(x-3)$.

The tangent line at $(3,-4)$ has slope $m=\frac{3}{4}$ and equation $y+4=\frac{3}{4}(x-3)$.

Adding these equations yields $2y=0$ and so $y=0$. Substituting $y=0$ into the second equation yields

$4=\frac{3}{4}(x-3)$

$\frac{16}{3}=x-3$

$x=\frac{25}{3}$

The tangent lines intersect at $\left(\frac{25}{3},0\right)$.

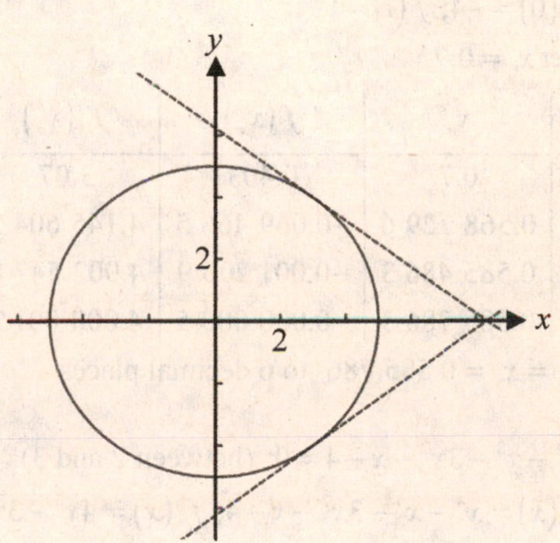

24.2 Newton's Method for Solving Equations

1. $x^2 - 5x + 1 = 0,\ 0 < x < 1,\ f(x) = x^2 - 5x + 1$

$f(0) = 1,\ f(1) = -3$, choose $x_1 = 0.5$

$f'(x) = 2x - 5$

$$x_2 = x_1 - \frac{f(x_1)}{f'(x_1)} = 0.5 - \frac{0.5^2 - 5(0.5) + 1}{2(0.5) - 5}$$

$= 0.1875$

$$x_3 = 0.1875 - \frac{(0.1875)^2 - 5(0.1875) + 1}{2(0.1875) - 5}$$

$= 0.208\ 614\ 864\ 9$

Using the quadratic formula,

$$x = \frac{(-5) - \sqrt{(-5)^2 - 4(1)(1)}}{2(1)}$$

$= 0.208\ 712\ 152\ 5$

Therefore, x_3 is correct to 3 decimal places.

5. $x^3 - 6x^2 + 10x - 4 = 0$ (between 0 and 1)

$f(x) = x^3 - 6x^2 + 10x - 4;$

$f'(x) = 3x^2 - 12x + 10;$

$f(0) = -4;\ f(1) = 1$

Let $x_1 = 0.7$

n	x_n	$f(x_n)$	$f'(x_n)$	$x_n - \frac{f(x_n)}{f'(x_n)}$
1	0.7	0.403	3.07	0.568 729 6
2	0.568 729 6	−0.069 466 5	4.145 604 5	0.585 486 3
3	0.585 486 3	−0.001 200 9	4.002 547 0	0.585 786 3
4	0.585 786 3	−0.000 000 5	4.000 001 2	0.585 786 4

$x_4 = x_5 = 0.585\ 786$ to 6 decimal places

9. $x^4 - x^3 - 3x^2 - x - 4 = 0$; (between 2 and 3)

$f(x) = x^4 - x^3 - 3x^2 - x - 4;\ f'(x) = 4x^3 - 3x^2 - 6x - 1$

$f(2) = -10;\ f(3) = 20.$

Let $x_1 = 2.3$

n	x_n	$f(x_n)$	$f'(x_n)$	$x_n - \frac{f(x_n)}{f'(x_n)}$
1	2.3	−6.3529	17.998	2.652 978 1
2	2.652 978 1	3.097 272 7	36.657 000 1	2.568 484 8
3	2.568 484 8	0.217 499 3	31.576 096	2.561 596 7
4	2.561 596 7	0.001 367 1	31.179 597	2.561 552 8

$x_4 = x_5 = 2.561\ 6$ to 4 decimal places

13. $2x^2 = \sqrt{2x+1}$ (the positive real solution)

$f(x) = 2x^2 - \sqrt{2x+1}$

$f'(x) = 4x - \dfrac{1}{\sqrt{2x+1}}$

Let $x_1 = 0.8$, since we can see that the graphs of $2x^2$ and $\sqrt{2x-1}$ intersect near $x = 0.8$.

n	x_n	$f(x_n)$	$f'(x_n)$	$x_n - \frac{f(x_n)}{f'(x_n)}$
1	0.8	−0.332 451 5	2.579 826 3	0.928 865 9
2	0.928 865 9	0.035 100 9	3.123 916 4	0.917 629 7
3	0.917 629 7	2.656×10^{-4}	3.076 632 0	0.917 543 3
4	0.917 543 3	1.569×10^{-8}	3.076 268 6	0.917 543 3

The positive root is approximately 0.917 543 3 to 7 decimal places.

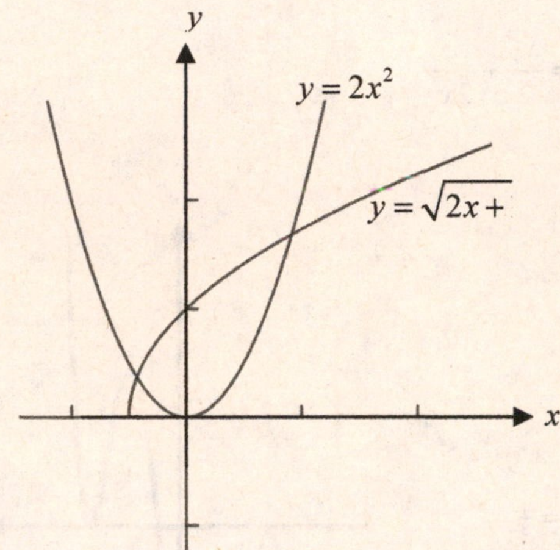

7. $f(x) = x^3 - 2x^2 - 5x + 4$. From the graph, one root lies between -1 and -2, a second between 0 and 1, and a third between 3 and 4

$f'(x) = 3x^2 - 4x - 5$

Let $x_1 = -1.7$

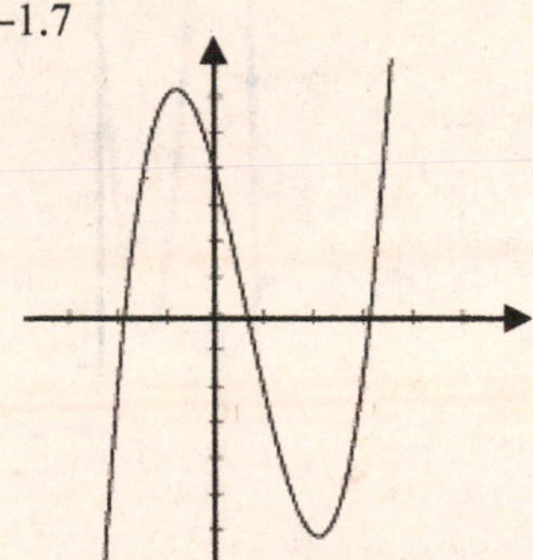

n	x_n	$f(x_n)$	$f'(x_n)$	$x_n = \frac{f(x_n)}{f'(x_n)}$
1	−1.7	1.807	10.47	−1.872 588 3
2	−1.872 588 3	−0.216 626 7	13.010 114 8	−1.855 937 7
3	−1.855 937 7	−0.002 107 4	12.757 265 2	−1.855 772 5
4	−1.855 772 5	$-2.065\ 005\ 0 \times 10^{-7}$	12.754 765 1	−1.855 772 5

Let $x_1 = 0.7$

1	0.7	−0.137	−6.33	0.678 357 0
2	0.678 357 0	$3.670\ 385\ 5 \times 10^{-5}$	−6.332 923 3	0.678 362 8

Let $x_1 = 3.1$

1	3.1	−0.929	11.43	3.181 277 3
2	3.118 127 73	0.048 760 8	12.636 467 2	3.177 418 6
3	3.177 418 6	$1.122\ 688\ 9 \times 10^{-4}$	12.578 292 6	3.177 409 7

The roots are −1.8558, 0.6784, and 3.1774.

21. $f(x) = x^2 - a$

$f'(x) = 2x$

$$x_2 = x_1 - \frac{f(x_1)}{f'(x_1)} = x_1 - \frac{x_1^2}{2x_1} = x_1 - \frac{x_1}{2} + \frac{a}{2x_1}$$

$x_2 = \frac{x_1}{2} + \frac{a}{2x_1}$. Similarly, $x_3 = \frac{x_2}{2}^2 + \frac{a}{2x_2}$ which generalizes to $x_{n+1} = \frac{x_n}{2} + \frac{a}{2x_n}$.

25. $f(x) = 3x^5 - x^4 - 12x^3 + 4x^2 + 12x - 4$

n	x_n	$f(x_n)$	$f'(x_n)$	$x_n = \frac{f(x_n)}{f'(x_n)}$
1	−1.4	−0.00832	−1.16	−1.407172414
2	−1.407172414	−0.002060671	−0.582677811	−1.410708968
3	−1.410708968	−0.000512822	−0.292000057	−1.412465208
4	−1.412465208	−0.000127917	−0.146165305	−1.413340365

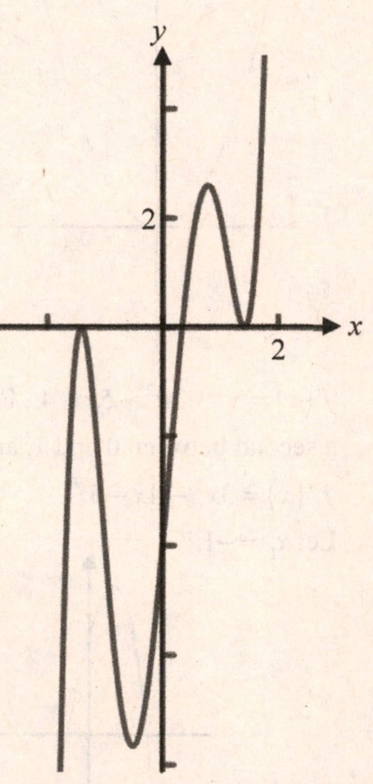

$f(x) = 3x^5 - x^4 - 12x^3 + 4x^2 + 12x - 4 = (3x-1)(x^2-2)$ has roots $x = \frac{1}{3}$ and $x = \pm\sqrt{2}$. Newton's method converges more slowly to the root $x = -\sqrt{2}$ because $f'(x)$ is also close to zero for this value of x.

29. $V = \frac{1}{6}\pi h(h^2 + 3r^2) = \frac{1}{6}\pi h^3 + \frac{1}{2}\pi r^2 h$

$180\ 000 = \frac{1}{6}\pi h^3 + 1800\pi h$

$f(h) = \frac{1}{6}\pi h^3 + 1800\pi h - 180\ 000$

$f'(h) = \frac{1}{2}\pi h^2 + 1800\pi$

From the figure, our first guess is $r/2 = 30$.
Evaluating, $f(30) = 3783$, $f(29) = -3239$.
So we choose $x_1 = 29.5$.

n	h_n	$f(h_n)$	$f'(h_n)$	$h_n - \frac{f(h_n)}{f'(h_n)}$
1	29.5	260.594 022 3	7021.852	29.462 888
2	29.462 888	0.063 794 79	7018.415	29.462 879

The root is $h = 29.462\ 9$

The height of the dome is $h = 29.5$ m

24.3 Curvilinear Motion

1. $x = 4t^2 \qquad y = 1 - t^2$

$v_x = \frac{dx}{dt} = 8t\big|_{t=2} = 16 \qquad \frac{dy}{dt} = -2t\big|_{t=2} = -4$

$v = \sqrt{16^2 + (-4)^2} = 16.5$

$\tan\theta = \frac{-4}{16},\ \theta = -14.0^\circ$

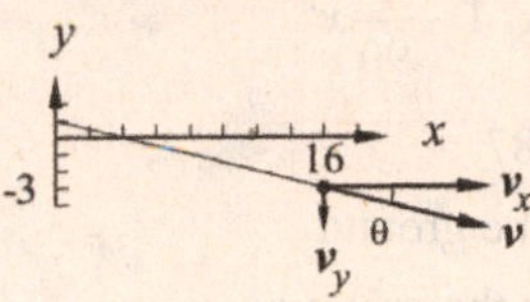

5. $x = t(2t+1)^2$

$\frac{dx}{dt} = t(2)(2t+1)(2) + (2t+1)^2(1)$

$= 12t^2 + 8t + 1 = v_x$

$y = 6(4t+3)^{-1/2}$

$\frac{dy}{dt} = 6\ -\frac{1}{2}\ (4t+3)^{-3/2}(4)$

$= \frac{-12}{(4t+3)^{3/2}} = v_y$

$v_x\big|_{t=0.5} = 8$

$v_y\big|_{t=0.5} = -1.0733$

$v = \sqrt{8^2 + (-1.0773)^2}$

$v = 8.07$

$v_x > 0, v_y < 0$, so θ is in QIV.

$\theta_{\text{ref}} = \tan^{-1}\frac{1.0733}{8} = 7.641^\circ$

Therefore, $\theta = 360^\circ - \theta_{\text{ref}} = 352.36^\circ$ (or $\theta = -7.64^\circ$)

Therefore, v is 8.07 at $\theta = 352.36^\circ$

To sketch the curve we tabulate a few points:

t	x	y
0	0	3.464
0.5	2	2.683
1	9	2.268

9. $x = t(2t+1)^2$

$v_x = 12t^2 + 8t + 1$

$a_x = 24t + 8;$

$a_x\big|_{t=0.5} = 20.0$

$y = \frac{6}{\sqrt{4t+3}}$

$v_y = -12(4t+3)^{-3/2}$

$a_y = \frac{72}{(4t+3)^{5/2}}$

$a_y\big|_{t=0.5} = 1.288$

$a = \sqrt{20.0^2 + 1.288^2}$

$a = 20.0$

$\theta = \tan^{-1}\frac{1.288}{20.0}$

$\theta = 3.68^\circ$

Therefore, a is 20.0 at $\theta = 3.68^\circ$.

13. $y = 2.0 + 0.80x - 0.20x^2,$

$\frac{dy}{dt} = v_y$

$= 0.80\frac{dx}{dt} - 0.40x\frac{dx}{dt}$

$v_y\big|_{(4.0,2.0)} = 0.80v_x - 0.40xv_x\big|_{(4.0,\ 2.0)}$

$= 0.80(5.0) - 0.40(4.0)(5.0)$

$= -4.0$

$v = \sqrt{(5.0)^2 + (-4.0)^2} = 6.40$ m/s

$v_x > 0, v_y < 0$, so θ is in QIV

$\tan\theta = \frac{-4.0}{5.0}$

$\theta = -39^\circ$ or $\theta = 360^\circ - 39^\circ = 321^\circ$

17. $x = 25t$

$v_x = \frac{dx}{dt} = 25$

$a_x = \frac{d^2x}{dt^2} = 0$

$y = 36t - 4.9t^2$

$v_y = \frac{dy}{dt}$

$= 36 - 9.8t$

$v_y\big|_{t=6.0} = 36 - 9.8(6.0)$

$= -22.8$

$v = \sqrt{25^2 + (-22.8)^2}$

$= 34 \text{ m/s}$

$v_x > 0, v_y < 0$, so θ_v is in QIV

$\theta_v = \tan^{-1}\frac{-22.8}{25}$

$= -42.4^\circ$

$a_y = \frac{d^2y}{dt^2}$

$= -9.8$

$a = \sqrt{0^2 + (-9.8)^2}$

$= 9.8 \text{ m/s}^2$

$a_x = 0, a_y < 0$, so $\theta_a = -90^\circ$

Acceleration is pointing vertically downward.

21. $y = 3.00 + x^{-1.50}$, at $t = 0$, $(x, y) = (1.00, 4.00)$.

$v_x = 1.20$ cm/s.

$\frac{dy}{dt} = -1.50x^{-2.50}\frac{dx}{dt}$

$= -1.50x^{-2.50}(1.20)$

$= -1.80x^{-2.50}$

At $t = 0.500$ s,

$x = 1.00 + 1.20(0.500) = 1.60$

$\frac{dy}{dt}\Big|_{x=1.60} = -1.8(1.60)^{-2.50}$

$= -0.556$

$v = \sqrt{1.20^2 + (-0.556)^2}$

$= 1.32$ cm/s

$v_x > 0, v_y < 0$, so θ is in QIV.

$\theta = \tan^{-1}\frac{-0.556}{1.20} = -24.9^\circ$

$\theta = -24.9^\circ$

$v_x = 1.20$

v

$v_y = -0.556$

25. $y = x - \frac{1}{90}x^3; v_x = x$; rocket hits the ground when $y = 0$.

$v_y = v_x - \frac{1}{30}x^2 v_x$

$v_y = x - \frac{1}{30}x^2(x) = x - \frac{1}{30}x^3$

$y = 0 = x - \frac{1}{90}x^3 = x\ \ 1 - \frac{1}{90}x^2$

$x = 0, x = \sqrt{90} = 9.487$

$x = 0$; rocket leaves the ground

$x = 9.487$; rocket hits the ground

$v_x = x$

$v_x\big|_{x=9.487} = 9.487$

$v_y = x - \frac{1}{30}x^3$

$v_y\big|_{x=9.487} = -18.98$

$v = \sqrt{90.0 + (18.98)^2} = 21.2$

$v_x > 0, v_y < 0$, so θ is in QIV.

$\theta_{\text{ref}} = \tan^{-1}\frac{18.97}{\sqrt{9.487}} = 63.4^\circ$

$\theta = 360^\circ - \theta_{\text{ref}} = 296.6^\circ$ (or $\theta = -63.4^\circ$)

Therefore, v is 21.2 km/min at $\theta = 296.6^\circ$.

29. $h = k\sqrt{x} = kx^{1/2}; v_x = 350$ m/s

$k = \frac{h}{\sqrt{x}}$

$= \frac{280}{\sqrt{400}}$

$= 14$

$h = 14x^{1/2}$

$\frac{dh}{dt} = 7x^{-1/2}v_x$

$\frac{dh}{dt}\Big|_{x=400} = \frac{7(350)}{20}$

$= 122.5$ m/s

$v = \sqrt{350^2 + 122.5^2}$

$v = 370.8$ m/s

$\theta = \tan^{-1}\frac{122.5}{350} = 19.3^\circ$

Therefore, v is 370 m/s at $\theta = 19^\circ$.

24.4 Related Rates

1. $E = 2.800T + 0.012T^2$

$$\frac{dE}{dt} = 2.800\frac{dT}{dt} + 0.024T\frac{dT}{dt}$$

$$\left.\frac{dE}{dt}\right|_{T=100^\circ C} = 2.800(1.00) + 0.024(100)(1.00)$$

$$\left.\frac{dE}{dt}\right|_{T=100^\circ C} = 5.20 \text{ V/min}$$

5. $x^2 + 3y^2 + 2y = 10$

$$2x\frac{dx}{dt} + 6y\frac{dy}{dt} + 2\frac{dy}{dt} = 0$$

$$\frac{dy}{dt} = \frac{-x\frac{dx}{dt}}{3y+1}$$

$$\left.\frac{dy}{dt}\right|_{(3,-1)} = \frac{(-3)(2)}{3(-1)+1}$$

$$= 3$$

9. $R = 4.000 + 0.003T^2;\ \frac{dT}{dt} = 0.100^\circ\text{C/s}$

$$\frac{dR}{dt} = 0 + 0.006T\frac{dT}{dt}$$

$$\left.\frac{dR}{dt}\right|_{T=150^\circ C} = 0 + 0.006(150)(0.100)$$

$$= 0.0900\ \Omega\text{/s}$$

13. $D = \sqrt{4.0 + x^2}$

$$\frac{dD}{dt} = \frac{x}{\sqrt{4.0+x^2}} \cdot \frac{dx}{dt}$$

$$\left.\frac{dD}{dt}\right|_{x=6.2} = \frac{6.2}{\sqrt{4+6.2^2}} \cdot 350$$

$$= 330 \text{ mi/h}$$

17. $B = \dfrac{k}{\left(r^2 + \left(\frac{l}{2}\right)^2\right)^{3/2}} = k\left(r^2 + \left(\frac{l}{2}\right)^2\right)^{-3/2}$

$$\frac{dB}{dt} = -\frac{3}{2}k\left(r^2 + \left(\frac{l}{2}\right)^2\right)^{-5/2} 2r\frac{dr}{dt}$$

$$\frac{dB}{dt} = \frac{-3kr\frac{dr}{dt}}{\left(r^2 + \left(\frac{l}{2}\right)^2\right)^{5/2}}$$

21. $A = \pi r^2;\ \frac{dr}{dt} = 0.020$ mm/mo;

$$\frac{dA}{dt} = 2\pi r\frac{dr}{dt}$$

$$\left.\frac{dA}{dt}\right|_{r=1.2} = 2\pi(1.2)(0.020)$$

$$= 0.15 \text{ mm}^2\text{/month}$$

25. $V = \frac{4}{3}\pi r^3$

$$\frac{dV}{dt} = \left(4\pi r^2\right)\frac{dr}{dt}$$

Also, with $k > 0$,

$$\frac{dV}{dt} = -kA$$

$$= -k\left(4\pi r^2\right)$$

Therefore,

$$\frac{dr}{dt} = -k$$

and radius decreases at a constant rate.

29. Let s be the length of the shadow and D be the distance from the top of the building to the end of the shadow. Using the Pythagorean theorem,

$$D^2 = 24^2 + s^2$$

When $s = 18$, $D^2 = 24^2 + 18^2 = 900$ so $D = 30$.

$$2D\frac{dD}{dt} = 2s\frac{ds}{dt}$$

$$60\frac{dD}{dt} = 36(0.18)$$

$$\frac{dD}{dt} = \frac{36(0.18)}{60} = 0.108$$

The distance D increases at 10.8 cm/min.

33. $w = 650\dfrac{6400}{6400+h}$

$$\frac{dw}{dt} = -\frac{650(6400)}{(6400+h)^2}\frac{dh}{dt}$$

$$\left.\frac{dw}{dt}\right|_{h=1200} = \frac{-650(6400)}{(6400+1200)^2}(6.0)$$

$$= -0.43 \text{ N/s}$$

37. Let x be the distance travelled by the jet going due east, and y be the distance travelled by the jet going north of east.

Since the second jet remains due north of the first jet, we have a right triangle and we can use the Pythagorean theorem. $x^2 + z^2 = y^2$

Taking the derivative of this expression,

$$2x\frac{dx}{dt} + 2z\frac{dz}{dt} = 2y\frac{dy}{dt}$$

$$\frac{dz}{dt} = \frac{1}{z}\left(y\frac{dy}{dt} - x\frac{dx}{dt}\right)$$

$$x\Big|_{t=(1/2)} = 1600\left(\frac{1}{2}\right) = 800 \text{ mi}$$

$$y\Big|_{t=(1/2)} = 1800\left(\frac{1}{2}\right) = 900 \text{ mi}$$

$$z = \sqrt{y^2 - x^2}$$

$$= \sqrt{900^2 - 800^2}$$

$$= 412.3 \text{ mi}$$

$$\frac{dx}{dt} = 1600; \frac{dy}{dt} = 1800 \text{ mi/h}$$

Substituting,

$$\frac{dz}{dt} = \frac{1}{z}\left(y\frac{dy}{dt} - x\frac{dx}{dt}\right)$$

$$\frac{dz}{dt}\Big|_{t=0.5} = \frac{1}{412.3}(900(1800) - 800(1600))$$

$$= 820 \text{ mi/h}$$

41. $\frac{dx}{dt} = -5.00$ ft/s; $x = 10.0$ ft

By similar triangles;

$$\frac{6.00}{d-x} = \frac{15.0}{d};$$

$$6.00d = 15.0(d-x)$$

$$9.00d = 15.0x$$

$$9.00\frac{dd}{dt} = 15.0\frac{dx}{dt}$$

$$\frac{dd}{dt} = \frac{15.0}{9.00}\frac{dx}{dt}$$

$$\frac{dd}{dt}\Big|_{x=10.00} = -8.33 \text{ ft/s}$$

The negative sign means the end of man's shadow is approaching the light post. Also, we note that $\frac{dd}{dt}$ does not depend on the value of x.

24.5 Using Derivatives in Curve Sketching

1. $f(x) = x^3 + 3x^2$

$f'(x) = 3x^2 + 6x = 3x(x+2)$ with $x = -2$, $x = 0$ as critical values.

If $x < -2$, $f'(x) = 3x(x+2) > 0$. $f(x)$ is increasing

If $-2 < x < 0$, $f'(x) = 3x(x+2) < 0$. $f(x)$ is decreasing

If $x > 0$, $f'(x) = 3x(x-4) > 0$. $f(x)$ is increasing

inc. $x < -2$, $x > 0$; dec. $-2 < x < 0$

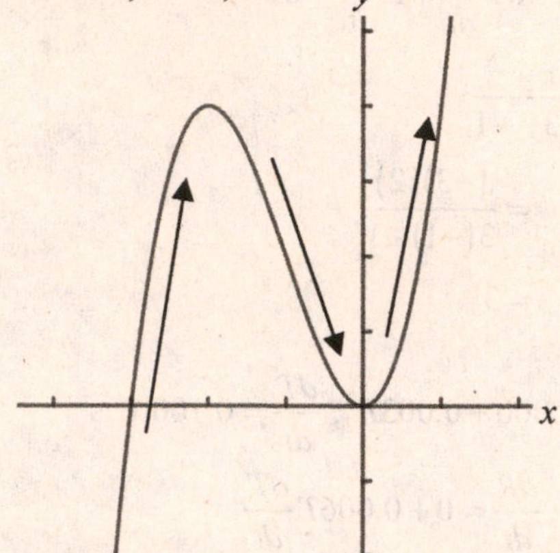

5. $y = x^2 + 2x$

$y' = 2x + 2$

$2x + 2 > 0$

$2x > -2$

$x > -1$; $f(x)$ increasing.

$2x + 2 < 0$

$2x < -2$

$x < -1$; $f(x)$ decreasing.

9. $y = x^2 + 2x$

$y' = 2x + 2$

$y' = 0$ at $x = -1$

$y'' = 2 > 0$ at $x = -1$ and $(-1, -1)$ is a relative minimum. Note also that y decreases for $x < -1$ and increases for $x > -1$.

13. $y = x^2 + 2x$

$y' = 2x + 2$

$y'' = 2$

Thus, $y'' > 0$ for all x. The graph is concave up for all x and has no points of inflection.

17. $y = x^2 + 2x$

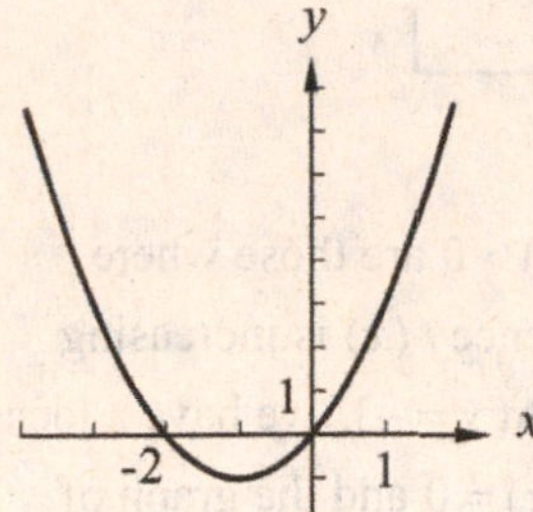

21. $y = 12x - 2x^2$

$y' = 12 - 4x$

$y' = 0$ at $x = 3$

For $x = 3, y = 12(3) - 2(3)^2 = 18$ and $(3, 18)$ is a critical point.

$12 - 4x < 0$ for $x > 3$ and the function decreases; $12 - 4x > 0$ for $x < 3$ and the function increases. Hence $(3, 18)$ is a maximum.

$y'' = -4$; thus $y'' < 0$ for all x. There are no inflection points.

The graph is concave down for all x, and (3, 18) is a maximum point.

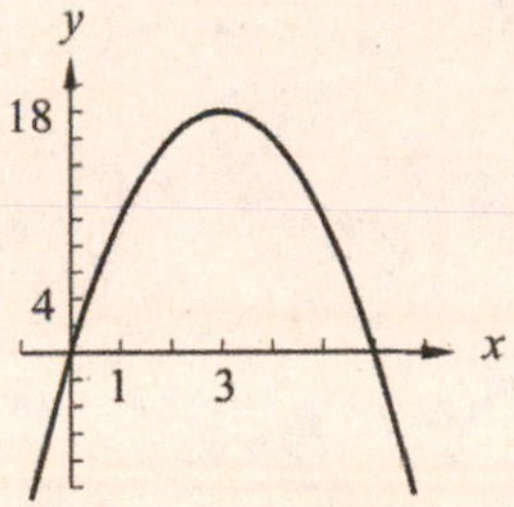

25. $y = x^3 + 3x^2 + 3x + 2$

$y' = 3x^2 + 6x + 3$

$= 3(x^2 + 2x + 1)$

$= 3(x+1)^2$

$= 0$ for $x = -1$

$(-1, 1)$ is a critical point.

$3(x+1)^2 > 0$ for all x so the function is always increasing

$y'' = 6x + 6$

$= 0$ for $x = -1$

$6x + 6 < 0$ for $x < -1$ and the graph is concave down.

$6x + 6 > 0$ for $x > -1$ and the graph is concave up.

Therefore $(-1, 1)$ is an inflection point.

Since there is no change in slope from positive to negative or vice versa, there are no maximum or minimum points.

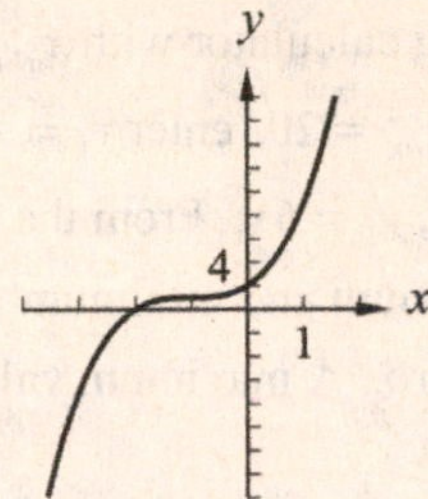

29. $y = 4x^3 - 3x^4 + 6$

$y' = 12x^2 - 12x^3 = 12x^2(1 - x) = 0$

$12x^2(1 - x) = 0$ for $x = 0$ and $x = 1$

$(0, 6)$ and $(1, 4)$ are critical points.

$12x^2 - 12x^3 > 0$ for $x < 0$ and the slope is positive.

$12x - 12x^3 > 0$ for $0 < x < 1$ and the slope is positive.

$12x - 12x^3 < 0$ for $x > 1$ and the slope is negative.

$y'' = 24x - 36x^2$

$= 24x - 36x^2$

$= 12x(2 - 3x)$

$= 0$ for $x = 0,\ x = \frac{2}{3}$

$(0, 6)$ and $\left(\frac{2}{3}, \frac{178}{27}\right)$ are possible inflection points.

$24x - 36x^2 < 0$ for $x < 0$ and the graph is concave down.

$24x - 36x^2 > 0$ for $0 < x < \frac{2}{3}$ and the graph is concave up.

$24x - 36x^2 < 0$ for $x > \frac{2}{3}$ and the graph is concave down.

$(1, 7)$ is a relative maximum point since $y' = 0$ at $(1, 7)$ and the slope is positive for $x < 1$ and negative for $x > 1$. $(0, 6)$ and $\left(\frac{2}{3}, \frac{178}{27}\right)$ are inflection pts since there is a concavity change.

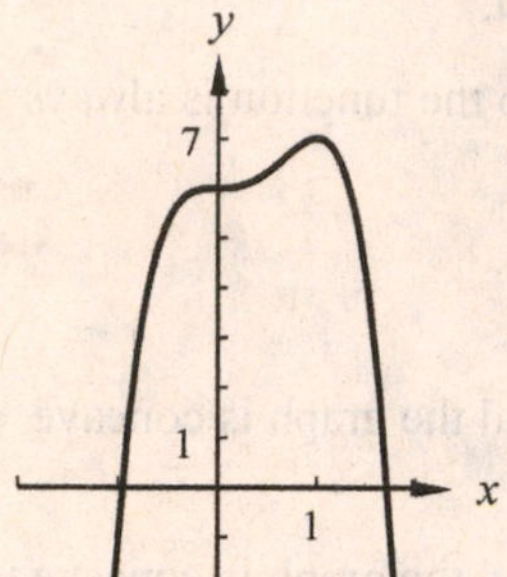

33. $y = x^3 - 12x$

$y' = 3x^2 - 12$

$y'' = 6x.$

On a graphing calculator with $x_{min} = -5$, $x_{max} = 5$, $y_{min} = -20$, $y_{max} = 20$, enter $y_1 = x^3 - 12x$, $y_2 = 3x^2 - 12$, $y_3 = 6x$. From the graph it is observed that the maximum and minimum values of y occur when y' is zero. A maximum value for y occurs

when $x = -2$, and a minimum value occurs when $x = 2$. An inflection point (change in curvature) occurs when y'' is zero. x is also zero at this point. Where $y' > 0$, y inc.; $y' < 0$, y dec. $y'' > 0$, y concave up; $y'' < 0$, y concave down, $y'' = 0$, y has inflection.

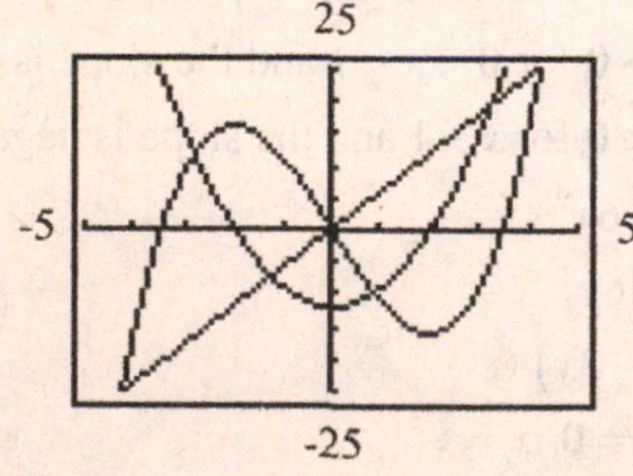

37. The left relative max is above the left relative min and below the right relative min.

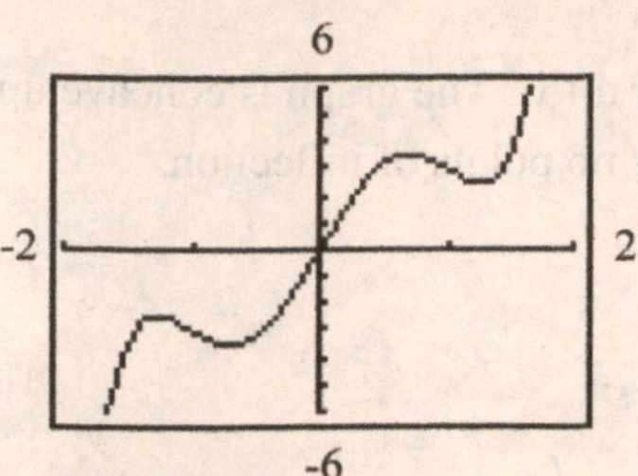

41. The points where $f'(x) > 0$ are those where $f(x)$ is increasing. Hence $f(x)$ is increasing from $x = -1$ to $x = 3$. At $x = -1$, we have a local minimum because $f'(x) = 0$ and the graph of $f'(x)$ is increasing (hence $f''(x) > 0$.) At $x = 3$, we have a local minimum because $f'(x) = 0$ and $f''(x) < 0$.

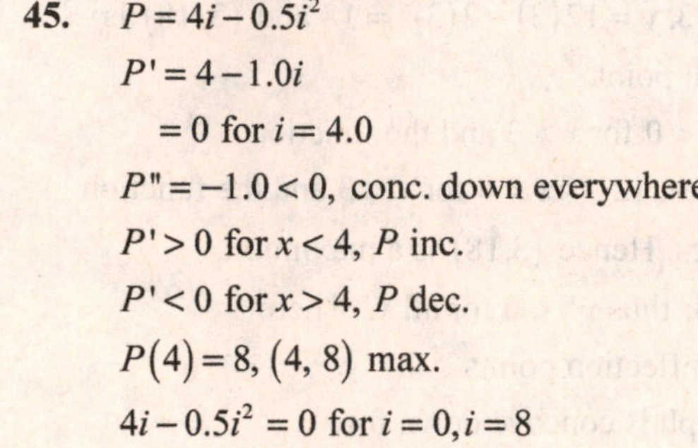

45. $P = 4i - 0.5i^2$

$P' = 4 - 1.0i$

$= 0$ for $i = 4.0$

$P'' = -1.0 < 0$, conc. down everywhere

$P' > 0$ for $x < 4$, P inc.

$P' < 0$ for $x > 4$, P dec.

$P(4) = 8$, $(4, 8)$ max.

$4i - 0.5i^2 = 0$ for $i = 0, i = 8$

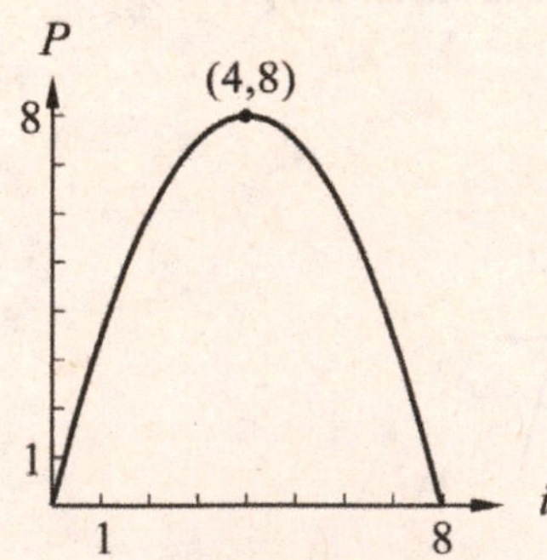

49. $R = 75 - 18i^2 + 8i^3 - i^4$

$R' = -36i + 24i^2 - 4i^3$

$= -4i(i^2 - 6i + 9)$

$= -4i(i-3)^2$

$R' = 0$ for $i = 0$ and $i = 3$

$(0, 75)$ and $(3, 48)$ are critical points.

$R' > 0$ for $i < 0$, $R' < 0$ for $0 < i < 3$

$R' < 0$ for $i > 3$

Max. at (0, 75), no max. or min. at (3, 48)

$R'' = -36 + 48i - 12i^2$

$= -12(i-1)(i-3)$

$(1, 64)$ and $(3, 48)$ are possible inflection points.

$R'' < 0$ for $i < 1$, concave down

$R'' > 0$ for $1 < i < 3$, concave up

$R'' < 0$ for $i > 3$, concave down

$(1, 64)$ and $(3, 48)$ are inflection points.

(From calculator graph, $R = 0$ for $i = -1.5$ and $i = 5.0$)

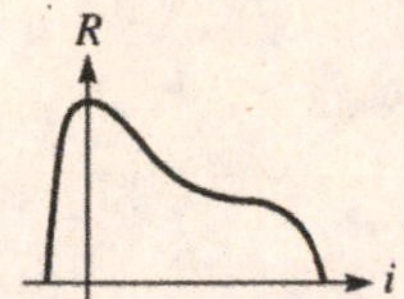

53. $V = x(8-2x)(12-2x), 0 < x < 4$

$V = 4(24x - 10x^2 + x^3)$

$= 4x^3 - 40x^2 + 96x$

$f'(x) = 4(24 - 20x + 3x^2)$

$= 0$

By quadratic solution:

$x = 1.57$ and $x = 5.10$ (reject)

$f''(x) = 4(-20 + 6x)$

$f''(1.57) < 0$, rel max. $(1.57, 67.6)$

$f''(5.10) > 0$, rel min. $(5.10, -20.2)$

$f''(x) = 0$; $x = \frac{20}{6}$, infl. $\frac{10}{3}$, 23.7

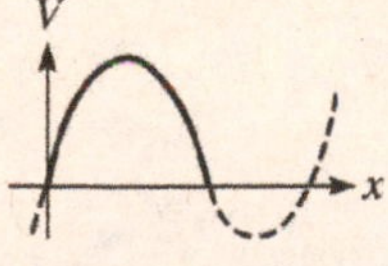

57. $f(-1) = 0$, root at $(-1, 0)$

$f(2) = 2$, point on curve $(2, 2)$

$f'(x) < 0$ for $x < -1$

Therefore, $f(x)$ decreasing for $x < -1$.

$f'(x) > 0$ for $x > -1$

Therefore, $f(x)$ increasing for $x > -1$.

$f''(x) < 0$ for $0 < x < 2$

Therefore, $f(x)$ concave down for $0 < x < 2$.

$f''(x) > 0$ for $x < 0$ or $x > 2$

Therefore, $f(x)$ concave up for $x < 0$.

Therefore, $f(x)$ concave up for $x > 2$.

Summary: $(-1, 0)$ min. point

Inflection point at y-intercept

$(2, 2)$ inflection point

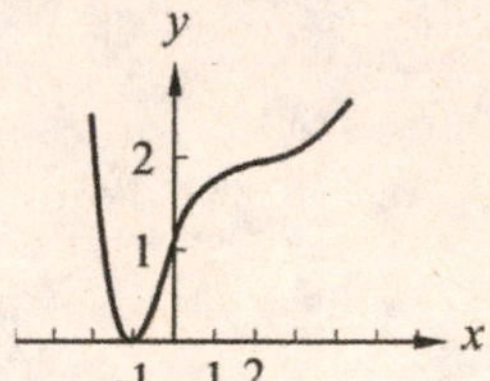

24.6 More on Curve Sketching

1. $y = x - \frac{4}{x}$

(1) Intercepts:

For $x = 0$, y is undefined which means curve is not continuous at $x = 0$ and there are no y-intercepts.

For $y = 0$, $0 = x - \frac{4}{x}$, $x^2 - 4 = 0$, $x = \pm 2$ are the x-intercepts.

(2) Symmetry: symmetric to the origin since $-y = -x - \frac{4}{-x}$ is the same as $y = x - \frac{4}{x}$

(3) Behaviour as x becomes large:

As $x \to \pm\infty$, $\frac{4}{x} \to 0$ and $y \to x$. $y = x$ is an asymptote.

(4) Vertical asymptotes:

y is undefined for $x = 0$. As $x \to 0^-$, $y \to \infty$ and as $x \to 0^+$, $y \to -\infty$. The x-axis is a vertical asymptote.

(5) Domain and range:

domain: $x \neq 0$, range: $-\infty < y < \infty$

(6) Derivatives:

$y' = 1 + \frac{4}{x^2} > 0$, y inc. for $x \neq 0$

$y'' = -\frac{8}{x^3} > 0$ for $x < 0$, y conc. up

$y'' = -\frac{8}{x^3} < 0$ for $x > 0$, y conc. down

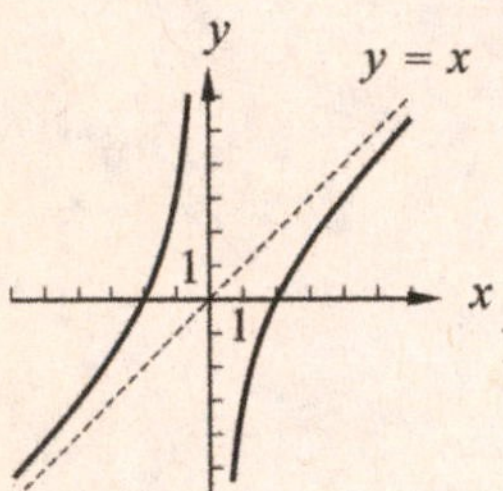

5. $y = x^2 + \frac{2}{x} = \frac{x^3 + 2}{x}$

(1) $\frac{2}{x}$ is undefined for $x = 0$, so the graph is not continuous at the y-axis; i.e., no y-intercept exists.

(2) $\frac{x^3 + 2}{x} = 0$ at $x = \sqrt[3]{-2} = -\sqrt[3]{2}$. There is an x-intercept at $\left(-\sqrt[3]{2}, 0\right)$.

(3) As $x \to \infty$, $x^2 \to \infty$ and $\frac{2}{x} \to 0$, so $x^2 + \frac{2}{x} \to \infty$.

(4) As $x \to 0$ through positive x, $x^2 \to 0$ and $\frac{2}{x} \to \infty$, so $x^2 + \frac{2}{x} \to \infty$.

(5) As $x \to -\infty$, $x^2 \to \infty$ and $\frac{2}{x} \to 0$ so $x^2 + \frac{2}{x} \to \infty$.

$x = 0$ is a vertical asymptote

(6) As $x \to 0$ through negative numbers, $x^2 \to 0$ and $\frac{2}{x} \to -\infty$, so $x^2 + \frac{2}{x} \to -\infty$.

(7) $y' = 2x - 2x^{-2} = 0$ at $x = 1$ and the slope is zero at $(1, 3)$.

(8) $y'' = 2 + 4x^{-3} = 0$ at $x = -\sqrt[3]{2}$ and $\left(-\sqrt[3]{2}, 0\right)$ is an inflection point.

(9) $y'' > 0$ at $x = 1$, so the graph is concave up and $(1, 3)$ is a relative minimum.

(10) Since $\left(-\sqrt[3]{2}, 0\right)$ is an inflection pt, $f''(x) < 0$ for $-\sqrt[3]{2} < x < 0$ and the graph is concave down. $f''(x) > 0$ for $x < -\sqrt[3]{2}$ and the graph is concave up. $f''(x) > 0$ for $x > 0$ and the graph is concave up.

(11) Not symmetrical about the x or y-axis nor about the origin.

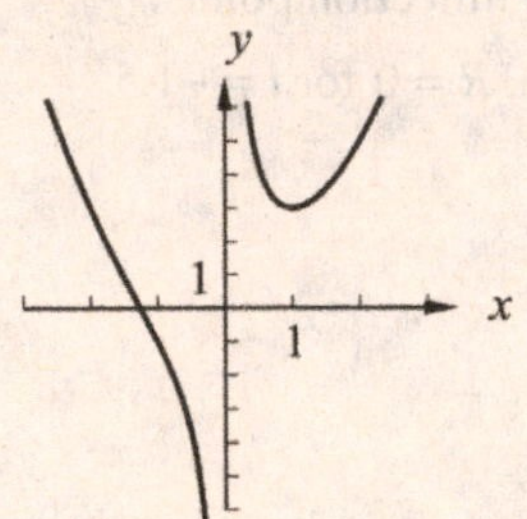

9. $y = \frac{x^2}{x+1}$

$= \frac{x^2 - 1 + 1}{x+1}$

$= x - 1 + \frac{1}{x+1}$

Intercepts:

(1) Function undefined at $x = -1$; not continuous at $x = -1$.

(2) At $x = 0$, $y = 0$. The origin is the only intercept.

(3) Behaviour as x becomes large: As $x \to \pm\infty$, $y \to x - 1$, so $y = x - 1$ is a slant asymptote.

Vertical asymptotes:

(4) As $x \to -1$ from the left, $x+1 \to 0$ through negative values and $\frac{x^2}{x+1} \to -\infty$ since $x^2 > 0$ for all x. As $x \to -1$ from the right, $x+1 \to 0$ through positive values and $\frac{x^2}{x+1} \to +\infty$. $x = -1$ is an asymptote.

Symmetry:

(5) The graph is not symmetrical about the y-axis or the x-axis or the origin.

Derivatives:

(6) $$y' = \frac{(x+1)(2x) - x^2(1)}{(x+1)^2} = \frac{x(x+2)}{(x+1)^2}$$

$y' = 0$ at $x = -2$, $x = 0$.

$(-2, -4)$ and $(0, 0)$ are critical points. Checking the derivative at $x = -3$, the slope is positive, and at $x = -1.5$ the slope is negative. $(-2, -4)$ is a relative maximum point. Checking the derivative at $x = -0.5$, the slope is negative and at $x = 1$ the slope is positive, so $(0, 0)$ is a relative minimum point.

$$y'' = \frac{(x+1)^2(2x+2) - (x^2+2x)(2(x+1))}{(x+1)^4} = \frac{2(x+1)(x^2+2x+1-x^2-2x)}{(x+1)^4} = \frac{2}{(x+1)^3}$$

$y'' \neq 0$, so no inflection points.

$y'' > 0$ for $x > -1$, concave up

$y'' < 0$ for $x < -1$, concave down

Int. $(0, 0)$, max. $(-2, -4)$, min $(0, 0)$, asym. $x = -1$

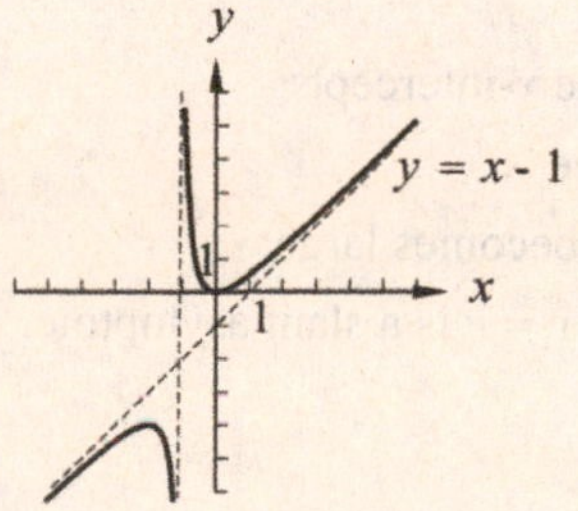

13. $y = \frac{4}{x} - \frac{4}{x^2}$

Intercepts:

(1) There are no y intercepts since $x = 0$ is undefined.

(2) $y = 0$ when $x = 1$ so $(1, 0)$ is an x-intercept,

Asymptotes:

(3) $x = 0$ is an asymptote; the denominator is 0.
$y = 0$ is an asymptote because as $x \to \pm\infty$, $y \to 0$.

Symmetry:

(4) Not symmetrical about the y-axis since $\frac{4}{x} - \frac{4}{x^2}$ is different from $\frac{4}{(-x)} - \frac{4}{(-x)^2}$

(5) Not symmetrical about the x-axis since $y = \frac{4}{x} - \frac{4}{x^2}$ is different from $-y = \frac{4}{x} - \frac{4}{x^2}$

(6) Not symmetrical about the origin since $y = \frac{4}{x} - \frac{4}{x^2}$ is different from $-y = \frac{4}{-x} - \frac{4}{(-x)^2}$

Derivatives:

(7) $y' = -4x^{-2} + 8x^{-3}$
$= 0$ at $x = 2$

(8) $y'' = 8x^{-3} - 24x^{-4}$
$= 0$ at $x = 3$

$y''(2) < 0$ so $(2, 1)$ is a relative maximum.

$y'' < 0$ (concave down) for $x < 3$ and $y'' > 0$ (concave up) for $x > 3$ so $\left(3, \frac{8}{9}\right)$ is an inflection.

Behaviour as x becomes large:

(9) As $x \to \infty$ or $-\infty$, $\frac{4}{x}$ and $-\frac{4}{x^2}$ each approach 0, so the x-axis is a horizontal asymptote.

As $x \to 0$, $\frac{4}{x} - \frac{4}{x^2} = \frac{4x-4}{x^2}$ approaches $-\infty$, through positive or negative values of x, so the y-axis is a vertical asymptote.

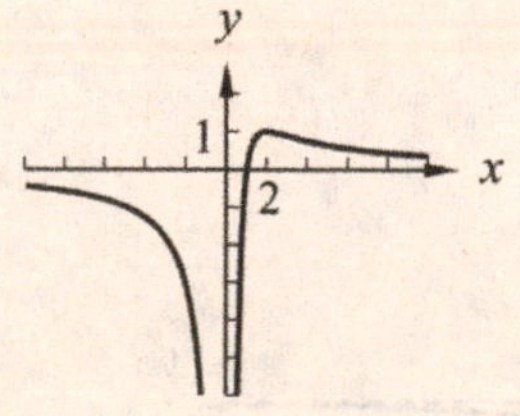

17. $y=\dfrac{9x}{9-x^2}$

(1) Intercept, $(0,\,0)$

(2) Vertical asymptotes at $x=-3,\ x=3$

(3) $y'=\dfrac{(9-x^2)(9)-(9x)(-2x)}{(9-x^2)^2}$

$=\dfrac{9(x^2+9)}{(9-x^2)^2}$

$\neq 0$ for all $x\neq \pm 3$

In fact, $y'>0$ for all $x\neq\pm3$, so the graph is always increasing.

(4)

$$y''=\frac{9\left[(9-x^2)^2(2x)-(x^2+9)(2)(9-x^2)(-2x)\right]}{(9-x^2)^4}$$

$$=\frac{9(9-x^2)(2x)\left[9-x^2+2x^2+18\right]}{(9-x^2)^4}$$

$$=\frac{18x(x^2+27)}{(9-x^2)^3}$$

$=0$ for $x=0$

Concavity changes at $x=0, x=\pm3$:

$y''>0$ for $x<-3$ and $0<x<3$ (concave up)

$y''<0$ for $-3<x<0$ and $x>3$ (concave down)

Therefore $(0,0)$ is an inflection point.

(5) Symmetry: There is symmetry to the origin.

(6) As $x\to+\infty$ and as $x\to-\infty$, $y\to 0$. Therefore, $y=0$ is an asymptote.

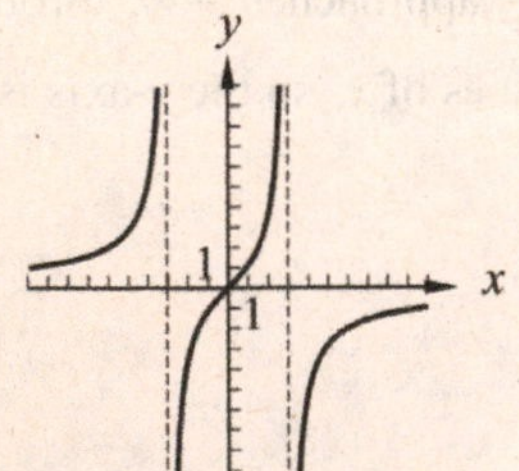

21.

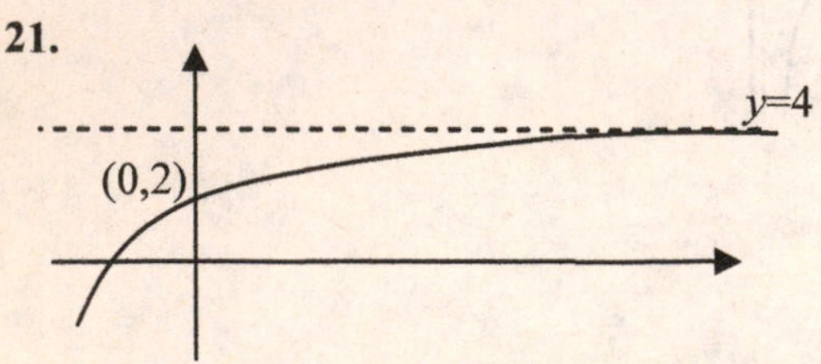

25. $C_T=f(C)=\dfrac{6C}{6+C}$

$f'(C)=\dfrac{36}{(6+C)^2}$

$\neq 0$ for all $C\geq 0$

$f''(C)=\dfrac{-72}{(6+C)^3}$

$\neq 0$ for all $C\geq 0$

$f'(C)\neq 0$; $f''(C)\neq 0$, therefore no max, no min, no infl. points

$f'(C)>0$ for all C, therefore C_T is increasing for all C.

$f''(C)<0$ for $C>-6$, therefore C_T is concave down for $C>-6$ (only values of $C\geq 0$ have physical significance.)

$C=0$, $C_T=0$, $(0,\,0)$ is the only intercept

No symmetry WRT axes or origin.

$\lim\limits_{x\to+\infty}\dfrac{6C}{6+C}=\lim\limits_{x\to+\infty}\dfrac{6}{\frac{6}{C}+1}=6$, therefore, horizontal asymptote at $C_T=6$.

Vertical asymptote at $C=-6$ (capacitance > 0)

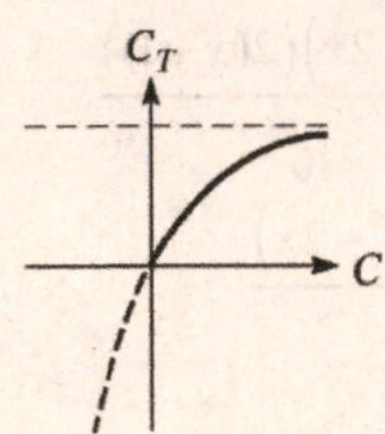

29. $v=k\left(r-\dfrac{1}{r^3}\right)$

$=r-\dfrac{1}{r^3}$ for $k=1$.

(1) Intercepts: For $r=0$, v is undefined. There is no v-intercept.

For $v=0=r-\frac{1}{r^3}$

$r^4=1$, $r=\pm1$ are the x-intercepts.

(2) Symmetry: none

(3) Behaviour as r becomes large:

As $r\to+\infty$, $v\to r$. $v=r$ is a slant asymptote.

(4) Vertical asymptotes: $r = 0$ is a vertical asymptote since v is undefined for $r = 0$.

(5) Domain and range: Domain is all $r \neq 0$ range is $-\infty < v < \infty$

(6) Derivatives:

$$v' = 1 + \frac{3}{r^4} > 0,\ v \text{ inc. } r \neq 0$$

$$v'' = -\frac{12}{r^5},\ v'' < 0,\ r > 0,\ v \text{ conc. down}$$

$$v'' = -\frac{12}{r^5},\ v'' > 0,\ r < 0,\ \text{conc. up.}$$

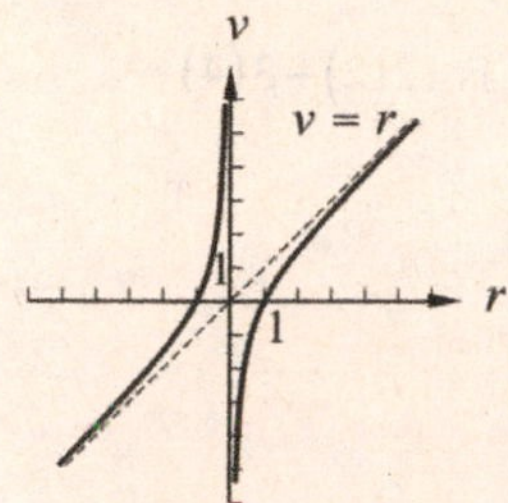

24.7 Applied Maximum and Minimum Problems

1. $A = xy,\ 2x + 2y = 2400 \quad x \quad y \quad 1200$

$$A = (1200 - y)y$$
$$= 1200y - y^2$$
$$A' = 1200 - 2y$$
$$= 0 \text{ for } y = 600$$
$$A'' = -2 < 0, \text{ so } y = 600 \text{ is max.}$$
$$x + 600 = 1200$$
$$x = 600$$
$$A_{max} = 600(600) = 360000 \text{ ft}^2$$

5. $P = 12I - 5.0I^2$

$$\frac{dP}{dI} = 12 - 10I$$
$$= 0 \text{ for } I = 1.2 \text{ A}$$
$$\frac{d^2P}{dI^2} = -10 < 0 \text{ for all } I, \text{ therefore max. power}$$

at $I = 1.2$ A

9. $S = 360A - 0.1A^3$, find maximum S.

$$S' = 360 - 0.3A^2$$
$$= 0$$
$$A^2 = 1200$$
$$A = 35 \text{ m}^2$$

$S'' = -0.6A < 0$ for all valid (positive) A so the graph is concave down and $A = 35 \text{ m}^2$ is a max. Maximum savings are

$$S = 360(35) - 0.1(35)^3 = \$8300.$$

13. $v = k\sqrt{\frac{l}{a} + \frac{a}{l}}$

$$\frac{dv}{dl} = \frac{k}{2\sqrt{\frac{l}{a} + \frac{a}{l}}} \quad \frac{1}{a} - \frac{a}{l^2}$$
$$= 0$$
$$l^2 = a^2$$
$$l = a$$

For $l < a, \frac{dv}{dl} < 0$, and for $l > a, \frac{dv}{dl} > 0$.

Therefore, v is minimum at $l = a$.

17. $xy = A$

Substitute $y = \frac{A}{x}$ into $P = 2x + 2y$:

$$P = 2x + 2\frac{A}{x}$$
$$= 2x + \frac{2A}{x}$$
$$\frac{dP}{dx} = 2 - \frac{2A}{x^2}$$
$$= 0 \text{ for } x = \sqrt{A}$$
$$\frac{d^2P}{dx^2} = \frac{4A}{x^3}\Big|_{x=\sqrt{A}} > 0, \text{ so min. at } (\sqrt{A}, \sqrt{A})$$

The dimensions of the chip are $\sqrt{A}$ by $\sqrt{A}$ for minimum perimeter.

21. Distance traveled from B is $16.0t$; $40.0-16.0t$ is side of a right triangle. Distance travelled from A is $18.0t$.

$$d=\sqrt{(40.0-16.0t)^2+(18.0t)^2}$$
$$=\sqrt{1600-1280t+580t^2}$$
$$y=d^2=1600-1280t+580t^2$$
$$\frac{dy}{dt}=-1280+1160t$$
$$=0 \text{ for } t=1.10$$
$$\frac{d^2y}{dt^2}=1160>0, \text{ so } y \text{ is minimized.}$$

Minimum will occur after $t=1.1$ h

25. $l+w+h\le 108$ in

$w=h=2r$

$l+4r=108$

Substitute $l=108-4r$ into $V=\pi r^2 l$:

$$V=\pi r^2(108-4r)$$
$$=108\pi r^2-4\pi r^3$$
$$\frac{dV}{dr}=216\pi r-12\pi r^2$$
$$=12\pi r(18-r)$$
$$=0 \text{ for } r=0 \text{ (reject) and } r=18 \text{ in}$$
$$\frac{d^2V}{dr^2}=216\pi-24\pi r$$

$\left.\frac{d^2V}{dr^2}\right|_{r=18}<0$, max. at $r=18$, $l=36$, with $V=36000\text{ in}^3$.

29. $A=\frac{1}{2}(a+b)h;\ a=6$ ft

Let $b=2x$

$$A=\frac{1}{2}(6+2x)\left(\sqrt{3^2-x^2}\right)$$
$$=(3+x)\left(3^2-x^2\right)^{1/2}$$
$$A'=(3+x)\ \frac{1}{2}\ \left(3^2-x^2\right)^{-1/2}(-2x)$$
$$+(1)\left(3^2-x^2\right)^{1/2}=0$$
$$A'=(9-x^2-3x-x^2)\left(9-x^2\right)^{-1/2}$$
$$A'=(3-2x)(3+x)\left(9-x^2\right)^{-1/2}$$

The only positive solution is $x=1.5$

$A'>0$ for $x<1.5$, and $A'>0$ for $x>1.5$, so area is maximized when $x=1.5$.

$b=2x=3$ ft

33. $y=6x^2-x^3$; maximum slope? What value of x gives a maximum $\frac{dy}{dx}$?

$$\frac{dy}{dx}=12x-3x^2;$$
$$y'=12x-3x^2$$
$$\frac{dy'}{dx}=12-6x$$
$$=0 \text{ for } x=2$$

$\frac{d^2y'}{dx^2}=-6<0$ for all x, therefore, $x=2$, max. $(2,16)$

Maximum slope at $x=2$ is $12(2)-3(4)$ $=24-12=12$.

37. $y+\sqrt{x^2-100}=50$

$$y=50-\sqrt{x^2-100}$$
$$L=2x+y$$
$$L(x)=2x+50-(x^2-100)^{1/2}$$
$$L'(x)=2-x(x^2-100)^{-1/2}=2-\frac{x}{\sqrt{x^2-100}}$$
$$2-\frac{x}{\sqrt{x^2-100}}=0$$
$$2\sqrt{x^2-100}=x$$
$$4x^2-400=x^2$$
$$x^2=\frac{400}{3}$$
$$x=\frac{20}{\sqrt{3}}$$

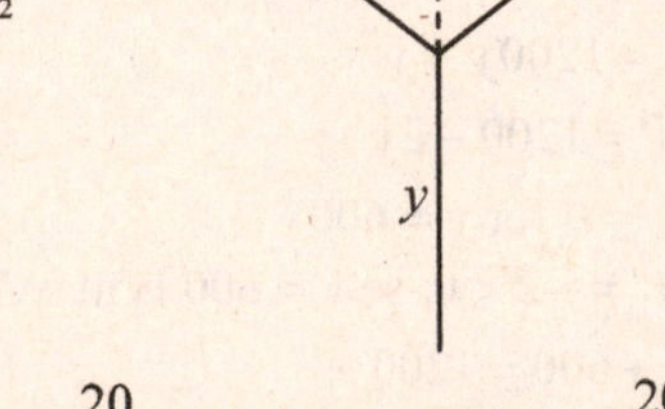

$L'(x)<0$ for $x<\frac{20}{\sqrt{3}}$ and $L'(x)>0$ for $x>\frac{20}{\sqrt{3}}$, so we have minimized $L(x)$.

$$y=50-\frac{10}{\sqrt{3}}$$
$$L=y+2x=50-\frac{10}{\sqrt{3}}+\frac{40}{\sqrt{3}}$$
$$L=50+\frac{30}{\sqrt{3}}=50+10\sqrt{3}$$

41. $E = \frac{100T(1-fT)}{T+f}\Big|_{f=0.25}$

$= \frac{100T(1-0.25T)}{T+0.25}$

$= \frac{400T-100T^2}{4T+1}$

$E' = -\frac{200(2T^2+T-2)}{(4T+1)^2}$

$= 0$ for $T = 0.78, -1.28 < 0$, reject

$E'' = \frac{-3400}{(4T+1)^3}$

$E''|_{T=0.78} = -48.6 < 0,$

E is maximum for $T = 0.78 =$ tangent of acute pitch angle of screw.

Acute pitch angle of screw $= \tan^{-1} 0.78 = 38^\circ$.

45. $V = (8.00-2x)(8.00-2x)x$

$= 64.0x - 32.0x^2 + 4x^3$

$\frac{dV}{dx} = 64.0 - 64.0x + 12x^2 = 4(16-16x+3x^2)$

$= 4(4-x)(4-3x)$

$= 0$ for $x = 4$ and $x = \frac{4}{3}$

$\frac{d^2V}{dx^2} = -64 + 24x$

$\frac{d^2V}{dx^2}\Big|_{x=4} > 0$, minimum volume

$\frac{d^2V}{dx^2}\Big|_{x=4/3} < 0$, maximum volume ($x = 4/3$ in)

49. Let C = total cost, and let x be the vertical distance from the loading area to the point P. Then,

$C = 50\ 000(10-x) + 80\ 000\sqrt{x^2+2.5^2}$

$= 500\ 000 - 50\ 000x + 80\ 000(x^2+6.25)^{1/2}$

$C' = -50\ 000 + 40\ 000(x^2+6.25)^{-1/2}(2x)$

$= -50\ 000 + 80\ 000x(x^2+6.25)^{-1/2}$

$= -50\ 000 + \frac{80\ 000x}{\sqrt{x^2+6.25}}$

$= \frac{-50\ 000\sqrt{x^2+6.25}+80\ 000x}{\sqrt{x^2+6.25}} = 0$

$-50\ 000\sqrt{x^2+6.25} + 80\ 000x = 0$

$\sqrt{x^2+6.25} = \frac{-80\ 000x}{-50\ 000}$

$\sqrt{x^2+6.25} = \frac{8x}{5}$; square both sides

$x^2 + 6.25 = \frac{64}{25}x^2$

$6.25 = \frac{64}{25}x^2 - x^2$

$6.25 = \frac{39}{25}x^2$

$x^2 = 6.25\left(\frac{25}{39}\right) = 4.00$

$x = 2.00$ mi; $10 - x = 8.00$ mi

The pipeline should turn 8.00 mi downstream from the refinery.

53. Let x be the number sold in excess of 1000.

Profit on first 1000 units at \$10.00/unit = \$10 000

Profit on x units over 1000 at

\$ 10.00 $x(0.2)$ /unit is $x[10.00\quad 0.02x]$

$P_T = 10\ 000 + x[10.00 - 0.02x]$

$= 10\ 000 + 10.00x - 0.02x^2$

$\frac{dP_T}{dx} = 10.00 - 0.04x = 0$

$x = 250$ units in excess of 1000

$\frac{d^2P_T}{dx^2} = -0.04 < 0$ for all x, therefore max. at $x = 250$.

Therefore, the before company should produce 1250 units per week to maximize profit.

24.8 Differentials and Linear Approximations

1. $s = \frac{4t}{t^3+4}$

$ds = \frac{(t^3+4)(4) - 4t(3t^2)}{(t^3+4)^2}dt$

$ds = \frac{-8t^3+16}{(t^3+4)^2}dt$

5. $y = x^4 + 3x$

$\frac{dy}{dx} = 4x^3 + 3$

$dy = (4x^3+3)dx$

9. $s = 3(t^2-5)^4$

$\frac{ds}{dt} = 12(t^2-5)^3(2t)$

$ds = 24t(t^2-5)^3\,dt$

13. $y = x(1-x)^3$

$dy = \left[x\cdot 3(1-x)^2(-1) + (1-x)^3\cdot 1\right]dx$

$dy = (1-x)^2(-3x+(1-x))dx$

$dy = (1-x)^2(1-4x)dx$

17. $y = f(x) = 7x^2 + 4x,$

$dy = f'(x)dx$

$= (14x+4)dx$

$\Delta y = f(x+\Delta x) - f(x)$

$= 7(4.2)^2 + 4(4.2) - (7\cdot 4^2 + 4\cdot 4)$

$= 12.28$

$dy = (14\cdot 4 + 4)(0.2) = 12$

21. $f(x) = x^2 + 2x;\ f'(x) = 2x+2$

$L(x) = f(a) + f'(a)(x-a)$

$= f(0) + f'(0)(x-0)$

$L(x) = 0^2 + 2\cdot 0 + (2\cdot 0 + 2)(x-0)$

$L(x) = 2x$

25. $C = 2\pi r,\ r = 6370,\ dr = 250$

$dC = 2\pi dr$

$= 2\pi(250)$

$= 1570$ km

29. $\lambda = \frac{k}{f},$

$685 = \frac{k}{4.38\times 10^{14}};$

$k = 3.00\times 10^{17}$ nm · H_z

$\frac{d\lambda}{df} = \frac{-k}{f^2}$

$d\lambda = \frac{-k}{f^2}df$

$d\lambda = \frac{-3.00\times 10^{17}}{(4.38\times 10^{14})^2}\cdot(0.20\times 10^{14})$

$d\lambda = -31.3$ nm

33. $r = k\sqrt{\lambda}$

$\frac{dr}{d\lambda} = \frac{k}{2\sqrt{\lambda}}$

$\frac{dr}{r} = \frac{1}{r}\cdot\frac{k\,d\lambda}{2\sqrt{\lambda}}$

$= \frac{1}{2}\cdot\frac{k}{k\sqrt{\lambda}}\cdot\frac{d\lambda}{\sqrt{\lambda}}$

$\frac{dr}{r} = \frac{1}{2}\cdot\frac{d\lambda}{\lambda}$

37. $y = f(x) = \sqrt{x},\ x = 4,\ \Delta x = dx = 0.05$

$\Delta y = f(x+\Delta x) - f(x) \approx dy = \frac{1}{2\sqrt{x}}dx$

$\sqrt{4.05} \approx \sqrt{4} + \frac{1}{2\sqrt{4}}(0.05)$

$\sqrt{4.05} \approx 2.0125$

41. $f(x) = \sqrt{2-x}$

$f'(x) = \frac{-1}{2\sqrt{2-x}}$

$L(x) = f(a) + f'(a)(x-a)$

$L(x) = f(1) + f'(1)(x-1)$

$= \sqrt{2-1} + \frac{-1}{2\sqrt{2-1}}(x-1)$

$L(x) = 1 - \frac{1}{2}(x-1)$

$= -\frac{1}{2}x + \frac{3}{2}$

$\sqrt{1.9} = f(0.1) \approx L(0.1)$

$= 1 - \frac{1}{2}(0.1-1) = 1.45$

Review Exercises

1. This is false. The tangent line has slope

$\left.\frac{dy}{dx}\right|_{x=1} = 6$ and so the normal line has slope

$-\frac{1}{6}$.

5. This is false.

The inflection points occur when $f''(x) = 0$.

Here,

$f''(x) = 24x^2 - 8$

which is 0 when $x = \pm\frac{1}{\sqrt{3}}$. Finally,

$y = f \;\; \pm\frac{1}{\sqrt{3}} \;\; = \frac{2}{9} - \frac{4}{3} = -\frac{10}{9}$.

9. $y = 3x - x^2$ at $(-1, -4)$;

$y' = 3 - 2x$

$y'|_{x=-1} = 3 - 2(-1)$

$= 5$

$m = 5$ for tangent line

$y - y_1 = 5(x - x_1)$

$y - (-4) = 5 \;\; x - (-1)$

$y + 4 = 5x + 5$

$5x - y + 1 = 0$

To graph:

$y - \frac{9}{4} = -\left(x^2 - 3x + \frac{9}{4}\right)$

This is a parabola opening downward, with vertex at $\left(\frac{3}{2}, \frac{9}{4}\right)$, with intercepts $(0,0)$ and $(0,3)$.

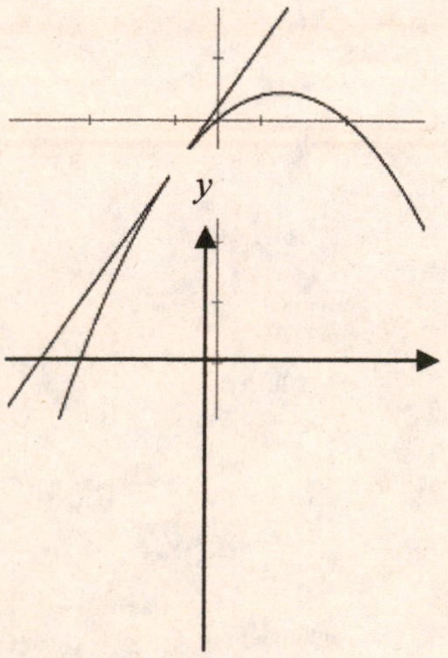

13. $y = \sqrt{x^2 + 3}$; $m_{\tan} = \frac{1}{2}$

$y = \left(x^2 + 3\right)^{1/2}$

$\frac{dy}{dx} = \frac{1}{2}\left(x^2 + 3\right)^{-1/2}(2x)$

$= \frac{x}{\sqrt{x^2 + 3}}$

$m_{\tan} = \frac{dy}{dx}$

$\frac{1}{2} = \frac{x}{\sqrt{x^2 + 3}}$

Squaring both sides,

$4x^2 = x^2 + 3$

$3x^2 = 3$

$x^2 = 1$

$x = 1$ and $x = -1$.

$x = 1$ checks in the equation, but $x = -1$ does not (it is an extraneous root), so the only point where the slope is $\frac{1}{2}$ is $(1, 2)$.

Therefore,

$y - 2 = \frac{1}{2}(x - 1)$

$y = \frac{1}{2}x + \frac{3}{2}$ is the equation of the tangent line.

17. $y = 0.5x^2 + x$

$v_y = \frac{dy}{dt} = \frac{x\,dx}{dt} + \frac{dx}{dt}$

Substitute $v_x = 0.5\sqrt{x}$:

$v_y = x\left(0.5\sqrt{x}\right) + 0.5\sqrt{x}$

Find v_y at $(2, 4)$:

$v_y|_{x=2} = 2\left(0.5\sqrt{2}\right) + 0.5\sqrt{2}$

$= 1.5\sqrt{2} = 2.12$

21. $x^3 - 3x^2 - x + 2 = 0$ (between 0 and 1)

$f(x) = x^3 - 3x^2 - x + 2$

$f'(x) = 3x^2 - 6x - 1$

$f(0) = 0^3 - 3(0^2) - 0 + 2 = 2$

$f(1) = 1^3 - 3(1^2) - 1 + 2 = -1$

The root is possibly closer to 1 than 0. Let $x_1 = 0.6$:

n	x_n	$f(x_n)$	$f'(x_n)$	$x_n - \frac{f(x_n)}{f'(x_n)}$
1	0.6	0.536	−3.52	0.752 272 7
2	0.752 272 7	−0.024 293 6	−3.815 893 6	0.745 906 3
3	0.745 906 3	−0.000 030 4	−3.806 309 2	0.745 898 3

$x_4 = x_3 = 0.745\ 9$

25. $y = 4x^2 + 16x$

(1) The graph is continuous for all x.

(2) The intercepts are $(0, 0)$ and $(-4, 0)$.

(3) As $x \to +\infty$ and $-\infty$, $y \to +\infty$.

(4) The graph is not symmetrical about either axis or the origin.

(5) $y' = 8x + 16$

$y' = 0$ at $x = -2$. $(-2, -16)$ is a critical point.

(6) $y'' = 8 > 0$ for all x; the graph is concave up and $(-2, -16)$ is a minimum.

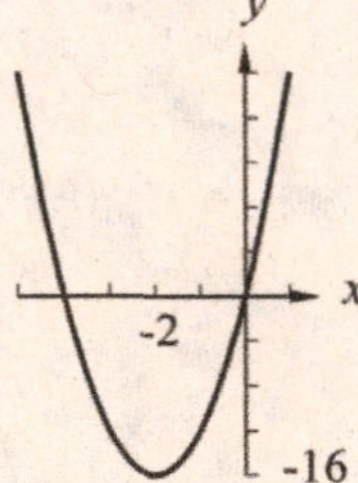

29. $y = x^4 - 32x$

(1) The graph is continuous for all x.

(2) The intercepts are $(0, 0)$ and $(2\sqrt[3]{4}, 0)$.

(3) As $x \to -\infty$, $y \to +\infty$; as $x \to +\infty$, $y \to +\infty$.

(4) The graph is not symmetrical about either axis or the origin.

(5) $y' = 4x^3 - 32 = 0$ for $x = 2$

(6) $y'' = 12x^2$; $y'' = 0$ at $x = 0$; $(0, 0)$ is a possible point of inflection. Since $f''(x) > 0$; the graph is concave up everywhere (except 0) and (0, 0) is not an inflection point. $(2, -48)$ is a minimum

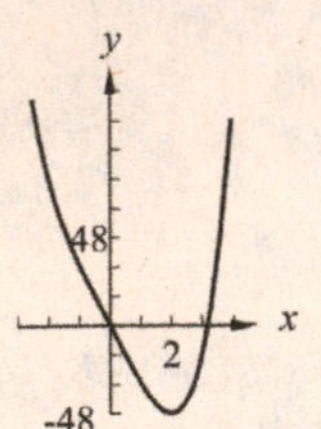

33. $y = f(x)$

$$= 2x^3 + \frac{5}{x^2}$$

$$dy = f'(x)dx$$

$$= \left(6x^2 - \frac{10}{x^3}\right)dx$$

37. $y = f(x) = 4x^3 - 12,\ x = 2,\ \Delta x = 0.1$

$$\Delta y - dy = f(x+\Delta x) - f(x) - f'(x)dx$$

$$= 4(2.1)^3 - 12 - \left(4(2)^3 - 12\right) - 12(2)^2(0.1)$$

$$= 0.244$$

41. $V = f(r) = \frac{4}{3}\pi r^3,\ r = 3.500,\ \Delta r = 0.012$

$$dV = f'(r)dr$$

$$= 4\pi r^2 dr$$

$$= 4\pi(3.500)^2(0.012)$$

$$dV = 1.85\ \text{m}^3$$

45. $Z = \sqrt{R^2 + X^2}$

$$= \left(R^2 + X^2\right)^{1/2}$$

$$dZ = \frac{1}{2}\left(R^2 + X^2\right)^{-1/2}(2R)dR$$

$$dZ = \frac{RdR}{\sqrt{R^2 + X^2}}$$

$$\text{relative error} = \frac{dZ}{Z} = \frac{RdR}{\sqrt{R^2 + X^2}} \cdot \frac{1}{Z}$$

$$= \frac{RdR}{\sqrt{R^2 + X^2}} \cdot \frac{1}{\sqrt{R^2 + X^2}}$$

$$= \frac{RdR}{R^2 + X^2}$$

49. $y = x^2 + 2$ and $y = 4x - x^2$

$y' = 2x;\qquad y' = 4 - 2x$

To find the point of intersection:

$2x = 4 - 2x$

$4x = 4$

$x = 1$

$y = 1 + 2 = 3$

The point (1, 3) belongs to both graphs; the slope of the tangent line is 2.

$y - y_1 = 2(x - x_1)$

$y - 3 = 2(x - 1)$

$y - 3 = 2x - 2$

$2x - y + 1 = 0$ is the equation of the tangent line.

53. (a) A counterexample is $f(x) = x^3$. Here, $f'(x) = 3x^2$ is zero at $x = 0$, but this is not a local extreme point.

(b) A counterexample is $f(x) = x^4$. Here, $f''(x) = 12x^2$. The concavity is upward on both sides of $x = 0$ and $f''(0) = 0$, but $x = 0$ is not a point of inflection.

57. $y = k\left(x^4 - 30x^3 + 1000x\right) = 0$. By inspection, $x = 0.0$ m is the first value of x where the deflection is zero. From the graph of $y = x^4 - 30x^3 + 1000x$, there is a zero between $x = 6$ and $x = 7$. Letting $x_1 = 6$, successive iterations of Newton's method give

$x_1 = 6$
$x_2 = 6.593\ 023\ 253$
$x_3 = 6.527\ 855\ 923$
$x_4 = 6.527\ 036\ 576$
$x_5 = 6.527\ 036\ 447$
$x_6 = 6.527\ 036\ 447$

$x = 6.527$ m is the second value where the defection is zero.

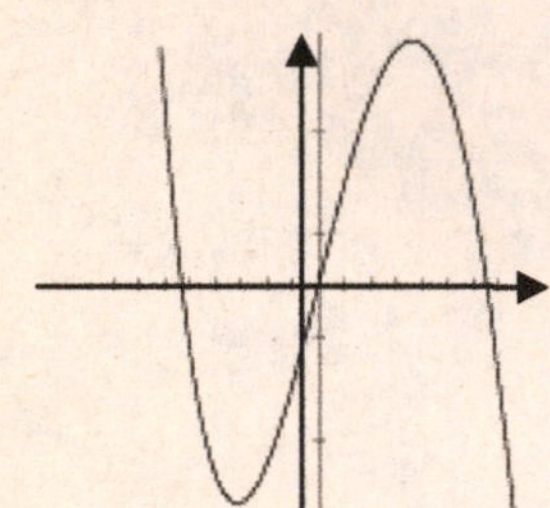

61.

$$D^2 = (4.0t)^2 + 250^2$$

$$2D\frac{dD}{dt} = 2t$$

$$\frac{dD}{dt} = \frac{t}{D} = \frac{(60)}{\sqrt{(4.0(60))^2 + 250^2}}$$

$$= 0.173 \text{ ft/s}$$

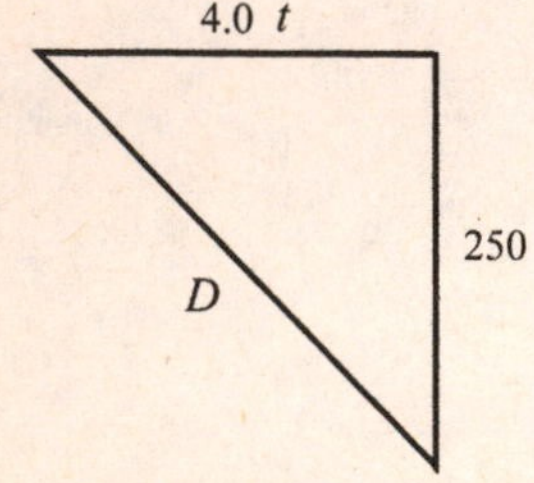

65. $y = x^2 - \dfrac{2}{x}$

(1) Intercepts: y is undefined for $x = 0$ no y-intercept. $y = 0$ gives $0 = x^2 - \dfrac{2}{x}$;

$x^3 = 2$;

$x = \sqrt[3]{2} \approx 1.26$

(2) Symmetry: none

(3) Behaviour as x becomes large: $y \to \infty$ as $x \to \pm\infty$. As $x \to \pm\infty$, $\frac{2}{x} \to 0$, so y increases as x^2.

(4) Vertical asymptotes: $x = 0$ is a vertical asymptote

(5) Domain: $x \neq 0$; Range: $-\infty < y < \infty$

(6) Derivatives:

$y' = 2x + \dfrac{2}{x^2}$

$= 0$ for $x = -1$

$y'' = 2 - \dfrac{4}{x^3}$

$y''|_{x=-1} > 0$, so $(-1, 3)$ is a min

$y'' = 0 = 2 - \dfrac{4}{x^3}$

$x^3 = 2$

$x = \sqrt[3]{2}$

$y'' < 0$ for $x < \sqrt[3]{2}$; $y'' > 0$ for $x > \sqrt[3]{2}$, so $\left(\sqrt[3]{2}, 0\right)$ is an inflection point

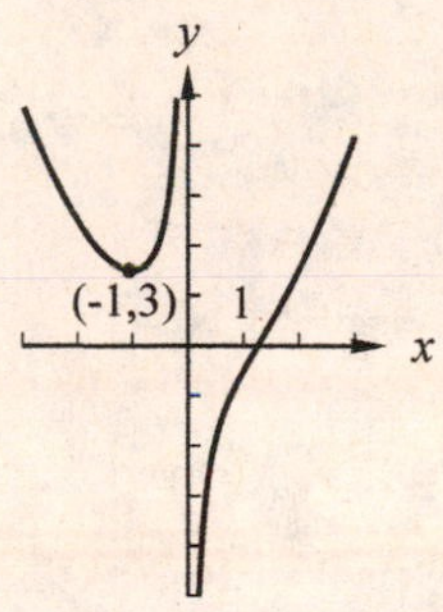

From the graph $y = x^2 - \dfrac{2}{x} > 0$ for $x < 0$ or $x > 1.26$

$x^2 - \dfrac{2}{x} > 0 \Rightarrow x^2 > \dfrac{2}{x}$ for $x < 0$ or $x > 1.26$

The graph of $x^2 > \dfrac{2}{x}$ is

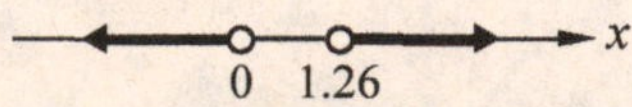

69. $f(0)=2 \Rightarrow y$-intercept is 2

$$\left.\begin{array}{l} f'(x)<0 \text{ for } x<0 \Rightarrow f \text{ is dec. for } x<0 \\ f'(x)>0 \text{ for } x>0 \Rightarrow f \text{ is inc. for } x>0 \end{array}\right\}(0,\ 2) \text{ min}$$

$f''(x)>0$ for all $x \Rightarrow f$ is conc. up for all x.

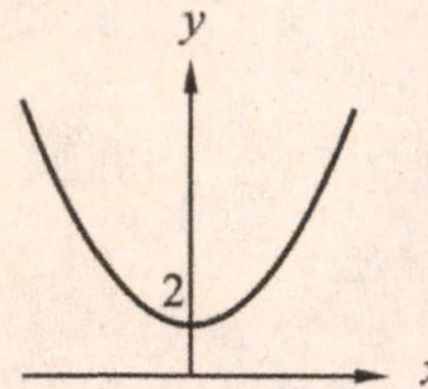

73. The bottom edge of the trapezoid is $z = 1500 - 3y - x$.
The area of the trapezoid is

$$A = \frac{y}{2}(x+z)$$

$$A(y) = \frac{y}{2}(1500-3y)$$

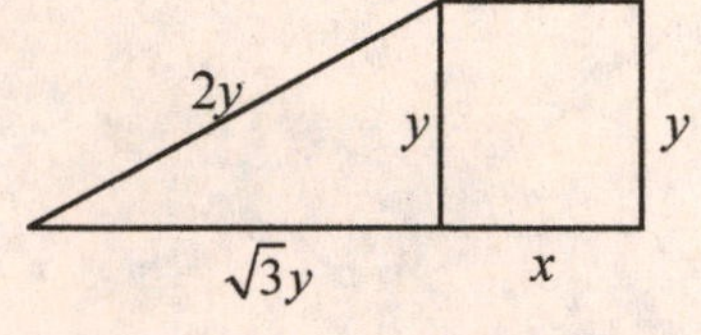

$$A(y) = 750y - \frac{3}{2}y^2$$

$$A'(y) = 750 - 3y$$

which is 0 when $y = 250$

$A''(y) = -3 < 0$ and so $y = 250$ ft. maximizes the area.

The bottom of the trapezoid is $x + \sqrt{3}y$ by dissecting the trapezoid into a right triangle and a rectangle. This right triangle has angles 30° and 60°, justifying the length of $\sqrt{3}y$.

Finally, $750 = 2x + 250\sqrt{3}$ and so $x = 158.5$ ft.
We have $x = 158.5$ ft and $y = 250$ ft.

77. $$y = \frac{300}{0.0005x^2+2} - 50,\ 0<x<100$$

$$y' = -300\left(0.0005x^2+2\right)^{-2}(0.001x)$$

$$y'\big|_{x=50} = -1.42$$

$y'<0$ for $0<x<100$ y is dec for 0 x 100

$$y' = \frac{-0.3x}{\left(0.0005x^2+2\right)^2}$$

$$y'' = \frac{\left(0.0005x^2+2\right)^2(-0.3) + 0.3x\left(2\left(0.0005x^2+2\right)^1(0.001x)\right)}{\left(0.0005x^2+2\right)^4}$$

$$= \frac{-0.00015x^2 - 0.6 + .0006x^2}{\left(0.0005x^2+2\right)^3} = 0$$

$0.00045x^2 - 0.6 = 0$

$x = 36.5,\ x = -36.5$ (reject)

$y(36.5) = 62.5$

$y'' < 0$ for $x < 36.5$; $y'' > 0$ for $x > 36.5$, so $x = 37$ is an inflection point.

y is conc. down for $x < 36.5$ and conc. up for $x > 36.5$.

$y(0) = 100$, y-intercept

$y = 0 \Rightarrow x = 89$, x-intercept

$$\text{L}(x) = -1.42(x-50) + y(50)$$
$$= -1.42x + 71 + 42$$
$$L(x) = -1.42x + 113$$

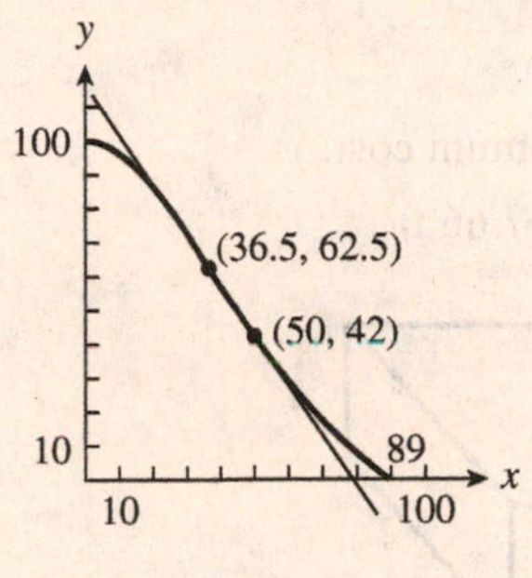

81.
$$\frac{1}{C_T} = \frac{1}{C_1} + \frac{1}{C_2}$$
$$= \frac{C_2 + C_1}{C_1 C_2}$$
$$C_T = \frac{C_1 C_2}{C_1 + C_2}$$
$$C_1 + C_2 = 12$$
$$C_T = \frac{C_1(12 - C_1)}{12}$$
$$\frac{dC_T}{dC_1} = \frac{1}{12}(12 - 2C_1)$$
$= 0$ for $C_1 = 6\,\mu\text{F}$, and $C_2 = 6\,\mu\text{F}$

$\dfrac{d^2C_T}{dC_1^2} = -\dfrac{1}{6} < 0$, so total capacitance is a maximum.

85. $t=\dfrac{\sqrt{16+x^2}}{3}+\dfrac{5-x}{5}$

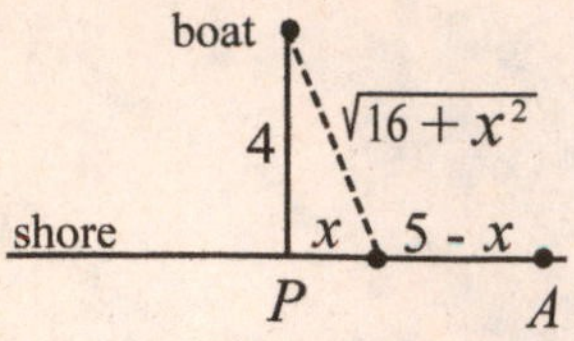

$$\frac{dt}{dx}=\frac{x}{3\sqrt{16+x^2}}-\frac{1}{5}=0$$

$$5x=3\sqrt{16+x^2}$$
$$25x^2=9\left(16+x^2\right)$$
$$=144+9x^2$$
$$16x^2=144$$
$$x^2=9$$
$x=3$, a minimum since

$$\frac{d^2t}{dx^2}=\frac{16}{3\left(16+x^2\right)^{3/2}}>0$$

The boat should land 3 km from P toward A.

89. $V=\dfrac{1}{3}\pi r^2 h$

$$=\frac{1}{3}\pi r^2\cdot r$$
$$=\frac{1}{3}\pi r^3$$
$$\frac{dV}{dt}=\pi r^2\frac{dr}{dt}$$

When $h=r=10.0$,

$$100.0=\pi(10.0)^2\frac{dr}{dt}$$
$$\frac{dr}{dt}=0.318\text{ ft/min}$$

93. $V=1350=0.75l^2h\Rightarrow h=\dfrac{1800}{l^2}$

$$c=0.75l^2(6)+2(0.75lh)(9)$$
$$+2(lh)(9)+0.75l^2(4.5)$$
$$c=4.5l^2+13.5l\left(\frac{1800}{l^2}\right)+18l\left(\frac{1800}{l^2}\right)+3.375l^2$$
$$c=7.875l^2+\frac{56700}{l}$$
$$\frac{dc}{dl}=15.75l-\frac{56700}{l^2}$$
$=0$ for $l^3=3600$ or $l=15.33$.

$$h=\frac{1800}{15.32^2}=7.66$$
$$w=0.75\,l=11.50$$

dimensions for minimum cost:

11.5 ft by 15.3 ft by 7.66 ft

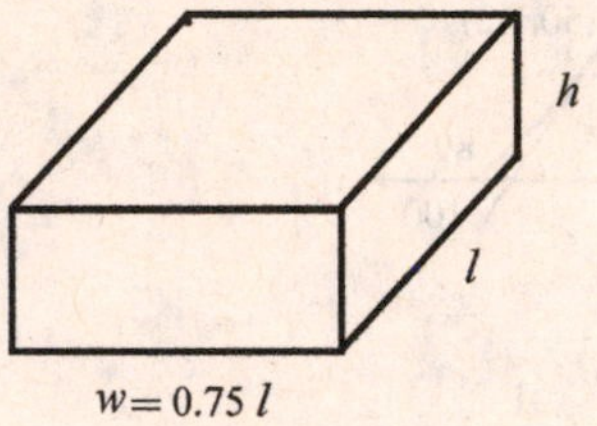

Chapter 25

Integration

25.1 Antiderivatives

1. $f(x)=12x^3$; power of x required is 4, $F(x)=ax^4$

$F'(x)=4ax^3=12x^3$

$4a=12$

$a=3$

$F(x)=3x^4$

5. $f(x)=3x^2$

The power of x required in the antiderivative is 3. Therefore, we must multiply by $\frac{1}{3}$. An antiderivative of $3x^2$ is $\frac{1}{3}(3x^3)=x^3$, or $a=1$.

9. The power of x required in the antiderivative of $f(x)=9\sqrt{x}=9x^{1/2}$ is $\frac{3}{2}$.

$\frac{d}{dx}x^{3/2}=\frac{3}{2}x^{1/2}$

Write $f(x)=9\cdot\frac{2}{3}\cdot\frac{3}{2}x^{1/2}$

The antiderivative of $9\sqrt{x}$ is $\frac{2}{3}\cdot 9x^{3/2}=6x^{3/2}$, or $a=6$.

13. $f(x)=5x^2$

The power of x required in the antiderivative is 3.

$\frac{d}{dx}ax^3=3ax^2$

$3a=5, a=\frac{5}{3}$

An antiderivative of $5x^2$ is $\frac{5}{3}x^3$.

17. $f(x)=2x^{3/2}-3x$

$F(x)=ax^{5/2}+bx^2$

$\frac{d}{dx}\left(ax^{5/2}+bx^2\right)=\frac{5}{2}ax^{3/2}+2bx$

$\frac{5}{2}a=2;\ a=\frac{4}{5}$

$2b=-3, b=-\frac{3}{2}$

An antiderivative of $2x^{3/2}-3x$ is $\frac{4}{5}x^{5/2}-\frac{3}{2}x^2$.

21. $f(x)=-\frac{7}{x^6}+\frac{1}{3^2}=-7x^{-6}+\frac{1}{9}$

$F(x)=ax^{-5}+bx$

$\frac{d}{dx}\left(ax^{-5}+bx\right)=-5ax^{-6}+b$

$-5a=-7$

$a=\frac{7}{5}$

$b=\frac{1}{9}$

An antiderivative of $-\frac{7}{x^6}+\frac{1}{3^2}$ is $\frac{7}{5}x^{-5}+\frac{1}{9}x$.

25. $f(x)=50x^{99}-39x^{-79}$

$F(x)=ax^{100}+bx^{-78}$

$\frac{d}{dx}\left(ax^{100}+bx^{-78}\right)=100ax^{99}-78bx^{-79}$

$100a=50, a=\frac{1}{2}$

$78b=39, b=\frac{1}{2}$

An antiderivative of $50x^{99}-39x^{-79}$ is $\frac{1}{2}x^{100}+\frac{1}{2}x^{-78}$.

29. $f(x)=6(2x+1)^5(2)$;

$F(x)=a(2x+1)^6$

$\frac{d}{dx}(2x+1)^6=6a(2x+1)^5(2)$

This is exactly $f(x)$, so $a=1$ and an antiderivative is $(2x+1)^6$.

33. $f(x)=4x^3(2x^4+1)^4$

$F(x)=a(2x^4+1)^5$

$\frac{d}{dx}\left(a(2x^4+1)^5\right)=5a(2x^4+1)^4(8x^3)$

$=10a\left(4x^3(2x^4+1)^4\right)$

$10a=1, a=\frac{1}{10}.$

An antiderivative is $\frac{1}{10}(2x^4+1)^5$.

Check:

$\frac{d}{dx}\frac{1}{10}(2x^4+1)^5=\frac{1}{10}\left[5(2x^4+1)^4 8x^3\right]$

$=\frac{1}{10}\left[40(2x^4+1)^4 x^3\right]$

$=4x^3(2x^4+1)^4$

37. $f(x)=(3x+1)^{1/3}$

$F(x)=(3x+1)^{4/3}$

$\frac{d}{dx}a(3x+1)^{4/3}=a\frac{4}{3}(3x+1)^{1/3}(3)$

$4a=1;\ a=\frac{1}{4}$

An antiderivative is $\frac{1}{4}(3x+1)^{4/3}$

Check:

$\frac{d}{dx}\frac{1}{4}(3x+1)^{4/3}=\frac{1}{4}\cdot\frac{4}{3}(3x+1)^{1/3}(3)=(3x+1)^{1/3}.$

41. $F(x)=(x+5)^3$ is the correct antiderivative of $f(x)=3(x+5)^2$ since $F'(x)=3(x+5)(1)=f(x)$

$F(x)=(2x+5)^3$ is not the correct antiderivative of $f(x)=3(2x+5)^2$ since

$F'(x)=3(2x+5)^2(2)=2\cdot 3(2x+5)^2$

$=2f(x)\neq f(x)$

$f(x)$ must include the derivative of the function inside the power as a factor.

25.2 The Indefinite Integral

1. $\int 8x\,dx=8\int x^1 dx=8\frac{x^{1+1}}{1+1}+C=4x^2+C$

5. $\int 2x\,dx=2\int x\,dx$

$u=x;\ du=dx;\ n=1$

$2\int x\,dx=2\left(\frac{x^{1+1}}{1+1}\right)+C$

$=x^2+C$

9. $\int 8x^{3/2}dx=8\int x^{3/2}dx$

$u=x;\ du=dx;\ n=\frac{3}{2}$

$\int 8x^{3/2}dx=\frac{8x^{(3/2)+1}}{\frac{3}{2}+1}+C$

$=\frac{8x^{5/2}}{\frac{5}{2}}+C$

$=\frac{16}{5}x^{5/2}+C$

13. $\int(x^7-3x^5)dx=\int x^7dx-3\int x^5dx$

$=\frac{x^8}{8}-3\frac{x^6}{6}+C$

$=\frac{1}{8}x^8-\frac{1}{2}x^6+C$

17. $\int\left(\frac{t^2}{2}-\frac{2}{t^2}\right)dt=\frac{t^{2+1}}{2(2+1)}-\frac{2t^{-2+1}}{-2+1}+C$

$=\frac{t^3}{6}+\frac{2}{t}+C$

21. $\int(2x^{-2/3}+3^{-2})dx=\int 2x^{-2/3}dx+\int 3^{-2}dx$

$=2\int x^{-2/3}dx+3^{-2}\int dx$

$=\frac{2}{\frac{1}{3}}(x^{1/3})+3^{-2}(x^1)+C$

$=6x^{1/3}+\frac{1}{9}x+C$

25. $\int (x^2-1)^5(2xdx)$;

$u = x^2 - 1;\ du = 2xdx;\ n = 5$

$$\int (x^2-1)^5(2xdx) = \frac{(x^2-1)^6}{6} + C$$

$$= \frac{1}{6}(x^2-1)^6 + C$$

29. $40(2\theta^5+5)^7\theta^4 d\theta = 4\ (2\theta^5+5)^7 \cdot (10\theta^4)d\theta$

$$= 4\ u^7 du$$

$$=4 \cdot \frac{u^8}{8} + C$$

$$= \frac{(2\theta^5+5)^8}{2} + C$$

$$= \frac{(2\theta^5+5)^8}{2} + C$$

$u = 2\theta^5 + 5, du = 10\theta^4, n = 7$

33. $\int \frac{4xdx}{\sqrt{6x^2+1}} = \int (6x^2+1)^{-1/2} 4xdx$

$u = 6x^2 + 1;\ du = 12xdx;\ n = -\frac{1}{2}$

$$\int (6x^2+1)^{-1/2} 4xdx = \frac{1}{3}\int (6x^2+1)^{-1/2}(12xdx)$$

$$= \frac{1}{3}\frac{(6x^2+1)^{1/2}}{1/2} + C$$

$$= \frac{2}{3}\sqrt{6x^2+1} + C$$

37. $\frac{dy}{dx} = 6x^2$

$dy = 6x^2 dx$

$y = \int 6x^2 dx = 6\int x^2 dx = \frac{6x^3}{3} + C = 2x^3 + C$

The curve passes through $(0, 2)$:

$2 = 2(0^3) + C$

$C = 2$

$y = 2x^3 + 2$

41. $\int 3x^2 dx = x^3 + C$

$\int 3x^2 dx \neq x^3$ since the constant of integration must be included.

45. $\int 3(2x+1)^2\, dx = \frac{3}{2}\int (2x+1)^2(2dx)$

$$= \frac{3}{2}\frac{(2x+1)^3}{3} + C$$

$$= \frac{(2x+1)^3}{2} + C$$

$\int 3(2x+1)^2\, dx \neq (2x+1)^3 + C$ because the factor of $\frac{1}{2}$ is missing.

49. $f'(x) = 4x - 5$

$f(x) = \int (4x-5)\, dx = 2x^2 - 5x + C$

$10 = f(-1) = 2(-1)^2 - 5(-1) + C = 7 + C;\ C = 3$

$f(x) = 2x^2 - 5x + 3$

53. $v(t) = 10 + t$

$s(t) = 10t + \frac{1}{2}t^2 + C$ where $C = 0$

since $v(0) = 0$

$s(t) = 10t + \frac{1}{2}t^2$

57. $\frac{dT}{dr} = -4500(r+1)^{-3}$

$dT = -4500(r+1)^{-3}\, dr$

$T = 2250(r+1)^{-2} + C$

$T = 2500^\circ\text{C}$ for $r = 0$:

$2500 = 2250(0+1)^{-2} + C$

$C = 250$

$T = 2250(r+1)^{-2} + 250$

61. $\frac{d^2y}{dx^2}=\frac{d(y')}{dx}=6$

$dy'=6dx$

$y'=6x+C$

$y'=8$ for $x=1$:

$8=6(1)+C$

$C=2$

$y'=\frac{dy}{dx}=6x+2$

$dy=(6x+2)\,dx$

$y=3x^2+2x+C_1$

The curve passes through $(1, 2)$:

$2=3(1)^2+2(1)+C_1$

$C_1=-3$

$y=3x^2+2x-3$

25.3 The Area Under a Curve

1. **(a)**

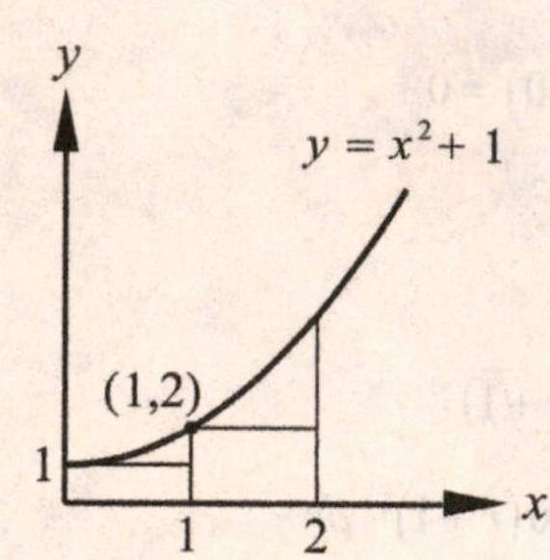

$A=1(1)+1(2)$

$A=3$

(b)

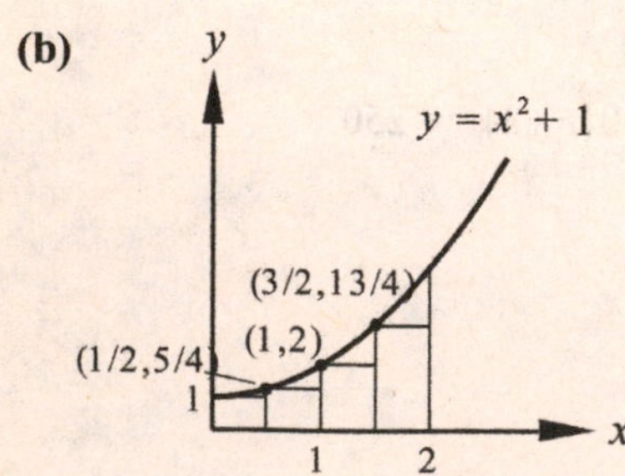

$A=\frac{1}{2}\left[1+\frac{5}{4}+2+\frac{13}{4}\right]$

$A=\frac{15}{4}$

5. $y=3x$, between $x=0$ and $x=3$

(a)

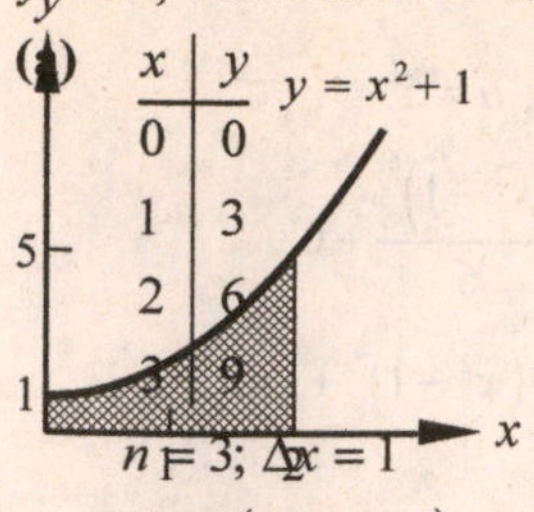

x	y
0	0
1	3
2	6
3	9

$n=3$; $\Delta x=1$

$A=1(0+3+6)=9$

(b)

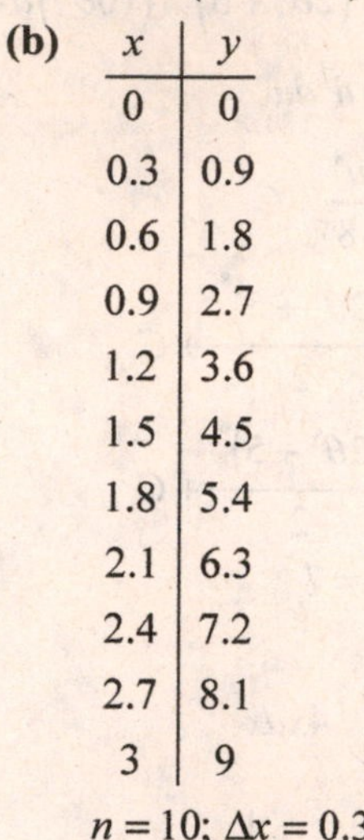

x	y
0	0
0.3	0.9
0.6	1.8
0.9	2.7
1.2	3.6
1.5	4.5
1.8	5.4
2.1	6.3
2.4	7.2
2.7	8.1
3	9

$n=10$; $\Delta x=0.3$

$A=0.3(0+0.9+1.8+2.7+3.6+4.5+5.4$
$+6.3+7.2+8.1)$
$=0.3(40.5)=12.2$

9. $y=4x-x^2$, between $x=1$ and $x=4$

(a) $n=6$, $\Delta x=0.5$

$A=0.5(3.00+3.75+3.75+3.00+1.75+0.00)$

$A=7.62$

x	y
1.0	3.00
1.5	3.75
2.0	4.00
2.5	3.75
3.0	3.00
3.5	1.75
4.0	0.00

(b) $n=10$, $\Delta x=0.3$

$A=0.3(3.00+3.51+3.84+3.96+3.75+3.36$
$+2.79+2.04+1.11)$

$A=8.21$

x	y
1.0	3.00
1.3	3.51
1.6	3.84
1.9	3.99
2.2	3.96
2.5	3.75
2.8	3.36
3.1	2.79
3.4	2.04
3.7	1.11
4.0	0.00

13. $y = \dfrac{1}{\sqrt{x+1}}$, between $x = 3$ and $x = 8$

(a) $n = 5,\ \Delta x = \dfrac{8-3}{5} = 1$

$A = \sum_{i=1}^{5} A_i = \sum_{i=1}^{5} y_i \Delta x$

$y_1 = f(4)$

$A = (0.447 + 0.408 + \cdots + 0.354 + 0.333)(1)$

$A = 1.92$

x	y
3	0.5
4	0.447
5	0.408
6	0.378
7	0.355
8	0.333

(b) $n = 10,\ \Delta x = \dfrac{8-3}{10} = 0.5$

$A = \sum_{i=1}^{10} A_i = \sum_{i=1}^{10} y_i \Delta x$

$y_1 = f(3.5)$

$A = (0.471 + 0.447 + \cdots + 0.343 + 0.333)(0.5)$

$A = 1.96$

x	y
3	0.5
3.5	0.471
4	0.447
4.5	0.426
5	0.408
5.5	0.392
6	0.378
6.5	0.365
7	0.354
7.5	0.343
8	0.333

17. $y = x^2$, between $x = 0$ and $x = 2$

$\int x^2 dx = \dfrac{x^3}{3} + C$

$F(x) = \dfrac{x^3}{3}$

$$A_{0,\,2} = \left[\int x^2 dx\right]_0^2 = F(2) - F(0) = \frac{8}{3} - 0 = \frac{8}{3}$$

21. $y = \dfrac{1}{x^2} = x^{-2}$, between $x = 1$ and $x = 5$

$\int x^{-2} dx = \dfrac{x^{-1}}{-1} + C$

$F(x) = \dfrac{x^{-1}}{-1}$

$$A_{1,5} = \left[\int x^{-2} dx\right]_1^5 = F(5) - F(1) = -\frac{1}{5} - (-1) = \frac{4}{5} = 0.8$$

25. $y = 3x$, $x = 0$ to $x = 3$, $n = 10$, $\Delta x = 0.3$

Using the table in 5(b) and $y_1 = f(0.3)$,

$$A = 0.3(0.9+1.8+2.7+3.6+4.5+5.4+6.3+7.2+8.1+9.0)$$

$$A = 14.85$$

$A_{inscribed}$	<	A_{exact}	<	$A_{circumscribed}$
12.15	<	13.5	<	14.85

$\frac{12.15+14.85}{2} = 13.5$ because the extra area above $y = 3x$ using circumscribed rectangles is the same as the omitted area under $y = 3x$ using inscribed rectangles.

25.4 The Definite Integral

1. $\int_1^4 (x^{-2}-1)dx = -\frac{1}{x} - x\Big|_1^4 = -\frac{1}{4} - 4 - \left(-\frac{1}{1} - 1\right)$

$\int_1^4 (x^{-2}-1)dx = -\frac{9}{4}$

5. $\int_1^4 x^{3/2}dx = \frac{2}{5}x^{5/2}\Big|_1^4 = \frac{64}{5} - \frac{2}{5} = \frac{62}{5}$

9. $u = 1-x$; $du = -dx$

$$\int_{-1.6}^{0.7}(1-x)^{1/3}dx = -\int_{-1.6}^{0.7}(1-x)^{1/3}(-dx)$$

$$= -\frac{3}{4}(1-x)^{4/3}\Big|_{-1.6}^{0.7}$$

$$= -\frac{3}{4}(0.2008 - 3.5752)$$

$$= 2.53$$

13. $\int_1^8 (\sqrt[3]{x}-2)dx = \int_1^8 x^{1/3}dx - 2\int_1^8 dx$

$$= \left(\frac{3}{4}x^{4/3} - 2x\right)\Big|_1^8$$

$$= (12-16) - \left(\frac{3}{4} - 2\right)$$

$$= -\frac{11}{4}$$

17. $u = 4 - x^2$; $du = -2xdx$

$$\int_{-2}^{-1} 12x(4-x^2)^3 dx = -6\int_{-2}^{-1}(4-x^2)^3(-2xdx)$$

$$= -\frac{3(4-x^2)^4}{2}\Bigg|_{-2}^{-1}$$

$$= -\left(\frac{243}{2} - 0\right) = -\frac{243}{2}$$

21. $u = 6x+1$; $du = 6dx$

$$\int_{2.75}^{3.25}\frac{dx}{\sqrt[3]{6x+1}} = \int_{2.75}^{3.25}(6x+1)^{-1/3}dx$$

$$= \frac{1}{6}\int_{2.75}^{3.25}(6x+1)^{-1/3}(6dx)$$

$$= \frac{1}{6}\cdot\frac{3}{2}(6x+1)^{2/3}\Big|_{2.75}^{3.25}$$

$$= \frac{1}{4}(6x+1)^{2/3}\Big|_{2.75}^{3.25}$$

$$= \frac{1}{4}(7.4904 - 6.7405) = 0.188$$

25. $\int_3^7 \sqrt{16t^2+8t+1}\,dt = \int_3^7 \sqrt{(4t+1)^2}\,dt$

$$= \int_3^7 (4t+1)dt$$

$$= \frac{4t^2}{2} + t\Big|_3^7$$

$$= 2(49) + 7 - 2(9) - 3 = 84$$

29. $u = 2x^2 - x + 1$; $du = (4x-1)dx$

$$\int_{-1}^{2}\frac{8x-2}{(2x^2-x+1)^3}dx$$

$$= \int_{-1}^{2}(8x-2)(2x^2-x+1)^{-3}dx$$

$$= 2\int_{-1}^{2}(4x-1)(2x^2-x+1)^{-3}dx$$

$$= \frac{2(2x^2-x+1)^{-2}}{-2}\Bigg|_{-1}^{2}$$

$$= -(2x^2-x+1)^{-2}\Big|_{-1}^{2}$$

$$= -\frac{1}{7^2} - -\frac{1}{4^2} = 0.0421$$

33. $u = z^2 + 4; du = 2zdz$

$$\int_{\sqrt{5}}^{3} 8z\sqrt[4]{z^4 + 8z^2 + 16}dz$$

$$= \int_{\sqrt{5}}^{3} 8z\left(\left(z^2+4\right)^2\right)^{1/4} dz$$

$$= 4\int_{\sqrt{5}}^{3}\left(z^2+4\right)^{1/2} 2zdz$$

$$= \frac{8}{3}\left(z^2+4\right)^{3/2}\Big|_{\sqrt{5}}^{3} = 52.8$$

37. $y^2 = 4x, \quad y > 0$

For $x = 1, \quad y^2 = 4(1) \quad y \quad 2$

For $x = 4, \quad y^2 = 4(4) \quad y \quad 4$

$$\int_{x=1}^{x=4} ydx = \int_1^4 2\sqrt{x}dx$$

$$= 2\cdot\frac{2}{3}x^{3/2}\Big|_1^4$$

$$= \frac{4}{3}\left(4^{3/2} - 1^{3/2}\right) = \frac{28}{3}$$

41. $$\int_{-1}^{1} t^{2k}dt = \frac{t^{2k+1}}{2k+1}\Big|_{-1}^{1} = \frac{1^{2k+1}}{2k+1} - \frac{(-1)^{2k+1}}{2k+1}$$

$$= \frac{1}{2k+1} + \frac{1}{2k+1}$$

$$\int_{-1}^{1} t^{2k}dt = \frac{2}{2k+1}$$

45. $\int_{-3}^{3} |x-1|\ dx = 8 + 2 = 10,$

the area of the two triangles depicted.

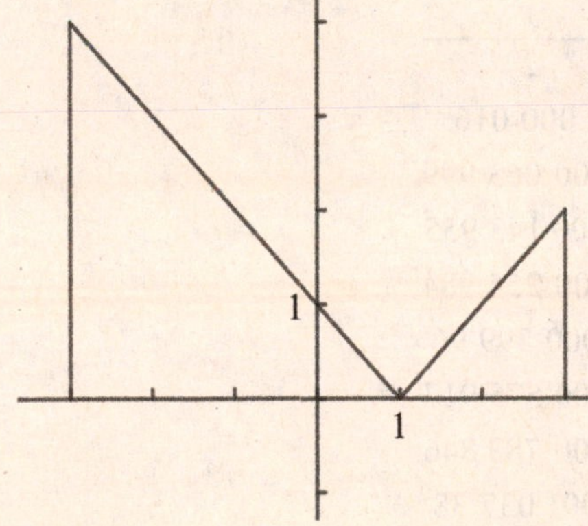

49. $$W = \int_0^{80} (1000 - 5x)\,dx$$

$$= \left(1000x - \frac{5}{2}x^2\right)\Big|_0^{80}$$

$$= 1000(80) - \frac{5}{2}(80)^2 - [0]$$

$$= 80\ 000 - 16\ 000 = 64\ 000\ \text{N}\cdot\text{m}$$

53. $$\frac{3N}{2E_F^{\ 3/2}}\int_0^{E_F} E^{3/2}dE = \frac{3N}{2E_F^{\ 3/2}}\cdot\frac{E^{5/2}}{\frac{5}{2}}\Big|_0^{E_F}$$

$$\frac{3N}{2E_F^{\ 3/2}}\int_0^{E_F} E^{3/2}dE = \frac{3N}{2E_F^{\ 3/2}}\cdot\frac{2E_F^{\ 5/2}}{5} = \frac{3NE_F}{5}$$

25.5 Numerical Integration: The Trapezoidal Rule

1. $\int_1^3 \frac{1}{x}dx,\ n = 2,\ h = \frac{b-a}{n} = \frac{3-1}{2} = 1$

x	y
1	1
2	$\frac{1}{2}$
3	$\frac{1}{3}$

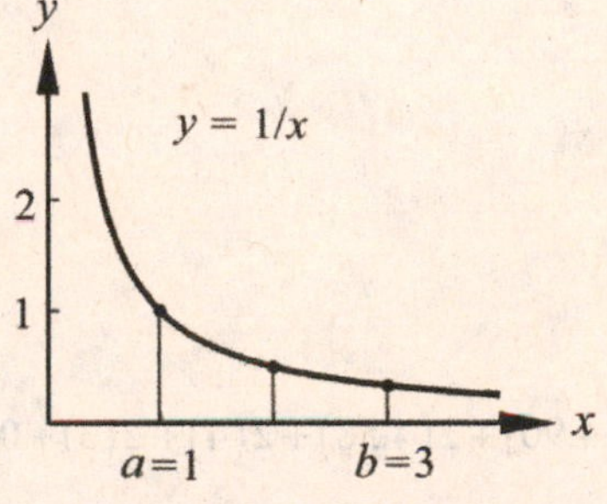

$$A = \frac{1}{2}\left(1 + 2\left(\frac{1}{2}\right) + \frac{1}{3}\right) = \frac{7}{6}$$

5. $\int_1^4 (1+\sqrt{x})dx;\ n=6;\ h=\frac{4-1}{6}=\frac{1}{2};\ \frac{h}{2}=\frac{1}{4}$

n	x_n	y_n
0	1	2
1	1.5	2.22
2	2	2.41
3	2.5	2.58
4	3	2.73
5	3.5	2.87
6	4	3

$$A_T=\frac{1}{4}[2+2(2.22)+2(2.41)+2(2.58)+2(2.73)+2(2.87)+3]$$

$$A=\frac{1}{4}(30.62)=7.66$$

$$A=\int_1^4 (1+x^{1/2})dx=\left(x+\frac{2}{3}x^{3/2}\right)\Bigg|_1^4$$

$$A=4+\frac{16}{3}-\left(1+\frac{2}{3}\right)=\frac{23}{3}=7.67$$

9. $\int_0^5 \sqrt{25-x^2}dx;\ n=5;\ h=\frac{5}{5}=1;\ \frac{h}{2}=\frac{1}{2}$

n	x_n	y_n
0	0	5
1	1	4.90
2	2	4.58
3	3	4
4	4	3
5	5	0

$$A_T=\frac{1}{2}[5+2(4.90)+2(4.58)+2(4)+2(3)+0]=19.0$$

13. $\int_0^4 2^x dx;\ n=12;\ h=\frac{4}{12}=\frac{1}{3};\ \frac{h}{2}=\frac{1}{6}$

x	y
0	1
$\frac{1}{3}$	1.260
$\frac{2}{3}$	1.587
1	2
$1\frac{1}{3}$	2.520
$1\frac{2}{3}$	3.175
2	4
$2\frac{1}{3}$	5.040
$2\frac{2}{3}$	6.350
3	8
$3\frac{1}{3}$	10.080
$3\frac{2}{3}$	12.699
4	16

$$A=\frac{1}{6}[1+2(1.260)+2(1.587)+2(2)+2(.520)+2(3.175)+2(4)+2(5.040)+2(6.350)+2(8)+2(10.080)+2(12.699)+16]$$

$$A=21.7$$

17. The approximate value is less than the exact value because the upper boundaries of all of the trapezoids are below the curve.

21. $L=2\int_0^{50}\sqrt{6.4\times10^{-7}x^2+1}dx;\ n=10;\ h=\frac{50}{10}=5;\ \frac{h}{2}=\frac{5}{2}$

x	y
0	2
5	2.000 016
10	2.000 063 999
15	2.000 143 955
20	2.000 255 984
25	2.000 399 96
30	2.000 575 917
35	2.000 783 846
40	2.001 037 38
45	2.001 295 58
50	2.001 599 361

$L = \frac{5}{2}[2 + 2(2.000\ 016$
$+ 2.000\ 063\ 999 + 2.000\ 143\ 995$
$+ 2.000\ 255\ 984 + 2.000\ 399\ 96 + 2.000\ 575\ 917$
$+ 2.000\ 783\ 846 + 2.001\ 023\ 838 + 2.001\ 295\ 58)$
$+ 2.001\ 599\ 361]$

$L = 100.026\ 791\ 7$ m

$L = 100.027$ m

25.6 Simpson's Rule

1. $\int_0^1 \frac{dx}{x+2}$, $n = 2$, $h = \frac{1-0}{2} = \frac{1}{2}$

x	y
0	$\frac{1}{2}$
$\frac{1}{2}$	$\frac{2}{5}$
1	$\frac{1}{3}$

$$\int_0^1 \frac{dx}{x+2} = \frac{\frac{1}{2}}{3}\left(\frac{1}{2} + 4\left(\frac{2}{5}\right) + \frac{1}{3}\right) = 0.406$$

5. $\int_1^4 \left(2x + \sqrt{x}\right) dx$; $n = 6$; $h = \frac{4-1}{6} = \frac{1}{2}$; $\frac{h}{3} = \frac{1}{6}$

$$A_S = \frac{1}{6}\left[3 + 4(4.22) + 2(5.41) + 4(6.58) + 2(7.73) + 4(8.87) + 10\right]$$

$$A_S = \frac{1}{6}(117.96) = 19.7$$

n	x_n	y_n
1	1	3
2	1.5	4.22
3	2	5.41
4	2.5	6.58
5	3	7.73
6	3.5	8.87
7	4	10

$$A = \int_1^4 \left(2x + \sqrt{x}\right) dx = 2\int_1^4 x\,dx + \int_1^4 x^{1/2}\,dx$$

$$= \left(x^2 + \frac{2}{3}x^{3/2}\right)\Bigg|_1^4 = 16 + \frac{16}{3} + -\left(1 + \frac{2}{3}\right)$$

$$= \frac{59}{3} = 19.7$$

9. $\int_1^5 \frac{dx}{x^2 + x}$; $n = 10$; $h = 0.4$; $\frac{h}{3} = \frac{0.4}{3}$

$$A_S = \frac{0.4}{3}\left[0.5000 + 4(0.2976) + 2(0.1984) + 4(0.1420) + 2(0.1068) + 4(0.0833) + 2(0.0668) + 4(0.0548) + 2(0.0458) + 4(0.0388) + 0.0333\right]$$

$$= \frac{0.4}{3}(3.8349) = 0.511$$

13. $h = 2$; $\frac{h}{3} = \frac{2}{3}$

x	y
2	0.670
4	2.34
6	4.56
8	3.67
10	3.56
12	4.78
14	6.87

$$\int_2^{14} y\,dx = \frac{2}{3}\left[0.67 + 4(2.34) + 2(4.56) + 4(3.67) + 2(3.56) + 4(4.78) + 6.87\right] = 44.6$$

17. $\bar{x} = 0.9129\int_0^3 x\sqrt{0.3 - 0.1}dx$; $n = 12$; $h = \frac{3}{12} = \frac{1}{4}$; $\frac{h}{3} = \frac{1}{12}$

n	x_n	y_n
1	0	0
2	0.25	0.131
3	0.50	0.25
4	0.75	0.356
5	1	0.447
6	1.25	0.523
7	1.50	0.581

n	x_n	y_n
8	1.75	0.619
9	2	0.632
10	2.25	0.616
11	2.5	0.559
12	2.75	0.435
13	3	0

$$\bar{x} = A_S = \frac{1}{12}\left[0 + 4(0.131) + 2(0.25) + 4(0.356) + 2(0.447) + 4(0.523) + 2(0.581) + 4(0.619) + 2(0.632) + 4(0.616) + 2(0.559) + 4(0.435) + (0)\right](0.9129)$$

$$= \frac{1}{12}(15.658)(0.9129) = 1.191$$

$\bar{x} = 1.19$ cm

Review Exercises

1. This is false.

By the chain rule,

$$\frac{d}{dx}\left(\frac{1}{6}(3x^2+1)^6\right)=\left(3x^2+1\right)^5\cdot 6x$$

which is not the integrand.

5. $\int\left(4x^3-x\right)dx=\int 4x^3dx-\int xdx$

$$=\frac{4x^4}{4}-\frac{x^2}{2}+C$$

$$=x^4-\frac{1}{2}x^2+C$$

9. $\int_1^4\left(\frac{\sqrt{x}}{2}+\frac{2}{\sqrt{x}}\right)dx=\frac{1}{2}\int_1^4 x^{1/2}dx+2\int_1^4 x^{-1/2}dx$

$$=\frac{1}{2}\frac{x^{3/2}}{\frac{3}{2}}+\frac{2x^{1/2}}{\frac{1}{2}}\Bigg|_1^4$$

$$=\frac{1}{3}x^{3/2}+4x^{1/2}\Big|_1^4$$

$$=\frac{1}{3}(4)^{3/2}+4(4)^{1/2}$$

$$-\frac{1}{3}(1)^{3/2}+4(1)^{1/2}=\frac{19}{3}$$

13. $\int_0^2 5x(4-x)\,dx=\int_0^2\left(20x-5x^2\right)dx$

$$=10x^2-\frac{5x^3}{3}\Bigg|_0^2$$

$$=10(2)^2-\frac{5\cdot 2^3}{3}-10(0)^2-\frac{5\cdot 0^3}{3}$$

$$=\frac{80}{3}$$

17. $\int_{-2}^{5}\frac{dx}{\sqrt[3]{x^2+6x+9}}=\int_{-2}^{5}\frac{dx}{\sqrt[3]{(x+3)^2}}$

$$=\int_{-2}^{5}(x+3)^{-2/3}dx$$

$$=3(x+3)^{1/3}\Big|_{-2}^{5}$$

$$=3\sqrt[3]{5+3}-3\sqrt[3]{-2+3}$$

$$=3\sqrt[3]{8}-3\sqrt[3]{1}$$

$$=3(2)-3=3$$

21. $\int 3(7-2x)^{3/4}dx=-\frac{3}{2}\int u^{3/4}du=-\frac{3}{2}\left(\frac{4}{7}\right)u^{7/4}+C$

$u=7-2x,\ du=-2dx$

$$\int 3(7-2x)^{3/4}dx=-\frac{6}{7}(7-2x)^{7/4}+C$$

25. $\int 4x^2\left(1-2x^3\right)^4dx=-\frac{2}{3}\int u^4\,du=-\frac{2}{3}\cdot\frac{u^5}{5}+C$

$u=1-2x^3,\ du=-6x^2dx$

$$\int 4x^2\left(1-2x^3\right)^4dx=-\frac{2}{15}\left(1-2x^3\right)^5+C$$

29. $\int_1^3\left(x^2+x+2\right)\left(2x^3+3x^2+12x\right)dx$

$$=\frac{1}{6}\int_1^3\left(2x^3+3x^2+12x\right)6\left(x^2+x+2\right)dx$$

$$=\frac{1}{6}\cdot\frac{1}{2}\left(2x^3+3x^2+12x\right)^2\Big|_1^3$$

$$=\frac{1}{12}\left(117^2-17^2\right)=\frac{3350}{3}$$

$u=2x^3+3x^2+12x$

$du=\left(6x^2+6x+12\right)dx=6\left(x^2+x+2\right)dx$

33. **(a)** $\int(1-2x)\,dx=\int dx-2\int x\,dx$

$$=x-x^2+C_1$$

(b) $\int(1-2x)\,dx=-\frac{1}{2}\int u\,du=-\frac{1}{2}\frac{u^2}{2}+C_2$

$u=1-2x,\ du=-2dx$

$$\int(1-2x)\,dx=-\frac{1}{4}(1-2x)^2+C_2$$

$$=-\frac{1}{4}\left(1-4x+4x^2\right)+C_2$$

$$=-\frac{1}{4}+x-x^2+C_2$$

$$=x-x^2+C_2-\frac{1}{4}$$

C_1 and C_2 are not equal, $C_1=C_2-\frac{1}{4}$. Indefinite integrals with the same integrand are only equal to within an arbitrary constant.

37. $${}_2^9(3x-2)^{1/2}\,dx = \frac{2}{9}(3x-2)^{3/2}\Big|_2^9$$
$$= \frac{2}{9}(125-8)$$
$$= \frac{234}{9}$$
$${}_2^6(3x-2)^{1/2}\,dx = \frac{2}{9}(3x-2)^{3/2}\Big|_2^6 = \frac{2}{9}(64-8) = \frac{112}{9}$$
$${}_6^9(3x-2)^{1/2}\,dx = \frac{2}{9}(3x-2)^{3/2}\Big|_6^9 = \frac{2}{9}(125-64) = \frac{122}{9}$$
$${}_2^6(3x-2)^{1/2}\,dx + {}_2^6(3x-2)^{1/2}\,dx = \frac{112}{9} + \frac{122}{9}$$
$$= \frac{234}{9}$$
$$= {}_2^9(3x-2)^{1/2}\,dx$$

41. $$\int_0^1 x^3 dx = \frac{x^4}{4}\Big|_0^1 = \frac{1}{4}$$
$$\int_1^2 (x-1)^3\,dx = \frac{(x-1)^4}{4}\Big|_1^2$$
$$= \frac{(2-1)^4}{4} - \frac{(1-1)^4}{4} = \frac{1}{4}$$

which shows $\int_0^1 x^3 dx = \int_1^2 (x-1)^3\,dx$

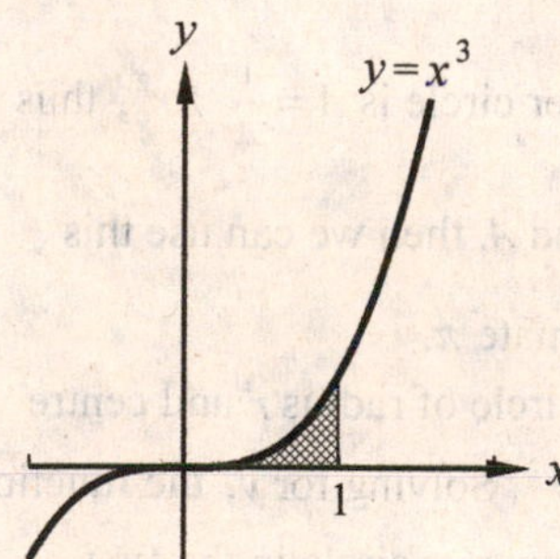

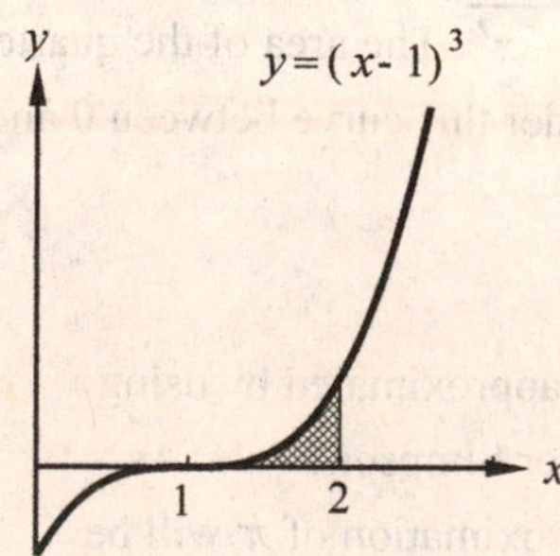

$y=(x-1)^3$ is $y=x^3$ shifted right one unit, so the areas are the same.

45. Since $f(x) > 0$, the graph is above the x-axis. Since $f''(x) < 0$, for $a \le x < b$, f is concave down for $a \le x \le b$.

$$\int_a^b f(x)\,dx > A_{\text{trapezoid}}$$

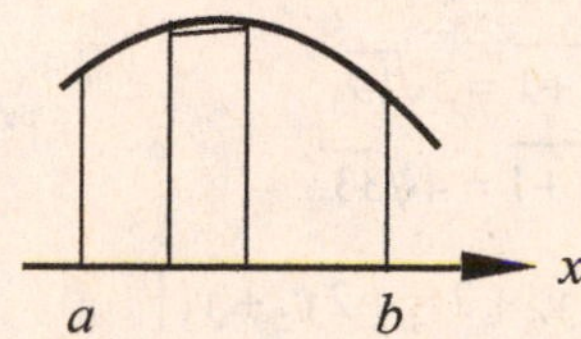

If a function is concave, the tops of the trapezoids are all below the curve. Therefore, the area under the curve is greater than the approximation by the trapezoidal rule.

49. $$h = \frac{3-1}{4} = \frac{1}{2}; \frac{h}{3} = \frac{1}{6}$$
$$x_0 = 1,\ f(1) = \frac{1}{2(1)-1} = 1,$$
$$x_1 = 1.5,\ f(1.5) = \frac{1}{2(1.5)-1} = \frac{1}{2},$$
$$x_2 = 2.0,\ f(2.0) = \frac{1}{2(2.0)-1} = \frac{1}{3},$$
$$x_3 = 2.5, f(2.5) = \frac{1}{2(2.5)-1} = \frac{1}{4},$$
$$x_4 = 3.0, f(3.0) = \frac{1}{2(3.0)-1} = \frac{1}{5}$$
$$\int_1^3 \frac{dx}{2x-1} \approx \frac{h}{3}[y_0 + 4y_1 + 2y_2 + 4y_3 + y_4]$$
$$= \frac{1}{6}\left[1 + \frac{4}{2} + \frac{2}{3} + \frac{4}{4} + \frac{1}{5}\right]$$
$$= \frac{73}{90} = 0.811$$

53. $y = x\sqrt[3]{2x^2+1},\ a=1,\ b=4,\ n=3.$

$h = \frac{b-a}{n} = \frac{4-1}{3} = 1, \frac{h}{2} = \frac{1}{2}$

$x_0 = 1,\ y_0 = 1\sqrt[3]{2\cdot 1^2+1} = \sqrt[3]{3},$

$x_1 = 2,\ y_1 = 2\sqrt[3]{2\cdot 2^2+1} = 2\sqrt[3]{9},$

$x_2 = 3,\ y_2 = 3\sqrt[3]{2\cdot 3^2+1} = 3\sqrt[3]{19},$

$x_3 = 4,\ y_3 = 4\sqrt[3]{2\cdot 4^2+1} = 4\sqrt[3]{33}$

$\int_1^4 x\sqrt[3]{2x^2+1}dx \approx \frac{1}{2}[y_0+2y_1+2y_2+y_3]$

$= \frac{1}{2}\left[\sqrt[3]{3}+4\sqrt[3]{9}+6\sqrt[3]{19}+4\sqrt[3]{33}\right] = 19.30$

57. $y = x\sqrt[3]{2x^2+1},\ a=1,\ b=4,\ n=6.$

$h = \frac{b-a}{n} = \frac{4-1}{6} = \frac{1}{2}, \frac{h}{3} = \frac{1}{6}$

$x_0 = 1,\ y_0 = 1\sqrt[3]{2\cdot 1^2+1} = \sqrt[3]{3},$

$x_1 = 1.5,\ y_1 = 1.5\sqrt[3]{2\cdot(1.5)^2+1} = 1.5\sqrt[3]{5.5},$

$x_2 = 2.0,\ y_2 = 2.0\sqrt[3]{2\cdot 2^2+1} = 2.0\sqrt[3]{9},$

$x_3 = 2.5,\ y_3 = 2.5\sqrt[3]{2\cdot(2.5)^2+1} = 2.5\sqrt[3]{13.5}$

$x_4 = 3.0,\ y_4 = 3.0\sqrt[3]{2(3.0)^2+1} = 3.0\sqrt[3]{19},$

$x_5 = 3.5, y_5 = 3.5\sqrt[3]{2(3.5)^2+1} = 3.5\sqrt[3]{25.5},$

$x_6 = 4.0,\ y_6 = 4.0\sqrt[3]{2(4.0)^2+1} = 4.0\sqrt[3]{33}$

$\int_1^4 x\sqrt[3]{2x^2+1}dx \approx \frac{h}{3}[y_0+4y_1+2y_2+4y_3+2y_4+4y_5+y_6]$

$\int_1^4 x\sqrt[3]{2x^2+1}dx \approx \frac{1}{6}\big[\sqrt[3]{3}+4(1.5)\sqrt[3]{5.5}+2(2.0)\sqrt[3]{9}$
$+4(2.5)\sqrt[3]{13.5}+2(3.0)\sqrt[3]{19}+4(3.5)\sqrt[3]{25.5}+4.0\sqrt[3]{33}\big]$

$\int_1^4 x\sqrt[3]{2x^2+1}dx \approx 19.04$

61. $h = \frac{4}{8} = \frac{1}{2}, \frac{h}{3} = \frac{1}{6}$

$A_S = \frac{h}{3}[y_0+4y_1+2y_2+4y_3+2y_4+4y_5$
$+2y_6+4y_7+y_8]$

x	y	x	y
0	5	$\frac{5}{2}$	6.915
$\frac{1}{2}$	6.121	3	6.646
1	6.646	$\frac{7}{2}$	6.121
$\frac{3}{2}$	6.915	4	5
2	7		

$A_S = \frac{1}{6}\big[5+4(6.121)+2(6.646)+4(6.915)+2(7)$
$+4(6.915)+2(6.646)+4(6.121)+5\big]$

$A_S = 25.8\text{ m}^2$

65. $\frac{dy}{dx} = k\left(2L^3-12Lx+2x^4\right)$

$dy = k\left(2L^3x^0-12Lx+2x^4\right)dx$

$y = \int k\left(2L^3x^0-12Lx+2x^4\right)dx$

$= k\int\left(2L^3x^0-12Lx+2x^4\right)dx$

$= k\left(2L^3x - \frac{12Lx^2}{2} + \frac{2x^5}{5}\right)+C$

$y=0$ for $x=0$;

$0 = k(0-0+0)+C$

$C=0$

$y = k\left(2L^3x-6Lx^2+\frac{2}{5}x^5\right)$

69. The area of a quarter circle is $A = \frac{1}{4}\cdot\pi r^2$, thus $\pi = \frac{4\cdot A}{r^2}$. If we find A, then we can use this formula to approximate π.
The equation of a circle of radius r and centre $(0,0)$ is $x^2+y^2=r^2$. Solving for y, the function that represents the quarter circle in the first quadrant is $y=\sqrt{r^2-x^2}$. The area of the quarter circle is the area under this curve between 0 and r. In other words,

$A = \int_0^r \sqrt{r^2-x^2}\,dx.$

The integral can be approximated by using the trapezoidal rule or Simpson's rule. As n increases, the approximation of π will be better. Either of these numerical procedures can be programmed easily by the student.

Chapter 26

Applications of Integration

26.1 Applications of the Indefinite Integral

1. $s = \int (v_0 - 32t)\,dt = v_0 t - 16t^2 + C$

$200 = v_0(0) - 16(0)^2 + C$

$200 = C$

$s = v_0 t - 16t^2 + 200$

$0 = v_0(2.5) - 16(2.5)^2 + 200$

$v_0 = -40$ ft/s

The initial velocity was 40 ft/s downward.

5. $\frac{ds}{dt} = -0.25$ m/s

$ds = -0.25dt$

$s = -0.25\int dt = -0.25t + C_1$

$t = 0,\ s = 8$; therefore, $C_1 = 8$

$s = -0.25t + 8.00$

$= 8.00 - 0.25t$

9. $a = -8.0t,$

$v = (-8.0t)dt = -4.0t^2 + C$

$64 = v(0) = -4.0(0)^2 + C;\ C = 64$

$v = -4.0t^2 + 64$

When $t = 4.0,\ v(t) = 0,$ so the fire engine travels for 4.0 seconds in order to stop completely.

$s = \left(-4.0t^2 + 64\right) dt$

$s = -\frac{4.0}{3}t^3 + 64t + D$

$0 = s(0) = -\frac{4.0}{3}(0)^3 + 64(0) + D;\ D = 0$

$s(t) = -\frac{4.0}{3}t^3 + 64t$

$s(4.0) = -\frac{4.0}{3}(4.0)^3 + 64(4.0) = \frac{512}{3} = 170.66$

The fire engine travels 170 feet before stopping.

13. $a = \frac{88.0 \text{ ft/s}}{3.60 \text{ s}} = \frac{220}{9}\ \frac{\text{ft}}{\text{s}^2}$

$v(t) = \frac{220}{9}t + C$

$0 = v(0) = \frac{220}{9}(0) + C; C = 0$

$v(t) = \frac{220}{9}t$

$s(t) = \frac{110}{9}t^2 + D$

$0 = s(0) = \frac{110}{9}(0)^2 + D; D = 0$

$s(t) = \frac{110}{9}t^2$

$s(3.60) = \frac{110}{9}(3.60)^2 = 158.4$

The Corvette goes 158 ft in this time.

17. $v = -32dt$

$= -32t + C$

$v = v_0;\ t = 0$

$C = v_0$

$v = -32t + v_0$

$s = (-32t + v_0)dt$

$= -16t^2 + v_0 t + C_1$

$s = 0,\ t = 0$

$C_1 = 0$

$s = -16t^2 + v_0 t$

$s = 90$ when $v = 0$

$90 = -16t^2 + v_0 t$

$0 = -32t + v_0$

Substitute $t = \frac{v_0}{32}$ into $90 = -16t^2 + v_0 t$:

$90 = -16\left(\frac{v_0}{32}\right)^2 + v_0\left(\frac{v_0}{32}\right)$

$\frac{16v_0^2}{32^2} = 90$

$v_0 = \sqrt{\frac{90(32)^2}{16}} = 76$ ft/s

21. $i = 0.230\mu\text{A} = 0.230 \times 10^{-6}\ \text{A}$

$t = 1.50\ \text{ms} = 1.50 \times 10^{-3}\ \text{s}$

$q = \ i\,dt$

$= \ 0.230 \times 10^{-6}\,dt$

$= 0.230 \times 10^{-6}t + C_1$

$q = 0,\ t = 0$

$C_1 = 0;$

$q = 0.230 \times 10^{-6}t$

Find q for $t = 1.50 \times 10^{-3}$ s:

$q = 0.230 \times 10^{-6}\left(1.50 \times 10^{-3}\right)$

$= 0.345 \times 10^{-9} = 0.345\ \text{nC}$

25. $C = 2.5\ \mu\text{F} = 2.5 \times 10^{-6}\ \text{F}$

$i = 25\ \text{mA} = 25 \times 10^{-3}\text{A}$

$t = 12\ \text{ms} = 12 \times 10^{-3}\ \text{s}$

$V_c = \dfrac{1}{C}\ \ i\,dt$

$= \dfrac{1}{2.5 \times 10^{-6}}\ \ 0.025dt$

$= \dfrac{1}{2.5 \times 10^{-6}}(0.025)t + C_1$

$= 1.0 \times 10^4 t + C_1$

$V_c = 0$ at $t = 0$ so $C_1 = 0$

$V_c = 1.0 \times 10^4 t$

Find V_c for $t = 0.012$ s:

$V_c = 1.0 \times 10^4\,(0.012)$

$= 120\ \text{V}$

29. $\omega = \dfrac{d\theta}{dt} = 16t + 0.5t^2$

$d\theta = \left(16t + 0.50t^2\right)dt$

$\theta = 8t^2 + \dfrac{0.50t^3}{3} + C$

$\theta = 0,\ t = 0$ so $C = 0$

$\theta = 8t^2 + \dfrac{0.50t^3}{3}$

Find θ for $t = 10.0$ s:

$\theta = 8(10.0)^2 + \dfrac{0.50}{3}(10.0)^3$

$= 970\ \text{rad}$

33. $V = \dfrac{1}{C}\ \ i\,dt$

If $i = 1$ mA $= 0.001$ A, $C = 1\ \mu\text{F} = 0.000001$ F then

$\Delta V = \dfrac{1}{0.000001}\ {}_0^{0.001}\ 0.001\,dt = 1000t\Big|_0^{0.001}$

$= 1 - 0 = 1$

26.2 Areas by Integration

1. $A = {}_1^3\ y\,dx = {}_1^3\ x^2\,dx$

$= \dfrac{x^3}{3}\Big|_1^3$

$A = \dfrac{(3)^3}{3} - \dfrac{(1)^3}{3}$

$A = \dfrac{26}{3}$

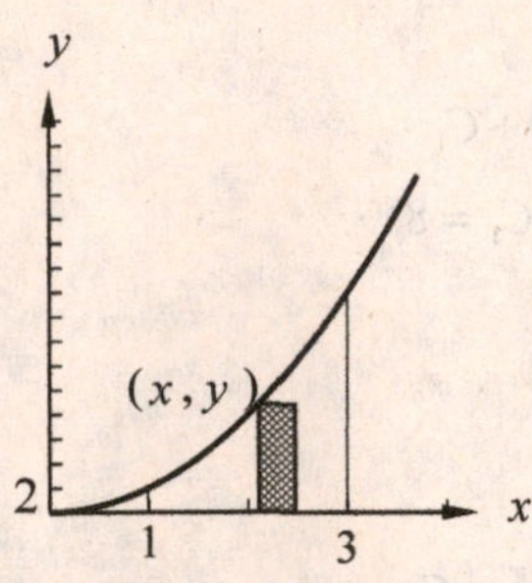

5. $y = 8 - 2x^2;\ y = 0,$

Intersections are at $x = -2, x = 2$

$A = \int_{-2}^{2}(8 - 2x^2)dx$

$= 8x - \dfrac{2}{3}x^3\Big|_{-2}^{2}$

$= \dfrac{64}{3}$

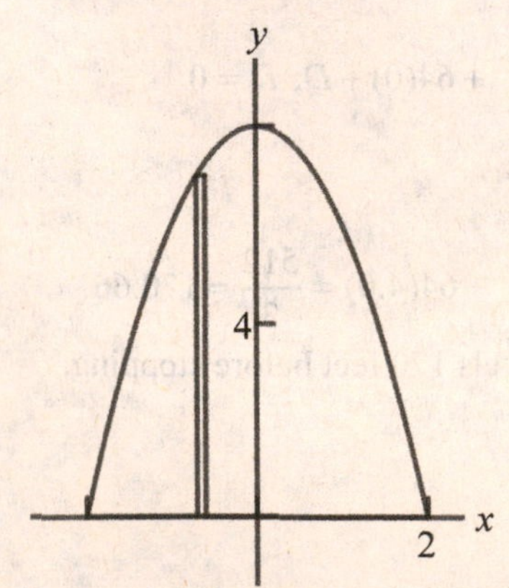

9. $y = 3x^{-2}$; $y = 0$, $x = 2$, $x = 3$

$$A = \int_2^3 3x^{-2}dy$$

$$= -3x^{-1}\Big|_2^3$$

$$= -\frac{3}{x}\Big|_2^3$$

$$= -1 - \left(-\frac{3}{2}\right) = \frac{1}{2}$$

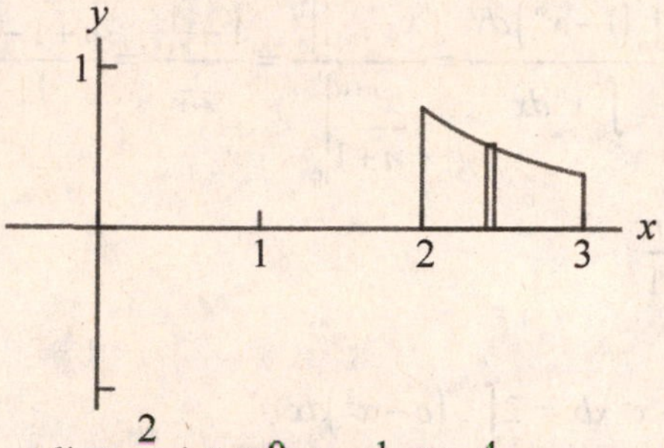

13. $y = \dfrac{2}{\sqrt{x}}$; $x = 0$, $y = 1$, $y = 4$

$$\sqrt{x} = \frac{2}{y}$$

$$x = \frac{4}{y^2}$$

$$A = {}_1^4 xdy$$

$$= 4\ {}_1^4 y^{-2}dy$$

$$= -4y^{-1}\Big|_1^4$$

$$= -\frac{4}{y}\Big|_1^4$$

$$= -\frac{4}{4} - \ -\frac{4}{1} \ = -1 + 4 = 3$$

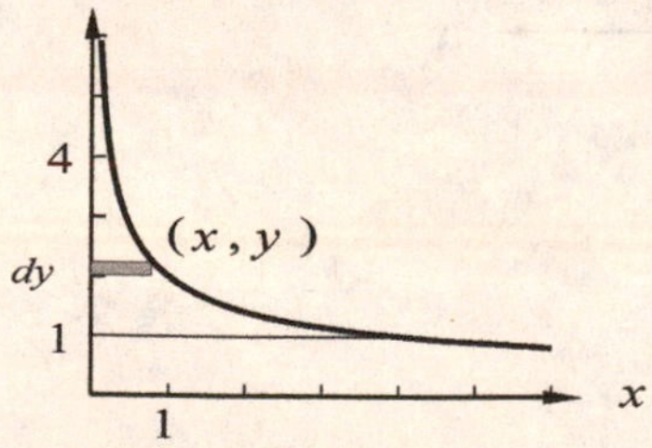

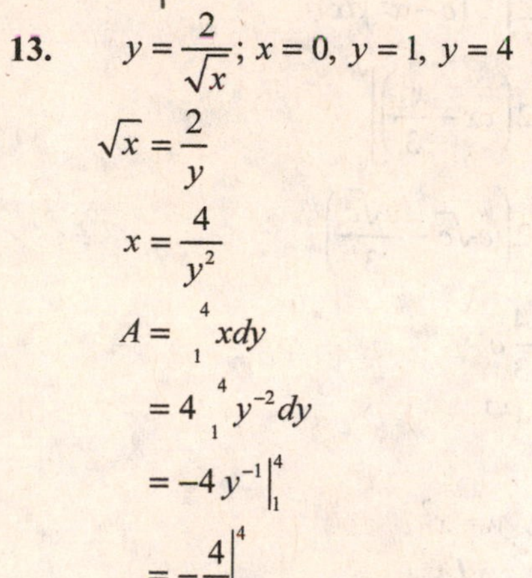

17. Intersections:

$$x - 4\sqrt{x} = 0$$

$$x\ \ 1 - \frac{4}{\sqrt{x}}\ \ = 0$$

$$x = 0 \quad \text{or} \quad x = 16$$

$$A = {}_0^4\left(0 - \left(x - 4\sqrt{x}\right)\right)dx$$

$$A = -\frac{x^2}{2} + \frac{4x^{3/2}}{\frac{3}{2}}\Bigg|_0^4$$

$$A = \frac{40}{3}$$

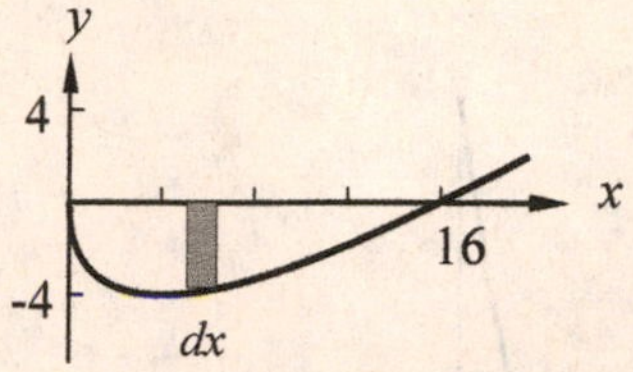

21. $y = x^4 - 8x^2 + 16$, $y = 16 - x^4$

Intersections:

$$x^4 - 8x^2 + 16 = 16 - x^4$$

$$2x^4 - 8x^2 = 0$$

$$2x^2\left(x^2 - 4\right) = 0$$

$$x = 0 \text{ or } x = 2 \text{ or } x = -2$$

$$A = 2\int_0^2\left(16 - x^4 - \left(x^4 - 8x^2 + 16\right)\right)dx$$

$$A = 2\int_0^2\left(8x^2 - 2x^4\right)dx$$

$$= 2\left(\frac{8}{3}x^3 - \frac{2}{5}x^5\right)\Bigg|_0^2 = \frac{256}{15}$$

$\int_{-2}^2 y_2 - y_1 dx = 2\int_0^2 y_2 - y_1 dx$ because of symmetry with respect to the y-axis.

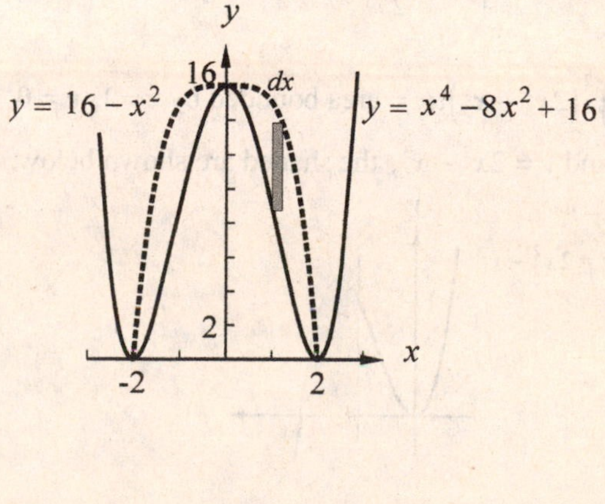

25. $y = x^5;\ x = -1,\ x = 2,\ y = 0$

Between -1 and 0, x^5 is below $y = 0$.

Between 0 and 2, x^5 is above $y = 0$.

We compute two integrals and sum.

$$A = \int_{-1}^{0}(0-y)\,dx + \int_{0}^{2}(y-0)\,dx$$

$$= -\int_{-1}^{0} x^5\,dx + \int_{0}^{2} x^5\,dx$$

$$= -\frac{x^6}{6}\Big|_{-1}^{0} + \frac{x^6}{6}\Big|_{0}^{2}$$

$$= 0 - \left(-\frac{1}{6}\right) + \frac{64}{6} - 0 = \frac{65}{6}$$

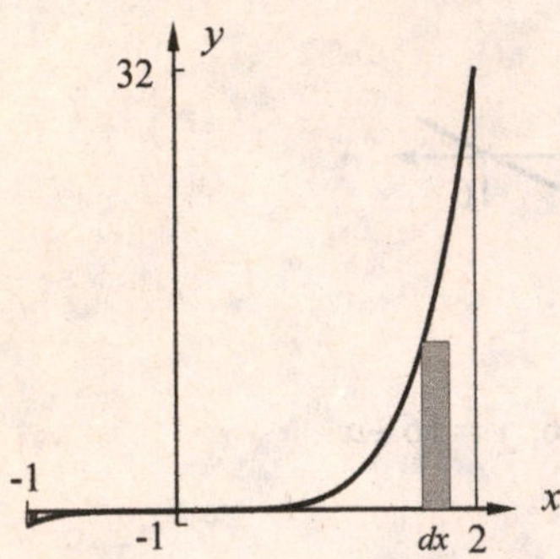

29. $\int_1^2 (2x^2 - x^3)\,dx$ = area bounded by $x = 1$, $y = 2x^2$, and $y = x^3$, the shaded area shown below.

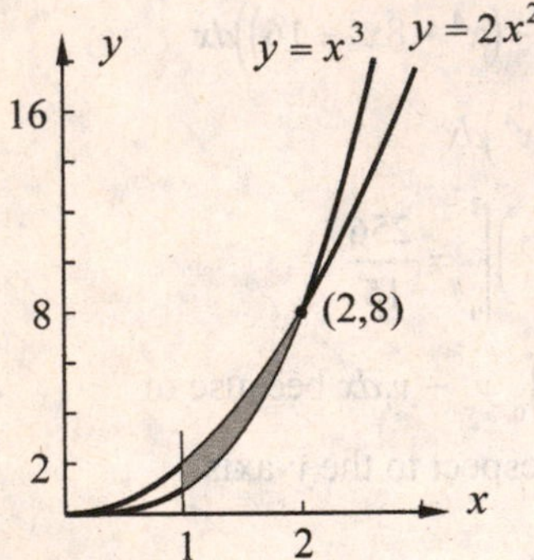

$\int_1^2 (2x^2 - x^3)\,dx$ = area bounded by $x = 1$, $y = 0$, and $y = 2x^2 - x^3$, the shaded are shown below.

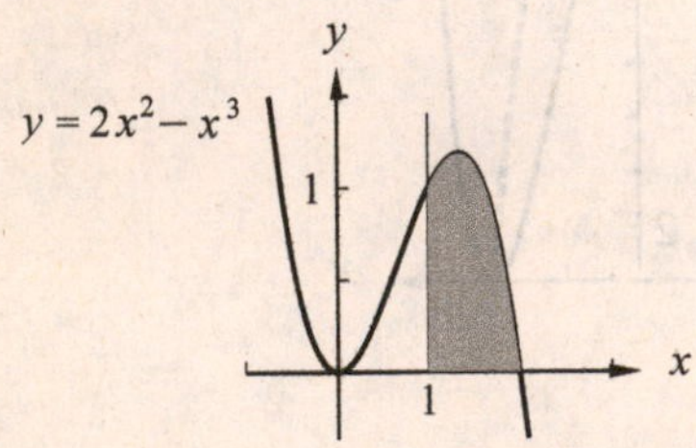

33.

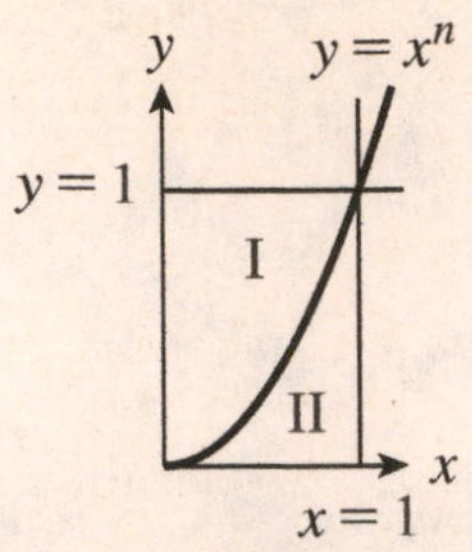

$$\frac{A_I}{A_{II}} = \frac{\int_0^1 (1-x^n)\,dx}{\int_0^1 x^n\,dx} = \frac{x - \frac{x^{n+1}}{n+1}\Big|_0^1}{\frac{x^{n+1}}{n+1}\Big|_0^1} = \frac{1 - \frac{1}{n+1}}{\frac{1}{n+1}} = \frac{n+1-1}{1}$$

$$\frac{A_I}{A_{II}} = \frac{n}{1}$$

37. $\int_0^2 (4-x^2)\,dx = 2\int_0^{\sqrt{c}} (c-x^2)\,dx$

$$4x - \frac{x^3}{3}\Big|_0^2 = 2\left(cx - \frac{x^3}{3}\right)\Big|_0^{\sqrt{c}}$$

$$8 - \frac{8}{3} = 2\left(c\sqrt{c} - \frac{c\sqrt{c}}{3}\right)$$

$$\frac{16}{3} = \frac{4}{3}c^{3/2}$$

$$c = 4^{2/3}$$

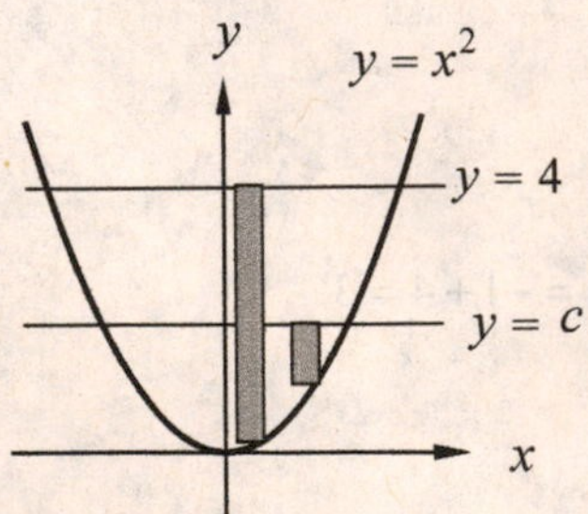

41. **(a)** $\int_0^2 \left(8x - x^4\right)dx = 4x^2 - \left.\frac{x^5}{5}\right|_0^2$

$= \frac{48}{5}$

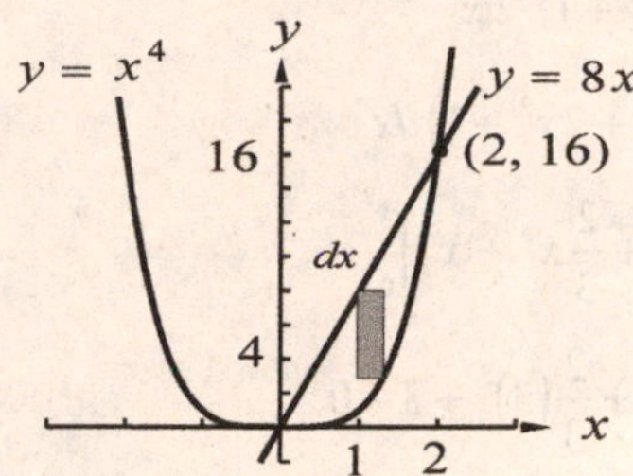

(b) $\int_0^{16} \sqrt[4]{y} - \frac{y}{8}\, dy = \frac{4}{5}y^{5/4} - \left.\frac{y^2}{16}\right|_0^{16}$

$= \frac{48}{5}$

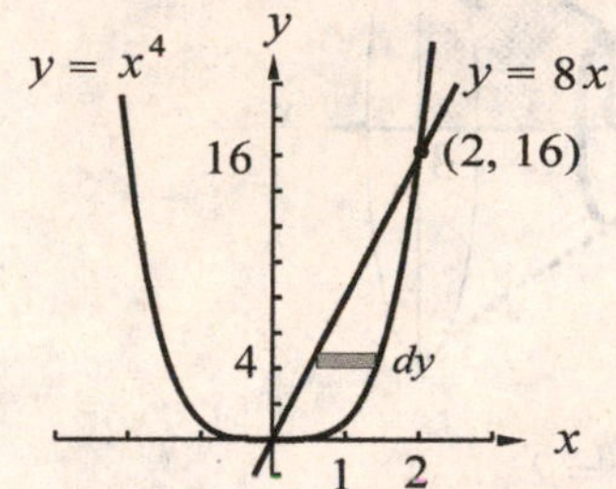

45. $v = 1 - 0.01\sqrt{2t+1}$; $t = 10$ s to $t = 100$ s

$$\Delta s = \int_{10}^{100}\left(1 - 0.01\sqrt{2t+1}\right)dt$$

$$= \int_{10}^{100} dt - 0.01\int_{10}^{100}(2t+1)^{1/2}\, dt$$

$$\Delta s = \int_{10}^{100} dt - \frac{0.01}{2}\int (2t+1)^{1/2}\, 2\, dt$$

$$= t - \frac{0.01}{2}\cdot\frac{2}{3}(2t+1)^{3/2}\Big|_{10}^{100}$$

$$= \left[t - \frac{0.01}{3}(2t+1)^{3/2}\right]_{10}^{100}$$

$$= 90.501 - 9.679$$

$\Delta s = 80.8$ km

change in position from $t = 10$ s to $t = 100$ s

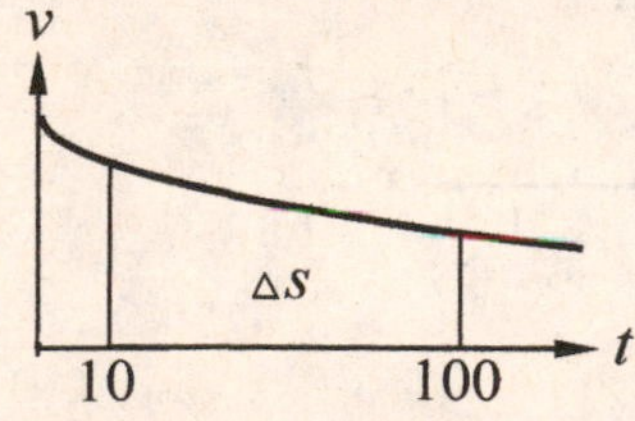

49. $y = 0.25x^4$ and $y = 12 - 0.25x^4$ intersect when

$0.25x^4 = 12 - 0.25x^4$

$0.50x^4 = 12$

$x^4 = 24$

$x = \sqrt[4]{24}$

$y = 0.25\left(\sqrt[4]{24}\right)^4 = 6$

$$A = 2\int_0^{\sqrt[4]{24}}\left(12 - 0.25x^4 - 0.25x^4\right)dx$$

$$A = 2\int_0^{\sqrt[4]{24}}\left(12 - 0.50x^4\right)dx$$

$$A = 2\left(12x - \frac{0.50x^5}{5}\right)\Bigg|_0^{\sqrt[4]{24}}$$

$A = 42.5\ \text{dm}^2$

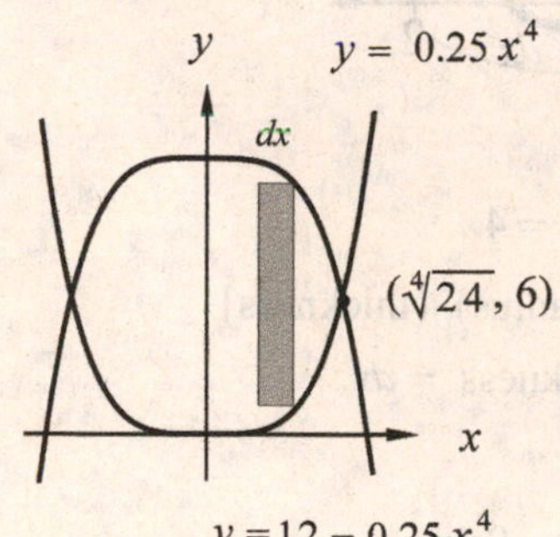

26.3 Volumes by Integration

1. $y = x^3, x = 2, y = 0$ about the x-axis.

$$V = \int_0^2 \pi y^2 dx = \pi\int_0^2 \left(x^3\right)^2 dx$$

$$V = \pi\left.\frac{x^7}{7}\right|_0^2 = \frac{128\pi}{7}$$

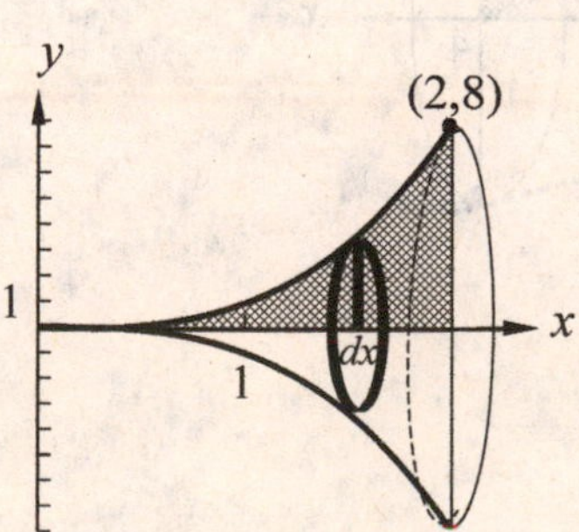

5. Shell: $dV = 2\pi(\text{radius})(\text{height})(\text{thickness})$

radius $= x$, height $= y$, thickness $= dx$

$$dV = 2\pi xy\,dx$$

$$V = \int_0^2 2\pi x(4-2x)\,dx$$

$$= 2\pi\left[2x^2 - \frac{2x^3}{3}\right]_0^2$$

$$= \frac{16\pi}{3}$$

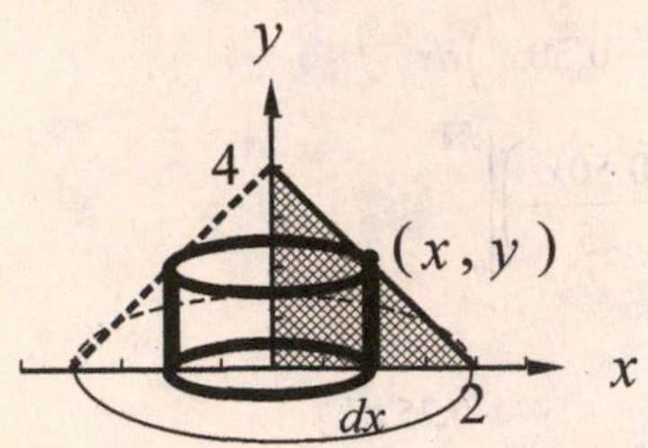

9. $y = 3\sqrt{x}, y = 0, x = 4$

Disc: $dV = \pi(\text{radius})^2(\text{thickness})$

radius $= y$, thickness $= dx$

$$dV = \pi y^2\,dx$$

$$V = \pi\,{}_0^4\, y^2\,dx = \pi\,{}_0^4\, 9x\,dx$$

$$= \pi\ \frac{9}{2}x^2\ \Big|_0^4$$

$$= \pi\ \frac{9}{2}(4)^2 - 0\ = 72\pi$$

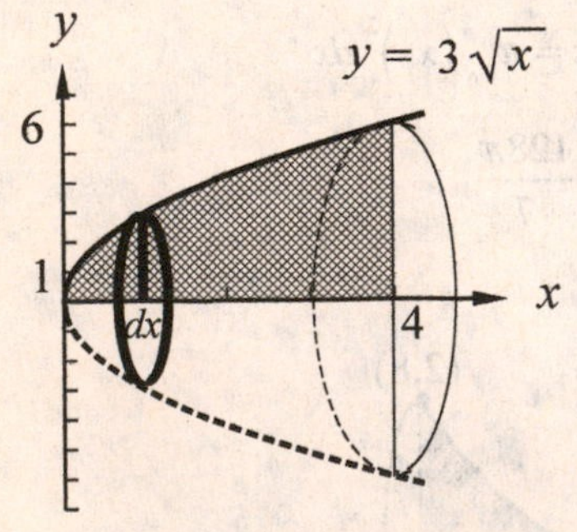

13. $y = x^2 + 1, x = 0, x = 3, y = 0$

Disc: $dV = \pi(\text{radius})^2(\text{thickness})$

radius $= y$, thickness $= dx$

$$dV = \pi y^2\,dx$$

$$V = \pi\,{}_0^3\,(x^2+1)^2\,dx$$

$$= \pi\,{}_0^3\,(x^4 + 2x^2 + 1)\,dx$$

$$= \pi\ \frac{1}{5}x^5 + \frac{2}{3}x^3 + x\ \Big|_0^3$$

$$= \pi\ \frac{1}{5}(3)^5 + \frac{2}{3}(3)^3 + 3 - 0$$

$$= \frac{348}{5}\pi$$

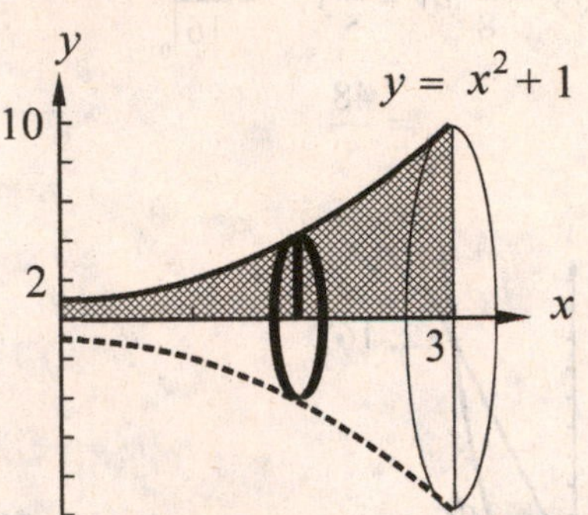

17. $y = 2x^{1/3}, x = 0, y = 2$

$x = \frac{1}{8}y^3; x^2 = \frac{1}{64}y^6$

Disc: $dV = \pi(\text{radius})^2(\text{thickness})$

radius $= x$, thickness $= dy$

$$dV = \pi x^2\,dy$$

$$V = \pi\,{}_0^2\,\tfrac{1}{64}y^6\,dy$$

$$= \frac{\pi}{7\cdot 64}y^7\Big|_0^2 = \frac{2\pi}{7}$$

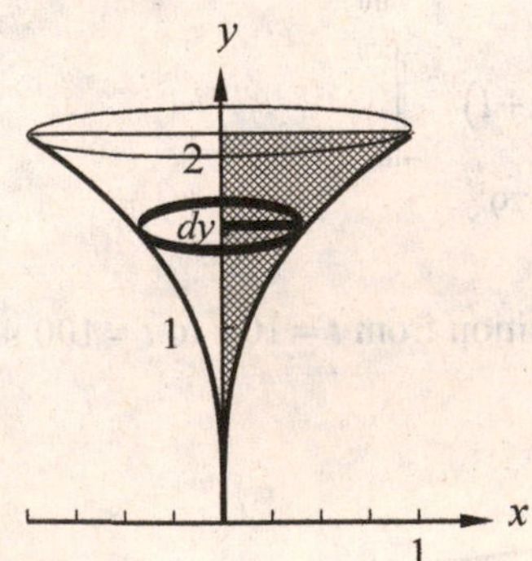

21. $x^2 - 4y^2 = 4, x = 3$

Shell: $dV = 2\pi(\text{radius})(\text{height})(\text{thickness})$

radius $= x$, height $= 2y$, thickness $= dx$

$$dV = 2\pi x(2y)dx$$

$$x^2 - 4y^2 = 4, y = \sqrt{\frac{x^2-4}{4}}$$

Intercept: $\frac{x^2-4}{4} = 0$

$$x = 2$$

$$V = 4\pi\ {}_2^3\, x\sqrt{\frac{x^2-4}{4}}dx$$

$$= \frac{4\pi}{2}\ {}_2^3\left(x^2-4\right)^{1/2} 2xdx$$

$$u = x^2 - 4, du = 2xdx$$

$$V = \pi \cdot \frac{2}{3}\left(x^2-4\right)^{3/2}\Big|_2^3$$

$$= \frac{2\pi}{3}\left(5^{3/2}\right) - 0$$

$$= \frac{10\sqrt{5}}{3}\pi$$

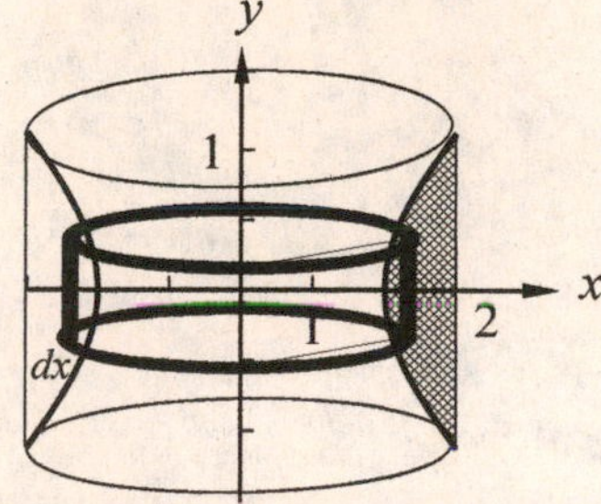

25. $y = \sqrt{4-x^2}$, Quad I

Shell: $dV = 2\pi(\text{radius})(\text{height})(\text{thickness})$

radius $= x$, height $= y$, thickness $= dx$

$$dV = 2\pi xydx$$

$$V = 2\pi\ {}_0^2\, x\sqrt{4-x^2}dx$$

$$u = 4 - x^2, du = -2xdx$$

$$V = -\pi\ {}_0^2\left(4-x^2\right)^{1/2}(-2xdx)$$

$$= -\pi\frac{2}{3}\left(4-x^2\right)^{3/2}\Big|_0^2$$

$$= -\frac{2\pi}{3}(0-8) = \frac{16\pi}{3}$$

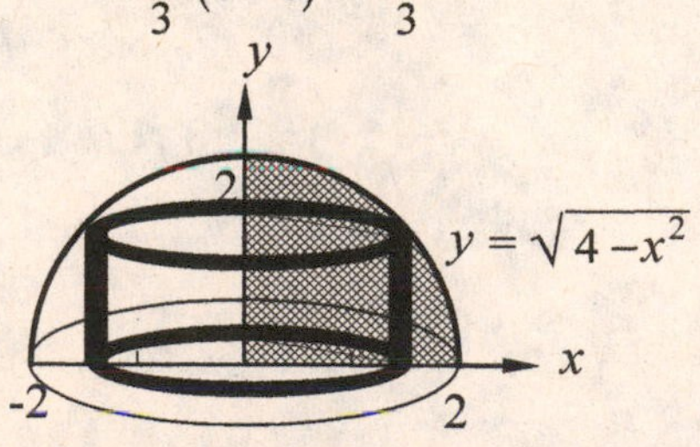

29. $y = 4 - x^2, y = 0$, rotated around $x = 2$

Shell: $dV = 2\pi(\text{radius})(\text{height})(\text{thickness})$

radius $= 2 - x$, height $= y$, thickness $= dx$

$$dV = 2\pi(2-x)ydx$$

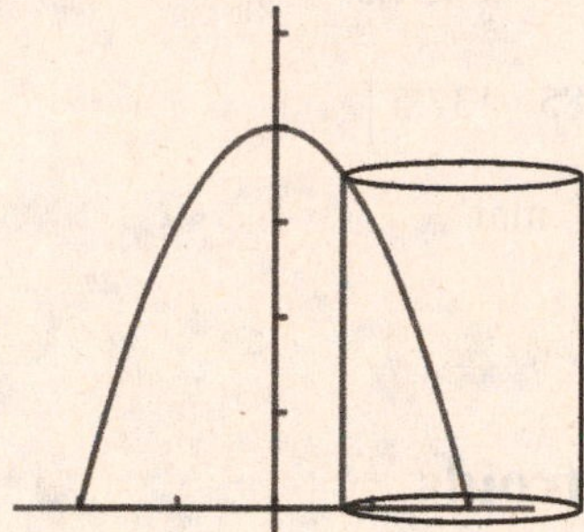

$$V = 2\pi\int_{-2}^{2}(2-x)(4-x^2)dx$$

$$= 2\pi\int_{-2}^{2}(8-4x-2x^2+x^3)dx$$

$$= 2\pi\left[8x - 2x^2 - \frac{2}{3}x^3 + \frac{1}{4}x^4\right]\Bigg|_{-2}^{2}$$

$$= 2\pi\left[\left(16-8-\frac{16}{3}+4\right)-\left(-16-8+\frac{16}{3}+4\right)\right]$$

$$= \frac{128}{3}\pi$$

33. The volume of the solid is the sum of elements of volume as we vary x. Each element has width dx and length and height $2y$, so

$dV = 4y^2dx = 4x\ dx$ (because $x = y^2$)

As we sum all the elements of volume we get

$$V = {}_0^{25}\, 4x\ dx = 2x^2\Big|_0^{25} = 1250 \text{ cm}^3$$

37. Equation of ellipse:

$$\frac{x^2}{137.5^2} + \frac{y^2}{85^2} = 1$$

$$85^2x^2 + 137.5y^2 = 85^2 \cdot 137.5^2$$

$$y^2 = 85^2 - \frac{85^2x^2}{137.5^2}$$

Disc: $dV = \pi y^2dx$

Because of symmetry, the volume is twice that from the first quadrant.

$$V = 2\left\{\int_0^{137.5} \pi\left(85^2 - \frac{85^2 x^2}{137.5^2}\right)dx\right\}$$

$$= 2\pi\left(85^2 x - \frac{85^2}{3\cdot 137.5^2}x^3\right)\Bigg|_0^{137.5}$$

$$= 2\pi\left(\frac{2}{3}\cdot 85^2\cdot 137.5\right)$$

$$= 4.2\times 10^6 \text{ mm}^3$$

26.4 Centroids

1. $y = |x|$ is symmetric with respect to y-axis $\Rightarrow \bar{x} = 0$

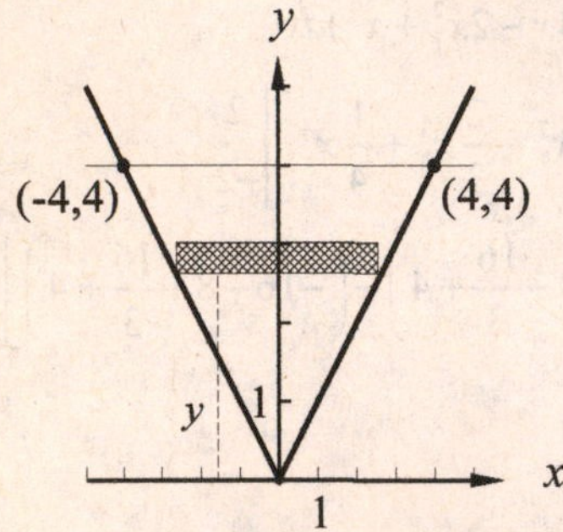

$$\bar{y} = \frac{\int_0^4 y(2x)dy}{\int_0^4 2xdy}$$

$$\bar{y} = \frac{\int_0^4 2y^2 dy}{\int_0^4 2ydy} = \frac{\left.\frac{y^3}{3}\right|_0^4}{\left.\frac{y^2}{2}\right|_0^4}$$

$$= \frac{\frac{4^3}{3}}{\frac{4^2}{2}} = \frac{8}{3}$$

$$(\bar{x}, \bar{y}) = \left(0, \frac{8}{3}\right)$$

5. $$M\overline{x} = m_1x_1 + m_2x_2 + m_3x_3 + m_4x_4$$

$$(31+24+15+84)\overline{x} = 31(-3.5)+24(0)+15(2.6)+84(3.7)$$

$$154\overline{x} = 241.3$$

$$\overline{x} = 1.6 \text{ cm}$$

42 g 24 g 15 g 84 g

-4 -3 -2 -1 0 1 2 3 4

-3.5 0 2.6 3.7

9. Break area into three ractangles.

First: center $(-1.00, -1.00)$; $A_1 = 4.00$

Second: center $(0, 0.50)$; $A_2 = 4.00$

Third: center $(2.50, 1.50)$; $A_3 = 3.00$

Taking moments with respect to the y-axis:

$4.00(-1.00) + 4.00(0) + 3.00(2.50) = (4.00 + 4.00 + 3.00)\overline{x}$

$\overline{x} = 0.32$

Taking moments with respect to the x-axis:

$4.00(-1.00) + 4.00(0.50) + 3.00(1.50) = 11.00\overline{y}$

$\overline{y} = 0.23$

$(0.32 \text{ in}, 0.23 \text{ in})$ is the centroid.

13. $y = 4 - x$, and axes

$x = 4 - y$

$$\overline{x} = \frac{\int_0^4 xy\,dy}{\int_0^4 y\,dx} = \frac{\int_0^4 x(4-x)dx}{\int_0^4 (4-x)dx} = \frac{\int_0^4 (4x-x^2)dx}{\int_0^4 (4-x)dx} = \frac{(2x^2-\frac{1}{3}x^3)\big|_0^4}{(4x-\frac{1}{2}x^2)\big|_0^4} = \frac{\frac{32}{3}}{8} = \frac{4}{3}$$

$$\overline{y} = \frac{\int_0^4 y(x)\,dy}{\int_0^4 x\,dx} = \frac{\int_0^4 y(4-y)dy}{\int_0^4 (4-y)dy} = \frac{\int_0^4 (4y-y^2)dy}{\int_0^4 y(4-y)dy} = \frac{(2y^2-\frac{1}{3}y^3)\big|_0^4}{(4y-\frac{1}{2}y^2)\big|_0^4} = \frac{4}{3}$$

The centroid is at $(\overline{x}, \overline{y}) = \left(\frac{4}{3}, \frac{4}{3}\right)$.

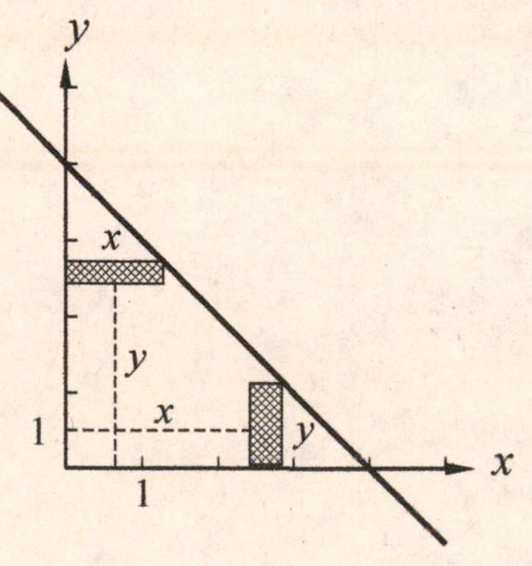

17. Right boundary is $x = \frac{1}{2}y - 1$

Left boundary is $x = \frac{y}{3} - \frac{2}{3}$

$$A = \int_2^8 \left[\left(\frac{1}{2}y - 1\right) - \left(\frac{y}{3} - \frac{2}{3}\right)\right] dy$$

$$= \int_2^8 \left(\frac{1}{6}y - \frac{1}{3}\right) dy$$

$$= \frac{1}{12}y^2 - \frac{1}{3}y \Big|_2^8$$

$$= \left(\frac{64}{12} - \frac{8}{3}\right) - \left(\frac{4}{12} - \frac{2}{3}\right)$$

$$= 3$$

The center of mass of the thin strip at a given value of y has $x-$ coordinate $\frac{5}{12}y - \frac{5}{6}$, so

$$\bar{x} = \frac{\int_2^8 \left(\frac{5}{12}y - \frac{5}{6}\right)\left(\frac{1}{6}y - \frac{1}{3}\right) dy}{3}$$

$$= \frac{5}{216}\int_2^8 \left(y^2 - 4y + 4\right) dy$$

$$= \frac{5}{216}\left(\frac{1}{3}y^3 - 2y^2 + 4y\right)\Big|_2^8$$

$$= \frac{5}{216}\left[\left(\frac{512}{3} - 128 + 32\right) - \left(\frac{8}{3} - 8 + 8\right)\right]$$

$$= \frac{5}{3}$$

$$\bar{y} = \frac{\int_2^8 y\left(\frac{1}{6}y - \frac{1}{3}\right) dy}{3}$$

$$= \frac{1}{3}\int_2^8 \left(\frac{1}{6}y^2 - \frac{1}{3}y\right) dy$$

$$= \frac{1}{3}\left(\frac{1}{18}y^3 - \frac{1}{6}y^2\right)\Big|_2^8$$

$$= \frac{1}{54}\left[(512 - 192) - (8 - 12)\right]$$

$$= 6$$

The centroid of mass is at $(\bar{x}, \bar{y}) = \left(\frac{5}{3}, 6\right)$.

21. Curves intersect when $\frac{x^2}{4p} = a$

$$x = \pm 2\sqrt{pa}$$

$$y = a$$

Region is symmetric with respect to $y-$axis, so $\bar{x} = 0$.

$$\bar{y} = \frac{\int y(2x)dy}{\int 2x\,dy}$$

$$= \frac{\int_0^a 2y(2\sqrt{py})dy}{\int_0^a 2(2\sqrt{py})dy}$$

$$\bar{y} = \frac{4\sqrt{p}\int_0^a y^{3/2}dy}{4\sqrt{p}\int_0^a y^{1/2}dy}$$

$$= \frac{\frac{2}{5}y^{5/2}\big|_0^a}{\frac{2}{3}y^{3/2}\big|_0^a}$$

$$= \frac{3}{5}\frac{a^{5/2}}{a^{3/2}} = \frac{3}{5}a$$

$$(\bar{x}, \bar{y}) = \left(0, \frac{3}{5}a\right)$$

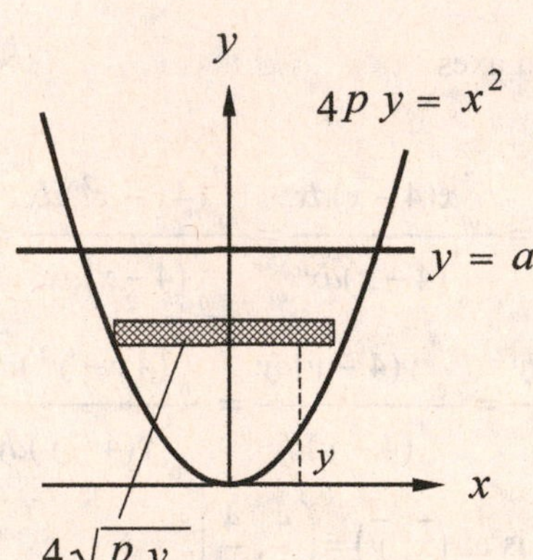

25. $y^2 = 4x,\ x = 0,\ y = 2,$

$y^2 = 4x,\ x = \dfrac{y^2}{4}$

Rotate about y-axis, $\overline{x} = 0$

$$\overline{y} = \frac{\int_0^2 y\left(\left(\frac{y^2}{4}\right)^2\right)dy}{\int_0^2\left(\left(\frac{y^2}{4}\right)^2\right)dy}$$

$$\overline{y} = \frac{\int_0^2\left(\frac{1}{16}y^5\right)dy}{\int_0^2\left(\frac{1}{16}y^4\right)dy} = \frac{\frac{1}{6}y^6\Big|_0^2}{\frac{1}{5}y^5\Big|_0^2}$$

$$= \frac{\frac{64}{6}}{\frac{32}{5}} = \frac{5}{3}$$

$$\left(\overline{x}, \overline{y}\right) = \left(0, \frac{5}{3}\right)$$

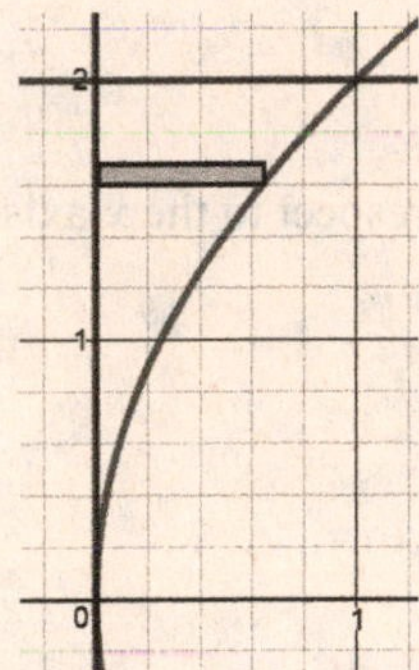

29. Triangle is area bounded by $y = \dfrac{3}{2}x$, x-axis, $x = 3$

The area of the sail is $A = \dfrac{1}{2}bh = \dfrac{1}{2}(3)\ \dfrac{9}{2}\ = \dfrac{27}{4}$

$$\overline{x} = \frac{\int_0^3 xy\,dx}{A} = \frac{\int_0^3 x\ \frac{3}{2}x\ dx}{\frac{27}{4}}$$

$$= \frac{\frac{3}{2}\int_0^3 x^2dx}{\frac{27}{4}} = \frac{\frac{x^3}{2}\Big|_0^3}{\frac{27}{4}} = \frac{\frac{27}{2}}{\frac{27}{4}} = 2$$

$x = \dfrac{2}{3}y$, so

$$\overline{y} = \frac{\int_0^{9/2} y(3-x)dy}{A} = \frac{\int_0^{9/2} y\ 3 - \frac{2}{3}y\ dy}{\frac{27}{4}}$$

$$= \frac{\int_0^{9/2} 3y - \frac{2}{3}y^2\ dy}{\frac{27}{4}} = \frac{\frac{3}{2}y^2 - \frac{2}{9}y^3\Big|_0^{9/2}}{\frac{27}{4}}$$

$$= \frac{\frac{243}{8} - \frac{81}{4}}{\frac{27}{4}} = \frac{81}{8}\cdot\frac{4}{27} = \frac{3}{2}$$

$$\left(\overline{x}, \overline{y}\right) = \left(2, \frac{3}{2}\right)$$

33. Bounded area: $y = -4x + 80,\ y = 60,\ y = 0,$ $x = 0, \overline{x} = 0$

$y = -4x + 80; 4x = 80 - y; x = 20 - \dfrac{1}{4}y$

$$\overline{y} = \frac{\int_0^{60} 400y - 10y^2 + \frac{1}{16}y^3\ dy}{\int_0^{60} 400 - 10y + \frac{1}{16}y^2\ dy}$$

$$= \frac{200y^2 - \frac{10}{3}y^3 + \frac{1}{64}y^4\Big|_0^{60}}{400y - 5y^2 + \frac{1}{48}y^3\Big|_0^{60}}$$

$$= \frac{720\,000 - 720\,000 + 202\,500}{24\,000 - 18\,000 + 4\,500}$$

$= 19.3\,\text{cm}$ from larger base

$$\left(\overline{x}, \overline{y}\right) = (0, 19.3)$$

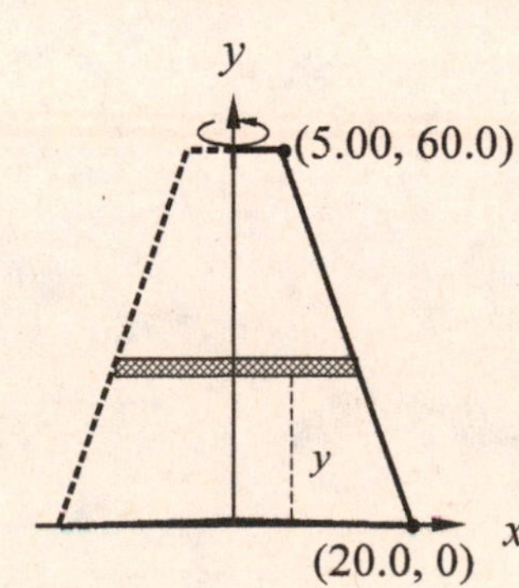

26.5 Moments of Inertia

1. $I_y = k\int_0^1 x^2 y\,dx$

$= k\int_0^1 4x^3\,dx$

$= k\,x^4\Big|_0^1 = k$

$m = k\int_0^1 4x\,dx$

$= k\left(2x^2\right)\Big|_0^1$

$= k\rho$

$R_y^2 = \frac{I_y}{m} = \frac{k}{2k} = \frac{1}{2}$

$R_y = \frac{\sqrt{2}}{2}$

5. $I = m_1x_1^2 + m_2x_2^2 + m_3x_3^2$

$I = 82.0(-3.80)^2 + 90.0(0.00)^2 + 62.0(5.50)^2$

$I = 3060\text{g}\cdot\text{cm}^2$

$I = MR^2$

$3060 = (82.0 + 90.0 + 62.0)R^2$

$R = 3.62\text{cm}$

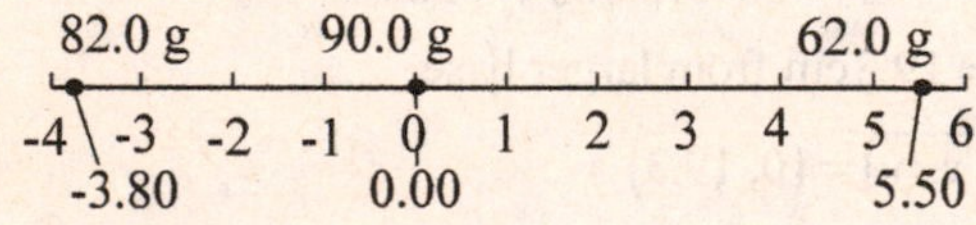

9. $y^2 = x,\ x = 9$, x-axis, with respect to the x-axis

Intersection: $(9,3)$

$I_x = k\int_0^3 y^2\left(9 - y^2\right)dy$

$= k\int_0^3 \left(9y^2 - y^4\right)dy$

$= k\left(3y^3 - \frac{1}{5}y^5\right)\Big|_0^3$

$= k\left(81 - \frac{243}{5}\right) = \frac{162}{5}k$

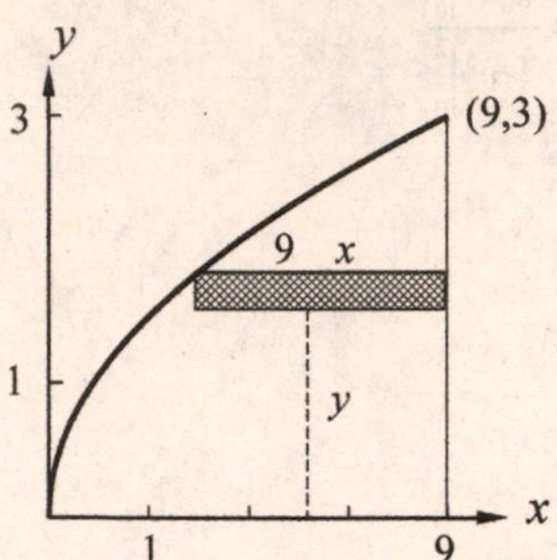

13. $y = x^2,\ x = 3$, x-axis, with respect to the x-axis

$x = \sqrt{y}$

Intersection: $(3,9)$

$I_x = k\int_0^9 y^2\left(3 - \sqrt{y}\right)dy$

$= k\int_0^9 \left(3y^2 - y^{5/2}\right)$

$= k\left(y^3 - \frac{2}{7}y^{7/2}\right)\Big|_0^9$

$= \frac{729}{7}k$

$m = k\int_0^9 \left(3 - \sqrt{y}\right)dy$

$= k\left(3y - \frac{2}{3}y^{3/2}\right)\Big|_0^9$

$= 9k$

$R_x^2 = \frac{\frac{729}{7}k}{9k}$

$R_x = \frac{9}{\sqrt{7}}$

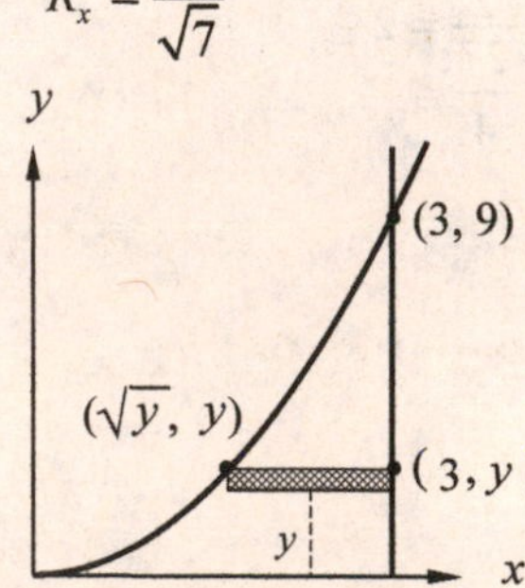

17. $y^2 = 4x,\ y = 2$, y-axis, rotated about x-axis

$$\frac{y^2}{4} = x$$

$$I_x = 2\pi k \int_0^2 xy^3\,dy$$

$$= 2\pi k \int_0^2 \frac{y^2}{4} y^3\,dy$$

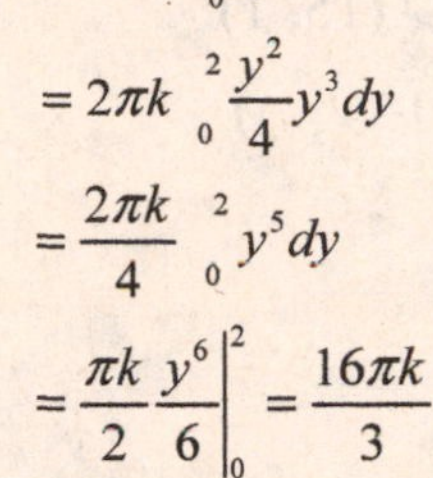

$$= \frac{2\pi k}{4} \int_0^2 y^5\,dy$$

$$= \frac{\pi k}{2} \frac{y^6}{6}\Big|_0^2 = \frac{16\pi k}{3}$$

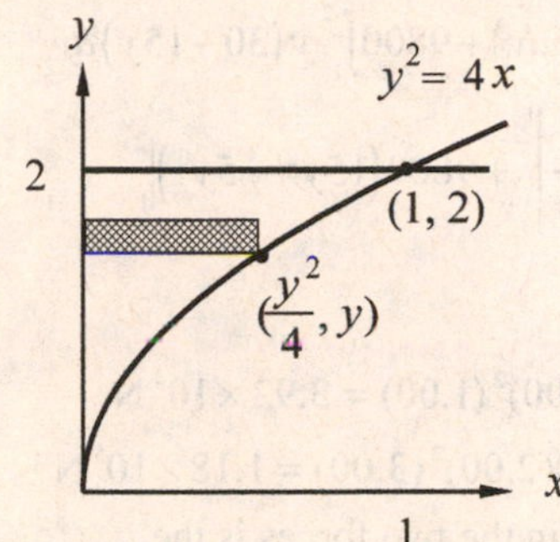

21. $y = \frac{3}{2}x;\ x = 3,\ y = 0$

$$I_y = k \int_0^3 x^2 \quad \frac{9}{2} - y \quad dx$$

$$= k \int_0^3 x^2 \quad \frac{9}{2} - \frac{3}{2}x \quad dx$$

$$= \frac{3k}{2} \int_0^3 \left(3x^2 - x^3\right) dx$$

$$= \frac{3k}{2} \quad x^3 - \frac{1}{4}x^4 \quad \Big|_0^3$$

$$= \frac{3k}{2} \quad 27 - \frac{81}{4}$$

$$= \frac{81}{8}k$$

25. $r = 0.600$ cm, $h = 0.800$ cm, $m = 3.00$ g

$$y = \frac{0.600}{0.800}x = 0.750x;\ x = 1.333y$$

$$I_x = 2\pi k \int_0^{0.600} (0.800 - 1.333y))y^3\,dy$$

$$= 2\pi k\,(0.200y^4 - 0.2667y^5)\,|_0^{0.600}$$

$$= 2\pi k\,(0.005\,181)$$

$$m = \frac{k}{3}\pi r^2 h,\ 2\pi k = \frac{6m}{r^2 h} = \frac{6(3.00)}{(0.600^2)(0.800)}$$

$$= 62.5 \text{ g/cm}^3$$

$$I_x = (62.5)(0.005\,181) = 0.324 \text{ g} \cdot \text{cm}^2$$

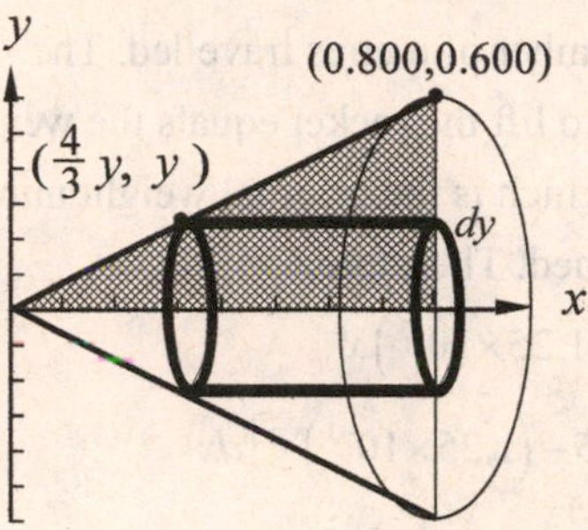

26.6 Other Applications

1. $F = 62.4 \int_2^6 5.00h\,dh$

$$= 312 \int_2^6 h\,dh$$

$$= 156h^2\Big|_2^6$$

$$= 156(36 - 4)$$

$$= 4990 \text{ lb}$$

5. $F = kx$

$$160 = k(0.080)$$

$$k = 2000 \text{ lb/in}$$

$$180 = 2000(x)$$

$$x = 0.09 \text{ cm}$$

$$W = \int_0^{0.09} 2000x\,dx$$

$$= 1000x^2\Big|_0^{0.09}$$

$$= 8.1 \text{ lb} \cdot \text{in}$$

9. $F = \dfrac{k}{x^2}$

$$W = k\int_{10}^{100} \frac{dx}{x^2}$$
$$= k\int_{10}^{100} x^{-2} dx$$
$$= -kx^{-1}\Big|_{10}^{100}$$
$$= k\frac{(-1)}{x}\Big|_{10}^{100}$$
$$= -k\left(\frac{1}{100} - \frac{1}{10}\right) = \frac{-k(1-10)}{100}$$
$$= \frac{9k}{100} = 0.09k \text{ ft}\cdot\text{lb}$$

13. Let x be the number of metres travelled. The work required to lift the rocket equals the weight of the rocket, which is the original weight minus the fuel consumed. Therefore,

$$f(x) = 32.5 - (1.25\times10^{-3})x$$
$$W = \int_0^{12\,000}\left[32.5 - (1.25\times10^{-3})x\right]dx$$
$$= 32.5x - 6.125\times10^{-4}x^2\Big|_0^{12\,000}$$
$$= 390000 - 88200$$
$$= 300000 \text{ ft}\cdot\text{ton}$$

17. $W = \int_{2.50}^{5.00} 62.4(7.00 - y)640\, dy$

$$= 279552y - 19968y^2\Big|_{2.50}^{5.00}$$
$$W = 324480 \text{ ft}\cdot\text{lb}$$

Hence, 324000 foot-pounds of work are done.

21. $F = k\int_a^b lh\, dh$

$$= 62.4\int_0^{2.5}(12.0x)dx$$
$$= 62.4(6.0x^2)\Big|_0^{2.5}$$
$$= 62.4(37.5 - 0) = 2340 \text{ lb}$$

25.

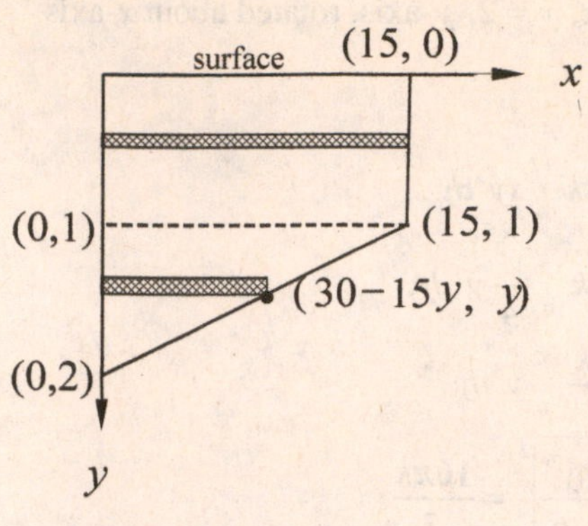

$$\gamma = 9800$$
$$F = 9800\int_0^1 y(15dy) + 9800\int_1^2 y(30-15y)dy$$
$$F = 9800(15)\frac{y^2}{2}\Big|_0^1 + 9800(15y^2 - 5y^3)\Big|_1^2$$
$$F = 172\,000 \text{ N}$$

29. $F_{\text{TOP}} = 9800(2.00)^2(1.00) = 3.92\times10^4 \text{ N}$

$F_{\text{BOTTOM}} = 9800(2.00)^2(3.00) = 1.18\times10^5 \text{ N}$

The difference in the two forces is the buoyant force.

33. $e = 0.768s - 0.000\,04s^3$, $s_1 = 30.0$ km/h, $s_2 = 90.0$ km/h

$$e_{av} = \int_{30.0}^{90.0}\frac{(0.768s - 0.000\,04s^3)}{60.0}ds$$
$$e = \frac{1}{60.0}\left(\frac{0.768s^2}{2} - 0.000\,01s^4\right)\Big|_{30.0}^{90.0}$$
$$= \frac{1}{60.0}\left\{\frac{0.768}{2}(90.0)^2 - 0.000\,01(90.0)^4 - \left(\frac{0.768}{2}(30.0)^2 - 0.000\,01(30.0)^4\right)\right\}$$
$$= \frac{1}{60.0}(2.454\times10^3 - 3.375\times10^2)$$
$$= 35.3\%$$

37. $y = \frac{r}{h}x$

$$\frac{dy}{dx} = \frac{r}{h}; \left(\frac{dy}{dx}\right)^2 = \frac{r^2}{h^2}$$

$$S = 2\pi \int_0^h \frac{r}{h} x \sqrt{1 + \frac{r^2}{h^2}} dx$$

$$= 2\pi \frac{r}{h} \frac{1}{h} \sqrt{h^2 + r^2} \int_0^h x\, dx$$

$$= \frac{2\pi r}{h^2} \sqrt{h^2 + r^2} \frac{x^2}{2} \Big|_0^h$$

$$= \frac{\pi r}{h^2} \sqrt{h^2 + r^2} x^2 \Big|_0^h$$

$$= \frac{\pi r}{h^2} \sqrt{h^2 + r^2} h^2$$

$$= \pi r \sqrt{h^2 + r^2}$$

Review Exercises

1. This is true.

Let v_0 be the initial velocity. Then $v(t) = v_0 + 8t$ and

$$s(t) = \int (v_0 + 8t) dt$$

$$= v_0 t + 4t^2 + C$$

The distance traveled in the second before a given time t is

$s(t) - s(t-1) = v_0 + 8t - 4$.

The distance traveled in the second after t is

$s(t+1) - s(t) = v_0 + 8t + 4$.

The difference between these two distances traveled is 8.

5. This is true.

The area element has area 4 dh and if it is h units deep, the force is $wh \cdot 4\, dh$; the given integral results.

9. $\frac{d^2s}{dt^2} = -32$

$$\frac{ds}{dt} = -32t + v_0$$

$$s = -16t^2 + v_0 t + s_0$$

at $t = 0$, $s_0 = 200$, $v_0 = 20$

$$s(t) = -16t^2 + 20t + 200$$

By the quadratic formula, $s(t) = 0$ when $t = 4.2$ seconds

13. $a_{\text{vert}} = -32$

$v_{\text{vert}} = -32t + 64$

$s_{\text{vert}} = -16t^2 + 64t$

We want to know at what times s_{vert} is between -3 and 3.
From the quadratic formula, $s_{\text{vert}} = -3$ when $t = 3.953$ and
$s_{\text{vert}} = 3$ when $t = 4.046$.
The corresponding horizontal positions of the ball are
$3.953(45) = 177.9$ and $4.046(45) = 182.07$. The ball will
reach the back of the endzone within 3.00 feet of the throwing
height.

17. $V_c = \dfrac{1}{C}\int i\,dt$

$V_c = \dfrac{i}{C}\int dt = \dfrac{it}{C} + V_0 = \dfrac{it}{C} + 0$

$V_c = \dfrac{12\times10^{-3}\left(25\times10^{-6}\right)}{5.5\times10^{-9}}$

$= 55 \text{ V}$

21. Intercept: $(5,0)$

$A = \int_0^5 \left((5-x)^{1/2} - (x-5)\right)dx$

$A = \left[-\dfrac{2}{3}(5-x)^{3/2} - \dfrac{1}{2}(x-5)^2\right]\Bigg|_0^5$

$= 5.046$

25. $A = \int_0^3 \left((x^2+1) - (x^3 - 2x^2 + 1)\right)dx$

$A = \int_0^3 \left(3x^2 - x^3\right)dx$

$A = x^3 - \dfrac{x^4}{4}\Bigg|_0^3$

$A = \dfrac{27}{4}$

29.

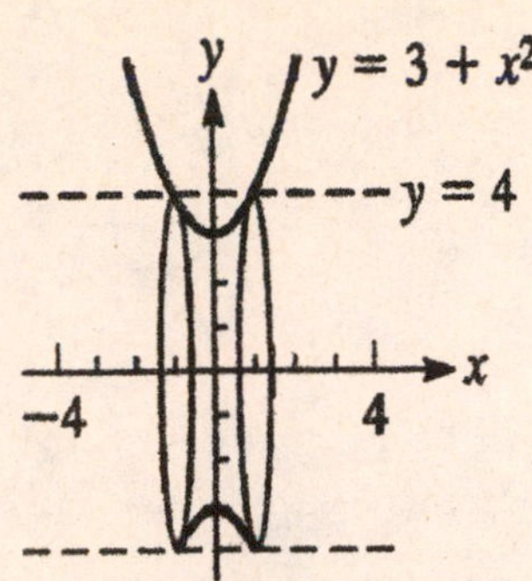

$$V = \pi\int_{-1}^{1} 4^2\,dx - \pi\int_{-1}^{1} y^2\,dx = \pi\int_{-1}^{1}\left(4^2 - y^2\right)dx$$
$$= \pi\int_{-1}^{1}\left[16-\left(3+x^2\right)^2\right]dx$$
$$= \pi\int_{-1}^{1}\left(16-9-6x^2-x^4\right)dx$$
$$= \pi\int_{-1}^{1}\left(7-6x^2-x^4\right)dx$$
$$= \pi\left(7x-2x^3-\frac{1}{5}x^5\right)\Big|_{-1}^{1}$$
$$= \pi\left(7-2-\frac{1}{5}\right)-\pi\left(-7+2+\frac{1}{5}\right) = \frac{24\pi}{5}+\frac{24\pi}{5}$$
$$= \frac{48\pi}{5}$$

33.

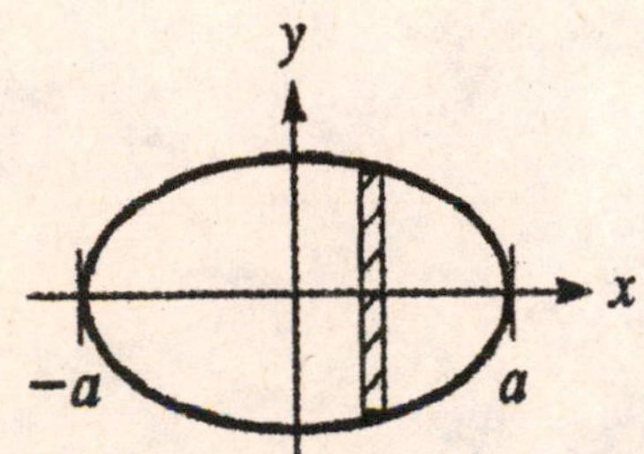

$$V = \pi\int_{-a}^{a} y^2\,dx = \pi\int_{-a}^{a}\left(\frac{a^2b^2-b^2x^2}{a^2}\right)dx = \pi\int_{-a}^{a}\left(b^2-\frac{b^2}{a^2}x^2\right)dx$$
$$= \pi\left(b^2x-\frac{b^2}{3a^2}x^3\right)\Big|_{-a}^{a}$$
$$= \pi\left[\left(ab^2-\frac{ab^2}{3}\right)-\left(-ab^2+\frac{ab^2}{3}\right)\right] = \pi\left(2ab^2-\frac{2ab^2}{3}\right)$$
$$= \pi\left(\frac{4ab^2}{3}\right) = \frac{4\pi ab^2}{3} = \frac{4}{3}\pi ab^2$$

37. $y^2 = x^3$ and $y = 3x$ intersect at (0, 0) and (9, 27).

$$y = x^{3/2}, x = y^{2/3}, x = \frac{y}{3}$$

$$\bar{x} = \frac{\int_0^9 x\left(3x - x^{3/2}\right)dx}{\int_0^9 \left(3x - x^{3/2}\right)dx}$$

$$= \frac{\int_0^9 \left(3x^2 - x^{5/2}\right)dx}{\int_0^9 \left(3x - x^{3/2}\right)dx}$$

$$\bar{x} = \frac{x^3 - \frac{2}{7}x^{7/2}\Big|_0^9}{\frac{3}{2}x^2 - \frac{2}{5}x^{5/2}\Big|_0^9} = \frac{30}{7}$$

$$\bar{y} = \frac{\int_0^{27} y\left(y^{2/3} - \frac{1}{3}y\right)dy}{\frac{243}{10}} = \frac{\int_0^{27}\left(y^{5/3} - \frac{y^2}{3}\right)dy}{\frac{243}{10}}$$

$$\bar{y} = \frac{\frac{3}{8}y^{8/3} - \frac{1}{9}y^3\Big|_0^{27}}{\frac{243}{10}} = \frac{45}{4}$$

$$\left(\bar{x}, \bar{y}\right) = \frac{30}{7}, \frac{45}{4}$$

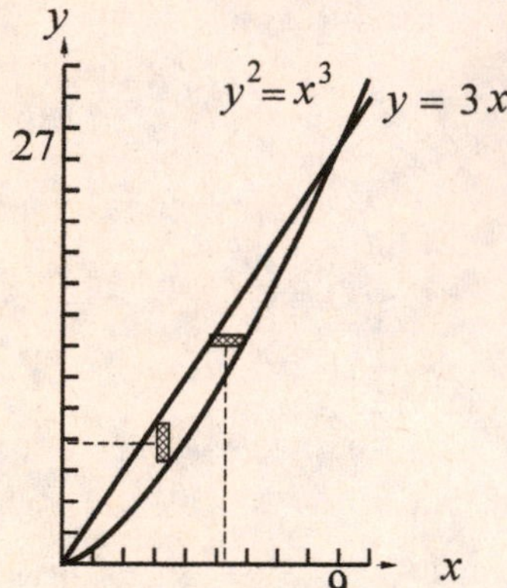

41. $I_y = k\int_0^2 x^2\left(\left(3x - x^2\right) - x\right)dx$

$$I_y = k\int_0^2 \left(2x^3 - x^4\right)dx$$

$$I_y = k\ \frac{x^4}{2} - \frac{x^5}{5}\ \Big|_0^2$$

$$I_y = \frac{8k}{5}$$

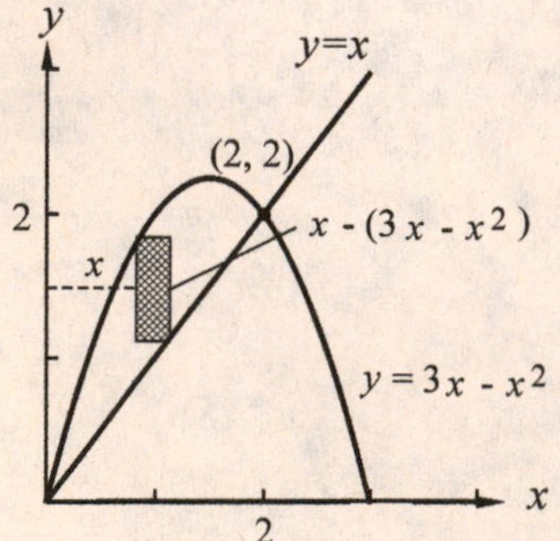

45. rope has a weight per unit length $= \frac{10}{100}$ lb/ft $= 0.1$ lb/ft

$$W = W_{\text{bucket}} + W_{\text{rope}}$$

$$W = 80(100) + \int_0^{100} 0.10(100 - x)\,dx$$

$$= 8000 + \left[10x - 0.05x^2\right]_0^{100}$$

$$= 8500 \text{ ft} \cdot \text{lb}$$

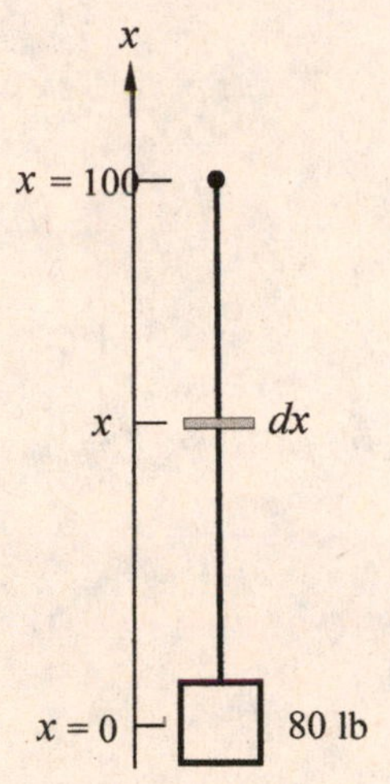

49. $a = -18$

$$v(t) = 28 - \int 18\,dt$$

$$= 28 - 18t$$

$$v(5) = 28 - 18t\big|_{t=5} = -62 \text{ ft/s}$$

After 5.0 s the velocity of the rock is 62 ft/s down the slope.

53.

$$\overline{x} = \frac{\int_0^3 x\left(3x^2 - x^3\right)dx}{\int_0^3 \left(3x^2 - x^3\right)dx}$$

$$= \frac{\int_0^3 \left(3x^3 - x^4\right)dx}{\int_0^3 \left(3x^2 - x^3\right)dx}$$

$$= \frac{\left.\frac{3}{4}x^4 - \frac{1}{5}x^5\right|_0^3}{\left.x^3 - \frac{1}{4}x^4\right|_0^3}$$

$$= \frac{243/20}{27/4}$$

$$= \frac{9}{5}$$

57. The parabola has equation

$x^2 = 4py$

Substituting $(2,4)$:

$4 = 4p(4)$

$4p = 1$

$x^2 = y$

$x = \sqrt{y}$

$V = \frac{1}{2}(4)(2)(6) + 6\int_0^4 \left(2\sqrt{y}\right)dy$

$V = 24 + 12\left.\frac{2}{3}y^{3/2}\right|_0^4$

$V = 88 \text{ ft}^3$

61. $T_{\text{ave}} = \frac{1}{24}\int_0^{24}\left(5.00\times10^{-4}t^4 - 0.0240t^3 + 0.300t^2 + 10.0\right)dt$

$T_{\text{ave}} = \frac{1}{24}\left. 1.00\times10^{-4}t^5 - 0.0060t^4 + 0.100t^3 + 10.0t \right|_0^{24}$

$= \frac{1}{24}(796.2624 - 1990.656 + 1382.4 + 240)$

$= 17.8° \text{ C}$

65. We place a coordinate system at the centre of the top of the float, with the y-axis perpendicular to the fixed diameter. Therefore, slices parallel to the y-axis have cross sections that are squares. We take those as our elements of volume.
An element of volume at x of thickness dx has area $(2y)^2 = 4\left(a^2 - x^2\right)$, so its volume is $4\left(a^2 - x^2\right)dx$.
The volume is the sum of all the elements of volume from $x = -a$ to $x = a$. Because of symmetry, we need only compute the integral from 0 to a.

$V = 2\int_0^a 4\left(a^2 - x^2\right)dx$

$= 8\left.\left(a^2x - \frac{x^3}{3}\right)\right|_0^a$

$= 8\left(\frac{2}{3}a^3\right) = \frac{16}{3}a^3$

Chapter 27

Differentiation of Transcendental Functions

27.1 Derivatives of the Sine and Cosine Functions

1. $r = \sin^2 2\theta^2$

$$\frac{dr}{d\theta} = 2\sin 2\theta^2 (2(2\theta))\cos 2\theta^2$$

$$\frac{dr}{d\theta} = 8\theta \sin 2\theta^2 \cos 2\theta^2$$

$$\frac{dr}{d\theta} = 4\theta \sin 4\theta^2$$

5. $y = 2\sin(2x^3 - 1)$

$$\frac{dy}{dx} = 2\cos(2x^3 - 1)(6x^2)$$

$$= 12x^2 \cos(2x^3 - 1)$$

9. $y = 3x + 2\cos(3x - \pi)$

$$\frac{dy}{dx} = 3 + 2[-\sin(3x - \pi)(3)]$$

$$= 3 - 6\sin(3x - \pi)$$

13. $y = 3\cos^3(5x + 2)$

$$\frac{dy}{dx} = 3(3)\cos^2(5x+2)[-\sin(5x+2)(5)]$$

$$\frac{dy}{dx} = -45\cos^2(5x+2)\sin(5x+2)$$

17. $y = 3x^3 \cos 5x$

$$\frac{dy}{dx} = 3[x^3(-5\sin 5x) + \cos 5x(3x^2)]$$

$$\frac{dy}{dx} = 9x^2 \cos 5x - 15x^3 \sin 5x$$

21. $y = \sqrt{1 + \sin 4x} = (1 + \sin 4x)^{1/2}$

$$\frac{dy}{dx} = \frac{1}{2}(1 + \sin 4x)^{-1/2}(4\cos 4x)$$

$$\frac{dy}{dx} = \frac{2\cos 4x}{\sqrt{1 + \sin 4x}}$$

25 $y = \dfrac{2\cos x^2}{3x - 1}$

$$\frac{dy}{dx} = \frac{(3x-1)(2)(-\sin x^2)(2x) - 2\cos x^2(3)}{(3x-1)^2}$$

$$\frac{dy}{dx} = \frac{-4x(3x-1)\sin x^2 - 6\cos x^2}{(3x-1)^2}$$

$$\frac{dy}{dx} = \frac{4x(1-3x)\sin x^2 - 6\cos x^2}{(3x-1)^2}$$

29. $s = \sin(3\sin 2t)$

$$\frac{ds}{dt} = \cos(3\sin 2t)\big(3\cos(2t)\big)(2)$$

$$\frac{ds}{dt} = 6\cos 2t \cos(3\sin 2t)$$

33. $p = \dfrac{1}{\sin s} + \dfrac{1}{\cos s} = (\sin s)^{-1} + (\cos s)^{-1}$

$$\frac{dp}{ds} = \frac{-\cos s}{\sin^2 s} + \frac{\sin s}{\cos^2 s}$$

37. 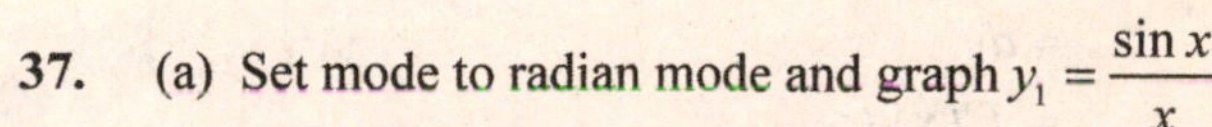
(a) Set mode to radian mode and graph $y_1 = \dfrac{\sin x}{x}$.

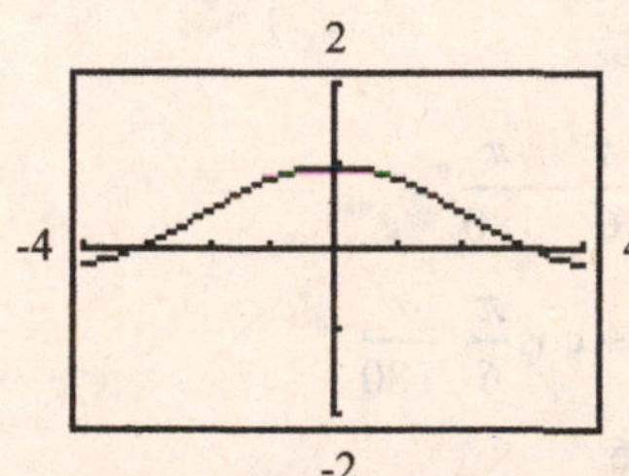

(b) Check the values, using a calculator. When $x = 0.5, y = 0.95885108$; when $x = 0.1$, $y = 0.99833417$; when $x = 0.05$, $y = 0.99958339$; when $x = 0.01$, $y = 0.99998333$; when $x = 0.001$, $y = 0.99999983$.

41. $y = \sin x$

$m_{\tan}\Big|_{x=0} = 1.0 \qquad m_{\tan}\Big|_{x=\pi/4} = 0.7 \qquad m_{\tan}\Big|_{x=\pi/2} = 0.0$

$m_{\tan}\Big|_{x=3\pi/4} = -0.7 \qquad m_{\tan}\Big|_{x=\pi} = -1.0 \qquad m_{\tan}\Big|_{x=5\pi/4} = -0.7$

$m_{\tan}\Big|_{x=3\pi/2} = 0.0 \qquad m_{\tan}\Big|_{x=7\pi/4} = 0.7 \qquad m_{\tan}\Big|_{x=2\pi} = 1.0$

Plot points: $(0, 1.0), (\frac{\pi}{4}, 0.7), (\frac{\pi}{2}, 0.0), (\frac{3\pi}{4}, -0.7),$
$(\pi, -1.0), (\frac{5\pi}{4}, -0.7), (\frac{3\pi}{2}, 0.0), (\frac{7\pi}{4}, 0.7), (2\pi, 1.0).$

Resulting curve is $y = \cos x$.

45. $\frac{d}{dx}\sin x = \cos x$

$\frac{d^2}{dx^2}\sin x = -\sin x$

$\frac{d^3}{dx^3}\sin x = -\cos x$

$\frac{d^4}{dx^4}\sin x = \sin x$

49. $x = 30° = \frac{\pi}{6}$

$dx = 1° = \frac{\pi}{180}$

$y = f(x) = \sin x$

$$\begin{aligned}\Delta y &= f(x+\Delta x) - f(x)\\ &\approx dy\\ &= f'(x)dx\\ &= \cos x dx\end{aligned}$$

$$\begin{aligned}\sin 31° &= \sin\left(\frac{\pi}{6} + \frac{\pi}{180}\right)\\ &\approx \sin\frac{\pi}{6} + \cos\frac{\pi}{6}\left(\frac{\pi}{180}\right)\end{aligned}$$

$\sin 31° \approx 0.515$

53. $V = 3.0\sin 188t\cos 188t$

$\frac{dV}{dt} = 3.0[\sin 188t(-\sin 188t)(188) + \cos 188t(\cos 188t)(188)]$

$\frac{dV}{dt} = 3.0(188)[-\sin^2 188t + \cos^2 188t]$

$\frac{dV}{dt}\Big|_{t=2\times10^{-3}} = 3.0(188)[-\sin^2 188(2\times10^{-3}) + \cos^2 188(2\times10^{-3})]$

$\frac{dV}{dt} = 412 \text{ V/s}$

57. $r = \dfrac{100}{1-\cos\theta} = 100(1-\cos\theta)^{-1}$

$\dfrac{dr}{d\theta} = -100(1-\cos\theta)^{-2}(\sin\theta)$

$\dfrac{dr}{d\theta} = \dfrac{-100\sin\theta}{(1-\cos\theta)^2}$

$\theta = 120° = \dfrac{2\pi}{3}$

$\left.\dfrac{dr}{d\theta}\right|_{\theta=120°} = \dfrac{-100\sin\dfrac{2\pi}{3}}{\left(1-\cos\dfrac{2\pi}{3}\right)^2}$

$= -38.5$ km

27.2 Derivatives of the Other Trigonometric Functions

1. $y = 3\sec^2 x^2$

$\dfrac{dy}{dx} = 3(2)(\sec x^2)\dfrac{d}{dx}(\sec x^2)$

$\dfrac{dy}{dx} = 6\sec x^2 \sec x^2 \tan x^2 (2x)$

$\dfrac{dy}{dx} = 12x\sec^2 x^2 \tan x^2$

5. $y = 5\cot(2\pi - 3\theta)$

$\dfrac{dy}{d\theta} = -5\csc^2(2\pi - 3\theta)\cdot(-3)$

$= 15\csc^2(2\pi - 3\theta)$

9. $y = 4x^3 - 3\csc\sqrt{2x+3}$

$\dfrac{dy}{dx} = 12x^2 - 3[-\csc\sqrt{2x+3}\cot\sqrt{2x+3}\cdot\dfrac{1}{2}(2x+3)^{-1/2}(2)]$

$\dfrac{dy}{dx} = 12x^2 + \dfrac{3\csc\sqrt{2x+3}\cot\sqrt{2x+3}}{\sqrt{2x+3}}$

13. $y = 2\cot^4\left(\dfrac{1}{2}x+\pi\right)$

$\dfrac{dy}{dx} = 2(4)\cot^3\left(\dfrac{1}{2}x+\pi\right)\left[-\csc^2\left(\dfrac{1}{2}x+\pi\right)\left(\dfrac{1}{2}\right)\right]$

$= -4\cot^3\left(\dfrac{1}{2}x+\pi\right)\csc^2\left(\dfrac{1}{2}x+\pi\right)$

17. $y = 3\csc^4(7x - \pi/2)$

$$\frac{dy}{dx} = 3 \cdot 4\csc^3(7x - \pi/2)[-\csc(7x - \pi/2)\cot(7x - \pi/2)(7)]$$

$$= -84\csc^4(7x - \pi/2)\cot(7x - \pi/2)$$

21. $y = 4\cos x \csc x^2$

$$\frac{dy}{dx} = 4[\cos x(-\csc x^2 \cot x^2 \cdot 2x) + \csc x^2(-\sin x)]$$

$$\frac{dy}{dx} = -4\csc x^2(2x\cos x \cot x^2 + \sin x)$$

25. $y = \dfrac{2\cos 4x}{1 + \cot 3x}$

$$\frac{dy}{dx} = \frac{(1 + \cot 3x)[-2\sin 4x(4)] - 2\cos 4x(-\csc^2 3x)(3)}{(1 + \cot 3x)^2}$$

$$\frac{dy}{dx} = \frac{-8\sin 4x(1 + \cot 3x) + 6\cos 4x \csc^2 3x}{(1 + \cot 3x)^2}$$

$$\frac{dy}{dx} = \frac{2(-4\sin 4x - 4\sin 4x \cot 3x + 3\cos 4x \csc^2 3x)}{(1 + \cot 3x)^2}$$

29. $r = \tan(\sin 2\pi\theta)$

$$\frac{dr}{d\theta} = \sec^2(\sin 2\pi\theta)\cos(2\pi\theta)(2\pi)$$

$$= 2\pi \cos 2\pi\theta \sec^2(\sin 2\pi\theta)$$

33. $x\sec y - 2y = \sin 2x$

$$x\sec y \tan y\frac{dy}{dx} + \sec y - 2\frac{dy}{dx} = 2\cos 2x$$

$$x\sec y \tan y\frac{dy}{dx} - 2\frac{dy}{dx} = 2\cos 2x - \sec y$$

$$\frac{dy}{dx}(x\sec y \tan y - 2) = 2\cos 2x - \sec y$$

$$\frac{dy}{dx} = \frac{2\cos 2x - \sec y}{x\sec y \tan y - 2}$$

37. $y = \tan 4x \sec 4x$

$$\frac{dy}{dx} = \tan 4x \cdot 4\sec 4x \tan 4x + \sec 4x \cdot 4\sec^2 4x$$

$$dy = 4\sec 4x(\tan^2 4x + \sec^2 4x)dx$$

41. (a)

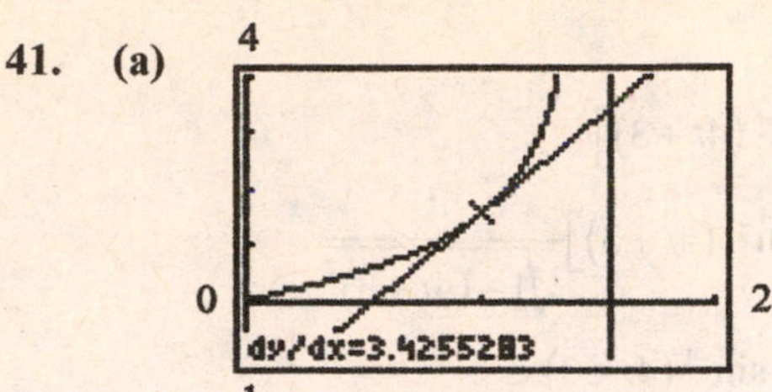

(b)

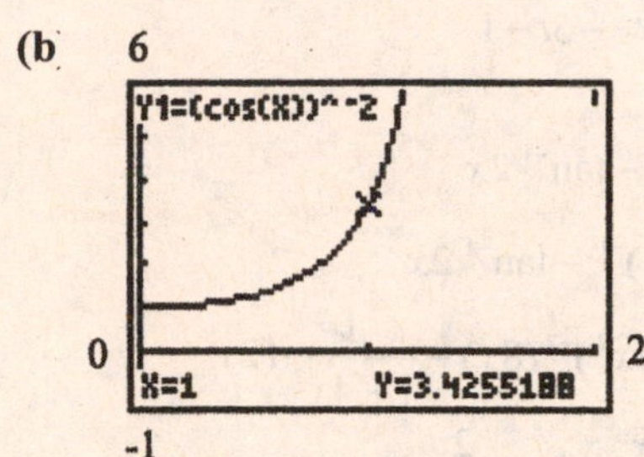

The values are the same to four decimal places.

45. $y = 2\cot 3x; x = \frac{\pi}{12};$

$$\frac{dy}{dx} = 2(-\csc^2 3x)(3)$$

$$= -6\csc^2 3x;$$

$$\left.\frac{dy}{dx}\right|_{x=\pi/12} = -6\csc^2 \frac{\pi}{4}$$

$$= -6(\sqrt{2})^2$$

$$= -12$$

49. $x = d\left(\sec\ (kL) - 1\right)$

$$x' = kd\left(\sec(kL)\tan(kL)\right)$$

53. $\theta = \frac{3t}{(2t+10)}$

$$h = 1000\tan\frac{3t}{2t+10};\ t = 5.0\ s$$

$$\frac{dh}{dt} = 1000\left[\sec^2\left(\frac{3t}{2t+10}\right)\right]\left(\frac{(2t+10)3 - 3t(2)}{(2t+10)^2}\right)$$

$$\left.\frac{dh}{dt}\right|_{t=5.0} = \left(\sec^2\frac{15}{20}\right)\left(\frac{30\,000}{400}\right)$$

$$= 140 \text{ ft/s}$$

27.3 Derivatives of the Inverse Trigonometric Functions

1. $y = \sin^{-1} x^2$

$$\frac{dy}{dx} = \frac{1}{\sqrt{1-(x^2)^2}}(2x) = \frac{2x}{\sqrt{1-x^4}}$$

5. $y = 3\sin^{-1} x^2$

$$\frac{dy}{dx} = 3\frac{1}{\sqrt{1-x^4}}(2x)$$
$$= \frac{6x}{\sqrt{1-x^4}}$$

9. $y = 6\cos^{-1}\sqrt{2-x}$

$$\frac{dy}{dx} = 6\cdot\frac{-1}{\sqrt{1-(2-x)}}\cdot\frac{1}{2}(2-x)^{-1/2}(-1)$$
$$\frac{dy}{dx} = \frac{3}{\sqrt{x-1}\sqrt{2-x}}$$
$$= \frac{3}{\sqrt{(x-1)(2-x)}}$$

13. $y = 6x\tan^{-1}\left(\frac{1}{x}\right)$

$$\frac{dy}{dx} = 6\tan^{-1}\left(\frac{1}{x}\right) + 6\frac{1}{1+\frac{1}{x^2}}\left(-\frac{1}{x^2}\right)$$
$$= 6\tan^{-1}\left(\frac{1}{x}\right) - \frac{6}{x^2+1}$$

17. $v = 0.4u\tan^{-1} 2u$

$$\frac{dv}{du} = 0.4u\cdot\frac{2}{1+4u^2} + 0.4\tan^{-1} 2u$$
$$\frac{dv}{du} = \frac{0.8u}{1+4u^2} + 0.4\tan^{-1} 2u$$

21. $y = x^{-3}\cos^{-1} x$

$$\frac{dy}{dx} = (-3)x^{-4}\cos^{-1} x - \frac{x^{-3}}{\sqrt{1-x^2}}$$
$$\frac{dy}{dx} = -\frac{3\cos^{-1} x}{x^4} - \frac{1}{x^3\sqrt{1-x^2}}$$

25. $u = \left[\sin^{-1}(4t+3)\right]^2$

$$\frac{du}{dt} = 2\left[\sin^{-1}(4t+3)\right]\frac{4}{\sqrt{1-(4t+3)^2}}$$
$$= \frac{2\sqrt{2}\sin^{-1}(4t+3)}{\sqrt{-2t^2-3t-1}}$$

29. $y = \frac{1}{1+4x^2} - \tan^{-1} 2x$

$$= (1+4x^2)^{-1} - \tan^{-1} 2x$$
$$\frac{dy}{dx} = -(1+4x^2)^{-2}(8x) - \frac{1}{1+4x^2}(2)$$
$$= \frac{-8x}{(1+4x^2)^2} - \frac{2}{1+4x^2}$$
$$= \frac{-8x - 2(1+4x^2)}{(1+4x^2)^2}$$
$$= \frac{-8x - 2 - 8x^2}{(1+4x^2)^2}$$
$$= \frac{-2(1+4x+4x^2)}{(1+4x^2)^2}$$
$$\frac{dy}{dx} = \frac{-2(1+2x)^2}{(1+4x^2)^2}$$

33. $2\tan^{-1} y + x^2 = 3x$

$$2\frac{1}{1+y^2}\frac{dy}{dx} + 2x = 3$$
$$\frac{dy}{dx} = \frac{(3-2x)(1+y^2)}{2}$$

37. $y = (\sin^{-1} x)^3$

$$\frac{dy}{dx} = 3(\sin^{-1} x)^2\frac{1}{\sqrt{1-x^2}}$$
$$dy = \frac{3(\sin^{-1} x)^2\,dx}{\sqrt{1-x^2}}$$

41. $y = x\tan^{-1}x$

$$\frac{dy}{dx} = \frac{x}{1+x^2} + \tan^{-1}x$$

$$\frac{d^2y}{dx^2} = \frac{1+x^2-x(2x)}{\left(1+x^2\right)^2} + \frac{1}{1+x^2}$$

$$\frac{d^2y}{dx^2} = \frac{1-x^2}{\left(1+x^2\right)^2} + \frac{1}{1+x^2} \times \frac{1+x^2}{1+x^2}$$

$$\frac{d^2y}{dx^2} = \frac{1-x^2+1+x^2}{\left(1+x^2\right)^2}$$

$$= \frac{2}{\left(1+x^2\right)^2}$$

45. $y = \tan^{-1}2x$

$$\frac{dy}{dx} = \frac{1}{1+(2x)^2}(2)$$

$$= \frac{2}{1+4x^2}$$

$$= 2\left(1+4x^2\right)^{-1}$$

$$\frac{d^2y}{dx^2} = 2(-1)\left(1+4x^2\right)^{-2}(8x)$$

$$= \frac{-16x}{\left(1+4x^2\right)^2}$$

49. $t = \frac{1}{\omega}\sin^{-1}\frac{A-E}{mE}$

$$= \frac{1}{\omega}\sin^{-1}\left(\frac{A-E}{E}\right)\left(\frac{1}{m}\right)$$

$$= \frac{1}{\omega}\sin^{-1}\left(\frac{A-E}{E}\right)m^{-1}$$

$$u = \left(\frac{A-E}{E}\right)m^{-1}; \frac{du}{dm} = -\left(\frac{A-E}{E}\right)m^{-2}$$

$$\frac{dt}{dm} = \frac{1}{\omega\sqrt{1-\left(\frac{A-E}{E}\right)^2 m^{-2}}}\left(\frac{-A+E}{Em^2}\right)$$

$$= \frac{E-A}{\omega Em^2\sqrt{1-\frac{(A-E)^2}{E^2m^2}}}$$

$$= \frac{E-A}{\omega Em^2\sqrt{\frac{E^2m^2-(A-E)^2}{E^2m^2}}}$$

$$\frac{dt}{dm} = \frac{E-A}{\omega m\sqrt{E^2m^2-(A-E)^2}}$$

53. $\tan\theta = \frac{h}{x}$

$$\theta = \tan^{-1}\frac{h}{x}$$

$$\frac{d}{dx}\frac{h}{x} = \frac{d}{dx}hx^{-1} = -hx^{-2} = -\frac{h}{x^2}$$

$$\frac{d\theta}{dx} = \frac{1}{1+\frac{h^2}{x^2}}\left(-\frac{h}{x^2}\right)$$

$$\frac{d\theta}{dx} = \frac{-h}{\frac{x^2+h^2}{x^2}\left(x^2\right)}$$

$$\frac{d\theta}{dx} = \frac{-h}{h^2+x^2}$$

27.4 Applications

1. Sketch the curve $y = \sin x - \frac{x}{2}, 0 \le x \le 2\pi.$

$x = 0 \quad y \quad 0, (0,0)$ is both x-intercept and y-intercept. Using Newton's method or zero feature on a graphing calculator, $(1.90, 0)$ is the only x-intercept for $0 \le x \le 2\pi$. For $x = 2\pi$,

$y = \sin 2\pi - \frac{2\pi}{2} = -\pi \quad (2\pi, \ \pi)$ is the right-hand endpoint.

No vertical asymptotes (no denominators that become zero).

$$\frac{dy}{dx} = \cos x - \frac{1}{2} = 0 \text{ for } x = \frac{\pi}{3}, \frac{5\pi}{3}$$

$$\frac{d^2y}{dx^2} = -\sin x$$

$$\left.\frac{d^2y}{dx^2}\right|_{x=\frac{\pi}{3}} = \frac{-\sqrt{3}}{2}, \text{ so max } \frac{\pi}{3}, 0.34$$

$$\left.\frac{d^2y}{dx^2}\right|_{x=\frac{5\pi}{3}} = \frac{\sqrt{3}}{2}, \text{ so min } \frac{5\pi}{3}, -3.48$$

$$\left.\frac{d^2y}{dx^2}\right|_{x=0,\pi} = 0 \text{ for } x = 0, \pi$$

$$\frac{d^2y}{dx^2} < 0 \text{ for } 0 < x < \pi$$

$$\frac{d^2y}{dx^2} > 0 \text{ for } x < 0, \pi < x < 2\pi$$

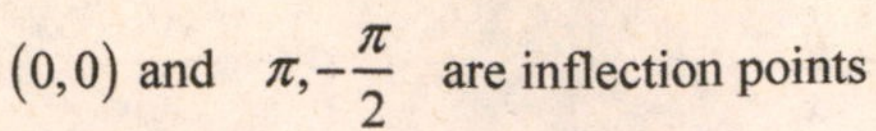

$(0,0)$ and $\pi, -\dfrac{\pi}{2}$ are inflection points

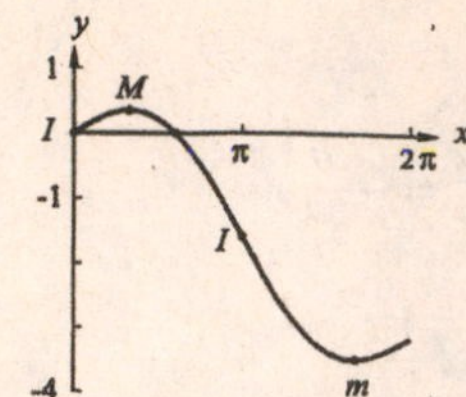

5. $y = \tan^{-1} x$

$\dfrac{dy}{dx} = \dfrac{1}{1+x^2} > 0$ for all x. Therefore, the graph is always increasing.

9.
$$y = x\sin^{-1} x$$
$$y\big|_{x=0.5} = 0.26179939$$
$$\frac{dy}{dx} = \frac{x}{\sqrt{1-x^2}} + \sin^{-1} x$$
$$\left.\frac{dy}{dx}\right|_{x=0.5} = 1.1009497$$
$$y - 0.26179939 = 1.1009497(x - 0.5)$$
$$y = 1.1x - 0.29$$

13. $y = 6\cos x - 8\sin x$; minimum value occurs when $f'(x) = 0$.

$$f'(x) = -6\sin x - 8\cos x = 0$$
$$\sin x = -\frac{8}{6}\cos x$$
$$\tan x = -\frac{8}{6} = -\frac{4}{3}$$

$\theta_{\text{ref}} = 0.927$ (a 3, 4, 5 triangle)

In QII: $\theta = \pi - 0.927 = 2.21$

In QIII: $\theta = \pi + 0.927 = 4.07$

$f''(x) = -6\cos x + 8\sin x$

$f''(2.21) > 0$, min; $f''(4.07) < 0$, max.

Minimum occurs where $x = 2.21$ rad.

$$f(2.21) = 6\left(\frac{-3}{5}\right) - 8\left(\frac{4}{5}\right)$$
$$= \frac{-18}{5} + \frac{-32}{5} = \frac{-50}{5} = -10$$

Minimum value of $y = -10$.

17.
$$y = 0.50\sin 2t + 0.30\cos t;$$
$$v = \frac{dy}{dt} = 1.00\cos 2t - 0.30\sin t$$
$$v\big|_{t=0.40} = 1.00\cos 0.80 - 0.30\sin 0.40$$
$$= 0.58 \text{ ft/s}$$
$$a = \frac{d^2y}{dt^2} = -2.00\sin 2t - 0.30\cos t$$
$$a\big|_{t=0.40} = -2.00\sin 0.80 - 0.30\cos 0.40$$
$$= -1.7 \text{ ft/s}^2$$

21.
$$v_x = \frac{dx}{dt}$$
$$= -19(6\pi)\sin 6\pi t$$
$$= 341$$
$$v_y = \frac{dy}{dt}$$
$$= 19(6\pi)\cos 6\pi t$$
$$= 111$$
$$v = \sqrt{v_x^2 + v_y^2}$$
$$v = \sqrt{(-19(6\pi)\sin 6\pi t)^2 + (19(6\pi)\cos 6\pi t)^2}$$
$$v\big|_{t=0.600} = 358 \text{ cm/s}$$
$$\tan\theta = \left.\frac{19(6\pi)\cos 6\pi t}{-19(6\pi)\sin 6\pi t}\right|_{t=0.600}$$
$$\theta = 0.314 \text{ rad}$$
$$= 0.314 \text{ rad}\cdot\frac{180^\circ}{\pi \text{ rad}}$$
$$= 18.0^\circ$$

25.
$$s = 16t^2$$
$$\tan\theta = \frac{200.0 - s}{100.0}$$
$$= \frac{200.0 - 16t^2}{100.0}$$
$$\theta = \tan^{-1}\left(\frac{200.0 - 16t^2}{100.0}\right) = \tan^{-1}(2 - 0.16t^2)$$
$$\frac{d\theta}{dt} = \frac{-0.32t}{1 + (2 - 0.16t^2)^2}$$
$$\left.\frac{d\theta}{dt}\right|_{t=1.0\text{ s}} = -0.073 \text{ rad/s}$$

29. $F = \dfrac{0.25w}{0.25\sin\theta + \cos\theta}$

$$\frac{dF}{d\theta} = \frac{-0.25w(0.25\cos\theta - \sin\theta)}{(0.25\sin\theta + \cos\theta)^2}$$

$$= 0$$

$$0.25\cos\theta - \sin\theta = 0$$

$$\tan\theta = 0.25$$

$$\theta = 0.245 \text{ rad}$$

$$= 0.245\cdot\frac{180}{\pi}$$

$$= 14.0^\circ$$

$$\frac{d^2F}{d\theta^2} = -0.25w\left[\frac{(0.25\sin\theta+\cos\theta)^2(-0.25\sin\theta-\cos\theta)}{(0.25\sin\theta+\cos\theta)^4} - \frac{(0.25\cos\theta-\sin\theta)^2(0.25\sin\theta+\cos\theta)}{(0.25\sin\theta+\cos\theta)^4}\right]$$

$\left.\dfrac{d^2F}{d\theta^2}\right|_{\theta=0.245} > 0$, so F is a minimum.

33. $V = 0.48(1.2 - \cos 1.26t)$

$$\frac{dV}{dt} = 0.48(1.26)\sin 1.26t$$

$$= 0$$

$$1.26t = 0, \pi$$

$$t = 0, \frac{\pi}{1.26}$$

$$\frac{d^2V}{dt^2} = 0.48(1.26)^2\cos 1.26t$$

$\left.\dfrac{d^2V}{dt^2}\right|_{t=\frac{\pi}{1.26}} < 0$, max at $t = \dfrac{\pi}{1.26}$

$$V_{max} = 0.48\ \ 1.2 - \cos 1.26\frac{\pi}{1.26} = 1.056$$

$$V_{max} = 1.06 \text{ L/s}$$

37.

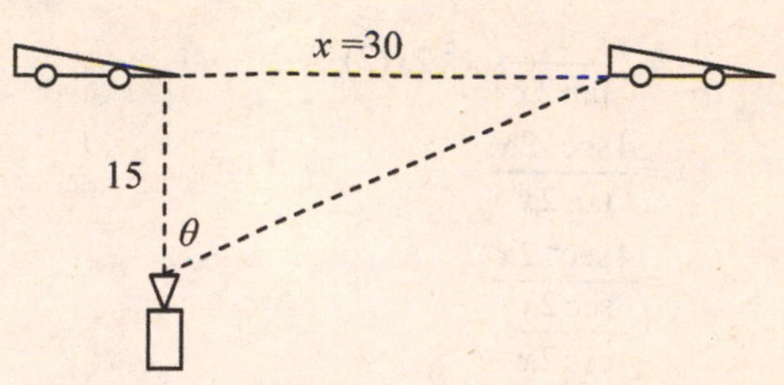

$$\tan\theta = \frac{x}{15}$$

$$\theta = \tan^{-1}\frac{x}{15}$$

$$\frac{d\theta}{dt} = \frac{1/15}{1+(x/15)^2}\frac{dx}{dt}$$

$$0.75 = \frac{1/15}{1+(30/15)^2}\frac{dx}{dt}$$

$$0.75 = \frac{1}{75}\frac{dx}{dt}$$

$$\frac{dx}{dt} = 56.25$$

The car is traveling 56 m/s.

27.5 Derivative of the Logarithmic Function

1. $y = \ln\cos 4x$

$$\frac{dy}{dx} = \frac{1}{\cos 4x}(-\sin 4x)(4)$$

$$\frac{dy}{dx} = -4\tan 4x$$

5. $y = 4\log_5(3-x)$

$$\frac{dy}{dx} = 4\frac{1}{3-x}(\log_5 e)(-1)$$

$$= \frac{4}{x-3}\log_5 e$$

9. $y = 2\ln\tan 2x$

$$\frac{dy}{dx} = 2\frac{1}{\tan 2x}\sec^2 2x(2)$$
$$= \frac{4\sec^2 2x}{\tan 2x}$$
$$= \frac{4\sec^2 2x}{\frac{\sec 2x}{\csc 2x}}$$
$$\frac{dy}{dx} = 4\sec 2x\csc 2x$$

13. $y = \ln\left(x - x^2\right)^3$

$$\frac{dy}{dx} = \frac{1}{\left(x - x^2\right)^3}\cdot 3\left(x - x^2\right)^2(1 - 2x)$$
$$= \frac{-6x + 3}{x - x^2}$$

17. $y = 3x\ln(6 - x)^2$

$$y = 6x\ln(6 - x)$$
$$\frac{dy}{dx} = 6x\cdot\frac{-1}{6 - x} + 6\ln(6 - x)$$
$$= \frac{6x}{x - 6} + 6\ln(6 - x)$$

21. $r = 0.5\ln[\cos(\pi\theta^2)]$

$$\frac{dr}{d\theta} = 0.5.\frac{1}{\cos(\pi\theta^2)}\cdot(-\sin(\pi\theta^2))\cdot 2\pi\theta$$
$$\frac{dr}{d\theta} = -\pi\theta\tan(\pi\theta^2)$$

25. $u = 3v\ln^2 2v$

$$\frac{du}{dv} = 3v\cdot 2\ln 2v\cdot\frac{1}{2v}\cdot 2 + 3\ln^2 2v$$
$$\frac{du}{dv} = 6\ln 2v + 3\ln^2 2v$$

29. $r = \ln\frac{v^2}{v + 2}$

$$= \ln v^2 - \ln(v + 2)$$
$$= 2\ln v - \ln(v + 2)$$
$$\frac{dr}{dv} = \frac{2}{v} - \frac{1}{v + 2}$$
$$= \frac{2v + 4 - v}{v(v + 2)}$$
$$\frac{dr}{dv} = \frac{v + 4}{v(v + 2)}$$

33. $\ln\frac{x}{y} = x$

$$\ln x - \ln y = x$$
$$\frac{d}{dx}(\ln x - \ln y) = \frac{d}{dx}(x)$$
$$\frac{1}{x} - \frac{1}{y}\frac{dy}{dx} = 1$$
$$\frac{1}{y}\frac{dy}{dx} = \frac{1}{x} - 1$$
$$\frac{dy}{dx} = \frac{y(1 - x)}{x}$$

37. $y = (1 + x)^{(1/x)}$

(a)

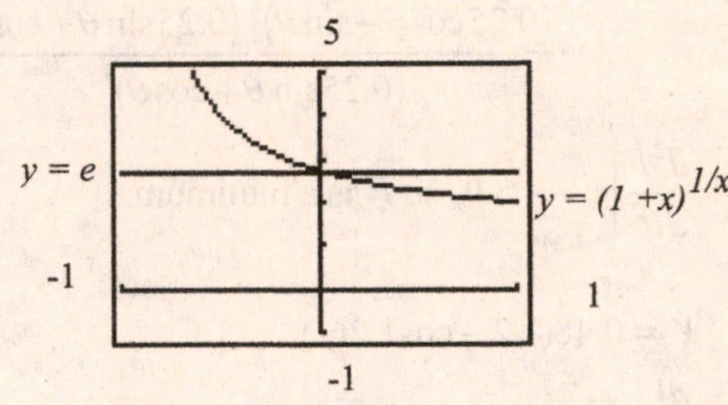

(b)

t	0.1	0.01	0.001	0.0001
y	2.5937	2.7048	2.7169	2.71815

41. $y = \ln\sqrt{1 - 4x^2}$; $x = 1/4$;

$$= \frac{1}{2}\ln\left(1 - 4x^2\right)$$
$$\frac{dy}{dx} = \frac{-8x}{2\left(1 - 4x^2\right)}$$
$$= -\frac{4x}{1 - 4x^2}$$
$$\left.\frac{dy}{dx}\right|_{x=1/4} = -\frac{4(1/4)}{1 - 4(1/4)^2}$$
$$= -\frac{4}{3}$$

45. $y = \tan^{-1} 2x + \ln(4x^2 + 1);$

$$m_{\tan} = \frac{dy}{dx}$$
$$= \frac{1}{1+4x^2}(2) + \frac{1}{4x^2+1}(8x)$$
$$= \frac{2}{1+4x^2} + \frac{8x}{4x^2+1}$$
$$= \frac{2+8x}{1+4x^2}$$

When $x = 0.625$,

$$m_{\tan} = \frac{2+8(0.625)}{1+4(0.625)^2}$$
$$= 2.73$$

49. $y_1 = \ln(x^2), x^2 \neq 0$

$$\frac{dy_1}{dx} = \frac{1}{x^2}(2x)$$
$$= \frac{2}{x}$$
$$\left.\frac{dy_1}{dx}\right|_{x=-1} = -2$$
$$y_2 = 2\ln x, x > 0$$
$$\frac{dy_2}{dx} = 2\ \frac{1}{x}$$
$$= \left.\frac{2}{x}\right|_{x=-1} \text{ is not defined since}$$
$$-1 < 0$$

For values of x where both functions are defined, the derivatives are the same.

53. $y = \ln \dfrac{1+\sqrt{1-x^2}}{x} - \sqrt{1-x^2}$

$$\frac{dy}{dx} = \frac{x}{1+\sqrt{1-x^2}} \cdot \frac{x\,\frac{-2x}{2\sqrt{1-x^2}} - \left(1+\sqrt{1-x^2}\right)}{x^2} + \frac{2x}{2\sqrt{1-x^2}}$$

$$\frac{dy}{dx} = -\frac{x}{1+\sqrt{1-x^2}}\ \frac{1+\sqrt{1-x^2}+\frac{x^2}{\sqrt{1-x^2}}}{x^2} + \frac{x}{\sqrt{1-x^2}}$$

$$\frac{dy}{dx} = -\frac{1}{1+\sqrt{1-x^2}} \cdot \frac{\sqrt{1-x^2}+1-x^2+x^2}{x\sqrt{1-x^2}} + \frac{x}{\sqrt{1-x^2}}$$

$$\frac{dy}{dx} = -\frac{1}{1+\sqrt{1-x^2}} \cdot \frac{1+\sqrt{1-x^2}}{x\sqrt{1-x^2}} + \frac{x^2}{x\sqrt{1-x^2}}$$

$$\frac{dy}{dx} = \frac{x^2-1}{x\sqrt{1-x^2}} = -\frac{\sqrt{1-x^2}}{x}$$

27.6 Derivative of the Exponential Function

1. $y = \ln \sin e^{2x}$

$\frac{dy}{dx} = \frac{1}{\sin e^{2x}}(\cos e^{2x})(2e^{2x})$

$\frac{dy}{dx} = 2e^{2x} \cot e^{2x}$

5. $y = 6e^{\sqrt{x}}$

$\frac{dy}{dx} = 6e^{\sqrt{x}} \cdot \frac{1}{2\sqrt{x}}$

$= \frac{3e^{\sqrt{x}}}{\sqrt{x}}$

9. $R = Te^{-3T}$

$\frac{dR}{dT} = T(e^{-3T})(-3) + (1)(e^{-3T})$

$= e^{-3T} - 3Te^{-T}$

$= e^{-3T}(1 - 3T)$

13. $r = \frac{2(e^{2s} - e^{-2s})}{e^{2s}}$

$r = 2(1 - e^{-4s})$

$\frac{dr}{ds} = 2(4e^{-4s})$

$\frac{dr}{ds} = 8e^{-4s}$

17. $y = \frac{2e^{3x}}{4x+3}$

$\frac{dy}{dx} = \frac{(4x+3)(2e^{3x})(3) - (2e^{3x})(4)}{(4x+3)^2}$

$\frac{dy}{dx} = \frac{(12x+9)(2e^{3x}) - 8e^{3x}}{(4x+3)^2}$

$= \frac{2e^{3x}(12x+5)}{(4x+3)^2}$

21. $y = (2e^{2x})^3 \sin x^2$

$= 8e^{6x} \sin x^2$

$\frac{dy}{dx} = 8e^{6x}(\cos x^2)(2x) + \sin x^2(8e^{6x})(6)$

$= 16e^{6x}(x \cos x^2 + 3\sin x^2)$

25. $y = e^{xy}$

$\frac{dy}{dx} = e^{xy} \quad y + x\frac{dy}{dx}$

$\left(1 - xe^{xy}\right)\frac{dy}{dx} = ye^{xy}$

$\frac{dy}{dx} = \frac{y}{1 - xe^{xy}}$

29. $I = \ln \sin 2e^{6t}$

$\frac{dI}{dt} = \frac{1}{\sin 2e^{6t}} \cdot \cos 2e^{6t} \cdot 12e^{6t}$

$= 12e^{6t} \cot 2e^{6t}$

33. **(a)** $e=e^x=2.7182818$ when $x=1.0000$. This is the slope of a tangent line to the curve $f(x)=e^x$ when $x=1.0000$. It is the curve $f'(x)=e^x$, since $\frac{de^x}{dx} = e^x$.

(b) $\frac{e^{1.0001} - e^{1.0000}}{0.0001} = 2.7184178$ This is the slope of a secant line through the curve.

$f(x) = e^x$ at $x = 1.0000$, where $\Delta x = 0.0001$.

$\lim_{\Delta x \to 0} \frac{e(x + \Delta x - e^x)}{\Delta x} = \frac{de^x}{dx} = e^x$

For $\Delta x = 0.0001$, the slope of the tangent line is approximately equal to the slope of the secant line.

37. $y = e^{-x/2} \cos 4x$

$\frac{dy}{dx} = -4(\sin 4x)\left(e^{-x/2}\right) - \frac{1}{2}e^{-x/2} \cos 4x$

$\left.\frac{dy}{dx}\right|_{x=0.625} = -1.46$

41.

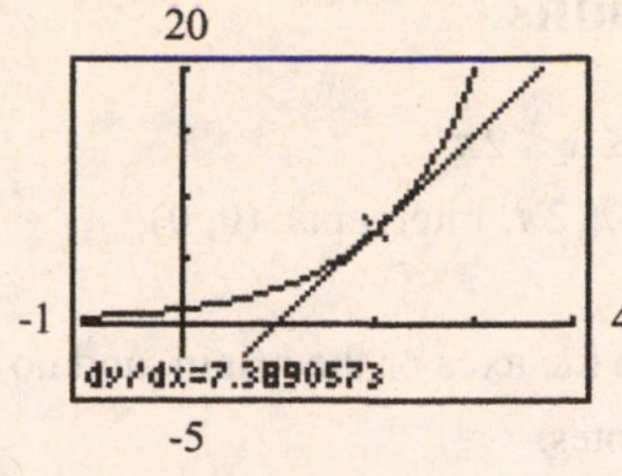

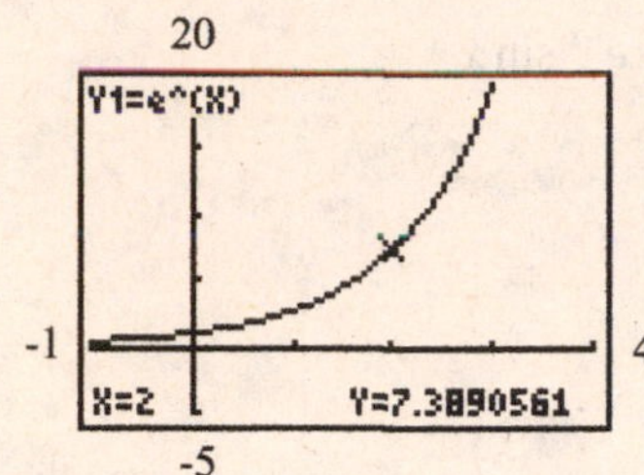

45. $y = ae^{mx}$

$y' = ame^{mx}$

$y'' = am^2e^{mx}$

If $y''+y'-6y=0$,

$$am^2e^{mx} + ame^{mx} - 6ae^{mx} = 0$$
$$m^2 + m - 6 = 0$$
$$(m+3)(m-2) = 0$$

The equation is satisfied for $m=-3, m=2$.

49. $R = e^{-0.0002t}; t = 1000\text{ h } (0 \le R \le 1)$

$$\frac{dR}{dt} = e^{-0.0002t}(-0.0002)$$
$$\frac{dR}{dt} = -0.0002e^{-0.0002t}$$
$$\left.\frac{dR}{dt}\right|_{t=1000} = -0.00016/\text{h}$$

53. $\cosh^2 u - \sinh^2 u$

$$= \frac{1}{4}(e^u + e^{-u})^2 - \frac{1}{4}(e^u - e^{-u})^2$$
$$= \frac{1}{4}(e^{2u} + 2e^0 + e^{-2u}) - \frac{1}{4}(e^{2u} - 2e^0 + e^{-2u})$$
$$= \frac{1}{4}\left(4e^0\right) = 1$$

27.7 L'Hospital's Rule

1.
$$\lim_{x\to\infty}\frac{3e^{2x}}{5\ln x} = \lim_{x\to\infty}\frac{\frac{d}{dx}3e^{2x}}{\frac{d}{dx}5\ln x} = \lim_{x\to\infty}\frac{6e^{2x}}{\frac{5}{x}} = \lim_{x\to\infty}\frac{6xe^{2x}}{5} = \infty$$

5.
$$\lim_{\theta\to 0}\frac{\tan\theta}{\theta} = \lim_{\theta\to 0}\frac{\frac{d}{d\theta}\tan\theta}{\frac{d}{d\theta}\theta} = \lim_{\theta\to 0}\frac{\sec^2\theta}{1} = 1$$

9.
$$\lim_{t\to\frac{\pi}{4}}\frac{1-\sin 2t}{\frac{\pi}{4}-t} = \lim_{t\to\frac{\pi}{4}}\frac{-2\cos 2t}{-1} = 0$$

13.
$$\lim_{x\to 0}\frac{\sin x - x}{x^3} = \lim_{x\to 0}\frac{\cos x - 1}{3x^2} = \lim_{x\to 0}\frac{-\sin x}{6x} = \lim_{x\to 0}\frac{-\cos x}{6} = -\frac{1}{6}$$

17.
$$\lim_{x\to 0} x\cot x = \lim_{x\to 0}\frac{x\cos x}{\sin x} = \lim_{x\to 0}\frac{\cos x}{\frac{\sin x}{x}} = 1$$

21.
$$\lim_{x\to 0}\frac{2\sin x}{5e^x} = \frac{2}{5}\frac{\lim_{x\to 0}\sin x}{\lim_{x\to 0}e^x} = \frac{2}{5}\cdot\frac{0}{1} = 0$$

25.
$$\lim_{x\to\infty}\frac{\ln x^3}{x^2} = \lim_{x\to\infty}\frac{3x^2/x^3}{2x} = \lim_{x\to\infty}\frac{3}{2x^2} = 0$$

29. $\lim\limits_{x\to 1}\dfrac{\ln x}{\sin 2\pi x}=\lim\limits_{x\to 1}\dfrac{1/x}{2\pi\cos 2\pi x}$

$=\dfrac{1}{2\pi}$

33. $\lim\limits_{x\to 2}\dfrac{x^3-8}{x^5-32}$

$=\lim\limits_{x\to 2}\dfrac{3x^2}{5x^4}$

$=\dfrac{3}{20}$

37. $\lim\limits_{\theta\to\frac{\pi}{2}^-}(\sec\theta-\tan\theta)=\lim\limits_{\theta\to\frac{\pi}{2}^-}\left(\dfrac{1}{\cos\theta}-\dfrac{\sin\theta}{\cos\theta}\right)$

$=\lim\limits_{\theta\to\frac{\pi}{2}^-}\left(\dfrac{1-\sin\theta}{\cos\theta}\right)$

$=\lim\limits_{\theta\to\frac{\pi}{2}^-}\dfrac{-\cos\theta}{-\sin\theta}$

$=0$

41. $y=\left(1+x^2\right)^{\frac{1}{x}}$

$\ln y=\dfrac{1}{x}\ln\left(1+x^2\right)$

$\lim\limits_{x\to+\infty}\ln y=\lim\limits_{x\to+\infty}\dfrac{\ln\left(1+x^2\right)}{x}$

$=\lim\limits_{x\to+\infty}\dfrac{\frac{2x}{1+x^2}}{1}$

$=\lim\limits_{x\to+\infty}\dfrac{2}{2x}=0$

$e^{\lim\limits_{x\to+\infty}\ln y}=e^0=1$

$\lim\limits_{x\to+\infty}\left(1+x^2\right)^{\frac{1}{x}}=1$

45. $\sin x$ varies between -1 and 1 as $x\to\infty$.

27.8 Applications

1. $y=e^{-x}\sin x,\ 0\le x\le 2\pi$

$y=0$ for $x=0,\pi,2\pi$. Intercepts: (0, 0), $(\pi,0),(2\pi,0)$

No symmetry to the axes or the origin, and no vertical asymptotes.

$\dfrac{dy}{dx}=e^{-x}\cos x-e^{-x}\sin x$

$=0$

$\sin x=\cos x$

$x=\dfrac{\pi}{4},\dfrac{5\pi}{4}$

$\dfrac{d^2y}{dx^2}=e^{-x}(-\sin x-\cos x)-e^{-x}(\cos x-\sin x)$

$\dfrac{d^2y}{dx^2}=e^{-x}(-2\cos x)$

$=0$ for $x=\dfrac{\pi}{2},\dfrac{3\pi}{2}$

$\left.\dfrac{d^2y}{dx^2}\right|_{x=\frac{\pi}{4}}=-2e^{-\pi/4}\cos\dfrac{\pi}{4}$

$=-0.64<0\Rightarrow\left(\dfrac{\pi}{4},0.322\right)$ is max

$\left.\dfrac{d^2y}{dx^2}\right|_{\frac{5\pi}{4}}=-2e^{-5\pi/4}\cos\dfrac{5\pi}{4}$

$=0.03>0\Rightarrow\left(\dfrac{5\pi}{4},-0.014\right)$ is a min

$\dfrac{d^2y}{dx^2}$ changes from negative to positive at $x=\dfrac{\pi}{2}$

$\left(\dfrac{\pi}{2},0.208\right)$ is an inflection point

$\dfrac{d^2y}{dx^2}$ changes from positive to negative at $x=\dfrac{3\pi}{2}$

$\left(\dfrac{3\pi}{2},-0.009\right)$ is an inflection point

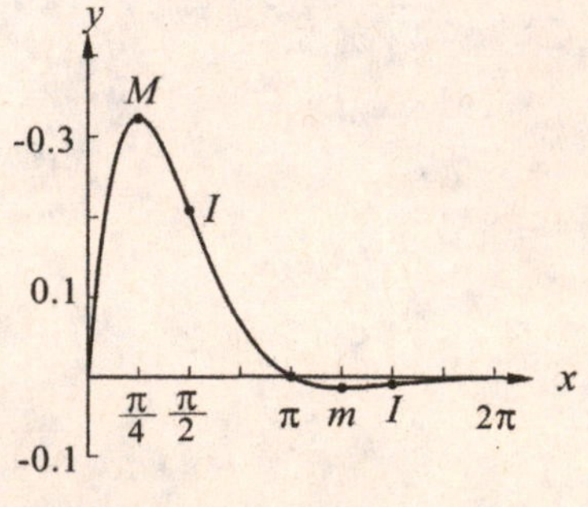

5. $y = 3xe^{-x} = \dfrac{3x}{e^x}$

$\dfrac{dy}{dx} = (3x)(-e^{-x}) + 3e^{-x}$

$= (3-3x)e^{-x}$

$\dfrac{d^2y}{dx^2} = (-3)(e^{-x}) + (3-3x)(-e^{-x})$

$= -3e^{-x} - 3e^{-x} + 3xe^{-x}$

$= (3x-6)e^{-x}$

(1) Intercept: $x = 0, y = 0$ (origin)

(2) Symmetry: None

(3) As $x \to +\infty$, $y = 0$, horizontal asymptote $y = 0$

As $x \to -\infty$, $y \to -\infty$

(4) Vertical asymptote: none

(5) Domain: all x, range to be determined

(6) Set $\dfrac{dy}{dx} = 0$

$e^{-x}(3-3x) = 0$ for $x = 1$

$f''(1) < 0$, so Max $\left(1, \dfrac{1}{2}\right)$

Set $\dfrac{d^2y}{dx^2} = 0$

$e^{-x}(3x-6) = 0$ for $x = 2$,

so inflection point at $\left(2, \dfrac{2}{e^2}\right)$

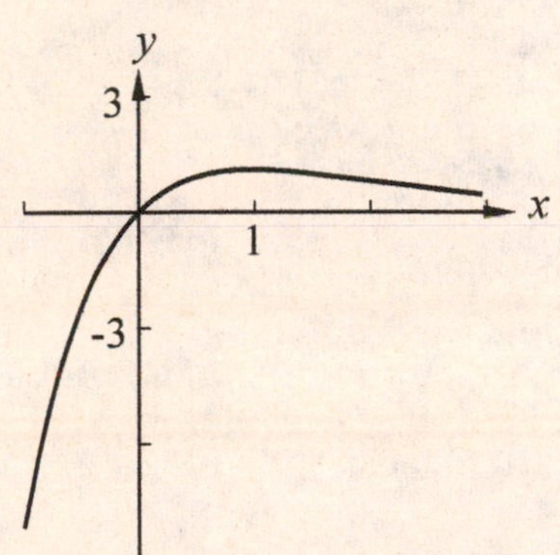

9. $y = 4e^{-x^2}$

$\dfrac{dy}{dx} = 4e^{-x^2}(-2x)$

$= \dfrac{-8x}{e^{x^2}}$

$\dfrac{d^2y}{dx^2} = \dfrac{e^{x^2}(-8) + 8x(2xe^{x^2})}{e^{2x^2}} = 8e^{-x^2}(2x^2-1)$

(1) Intercepts: $x = 0, y = 4, (0,4)$ intercept.

(2) Symmetry: yes, with respect to y-axis.

(3) As $x \to \pm\infty$, $y \to 0$ positively; x-axis is a horizontal asymptote.

(4) No vertical asymptote.

(5) Domain: all x; range to be determined.

(6) $\dfrac{dy}{dx} = \dfrac{-8x}{e^{x^2}} = 0$ for $x = 0$

$\left.\dfrac{d^2y}{dx^2}\right|_{x=0} < 0$, so max. at $(0,4)$

Range: $0 < y \le 4$

$\dfrac{d^2y}{dx^2} = 2x^2 - 1 = 0$ for $x = \pm\sqrt{\dfrac{1}{2}} = \pm\dfrac{\sqrt{2}}{2}$;

$\left(-\dfrac{\sqrt{2}}{2}, \dfrac{4}{\sqrt{e}}\right), \left(\dfrac{\sqrt{2}}{2}, \dfrac{4}{\sqrt{e}}\right)$ are inflection points.

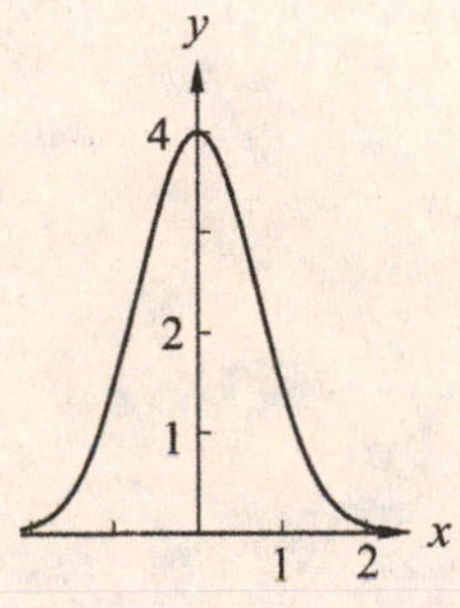

13. $y = \frac{1}{2}(e^x - e^{-x})$

$\frac{dy}{dx} = \frac{1}{2}(e^x + e^{-x})$

$\frac{d^2y}{dx^2} = \frac{1}{2}(e^x - e^{-x})$

(1) Intercepts: $x = 0, y = 0, (0, 0)$

(2) Symmetry with respect to the origin.

(3) $x \to +\infty, y \to +\infty, x \to -\infty, y \to -\infty$

(4) No vertical asymptote.

(5) Domain: all x; range: all y.

(6) $\frac{dy}{dx} = \frac{e^x + e^{-x}}{2} > 0$ for all x, so the function is always increasing and there are no maxima or minima.

$\frac{d^2y}{dx^2} = \frac{e^x - e^{-x}}{2}$

$= 0$

$e^x = e^{-x}$

$x = -x$

$x = 0$, therefore inflection point at (0,0)

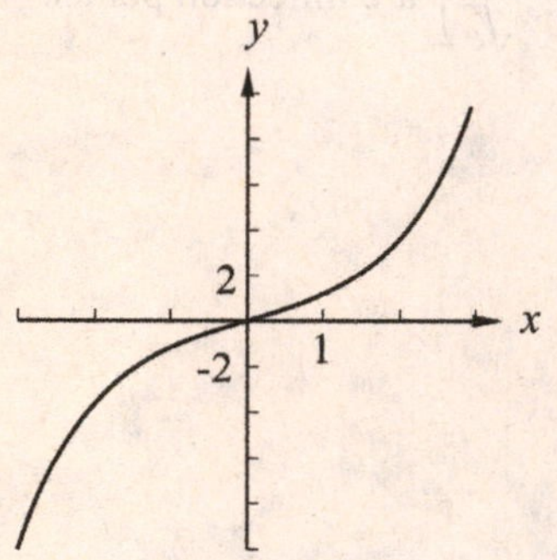

17. $y = x^2 \ln x$

$\frac{dy}{dx} = x^2 \frac{1}{x} + (\ln x)(2x)$

$= x + 2x \ln x;$

$\left.\frac{dy}{dx}\right|_{x=1} = 1 + 2\ln 1$

$= 1 + 2(0) = 1$

Slope is $1, x = 1, y = 0$; using slope intercept form of the equation and substituting gives $0 = 1(1) + b$ or $b = 1$. The equation is $y = (1)x - 1$ or $y = x - 1$.

21. $f(x) = x^2 - 3 + \ln 4x$

$f'(x) = 2x + \frac{1}{x}$

$f(1) = -0.6137$

$f(2) = 3.079$

Choose $x_1 = 1.5$

n	x_n	$f(x_n)$	$f'(x_n)$	$x_n - \frac{f(x_n)}{f'(x_n)}$
1	1.5	1.041 759 4	3.666 666 7	1.215 883 8
2	1.215 883 8	0.060 138 9	3.254 214 6	1.197 403 4
3	1.197 403 5	0.000 224 8	3.229 947 3	1.197 333 8
4	1.197 333 8	3.156×10^{-9}	3.229 856 6	1.197 333 8

The root is 1.197 333 8.

25. $P = 100e^{-0.005t}; t = 100$ days;

$$\frac{dP}{dt} = 100e^{-0.005t}(-0.005)$$

$$= -0.05e^{-0.005t}$$

$$\left.\frac{dP}{dt}\right|_{t=100} = -0.303 \text{ W/day}$$

29. $\ln p = \frac{a}{T} + b \ln T + c$

$$\frac{1}{p}\frac{dp}{dT} = -\frac{a}{T^2} + \frac{b}{T}$$

$$\frac{dp}{dT} = \left(-\frac{a}{T^2} + \frac{b}{T}\right)p$$

33. $y = \ln \sec x; -1.5 \le x \le 1.5;$

$\frac{dy}{dx} = \frac{1}{\sec x} \cdot \sec x \tan x$

$= \tan x = 0$ at $x = 0$;

$x = 0$ is a critical value

Multiples of 2π are also critical values, but they are not part of the path of the roller mechanism.

$\frac{d^2y}{dx^2} = \sec^2 x$

$\sec^2(0) = 1$ so the curve is concave up and there is a minimum point at $x = 0, y = \ln \sec 0 = \ln 1 = 0$

recurring at multiples of $x = 2\pi; (2\pi, 0), (4\pi, 0), \ldots$

$(0,0)$ is an intercept.

Asymptotes occur where $\frac{dy}{dx} = \tan x$ is undefined.
These values are odd multiples of $\frac{\pi}{2}; -\frac{\pi}{2}, \frac{\pi}{2}, \frac{3\pi}{2} \cdots$
The graph is periodic, but the path of the roller mechanism corresponds only to the part of the first cycle, with $-1.5 \le x \le 1.5$,

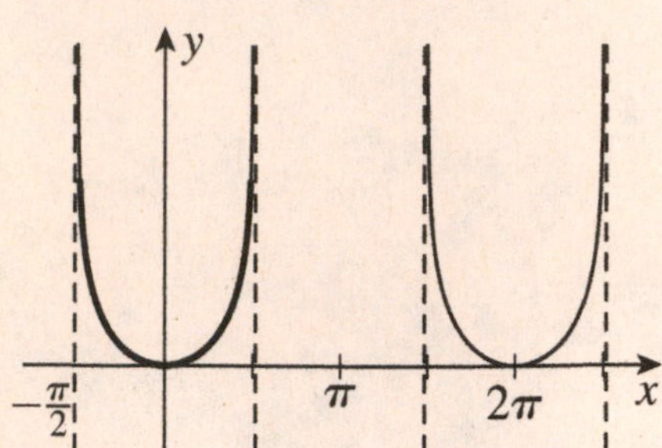

37. $y = \frac{1}{\sqrt{2\pi}} e^{-x^2/2}$

$$\frac{dy}{dx} = -\frac{1}{\sqrt{2\pi}} x e^{-x^2/2}$$

$$\frac{d^2y}{dx^2} = -\frac{1}{\sqrt{2\pi}}\left(e^{-x^2/2} - x^2 e^{-x^2/2}\right)$$

$$\frac{d^2y}{dx^2} = -\frac{e^{-x^2/2}}{\sqrt{2\pi}}\left(1 - x^2\right)$$

The locations where $\frac{d^2y}{dx^2} = 0$ are $x = \pm 1$.

For $x < -1$, $\frac{d^2y}{dx^2} > 0$. For $-1 < x < 1$, $\frac{d^2y}{dx^2} < 0$.

For $x > 1$, $\frac{d^2y}{dx^2} > 0$. Thus, $x = \pm 1$ correspond to inflection points.

41. $s = kx^2 \ln\frac{1}{x} = k[x^2(\ln 1 - \ln x)] = -kx^2 \ln x$

$$\frac{ds}{dx} = -k\left(x^2 \frac{1}{2} + \ln x 2x\right) = -k(x + 2x\ln x)$$

$$\frac{ds}{dx} = -kx(1 + 2\ln x) = 0$$

The critical value occurs when

$\ln x = -\frac{1}{2}; x = e^{-1/2} = \frac{1}{\sqrt{e}} = 0.607$

($x = 0$ is not in the domain.)

For $x < e^{-1/2}, \frac{ds}{dx} > 0$. For $x > e^{-1/2}, \frac{ds}{dx} < 0$.

Therefore, $x = e^{-1/2}$ corresponds to a maximum.

45. $a = 2w^2(\cos\theta + 0.5\cos 2\theta)$

$$\frac{da}{d\theta} = 2w^2(-\sin\theta - \sin 2\theta)$$

$$\frac{da}{d\theta} = -2w^2(\sin\theta + 2\sin\theta\cos\theta)$$

$$\frac{da}{d\theta} = -2w^2\sin\theta(1 + 2\cos\theta)$$

Critical values occur when $\sin\theta = 0$, i.e. when $\theta = 0, \theta = \pi$,

or when $\cos\theta = -\frac{1}{2}$, i.e. when $\theta = \frac{2\pi}{3}, \theta = \frac{4\pi}{3}$.

$$\frac{d^2a}{d\theta^2} = -2w^2(\cos\theta + 2\cos 2\theta)$$

$\left.\frac{d^2a}{d\theta^2}\right|_{\theta=0} = -6w^2 < 0$; acceleration is locally maximized at $\theta=0$

$\left.\frac{d^2a}{d\theta^2}\right|_{\theta=2\pi/3} = 3w^2 > 0$; acceleration is locally minimized at $\theta=\frac{2\pi}{3}$

$\left.\frac{d^2a}{d\theta^2}\right|_{\theta=\pi} = -6w^2 < 0$; acceleration is locally maximized at $\theta=\pi$

$\left.\frac{d^2a}{d\theta^2}\right|_{\theta=4\pi/3} = 3w^2 > 0$; acceleration is locally minimized at $\theta=\frac{4\pi}{3}$

At $\theta = 0, a = 3w^2$.

At $\theta = \frac{2\pi}{3}, a = -3w^2$.

At $\theta = \pi, a = -w^2$.

At $\theta = \frac{4\pi}{3}, a = -3w^2$.

The absolute maximum acceleration occurs at $\theta = 0$

and the absolute minimum acceleration occurs at $\theta = \frac{2\pi}{3}$

and $\theta = \frac{4\pi}{3}$.

Review Exercises

1. This is false. An extra factor of 2 is needed from the second application of the chain rule to cos $2x$.

5. This is true using the properties of logarithms. If $y = \ln(2x) = \ln 2 + \ln x$, then $\frac{dy}{dx} = \frac{1}{x}$.

9. $y = 3\cos(4x-1)$;

$$\frac{dy}{dx} = [-3\sin(4x-1)][4]$$
$$= -12\sin(4x-1)$$

13. $y = \csc^2(3x+2)$;

$$\frac{dy}{dx} = 2\csc(3x+2)[-\csc(3x+2)\cot(3x+2)](3)$$
$$= -6\csc^2(3x+2)\cot(3x+2)$$

17. $y = \left(e^{x-3}\right)^2$

$$\frac{dy}{dx} = 2\left(e^{x-3}\right)\left(e^{x-3}\right)(1)$$
$$= 2e^{2(x-3)}$$

21. $y = 10\tan^{-1}\left(\frac{x}{5}\right)$

$$\frac{dy}{dx} = 10\left[\frac{1}{1+\left(\frac{x}{5}\right)^2}\right]\frac{1}{5}$$
$$= \frac{2}{1+\left(\frac{x}{5}\right)^2}$$
$$= \frac{50}{25+x^2}$$

25. $y = \sqrt{\csc 4x + \cot 4x} = (\csc 4x + \cot 4x)^{1/2}$

$$\frac{dy}{dx} = \frac{1}{2}(\csc 4x + \cot 4x)^{-1/2}$$
$$\times\left(-4\csc 4x\cot 4x - 4\csc^2 4x\right)$$
$$= \frac{1}{2}(\csc 4x + \cot 4x)^{-1/2}(-4\csc 4x)$$
$$\times(\csc 4x + \cot 4x)$$
$$= -2\csc 4x(\csc 4x + \cot 4x)^{1/2}$$
$$= (-2\csc 4x)\sqrt{\csc 4x + \cot 4x}$$

29. $y = \dfrac{\cos^2 x}{e^{3x} + \pi^2}$

$$\frac{dy}{dx} = \frac{(e^{3x} + \pi^2)[2\cos x(-\sin x)] - (\cos^2 x)(e^{3x})(3)}{(e^{3x} + \pi^2)^2}$$

$$= \frac{(e^{3x} + \pi^2)[-2\sin x \cos x] - 3e^{3x}\cos^2 x}{(e^{3x} + \pi^2)^2}$$

$$= \frac{-\cos x\left[(e^{3x} + \pi^2)(2\sin x) + 3e^{3x}\cos x\right]}{(e^{3x} + \pi^2)^2}$$

$$= \frac{-\cos x\left[2e^{3x}\sin x + 2\pi^2 \sin x + 3e^{3x}\cos x\right]}{(e^{3x} + \pi^2)^2}$$

33. $y = \dfrac{\ln(\csc x^2)}{x}$

$$y = x^{-1}\ln(\csc x^2)$$

$$\frac{dy}{dx} = -x^{-2}\ln(\csc x^2) + x^{-1}\frac{1}{\csc x^2}(-\csc x^2 \cot x^2)(2x)$$

$$= -x^{-2}\ln(\csc x^2) - 2\cot x^2$$

37. $L = 0.1e^{-2t}\sec(\pi t)$

$$\frac{dL}{dt} = 0.1e^{-2t}\sec(\pi t)\tan(\pi t)\cdot\pi + \sec(\pi t)\cdot 0.1(-2)e^{-2t}$$

$$\frac{dL}{dt} = 0.1\pi e^{-2t}\sec(\pi t)\tan(\pi t) - 0.2e^{-2t}\sec(\pi t)$$

$$\frac{dL}{dt} = e^{-2t}\sec(\pi t)\cdot\left[0.1\pi\tan(\pi t) - 0.2\right]$$

41. $\tan^{-1}\dfrac{y}{x} = x^2 e^y$

$$u = \frac{y}{x} = yx^{-1};\ \frac{du}{dx} = -yx^{-2} + x^{-1}\frac{dy}{dx}$$

$$\frac{1}{1 + (yx^{-1})^2}\left(-yx^{-2} + x^{-1}\frac{dy}{dx}\right) = x^2 e^y \frac{dy}{dx} + 2xe^y$$

$$\frac{\frac{-y}{x^2} + \frac{1}{x}\frac{dy}{dx}}{1 + y^2x^{-2}} = x^2 e^y \frac{dy}{dx} + 2xe^y$$

$$\frac{-y}{x^2} + \frac{1}{x}\frac{dy}{dx} = \left(x^2 e^y \frac{dy}{dx} + 2xe^y\right)(1 + y^2x^{-2})$$

$$\frac{-y}{x^2} + \frac{1}{x}\frac{dy}{dx} = x^2 e^y \frac{dy}{dx} + 2xe^y + y^2 e^y \frac{dy}{dx} + 2x^{-1}y^2e^y$$

$$\frac{1}{x}\frac{dy}{dx} - x^2 e^y \frac{dy}{dx} - y^2 e^y \frac{dy}{dx} = 2xe^y + 2x^{-1}y^2e^y + \frac{y}{x^2}$$

$$\frac{dy}{dx}\left(\frac{1}{x}-x^2e^y-y^2e^y\right)=2xe^y+\frac{2y^2ey}{x}+\frac{y}{x^2}$$

$$\frac{dy}{dx}\left(\frac{1-x^3e^y-xy^2e^y}{x}\right)=\frac{2x^3e^y+2xy^2e^y+y}{x^2}$$

$$\frac{dy}{dx}=\frac{2x^3e^y+2xy^2e^y+y}{x-x^4e^y-x^2y^2e^y}$$

45. $e^x \ln xy + y = e^x$

Using implicit differentiation,

$$\frac{d}{dx}\left(e^x \ln xy + y\right)=\frac{d}{dx}\left(e^x\right)$$

$$e^x \ln xy + e^x \quad \frac{y+x\frac{dy}{dx}}{xy} \quad +\frac{dy}{dx}=e^x$$

$$e^x xy \ln xy + e^x \quad y + x\frac{dy}{dx} \quad + xy\frac{dy}{dx}=e^x xy$$

$$\left(xe^x+xy\right)\frac{dy}{dx}=e^x\left(xy-xy\ln xy-y\right)$$

$$\frac{dy}{dx}=\frac{xy-xy\ln xy-y}{xe^x+xy}$$

49. $y = x - \cos 0.5x$

$y = x-\cos 0.5x\big|_{x=0} = -1 \Rightarrow y\text{-int: } (0, -1)$

$0 = x-\cos 0.5x \Rightarrow x = 0.9 \Rightarrow x\text{-int: } (0, 0.9)$

$\frac{dy}{dx}=1+0.5\sin 0.5x$

$\quad =0$

$\sin x = -2$, no solution no critical points

$\frac{dy}{dx}>0$ for all x, so the function is always increasing

$\frac{d^2y}{dx^2}=0.25\cos 0.5x$

$\quad =0$ for $x=\pi+k(2\pi)$

$\frac{d^2y}{dx^2}=$ changes sign at $x=\pi+k(2\pi)$

$x \quad \pi \quad k(2\pi)$ are inflection points.

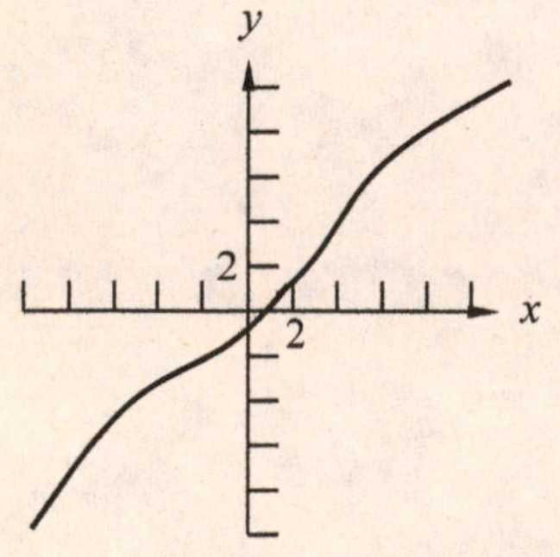

53. $y = 4\cos^2(x^2)$

$$\frac{dy}{dx} = 2\left[4\cos(x^2)\right]\left[-\sin(x^2)\right](2x)$$
$$= -16x\cos x^2 \sin x^2$$
$$\left.\frac{dy}{dx}\right|_{x=1} = -16\cos(1^2)\sin(1^2)$$
$$= -16(0.5403)(0.8415)$$
$$= -7.27$$
$$f(1) = 4\cos^2(1^2) = 4(0.5403)^2 = 1.168$$
$$y = -7.27x + b$$
$$1.168 = -7.27(1) + b$$
$$b = 8.44$$
$$y = -7.27x + 8.44$$
$$7.27x + y - 8.44 = 0$$

57. $$\lim_{x\to 0}\frac{\sin 2x}{\sin 3x} = \lim_{x\to 0}\frac{2\cos 2x}{3\cos 3x} = \frac{2}{3}$$

61. $$\lim_{x\to\infty}\frac{\ln x}{\sqrt[3]{x}} = \lim_{x\to\infty}\frac{\frac{1}{x}}{\frac{1}{3}x^{-\frac{2}{3}}}$$
$$= 3\lim_{x\to\infty}\frac{1}{\sqrt[3]{x}}$$
$$= 0$$

65. $y = \sin 3x$

$$\frac{dy}{dx} = 3\cos 3x$$
$$\frac{d^2y}{dx^2} = -9\sin 3x$$
$$= -9y$$

69. $y = e^x - 2e^{-x}$

$$\frac{dy}{dx} = e^x + 2e^{-x}$$
$$\frac{d^2y}{dx^2} = e^x - 2e^{-x} > 0$$
$$e^{2x} > 2$$
$$2x > \ln 2$$
$$x > \frac{1}{2}\ln 2 = 0.347$$

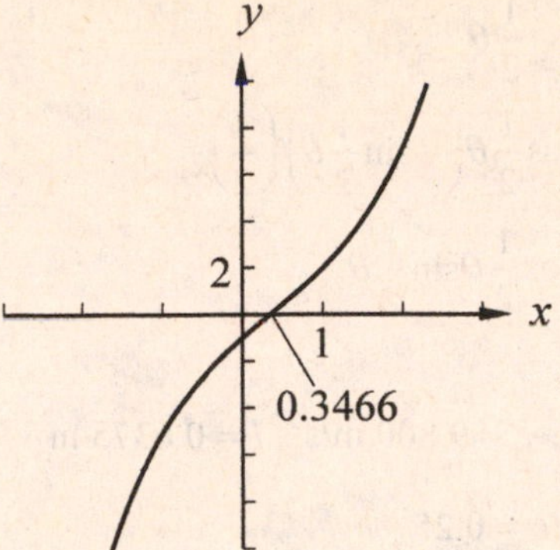

$y = e^x - 2e^{-x}$ is concave up for $x > \frac{1}{2}\ln 2$

73. $\ln r = \ln a + \ln\cos\theta - \frac{v}{w}\ln(\sec\theta + \tan\theta)$

$$r = e^{\ln a + \ln\cos\theta - \frac{v}{w}\ln(\sec\theta+\tan\theta)}$$
$$\frac{1}{r}\frac{dr}{d\theta} = -\frac{\sin\theta}{\cos\theta} - \frac{v}{w}\frac{\sec\theta\tan\theta + \sec^2\theta}{\sec\theta + \tan\theta}$$
$$\frac{dr}{d\theta} = -r\left(\cot\theta + \frac{v}{w}\sec\theta\right)$$
$$\frac{dr}{d\theta} = -e^{\ln a + \ln\cos\theta - \frac{v}{w}\ln(\sec\theta+\tan\theta)}\left(\cot\theta + \frac{v}{w}\sec\theta\right)$$

77. $\mu = \tan\theta,\ 18^\circ\left(\frac{\pi}{180^\circ}\right) = \frac{\pi}{10},\ \Delta\theta = 2^\circ\left(\frac{\pi}{180^\circ}\right) = \frac{\pi}{90}$

$$\Delta\mu \approx d\mu = \sec^2\theta d\theta$$
$$= \sec^2\frac{\pi}{10}\left(\frac{\pi}{90}\right)$$
$$= 0.03859$$

81. $y = 0.75\left(\sec\sqrt{0.15t} - 1\right)$

$$\frac{dy}{dt} = 0.75\sec\sqrt{0.15t}\tan\sqrt{0.15t}\,\frac{0.15}{2\sqrt{0.15t}}$$
$$\left.\frac{dy}{dt}\right|_{t=5.0} = 0.12 \text{ cm}$$

85. $W = 25\sin^2 2t$

$$P = \frac{dW}{dt} = 50\sin 2t\cos 2t(2)$$
$$P = 100\sin 2t\cos 2t$$

89. $I = kE_0^2 \cos^2 \frac{1}{2}\theta$

$$\frac{dI}{d\theta} = 2kE_0^2 \cos\frac{1}{2}\theta\left(-\sin\frac{1}{2}\theta\right)\left(\frac{1}{2}\right)$$

$$\frac{dI}{d\theta} = -kE_0^2 \cos\frac{1}{2}\theta \sin\frac{1}{2}\theta$$

93. $w = \sqrt{\frac{g}{l\cos\theta}}$, $g = 9.800 \text{ m/s}^2$, $l = 0.6375 \text{ m}$

$\theta = 32.50^\circ$, $d\theta = 0.25^\circ$

$$\Delta w \approx dw = \frac{1}{2}\sqrt{\frac{g}{l\cos\theta}}\tan\theta d\theta$$

$$\Delta w \approx \frac{1}{2}\sqrt{\frac{9.800}{0.6375\cos 32.5^\circ}}\tan 32.5^\circ\left(\frac{0.25\pi}{180^\circ}\right)$$

$\Delta w \approx 0.005934$ rad/s

97. $V = 32000e^{-0.14t}$

$$\frac{dV}{dt} = -4480e^{-0.14t}$$

$$\left.\frac{dV}{dt}\right|_{t=1} = -3900 \text{ \$/yr}$$

$$\left.\frac{dV}{dt}\right|_{t=5} = -2200 \text{ \$/yr}$$

101.

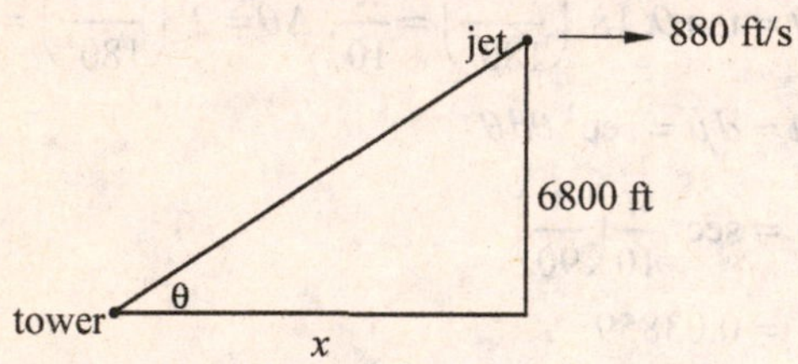

$$\theta = \tan^{-1}\frac{6800}{x}, \quad \frac{d\theta}{dt} = \frac{d\theta}{dx}\frac{dx}{dt}$$

$$\frac{d\theta}{dt} = \frac{1}{1+\left(\frac{6800}{x}\right)^2}\left(\frac{-6800}{x^2}\right)\frac{dx}{dt}$$

When $\theta = 13^\circ, x = \frac{6800}{\tan 13^\circ} = 29454$ ft

$$\left.\frac{d\theta}{dt}\right|_{\substack{x=29454 \\ dx/dt=880}} = -0.0065486205$$

When the angle of elevation of the jet from the contral tower is 13°, the angle of elevation is changing -0.0065 rad/s.

105. $x = r(\theta - \sin\theta)$; $y = r(1-\cos\theta)$; $r = 5.500$ cm;

$\frac{d\theta}{dt} = 0.12$ rad/s; $\theta = 35^\circ$

$$35^\circ = 35\frac{\pi}{180} \text{ rad}$$
$$= 0.6109 \text{ rad}$$

horizontal component of velocity,

$$v_x = \frac{dx}{dt} = \frac{d\left[5.5(\theta - \sin\theta)\right]}{dt}$$

$$= 5.5\left(\frac{d\theta}{dt} - \cos\theta\frac{d\theta}{dt}\right)$$

$$v_x\Big|_{\substack{\theta=0.6109 \\ d\theta/dt=0.12}} = 5.5(0.12 - 0.12\cos 0.6109)$$

$$= 0.119 \text{ cm/s}$$

vertical component of velocity,

$$v_y = \frac{dy}{dt} = \frac{d\left[5.5(1-\cos\theta)\right]}{dt}$$

$$= 5.5\sin\theta\frac{d\theta}{dt}$$

$$v_y\Big|_{\substack{\theta=0.6109 \\ d\theta/dt=0.12}} = (5.5)(0.12)\sin 0.6109$$

$$= 0.379$$

$$v = \sqrt{0.119^2 + 0.379^2}$$
$$= 0.397 \text{ cm/s}.$$

$$\theta = \tan^{-1}\frac{0.379}{0.119}$$
$$= 1.27$$
$$= 1.27 \cdot \frac{180}{\pi}$$
$$= 72.6^\circ$$

109. $\cos\theta = \dfrac{y}{4.0}$; $y = 4.0\cos\theta$

$\sin\theta = \dfrac{x}{4.0}$; $x = 4.0\sin\theta$

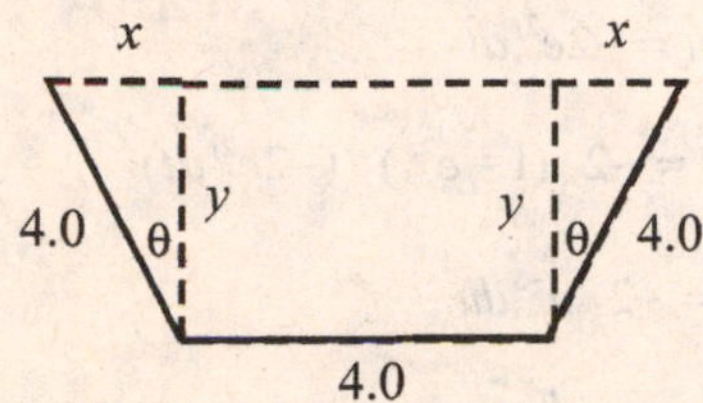

$A = 4.0y + xy$
$= 16\cos\theta + 16\sin\theta\cos\theta$
$= 16\cos\theta(1+\sin\theta)$

$\dfrac{d^2A}{d\theta^2} = -16(4\sin\theta\cos\theta + \cos\theta)$
$= -16\cos\theta(4\sin\theta + 1)$

(1)Domain: $0 \le \theta \le \dfrac{\pi}{2}$

(2)A – intercept at $\theta = 0$. Point is (0,16)

θ – intercept at $A = 0$. Point is $(\frac{\pi}{2}, 0)$

(3) Critical values

$\dfrac{dA}{d\theta} = -16\sin\theta + 16\cos^2\theta - 16\sin^2\theta$
$= -16(2\sin^2\theta + \sin\theta - 1)$
$= -16(2\sin\theta - 1)(\sin\theta + 1)$

$\dfrac{dA}{d\theta} = 0$ when $\sin\theta = \dfrac{1}{2}$ or $\sin\theta = -1$

($\sin\theta = -1, \theta = \dfrac{3\pi}{2}$ discarded, out of domain).

(4) Second derivative:

$\dfrac{d^2A}{d\theta^2} = -16(4\sin\theta\cos\theta + \cos\theta)$
$= -16\cos\theta(4\sin\theta + 1)$

$\dfrac{d^2A}{d\theta^2} < 0$ when $\sin\theta = \dfrac{1}{2}, \theta = \dfrac{\pi}{6}$ and so this maximizes the area. Point is $(\frac{\pi}{6}, 12\sqrt{3})$.

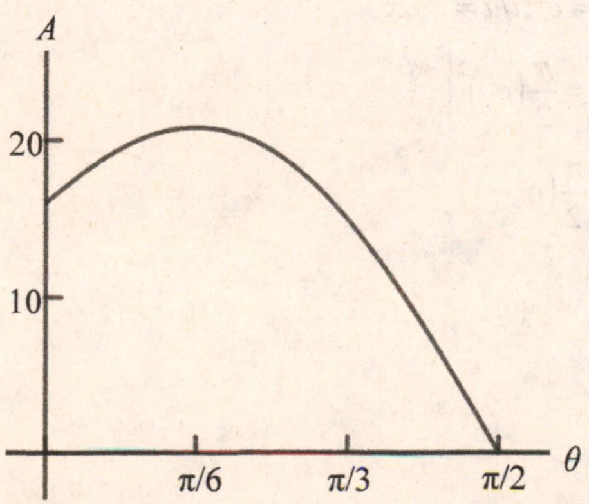

113.

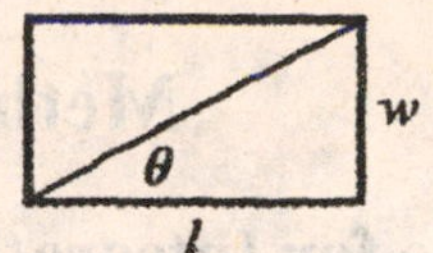

$P = 2l + 2w = 40$
$l + w = 20$
$l = 20 - w$

(a) To use an algebraic function, one dimension must be expressed in terms of the other one. We use l in terms of w:

$A = lw$
$= (20 - w)w$
$= 20w - w^2$

$\dfrac{dA}{dw} = 20 - 2w$
$= 0$ for $w = 10$ and $l = 10$

$\dfrac{d^2A}{dw^2} = -2 < 0$, so area is maximized.

The area is maximum when the rectangle is a square.

(b) To use trigonometry, both dimensions must be expressed in terms of the angle made with the diagonal.

$\tan\theta = \dfrac{w}{l}$
$= \dfrac{w}{20 - w}$ from which

$w = \dfrac{20\tan\theta}{1+\tan\theta}$ and $l = 20 - \dfrac{20\tan\theta}{1+\tan\theta}$

$A = wl$
$= \dfrac{20^2\tan\theta}{1+\tan\theta} - \dfrac{20^2\tan^2\theta}{(1+\tan\theta)^2}$

taking $\dfrac{dA}{d\theta} = 0$ gives a maximum for $\theta = 45^\circ$; again, a square.

Chapter 28

Methods of Integration

28.1 The Power Rule for Integration

1. $u = \cos x, du = -\sin x dx$

In order to be able to integrate, $\sin x$ must be changed to $-\sin x$, introducing an additional $-$ sign:

$$\cos^3 x \sin x\, dx = - \cos^3 x(-\sin x\, dx)$$

$$= \frac{1}{4}\cos^4 x + C$$

5. $u = \sin x;\ du = \cos x\, dx$

$$\int \sin^4 x \cos x\, dx = \frac{1}{5}\sin^5 x + C$$

9. $\int 4\tan^2 x \sec^2 x\, dx = 4\int \tan^2 x \sec^2 x\, dx$

$u = \tan x;\ du = \sec^2 x\, dx$

$$4\int \tan^2 x \sec^2 x dx = 4(\frac{1}{3}\tan^3 x) + C$$

$$= \frac{4}{3}\tan^3 x + C$$

13. $u = \sin^{-1} x;\ du = \frac{1}{\sqrt{1-x^2}}dx = \frac{dx}{\sqrt{1-x^2}}$

$$\int (\sin^{-1} x)^3\left(\frac{dx}{\sqrt{1-x^2}}\right) = \frac{1}{4}(\sin^{-1} x)^4 + C$$

17. $u = \ln(x+1);\ du = \frac{1}{x+1}(1)dx = \frac{dx}{x+1}$

$$\int [\ln(x+1)]^2 \frac{dx}{x+1} = \frac{1}{3}[\ln(x+1)]^3 + C$$

21. $u = 4 + e^x;\ du = e^x dx;$

$$\int 3(4+e^x)^3 e^x dx = \frac{3}{4}(4+e^x)^4 + C$$

23. $u = 1 - e^{2t}, du = -2e^{2t} dt$

$$\frac{4e^{2t}}{(1-e^{2t})^3}dt = -2\ (1-e^{2t})^{-3}(-2e^{2t} dt)$$

$$= -2\ u^{-3} du$$

$$= -2 \cdot \frac{u^{-3+1}}{-3+1} + C$$

$$= -2 \cdot \frac{1}{-2(u^2)} + C$$

$$= \frac{1}{(1-e^{2t})^2} + C$$

27. $u = 1 + \cot x, du = -\csc^2 x dx$

$${}_{\pi/6}^{\pi/4}(1+\cot x)^2(\csc^2 x\, dx)$$

$$= -{}_{\pi/6}^{\pi/4}(1+\cot x)^2(-\csc^2 x\, dx)$$

$$= -\frac{(1+\cot x)^3}{3}\Bigg|_{\pi/6}^{\pi/4}$$

$$= -\frac{(1+\cot\frac{\pi}{4})^3}{3} + \frac{(1+\cot\frac{\pi}{6})^3}{3}$$

$$= 2\sqrt{3} + \frac{2}{3}$$

$$= 4.13$$

31. $\int \frac{dx}{x\ln^2 x} = \int \ln^{-2} x\left(\frac{1}{x}dx\right) = \int u^{-2} du$

with $u = \ln x,\ du = \frac{1}{x}dx,\ n = -2$

35. Using disks,

$$V = \int_0^2 \pi\left(e^x\right)^2 dx = \int_0^2 \pi e^x\left(e^x dx\right)$$

$u = e^x;\ du = e^x dx$

$$V = \frac{\pi}{2}\left(e^x\right)^2\Big|_0^2$$

$$= \frac{\pi}{2}\left(e^4 - 1\right)$$

39. $\frac{dy}{dx} = m = \frac{(\ln x)^2}{x}$; passes through $(1, 2)$

$dy = \frac{(\ln x)^2}{x} dx$

$y = \int (\ln x)^2 \frac{dx}{x}$

$u = \ln x, du = \frac{dx}{x}; n = 2$

$y = \frac{1}{3}(\ln x)^3 + C$

Substitute $x = 1, y = 2$:

$2 = \frac{1}{3}(\ln 1)^3 + C$; therefore, $C = 2$

Therefore, $y = \frac{1}{3}(\ln x)^3 + 2$.

43. $i = 3(1 - e^{-t})^2 e^{-t}$; $\frac{dq}{dt} = 3(1 - e^{-t})^2 e^{-t}$;

$t = 0, q = 0$

$dq = 3(1 - e^{-t})^2 e^{-t} dt$;

$q = 3\int (1 - e^{-t})^2 e^{-t} dt$

$n = 2$; $u = 1 - e^{-t}$; $du = e^{-t} dt$

$q = \frac{3(1 - e^{-t})^3}{3} + C$

$= (1 - e^{-t})^3 + C$

Substitute $t = 0, q = 0$:

$0 = (1 - 1)^3 + C$; therefore, $C = 0$

Therefore, $q = (1 - e^{-t})^3$.

28.2 The Basic Logarithmic Form

1. $u = 2x + 1, du = 2dx$,

$\int \frac{dx}{2x+1} = \frac{1}{2}\int \frac{2dx}{2x+1} = \frac{1}{2}\ln|2x+1| + C$

5. $\int \frac{2x\,dx}{4 - 3x^2}$; $u = 4 - 3x^2$; $du = -6x\,dx$

$-\frac{1}{3}\int \frac{-6x\,dx}{4 - 3x^2} = -\frac{1}{3}\ln|4 - 3x^2| + C$

9. $0.4\int \frac{\csc^2 2\theta\,d\theta}{\cot 2\theta}$; $u = \cot 2\theta$; $du = -2\csc^2 2\theta\,d\theta$

$-\frac{0.4}{2}\int \frac{-2\csc^2 2\theta\,d\theta}{\cot 2\theta} = -0.2\ln|\cot 2\theta| + C$

13. $u = 1 - e^{-x}$; $du = e^{-x} dx$

$\int \frac{e^{-x} dx}{1 - e^{-x}} = \ln|1 - e^{-x}| + C$

17. $u = 1 + 4\sec x$; $du = 4\sec x \tan x\,dx$

$\int \frac{8\sec x \tan x\,dx}{1 + 4\sec x} = 2\int \frac{4\sec x \tan x\,dx}{1 + 4\sec x}$

$= 2\ln|1 + 4\sec x| + C$

21. $u = \ln r$; $du = \frac{dr}{r}$

$\int \frac{dx}{r\ln r} = \int \frac{1}{\ln r} \cdot \frac{dr}{r}$

$= \ln|\ln r| + C$

25. $n = -\frac{1}{2}$; $u = 1 - 2x$; $du = -2dx$

$\int \frac{16dx}{\sqrt{1 - 2x}} = -8\int (1 - 2x)^{-1/2}(2dx)$

$= -8(1 - 2x)^{1/2}(2) + C$

$= -16\sqrt{1 - 2x} + C$

29. $u = 4 + \tan 3x$; $du = 3\sec^2 3x\,dx$

$\int_0^{\pi/12} \frac{\sec^2 3x}{4 + \tan 3x} dx = \frac{1}{3}\int_0^{\pi/12} \frac{3\sec^2 3x\,dx}{(4 + \tan 3x)}$

$= \frac{1}{3}\ln|4 + \tan 3x|_0^{\pi/12}$

$= \frac{1}{3}(\ln 5 - \ln 4)$

$= \frac{1}{3}\ln\frac{5}{4}$

$= 0.0744$

33.

$$\begin{array}{r} 1 \\ x+4\overline{)x-4} \\ -x-4 \\ \hline -8 \end{array}$$

$\int \frac{x - 4}{x + 4} dx = \int dx - \int \frac{8}{x + 4} dx$

$= x - 8\ln|x + 4| + C$

37. $m = \frac{dy}{dx} = \frac{\sin x}{3+\cos x}; y = \int \frac{1}{3+\cos x} \times \sin x\, dx$

$y = -\int \frac{1}{3+\cos x}(-\sin x)\, dx;$

$u = 3+\cos x; du = -\sin x\, dx$

$y = -\int \frac{1}{u} du$

$= -\ln|u| + C$

$= -\ln(3+\cos x) + C$

Substitute $x = \frac{\pi}{3}, y = 2:$

$2 = -\ln\ 3+\cos\frac{\pi}{3}\ +C$

$2 = -\ln(3+0.5)+C$

$C = 2+\ln 3.5$

$y = -\ln(3+\cos x)+\ln 3.5+2;$ substituting for C

$y = \ln\frac{3.5}{3+\cos x}+2$

41. $v(t) = \frac{12.0}{0.200t+1}$

$s = \int_0^3 \frac{12.0}{0.200t+1} dt$

$u = 0.200t+1, du = 0.200dt$

$s = 60\int_0^3 \frac{0.200dt}{0.200t+1}$

$= 60\ln|0.200t+1|_0^3$

$= 60\ln 1.6 - 60\ln 1$

$= 28.2$ km

45. $t = 8.72\int_{175}^{425} \frac{1}{T-20.0} dT$

$= 8.72\ln|T-20.0|_{175}^{425}$

$= 8.72(\ln 405 - \ln 155)$

$= 8.38$ min

49. $y = \frac{50}{x^2+20}; x = 3.00; A = 6.61\text{ m}^2$

$u = x^2+20; du = 2x\,dx$

$\overline{x} = \frac{\int_0^3 x(y)dx}{6.61}$

$\overline{x} = \frac{1}{6.61}\int_0^3 x\frac{50}{x^2+20}dx$

$= \frac{25}{6.61}\int_0^3 \frac{2x\,dx}{x^2+20}$

$= \frac{25}{6.61}\ln\left|x^2+20\right|_0^3$

$\overline{x} = \frac{25}{6.61}(\ln 29 - \ln 20) = 1.41$ m

28.3 The Exponential Form

1. $u = x^3, du = 3x^2dx$

$\int x^2e^{x^3}dx = \frac{1}{3}\int e^{x^3}(3x^2dx)$

$= \frac{1}{3}e^{x^3}+C$

5. $u = 2x+5; du = 2dx$

$\int 4e^{2x+5}dx = 2\int e^{2x+5}(2dx)$

$= 2e^{2x+5}+C$

9. $y = x^3; du = 3x^2dx$

$6x^2e^{x^3}dx = 6\ e^{x^3}(x^2dx)$

$= \frac{6}{3}\ e^{x^3}(3x^2dx)$

$= 2e^{x^3}+C$

13. $u = 2\sec\theta; du = 2\sec\theta\tan\theta\, d\theta$

$\int 14(\sec\theta\tan\theta)\, e^{2\sec\theta}d\theta = 7\int e^{2\sec\theta}2\sec\theta\tan\theta\, d\theta$

$= 7e^{2\sec\theta}+C$

17. $u = 2x;\ du = 2dx$

$$\int_1^3 3e^{2x}(e^{-2x} - 1) = 3\int(e^0 - e^{2x})dx$$
$$= 3\int dx - \frac{3}{2}\int e^{2x}(2dx)$$
$$= 3x - \frac{3}{2}e^{2x}\Big|_1^3$$
$$= 9 - \frac{3}{2}e^6 - 3 - \frac{3}{2}e^2$$
$$= 6 - \frac{3}{2}(e^6 - e^2) = -588$$

21. $u = \tan^{-1} 2x;\ du = \frac{2}{4x^2+1}dx$

$$\int \frac{e^{\tan^{-1} 2x}dx}{8x^2+2} = \frac{1}{4}\int e^{\tan^{-1} 2x}\frac{2dx}{4x^2+1}$$
$$= \frac{1}{4}e^{\tan^{-1} 2x} + C$$

25. $u = \cos^2 x;$

$du = 2\cos x(-\sin x)dx$

$= -2\sin x \cos x\, dx$

$= -2\sin 2x\, dx$

$$\int_0^\pi (\sin 2x)e^{\cos^2 x}dx = -\frac{1}{2}\int_0^\pi e^{\cos^2 x}(-2\sin 2x\, dx)$$
$$= -\frac{1}{2}e^{\cos^2 x}\Big|_0^\pi$$
$$= -\frac{1}{2}[e^{(\cos\pi)^2} - e^{(\cos 0)^2}]$$
$$= -\frac{1}{2}[e - e] = 0$$

29. $A = \int_0^2 3e^x dx$

$$= 3e^x\Big|_0^2$$
$$= 3e^2 - 3e^0 = 19.2$$

33. $y = e^{x^2},\ x = 1, y = 0, x = 2$

Rotated about y-axis

Generates a shell of volume dV

$dV = 2\pi rh\, dx;\ r = x;\ h = y$

$V = 2\pi\int_1^2 xy\, dx = 2\pi\int_1^2 xe^{x^2}dx$

$u = x^2;\ du = 2x\, dx$

$V = \pi\int_1^2 e^{x^2}(2x\, dx)$

$= \pi e^{x^2}\Big|_1^2$

$= \pi(e^4 - e^1)$

$V = 163 \text{ units}^3$

37. To show $\int b^u du = \frac{b^u}{\ln b} + C;\ (b > 0;\ b \neq 1)$

Eq. 27.15: $\frac{d}{dx}b^u = b^u \ln b\frac{du}{dx}$

This means that the differential of b^u is

$d(b^u) = b^u \ln b(du),$

and $\frac{d(b^u)}{\ln b} = b^u du$

Reversing the process,

$$\int b^u du = \frac{1}{\ln b}b^u + C$$

41. $qe^{t/RC} = \frac{E}{R}\int e^{t/RC}dt$

$u = \frac{t}{RC};\ du = \frac{1}{RC}dt$

$qe^{t/RC} = RC\cdot\frac{E}{R}\int e^{t/RC}\ \frac{1}{RC}\ dt$

$qe^{t/RC} = EC(e^{t/RC}) + C_1$

Substitute $q=0$, $t = 0$:

$0 = EC + C_1$

$C_1 = -EC$

$qe^{t/RC} = EC(e^{t/RC}) - EC$

$q = EC - \frac{EC}{e^{t/RC}}$

$q = EC(1 - e^{-t/RC})$

28.4 Basic Trigonometric Forms

1. $u = x^3, du = 3x^2 dx$

If we change $x\,dx$ to $3x^2 dx$, then

$\sec^2 x^3 (3x^2 dx) = \tan x^3 + C$

5. $u = 3\theta; du = 3\,d\theta$

$$\int \sec^2 3\theta\, d\theta = \frac{1}{3}\int \sec^2 3\theta(3d\theta)$$

$$= \frac{1}{3}\tan 3\theta + C$$

9. $u = x^3;\ du = 3x^2 dx$

$${}_{0.5}^{1} x^2 \cot x^3 dx = \frac{1}{3}\ {}_{0.5}^{1} \cot x^3 (3x^2 dx)$$

$$= \frac{1}{3}\ln\left|\sin x^3\right|\Big|_{0.5}^{1}$$

$$= \frac{1}{3}(\ln|\sin 1| - \ln\left|\sin\frac{1}{8}\right|$$

$$= 0.6365$$

13. $u = \frac{1}{x} = x^{-1}; du = -1x^{-2}dx = -\frac{dx}{x^2}$

$$\int \frac{\sin(\frac{1}{x})}{2x^2}dx = -\frac{1}{2}\int \sin\left(\frac{1}{x}\right)\left(-\frac{dx}{x^2}\right)$$

$$= -\frac{1}{2}\left[-\cos\left(\frac{1}{x}\right)\right] + C$$

$$= \frac{1}{2}\cos\left(\frac{1}{x}\right) + C$$

17. $u = e^x, du = e^x dx$

$$\int \frac{e^x}{\sin(e^x)}dx = \int \csc(e^x)\cdot e^x dx$$

$$= \ln\left|\csc\left(e^x\right) - \cot\left(e^x\right)\right| + C$$

21. $u = 3x, du = 3dx$

$$\cos 3x \sec^2 3x\, dx = \frac{1}{3}\quad \sec 3x \cdot 3\,dx$$

$$= \frac{1}{3}\ln|\sec 3x + \tan 3x| + C$$

25. $\sin 3x\ \ \frac{1}{\sin 3x} + \frac{1}{\cos 3x}\ \ = 1 + \tan 3x;$

$u = 3x;\ du = 3\,dx$

$${}_{0}^{\pi/9} \sin 3x(\csc 3x + \sec 3x)dx$$

$$= \int_0^{\pi/9}(1 + \tan 3x)dx$$

$$= \int_0^{\pi/9} dx + \int_0^{\pi/9}\tan 3x\, dx$$

$$= \int_0^{\pi/9} dx + \frac{1}{3}\int_0^{\pi/9}\tan 3x(3\,dx)$$

$$= \left(x - \frac{1}{3}\ln|\cos 3x|\right)\Bigg|_0^{\pi/9}$$

$$= \frac{\pi}{9} - \frac{1}{3}\ln\left|\cos\frac{\pi}{3}\right| - \left(0 - \frac{1}{3}\ln|\cos 0|\right)$$

$$= \frac{\pi}{9} - \frac{1}{3}\ln\left(\frac{1}{2}\right) = \frac{\pi}{9} + \frac{1}{3}\ln 2$$

$$= 0.580$$

29. $\frac{dx}{1 + \sin x} = \frac{dx}{1 + \sin x}\cdot\frac{1 - \sin x}{1 - \sin x}$

$$= \frac{(1 - \sin x)}{1 - \sin^2 x}dx$$

$$= \frac{1 - \sin x}{\cos^2 x}dx$$

$$= \frac{1}{\cos^2 x}dx + \frac{-\sin x\, dx}{\cos^2 x}$$

$$= (\sec^2 x\, dx - \sec x \tan x)dx$$

$$= \tan x - \sec x + C$$

33. $y = \sec x;\ x = 0;\ x = \frac{\pi}{3}$

$y = 0$; rotated about x-axis, discs.

$dV = \pi r^2 dx;\ r = y$

$$V = {}_{0}^{\pi/3}\ \pi y^2 dx$$

$$= \pi\ {}_{0}^{\pi/3} \sec^2 x\, dx$$

$$= \pi \tan x\Big|_0^{\pi/3}$$

$$= \pi\sqrt{3} = 5.44$$

37. $y = \tan x^2;\ y = 0,\ x = 1$

$$\overline{x} = \frac{\int_0^1 xy\,dx}{0.3984} = \frac{\int_0^1 (\tan x^2)x\,dx}{0.3984}$$

$u = x^2;\ du = 2x\,dx$

$$\overline{x} = \frac{\frac{1}{2}\int_0^1 \tan x^2 (2x\,dx)}{0.3984}$$

$$= \frac{-\frac{1}{2}\ln\left|\cos x^2\right|\Big|_0^1}{0.3984}$$

$$= \frac{-\frac{1}{2}(\ln \cos 1 - \cos 0)}{0.3984} = \frac{0.3078}{0.3984}$$

$$= 0.7726 \text{ m}$$

28.5 Other Trigonometric Forms

1. $\int \sin^2 3x\,dx = \int \frac{1}{2}(1 - \cos 6x)\,dx$

$$= \frac{1}{2}\int dx - \frac{1}{12}\int \cos 6x(6\,dx)$$

$$= \frac{x}{2} - \frac{1}{12}\sin 6x + C$$

5. $u = 2x; du = 2dx; u = \cos 2x; du = -2\sin 2x\,dx$

$$\int \sin^3 2x\,dx = \int \sin^2 2x \sin 2x\,dx$$

$$= \int (1 - \cos^2 2x)\sin 2x\,dx$$

$$= \int \sin 2x\,dx - \int \cos^2 2x \sin 2x\,dx$$

$$= \frac{1}{2}\int \sin 2x(2dx) + \frac{1}{2}\int \cos^2 2x(-2\sin 2x)dx$$

$$= -\frac{1}{2}\cos 2x + \frac{1}{2}\frac{\cos^3 2x}{3} + C$$

$$= -\frac{1}{2}\cos 2x + \frac{1}{6}\cos^3 2x + C$$

9. $\int_0^{\pi/2} \sin^4 x \cos^3 x\,dx = \int_0^{\pi/2} \sin^4 x \cos^2 x \cdot \cos x\,dx$

$$= \int_0^{\pi/2} \sin^4 x(1-\sin^2 x)\cdot \cos x\,dx$$

$$= \int_0^{\pi/2} (\sin^4 x - \sin^6 x)\cdot \cos x\,dx$$

$$= \frac{1}{5}\sin^5 x - \frac{1}{7}\sin^7 x \Big|_0^{\pi/2}$$

$$= \frac{1}{5} - \frac{1}{7}$$

$$= \frac{2}{35}$$

13. $\int \sin^3 2\theta \cos^2 2\theta\, d\theta = -\frac{1}{2}\int \sin^2 2\theta \cos^2 2\theta \cdot -2\sin 2\theta\, d\theta$

$$= -\frac{1}{2}\int \left(1-\cos^2 2\theta\right)\cos^2 2\theta \cdot -2\sin 2\theta\, d\theta$$

$$= -\frac{1}{2}\int \cos^2 2\theta \cdot -2\sin 2\theta\, d\theta + \frac{1}{2}\int \cos^4 2\theta \cdot -2\sin 2\theta\, d\theta$$

$$= -\frac{1}{6}\cos^3 2\theta + \frac{1}{10}\cos^5 2\theta + C$$

17. $\int_0^{\pi/4} \frac{\tan x}{\cos^4 x}dx = \int_0^{\pi/4} \tan x \sec^4 x\,dx$

$u = \tan x; du = \sec^2 x\,dx$

$\int_0^{\pi/4} \tan x \sec^4 x\,dx = \int_0^{\pi/4} \tan x \sec^2 x(1+\tan^2 x)dx$

$$= \int_0^{\pi/4} (\tan x)^1 \sec^2 x\,dx + \int_0^{\pi/4} \tan^3 x \sec^2 x\,dx$$

$$= \frac{1}{2}(\tan x)^2 + \frac{1}{4}(\tan x)^4 \Big|_0^{\pi/4}$$

$$= \frac{1}{2}(1)^2 + \frac{1}{4}(1)^4 = \frac{3}{4}$$

21. $\int 0.5 \sin s \sin 2s\, ds = \int \sin^2 s \cos s\, ds$

$$= \frac{\sin^3 s}{3} + C$$

25. $\frac{1-\cot x}{\sin^4 x}dx$

$= (1-\cot x)\csc^4 x\, dx$

$= (1-\cot x)\csc^2 x(1+\cot^2 x)dx$

$= (1-\cot x+\cot^2 x-\cot^3 x)\csc^2 x\, dx$

$= \csc^2 x\, dx - \cot x\csc^2 x\, dx + \cot^2 x\csc^2 x\, dx - \cot^3 x\csc^2 x\, dx$

$= \csc^2 x\, dx + \cot x(-\csc^2 x\, dx) - \cot^2 x(-\csc^2 x\, dx) + \cot^3 x(-\csc^2 x\, dx)$

$= -\cot x + \frac{\cot^2 x}{2} - \frac{\cot^3 x}{3} + \frac{\cot^4 x}{4} + C$

$= \frac{1}{4}\cot^4 x - \frac{1}{3}\cot^3 x + \frac{1}{2}\cot^2 x - \cot x + C$

29. $\sec^6 x\, dx = \sec^4 x(1+\tan^2 x)dx$

$= \sec^4 x\, dx + \sec^4 x\tan^2 x\, dx$

$= \sec^2 x(1+\tan^2 x)dx + \sec^2 x(1+\tan^2 x)\tan^2 x dx$

$= \sec^2 x\, dx + \tan^2 x\sec^2 x\, dx + \tan^2 x\sec^2 x\, dx + \tan^4 x\sec^2 x\, dx$

$= \tan x + \frac{2}{3}\tan^3 x + \frac{1}{5}\tan^5 + C$

33. $u = e^{-x}, du = -e^{-x}$

$\frac{\sec e^{-x}}{e^x}dx = - \sec u du$

$= -\ln|\sec u + \tan u|$

$= -\ln\left|\sec e^{-x} + \tan e^{-x}\right| + C$

37. Rotate about x-axis, using discs

$V = \pi {}_0^\pi y^2 dx$

$= \pi {}_0^\pi \sin^2 x\, dx$

$= \pi {}_0^\pi \frac{1}{2}(1-\cos 2x)dx$

$= \frac{\pi}{2} {}_0^\pi dx - \frac{\pi}{2} {}_0^\pi \cos 2x\, dx$

$= \frac{\pi}{2}x - \frac{\pi}{2}\times\frac{1}{2}\sin 2x\Big|_0^\pi$

$= \frac{\pi^2}{2} - \frac{\pi}{4}\sin 2\pi - 0 + \frac{\pi}{4}\sin 0$

$= \frac{1}{2}\pi^2 = 4.93$

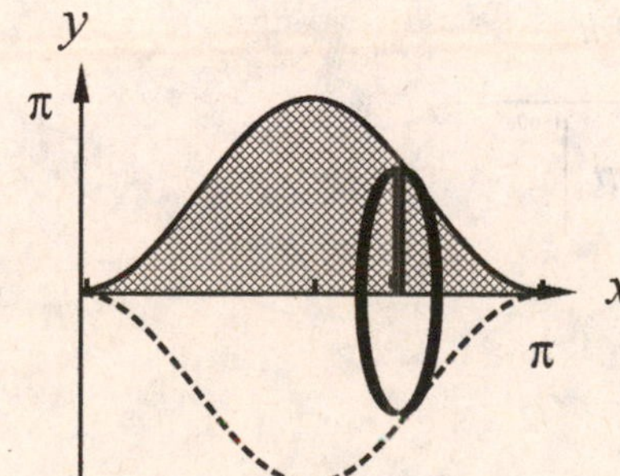

41. $\sin x \cos x\, dx; u = \sin x; du = \cos x\, dx$

$$u\, du = \frac{1}{2}u^2 + C = \frac{1}{2}\sin^2 x + C_1$$

Let $u = \cos x; du = -\sin x\, dx$

$$-\quad \cos x(-\sin x)dx = -\frac{1}{2}\cos^2 x + C_2$$

$$\frac{1}{2}\sin^2 x + C_1 = \frac{1}{2}(1 - \cos^2 x + C_2)$$

$$= \frac{1}{2} - \frac{1}{2}\cos^2 x + C_1$$

$$-\frac{1}{2}\cos^2 x + C_2 = \frac{1}{2} - \frac{1}{2}\cos^2 x + C_1$$

$$C_2 = C_1 + \frac{1}{2}$$

45. $v = \dfrac{\sin^3 t}{3} + 6$

$$s = \quad \frac{\sin^3 t}{3}dt + \quad 6dt$$

$$s = \frac{1}{3}\quad \sin^2 t \sin t dt + \quad 6dt$$

$$s = \frac{1}{3}\quad \left(1 - \cos^2 t\right)\sin t dt + 6 \quad dt$$

$$s = \frac{1}{3}\quad \sin t dt - \frac{1}{3}\quad \cos^2 t \sin t dt + 6 \quad dt$$

$$s = \frac{1}{3}\cos t + \frac{1}{9}\cos^3 t + 6t + s_0$$

$$s(0) = 0 = -\frac{1}{3}\cos 0 + \frac{1}{9}\cos^3 0 + 6(0) + s_0 \qquad s_0 \quad \frac{2}{9}$$

$$s = -\frac{1}{3}\cos t + \frac{1}{9}\cos^3 t + 6t + \frac{2}{9}$$

49. $V_{\text{rms}} = \sqrt{\dfrac{1}{T}\ {}_0^T\ y^2 dt}$

$$V_{\text{rms}} = \sqrt{\frac{1}{1/60.0}\ {}_0^{1/60.0}(340 \sin 120\pi t)^2\, dt}$$

$$= \sqrt{60}\sqrt{{}_0^{1/60.0}\ 340^2\ \frac{1 - \cos 240\pi t}{2}\, dt}$$

$$= \sqrt{60}\sqrt{\frac{340^2}{2}\quad t - \frac{1}{240\pi}\sin 240\pi t\ \Big|_0^{1/60.0}}$$

$$= 240 \text{ V}$$

28.6 Inverse Trigonometric Forms

1. Change dx to $-x\,dx$, then

$$\frac{-x\,dx}{\sqrt{9-x^2}} = \frac{1}{2}\left(9-x^2\right)^{-1/2}(-2x\,dx)$$

$$= \frac{1}{2}\frac{1\left(9-x^2\right)^{1/2}}{\frac{1}{2}} + C$$

$$= \sqrt{9-x^2} + C$$

5. $a = 8; u = x; du = dx;$

$$\int \frac{12dx}{64+x^2} = \frac{3}{2}\tan^{-1}\frac{x}{8} + C$$

9.

$$^{2}_{0}\frac{3e^{-t}dt}{1+9e^{-2t}}$$

$$= -\,^{2}_{0}\frac{-3e^{-t}dt}{1+(3e^{-t})^2} \text{ which has form } \frac{dx}{1+x^2}$$

$$= -\tan^{-1}\left(3e^{-t}\right)\Big|_0^2$$

$$= -\tan^{-1}\frac{3}{e^2} + \tan^{-1}3$$

$$= 0.863$$

13. $u = 9x^2 + 16; du = 18x\,dx$

$$\frac{8x\,dx}{9x^2+16} = \frac{8}{18}\frac{18x\,dx}{9x^2+16}$$

$$= \frac{4}{9}\ln\left|9x^2+16\right| + C$$

17. $a = 1; u = e^x; du = e^x dx$

$$\int \frac{2e^x dx}{\sqrt{1-e^{2x}}} = 2\sin^{-1}(e^x) + C$$

21. $a = 2; u = x+2; du = dx$

$$\frac{4dx}{\sqrt{-4x-x^2}} = \frac{4dx}{\sqrt{4-(x+2)^2}}$$

$$= 4\frac{dx}{\sqrt{4-(x+2)^2}}$$

$$= 4\sin^{-1}\frac{x+2}{2} + C$$

25. $a = 2; u = x; du; = dx; n = \frac{1}{2}; u = 4 - x^2;$

$du = -2x\,dx$

$$\frac{2-x}{\sqrt{4-x^2}}dx = \frac{2dx}{\sqrt{4-x^2}} - \frac{x\,dx}{\sqrt{4-x^2}}$$

$$= 2\frac{dx}{\sqrt{4-x^2}} + \frac{1}{2}\frac{(-2x\,dx)}{\left(4-x^2\right)^{1/2}}$$

$$= 2\sin^{-1}\frac{x}{2} + \frac{1}{2}\left(4-x^2\right)^{1/2}\cdot 2 + C$$

$$= 2\sin^{-1}\frac{x}{2} + \sqrt{4-x^2} + C$$

29.

$$\frac{x^2+3x^5}{1+x^6}dx = \frac{x^2}{1+x^6}dx + \frac{3x^5}{1+x^6}dx$$

$$\frac{x^2}{1+x^6}dx = \frac{1}{3}\frac{3x^2}{1+\left(x^3\right)^2}dx$$

$$= \frac{1}{3}\tan^{-1}x^3$$

$$\frac{3x^5}{1+x^6}dx = \frac{1}{2}\frac{6x^5dx}{1+x^6}$$

$$= \frac{1}{2}\ln\left(1+x^6\right)$$

$$\frac{x^2+3x^5}{1+x^6}dx = \frac{1}{3}\tan^{-1}x^3 + \frac{1}{2}\ln\left(1+x^6\right) + C$$

33. **(a)** General power, $u^{-1/2}du$ where $u = 4 - 9x^2$.
$du = -18x\,dx$; numerator can fit du of denominator.
Square root becomes $-1/2$ power.
Does not fit inverse sine form.

(b) Inverse sine; $a = 2; u = 3x; du = 3dx$

(c) Logarithmic; $u = 4 - 9x; du = -9dx$

37. $y = \frac{1}{1+x^2};$

$$A = \,^{2}_{0}\frac{1}{1+x^2}dx;$$

$a = 1; u = x; du = dx$

$$A = \frac{1}{1}\tan^{-1}\frac{x}{1}\Big|_0^2$$

$$= \tan^{-1}2 - \tan^{-1}0$$

$$= 1.11$$

41. $a = d; u = x; du = dx$

$$kd \quad \frac{dx}{d^2 + x^2} = kd \cdot \frac{1}{d} \tan^{-1} \frac{x}{d} + C$$

$$= k \tan^{-1} \quad \frac{x}{d} + C$$

45. $y = \frac{1}{1 + x^6}$; x-axis, $x = 1; x = 2$, WRT y-axis

$$I_y = \rho \, {}_1^2 \, x^2 y\, dx$$

$$= \rho \, {}_1^2 \, x^2 \frac{1}{1 + x^6} dx$$

$a = 1; u = x^3; du = 3x^2 dx$

$$I_y = \frac{\rho}{3} \, {}_1^2 \, \frac{3x^2 dx}{1 + x^6}$$

$$= \frac{\rho}{3} \tan^{-1} x^3 \Big|_{10}^{2}$$

$$I_y = \frac{\rho}{3}\left(\tan^{-1} 8 - \tan^{-1} 1\right)$$

$$= 0.220\rho$$

28.7 Integration by Parts

1. $u = \sqrt{1 - x}, dv = x dx$

$$du = \frac{-1}{2\sqrt{1 - x}} dx, v = \frac{x^2}{2}$$

$$x\sqrt{1 - x} dx = \frac{x^2}{2}\sqrt{1 - x} - \quad \frac{x^2}{2} \frac{-dx}{2\sqrt{1 - x}}$$

$$= \frac{x^2}{2}\sqrt{1 - x} + \frac{1}{4} \quad \frac{x^2 dx}{\sqrt{1 - x}}$$

No, the substitution $u = \sqrt{1 - x}, dv = x\,dx$ does not work since $\frac{x^2 dx}{\sqrt{1 - x}}$ is more complex than $x\sqrt{1 - x}dx$.

5. $4xe^{2x} dx$;

$u = x$; $du = dx$; $dv = e^{2x} dx$

$$v = \frac{1}{2} \quad e^{2x}(2dx) = \frac{1}{2} e^{2x}$$

$$4xe^{2x} dx = \frac{4}{2} xe^{2x} - \frac{4}{2} \quad e^{2x} dx$$

$$= 2xe^{2x} - \quad e^{2x}(2\,dx)$$

$$= 2xe^{2x} - e^{2x} + C$$

9. $2 \tan^{-1} x\, dx$;

$u = \tan^{-1} x; du = \frac{1}{1 + x^2} dx$;

$dv = dx; v = x$

$$2 \quad \tan^{-1} x\, dx = 2 \quad x \tan^{-1} x - \quad \frac{x\,dx}{1 + x^2}$$

$u = x^2; du = 2xdx$

$$= 2 \quad x \tan^{-1} x - \frac{1}{2} \quad \frac{2x\,dx}{1 + x^2}$$

$$= 2x \tan^{-1} x - \ln\left(1 + x^2\right) + C$$

13. $x \ln x\, dx$;

$u = \ln x; du = \frac{dx}{x}; dv = x\,dx; v = \frac{x^2}{2}$

$$x \ln x\, dx = \frac{1}{2} x^2 \ln x - \frac{1}{2} \quad x^2 \frac{dx}{x}$$

$$= \frac{1}{2} x^2 \ln x - \frac{1}{2} \quad x\,dx$$

$$= \frac{1}{2} x^2 \ln x - \frac{1}{2} \cdot \frac{x^2}{2} + C$$

$$= \frac{1}{2} x^2 \ln x - \frac{1}{4} x^2 + C$$

17. ${}_0^{\pi/2} e^x \cos x\, dx;$

We perform the integral as an indefinite integral first.

$u = e^x ;= e^z dx; dv = \cos x\, dx; v = \sin x$

$e^x \cos x\, dx = e^x \sin x - \ \sin x e^x\, dx$

$u = e^x ; du = e^x dx; dv = \sin x\, dx; v = -\cos x$

$e^x \cos x\, dx = e^x \sin x - \left(-e^x \cos x + \ e^x \cos dx\right)$

$e^x \cos dx = e^x \sin x + e^x \cos x - \ e^x \cos x\, dx$

$2\ e^x \cos x\, dx = e^x \sin x + e^x \cos x + C$

Now we evaluate the definite integral:

$$ {}_0^{\pi/2} e^x \cos x dx = \frac{1}{2} e^x \left(\sin + \cos x\right)\Big|_0^{\pi/2}$$

$$= \frac{1}{2} e^{\pi/2}(1+0) - \frac{1}{2} e^0 (0+1)$$

$$= \frac{1}{2}\left(e^{\pi/2} - 1\right) = 1.91$$

21. $xe^x\, dx;$

$u = x, du = dx; dv = e^x dx, v = e^x$

$xe^x\, dx = xe^x - \ e^x dx$

$= xe^x - e^x + C$

${}_0^{\ln 2} xe^x\, dx = xe^x - e^x \Big|_0^{\ln 2}$

$= (2\ln 2 - \ln 2) - (0 - 1)$

$= \ln 2 + 1$

25. $u = \cos(\ln x), \qquad dv = dx$

$v = x$

$du = -\sin(\ln x) \cdot \frac{1}{x} dx$

$\cos(\ln x) dx = x\cos(\ln x) + \ x\sin(\ln x)\frac{1}{x} dx$

$u = \sin(\ln x), \qquad dv = dx$

$v = x$

$du = \cos(\ln x) \cdot \frac{1}{x} dx$

$\cos(\ln x) dx = x\cos(\ln x) + x\sin(\ln x)$

$- \ x\cos(\ln x) \cdot \frac{1}{x} dx$

$2\ \cos(\ln x) dx = x\cos(\ln x) + x\sin(\ln x)$

$\cos(\ln x) dx = \frac{x}{2}\left(\cos(\ln x) + \sin(\ln x)\right) + C$

29. $A = {}_0^2 xe^{-x} dx;$

$u = x;\ du = dx;\ dv = e^{-x} dx; v = \ e^{-x} dx = -e^{-x}$

$A = -xe^{-x}\Big|_0^2 - {}_0^2 -e^{-x} dx$

$= -xe^{-x} - e^{-x}\Big|_0^2$

$= 2e^{-2} - e^{-2} - (0 - 1)$

$= 1 - \frac{3}{e^2} = 0.594$

33. $\bar{x} = \dfrac{{}_a^b xydx}{{}_a^b ydx}$

$= \dfrac{{}_0^{\pi/2} x(\cos x) dx}{{}_0^{\pi/2} \cos x dx}$

Let $u = x$; $du = dx$; $dv = \cos x$; $v = \sin x$

$\bar{x} = \dfrac{x \sin x\big|_0^{\pi/2} - {}_0^{\pi/2} \sin x dx}{\sin x\big|_0^{\pi/2}}$

$= \dfrac{x \sin x\big|_0^{\pi/2} - (-\cos x)\big|_0^{\pi/2}}{1}$

$= x\sin x + \cos x\big|_0^{\pi/2}$

$= \frac{\pi}{2} - 1 = 0.571$

37. $x\sin x\, dx;$

$u = x, du = dx; dv = \sin x\, dx, v = -\cos x$

$x\sin x\, dx = -x\cos x + \ \cos x\, dx$

$= -x\cos x + \sin x + C$

(a) $A = {}_0^{\pi} x\sin x\, dx = -x\cos x + \sin x\big|_0^{\pi} = \pi$

(b) $A = {}_{\pi}^{2\pi} -x\sin x\, dx = x\cos x - \sin x\big|_{\pi}^{2\pi} = 3\pi$

(The sign changes because the curve passes below the x – axis on this interval.)

(c) $A = {}_{2\pi}^{3\pi} x\sin x\, dx = -x\cos x + \sin x\big|_{2\pi}^{3\pi} = 5\pi$

The area increases by 2π for each successive interval.

41. $i = e^{-2t}\cos t = \dfrac{dy}{dt}; t = 0, q = q_0 = 0$

$q = \ e^{-2t}\cos t\ dt;$

$u = e^{-2t}; du = -2e^{-2t}dt; dv = \cos x\,dt; v = \sin t$

$q = \ e^{-2t}\cos t\,dt = e^{-2t}\sin t + 2\ \ e^{-2t}\sin t\,dt;$

$u = e^{-2t}; du = -2e^{-2t}dt; dv = \sin t\,dt$

$v = -\cos t$

$\ e^{-2t}\cos t\,dt = e^{-2t}\sin t + 2\left(-e^{-2t}\cos t - \ \ 2e^{-2t}\cos t\ dt\right)$

$\ e^{-2t}\cos t\,dt = e^{-2t}\sin t - 2e^{-2t}\cos t - 4\ \ e^{-2t}\cos t\,dt$

$5\ \ e^{-2t}\cos t\,dt = e^{-2t}(\sin t - 2\cos t)$

$\ e^{-2t}\cos t\,dt = \dfrac{1}{5}e^{-2t}(\sin t - 2\cos t) + C$

Substitute $t = 0, q = 0$:

$0 = \dfrac{1}{5}(1)(\sin 0 - 2\cos 0) + C$; therefore, $C = \dfrac{2}{5}$

$q = \dfrac{1}{5}e^{-2t}(\sin t - 2\cos t) + \dfrac{2}{5}$

$= \dfrac{1}{5}\ e^{-2t}(\sin t - 2\cos t) + 2$

28.8 Integration by Trigonometric Substitution

1. Delete the x^2 before the radical in the denominator.

$$\int \frac{dx}{\sqrt{1-x^2}} = \sin^{-1} x + C$$

5. $\displaystyle\int \frac{dx}{x^2\sqrt{x^2+1}}$.

Let $x = \tan\theta$, $dx = \sec^2\theta\,d\theta, \sqrt{x^2+1} = \sec\theta$

$$\int \frac{dx}{x^2\sqrt{x^2+1}} = \int \frac{\sec^2\theta\,d\theta}{\tan^2\theta\sec\theta}$$

$$= \int \frac{d\theta}{\tan\theta}$$

$$= \int \cot\theta\csc\theta\,d\theta$$

9. Let $x = \sin\theta$; $dx = \cos\theta\,d\theta, \sqrt{1-x^2} = \cos\theta$

$$\frac{\sqrt{1-x^2}}{x^2}dx = \frac{\cos\theta}{\sin^2\theta}\cos\theta\,d\theta$$

$$= \frac{\cos^2\theta}{\sin^2\theta}d\theta$$

$$= \cot^2\theta\,d\theta$$

$$= \left(\csc^2\theta - 1\right)d\theta$$

$$= \csc^2\theta\,d\theta - \ d\theta$$

$$= -\cot\theta - \theta + C$$

$$= \frac{-\sqrt{1-x^2}}{x} - \sin^{-1} x + C$$

13. Let $z = 3\tan\theta$; $dz = 3\sec^2\theta\,d\theta, \sqrt{z^2+9} = \sec\theta$

$$\frac{6dz}{z^2\sqrt{z^2+9}} = 6\ \frac{3\sec^2\theta\,d\theta}{9\tan^2\theta(3\sec\theta)}$$

$$= \frac{6}{9}\ \frac{\sec\theta\,d\theta}{\tan^2\theta}$$

$$= \frac{6}{9}\ \frac{\cos\theta\,d\theta}{\sin^2\theta}$$

$$= \frac{6}{9}\ \csc\theta\cot\theta\,d\theta$$

$$= -\frac{6}{9}\csc\theta + C$$

$$\tan\theta = \frac{z}{3}; \csc\theta = \frac{\sqrt{z^2+9}}{z}$$

$$\frac{-6}{9}\csc\theta + C = -\frac{2\sqrt{z^2+9}}{3z} + C$$

17. $\ {}_0^{0.5} \dfrac{x^3dx}{\sqrt{1-x^2}}$,

$x = \sin\theta$; $dx = \cos\theta\,d\theta, \sqrt{1-x^2} = \cos\theta$

$$\frac{\sin^3\theta\cos\theta\,d\theta}{\cos\theta} = \ \sin^3\theta\,d\theta$$

$$= \ \sin\theta\sin^2\theta\,d\theta$$

$$= \ \sin\theta(1-\cos^2\theta)d\theta$$

$$= \ \sin\theta\,d\theta - \ \cos^2\theta\sin\theta\,d\theta$$

$$= -\cos\theta + \frac{\cos^3\theta}{3} + C$$

$$\ {}_0^{0.5} \frac{x^3dx}{\sqrt{1-x^2}} = -\sqrt{1-x^2} + \frac{1}{3}\left(\sqrt{1-x^2}\right)^3\Big|_0^{0.5}$$

$$= -\sqrt{1-0.5^2} + \frac{1}{3}\left(\sqrt{1-0.5^2}\right)^3 + \sqrt{1} - \frac{1}{3}\sqrt{1}$$

$$= 0.017$$

21. $\dfrac{dy}{y\sqrt{4y^2-9}};$

$2y = 3\sec\theta; y = \dfrac{3}{2}\sec\theta, dy = \dfrac{3}{2}\sec\theta\tan\theta\, d\theta$

$\sqrt{4y^2-9} = 3\tan\theta$

$$\frac{\frac{3}{2}\sec\theta\tan\theta\, d\theta}{\frac{3}{2}\sec\theta\cdot 3\tan\theta} = \frac{1}{3}\ d\theta$$

$$= \frac{1}{3}\theta + C$$

$$= \frac{1}{3}\sec^{-1}\frac{2}{3}y + C$$

$$\int_{2.5}^{3}\frac{dy}{y\sqrt{4y^2-9}} = \frac{1}{3}\sec^{-1}\ \frac{2y}{3}\Bigg|_{2.5}^{3}$$

$$= \frac{1}{3}\cos^{-1}\ \frac{3}{2y}\Bigg|_{2.5}^{3}$$

$$= 0.0400$$

25. $\displaystyle\int\frac{2dx}{\sqrt{e^{2x}-1}}$

$a = 1; e^x = \sec\theta; e^x dx = \sec\theta\tan\theta d\theta$

$dx = \dfrac{\sec\theta\tan\theta\, d\theta}{e^x} = \dfrac{\sec\theta\tan\theta\, d\theta}{\sec\theta} = \tan\theta\, d\theta$

$\sqrt{e^{2x}-1} = \tan\theta$

$$2\int\frac{\tan\theta\, d\theta}{\tan\theta} = 2\int d\theta$$

$$= 2\theta + C$$

$$= 2\sec^{-1} e^x + C$$

$$\ln\left|\frac{x(x-2)}{(x-1)^2}\right|$$

29.

$A = \int_0^2 \sqrt{4-x^2}\, dx,$

$x = 2\sin\theta, dx = 2\cos\theta\, d\theta, \sqrt{4-x^2} = 2\cos\theta$

$$\int\sqrt{4-x^2}\,dx = 4\int\cos^2\theta\, d\theta$$

$$= 4\int\frac{1+\cos 2\theta}{2}d\theta$$

$$= 2\theta + \int\cos 2\theta(2d\theta)$$

$$= 2\theta + \sin 2\theta = 2\theta + 2\sin\theta\cos\theta$$

$$= 2\sin^{-1}\frac{x}{2} + 2\left(\frac{x}{2}\right)\left(\frac{\sqrt{4-x^2}}{2}\right)$$

$$A = \int_0^2\sqrt{4-x^2}\,dx$$

$$= 2\sin^{-1}\frac{x}{2} + \frac{1}{2}x\sqrt{4-x^2}\Bigg|_0^2$$

$$A = \pi$$

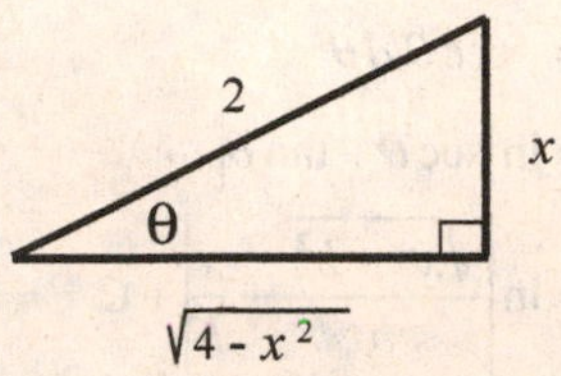

33. The runner from Exercise 41 of Section 28.2 has position $s_1(t) = 60\ln|0.200t+1|$.

The runner in Exercise 39 of Section 28.3 has position

$s_2(t) = 72 - 72e^{-t/6.00}$.

The runner from this exercise has velocity

$$v_3 = \frac{48}{0.25t^2+4} = \frac{192}{t^2+16}$$

$a = 4, t = 4\tan\theta, dt = 4\sec^2\theta d\theta, \sqrt{t^2+16} = 4\sec\theta$

$$s_3 = \int\frac{192}{t^2+16}dt = \int\frac{192\cdot 4\sec^2\theta}{16\sec^2\theta}d\theta$$

$$= \int 48 d\theta$$

$$= 48\tan^{-1}\left(\frac{t}{4}\right) + C$$

$s(0) = 0$ implies $C = 0$

$$s_3 = 48\tan^{-1}\left(\frac{t}{4}\right)$$

(a) After 3 h,

$s_1 = 60\ln 1.6 = 28.20$

$s_2 = 72 - 72e^{-0.5} = 28.32$

$s_3 = 48\tan^{-1} 0.75 = 30.89$

and so the runner from this exercise is ahead.

(b) After 4 h,

$$s_1 = 60\ln 1.8 = 35.37$$

$$s_2 = 72 - 72e^{-2/3} = 35.03$$

$$s_3 = 48\tan^{-1} 1 = 37.70$$

The runner from this exercise is in first, the runner from section 28.2 is in second, the runner from section 28.3 is in third.

37. $V = kQ\ {}_{-a}^{\ a}\dfrac{dx}{\sqrt{b^2+x^2}}$

$x = b\tan\theta; dx = b\sec^2\theta\, d\theta, \sqrt{b^2+x^2} = \sec\theta$

$$\frac{dx}{\sqrt{b^2+x^2}} = \frac{b\sec^2\theta\, d\theta}{b\sec\theta}$$

$$= \sec\theta\, d\theta$$

$$= \ln\left|\sec\theta + \tan\theta\right|$$

$$= \ln\left|\frac{\sqrt{x^2+b^2}}{b} + \frac{x}{b}\right| + C$$

$$V = kQ\ln\left|\frac{\sqrt{x^2+b^2}+x}{b}\right|\Bigg|_{-a}^{a}$$

$$= kQ\left[\ln\left|\frac{\sqrt{a^2+b^2}+a}{b}\right| - \ln\left|\frac{\sqrt{a^2+b^2}-a}{b}\right|\right]$$

$$= kQ\left(\ln\left|\sqrt{a^2+b^2}+a\right| - \ln|b| - \ln\left|\sqrt{a^2+b^2}-a\right| + \ln|b|\right)$$

$$V = kQ\ \ln\left|\frac{\sqrt{a^2+b^2}+a}{\sqrt{a^2+b^2}-a}\right|$$

41. $u = (x-4)^{2/3} \Rightarrow x = u^{3/2} - 4$

$$du = \frac{2}{3}(x-4)^{-1/3}\,dx$$

$$\frac{3}{2}u^{1/2}du = dx$$

$$\int x(x-4)^{2/3}\,dx = \int\left(u^{3/2}-4\right)(u)\left(\frac{3}{2}u^{1/2}du\right)$$

$$= \frac{3}{2}\int\left(u^3 - 4u^{3/2}\right)du$$

$$= \frac{3}{2}\left[\frac{u^4}{4} - \frac{4u^{5/2}}{5/2}\right] + C$$

$$= \frac{3}{8}(x-4)^{8/3} + \frac{12}{5}(x-4)^{5/3} + C$$

28.9 Integration by Partial Fractions: Nonrepeated Linear Factors

1. $\dfrac{10-x}{x^2+x-2}=\dfrac{10-x}{(x-1)(x+2)}=\dfrac{A}{x-1}+\dfrac{B}{x+2}$

$10-x=A(x+2)+B(x-1)$

for $x=-2$:

$10-1=A(1+2)+B(1-1)$

$9=3A$

$A=3$

for $x=1$:

$10-1=A(1+2)+B(1-1)$

$9=3A$

$A=3$

$\dfrac{10-x}{x^2+x-2}=\dfrac{3}{x-1}+\dfrac{-4}{x+2}$

5. $\dfrac{x^2-6x-8}{x^3-4x}=\dfrac{x^2-6x-8}{x(x^2-4)}$

$=\dfrac{x^2-6x-8}{x(x+2)(x-2)}$

$=\dfrac{A}{x}+\dfrac{B}{x+2}+\dfrac{C}{x-2}$

8. $\int\dfrac{x+2}{x(x+1)}dx=\int\dfrac{2}{x}dx-\int\dfrac{dx}{x+1}$

$=2\ln|x|-\ln|x+1|+C$

13. $\int_0^1\dfrac{2t+4}{3t^2+5t+2}dt$

$=\int_0^1\dfrac{8}{3t+2}dt-\int_0^1\dfrac{2}{t+1}dt$

$=\dfrac{8\ln(3t+2)}{3}\Big|_0^1-2\ln(t+1)\Big|_0^1$

$=\dfrac{8\ln 5}{3}-\dfrac{8\ln 2}{3}-2\ln 2+2\ln 1$

$=1.06$

17. $\dfrac{12x^2-4x-2}{4x^3-x}dx$

$=\dfrac{12x^2-4x-2}{x(4x^2-1)}dx=\dfrac{12x^2-4x-2}{x(2x+1)(2x-1)}dx$

$=2\dfrac{dx}{x}+\dfrac{3}{2x+1}dx-\dfrac{1}{2x-1}dx$

$=2\ln|x|+\dfrac{3\ln|2x+1|}{2}-\dfrac{\ln|2x-1|}{2}+C$

$=\dfrac{4\ln|x|}{2}+\dfrac{3\ln|2x+1|}{2}-\dfrac{\ln|2x-1|}{2}+C$

$=\dfrac{1}{2}\ln\left|\dfrac{x^4(2x+1)^3}{2x-1}\right|+C$

21. $\int\dfrac{dV}{(V^2-4)(V^2-9)}$

$=\int\dfrac{dV}{(V-2)(V+2)(V+3)(V-3)}$

$=\int\dfrac{\frac{1}{30}}{V-3}dV-\int\dfrac{\frac{1}{30}dV}{V+3}-\int\dfrac{\frac{1}{20}dV}{V-2}+\int\dfrac{\frac{1}{20}dV}{V+2}$

$=\dfrac{1}{30}\ln|V-3|-\dfrac{1}{30}\ln|V+3|$

$-\dfrac{1}{20}\ln|V-2|+\dfrac{1}{20}\ln|V+2|+C$

$=\dfrac{2}{60}\ln|V-3|-\dfrac{2}{60}\ln|V+3|-\dfrac{3}{60}\ln|V-2|$

$+\dfrac{3}{60}\ln|V+2|+C$

$=\dfrac{1}{60}\ln\left|\dfrac{(V+2)^3(V-3)^2}{(V-2)^3(V+3)^2}\right|+C$

25. $\dfrac{1}{u(a+bu)}=\dfrac{A}{u}+\dfrac{B}{a+bu}$

$$1=A(a+bu)+Bu$$

for $u=0$: $1=aA$

$$A=\frac{1}{a}$$

for $u=\frac{a}{b}$: $1=B\left(-\frac{a}{b}\right)$

$$B=-\frac{b}{a}$$

$$\int\frac{du}{u(a+bu)}$$

$$=\int\frac{\frac{1}{a}}{u}du+\int\frac{-\frac{b}{a}}{a+bu}du$$

$$=\frac{1}{a}\ln|u|-\frac{1}{a}\int\frac{b\,du}{a+bu}$$

$$=\frac{1}{a}\ln|u|-\frac{1}{a}\ln|a+bu|+C$$

$$=-\frac{1}{a}(-\ln|u|+\ln|a+bu|+C$$

$$=-\frac{1}{a}\ln\frac{|a+bu|}{|u|}+C$$

$$=\frac{1}{a}\ln\left|\frac{a+bu}{u}\right|+C$$

29. $u=\sin\theta,\ du=\cos\theta\,d\theta$

$$\int\frac{\cos\theta\,d\theta}{\sin^2\theta+2\sin\theta-3}=\int\frac{du}{u^2+2u-3}=\int\frac{du}{(u+3)(u-1)}$$

$$1=A(u-1)+B(u+3)$$

$u=1,\quad 1=B(1+3)\Rightarrow B=\frac{1}{4}$

$u=-3,\quad 1=A(-3-1)\Rightarrow A=-\frac{1}{4}$

$$\int\frac{\cos\theta\,d\theta}{\sin^2\theta+2\sin\theta-3}=\int\frac{-\frac{1}{4}du}{u+3}+\int\frac{\frac{1}{4}du}{u-1}$$

$$=-\frac{1}{4}\ln|u+3|+\frac{1}{4}\ln|u-1|+C$$

$$=-\frac{1}{4}\ln|\sin\theta+3|+\frac{1}{4}\ln|\sin\theta-1|+C$$

33. $\frac{3x+5}{x^2+5x}=\frac{3x+5}{x(x+5)}=\frac{A}{x}+\frac{B}{x+5}$

$=\frac{A(x+5)+Bx}{x(x+5)}$

$3x+5=A(x+5)+Bx$

$x=0, \quad 5=5A, \quad A=1$

$x=-5, \ -10=-5B, \quad B=2$

$y=\int\frac{3x+5}{x^2+5x}dx$

$y=\int\frac{dx}{x}+\int\frac{2}{x+5}dx$

$y=\ln|x|+2\ln|x+5|+C$

Substitute $x=1, y=0$:

$0=\ln|1|+2\ln|6|+C, \ C=-2\ln 6$

$y=\ln|x|+2\ln|x+5|-2\ln 6$

$y=\ln\frac{|x|(x+5)^2}{36}$

28.10 Integration by Partial Fractions: Other Cases

1. $\frac{2}{x(x+3)^2}=\frac{A}{x}+\frac{B}{x+3}+\frac{C}{(x+3)^2}$

5. $\int\frac{1}{x^2(x+1)}dx=\int\frac{dx}{x+1}-\int\frac{dx}{x}+\int\frac{dx}{x^2}$

$=\ln|x+1|-\ln|x|-\frac{1}{x}+C$

$=\ln\left|\frac{x+1}{x}\right|-\frac{1}{x}+C$

9. $\int\frac{2\,dx}{x^2(x^2-1)}=\int\frac{2\,dx}{x^2(x-1)(x+1)}$

$=\int\frac{dx}{x-1}-\int\frac{dx}{x+1}-\int\frac{2\,dx}{x^2}$

$=\ln|x-1|-\ln|x+1|+\frac{2}{x}+C$

$=\ln\left|\frac{x-1}{x+1}\right|+\frac{2}{x}+C$

13. $\int \frac{x^3-2x^2-7x+28}{(x+1)^2(x-3)^2}dx = \int \frac{dx}{(x-3)^2} + \int \frac{2\,dx}{(x+1)^2} + \int \frac{dx}{x+1}$

$$= \frac{-1}{x-3} - \frac{2}{x+1} + \ln|x+1| + C$$

17. $\int \frac{5x^2-3x+2}{x^3-2x^2}dx = \int \frac{5x^2-3x+2}{x^2(x-2)}dx$

$$= \int \left(\frac{1}{x} - \frac{1}{x^2} + \frac{4}{x-2}\right)dx$$

$$= \ln|x| + \frac{1}{x} + 4\ln|x-2| + C$$

21. $\frac{10x^3+40x^2+22x+7}{(4x^2+1)(x^2+6x+10)} = \frac{Ax+B}{4x^2+1} + \frac{Cx+D}{x^2+6x+10}$

$$10x^3+40x^2+22x+7$$

$$= (Ax+B)(x^2+6x+10) + (Cx+D)(4x^2+1)$$

$$= Ax^3+6Ax^2+10Ax+Bx^2+6Bx+10B+4Cx^3+Cx+4Dx^2+D$$

$$= (A+4C)x^3+(6A+B+4D)x^2+(10A+6B+C)x+10B+D$$

$$\left.\begin{array}{lll}(1) & A+4C & =10\\ (2) & 6A+B+4D & =40\\ (3) & 10A+6B+C & =22\\ (4) & 10B+D & 7\end{array}\right\} A=2, B=0, C=2, D=7$$

$$\int \frac{10x^3+40x^2+22x+7}{(4x^2+1)(x^2+6x+10)}dx = \int \frac{2x}{4x^2+1}dx + \int \frac{2x+7}{x^2+6x+10}dx$$

$$\int \frac{2x}{4x^2+1}dx = \frac{1}{4}\int \frac{8x}{4x^2+1}dx$$

$$= \frac{\ln(4x^2+1)}{4}$$

$$\int \frac{2x+7}{x^2+6x+10}dx = \int \frac{2x+7}{x^2+6x+9+1}dx$$

$$= \int \frac{2x+7}{(x+3)^2+1}dx$$

$u = x+3, du = dx, x = u-3$

$$\int \frac{2x+7}{x^2+6x+10}dx = \int \frac{2(u-3)+7}{u^2+1}du$$
$$= \int \frac{2u-6+7}{u^2+1}du$$
$$= \int \frac{2u}{u^2+1}du$$
$$= \int \frac{2u}{u^2+1}du + \int \frac{1}{u^2+1}du$$
$$= \ln\left(u^2+1\right) + \tan^{-1} u + C$$
$$= \ln\left(x^2+6x+10\right) + \tan^{-1}(x+3) + C$$

$$\int \frac{10x^3+40x^2+22x+7}{\left(4x^2+1\right)\left(x^2+6x+10\right)}dx$$
$$= \frac{\ln\left(4x^2+1\right)}{4} + \ln\left(x^2+6x+10\right) + \tan^{-1}(x+3) + C$$

25. $\dfrac{x}{(x-2)^3} = \dfrac{A}{(x-2)} + \dfrac{B}{(x-2)^2} + \dfrac{C}{(x-2)^3}$

$$= \frac{A(x-2)^2 + B(x-2) + C}{(x-2)^3}$$

$x = A\left(x^2-4x+4\right) + B(x-2) + C$

$x = Ax^2 - 4Ax + 4A + Bx - 2B + C$

$x^2:\ A = 0$

$x:\ 1 = -4A + B \quad B \quad 1$

Constant: $0 = 4A - 2B + C$

$0 = -2 + C$

$C = 2$

$$\frac{xdx}{(x-2)^3} = \frac{dx}{(x-2)^2} + \frac{2dx}{(x-2)^3}$$
$$= \frac{(x-2)^{-2+1}}{-2+1} + 2\frac{(x-2)^{-3+1}}{-3+1} + C$$
$$= \frac{-1}{(x-2)} - \frac{1}{(x-2)^2} + C$$
$$= \frac{-1(x-2)-1}{(x-2)^2} + C$$
$$= \frac{1-x}{(x-2)^2} + C$$

29. $V = \int_0^3 \pi \cdot \left(\dfrac{x}{(x+3)^2}\right)^2 dx = \dfrac{\pi}{72} = 0.0436$

33. $\bar{x} = \dfrac{\int_1^2 x \cdot \dfrac{4}{x^3+x} dx}{\int_1^2 \dfrac{4}{x^3+x} dx} = \dfrac{\pi - 4\tan^{-1}\frac{1}{2}}{2\ln\frac{8}{5}}$

$\bar{x} = 1.370$

34. $y = \int \dfrac{29x^2+36}{4x^4+9x^2} dx = \dfrac{13}{6}\tan^{-1}\dfrac{2x}{3} - \dfrac{4}{x} + C$

$5 = \dfrac{13}{6}\tan^{-1}\dfrac{2(1)}{3} - \dfrac{4}{1} + C \Rightarrow C$

$= 9 - \dfrac{13}{6}\tan^{-1}\dfrac{2}{3}$

$y = \dfrac{13}{6}\tan^{-1}\dfrac{2x}{3} - \dfrac{4}{x} + 9 - \dfrac{13}{6}\tan^{-1}\dfrac{2}{3}$

28.11 Integration by Use of Tables

1. $\int \dfrac{xdx}{(2+3x)^2}$ is formula 3 with $u = x$, $du = dx$, $a = 2$, $b = 3$.

5. $u = x^2$, $du = 2xdx$

$\int \dfrac{xdx}{(4-x^4)^{3/2}} = \dfrac{1}{2}\int \dfrac{2xdx}{\left(2^2-(x^2)^2\right)^{3/2}} = \dfrac{1}{2}\int \dfrac{du}{(2^2-u^2)^{3/2}}$

which is formula 25 with $a = 2$.

9. Formula 1; $u = x$; $a = 2$; $b = 5$; $du = dx$

$\dfrac{3xdx}{2+5x} = 3 \quad \dfrac{xdx}{2+5x}$

$= 3 \quad \dfrac{1}{25}(2+5x) - 2\ln|2+5x| \quad + C$

$= \dfrac{3}{25} \quad 2 \quad 5x \quad 2\ln|2 \quad 5x| \quad C$

13. Formula 24; $u = y$, $du = dy, a = 2$

$\int \dfrac{8dy}{(y^2+4)^{3/2}} = \dfrac{2y}{\sqrt{y^2+4}} + C$

17. Formula 17; $u = 2x$; $du = 2dx$; $a = 3$

$\dfrac{\sqrt{4x^2-9}}{x} dx = \dfrac{\sqrt{(2x)^2-3^2}}{2x} 2dx$

$= \sqrt{4x^2-9} - 3\sec^{-1} \quad \dfrac{2x}{3} \quad + C$

21. Formula 52; $u = r^2; du = 2r\,dr$

$$6\ \tan^{-1} r^2 (r\,dr)$$
$$= 3\ \tan^{-1} r^2 (2r\,dr)$$
$$= 3\ r^2 \tan^{-1} r^2 - \frac{1}{2}\ln(1+r^4)\ + C$$
$$= 3r^2 \tan^{-1} r^2 - \frac{3}{2}\ln(1+r^4) + C$$

25. Formula 11; $u = 2x$; $du = 2dx$; $a = 1$

$$\frac{dx}{2x\sqrt{x^2+\frac{1}{4}}} = \frac{2dx}{2x\sqrt{(2x)^2+1^2}}$$
$$= -\ln\ \frac{1+\sqrt{4x^2+1}}{2x}\ + C$$

29. Formula 40; a=1; u=x; $du = dx$; $b = 5$

$$\int_0^{\pi/12} \sin\theta\cos 5\theta\, d\theta = -\frac{\cos(-4\theta)}{2(-4)} - \frac{\cos 6\theta}{12}\Bigg|_0^{\pi/12}$$
$$= \frac{1}{8}\cos 4\theta - \frac{1}{12}\cos 6\theta\Bigg|_0^{\pi/12}$$
$$= 0.0208$$

33. $u = x^2, du = 2x\,dx, u^2 = x^4$

$$\frac{2xdx}{(1-x^4)^{3/2}} = \frac{du}{(1-u^2)^{3/2}}$$

Formula 25; $a = 1$

$$\frac{2xdx}{(1-x^4)^{3/2}} = \frac{u}{\sqrt{1-u^2}} + C;$$
$$\frac{2x\,dx}{(1-x^4)^{3/2}} = \frac{x^2}{\sqrt{1-x^4}} + C$$

37. Formula 46; $u = x^2$; $du = 2x\,dx$; $n = 1$

$$x^3 \ln x^2 dx = \frac{1}{2}\ \ x^2 \ln x^2 (2x\,dx)$$
$$= \frac{1}{2}\ (x^2)^2\ \ \frac{\ln x^2}{2} - \frac{1}{4}\ \ + C$$
$$= \frac{1}{2}\ \frac{x^4}{2}\ \ \ln x^2 - \frac{1}{2}\ \ + C$$
$$= \frac{1}{4}x^4\ \ \ln x^2 - \frac{1}{2}\ \ + C$$

41. Let $u = t^3$, $du = 3t^2 dt$, formula 19, $a = 1$

$$t^2\left(t^6+1\right)^{3/2} dt = \frac{1}{3}\left(u^2+1\right)^{3/2} du,$$

$$= \frac{1}{3}\;\frac{u}{4}\left(u^2+1\right)^{3/2} + \frac{3u}{8}\sqrt{u^2+1} + \frac{3}{8}\ln\left(u+\sqrt{u^2+1}\right)$$

$$= \frac{1}{12}t^3\left(t^6+1\right)^{3/2} + \frac{t^3}{8}\sqrt{t^6+1} + \frac{1}{8}\ln\left(t^3+\sqrt{t^6+1}\right) + C$$

45. From Exercise 35 of Section 26.6;

$$s = \int_a^b \sqrt{1+\left(\frac{dy}{dx}\right)^2}\, dx;$$

$$y = 0.000370x^2;\ \frac{dy}{dx} = 0.000740x$$

$$s = \int_{-1140}^{1140} \sqrt{1+(0.000740x)^2}\, dx$$

$$= 2\int_0^{1140} \sqrt{(0.000740x)^2+1}\, dx$$

Formula 14; u=0.000740x; du = 0.000740 $dx, a = 1$

$$s = 2\left[\frac{0.000740x}{2}\sqrt{(0.000740x)^2+1} + \frac{1}{2}\ln(0.000740x+\sqrt{(0.000740x)^2+1})\right]\Bigg|_0^{1140}$$

= 2530 m

49. $F = \gamma \int_0^3 lhdh = \gamma \int_0^3 x(3-y)dy = \gamma \int_0^3 \frac{3-y}{\sqrt{1+y}}dy$

Power rule and formula #6 with $a = 1, b = 1, u = y$

$$\frac{3-y}{\sqrt{1+y}}dy = 3\int \frac{dy}{\sqrt{1+y}} - \int \frac{ydy}{\sqrt{1+y}}$$

$$= 3\frac{(1+y)^{1/2}}{\frac{1}{2}} - \frac{-2(2-y)\sqrt{1+y}}{3(1)^2} + C$$

$$F = \gamma \int_0^3 \frac{3-y}{\sqrt{1+y}}dy$$

$$= \gamma\; 6(1+y)^{1/2} + \frac{2}{3}(2-y)(1+y)^{1/2}\Big|_0^3$$

$$= \gamma\; 6(2) + \frac{2}{3}(-1)(2) - 6(1) - \frac{2}{3}(2)(1)$$

$$F = \lambda\; 12 - \frac{4}{3} - 6 - \frac{4}{3}$$

$$= \frac{10\lambda}{3}$$

$$= \frac{10(9800)}{3}$$

$$= 32.7 \text{ kN}$$

Review Exercises

1. This is false. A factor of $\frac{1}{2}$ is missing.

5. This is true.

9. $u = -8x,\ du = -8dx$

$$\int e^{-8x}dx = -\frac{1}{8}\int e^{-8x}(-8dx)$$
$$= -\frac{1}{8}e^{-8x} + C$$

13. $\int_0^{\pi/2}\frac{4\cos\theta\, d\theta}{1+\sin\theta} = 4\int_0^{\pi/2}\frac{\cos\theta\, d\theta}{1+\sin\theta};$

$u = 1+\sin\theta, du = \cos\theta$

$$= 4\ln(1+\sin\theta)\Big|_0^{\pi/2}$$
$$= 2.77$$

17. $\int_0^{\pi/2}\cos^3 2\theta\, d\theta = \int_0^{\pi/2}\cos^2 2\theta\cos 2\theta\, d\theta$

$$= \int_0^{\pi/2}\left(1-\sin^2 2\theta\right)\cos 2\theta\, d\theta$$
$$= \int_0^{\pi/2}\cos 2\theta\, d\theta - \int_0^{\pi/2}\sin^2 2\theta\cos 2\theta\, d\theta$$
$$= \frac{1}{2}\int_0^{\pi/2}\cos 2\theta(2d\theta) - \frac{1}{2}\int_0^{\pi/2}\sin^2 2\theta\cos 2\theta(2d\theta)$$
$$= \frac{1}{2}\left[\sin 2\theta - \frac{1}{3}\sin^3 2\theta\right]_0^{\pi/2}$$
$$= \frac{1}{2}\left[\left(\sin\pi - \frac{1}{3}\sin^3\pi\right) - \left(\sin 0 - \frac{1}{3}\sin^3 0\right)\right]$$
$$= \frac{1}{2}(0) = 0$$

21. $(\sin t + \cos t)^2 \cdot \sin t\, dt$

$$= \left(\sin^2 t + 2\sin t\cos t + \cos^2 t\right)\cdot \sin t dt$$
$$= (1 + 2\sin t\cos t)\cdot \sin t dt$$
$$= \left(\sin t + 2\sin^2 t\cos t\right)dt$$
$$= \sin t\, dt + 2\ \sin^2 t(\cos t\, dt)$$
$$= -\cos t + \frac{2\sin^3 t}{3} + C$$

$(u = \sin t, du = \cos t dt)$

25. $\int 6\sec^4 3x dx = \int 6\sec^2 3x\sec^2 3x dx$

$$= \int 6\left(1+\tan^2 3x\right)\sec^2 3x dx$$
$$= 2\int \sec^2 3x(3dx) + 2\int \tan^2 3x\sec^2 3x(3dx)$$
$$= 2\tan 3x + 2\frac{\tan^3 3x}{3} + C$$
$$= \frac{2}{3}\tan^3 3x + 2\tan 3x + C$$

29. $u = x^2, du = 2xdx$

$$\frac{3xdx}{4+x^4} = 3\ \frac{xdx}{4+x^4} = 3\ \frac{1}{2^2+\left(x^2\right)^2}xdx$$
$$= \frac{3}{2}\ \frac{1}{2^2+\left(x^2\right)^2}2xdx$$
$$= \frac{3}{2}\ \frac{1}{2}\tan^{-1}\frac{x^2}{2}\ + C$$
$$= \frac{3}{4}\tan^{-1}\frac{x^2}{2} + C$$

33. $u = e^{2x}+1,\ du = e^{2x}(2dx)$

$$\frac{e^{2x}dx}{\sqrt{e^{2x}+1}} = \frac{1}{2}\ \left(e^{2x}+1\right)^{-1/2}e^{2x}(2dx)$$
$$= \frac{1}{2}\left(e^{2x}+1\right)^{1/2}(2) + C$$
$$= \sqrt{e^{2x}+1} + C$$

37. $\ _0^{\pi/6}3\sin^2 3\phi d\phi = \ _0^{\pi/6}3\cdot\frac{(1-\cos 6\phi)}{2}d\phi$

$$= \ _0^{\pi/6}\frac{3}{2}d\phi - \frac{1}{4}\ _0^{\pi/6}\cos 6\phi(6d\phi)$$
$$= \frac{3}{2}\phi\Big|_0^{\pi/6} - \frac{1}{4}\sin 6\phi\Big|_0^{\pi/6}$$
$$= \frac{3}{2}\ \frac{\pi}{6} - 0\ - \frac{1}{4}[\sin\pi - \sin 0] = \frac{\pi}{4}$$

41. $$\frac{3u^2-6u-2}{u^2(3u+1)}=\frac{Au+B}{u^2}+\frac{C}{3u+1}$$

$$=\frac{(Au+B)(3u+1)+Cu^2}{u^2(3u+1)}$$

$$3u^2-6u-2=3Au^2+Au+3Bu+B+Cu^2$$

$$3u^2-6u-2=(3A+C)u^2+(A+3B)u+B$$

(1) $3A+C=3,\ 3(0)+C=3,\ C=3$

(2) $A+3B=-6;\ A+3(-2)=-6,\ A=0$

(3) $\quad B=-2$

$$\int\frac{3u^2-6u-2}{u^2(3u+1)}du=\int\frac{-2}{u^2}du+\int\frac{3}{3u+1}du$$

$$=\frac{2}{u}+\ln|3u+1|+C$$

45. $u=\ln x, du=\frac{dx}{x}$

$$\int_1^e 3\cos(\ln x)\cdot\frac{dx}{x}=3\sin(\ln x)\Big|_1^e$$

$$=3\sin(\ln e)-3\sin(\ln 1)$$

$$=3\sin(1)-3\sin(0)$$

$$=3\sin 1=2.52$$

49. $u=\cos x,\ du=-\sin x dx$

$$\int\frac{\sin x\cos^2 x}{5+\cos^2 x}dx=-\int\frac{u^2du}{5+u^2}$$

$$=-\int\left(1-\frac{5}{5+u^2}\right)du$$

$$=-u+\frac{5}{\sqrt{5}}\tan^{-1}\frac{u}{\sqrt{5}}+C$$

$$=-\cos x+\frac{5}{\sqrt{5}}\tan^{-1}\frac{\cos x}{\sqrt{5}}+C$$

53. $\int e^{\ln 4x}dx=\int 4xdx=2x^2+C$

$\int \ln e^{4x}dx=\int 4x\,dx=2x^2+C$

The integrals are the same because the functions are the same.

57. Use the general power formula.

Let $u = e^x + 1, du = e^x dx, n = 2.$

$$
\begin{aligned}
e^x\left(e^x+1\right)^2 dx &= u^2 du \\
&= \frac{u^3}{3} + C_1 \\
&= \frac{(e^x+1)^3}{3} + C_1 \\
&= \frac{e^{3x}}{3} + e^{2x} + e^x + \frac{1}{3} + C_1
\end{aligned}
$$

$$
\begin{aligned}
\int e^x\left(e^x+1\right)^2 dx &= \int e^x\left(e^{2x}+2e^x+1\right)dx \\
&= \int\left(e^{3x}+2e^{2x}+e^x\right)dx \\
&= \int e^{3x}dx + \int 2e^{2x}dx + \int e^x dx \\
&= \frac{1}{3}\int e^{3x}(3dx) + 2\left(\frac{1}{2}\right)\int e^{2x}(2dx) + \int e^x dx \\
&= \frac{1}{3}e^{3x} + e^{2x} + e^x + C_2
\end{aligned}
$$

The two integrals differ by a constant, with $C_2 = C_1 + \frac{1}{3}$

61. (a)

$$
\begin{aligned}
u &= x^2 + 4 \\
du &= 2xdx \\
\frac{x}{\sqrt{x^2+4}}dx &= \frac{\frac{1}{2}du}{\sqrt{u}} \\
&= u^{1/2} + C \\
&= \sqrt{x^2+4} + C
\end{aligned}
$$

(b)

$$
\begin{aligned}
x &= 2\tan\theta \\
dx &= 2\sec^2\theta d\theta \\
\sqrt{x^2+4} &= \sec\theta
\end{aligned}
$$

$\sqrt{x^2+4}$ x θ   2

$$
\begin{aligned}
\int\frac{xdx}{\sqrt{x^2+4}} &= \int\frac{2\tan\theta\left(2\sec^2\theta d\theta\right)}{\sec\theta} \\
&= \int 2\tan\theta\sec\theta d\theta \\
&= 2\sec\theta + C \\
&= 2\frac{\sqrt{x^2+4}}{2} + C \\
&= \sqrt{x^2+4} + C
\end{aligned}
$$

The u-substitution is simpler than the trig substitution.

65. **(a)** $\sin 2x\ dx = -\frac{1}{2}\cos 2x + C_1$

using $u = 2x, du = 2\ dx.$

$\sin 2x\ dx = \ 2\sin x \cos x\, dx$

(b) $= \sin^2 x + C_2$

using $u = \sin x, du = \cos x\ dx.$

(c) $\sin 2x\ dx = \ 2\sin x \cos x\, dx$

Parts: $u = \sin x, dv = \cos x dx; du = \cos x dx, v = \sin x$

$2\sin x\cos x\, dx = 2\sin^2 x - \ 2\sin x\cos x dx$

$4\sin x\cos x\, dx = 2\sin^2 x + 2C_3$

$\sin 2x\ dx = \sin^2 x + C_3$

Since $-\frac{1}{2}\cos 2x = \sin^2 x - \frac{1}{2}$

$C_1 = C_2 + \frac{1}{2}; C_2 = C_3$

69. $x^2 + y^2 = 5^2;$

$y = \sqrt{25 - x^2};$

$A = 2\int_3^5 \sqrt{25 - x^2}\ dx$

$A = 2\left[\frac{x}{2}\sqrt{25-x^2} + \frac{25}{2}\sin^{-1}\frac{x}{5}\right]_3^5$ Formula 15 in table of integrals.

$A = 2\left[\frac{5}{2}\sqrt{0} + \frac{25}{2}\sin^{-1}1\right] - 2\left[\frac{3}{2}\sqrt{16} + \frac{25}{2}\sin^{-1}\frac{3}{5}\right]$

$= 2\left[\frac{25}{2}\left(\frac{\pi}{2}\right)\right] - 2\left[6 + \frac{25}{2}(0.6435)\right]$

$= 2[19.63] - 2[14.04] = 11.2$

73. $y = xe^x,\ y = 0,\ x = 2$

shells: $V = \int_0^2 2\pi xy\, dx = 2\pi\int_0^2 x\left(xe^x\right)dx$

$V = 2\pi \ {}_0^2\ x^2e^x dx$, formula 45

$V = 2\pi \ \ e^x\left(x^2 - 2x + 2\right)\ \Big|_0^2$

$V = 2\pi \ \ e^2\left(4 - 4 + 2\right) - e^0\left(0 - 0 + 2\right)$

$V = 2\pi \ \ 2e^2 - 2 \ \ = 4\pi \ \ e^2 - 1$

$V = 80.3$

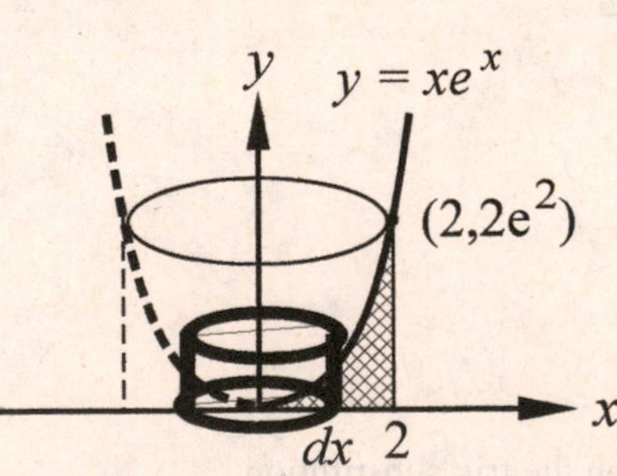

77. $y = \ln \sin x,\ x = 0.5 \text{ to } x = 2.5$

$$\frac{dy}{dx} = \frac{1}{\sin x}(\cos x) = \cot x$$

$$\sqrt{1 + \frac{dy}{dx}^{2}} = \sqrt{1 + \cot^2 x}$$

$$= \sqrt{\csc^2 x}$$

$$= \csc x$$

$$L = {}_{0.5}^{2.5} \csc x\, dx = \ln\left|\frac{\sin x}{\cos x + 1}\right|\Bigg|_{0.5}^{2.5}$$

$L = 2.47$

81. $\Delta S = \int (c_v / T)\, dT = \int \frac{a + bT + cT^2}{T} dT$

$$= \int \frac{a}{T} dT + \int b dT + \int cT\, dT$$

$$= a \ln T + bT + \frac{1}{2} cT^2 + C$$

85.

$$\int \frac{dv}{32 - 0.5v} = \int dt$$

$$-\frac{1}{0.5} \ln|32 - 0.5v| = t + C$$

$$-\frac{1}{0.5} \ln|32 - 0.5(0)| = 0 + C$$

$$C = -\frac{1}{0.5} \ln|32|$$

$$-\frac{1}{0.5} \ln|32 - 0.5v| = t - \frac{1}{0.5} \ln(32)$$

$$\frac{1}{0.5} \ln(32) - \frac{1}{0.5} \ln|32 - 0.5v| = t$$

$$\ln \frac{32}{|32 - 0.5v|} = 0.5t$$

$$\frac{32}{|32 - 0.5v|} = e^{0.5t}$$

$$32e^{-0.5t} = |32 - 0.5v|$$

$$32 - 0.5v = \pm 32e^{-0.5t}$$

$$0.5v = 32\left(1 \pm e^{-0.5t}\right)$$

$v = 64\left(1 \pm e^{-0.5t}\right)$ where the negative must be chosen to satisfy $v = 0$ at $t = 0$

$$v = 64\left(1 - e^{-0.5t}\right)$$

89. $V = {}_{2}^{4} \pi e^{-0.2x} dx = -5\pi e^{-0.2x}\Big|_{2}^{4}$

$= 3.47 \text{ cm}^3$

93. $r = 1.5x^{2/3} \quad x \quad \frac{r}{1.5}^{3/2}$

$$\frac{d}{dr}\ 16.00 - \frac{r}{1.5}^{3/2} = -\frac{r}{1.5}^{1/2}$$

$$S = {}_{0}^{1.5(16.00^{2/3})} 2\pi r \sqrt{1 + \frac{r}{1.5}}\, dr$$

Formula 5 with $u = r, du = dr, a = 1, b = \frac{1}{1.5}$:

$$r\sqrt{1 + \frac{r}{1.5}} dr = \frac{2(2 - \frac{3}{1.5}r)(1 + \frac{1}{1.5}r)^{3/2}}{15\left(\frac{1}{1.5}\right)^2}$$

When $x = 16, r = 1.5(16^{2/3}) = 9.524$

$$S = 2\pi\ - \frac{2(2 - \frac{3}{1.5}r)(1 + \frac{1}{1.5}r)^{3/2}}{15\left(\frac{1}{1.5}\right)^2}\Bigg|_{0}^{9.524}$$

$= (640.25 - 3.77)$

$S = 636 \text{ ft}^2$

Chapter 29

Partial Derivatives and Double Integrals

29.1 Functions of Two Variables

1. $f(x, y) = 3x^2 + 2xy - y^3$

$f(-2, 1) = 3(-2)^2 + 2(-2)(1) - (1)^3$

$= 7$

5. From geometry:

$A = 2\pi rh + 2\pi r^2;$

$V = \pi r^2 h$ so write h as a function of V:

$h = \frac{V}{\pi r^2}$

Substitute in A:

$A = 2\pi r \quad \frac{V}{\pi r^2} \quad + 2\pi r^2$

$= \frac{2V}{r} + 2\pi r^2$

9. $f(x, y) = 2x - 6y$

$f(0, -4) = 2(0) - 6(-4) = 24$

$f(-3, 2) = 2(-3) - 6(2) = -18$

13. $Y(y, t) = \frac{2 - 3y}{t - 1} + 2y^2 t$

$Y(2, -1) = \frac{2 - 3(2)}{(-1) - 1} + 2(2)^2(-1)$

$= 2 - 8$

$= -6$

$Y(y, 2) = \frac{2 - 3y}{2 - 1} + 2y^2(2)$

$= 2 - 3y + 4y^2$

17. $H(p, q) = p - \frac{p - 2q^2 - 5q}{p + q}$

$H(p, q + k)$

$= p - \frac{p - 2(q + k)^2 - 5(q + k)}{p + q + k}$

$= \frac{p(p + q + k) - p + 2(q^2 + 2kq + k^2) + 5(q + k)}{p + q + k}$

$= \frac{p^2 + pq + pk - p + 2q^2 + 4kq + 2k^2 + 5q + 5k}{p + q + k}$

21. $f(x, y) = xy + x^2 - y^2$

$f(x, x) - f(x, 0)$

$= x(x) + x^2 - x^2 - [x(0) + x^2 - 0^2]$

$= x^2 + x^2 - x^2 - x^2$

$= 0$

25. $f(x, y) = \frac{\sqrt{y}}{2x}$; considering $\sqrt{y}, y \geq 0$ for real values of $f(x, y)$; considering $2x, x \neq 0$ to avoid division by zero. Thus, $y < 0$ and $x = 0$ are not permissible.

29. $M = mv$

$M = (0.160)(45.0) = 7.20$ kg m/s

33. $p = \frac{nRT}{V}$; $n = 3$ mol, $R = 8.31$ J/mol·K

$T = 300$ K, $V = 50$ m^3

$p = \frac{3(8.31)(300)}{50} = 150$ Pa

37. $i = \frac{6.0 \sin 0.01t}{R + 0.12}$; $t = 0.75$ s, $R = 1.5\ \Omega$,

$i = \frac{6.0 \sin 0.01(0.75)}{1.5 + 0.12}$

$= \frac{6.0 \sin 0.0075}{1.62} = 0.0278$ A

41. $p = 2\ell + 2w; \ell = \frac{p - 2w}{2}$

$A = \ell w = \frac{p - 2w}{2} w = \frac{pw - 2w^2}{2}$

$p = 250$ cm, $w = 55$ cm

$A = \frac{250(55) - 2(55)^2}{2} = 3850$ cm^2

29.2 Curves and Surfaces in Three Dimensions

1. $3x - y + 2z + 6 = 0$

Intercepts: $(0, 0, -3)$, $(0, 6, 0)$, $(-2, 0, 0)$

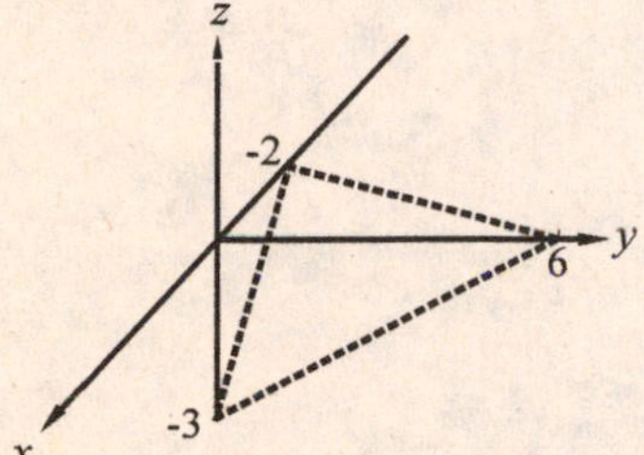

5. $x + y + 2z - 4 = 0$; plane

Intercepts: $(4, 0, 0)$, $(0, 4, 0)$, $(0, 0, 2)$

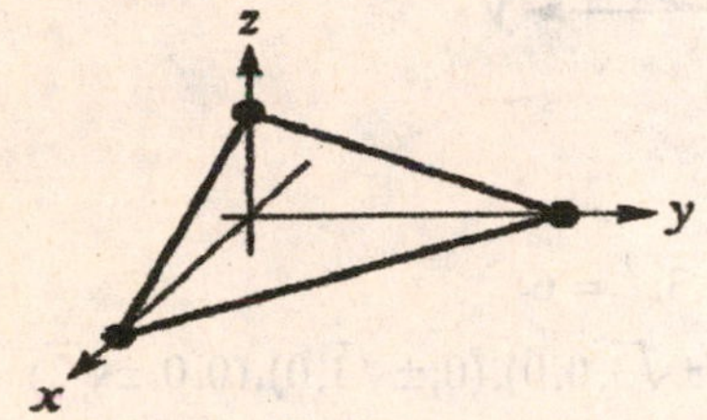

9. $z = y - 2x - 2$; plane

Intercepts: $(-1, 0, 0)$, $(0, 2, 0)$, $(0, 0, -2)$

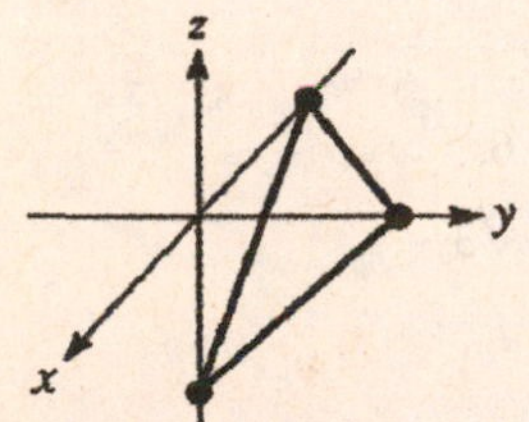

13. $x^2 + y^2 + z^2 = 4$

Intercepts: $(\pm 2, 0, 0), (0, \pm 2, 0), (0, 0, \pm 2)$

Traces:

yz-plane: $y^2 + z^2 = 4$, circle, $r = 2$

xz-plane: $x^2 + z^2 = 4$, circle, $r = 2$

xy-plane: $x^2 + y^2 = 4$, circle, $r = 2$

The surface is a sphere with radius 2.

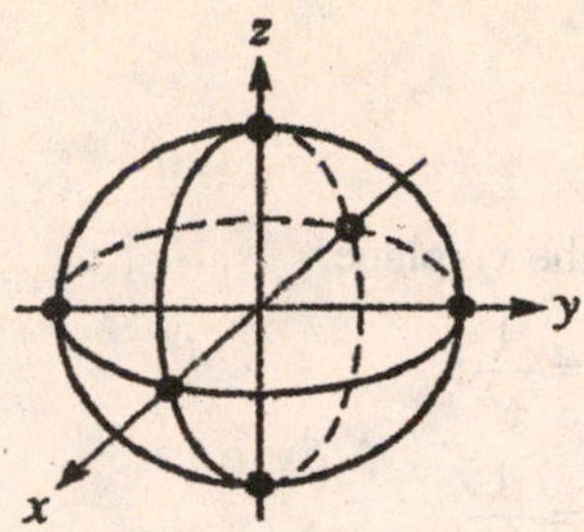

17. $z = 2x^2 + y^2 + 2$

Intercepts: No x-intercept, no y-intercept, (0,0,2)

Traces:

yz-plane:: $z = y^2 + 2$; parabola, $V(0,0,2)$

xz-plane: $z = 2x^2 + 2$, parabola, $V(0,0,2)$

xy-plane: No trace, $(2x^2 + y^2 + 2 \neq 0)$

Section: For $z = 4, 2x^2 + y^2 = 2$, ellipse

The surface is an elliptical paraboloid.

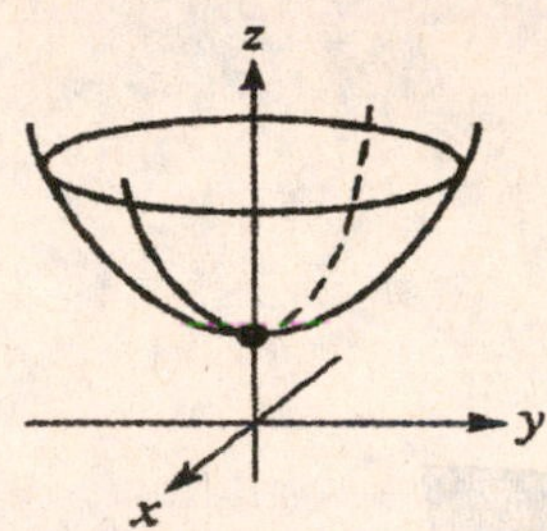

21. $x^2 + y^2 = 16$

Intercepts: $(\pm 4, 0, 0), (0, \pm 4, 0)$, no z-intercept.

Traces and sections:

Since z is not present in the equation, the trace and sections are circles $x^2 + y^2 = 16$, with $r = 4$, for all z. This is a cylindrical surface.

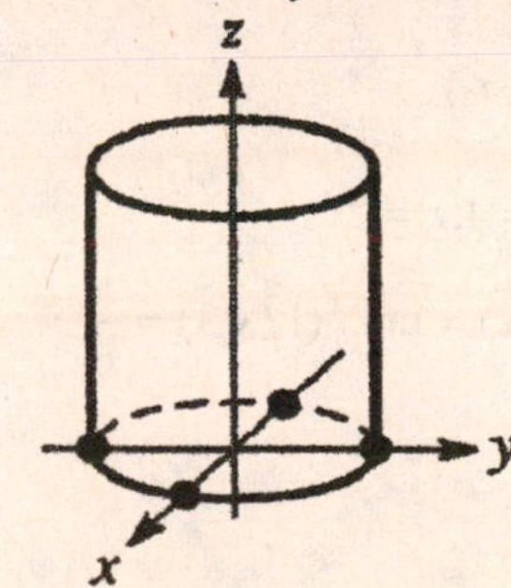

25. $z = \dfrac{1}{x^2 + y^2}$

Intercepts: none

Traces: None in the xy-plane;

in the xz-plane, $z = \dfrac{1}{x^2}$;

in the yz-plane, $z = \dfrac{1}{y^2}$.

Sections: For $z > 0$, circles of radius $\dfrac{1}{\sqrt{z}}$.

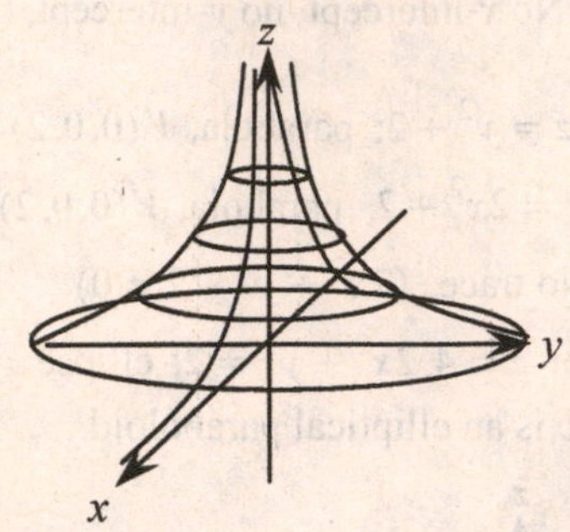

29. $z = y^4 - 4y^2 - 2x^2$

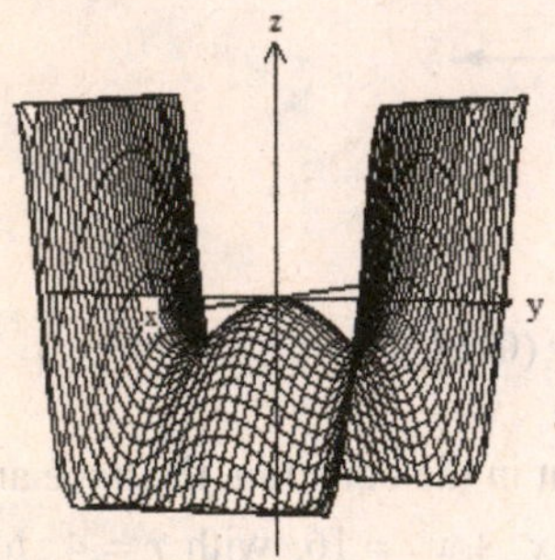

33. **(a)** $\left(\sqrt{3}, 1, 7\right)$

$r^2 = x^2 + y^2 = 4; r = 2$

$\theta = \tan^{-1}(y/x) = \tan^{-1}(1/\sqrt{3}) = \dfrac{\pi}{6}$

$z = z = 7$

$2, \dfrac{\pi}{6}, 7$

(b) $(0,4,1)$

$r^2 = x^2 + y^2 = 16; r = 4$

$\theta = \tan^{-1}(y/x)$; when $x = 0, y > 0, \theta = \dfrac{\pi}{2}$

$z = z = 1$

$4, \dfrac{\pi}{2}, 1$

37. $r^2 = 4z$

$x^2 + y^2 = 4z$

41. $2x^2 + 2y^2 + 3z^2 = 6$

Intercepts: $(\pm\sqrt{3},0,0), (0,\pm\sqrt{3},0), (0,0,\pm\sqrt{2})$

Traces:

yz-plane: $2y^2 + 3z^2 = 6$,

ellipse, $a = \sqrt{3}, b = \sqrt{2}$

xz-plane: $2x^2 + 3z^2 = 6$,

ellipse, $a = \sqrt{3}, b = \sqrt{2}$

xy-plane: $2x^2 + 2y^2 = 6$,

circle, $x^2 + y^2 = 3, r = \sqrt{3}$

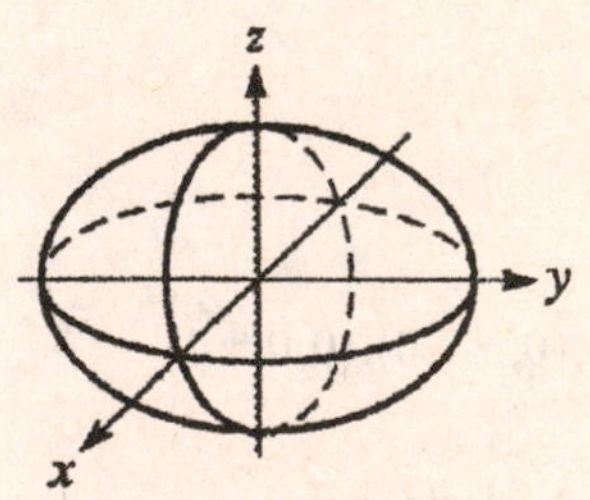

45. $x^2 + y^2 - 2y = 0$

Since z does not appear in the equation, all the traces and sections are circles,

$$x^2 + y^2 - 2y + 1 = 1$$
$$x^2 + (y-1)^2 = 1$$

with center $(0, 1, z)$ for z and $r = 1$. This is a cylindrical surface.

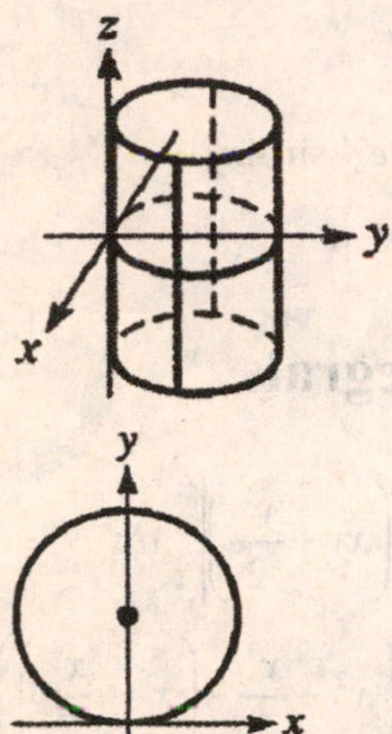

29.3 Partial Derivatives

1. $z = \dfrac{x \ln y}{y^2 + 1}$

$$\frac{\partial z}{\partial x} = \frac{\ln y}{y^2 + 1}$$

$$\frac{\partial z}{\partial y} = \frac{x\left(y^2+1\right)1/y - 2xy \ln y}{\left(y^2+1\right)^2}$$

$$\frac{\partial z}{\partial y} = \frac{x\left(y^2+1\right) - 2xy^2 \ln y}{y\left(y^2+1\right)^2}$$

5. $z = e^{3x} - \sin y$

$$\frac{\partial z}{\partial x} = 3e^{3x}$$

$$\frac{\partial z}{\partial y} = -\cos y$$

9. $f(x, y) = xe^{-2y}$

$$\frac{\partial f}{\partial x} = e^{-2y}$$

$$\frac{\partial f}{\partial y} = xe^{-2y}(-2) = -2xe^{2y}$$

13. $\phi = r\sqrt{1 + 2rs}$

$$\frac{\partial \phi}{\partial r} = r\left(\frac{1}{2}\right)(1+2rs)^{-1/2}(2s) + (1+2rs)^{1/2}$$

$$= \frac{rs}{(1+2rs)^{1/2}} + (1+2rs)^{1/2}$$

$$= \frac{1+3rs}{\sqrt{1+2rs}}$$

$$\frac{\partial \phi}{\partial s} = r\left(\frac{1}{2}\right)(1+2rs)^{-1/2}(2r)$$

$$= \frac{r^2}{\sqrt{1+2rs}}$$

17. $z = \sin x^2 y$

$$\frac{\partial z}{\partial x} = (\cos x^2 y)(2xy)$$

$$= 2xy \cos x^2 y$$

$$\frac{\partial z}{\partial y} = (\cos x^2 y)(x^2)$$

$$= x^2 \cos x^2 y$$

21. $f(x, y) = \dfrac{2\sin^3 2x}{1 - 3y}$

$$\frac{\partial f}{\partial x} = \frac{2(3)(\sin^2 2x)(\cos 2x)(2)}{1-3y}$$

$$= \frac{12 \sin^2 2x \cos 2x}{1 - 3y}$$

$$\frac{\partial f}{\partial y} = -(2\sin^3 2x)(1-3y)^{-2}(-3)$$

$$= \frac{6\sin^3 2x}{(1-3y)^2}$$

25. $f(x, y) = e^z \cos xy + e^{-2x} \tan y$

$$\frac{\partial f}{\partial x} = e^x[(-\sin xy)(y)] + (\cos xy)e^x + e^{-2x} \tan y(-2)$$

$$= e^x(\cos xy - y \sin xy) - 2e^{-2x} \tan y$$

$$\frac{\partial f}{\partial y} = e^x[-(\sin xy)(x)] + e^{-2x} \sec^2 y$$

$$= -xe^x \sin xy + e^{-2x} \sec^2 y$$

29. $z = x\sqrt{x^2 - y^2}$

$$\frac{\partial z}{\partial x} = x\frac{1}{2\sqrt{x^2 - y^2}}(2x) + \sqrt{x^2 - y^2}$$

$$\left.\frac{\partial z}{\partial x}\right|_{(5,3,20)} = \frac{(5)^2}{\sqrt{5^2 - 3^2}} + \sqrt{5^2 - 3^2} = \frac{41}{4}$$

33. $z = \frac{x}{y} + e^x \sin y$

$$\frac{\partial z}{\partial x} = \frac{1}{y} + e^x \sin y, \frac{\partial z}{\partial y} = \frac{-x}{y^2} + e^x \cos y$$

$$\frac{\partial^2 z}{\partial x^2} = e^x \sin y, \frac{\partial^2 z}{\partial y^2} = \frac{2x}{y^3} - e^x \sin y$$

$$\frac{\partial^2 z}{\partial x \partial y} = \frac{\partial^2 z}{\partial y \partial x} = -\frac{1}{y^2} + e^x \cos y$$

37. $y = \sin(\pi x)\sin(\pi t / 2)$

$$\frac{\partial y}{\partial x} = \pi \cos(\pi x)\sin(\pi t / 2)$$

$$\frac{\partial y}{\partial t} = \frac{\pi}{2}\sin(\pi x)\cos(\pi t / 2)$$

41. $\frac{1}{R_T} = \frac{1}{R_1} + \frac{1}{R_2}$

$$R_T = \frac{R_1 R_2}{R_1 + R_2}$$

$$\frac{\partial R_T}{\partial R_1} = \frac{(R_1 + R_2)(R_2) - R_1 R_2 (1)}{(R_1 + R_2)^2}$$

$$= \frac{R_2^2}{(R_1 + R_2)^2}$$

45. $d = \sqrt{x^2 + y^2}$

$$\frac{\partial d}{\partial x} = \frac{1}{2}\left(x^2 + y^2\right)^{-1/2}(2x)$$

$$= \frac{x}{\sqrt{x^2 + y^2}}$$

For $x = 6.50$ ft, $y = 4.75$ ft

$$\left.\frac{\partial d}{\partial x}\right|_{\substack{x=6.50 \\ y=4.75}} = \frac{6.50}{\sqrt{6.50^2 + 4.75^2}} = 0.807$$

49. $u(x,t) = 5e^{-t} \sin 4x$

$$\frac{\partial u}{\partial x} = 5e^{-t}(4\cos 4x) = 20e^{-t} \cos 4x$$

$$\frac{\partial^2 u}{\partial x^2} = 20e^{-t}(-4\sin 4x) - 80e^{-t} \sin 4x$$

$$\frac{\partial u}{\partial t} = -5e^{-t} \sin 4x$$

$$\frac{\partial u}{\partial t} = k\frac{\partial^2 u}{\partial x^2}, k = \frac{1}{16}$$

$$-5e^{-t} \sin 4x = \frac{1}{16}(-80e^{-t} \sin 4x)$$

29.4 Double Integrals

1. $$\int_0^1 \int_{x^2}^x (x+y)\,dy\,dx = \int_0^1 \left(xy + \frac{y^2}{2} \right)\Bigg|_{x^2}^{x} dx$$

$$= \int_0^1 \left(x^2 + \frac{x^2}{2} - \left(x^3 + \frac{x^4}{2} \right) \right) dx$$

$$= \int_0^1 \left(-\frac{x^4}{2} - x^3 + \frac{3x^2}{2} \right) dx$$

$$= -\frac{x^5}{10} - \frac{x^4}{4} + \frac{3x^2}{6}\Bigg|_0^1 = \frac{3}{20}$$

5. $$\int_2^4 \int_0^1 xy^2\,dx\,dy = \int_2^4 y^2 \left(\frac{1}{2}x^2 \right)\Bigg|_0^1 dy$$

$$= \int_2^4 y^2 \left(\frac{1}{2} - 0 \right) dy$$

$$= \frac{1}{2}\int_2^4 y^2\,dy$$

$$= \frac{1}{6}y^3\Big|_2^4$$

$$= \frac{1}{6}(64 - 8) = \frac{28}{3}$$

9. $$\int_0^1 \int_0^{\sqrt{1-x^2}} y\,dy\,dx = \int_0^1 \frac{1}{2}y^2\Bigg|_0^{\sqrt{1-x^2}} dx$$

$$= \int_0^1 \frac{1}{2}(1 - x^2)\,dx$$

$$= \frac{1}{2}x - \frac{1}{6}x^3\Bigg|_0^1$$

$$= \frac{1}{2} - \frac{1}{6}$$

$$= \frac{1}{3}$$

13. $\int_1^e \int_1^y \frac{1}{x}\,dx\,dy = \int_1^e \ln x\Big|_1^y\,dy$

$$= \int_1^e \ln y\,dy$$
$$= y(\ln y - 1)\Big|_1^e \text{ (Parts or Formula 46)}$$
$$= e(\ln e - 1) - (\ln 1 - 1)$$
$$= e(0) + 1 = 1$$

17.

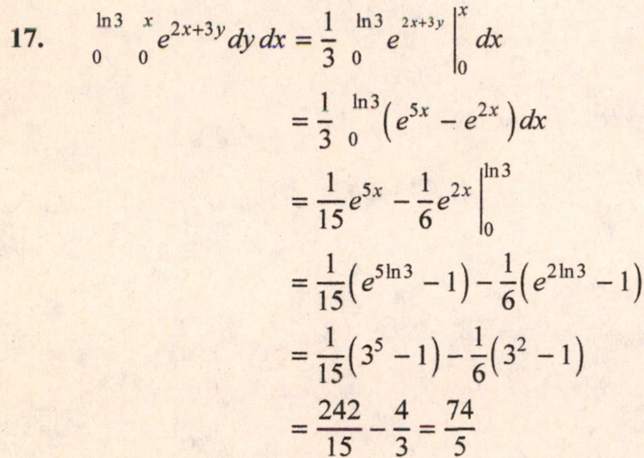

$$\int_0^{\ln 3}\int_0^x e^{2x+3y}\,dy\,dx = \frac{1}{3}\int_0^{\ln 3} e^{2x+3y}\Big|_0^x\,dx$$
$$= \frac{1}{3}\int_0^{\ln 3}\left(e^{5x} - e^{2x}\right)dx$$
$$= \frac{1}{15}e^{5x} - \frac{1}{6}e^{2x}\Big|_0^{\ln 3}$$
$$= \frac{1}{15}\left(e^{5\ln 3} - 1\right) - \frac{1}{6}\left(e^{2\ln 3} - 1\right)$$
$$= \frac{1}{15}\left(3^5 - 1\right) - \frac{1}{6}\left(3^2 - 1\right)$$
$$= \frac{242}{15} - \frac{4}{3} = \frac{74}{5}$$

21. The volume is four times the volume in the first octant.

The trace in the xy-plane is $x^2 + y^2 = 4$, so x goes from 0 to $\sqrt{4 - y^2}$. This curve crosses the y-axis at $y = 2$, so y goes from 0 to 2.

$$V = 4\int_0^2\int_0^{\sqrt{4-y^2}} z\,dx\,dy$$
$$= 4\int_0^2\int_0^{\sqrt{4-y^2}}\left(4 - x^2 - y^2\right)dx\,dy$$
$$= \int_0^2\left(16x - \frac{4}{3}x^3 - 4y^2x\right)\Big|_0^{\sqrt{4-y^2}}dy$$
$$= \int_0^2\left(16\sqrt{4-y^2} - \frac{4}{3}\left(4-y^2\right)^{3/2} - 4y^2\sqrt{4-y^2}\right)dy$$
$$= \frac{1}{3}\int_0^2\left(32 - 8y^2\right)\sqrt{4-y^2}\,dy$$
$$= \frac{8}{3}\int_0^2\left(4-y^2\right)^{3/2}dy$$
$$= \frac{8}{3}\left[\frac{y}{4}\left(4-y^2\right)^{3/2} + \frac{3(4)y}{8}\left(4-y^2\right)^{1/2} + \frac{3(16)}{8}\sin^{-1}\frac{y}{2}\right]_0^2 \text{ (Formula 20)}$$
$$= \frac{8}{3}\left[0 + 0 + 6\sin^{-1}1 - 0 - 0 - 0\right]$$
$$= \frac{8}{3}(6)\left(\frac{\pi}{2}\right) = 8\pi$$

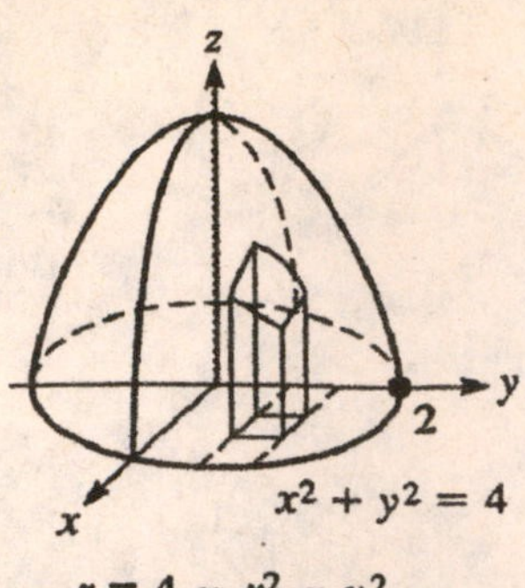

25. The trace in the xy-plane is $x^2 + y^2 = 9$, so y goes from 0 to $\sqrt{9-x^2}$. This curve crosses the x-axis at $x = 3$, so x goes from 0 to 3.

$$V = \int_0^3 \int_0^{\sqrt{9-x^2}} z\,dy\,dx$$

$$= \int_0^3 \int_0^{\sqrt{9-x^2}} (x+y)\,dy\,dx$$

$$= \int_0^3 \left(xy + \frac{1}{2}y^2 \right)\Big|_0^{\sqrt{9-x^2}}$$

$$= \int_0^3 \left[x\sqrt{9-x^2} + \frac{1}{2}\left(9-x^2\right) \right] dx$$

$$= -\frac{1}{3}\left(9-x^2\right)^{3/2} + \frac{9}{2}x - \frac{1}{6}x^3 \Big|_0^3$$

$$= 0 + \frac{27}{2} - \frac{27}{6} + \frac{1}{3}(27) - 0 + 0 = 18$$

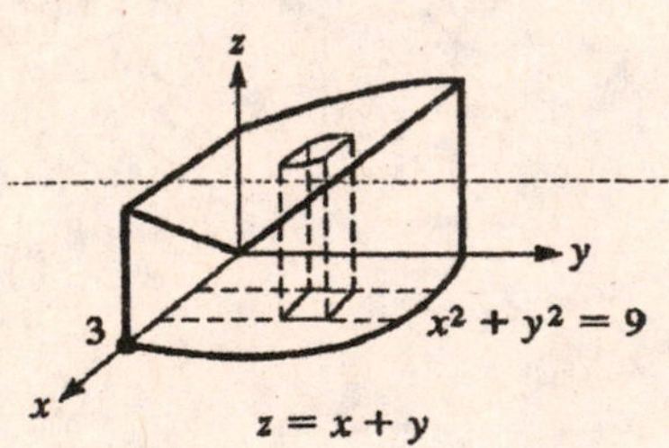

29. Set up the origin at the back lower corner. The volume is bounded by the planes $x = 0, x = 12, y = 0, z = 0$ and the plane parallel with the x-axis containing the points (0,5,0) and (0,0,10); this latter plane has equation $2y + z = 10$.
The double integral is then

$$\int_0^{12}\int_0^5 z\,dy\,dx = \int_0^{12}\int_0^5 (10-2y)\,dy\,dx$$
$$= \int_0^{12}\left(10y - y^2\right)\Big|_0^5\,dx$$
$$= \int_0^{12} 25\,dx$$
$$= 25x\Big|_0^{12}$$
$$= 300 \text{ cm}^3$$

33. The region of integration is bounded by $x = 0, x = 1, y = 2x,$ and $y = 2$. We rewrite the third condition as $x = \dfrac{1}{2}y$ and note that $y = 0$ is the lower bound on y. Thus, the integral becomes

$$\int_0^2\int_0^{y/2} f(x,y)\,dx\,dy$$

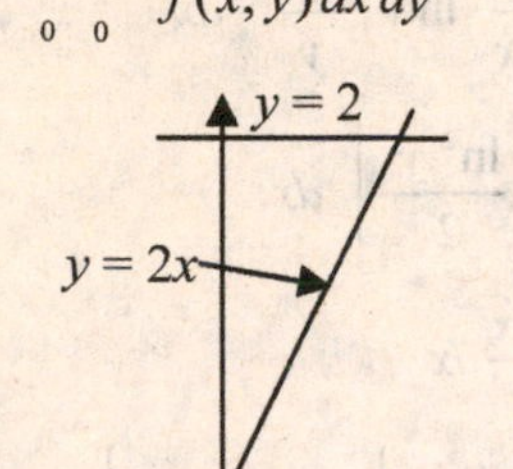

Review Exercises

1. This is false:

$$f(y^2,x) = \frac{2(y^2)^2x - x^2}{2(y^2)x} = \frac{2xy^4 - x^2}{2xy^2}.$$

The denominator is $2xy^2$, not $2x^2y$.

5. $f(x, y) = 4xy^3 - y^2$

$f(-4, 1) = 4(-4)(1)^3 - 1^2 = -17$

$f(1,-2) = 4(1)(-2)^3 - (-2)^2 = -36$

9. 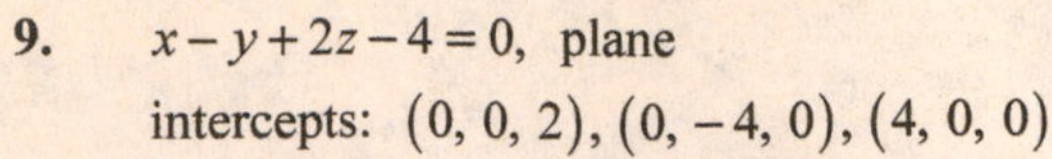
$x - y + 2z - 4 = 0$, plane
intercepts: $(0, 0, 2), (0, -4, 0), (4, 0, 0)$

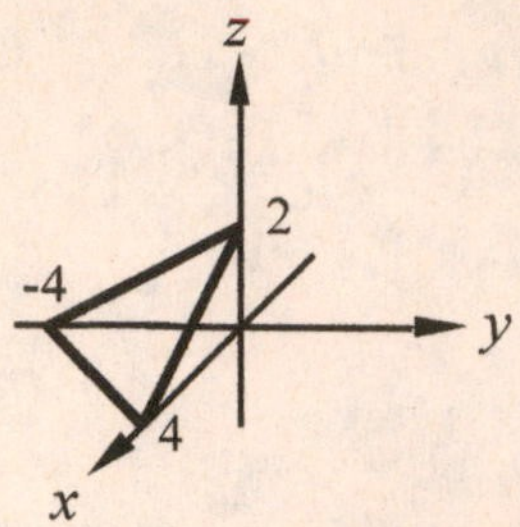

13. $z = 5x^3y^2 - 2xy^4$

$$\frac{\partial z}{\partial x} = 5y^2\left(3x^2\right) - 2y^4(1)$$
$$= 15x^2y^2 - 2y^4$$
$$\frac{\partial z}{\partial y} = 5x^3(2y) - 2x\left(4y^3\right)$$
$$= 10x^3y - 8xy^3$$

17. $z = \dfrac{e^{2y}}{x+y}$

$$\frac{\partial z}{\partial x} = -\frac{e^{2y}}{(x+y)^2}$$
$$\frac{\partial z}{\partial y} = \frac{2e^{2y}(x+y) - e^{2y}}{(x+y)^2}$$
$$= \frac{e^{2y}(2(x+y)-1)}{(x+y)^2}$$

21. $z = \cos^{-1}\sqrt{x+y}$

$$\frac{\partial z}{\partial x} = -\frac{1}{\sqrt{1-(x+y)}}\left(\frac{1}{2}\right)(x+y)^{-1/2}(1)$$
$$= -\frac{1}{2\sqrt{(x+y)(1-x-y)}}$$
$$\frac{\partial z}{\partial y} = -\frac{1}{\sqrt{1-(x+y)}}\left(\frac{1}{2}\right)(x+y)^{-1/2}(1)$$
$$= -\frac{1}{2\sqrt{(x+y)(1-x-y)}}$$

25. $r = 4e^s\cos 2t - 2te^{-s}$

$$\frac{\partial r}{\partial s} = 4e^s\cos 2t + 2te^{-s}$$
$$\frac{\partial^2 r}{\partial s^2} = 4e^s\cos 2t - 2te^{-s}$$
$$\frac{\partial r}{\partial t} = -8e^s\sin 2t - 2e^{-s}$$
$$\frac{\partial^2 r}{\partial t^2} = -16e^s\cos 2t$$
$$\frac{\partial^2 r}{\partial s\,\partial t} = \frac{\partial^2 r}{\partial t\,\partial s} = -8e^s\sin 2t + 2e^{-s}$$

29. $\int_0^3\int_1^x (x+2y)\,dy\,dx = \int_0^3 \left(xy+y^2\right)\Big|_1^x dx$

$$= \int_0^3\left(x^2+x^2-x-1\right)dx$$
$$= \int_0^3\left(2x^2-x-1\right)dx$$
$$= \frac{2}{3}x^3 - \frac{1}{2}x^2 - x\Big|_0^3$$
$$= \frac{2}{3}(27) - \frac{1}{2}(9) - 3 - 0 = \frac{21}{2}$$

33. $\int_1^e\int_1^x \frac{\ln y}{xy}\,dy\,dx = \int_1^e\int_1^x \frac{1}{x}\ \ln y\ \frac{dy}{y}\ dx$

$$= \int_1^e \frac{1}{x}\ \frac{\ln^2 y}{2}\Bigg|_1^x dx$$
$$= \int_1^e \frac{\ln^2 x}{2x}dx$$
$$= \frac{1}{6}\ln^3 x\Big|_1^e = \frac{1}{6}\ln^3 e - 0 = \frac{1}{6}$$

37. $z = e^{x+y}$

Intercepts: $(0, 0, 1)$

Traces:

$z > 0$ for all x and y since z is defined by an exponential.

In xy-plane: $z = e^x$

In yz-plane: $z = e^y$

Section: $z = 2 = e^{x+y}$

$\ln 2 = x + y$ (a line)

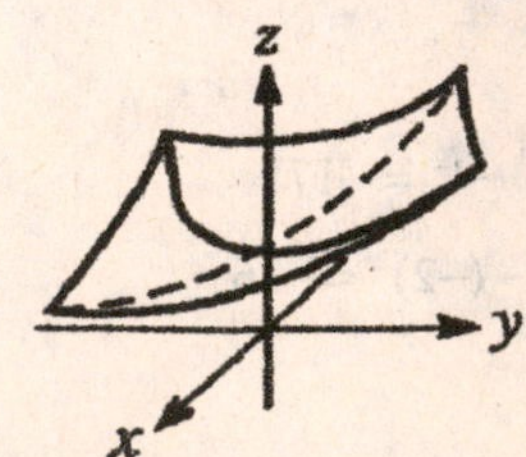

41. $V = \dfrac{rE}{r+R}$

$$\frac{\partial V}{\partial r} = \frac{E(r+R) - rE}{(r+R)^2}$$
$$= \frac{ER}{(r+R)^2}$$
$$\frac{\partial V}{\partial R} = -rE(r+R)^{-2}(1)$$
$$= \frac{-rE}{(r+R)^2}$$

45. $T = 2\pi\sqrt{\dfrac{\ell}{g}}$

$$\frac{\partial T}{\partial \ell} = 2\pi\sqrt{\frac{1}{g}}\ \frac{1}{2}\ell^{-1/2}$$
$$= \frac{\pi}{\sqrt{g\ell}}$$
$$\frac{T}{2\ell} = \frac{2\pi\sqrt{\ell/g}}{2\ell} = \frac{\pi}{\sqrt{g\ell}}$$
$$\frac{\partial T}{\partial \ell} = \frac{T}{2\ell}$$

49. $M = \dfrac{k}{\pi\left(b^2 - a^2\right)} \int_0^{2\pi} \int_a^b r^2\, dr\, d\theta$

$$= \frac{k}{\pi\left(b^2 - a^2\right)} \int_0^{2\pi} \frac{1}{3} r^3 \Big|_{r=a}^{r=b} d\theta$$
$$= \frac{k}{\pi\left(b^2 - a^2\right)} \int_0^{2\pi} \frac{b^3 - a^3}{3}\, d\theta$$
$$= \frac{2k\left(b^3 - a^3\right)}{3\left(b^2 - a^2\right)}$$
$$= \frac{2k\left(b^2 + ab + a^2\right)}{3(b+a)}$$

53. The region is bounded by the vertical plane $y = 0$, the vertical plane $x = 0$, the cylinder $x^2 + y^2 = 16$, and is under the plane $z = 8 - x$.

$$V = \int_0^4 \int_0^{\sqrt{16-x^2}} z\, dy\, dz$$
$$= \int_0^4 \int_0^{\sqrt{16-x^2}} (8-x)\, dy\, dz$$
$$= \int_0^4 (8-x)\, y\Big|_0^{\sqrt{16-x^2}} dx$$
$$= \int_0^4 (8-x)\sqrt{16-x^2}\, dx$$
$$= 8\int_0^4 \sqrt{16-x^2}\, dx - \int_0^4 x\sqrt{16-x^2}\, dx$$

(formula 15 and the power rule)

$$= 8\left[\frac{x}{2}\sqrt{16-x^2} + \frac{16}{2}\sin^{-1}\frac{x}{4}\right] + \frac{1}{3}\left(16-x^2\right)^{3/2}\Big|_0^4$$
$$= 8\left[8\sin^{-1}1\right] - \frac{1}{3}(16)^{3/2}$$
$$= 64\sin^{-1}1 - \frac{64}{3}$$
$$= 64\left(\frac{\pi}{2}\right) - \frac{64}{3} = 32\left(\pi - \frac{2}{3}\right)$$
$$= 79.2$$

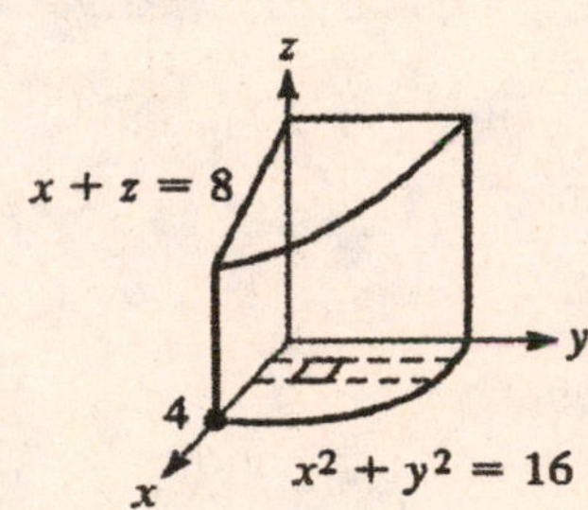

Chapter 30

Expansion of Functions in Series

30.1 Infinite Series

1. $\sum_{u=1}^{\infty} 0.5^n = 0.5 + 0.5^2 + 0.5^3 + 0.5^4 + \cdots + 0.5^n + \cdots$

$S_1 = 0.5, S_2 = 0.75, S_3 = 0.875, S_4 = 0.9375.$

The series now converges

5. $a_n = \frac{n}{n+1}; n = 0, 1, 2, 3, \ldots$

$a_0 = \frac{0}{0+1} = \frac{0}{1} = 0$ $\quad a_2 = \frac{2}{2+1} = \frac{2}{3}$

$a_1 = \frac{1}{1+1} = \frac{1}{2}$ $\quad a_3 = \frac{3}{3+1} = \frac{3}{4}$

9. $a_n = \cos\frac{n\pi}{2}, n = 0, 1, 2, 3, \ldots$

$a_0 = \cos\frac{0 \cdot \pi}{2} = 1$

$a_1 = \cos\frac{1 \cdot \pi}{2} = 0$

$a_2 = \cos\frac{2 \cdot \pi}{2} = -1$

$a_3 = \cos\frac{3 \cdot \pi}{2} = 0$

(a) $1, 0, -1, 0$

(b) $1 + 0 - 1 + 0 \ldots$

13. $\frac{1}{2\times3} - \frac{1}{3\times4} + \frac{1}{4\times5} - \frac{1}{5\times6} + \cdots$

$n = 1,\ a_1 = \frac{1}{(1+1)(1+2)} = \frac{1}{2\times3}$

$n = 2,\ a_2 = \frac{-1}{(2+1)(2+2)} = \frac{-1}{3\times4}$

$a_n = \frac{(-1)^{n+1}}{(n+1)(n+2)}$

17. $1 + \frac{1}{2} + \frac{2}{3} + \frac{3}{4} + \frac{4}{5} + \ldots$

$S_0 = 1; S_1 = 1 + \frac{1}{2} = \frac{3}{2} = 1.5$

$S_2 = 1 + \frac{1}{2} + \frac{2}{3} = \frac{13}{16} = 2.1666667$

$S_3 = 1 + \frac{1}{2} + \frac{2}{3} + \frac{3}{4} = \frac{35}{12} = 2.9166667$

$S_4 = 1 + \frac{1}{2} + \frac{2}{3} + \frac{3}{4} + \frac{4}{5} = \frac{223}{60} = 3.7166667$

The partial sums do not seem to converge, so the series appears to be divergent.

21. $\sum_{n=1}^{\infty} \frac{2n+1}{n^2(n+1)^2}.$

First five terms:

$a_1 = \frac{3}{4}; a_2 = \frac{5}{36}; a_3 = \frac{7}{144}; a_4 = \frac{9}{400}; a_5 = \frac{11}{900}$

First five partial sums:

$S_1 = 0.75;$

$S_2 = \frac{3}{4} + \frac{5}{36} = 0.8888889$

$S_3 = \frac{3}{4} + \frac{5}{36} + \frac{7}{144} = 0.9375000$

$S_4 = \frac{3}{4} + \frac{5}{36} + \frac{7}{144} + \frac{9}{400} = 0.9600000$

$S_5 = \frac{3}{4} + \frac{5}{36} + \frac{7}{144} + \frac{9}{400} + \frac{11}{900} = 0.9722222$

Convergent, converging to 1 (approx. sum)

25. $1 + 2 + 4 + \cdots + 2^n + \cdots; n = 0, 1, 2, 3, \cdots$

$S_0 = 1$

$S_1 = 3$

$S_2 = 7$

$S_3 = 15$

$S_n = 2^{n+1} - 1$

$\lim_{n\to\infty} Sn = \lim_{n\to\infty}(2^{n+1} - 1) = \infty$, divergent

Also, it is a geometric series with $r = 2 > 1$, so the series is divergent.

29. $10 + 9 + 8.1 + 7.29 + 6.561 + \cdots + 10(0.9)^n + \cdots;$

$n = 0, 1, 2, 3, \ldots$

$a = 10, r = 0.9 < 1$, so the series is convergent.

$$S = \frac{10}{1 - 0.9} = 100$$

33. $\sum_{n=0}^{\infty} (x-4)^n$ is a GS with $a_1 = 1$, $r = x - 4$ which, converges for

$$|x - 4| < 1$$
$$-1 < x - 4 < 1$$
$$3 < x < 5$$

37. $a_n = \dfrac{e^n}{n^3}$

Letting $n \to \infty$, using L'Hospital's rule successively,

$$\lim_{n\to\infty} \frac{e^n}{n^3} = \lim_{n\to\infty} \frac{e^n}{3n^2} = \lim_{n\to\infty} \frac{e^n}{6n} = \lim_{n\to\infty} \frac{e^n}{6} = \infty$$

41. $S_n = \dfrac{a_1(1 - r^n)}{(1 - r)}$; $r \neq 1$; geometric series

Series: $\frac{1}{2} + \frac{1}{4} + \frac{1}{8} + \cdots;\ a_n = \frac{1}{2^n}, a = \frac{1}{2}, r = \frac{1}{2}$

$$f(x) = \frac{a_1(1 - r^x)}{(1 - r)};\ f(x) = \frac{(1 - r^x)}{(1 - \frac{1}{2})} = (1 - r^x)$$

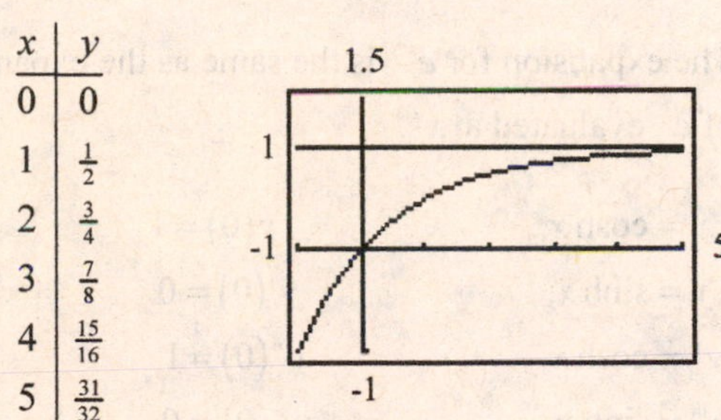

x	y
0	0
1	$\frac{1}{2}$
2	$\frac{3}{4}$
3	$\frac{7}{8}$
4	$\frac{15}{16}$
5	$\frac{31}{32}$

The infinite series approaches 1.

45. $\sum_{n=0}^{\infty} x^n = 1 + x + x^2 + \cdots + x^n + \cdots$

For $|x| < 1$, $a_1 = 1, r = x$, and the series converges because $|r| < 1$.

$$S = \frac{1}{1 - x}$$

$$\sum_{n=0}^{\infty} x^n = \frac{1}{1 - x}$$

30.2 Maclaurin Series

1. $f(x) = \dfrac{2}{2 + x},\ f(0) = 1$

$$f'(x) = \frac{-2}{(2 + x)^2},\ f'(0) = -\frac{1}{2}$$

$$f''(x) = \frac{4}{(2 + x)^3},\ f''(0) = \frac{1}{2}$$

$$f'''(x) = \frac{-12}{(2 + x)^4},\ f'''(0) = -\frac{3}{4}$$

$$f(x) = \frac{2}{2 + x} = 1 - \frac{1}{2}x + \frac{1}{4}x^2 - \frac{1}{8}x^3 + \cdots$$

5. $f(x) = \cos x \qquad f(0) = 1$

$f'(x) = -\sin x \qquad f'(0) = 0$

$f''(x) = -\cos x \qquad f''(0) = -1$

$f'''(x) = \sin x \qquad f'''(0) = 0$

$f^{iv}(x) = \cos x \qquad f^{iv}(0) = 1$

$$f(x) = \cos x = f(0) + f''(0)\frac{x^2}{2!} + f^{iv}(0)\frac{x^4}{4!} - \cdots$$

$$\cos x = 1 - 1\frac{x^2}{2} + 1\frac{x^4}{24} - \cdots$$

$$\cos x = 1 - \frac{1}{2}x^2 + \frac{1}{24}x^4 - \cdots$$

9. $f(x) = e^{-2x} \qquad f(0) = 1$

$f'(x) = -2e^{-2x} \qquad f'(0) = -2$

$f''(x) = 4e^{-2x} \qquad f''(0) = 4$

$$e^{-2x} = 1 - 2x + 4\frac{x^2}{2} - \cdots$$
$$= 1 - 2x + 2x^2 - \cdots$$

13. $f(x) = \dfrac{1}{(1 - x)} \qquad f(0) = 1$

$$f'(x) = \frac{1}{(1 - x)^2} \qquad f'(0) = 1$$

$$f''(x) = \frac{2}{(1 - x)^3} \qquad f''(0) = 2$$

$$\frac{1}{(1 - x)} = 1 + x + \frac{2x^2}{2} + \cdots$$
$$= 1 + x + x^2 + \cdots$$

17. $f(x)=\cos^2 x \qquad f(0)=1$
$f'(x)=-2\sin x\cos x \qquad f'(0)=0$
$f''(x)=2-4\cos^2 x \qquad f''(0)=-2$
$f'''(x)=8\sin x\cos x \qquad f'''(0)=0$
$f^{iv}(x)=16\cos^2 x-8 \qquad f^{iv}(0)=8$

$$\cos^2 x=1-2\frac{x^2}{2!}+8\frac{x^4}{4!}+\cdots$$
$$=1-x^2+\frac{1}{3}x^4+\cdots$$

21. $f(x)=\tan^{-1}x$

$f'(x)=\frac{1}{1+x^2}=(1+x^2)^{-1}$

$f''(x)=-(1+x^2)^{-2}2x=-2x(1+x^2)^{-2}$

$f'''(x)=-2x[-2(1+x^2)^{-3}(2x)]+(1+x^2)^{-2}(-2)$

$f(0)=0$
$f'(0)=1$
$f''(0)=0$
$f'''(0)=-2$

$$f(x)=0+1x+\frac{0x^2}{2!}-\frac{2x^3}{3!}+\cdots$$
$$=x-\frac{1}{3}x^3+\cdots$$

25. $f(x)=\ln\cos x$

$f'(x)=-\frac{1}{\cos x}\sin x=-\tan x$

$f''(x)=-\sec^2 x$

$f'''(x)=-2\sec x\sec x\tan x=-2\sec^2 x\tan x$

$f^{iv}(x)=-2\sec^2 x\sec^2 x-2\tan x(2\sec x\sec x\tan x)$

$f(0)=\ln 1=0$
$f'(0)=0$
$f''(0)=-1$
$f'''(0)=0$
$f^{iv}(0)=-2-0=-2$

$$f(x)=0+0x-\frac{1x^2}{2!}+\frac{0x^3}{3!}-\frac{2x^4}{4!}+\cdots$$
$$=-\frac{1}{2}x^2-\frac{1}{12}x^4-\cdots$$

29.
$$\begin{array}{r|l} & 1+x+x^2+\ldots \\ 1-x & 1 \\ & \underline{1-x} \\ & x \\ & \underline{x-x^2} \\ & x^2 \\ & x^2-x^3 \\ & \vdots \end{array}$$

which reflects the answer in Exercise 13.

33. (a) $f(x)=e^x \qquad f(0)=1$
$f'(x)=e^x \qquad f'(0)=1$
$f''(x)=e^x \qquad f''(0)=1$

$$e^x=1+x+\frac{1}{2}x^2+\cdots$$

(b) $f(x)=e^{x^2} \qquad f(0)=1$
$f'(x)=2xe^{x^2} \qquad f'(0)=0$
$f''(x)=2e^{x^2}(2x^2+1) \qquad f''(0)=2$
$f'''(x)=4xe^{x^2}(2x^2+3) \qquad f'''(0)=0$
$f^{iv}(x)=4e^{x^2}(4x^4+12x^2+3) \quad f^{iv}(0)=12$

$$e^{x^2}=1+\frac{2x^2}{2}+\frac{12x^4}{4!}+\cdots$$
$$=1+x^2+\frac{1}{2}x^4+\cdots$$

The expansion for e^{x^2} is the same as the expansion of e^x evaluated at x^2.

37. $y=\cosh x, \qquad y(0)=1$
$y=\sinh x, \qquad y'(0)=0$
$y''=\cosh x, \qquad y''(0)=1$
$y'''=\sinh x, \qquad y'''(0)=0$
$y^{iv}=\cosh x \qquad y^{iv}(0)=1$

$$\cosh x=1+0\cdot x+1\frac{x^2}{2!}+0\cdot\frac{x^3}{3!}+\cdots$$
$$=1+\frac{x^2}{2!}+\frac{x^4}{4!}+\cdots$$

41. $y = 4e^{-0.2t}\cos t,\ y(0) = 4$

$$y' = -0.8e^{-0.2t}\cos t - 4.0e^{-0.2t}\sin t$$
$$y'(0) = -0.8$$
$$y'' = -3.84e^{-0.2t}\cos t + 1.6e^{-0.2t}\sin t$$
$$y''(0) = -3.84$$

$$y = 4e^{-0.2t}\cos t = 4 - 0.8t - 3.84\frac{t^2}{2!} + \cdots$$
$$= 4 - 0.8t - 1.92t^2 + \cdots$$

30.3 Operations with Series

1. $e^x = 1 + x + \frac{x^2}{2!} + \frac{x^3}{3!} + \cdots$

$$e^{2x^2} = 1 + 2x^2 + \frac{(2x^2)^2}{2!} + \frac{(2x^2)^3}{3!} + \cdots$$
$$e^{2x^2} = 1 + 2x^2 + 2x^4 + \frac{4}{3}x^6 + \cdots$$

5. $f(x) = \sin\ \frac{1}{2}x\ ;$

$$g(x) = \sin x = x - \frac{x^3}{3!} + \frac{x^5}{5!} - \frac{x^7}{7!} + \cdots$$
$$f(x) = g\ \frac{1}{2}x$$
$$= \frac{1}{2}x - \frac{\left(\frac{1}{2}x\right)^3}{3!} + \frac{\left(\frac{1}{2}x\right)^5}{5!} - \frac{\left(\frac{1}{2}x\right)^7}{7!} + \cdots$$
$$= \frac{1}{2}x - \frac{x^3}{2^3 3!} + \frac{x^5}{2^5 5!} - \frac{x^7}{2^7 7!} + \cdots$$

9. $f(x) = \ln\left(1 + x^2\right)$

$$g(x) = \ln(1 + x)$$
$$= x - \frac{x^2}{2} + \frac{x^3}{3} - \frac{x^4}{4} + \cdots$$
$$\ln\left(1 + x^2\right) = g(x^2)$$
$$= x^2 - \frac{\left(x^2\right)^2}{2} + \frac{\left(x^2\right)^3}{3} - \frac{\left(x^2\right)^4}{4} + \cdots$$
$$= x^2 - \frac{1}{2}x^4 + \frac{1}{3}x^6 - \frac{1}{4}x^8 + \cdots$$

13. $\int_0^{0.5} e^{-\sqrt{x}}\,dx = \int_0^{0.5} 1 - \sqrt{x} + \frac{\left(\sqrt{x}\right)^2}{2}\,dx$

$$= \int_0^{0.5} 1 - x^{1/2} + \frac{1}{2}x\ dx$$
$$= x - \frac{2}{3}x^{3/2} + \frac{1}{4}x^2\ \Big|_0^{0.5}$$
$$= 0.5 - \frac{2}{3}(0.5)^{3/2} + \frac{1}{4}(0.5)^2$$
$$= 0.327$$

17. $f(x) = \frac{2}{1 - x^2} = \frac{1}{1 + x} + \frac{1}{1 - x}$

$$f(x) = 1 - x + x^2 - x^3 + x^4 - x^5 + x^6 - x^7 + x^8 + \cdots$$
$$+ 1 + x + x^2 + x^3 + x^4 + x^5 + x^6 + x^7 + x^8 + \cdots$$
$$f(x) = 2\left(1 + x^2 + x^4 + x^6 + \cdots\right)$$

21. $x^2\ln(1 - x)^2$

$$= x^2\left(2\ln(1 - x)\right)$$
$$= 2x^2\ln\left(1 + (-x)\right)$$
$$= 2x^2\ \ (-x) - \frac{(-x)^2}{2} + \frac{(-x)^3}{3} - \frac{(-x)^4}{4} + \cdots$$
$$= 2x^2\ \ -x - \frac{x^2}{2} - \frac{x^3}{3} - \frac{x^4}{4} - \cdots$$
$$= -2x^3 - x^4 - \frac{2}{3}x^5 - \frac{1}{2}x^6 - \cdots$$

25. $e^x = 1 + x + \frac{x^2}{2!} + \frac{x^3}{3!} + \frac{x^4}{4!} + \cdots$

$$\frac{d}{dx}e^x = 0 + 1 + \frac{2x}{2!} + \frac{3x^2}{6} + \frac{4x^3}{24} + \cdots$$
$$= 1 + x + \frac{x^2}{2} + \frac{x^3}{6} + \cdots$$
$$= 1 + x + \frac{x^2}{2!} + \frac{x^3}{3!} + \cdots$$
$$= e^x$$

29. $y = 4e^{-0.2t}\cos t$

$$= 4\left[1 - 0.2t + 0.02t^2 - \cdots\right]$$

$$\cdot\left[1 - \frac{1}{2}t^2 + \frac{1}{24}t^4 - \frac{1}{720}t^6 + \cdots\right]$$

Considering through the t^3 terms in the product:

$$y = 4\left[1 - \frac{1}{2}t^2 - 0.2t + 0.1t^3 + .02t^2 + \cdots\right]$$

$$= 4 - 0.8t - 1.92t^2 + .4t^3 + \cdots$$

33. $\lim\limits_{x\to 0}\dfrac{\sin x - x}{x^3}$

$$= \lim_{x\to 0}\frac{x - \frac{1}{6}x^3 + \frac{1}{120}x^5 - \frac{1}{5040}x^7 + \cdots - x}{x^3}$$

$$= \lim_{x\to 0} -\frac{1}{6} + \frac{1}{120}x^2 - \frac{1}{5010}x^4 + \cdots$$

$$= -\frac{1}{6}$$

37. $\lim\limits_{x\to 0}\dfrac{\ln(1+x^2)}{x^2} = \lim\limits_{x\to 0}\dfrac{1}{x^2}\ \dfrac{x^2}{1} - \dfrac{x^4}{2} + \dfrac{x^6}{3} - \ldots$

$$= \lim_{x\to 0}\ 1 - \frac{x^2}{2} + \frac{x^4}{3} - \ldots$$

$$= 1$$

Using L'Hospital's rule,

$$\lim_{x\to 0}\frac{\ln(1+x^2)}{x^2} = \lim_{x\to 0}\frac{2x/(1+x^2)}{2x}$$

$$= \lim_{x\to 0}\frac{1}{1+x^2}$$

$$= 1$$

45. $y_1 = \ln(1+x)$,

$y_2 = x$,

$y_3 = x - \frac{1}{2}x^2$,

$y_4 = x - \frac{1}{2}x^2 + \frac{1}{3}x^3$

1
y2
y4
y1
y3
-1
1
-1

30.4 Computations by Use of Series Expansions

1. $e^x = 1 + x + \dfrac{x^2}{2!} + \cdots$

$$e^{-0.1} = 1 + (-0.1) + \frac{(-0.1)^2}{2!} + \cdots$$

$$e^{-0.1} = 0.905$$

5. $\sin 0.1, (2\text{ terms});$

$$\sin x = x - \frac{x^3}{3!}$$

$$\sin 0.1 = 0.1 - \frac{(0.1)^3}{6}$$

$$= 0.0998333$$

$(\sin 0.1 = 0.0998334$ from calculator$)$

9. $\cos\pi^\circ$, (2 terms); $\pi^\circ = \dfrac{\pi^2}{180}$ radians

$$\cos x = 1 - \frac{x^2}{2!}$$

$$\cos\pi^\circ = 1 - \frac{\left(\frac{\pi^2}{180}\right)^2}{2}$$

$$= 0.9984967733$$

$(\cos\pi^\circ = 0.9984971499$ from calculator$)$

13. $\sin 0.3625, (3\text{ terms})$

$$\sin x = x - \frac{x^3}{3!} + \frac{x^5}{5!}$$

$$\sin 0.3625 = 0.3625 - \frac{(0.3625)^3}{6} + \frac{(0.3625)^5}{5!}$$

$$= 0.35461303$$

$(\sin 0.3625 = 0.35461287$ from calculator$)$

17. 1.032^6, 3 terms

$$(1+x)^6 = 1 + 6x + 15x^2$$

$$(1.032) = 1 + 6(0.032) + 15(0.032)^2$$

$$= 1.20736$$

$((1.032)^6 = 1.20803$, from calculator$)$

21. $\sqrt{1+x}=(1+x)^{1/2}$

$$(1+x)^n=1+nx+\frac{n(n-1)}{2!}x^2$$

$$(1+x)^{1/2}=1+\frac{1}{2}x+\frac{\frac{1}{2}\ -\frac{1}{2}}{2!}x^2$$

$$=1+\frac{1}{2}x-\frac{1}{8}x^2$$

$$\sqrt{1.1076}=\sqrt{1+0.1076}=(1+0.1076)^{1/2}$$

$$(1+0.1076)^{1/2}=1+\frac{1}{2}(0.1076)-\frac{1}{8}(0.1076)^2$$

$$=1.052353$$

$\left(\sqrt{1.1076}=1.052426 \text{ from calculator}\right)$

25. From Exercise 5,

$$\sin(0.1)=0.1-\frac{0.1^3}{6}=0.0998$$

The maximum possible error is the value of the first term omitted,

$$\frac{x^5}{5!}=\left|\frac{0.1^5}{120}\right|=8.3\times10^{-8}$$

29. $(1+x)^n=1+nx+\dfrac{n(n-1)}{2!}x^2$

$$\sqrt{3.92}=2(1+(-0.02))^{1/2}$$

$$=2\left[1+\frac{1}{2}(-0.02)+\frac{\frac{1}{2}\left(\frac{1}{2}-1\right)}{2!}(-0.02)^2\right]$$

$$=1.9799$$

33. $e^x=1+x+\dfrac{x^2}{2}+\dfrac{x^3}{3!}+\dfrac{x^4}{4!}+\cdots>1+x+\dfrac{x^2}{2}$

for $x>0$ since the terms of the expansion for e^x after those on right-hand side of the inequality have a positive value.

37. $f(t)=\dfrac{E}{R}\left(1-e^{-Rt/L}\right);$

$$e^x=1+x+\frac{x^2}{2}+\cdots$$

$$e^{-Rt/L}=1-\frac{Rt}{L}+\frac{R^2t^2}{2L^2}+\cdots$$

$$i=\frac{E}{R}\ 1-\ 1-\frac{Rt}{L}+\frac{R^2t^2}{2L^2}\quad=\frac{E}{L}\ t-\frac{Rt^2}{2L}$$

The approximation will be valid for small values of t.

30.5 Taylor Series

1. $f(x)=x^{1/2}, f(1)=1$

$$f'(x)=\frac{1}{2x^{1/2}}, f'(1)=\frac{1}{2}$$

$$f''(x)=-\frac{1}{4x^{3/2}}, f''(1)=-\frac{1}{4}$$

$$f'''(x)=\frac{3}{8x^{5/2}}, f'''(1)=\frac{3}{8}$$

$$\sqrt{x}=1+\frac{1}{2}(x-1)+\frac{-\frac{1}{4}(x-1)^2}{2!}+\frac{\frac{3}{8}(x-1)^3}{3!}+\cdots$$

$$\sqrt{x}=1+\frac{1}{2}(x-1)-\frac{1}{8}(x-1)^2+\frac{1}{16}(x-1)^3-\cdots$$

5. $\sqrt{4.3}$

$$\sqrt{x}=2+\frac{(x-4)}{4}-\frac{(x-4)^2}{64}+\frac{(x-4)^3}{512}$$

$$\sqrt{4.3}=2+\frac{(4.3-4)}{4}-\frac{(4.3-4)^2}{64}+\frac{(4.3-4)^3}{512}$$

$$=2.074;$$

$(\sqrt{4.3}=2.0736 \text{ from calculator})$

9. $\sin x=\dfrac{1}{2}+\dfrac{\sqrt{3}}{2}\ x-\dfrac{\pi}{6}\ -\dfrac{1}{4}\ x-\dfrac{\pi}{6}^{\ 2}$

$$\sin\frac{9\pi}{60}=\frac{1}{2}+\frac{\sqrt{3}}{2}\ -\frac{\pi}{60}$$

$$-\frac{1}{4}\ -\frac{\pi}{60}^{\ 2}$$

$$=0.45399;$$

$\left(\sin\dfrac{9\pi}{60}=0.4539904997 \text{ from calculator}\right)$

13. $\sin x; a = \frac{\pi}{3}$

$f(x) = \sin x \qquad f \ \frac{\pi}{3} \ = \frac{\sqrt{3}}{2}$

$f'(x) = \cos x \qquad f' \ \frac{\pi}{3} \ = \frac{1}{2}$

$f''(x) = -\sin x \qquad f'' \ \frac{\pi}{3} \ = -\frac{\sqrt{3}}{2}$

$$\sin x = \frac{\sqrt{3}}{2} + \frac{1}{2} \ x - \frac{\pi}{3} \ - \frac{\sqrt{3}}{2}\frac{1}{2!} \ x - \frac{\pi}{3} \ ^2 - \cdots$$

$$= \frac{1}{2} \ \sqrt{3} + \ x - \frac{\pi}{3} \ - \frac{\sqrt{3}}{2!} \ x - \frac{\pi}{3} \ ^2 + \cdots$$

17. $\tan x; a = \frac{\pi}{4}$

$f(x) = \tan x \qquad f \ \frac{\pi}{4} \ = 1$

$f'(x) = \sec^2 x \qquad f' \ \frac{\pi}{4} \ = (\sqrt{2})^2 = 2$

$f''(x) = 2\sec x \sec x \tan x = 2\sec^2 x \tan x$

$f''(x) = 2(\sqrt{2})^2(1) = 4$

$$\tan x = 1 + 2 \ x - \frac{\pi}{4} \ + \frac{4\left(x - \frac{\pi}{4}\right)^2}{2!} + \cdots$$

$$= 1 + 2 \ x - \frac{\pi}{4} \ + 2 \ x - \frac{\pi}{4} \ ^2 + \cdots$$

21. $f(x) = \frac{1}{x+2}, \ f(3) = \frac{1}{5}$

$f'(x) = -\frac{1}{(x+2)^2}, \ f'(3) = -\frac{1}{25}$

$f''(x) = \frac{2}{(x+2)^3}, \ f''(3) = \frac{2}{125}$

$$\frac{1}{x+2} = \frac{1}{5} - \frac{1}{25}(x-3) + \frac{1}{125}(x-3)^2$$

25. $\sqrt{9.3}; a = 9$

$f(x) = \sqrt{x} \qquad f'(9) = 3$

$f'(x) = \frac{1}{2\sqrt{x}} \qquad f'(9) = \frac{1}{6}$

$f''(x) = -\frac{1}{4x^{3/2}} \qquad f''(9) = -\frac{1}{108}$

$$\sqrt{x} = 3 + \frac{1}{6}(x-9) - \frac{1}{108}\frac{(x-9)^2}{2!}$$

$$\sqrt{9.3} = 3 + \frac{1}{6}(0.3) - \frac{1}{108}\frac{(0.3)^2}{2} = 3.0496$$

29. $\sin x = \frac{1}{2} \ \sqrt{3} + \ x - \frac{\pi}{2} \ - \frac{\sqrt{3}}{2} \ x - \frac{\pi}{3} \ ^2 ; a = \frac{\pi}{3}$

$$61° = 60° + 1° = \frac{\pi}{3} + \frac{\pi}{180}$$

$$\sin 61° = \frac{1}{2} \ \sqrt{3} + \frac{\pi}{180} - \frac{\sqrt{3}}{2} \ \frac{\pi}{180} \ ^2 = 0.87462$$

33. Expand $f(x) = 2x^3 + x^2 - 3x + 5$ about $x = 1$

$f'(x) = 6x^2 + 2x - 3$

$f''(x) = 12x + 2$

$f'''(x) = 12$

$$f(x) = f(1) + f'(1)(x-1)\frac{f''(1)(x-1)^2}{2!} + \frac{f'''(1)(x-1)^2}{3!}$$

$$= 2(1)^3 + 1^2 - 3(1) + 5 + \left(6(1)^2 + 2(1) - 3\right)(x-1) + \frac{(12(1)+2)(x-1)^2}{2!} + \frac{12(x-1)^3}{3!}$$

$$= 5 + 5(x-1) + 7(x-1)^2 + 2(x-1)^3$$

37. $i = 6\sin \pi t, \qquad i(\pi/2) = 6\sin \pi^2/2$

$i' = 6\pi \cos \pi t, \qquad i'(\pi/2) = 6\pi \cos \pi^2/2$

$i'' = -6\pi^2 \sin \pi t, \quad i''(\pi/2) = -6\pi^2 \sin \pi^2/2$

$$i = 6\sin\frac{\pi^2}{2} + 6\pi\cos\frac{\pi^2}{2}\left(t - \frac{\pi}{2}\right) - 3\pi^2\sin\frac{\pi^2}{2}\left(t - \frac{\pi}{2}\right)^2 + \cdots$$

41. $f(x) = \frac{1}{x}$; $x = 0$ to $x = 4$

(a) $y_1 = \frac{1}{x}$

(b) $y_2 = \frac{1}{2} - \frac{1}{4}(x-2)$

Graph in part (b) will fit the graph in part (a) well for values of x close to $x = 2$.

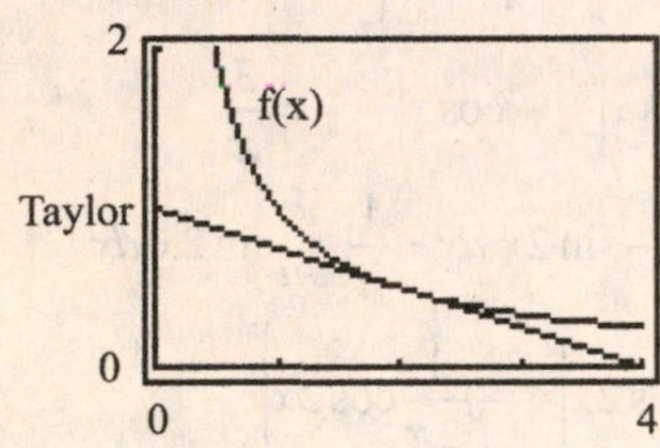

30.6 Introduction to Fourier Series

1. $f(x) = \begin{cases} -2, -\pi \le x < 0 \\ 2, \quad 0 \le x < \pi \end{cases}$

$$a_0 = \frac{1}{2\pi}\int_{-\pi}^{0}(-2)dx + \frac{1}{2\pi}\int_{0}^{\pi}2dx = 0$$

$$a_n = \frac{1}{\pi}\int_{-\pi}^{0}-2\cos nx\,dx + \frac{1}{\pi}\int_{0}^{\pi}2\cos nx\,dx = 0$$

$$b_n = \frac{1}{\pi}\int_{-\pi}^{0}-2\sin nx\,dx + \frac{1}{\pi}\int_{0}^{\pi}2\sin nx\,dx$$

$$= \frac{4}{\pi}\left(\frac{1}{n} - \frac{\cos(\pi n)}{n}\right)\Bigg|_0^{\pi}$$

$$b_n = \frac{4}{\pi}(1 - \cos \pi n)\Big|_0^{\pi} = \begin{cases} \frac{8}{n\pi}, n \text{ odd} \\ 0, n \text{ neven} \end{cases}$$

$$b_1 = \frac{8}{\pi}, b_3 = \frac{8}{3\pi}, b_5 = \frac{8}{5\pi}$$

$$f(x) = \frac{8}{\pi}\sin x + \frac{8}{3\pi}\sin 3x + \frac{8}{5\pi}\sin 5x + \cdots$$

$$f(x) = \frac{8}{\pi}\left(\sin x + \frac{1}{3}\sin 3x + \frac{1}{5}\sin 5x + \cdots\right)$$

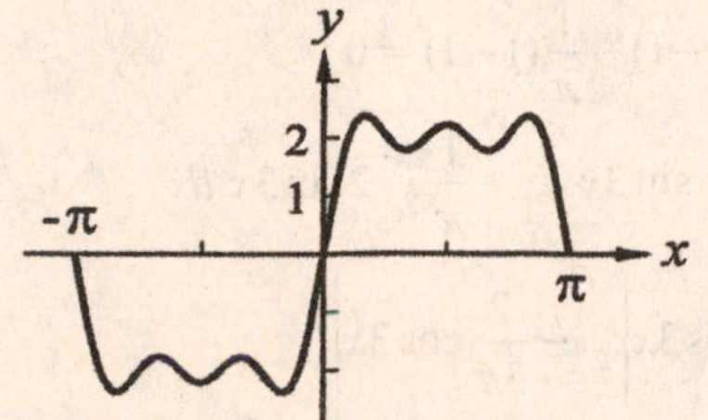

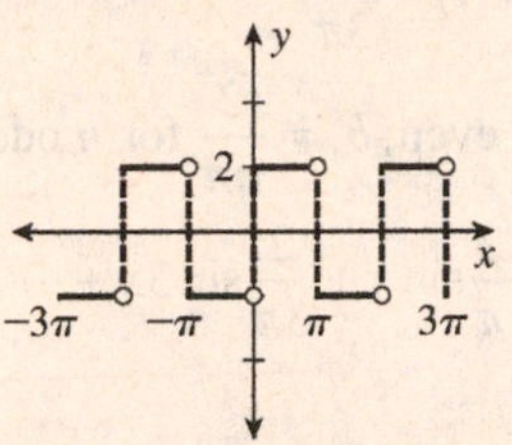

5. $f(x) = \begin{cases} 1 & -\pi \le x < 0 \\ 2 & 0 \le x < \pi \end{cases}$

$$a_0 = \frac{1}{2\pi}\int_{-\pi}^{0}1\,dx + \frac{1}{2\pi}\int_{0}^{\pi}2\,dx$$

$$= \frac{x}{2\pi}\Bigg|_{-\pi}^{0} + \frac{2x}{2\pi}\Bigg|_{0}^{\pi}$$

$$= 0 + \frac{\pi}{2\pi} + \frac{2\pi}{2\pi} - 0 = \frac{1}{2} + 1 = \frac{3}{2}$$

$$a_1 = \frac{1}{\pi}\int_{-\pi}^{0}1\cos x\,dx + \frac{1}{\pi}\int_{0}^{\pi}2\cos x\,dx$$

$$= \frac{1}{\pi}\sin x\Bigg|_{-\pi}^{0} + \frac{2}{\pi}\sin x\Bigg|_{0}^{\pi}$$

$$= \frac{1}{\pi}(0-0) + \frac{2}{\pi}(0-0) = 0$$

$a_n = 0$ since $\sin n\pi = 0$

$$b_1 = \frac{1}{\pi}\int_{-\pi}^{0}1\sin x\,dx + \frac{1}{\pi}\int_{0}^{\pi}2\sin x\,dx$$

$$= -\frac{1}{\pi}\cos x\Bigg|_{-\pi}^{0} - \frac{2}{\pi}\cos x\Bigg|_{0}^{\pi}$$

$$= -\frac{1}{\pi}(1+1) - \frac{2}{\pi}(-1-1)$$

$$= -\frac{2}{\pi} + \frac{4}{\pi} = \frac{2}{\pi}$$

$$b_2 = \frac{1}{\pi}\int_{-\pi}^{0} 1\sin 2x\,dx + \frac{1}{\pi}\int_{0}^{\pi} 2\sin x\,dx$$

$$= -\frac{1}{2\pi}\cos 2x\Big|_{-\pi}^{0} - \frac{1}{\pi}\cos 2x\Big|_{0}^{\pi}$$

$$= -\frac{1}{2\pi}(1-1) - \frac{1}{\pi}(1-1) = 0$$

$$b_3 = \frac{1}{\pi}\int_{-\pi}^{0} \sin 3x\,dx + \frac{1}{\pi}\int_{0}^{\pi} 2\sin 3x\,dx$$

$$= -\frac{1}{3\pi}\cos 3x\Big|_{-\pi}^{0} - \frac{2}{3\pi}\cos 3x\Big|_{0}^{\pi}$$

$$= -\frac{1}{3\pi}(1+1) - \frac{2}{3\pi}(-1-1) = \frac{2}{3\pi}$$

Therefore, $b_n = 0$ for n even; $b_n = \frac{2}{n\pi}$ for n odd.

Therefore, $f(x) = \frac{3}{2} + \frac{2}{\pi}\sin x + \frac{2}{3\pi}\sin 3x + \cdots$

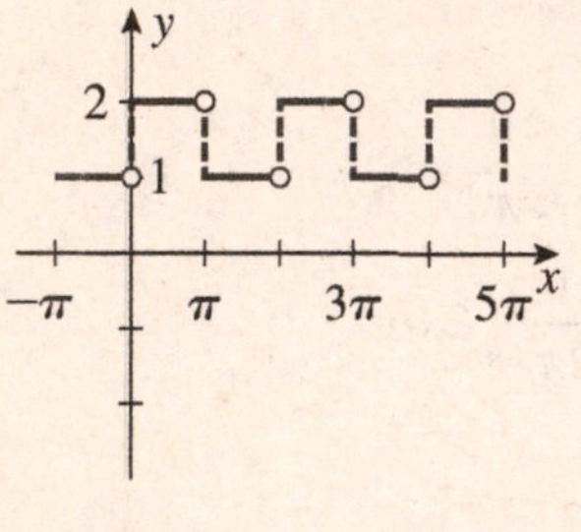

9. $$f(x) = \begin{cases} -1 & -\pi \le x < 0 \\ 0 & 0 \le x < \frac{\pi}{2} \\ 1 & \frac{\pi}{2} \le x < \pi \end{cases}$$

$$a_0 = \frac{1}{2\pi}\int_{-\pi}^{0} -dx + \frac{1}{2\pi}\int_{\pi/2}^{\pi} dx$$

$$= -\frac{1}{2\pi}x\Big|_{-\pi}^{0} + \frac{1}{2\pi}x\Big|_{\pi/2}^{\pi}$$

$$= -\frac{1}{2\pi}\ \pi - \ \pi - \frac{\pi}{2}\ = -\frac{1}{4}$$

$$a_1 = \frac{1}{\pi}\int_{-\pi}^{0} -\cos x\,dx + \frac{1}{\pi}\int_{\pi/2}^{\pi} \cos x\,dx$$

$$= -\frac{1}{\pi}\sin x\Big|_{-\pi}^{0} + \frac{1}{\pi}\sin x\Big|_{\pi/2}^{\pi}$$

$$= -\frac{1}{\pi}(\sin x\,|_{-\pi}^{0} - \sin x\,|_{\pi/2}^{\pi}) = -\frac{1}{\pi}$$

$$a_2 = \frac{1}{\pi}\int_{-\pi}^{0} -\cos 2x\,dx + \frac{1}{\pi}\int_{\pi/2}^{\pi} \cos 2x\,dx$$

$$= -\frac{1}{2\pi}\sin 2x\Big|_{-\pi}^{0} + \frac{1}{2\pi}\sin 2x\Big|_{\pi/2}^{\pi}$$

$$= -\frac{1}{2\pi}\left(\sin 2x\,|_{-\pi}^{0} - \sin 2x\,|_{\pi/2}^{\pi}\right) = 0$$

$$a_3 = \frac{1}{\pi}\int_{-\pi}^{0} -\cos 3x\,dx + \frac{1}{\pi}\int_{\pi/2}^{\pi} \cos 3x\,dx$$

$$= -\frac{1}{3\pi}\sin 3x\Big|_{-\pi}^{0} + \frac{1}{3\pi}\sin 3x\Big|_{\pi/2}^{\pi}$$

$$= -\frac{1}{3\pi}\ \sin 3x\Big|_{-\pi}^{0} - \sin 3x\Big|_{\pi/2}^{\pi}\ = \frac{1}{3\pi}$$

Therefore, $a_n = \pm\frac{1}{n\pi}$ for n odd; $a_n = 0$ for n even

$$b_1 = \frac{1}{\pi}\int_{-\pi}^{0} -\sin x\,dx + \frac{1}{\pi}\int_{\pi/2}^{\pi} \sin x\,dx$$

$$= \frac{1}{\pi}\cos x\Big|_{-\pi}^{0} - \frac{1}{\pi}\cos x\Big|_{\pi/2}^{\pi}$$

$$= \frac{1}{\pi}\ \cos x\Big|_{-\pi}^{0} - \cos x\Big|_{\pi/2}^{\pi}\ = \frac{3}{\pi}$$

$$b_2 = \frac{1}{\pi}\int_{-\pi}^{0} -\sin 2x\,dx + \frac{1}{\pi}\int_{\pi/2}^{\pi} \sin 2x\,dx$$

$$= \frac{1}{2\pi}\cos 2x\Big|_{-\pi}^{0} - \frac{1}{2\pi}\cos 2x\Big|_{\pi/2}^{\pi}$$

$$= \frac{1}{2\pi}\ \cos x\Big|_{-\pi}^{0} - \cos 2x\Big|_{\pi/2}^{\pi}\ = -\frac{1}{\pi}$$

$$b_3 = \frac{1}{\pi}\int_{-\pi}^{0} -\sin 3x\,dx + \frac{1}{\pi}\int_{\pi/2}^{\pi} \sin 3x\,dx$$

$$= \frac{1}{3\pi}\cos 3x\Big|_{-\pi}^{0} - \frac{1}{3\pi}\cos 3x\Big|_{\pi/2}^{\pi}$$

$$= \frac{1}{3\pi}\left(\cos 3x\Big|_{-\pi}^{0} - \cos 3x\Big|_{\pi/2}^{\pi}\right)$$

$$= \frac{1}{\pi}$$

$$b_n = (-1)^n\frac{1}{\pi} \text{ for } n > 1$$

$$f(x) = -\frac{1}{4} - \frac{1}{\pi}\cos x + \frac{1}{3\pi}\cos 3x - \cdots$$
$$+ \frac{3}{\pi}\sin x - \frac{1}{\pi}\sin 2x + \frac{1}{\pi}\sin 3x - \cdots$$

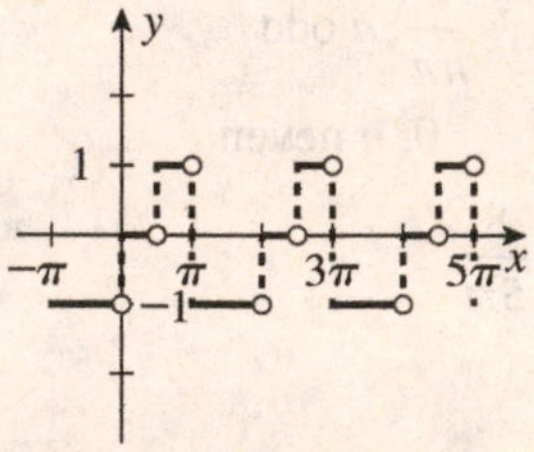

13. $f(x)=e^x, -\pi\le x<\pi$

$$a_0=\frac{1}{2\pi}\int_{-\pi}^{\pi}e^x\,dx=\frac{e^\pi-e^{-\pi}}{2\pi}$$

$$a_1=\frac{1}{\pi}\int_{-\pi}^{\pi}e^x\cos x\,dx=-\frac{e^\pi-e^{-\pi}}{2\pi}$$

$$a_2=\frac{1}{\pi}\int_{-\pi}^{\pi}e^x\cos 2x\,dx=\frac{e^\pi-e^{-\pi}}{5\pi}$$

$$b_1=\frac{1}{\pi}\int_{-\pi}^{\pi}e^x\sin x\,dx=-\frac{e^\pi-e^{-\pi}}{2\pi}$$

$$b_2=\frac{1}{\pi}\int_{-\pi}^{\pi}e^x\sin 2x\,dx=-\frac{2(e^\pi-e^{-\pi})}{5\pi}$$

$$e^x=\frac{e^\pi-e^{-\pi}}{2\pi}-\frac{e^\pi-e^{-\pi}}{2\pi}\cos x+\frac{e^\pi-e^{-\pi}}{5\pi}\cos 2x+\cdots+\frac{e^\pi-e^{-\pi}}{2\pi}\sin x-\frac{2(e^\pi-e^{-\pi})}{5\pi}\sin 2x+\cdots$$

$$e^x=\frac{e^\pi-e^{-\pi}}{\pi}\left(\frac{1}{2}-\frac{1}{2}\cos x+\frac{1}{5}\cos 2x+\cdots+\frac{1}{2}\sin x-\frac{2}{5}\sin 2x+\cdots\right)$$

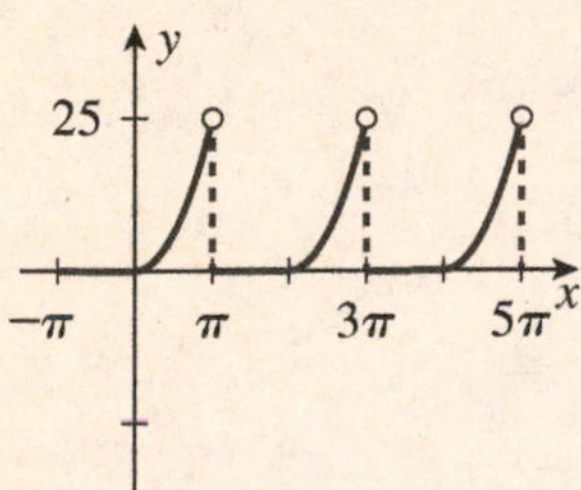

17. $f(x)=\begin{cases}1, -\pi\le x<0\\ 2, 0\le x<\pi\end{cases}$

Graph $y_1=\frac{3}{2}+\frac{2}{\pi}\sin x+\frac{2}{3\pi}\sin 3x$

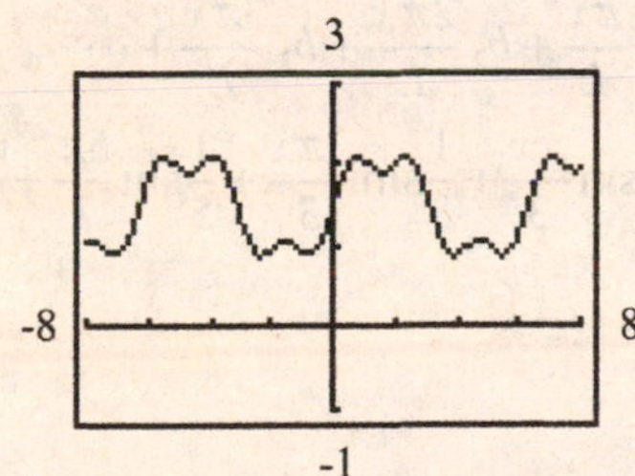

21. $F(t)=\begin{cases}0, -\pi\le t<0\\ t^2+t,\ 0<t<\pi\end{cases}$

$$a_0=\frac{1}{2\pi}\int_0^{\pi}(t^2+t)dt=\frac{\pi^2}{6}+\frac{\pi}{4}$$

$$a_1=\frac{1}{\pi}\int_0^{\pi}(t^2+t)\cos t\,dt=-\frac{2}{\pi}-2$$

$$a_2=\frac{1}{\pi}\int_0^{\pi}(t^2+t)\cos 2t\,dt=\frac{1}{2}$$

$$a_3=\frac{1}{\pi}\int_0^{\pi}(t^2+t)\cos 3t\,dt=\frac{-2-2\pi}{9\pi}$$

$$b_1=\frac{1}{\pi}\int_0^{\pi}(t^2+t)\sin t\,dt=\pi-\frac{4}{\pi}+1$$

$$b_2=\frac{1}{\pi}\int_0^{\pi}(t^2+t)\sin 2t\,dt=\frac{-\pi-1}{2}$$

$$b_3=\frac{1}{\pi}\int_0^{\pi}(t^2+t)\sin 3t\,dt=\frac{\pi}{3}-\frac{4}{27\pi}+\frac{1}{3}$$

$$F(t)=\frac{\pi^2}{6}+\frac{\pi}{4}-\left(\frac{2}{\pi}+2\right)\cos t+\frac{1}{2}\cos 2t-\left(\frac{2+2\pi}{9\pi}\right)\cos 3t+\cdots+\left(\pi-\frac{4}{\pi}+1\right)\sin t-\left(\frac{\pi+1}{2}\right)\sin 2t+\left(\frac{\pi}{3}-\frac{4}{27\pi}+\frac{1}{3}\right)\sin 3t+\cdots$$

30.7 More About Fourier Series

1. $f(x)\begin{cases}2 & -\pi\le x-\frac{\pi}{2}, \frac{\pi}{2}\le x<\pi\\ 3 & -\frac{\pi}{2}\le x<\frac{\pi}{2}\end{cases}$

From Example 3,

$$f_1(x)=\frac{1}{2}+\frac{2}{\pi}\left(\cos x-\frac{\cos 3x}{3}+\frac{\cos 5x}{5}\right)-\cdots$$

$$f(x)=f_1(x)+2$$

$$=\frac{5}{2}+\frac{2}{\pi}\left(\cos x-\frac{\cos 3x}{3}+\frac{\cos 5x}{5}-\cdots\right)$$

5. $f(x)=\begin{cases}5 & -3\le x<0\\ 0 & 0\le x<3\end{cases}$

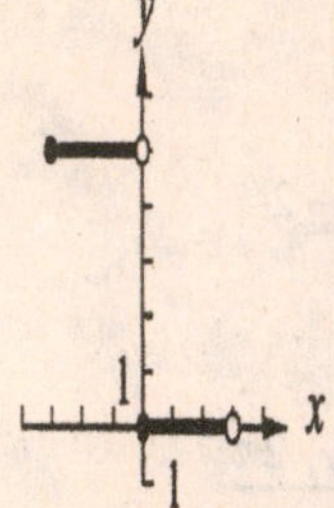

from the graph $f(x)$ is neither odd nor even. (It is, however, a shift of an odd function.)

9. $f(x)=|x| \quad -4\le x<4$

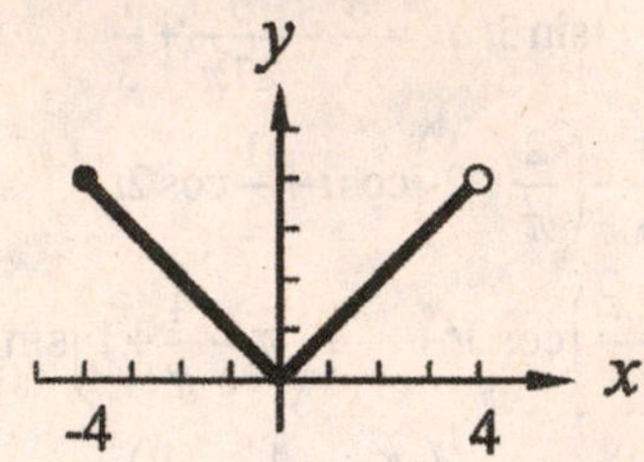

is even from the graph.

13. From the graph, $f(x)=2-x,\ -4\le x<4$ is not odd or even. However, $f(x)$ is the same as the function $f_1(x)=-x$ shifted up 2 units. Since $f_1(x)$ is odd, its expansion contains only sine terms. Therefore, the expansion of $f(x)$ contains only sine terms and the constant term 2.

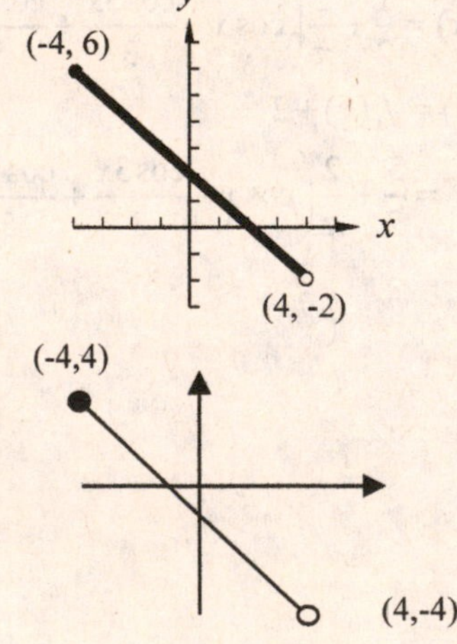

17. $f(x)=\begin{matrix}5 & -3\le x<0\\ 0 & 0\le x<3\end{matrix}$

period $=6=2L,\ L=3$

Although $f(x)$ is neither odd nor even, it is the same as the function

$f_1(x)=\begin{matrix}\frac{5}{2} & -3\le x<0\\ -\frac{5}{2} & 0\le x<3\end{matrix}$

shifted up $\frac{5}{2}$ units. Since $f_1(x)$ is odd, the expansion of $f(x)$ will have only sine terms plus the constant $\frac{5}{2}$. Therefore,

$$a_0=\frac{5}{2}$$

$$a_n=0$$

$$b_n=\frac{1}{L}\int_{-L}^{L} f(x)\sin\frac{n\pi x}{L}dx$$

$$=\frac{1}{3}\int_{-3}^{0} 5\sin\frac{n\pi x}{3}dx+\frac{1}{3}\int_{0}^{3} 0\cdot\sin\frac{2\pi x}{3}dx$$

$$b_n=\frac{5\cos(n\pi)-5}{n\pi}=\frac{5}{\pi}\ \frac{\cos(n\pi)-1}{n}$$

n	b_n
1	$\frac{5}{\pi}\cdot(-2)=\frac{-10}{\pi}$
2	0
3	$\frac{5}{\pi}\left(-\frac{2}{3}\right)=\frac{-10}{3\pi}$
4	0
5	$\frac{5}{\pi}\left(-\frac{2}{5}\right)=\frac{-10}{5\pi}$

$$f(x)=a_0+a_1\cos\frac{\pi x}{L}+a_2\cos\frac{2\pi x}{L}+a_3\cos\frac{3\pi x}{L}+\cdots$$

$$+b_1,\sin\frac{\pi x}{L}+b_2\frac{2\pi x}{L}+b_3\frac{3\pi x}{L}+\cdots$$

$$f(x)=\frac{5}{2}-\frac{10}{\pi}\left(\sin\frac{\pi x}{3}+\frac{1}{3}\sin\frac{3\pi x}{3}+\frac{1}{5}\sin\frac{5\pi}{3}+\cdots\right)$$

21. $f(x) = \begin{matrix} -x & -4 \le x < 0 \\ x & 0 \le x < 4 \end{matrix}$

period $= 2L = 8$, so $L = 4$

The function is even, and therefore, its Fourier expansion contains only cosine terms and possibly a constant. Therefore $b_n = 0$ and

$$a_0 = \frac{1}{8}\int_{-4}^{0} -x dx + \frac{1}{8}\int_{0}^{4} x dx$$

$$= -\frac{1}{16}x^2\Big|_{-4}^{0} + \frac{1}{16}x^2\Big|_{0}^{4}$$

$$= 2$$

$$a_n = \frac{1}{4}\int_{-4}^{0} -x\cos\frac{n\pi x}{4}dx + \frac{1}{4}\int_{0}^{4} x\cos\frac{n\pi x}{4}dx$$

$$= 2\left(\frac{1}{4}\int_{0}^{4} x\cos\frac{n\pi x}{4}dx\right)$$

$$= \frac{1}{2}\frac{16}{(n\pi)^2}\int_{0}^{4}\frac{n\pi}{4}x\cos\frac{n\pi}{4}x\frac{n\pi}{4}dx$$

(formula 48 from appendix)

$$= \frac{8}{(n\pi)^2}\left(\cos\frac{n\pi x}{4} + \frac{n\pi x}{4}\sin\frac{n\pi x}{4}\right)\Big|_{0}^{4}$$

$$= \frac{8}{(n\pi)^2}(\cos n\pi + n\pi\sin n\pi - 1)$$

$$= -\frac{8}{(n\pi)^2}(1 - \cos n\pi)$$

$$a_1 = -\frac{16}{\pi^2}; a_2 = 0; a_3 = -\frac{16}{9\pi^2}$$

Therefore,

$$f(x) = 2 - \frac{16}{\pi^2}\cos\frac{\pi x}{4} - \frac{16}{9\pi^2}\cos\frac{3\pi x}{4} + \cdots$$

$$= 2 - \frac{16}{\pi^2}\left(\cos\frac{\pi x}{4} + \frac{1}{9}\cos\frac{3\pi x}{4}\cdots\right)$$

25. Expand $f(x) = x^2$ in a half-range cosine series for $0 \le x < 2$.

$$a_0 = \frac{1}{L}\int_{0}^{L} f(x)dx = \frac{1}{2}\int_{0}^{2} x^2 dx$$

$$= \frac{1}{2}\frac{x^3}{3}\Big|_{0}^{2}$$

$$= \frac{1}{6}(2^3 - 0) = \frac{4}{3}$$

$$a_n = \frac{2}{L}\int_{0}^{L} f(x)\cos\frac{n\pi x}{L}dx, (n = 1, 2, 3, \ldots)$$

$$a_n = \frac{2}{2}\int_{0}^{2} x^2\frac{n\pi x}{2}dx$$

$$= \left(\frac{2x}{\frac{n^2\pi^2}{4}}\cos\frac{n\pi x}{2} + \left(\frac{x^2}{\frac{n\pi}{2}} - \frac{2}{\frac{n^3\pi^3}{8}}\right)\sin\frac{n\pi x}{2}\right)\Big|_{0}^{2}$$

$$a_n = \left(\frac{8x}{n^2\pi^2}\cos\frac{n\pi x}{2} + \left(\frac{2x^2}{n\pi} - \frac{16}{n^3\pi^3}\right)\sin\frac{n\pi x}{2}\right)\Big|_{0}^{2}$$

$$a_n = \frac{16}{n^2\pi^2}\cos n\pi + \left(\frac{8}{n\pi} - \frac{16}{n^3\pi^3}\right)\sin n\pi$$

$$a_n = \frac{16}{n^2\pi^2}\cos n\pi$$

$$a_1 = \frac{16}{1^2\pi^2}\cos\pi = \frac{-16}{\pi^2}$$

$$a_2 = \frac{16}{2^2\pi^2}\cos 2\pi = \frac{4}{\pi^2}$$

$$a_3 = \frac{16}{3^2\pi^2}\cos 3\pi = \frac{-16}{9\pi^2}$$

$$f(x) = \frac{4}{3} - \cos\frac{\pi x}{2} + \frac{4}{\pi^2}\cos\frac{2\pi x}{2} - \frac{16}{9\pi^2}\cos\frac{3\pi x}{2} + \cdots$$

$$f(x) = \frac{4}{3} - \frac{16}{\pi^2}\left(\cos\frac{\pi x}{2} - \frac{1}{4}\cos\pi x + \frac{1}{9}\cos\frac{3\pi x}{2} - \cdots\right)$$

Review Exercises

1. This is true. The series is geometric with $r = \frac{2}{3}$.

5. This is false. We have the Taylor series

$$f(x) = f(2) + f'(2)(x-2) + \frac{f''(2)}{2!}(x-2)^2 + \frac{f'''(2)}{3!}(x-2)^3 + \ldots$$

9. $F(x) = \sin x = x - \frac{x^3}{3!} + \frac{x^5}{5!} - \cdots$

$$f(x) = \sin 2x^2$$
$$= F(2x^2)$$
$$= 2x^2 - \frac{4}{3}x^6 + \frac{4}{15}x^{10} - \cdots$$

13. $f(x) = \sin^{-1} x \qquad f(0) = 0$

$$f'(x) = \frac{1}{\sqrt{1-x^2}} \qquad f'(0) = 1$$

$$f''(x) = \frac{x}{\left(1-x^2\right)^{3/2}} \qquad f''(0) = 0$$

$$f'''(x) = \frac{\left(1-x^2\right)^{3/2} - x(3/2)\left(1-x^2\right)^{1/2}(-2x)}{\left(1-x^2\right)^3}$$

$$= \frac{2x^2+1}{\left(1-x^2\right)^{5/2}} \qquad f'''(0) = 1$$

$$f^{iv}(x) = \frac{\left(1-x^2\right)^{5/2}(4x) - \left(2x^2+1\right)(5/2)\left(1-x^2\right)^{3/2}(-2x)}{\left(1-x^2\right)^5}$$

$$= \frac{6x^3+9x}{\left(1-x^2\right)^{7/2}} \qquad f^{iv}(0) = 0$$

$$f^{v}(x) = \frac{\left(1-x^2\right)^{7/2}\left(18x^2+9\right) - \left(6x^3+9x\right)(7/2)\left(1-x^2\right)^{5/2}(-2x)}{\left(1-x^2\right)^7}$$

$$= \frac{24x^4+72x^2+9}{\left(1-x^2\right)^{9/2}} \qquad f^{v}(0) = 9$$

$$f(x) = x + \frac{x^3}{3!} + 9\,\frac{x^5}{5!} + \cdots = x + \frac{1}{6}x^3 + \frac{3}{40}x^5 + \cdots$$

17. $e^x = 1 + x + \frac{x^2}{2} + \cdots$ Let $x = -0.2$

$$e^{-0.2} = 1 - 0.2 + \frac{(-0.2)^2}{2} + \cdots = 0.82$$

21. $(1+x)^{-1} = 1 - x + x^2 + \cdots$

$$1.086^{-1} = 1 - 0.086 + (0.086)^2$$
$$= 0.921$$

25. $f(x) = \tan x;\ a = \frac{\pi}{4}$ $\quad f\left(\frac{\pi}{4}\right) = 1$

$f'(x) = \sec^2 x = 1 + \tan^2 x \quad f'\left(\frac{\pi}{4}\right) = 2$

$f''(x) = 2\tan x \sec^2 x \quad f''\left(\frac{\pi}{4}\right) = 4$

$$f(x) = 1 + 2\left(x - \frac{\pi}{4}\right) + \frac{4\left(x - \frac{\pi}{4}\right)^2}{2!} + \cdots$$

$$\tan 43.62^\circ = \tan\left(45^\circ - 1.38^\circ\right)$$
$$= \tan\left(\frac{\pi}{4} - \frac{1.38\pi}{180}\right)$$
$$= 1 + 2\left(\frac{\pi}{4} - \frac{1.38\pi}{180} - \frac{\pi}{4}\right) + 2\left(\frac{\pi}{4} - \frac{1.38\pi}{180} - \frac{\pi}{4}\right)^2 + \cdots$$
$$= 1 + 2(-0.0240855) + 2(0.0005801)$$
$$= 0.953$$

29. $\int_{0.1}^{0.2} \frac{1 - \frac{x^2}{2} + \frac{x^4}{24} + \cdots}{\sqrt{x}}\, dx$

$$= \int_{0.1}^{0.2}\left(x^{-1/2} - \frac{x^{3/2}}{2} + \frac{x^{7/2}}{24} + \cdots\right)dx$$
$$= \int_{0.1}^{0.2} x^{-1.2}\,dx - \frac{1}{2}\int_{0.1}^{0.2} x^{3/2}\,dx + \frac{1}{24}\int_{0.1}^{0.2} x^{7/2}\,dx$$
$$= 2x^{1/2} - \frac{1}{5}x^{5/2} + \frac{1}{108}x^{9/2} + \cdots\Big|_{0.1}^{0.2}$$
$$= 0.259$$

33. $f(x)=\sin^{-1}x,\ a=\frac{1}{2}$

$$f(a)=f\left(\frac{1}{2}\right)$$
$$=\sin^{-1}\frac{1}{2}$$
$$=\frac{\pi}{6}$$

$$f'(x)=\frac{1}{\sqrt{1-x^2}}=(1-x^2)^{-1/2}, f'\left(\frac{1}{2}\right)=\frac{2}{\sqrt{3}}$$

$$f''(x)=-\frac{1}{2\left(1-x^2\right)^{3/2}}, f''\left(\frac{1}{2}\right)=-\frac{4}{3\sqrt{3}}$$

$$f(x)=f(a)+f'(a)(x-a)+f''(a)\frac{(x-a)^2}{2!}+\cdots$$

$$f(x)=\frac{\pi}{6}+\frac{2}{\sqrt{3}}\left(x-\frac{1}{2}\right)-\frac{4}{3\sqrt{3}}\frac{(x-\frac{1}{2})^2}{2!}+\cdots$$

$$\sin^{-1}(x)=\frac{\pi}{6}+\frac{2}{\sqrt{3}}\left(x-\frac{1}{2}\right)-\frac{2}{3\sqrt{3}}\left(x-\frac{1}{2}\right)^2+\cdots$$

37. $f(x)=\begin{matrix}\pi-1, & -4\le x<0\\ \pi+1, & 0\le x<4\end{matrix}$ is Example 7 in 30.7

shifted up $\pi-1$ units

$$f(x)=\pi-1+1+\frac{4}{\pi}\sin\frac{\pi x}{4}+\frac{4}{3\pi}\sin\frac{3\pi x}{4}+\cdots$$

$$f(x)=\pi+\frac{4}{\pi}\ \sin\frac{\pi x}{4}+\frac{1}{3}\sin\frac{3\pi x}{4}+\cdots$$

41. $f(x)=x$

$-2\le x<2$, period $=4, L=2$;

The function is odd, so the series consist only of sine terms.

$$b_n=\frac{1}{2}\,{}_{-2}^{\;2}\,x\sin\frac{n\pi x}{2}dx$$

$$=\frac{1}{2}\ \frac{2}{n\pi}\ ^{2}\ \sin\frac{n\pi x}{2}-\frac{n\pi x}{2}\cos\frac{n\pi x}{2}\ \Big|_{-2}^{2}$$

$$=\frac{2}{n^2\pi^2}[\sin n\pi-n\pi\cos n\pi-\sin(-n\pi)-n\pi\cos(-n\pi)]$$

$$=\frac{2}{n^2\pi^2}(-n\pi\cos n\pi-n\pi\cos n\pi)$$

$$=\frac{-4}{n\pi}\cos n\pi;$$

$$b_1 = -\frac{4}{\pi}\cos\pi = \frac{4}{\pi},$$
$$b_2 = -\frac{4}{2\pi}\cos 2\pi = -\frac{2}{\pi},$$
$$b_3 = \frac{-4}{3\pi}\cos 3\pi = \frac{4}{3\pi}$$

$$f(x) = \frac{4}{\pi}\left(\sin\frac{\pi x}{2} - \frac{1}{2}\sin\pi x + \frac{1}{3}\sin\frac{3\pi x}{2} - \cdots\right)$$

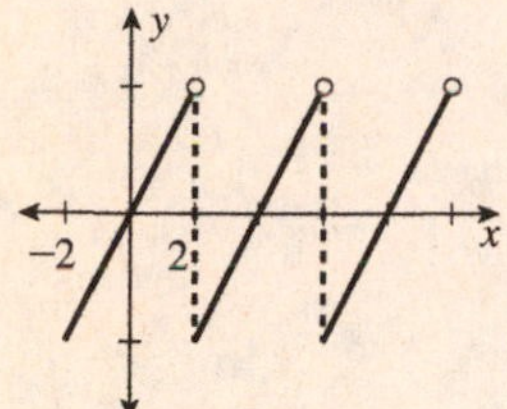

45. It is a geometric series for which $|r| < 1 = 0.75$. Therefore, the series converges.

$$S = \frac{64}{1-0.75} = 256$$

49. $f(x) = \tan x, \qquad f\left(\frac{\pi}{4}\right) = 1$

$f'(x) = 1 + \tan^2 x, \qquad f'\left(\frac{\pi}{4}\right) = 2$

$f''(x) = 2\tan x\left(+\tan^2 x\right), \qquad f''(x) = 4$

$$f(x) = \tan x = 1 + 2\left(x - \frac{\pi}{4}\right) + \frac{4\left(x - \frac{\pi}{4}\right)^2}{2!} + \cdots$$
$$= 1 + 2\left(x - \frac{\pi}{4}\right) + 2\left(x - \frac{\pi}{4}\right)^2 + \cdots$$

53. $\cos x = 1 - \frac{x^2}{2!} + \frac{x^4}{4!} - \cdots$

$$\lim_{x\to 0}\frac{1-\cos x}{x^2} = \lim_{x\to 0}\frac{1 - \left(1 - \frac{x^2}{2} + \frac{x^4}{4!} - \cdots\right)}{x^2}$$
$$= \lim_{x\to 0}\frac{1}{x^2}\ \frac{x^2}{2!} - \frac{x^4}{4!} + \cdots$$
$$= \lim_{x\to 0}\ \frac{1}{2!} - \frac{x^2}{4!} + \cdots$$
$$= \frac{1}{2}$$

57. $\sin^2 x = \frac{1}{2}(1-\cos 2x)$

$$= \frac{1}{2}\left(1-\left(1-\frac{(2x)^2}{2!}+\frac{(2x)^4}{4!}-\frac{(2x)^6}{6!}+\cdots\right)\right)$$

$$= \frac{1}{2}\left(1-1+2x^2-\frac{2}{3}x^4+\frac{4}{45}x^6-\cdots\right)$$

$$= x^2-\frac{1}{3}x^4+\frac{2}{45}x^6-\cdots$$

61. $\ln(1+x) = x-\frac{1}{2}x^2+\frac{1}{3}x^3-\frac{1}{4}x^4+\ldots$

$\ln(x) = \ln(1+(x-1))$

$$= (x-1)-\frac{1}{2}(x-1)^2+\frac{1}{3}(x-1)^3-\frac{1}{4}(x-1)^4+\ldots$$

which are precisely the terms obtained by expanding $\ln(x)$ directly with $a=1$.

65. $e^x = 1+x+\frac{x^2}{2!}+\frac{x^3}{3!}$

$$e^{0.9} = 1+(0.9)+\frac{0.9^2}{2}+\frac{0.9^3}{6} = 2.4265$$

$$e^x = e\left[1+(x-1)+\frac{(x-1)^2}{2}\right]$$

$$e^{0.9} = e\left[1+(0.9-1)+\frac{(0.9-1)^2}{2}\right] = 2.4600$$

$e^{0.9} = 2.459603$ directly from the calculator.

69. $\tan^{-1} x = \int \frac{1}{1+x^2}dx$

$$= \int\left(1-x^2+x^4-x^6+\cdots\right)dx$$

$$\tan^{-1} x = x-\frac{x^3}{3}+\frac{x^5}{5}-\frac{x^7}{7}+\cdots$$

73. $N = N_0 e^{-\lambda t}\cdot N$

$$= N_0\left[1+(-\lambda t)+\frac{(-\lambda t)^2}{2!}+\frac{(-\lambda t)^3}{3!}+\cdots\right]$$

$$= N_0\left[1-\lambda t+\frac{\lambda^2 t^2}{2}-\frac{\lambda^3 t^3}{6}+\cdots\right]$$

77. $\dfrac{N_0}{1-e^{-k/T}} = N_0\left(\dfrac{1}{1-e^{-k/T}}\right);$

Let $x = e^{-k/T}$

$$N_0\left(\frac{1}{1-e^{-k/T}}\right) = N_0\left(\frac{1}{1-x}\right)$$

The Maclaurin's expansion for $f(x) = \dfrac{1}{1-x}$ is

$$f(x) = \frac{1}{1-x}; \qquad f(0) = 1$$

$$f'(x) = \frac{1}{(1-x)^2}; \qquad f'(0) = 1$$

$$f''(x) = \frac{2}{(1-x)^3}; \qquad f''(0) = 2$$

$$f(x) = 1 + x + \frac{2x^2}{2!} + \cdots = 1 + x + x^2 + \cdots$$

Substituting $e^{-k/T}$ for x:

$$f\left(e^{-k/T}\right) = 1 + e^{-k/T} + e^{-2k/T} + \cdots$$

Therefore,

$$\frac{N_0}{1-e^{-k/T}} = N_0\left(1 + e^{-k/T} + e^{-2k/T} + \cdots\right)$$

81. $\sin x = x - \dfrac{1}{6}x^3 + \dfrac{1}{120}x^5 - \dfrac{1}{5040}x^7 + \dfrac{1}{362\,880}x^9 + \cdots$

$\cos x = 1 - \dfrac{1}{2}x^2 + \dfrac{1}{24}x^4 - \dfrac{1}{720}x^6 + \dfrac{1}{40\,320}x^8 + \cdots$

(a) Use the double angle trigonometric identity, and the cosine series evaluated at $2x$:

$$\sin^2 x = \frac{1-\cos 2x}{2}$$

$$= \frac{1}{2}\ 1 - 1 - \frac{1}{2}(2x)^2 + \frac{1}{24}(2x)^4 - \frac{1}{720}(2x)^6 + \frac{1}{40\,320}(2x)^8 \ + \cdots$$

$$\sin^2 x = \frac{1-\cos 2x}{2}$$

$$= \frac{1}{2}\ 1 - 1 - \frac{1}{2}(2x)^2 + \frac{1}{24}(2x)^4 - \frac{1}{720}(2x)^6 + \frac{1}{40\,320}(2x)^8 \ + \cdots$$

$$\sin^2 x = x^2 - \frac{1}{3}x^4 + \frac{2}{45}x^6 - \frac{1}{315}x^8 + \cdots$$

(b) Use two trigonometric identities and the cosine series evaluated at $2x$:

$$\sin^2 x = 1-\cos^2 x = 1-\frac{1+\cos 2x}{2} = 1-\frac{1}{2}(1+1-\frac{1}{2}(2x)^2$$

$$+\frac{1}{24}(2x)^4-\frac{1}{720}(2x)^6+\frac{1}{40\,320}(2x)^8+\cdots$$

$$\sin^2 x = 1-\frac{1}{2}(2-\frac{1}{2}(2x)^2+\frac{1}{24}(2x)^4-\frac{1}{720}(2x)^6$$

$$+\frac{1}{40\,320}(2x)^8+\cdots$$

$$\sin^2 x = x^2-\frac{1}{3}x^4+\frac{2}{45}x^6-\frac{1}{315}x^8+\cdots$$

(c) Square the sine series term by term:

$$\sin^2 x = \quad x-\frac{1}{6}x^3+\frac{1}{120}x^5-\frac{1}{5040}x^7+\cdots \quad ^2$$

$$\sin^2 x = x^2-\frac{1}{3}x^4+\frac{2}{45}x^6-\frac{1}{315}x^8+\cdots$$

(d) Use a trigonometric identity and square the cosine series term by term:

$$\sin^2 x = 1-\cos^2 x$$

$$=1-\left(1-\frac{1}{2}x^2+\frac{1}{24}x^4-\frac{1}{720}x^6+\frac{1}{40\,320}x^8+\cdots\right)^2$$

$$\sin^2 x = 1-\cos^2 x$$

$$=1-\left(1-\frac{1}{2}x^2+\frac{1}{24}x^4-\frac{1}{720}x^6+\frac{1}{40\,320}x^8+\cdots\right)^2$$

$$\sin^2 x = 1-\left(1-x^2+\frac{1}{3}x^4-\frac{2}{45}x^6+\frac{1}{315}x^8+\cdots\right)$$

$$\sin^2 x = x^2-\frac{1}{3}x^4+\frac{2}{45}x^6-\frac{1}{315}x^8+\cdots$$

(e) Multiply the sine and cosine series term by term and integrate term by term:

$$\sin x\cos x = x-\frac{x^3}{2}+\frac{x^5}{24}-\frac{x^3}{6}+\frac{x^5}{12}+\frac{x^5}{120}\cdots$$

$$= x-\frac{2}{3}x^3+\frac{2}{15}x^5+\cdots$$

$$\sin^2 x = 2 \quad x-\frac{2}{3}x^3+\frac{2}{15}x^5+\cdots dx$$

$$= 2 \quad \frac{x^2}{2}-\frac{2}{3}\cdot\frac{x^4}{4}+\frac{2}{15}\cdot\frac{x^6}{6}+\cdots$$

$$= x^2-\frac{1}{3}x^4+\frac{2}{45}x^6+\cdots$$

Chapter 31

Differential Equations

31.1 Solutions of Differential Equations

1. $y = c_1e^{-x} + c_2e^{2x}$

$\frac{dy}{dx} = -c_1e^{-x} + 2c_2e^{2x}$

$\frac{d^2y}{dx^2} = c_1e^{-x} + 4c_2e^{2x}$

$$\begin{aligned}\frac{d^2y}{dx^2} - \frac{dy}{dx} &= c_1e^{-x} + 4c_2e^{2x} - (-c_1e^{-x} + 2c_2e^{2x}) \\ &= c_1e^{-x} + 4c_2e^{2x} + c_1e^{-x} - 2c_2e^{2x} \\ &= 2c_1e^{-x} + 2c_2e^{2x} \\ &= 2(c_1e^{-x} + c_2e^{2x}) \\ &= 2y\end{aligned}$$

$y = 4e^{-x}$

$\frac{dy}{dx} = -4e^{-x}$

$\frac{d^2y}{dx^2} = 4e^{-x}$

$$\begin{aligned}\frac{d^2y}{dx^2} - \frac{dy}{dx} &= 4e^{-x} - (-4e^{-x}) \\ &= 8e^{-x} \\ &= 2(4e^{-x}) \\ &= 2y\end{aligned}$$

5. $y'' + 3y' - 4y = 3e^x$

$y = c_1e^x + c_2e^{-4x} + \frac{3}{5}xe^x$

$$\begin{aligned}y' &= c_1e^x - 4c_2e^{-4x} + \frac{3}{5}(xe^x + e^x) \\ &= \left(c_1 + \frac{3}{5}\right)e^x - 4c_2e^{-4x} + \frac{3}{5}xe^x\end{aligned}$$

$$\begin{aligned}y'' &= \left(c_1 + \frac{3}{5}\right)e^x - 16c_2e^{-4x} + \frac{3}{5}(xe^x + e^x) \\ &= \left(c_1 + \frac{6}{5}\right)e^x + 16c_2e^{-4x} + \frac{3}{5}xe^x\end{aligned}$$

Substitute y, y', y'' into the differential equation.

$$c_1 + \frac{6}{5}\ e^x + 16c_2e^{-4x} + \frac{3}{5}xe^x$$

$$+3\quad c_1 + \frac{3}{5}\ e^x - 4c_2e^{-4x} + \frac{3}{5}xe^x$$

$$-4\ \ c_1e^x + c_2e^{-4x} + \frac{3}{5}xe^x\ = 3e^x$$

$$c_1 + \frac{6}{5} + 3c_1 + \frac{9}{5} - 4c_1\ \ e^x + \left(16c_2 - 12c_2 - 4c_2\right)e^{-4x}$$

$$+\ \ \frac{3}{5} + \frac{9}{5} - \frac{12}{5}\ \ xe^x = 3e^x$$

$3e^x = 3e^x$

Since y is a solution to the second-order equation, and there are two arbitrary constants, it is the general solution.

9. $y = 3\cos 2x, y' = -6\sin 2x,\ y'' = -12\cos 2x$

Substitute y and y' into the differential equation.

$$\begin{aligned}y'' + 4y &= 0 \\ -12\cos 2x + 4(3\cos 2x) &= 0 \\ 0 &= 0\end{aligned}$$

$y = c_1 \sin 2x + c_2 \cos 2x; y' = 2c_1 \cos 2x - 2c_2 \sin 2x$

$y'' = -4c_1 \sin 2x - 4c_2 \cos 2x$

Substitute y and y' into the differential equation.

$$\begin{aligned}-4c_1 \sin 2x - 4c_2 \cos 2x + 4(c_1 \sin 2x + c_2 \cos 2x) &= 0 \\ 0 &= 0\end{aligned}$$

Both functions are solutions to the differential equation.

13. $y = 2 + x - x^3$

$\frac{dy}{dx} = 1 - 3x^2$

This is the original equation, so the function is a solution to the differential equation.

17. $y''+9y=4\cos x;$

$2y=\cos x;\ y=\frac{1}{2}\cos x;$

$y'=-\frac{1}{2}\sin x;\ y''=-\frac{1}{2}\cos x$

Substitute y and y''.

$-\frac{1}{2}\cos x+9\ \ \frac{1}{2}\cos x\ =4\cos x$

$-\frac{1}{2}\cos x+9\ \ \frac{1}{2}\cos x\ =4\cos x$

$\frac{8}{2}\cos x=4\cos x$

$4\cos x=4\cos x$ identity

The function is a solution to the differential equation.

21. $x\frac{d^2y}{dx^2}+\frac{dy}{dx}=0;$

$y=c_1\ln x+c_2$

$\frac{dy}{dx}=\frac{c_1}{x}=c_1x^{-1}$

$\frac{d^2y}{dx^2}=-c_1x^{-2}=-\frac{c_1}{x^2}$

Substitute $\frac{dy}{dx}$ and $\frac{d^2y}{dx^2}$.

$x\ \ \frac{-c_1}{x^2}\ \ +c_1x^{-1}=0$

$-\frac{c_1}{x}+\frac{c_1}{x}=0$

$0=0$ identity

The function is a solution to the differential equation.

25. $y=c_1e^x+c_2e^{2x}+\frac{3}{2}$

$y'=c_1e^x+2c_2e^{2x}$

$y''=c_1e^x+4c_2e^{2x}$

$y''-3y'+2y=3$

$c_1e^x+4c_2e^{2x}-3\left(c_1e^x+2c_2e^{2x}\right)$

$+2\left(c_1e^x+c_2e^{2x}+\frac{3}{2}\right)=3$

$c_1e^x+4c_2e^{2x}-3c_1e^x-6c_2e^{2x}$

$+2c_1e^x+2c_2e^{2x}+3=3$

$3=3$ identity

The function is a solution to the differential equation.

29. $(y')^2+xy'=y$

$y=cx+c^2$

$y'=c$

Substitute y'.

$(c)^2+x(c)=y$

$c^2+cx=y$

$y=y$ identity

The function is a solution to the differential equation.

33. Differential equation: $y''-2y'-3y=-4e^x$

Proposed solution: $y=e^x+e^{2x}$

$y'=e^x+2e^{2x}$

$y''=e^x+4e^{2x}$

$y''-2y'-3y$

$=\left(e^x+4e^{2x}\right)-2\left(e^x+2e^{2x}\right)-3\left(e^x+e^{2x}\right)$

$=-4e^x-3e^{2x}$

and so $y=e^x+e^{2x}$ is not a solution to $y''-2y'-3y=-4e^x$.

37. $N=N_0e^{kt}$

$\frac{dN}{dt}=N_0ke^{kt}=kN$

and so $N=N_0e^{kt}$ is a solution. Since there is one independent constant k and the equation is first-order, this is a general solution.

31.2 Separation of Variables

1. $2xy\,dx + (x^2+1)dy = 0$

$$\frac{2x\,dx}{x^2+1} + \frac{dy}{y} = 0$$

$$\ln(x^2+1) + \ln y = \ln c$$

$$\ln(y(x^2+1)) = \ln c$$

$$y(x^2+1) = c$$

5. $\dfrac{dy}{20+y} = k\,dt$

Integrate:

$$\ln|20+y| = kt + c$$

9. $p^{-1/2}\,dp = x^{-1/2}\,dx$

Integrate:

$$2p^{1/2} = 2x^{1/2} + c$$

13. $2x\,dx + dy = 0$

Integrate:

$$x^2 + y = c$$

$$y = c - x^2$$

17. $\dfrac{dV}{dP} = \dfrac{-V}{P^2}$

Multiply by $\frac{dP}{V}$ and integrate.

$$\frac{dV}{V} = -\frac{dP}{P^2}$$

$$\ln V = \frac{1}{P} + c$$

21. $dy + \ln xy\,dx = (4x + \ln y)dx$

$$dy + \ln x dx + \ln y dx = 4x\,dx + \ln y\,dx$$

$$dy + (\ln x - 4x)\,dx = 0$$

Integrate

$$y + x\ln x - x - 2x^2 = c$$

$$y = 2x^2 + x - x\ln x + c$$

25. $e^{x+y}\,dx + dy = 0$

Use properties of exponents to rewrite as

$$e^x e^y dx + dy = 0$$

Divide by e^y and integrate.

$$e^x dx + \frac{dy}{e^y} = 0$$

$$e^x dx + e^{-y} dy = 0$$

$$e^x - e^{-y} = c$$

29. $x\dfrac{dy}{dx} = y^2 + y^2\ln x$

$$x\frac{dy}{dx} = y^2(1+\ln x)$$

Multiply by $\frac{dx}{xy^2}$ and integrate.

$$\frac{dy}{y^2} = \frac{1+\ln x}{x}dx$$

$$-\frac{1}{y} = \frac{1}{2}(1+\ln x)^2 + \frac{c}{2}$$

$$-2 = y(1+\ln x)^2 + cy$$

$$y(1+\ln x)^2 + cy + 2 = 0$$

33. $yx^2\,dx = y dx - x^2 dy$

Divide by yx^2 and integrate.

$$dx = \frac{dx}{x^2} - \frac{dy}{y}$$

$$x + c = \frac{x^{-1}}{-1} - \ln y$$

$$x + \frac{1}{x} + \ln y = c$$

$$x^2 + 1 + x\ln y + cx = 0$$

37. $\frac{dy}{dx}=(1-y)\cos x; x=\frac{\pi}{6}$ when $y=0$

Multiply by $\frac{dx}{1-y}$ and integrate.

$$\frac{1}{1-y}dy=\cos x\,dx$$
$$-\ln(1-y)=\sin x+c$$
$$\sin x+\ln(1-y)=c$$

Substitute $x=\frac{\pi}{6}, y=0$

$$\sin\frac{\pi}{6}+\ln 1=c$$
$$c=\frac{1}{2}$$
$$\sin x+\ln(1-y)=\frac{1}{2}$$
$$2\ln(1-y)=1-2\sin x$$

41. $\frac{dP}{dt}=0.1P-20$

$$\frac{dP}{0.1P-20}=dt$$
$$10\ln|0.1P-20|=t+c$$

Let $A=e^{c/10}$

If $P>200$,

$$0.1P-20=e^{(t+c)/10}$$
$$P=200+10Ae^{t/10}$$

If $P<200$,

$$20-0.1P=e^{(t+c)/10}$$
$$P=200-10Ae^{t/10}$$

Finally, if $P=200$, $\frac{dP}{dt}=0$

and the population remains at 200.

31.3 Integrating Combinations

1. $x\,dy+y\,dx+2xy^2dy=0$

$$\frac{x\,dy+y\,dx}{xy}+2y\,dy=0$$
$$\frac{d(xy)}{xy}+2y\,dy=0$$
$$\ln xy+y^2=c$$

5. $y\,dx-x\,dy+x^3dx=2dx$

$$x\,dy-y\,dx-x^3dx=-2dx$$
$$\frac{(x\,dy-y\,dx)}{x^2}-x\,dx=-\frac{2dx}{x^2}$$
$$d\frac{y}{x}-xdx=-\frac{2dx}{x^2}$$
$$\frac{y}{x}-\frac{1}{2}x^2=2x^{-1}+c_1$$
$$y-\frac{1}{2}x^3=2+c_1x$$
$$2y-x^3=4+2c_1x$$
$$x^3-2y=cx-4$$

9. $\sin x\,dy=(1-y\cos x)\,dx$

$$\sin x\,dy+y\cos x\,dx=dx$$
$$d(y\sin x)=dx$$
$$y\sin x=x+c$$

13. $\tan(x^2+y^2)dy+x\,dx+y\,dy=0$

$$dy+\frac{x\,dx+y\,dy}{\tan(x^2+y^2)}=0$$

Multiply both sides by 2 so that the numerator of the quotient is the differential of x^2+y^2.

$$2dy+\cot(x^2+y^2)(2x\,dx+2y\,dy)=0$$
$$2dy+\cot(x^2+y^2)d(x^2+y^2)=0$$
$$2y+\ln\sin(x^2+y^2)=c$$

17. $10xdy+5ydx+3ydy=0$

$$5(2xdy+ydx)+3ydy=0$$

Multiply by y.

$$5(2xydy+y^2dx)+3y^2dy=0$$
$$5d(xy^2)+3y^2dy=0$$
$$5xy^2+y^3=c$$

21. $ydx - xdy = y^3dx + y^2xdy; x = 2, y = 4$

$$\frac{ydx - xdy}{y^2} = ydx + xdy$$

$$d\frac{x}{y} = d(xy)$$

$$\frac{x}{y} = xy + c$$

Substitute $x = 2, y = 4$:

$$\frac{2}{4} = 2(4) + c$$

$$c = -\frac{15}{2}$$

$$\frac{x}{y} = xy - \frac{15}{2}$$

$$2x = 2xy^2 - 15y$$

25. $e^{-x}dy - 2y\,dy = ye^{-x}\,dx$

$$e^{-x}dy - ye^{-x}\,dx = 2y\,dy$$

$$d\left(ye^{-x}\right) = d\left(y^2\right)$$

$$ye^{-x} = y^2 + c$$

31.4 The Linear Differential Equation of the First Order

1. $dy + \left(\frac{2}{x}\right)ydx = 3dx$

$$e^{\int \frac{2}{x}dx} = e^{2\ln x}$$

$$= e^{\ln x^2}$$

$$= x^2$$

$$yx^2 = \int 3x^2dx + c$$

$$yx^2 = x^3 + c$$

$$y = x + cx^{-2}$$

5. $dy + 2ydx = 2e^{-4x}dx; P = 2, Q = 2e^{-4x}$

$$e^{\int 2dx} = e^{2x}$$

$$ye^{2x} = \int 2e^{-4x}e^{2x}dx$$

$$= -\int e^{-2x}(-2dx)$$

$$= -e^{-2x} + c$$

$$y = -e^{-2x}e^{-2x} + ce^{-2x}$$

$$y = -e^{-4x} + ce^{-2x}$$

9. $dy = 3x^2(2-y)dx$

$$dy = 6x^2dx - 3x^2ydx$$

$$dy + 3x^2y = 6x^2dx; P = 3x^2, Q = 6x^2$$

$$e^{3x^2dx} = e^{x^3}$$

$$ye^{x^3} = 6x^2e^{x^3}dx + c$$

$$= 2e^{x^3} + c$$

$$y = ce^{-x^3} + 2$$

Note: The equation can also be solved by separation of variables.

13. $dr + r\cot\theta d\theta = d\theta$

$$dr + \cot\theta rd\theta = d\theta; P = \cot\theta, Q = 1$$

$$e^{\int\cot\theta d\theta} = e^{\ln\sin\theta} = \sin\theta$$

$$r\sin\theta = \int\sin\theta d\theta + c$$

$$r\sin\theta = -\cos\theta + c$$

$$r = -\frac{\cos\theta}{\sin\theta} + \frac{c}{\sin\theta}$$

$$r = -\cot\theta + c\csc\theta$$

17. $y' + y = x + e^x$

$$dy + y\,dx = \left(x + e^x\right)dx; P = 1, Q = x + e^x$$

$$ye^{\int dx} = \int\left(x + e^x\right)e^{\int dx}dx$$

$$ye^x = \int\left(xe^x + e^{2x}\right)dx$$

$$ye^x = xe^x - e^x + \frac{1}{2}e^{2x} + c$$

$$y = x - 1 + \frac{1}{2}e^x + ce^{-x}$$

21. $y' = x^3(1-4y)$

$$\frac{dy}{dx} = x^3 - 4x^3y$$

$$dy + 4x^3ydx = x^3dx; P = 4x^3, Q = x^3$$

$$e^{4\int x^3dx} = e^{x^4}$$

$$ye^{x^4} = \int x^3e^{x^4}dx + c$$

$$= \frac{1}{4}\int e^{x^4}4x^3dx + c$$

$$= \frac{1}{4}e^{x^4} + c$$

$$y = \frac{1}{4} + ce^{-x^4}$$

25. $\sqrt{1+x^2}\,dy + x(1+y)\,dx = 0$

$\sqrt{1+x^2}\,dx + x dx + xy\,dx = 0$

$dy + \frac{x}{\sqrt{1+x^2}}y\,dx = -\frac{x}{\sqrt{1+x^2}}dx, \quad P = \frac{x}{\sqrt{1+x^2}}$

$e^{\int \frac{x}{\sqrt{1+x^2}}dx} = e^{\sqrt{1+x^2}}$

$ye^{\sqrt{1+x^2}} = \int \frac{-x}{\sqrt{1+x^2}}e^{\sqrt{1+x^2}}\,dx + c$

$ye^{\sqrt{1+x^2}} = -e^{\sqrt{1+x^2}} + c$

$y = -1 + ce^{-\sqrt{1+x^2}}$

29. $y' = 2(1-y)$

(1) Solve by separation of variables.

$\frac{dy}{1-y} = 2dx$

$-\ln(1-y) = 2x - \ln c$

$\ln\frac{c}{1-y} = 2x$

$c = (1-y)e^{2x}$

$1 - y = ce^{-2x}$

$y = 1 - ce^{-2x}$

(2) Solve as a first-order equation.

$dy = 2dx - 2ydx$

$dy + 2ydx = 2dx; P = 2, Q = 2$

$e^{\int 2dx} = e^{2x}$

$ye^{2x} = \int 2e^{2x}dx + c$

$ye^{2x} = e^{2x} + c$

$y = 1 + ce^{-2x}$

33. $\frac{dy}{dx} + 2y\cot x = 4\cos x; x = \frac{\pi}{2}, y = \frac{1}{3};$

$dy + 2y\cot x dx = 4\cos x dx; P = 2\cot x; Q = 4\cos x$

$e^{\int 2\cot x dx} = e^{2\ln\sin x}$

$= e^{\ln(\sin x)^2}$

$= \sin^2 x$

$y\sin^2 x = \int 4\cos x(\sin x)^2\,dx + c$

$= \frac{4(\sin x)^3}{3} + c$

$y = \frac{4}{3}\sin x + c(\csc^2 x)$

Substitute $x = \frac{\pi}{2}, y = \frac{1}{3}$

$\frac{1}{3} = \frac{4}{3}\sin\frac{\pi}{2} + c$

$c = -1$

$y = \frac{4}{3}\sin x - \csc^2 x$

37. $y' + P(x)y = Q(x)y^2$

$dy + P(x)y\,dx = Q(x)y^2\,dx$, which is not linear because $Q(x)y^2$ is not a function of x only.

Let $u = \frac{1}{y}$, so $dy = -y^2\,du$, then

$dy + P(x)y\,dx = Q(x)y^2dx$ is

$-y^2du + P(x)y\,dx = Q(x)y^2\,dx$

$du - P(x)u\,dx = -Q(x)\,dx$, which is linear.

41. $\frac{dP}{dt} = 0.02P - 0.3e^{0.05t}$

As a linear equation of the first order:

$dP - 0.02P\,dt = -0.3e^{0.05t}\,dt$

$P = -0.02, Q = -0.3e^{0.05t}$

$e^{\int P dt} = e^{-0.02t}$

$Pe^{-0.02t} = \int -0.3e^{0.05t}e^{-0.02t}\,dt$

$Pe^{-0.02t} = -\int 0.3e^{0.03t}\,dt$

$Pe^{-0.02t} = -10e^{0.03t} + c$

$P = ce^{0.02t} - 10e^{0.05t}$

At $t = 0, P = 30$, so $30 = c - 10$, or $c = 40$.

We have

$P(t) = 40e^{0.02t} - 10e^{0.05t}$.

The population reaches a maximum at $t = 15.667$ years and is zero at $t = 46.21$ years.

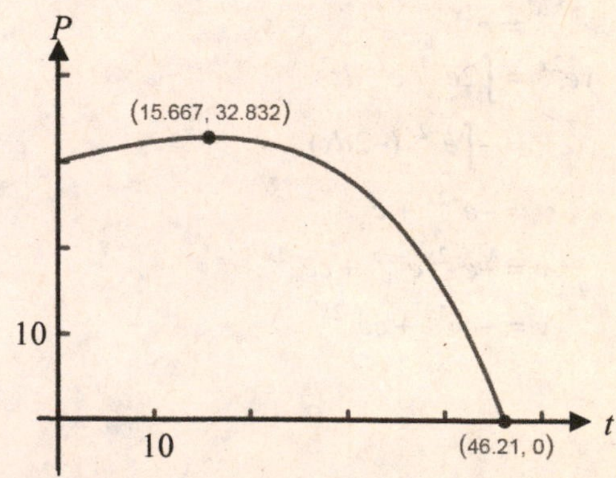

31.5 Numerical Solutions of First-order Equations

1. $\frac{dy}{dx} = x+1$

x	y	$x+1$	dy	y(correct)
0.0	1.00	1.0	0.20	1.00
0.2	1.20	1.2	0.24	1.22
0.4	1.44	1.4	0.28	1.48
0.6	1.72	1.6	0.32	1.78
0.8	2.04	1.8	0.36	2.12
1.0	2.40	2.0	0.40	2.50

$y = \frac{1}{2}x^2 + x + c$

$y = 1$ when $x = 0$

$c = 1$

$y = \frac{1}{2}x^2 + x + 1$

5.

x	y Approximate	y Exact
0.0	1	1
0.1	$1.0+(0.0+1)(0.1)=1.10$	1.105
0.2	$1.10+(0.1+1)(0.1)=1.21$	1.220
0.3	$1.21+(0.2+1)(0.1)=1.33$	1.345
0.4	$1.33+(0.3+1)(0.1)=1.46$	1.480
0.5	$1.46+(0.4+1)(0.1)=1.60$	1.625
0.6	$1.60+(0.5+1)(0.1)=1.75$	1.780
0.7	$1.75+(0.6+1)(0.1)=1.91$	1.945
0.8	$1.91+(0.7+1)(0.1)=2.08$	2.120
0.9	$2.08+(0.8+1)(0.1)=2.26$	2.305
1.0	$2.26+(0.9+1)(0.1)=2.45$	2.500

9. $\frac{dy}{dx} = xy+1,\ x=0$ to $x=0.4,\ \Delta x = 0.1,\ (0, 0)$

x	y	y to 4 Places
0	0	0
0.1	0.1003339594	0.1003
0.2	0.20268804	0.2027
0.3	0.3091639819	0.3092
0.4	0.42203172548	0.4220

13. $\frac{dy}{dx} = \cos(x+y),\ x=0$ to $x=0.6,\ \Delta x = 0.1,\ \left(0, \frac{\pi}{2}\right)$

x	y
0	$\frac{\pi}{2} = 1.5708$
0.1	1.5660
0.2	1.5521
0.3	1.5302
0.4	1.5011
0.5	1.4656
0.6	1.4244

17. $\frac{di}{dt} = 2i = \sin t$

t	i	$\sin t - 2i$	di
0.0	0.0000	0.0000	0.0000
0.1	0.0000	0.0998	0.0100
0.2	0.0100	0.1787	0.0179
0.3	0.0279	0.2398	0.0240
0.4	0.0518	0.2857	0.0286
0.5	0.0804	0.3186	0.0319

$i = 0.0804$ A for $t = 0.5$ s

$di + 2i\,dt = \sin t\,dt;\ e^{\int 2dt} = e^{2t}$

$ie^{2t} = \int e^{2t}\sin t\,dt = \frac{e^{2t}(2\sin t - \cos t)}{4+1} + c$

$i = \frac{1}{5}(2\sin t - \cos t) + ce^{-2t}$ (Formula 49)

$i = 0$ for $t = 0$, $0 = \frac{1}{5}(0-1) + c$, $c = \frac{1}{5}$

$i = \frac{1}{5}\left(2\sin t - \cos t + e^{-2t}\right)$

$i = 0.0898$ A for $t = 0.5t$

31.6 Elementary Applications

1. $y^2 = cx$

$2y\frac{dy}{dx} = c = \frac{y^2}{x}$

$\frac{dy}{dx} = \frac{y}{2x}$ for slope of any member of family

$\frac{dy}{dx} = -\frac{2x}{y}$ for slope of orthogonal trajectories.

$y\,dy = -2x\,dx$

$\frac{y^2}{2} = -x^2 + \frac{k}{2}$

$y^2 + 2x^2 = k$

5. $\frac{dy}{dx} = \frac{2x}{y};$

$y\,dy = 2x\,dx$

$\frac{1}{2}y^2 = x^2 + c$

Substitute $x = 2, y = 3:$

$\frac{1}{2}(9) = 4 + c$

$c = 0.5$

$\frac{1}{2}y^2 = x^2 + 0.5$

$y^2 = 2x^2 + 1$

$y = \pm\sqrt{2x^2 + 1}$

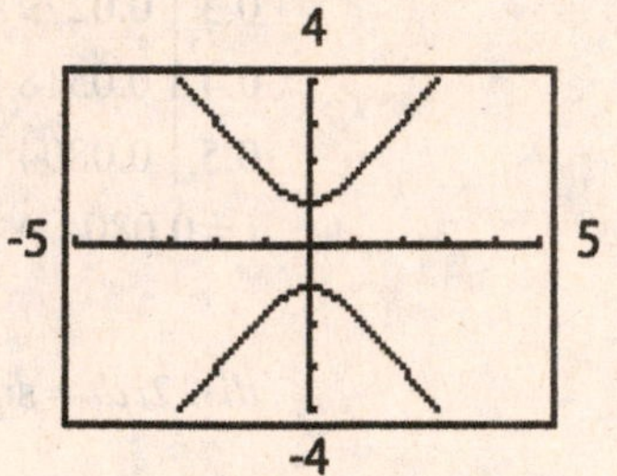

9. $\frac{dy}{dx} = ce^x$

$y = ce^x$, so $c = \frac{y}{e^x}$

Substitute for k in the equation for the derivative.

$\frac{dy}{dx} = \frac{y}{e^x}e^x = y$

$\left.\frac{dy}{dx}\right|_{OT} = -\frac{1}{y}$

$y\,dy = -dx$

Integrating,

$\frac{y^2}{2} = -x + \frac{k}{2}$

$y^2 = k - 2x$

The orthogonal trajectories to the exponential family are parabolas.

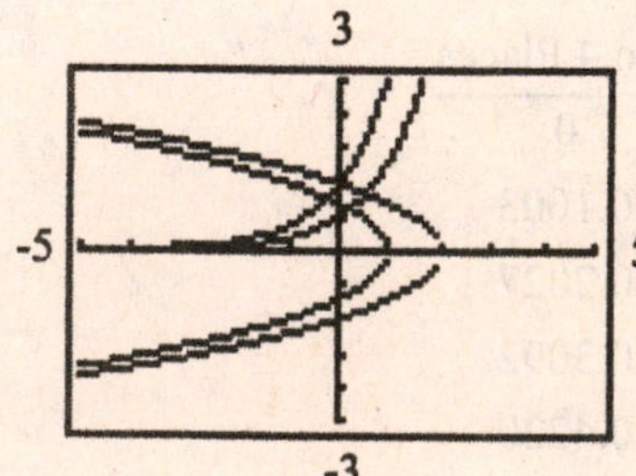

13. From Example 3,

$N = N_0(0.5)^{t/5.27}$

When $t = 2.00$:

$N = N_0(0.5)^{2.00/5.27}$

$= N_0(0.769)$

76.9% of the initial amount remains

17. $\frac{dN}{dt} = r - kN$

Solving by separation of variables:

$\frac{dN}{r - kN} = dt$

$-\frac{1}{k}\ln(r - kN) = t + c$

$N = 0$ for $t = 0$

$c = -\frac{1}{k}\ln r$

$-\frac{1}{k}\ln(r - kN) = t - \frac{1}{k}\ln r$

$\ln\frac{r - kN}{r} = -kt$

$r - kN = re^{-kt}$

$N = \frac{r}{k}\left(1 - e^{-kt}\right)$

21. $\frac{dP}{dt} = kP + I$

Solving by separation of variables:

$\frac{kdP}{kP + I} = k\,dt$

$\ln(kP + I) = kt + c$

Substitute $t = 0, P = 35.9, I = 0.24, k = 0.008$:

$\ln(0.008(35.9) + 0.24) = 0.008(0) + c$

$c = \ln 0.5272$

Find P for $t = 2025 - 2015 = 10$:

$\ln(0.008P + 0.24) = 0.008t + \ln 0.5272$

$\ln(0.008P + 0.24) = 0.008(10) + \ln 0.5272$

$\ln(0.008P + 0.24) = -0.5602$

$0.008P + 0.24 = e^{-0.5602}$

$P = 41.4$ million

25. $\frac{dM}{dt} = KM$

Solve by separation of variables:

$\frac{dM}{M} = K\,dt$

$\ln M = Kt + c$

Substitute $M = 5250, t = 0$:

$\ln 5250 = c$

$\ln M = Kt + \ln 5250;$

Substitute $M = 5460, t = 2.00$:

$\ln 5460 = 2.00K + \ln 5250$

$K = \frac{1}{2}\ln\frac{5460}{5250} = 0.0196$

$\ln M = 0.0196t + \ln 5250$

$\ln\frac{M}{5250} = 0.0196t;$

$\frac{M}{5250} = e^{0.0196t}$

$M = 5250e^{0.0196t}$

29. $\frac{dP}{dt} = kP$

$\frac{dP}{P} = k\,dt$

Solving by separation of variables:

$\ln P = kt + c$

Substitute $t = 0, P = P_0$ (the original investment):

$\ln P_0 = c$

$\ln P = kt + \ln P_0$

$\ln\frac{P}{P_0} = kt$

$\frac{P}{P_0} = e^{kt}$

Using $k = 0.04$ (the interest rate):

$P = P_0e^{0.04t}$

Find P for $P_0 = \$1000, t = 1$:

$P = 1000e^{0.04(1)} = \$1040.81$

33. See Example 4:

$i = \frac{E}{R}\left(1 - e^{-(R/L)t}\right)$

$\lim_{i\to\infty}\left(\frac{E}{R} - \frac{E}{R}e^{-(R/L)t}\right) = \frac{E}{R} - 0 = \frac{E}{R}$

37. $Ri+\frac{q}{C}=0$

Use $i=\frac{dq}{dt}$:

$R\frac{dq}{dt}+\frac{q}{C}=0$

Solve by separation of variables:

$\frac{dq}{q}+\frac{1}{RC}dt=0$

$\ln q=-\frac{1}{RC}t+c$

Let $q=q_0$ when $t=0$, so $\ln q_0=c$

$\ln q=-\frac{1}{RC}t+\ln q_0$

$\ln q-\ln q_0=-\frac{1}{RC}t$

$\ln\frac{q}{q_0}=-\frac{1}{RC}t$

$\frac{q}{q_0}=e^{(-1/RC)t}$

$q=q_0e^{-t/RC}$

41. $8\text{ mi/hr}=\frac{44\text{ ft/s}}{30\text{ mi/h}}\times 8\text{ mi/h}=11.73\text{ ft/s}$

$10\frac{dv}{dt}=20-2v$

$5dv=10dt-v\ dt$

$\frac{5dv}{10-v}=dt$

$-5\ln|10-v|=t+c$

$v=11.73$ ft/s when $t=0$

$c=-5\ln 1.73=-2.74$

$-5\ln|10-v|=t-2.74$

$|10-v|=e^{(2.74-t)/5}$

Since $v_0>10$,

$v=10+e^{(2.74-t)/5}$

For $t=3.0\text{ min}=180\text{ s}$

$v=10+e^{(2.74-180)/5}=10\text{ ft/s}$

45. $\frac{dp}{dh}=kp,\ h=0,\ p=15\text{ lb/in}^2$

$h=9800\text{ ft}, p=10\text{ lb/in}^2$

$\int\frac{dp}{p}=\int k\ dh$

$\ln p=kh+c$

Substitute $p=15, h=0$:

$\ln 15=c$

$\ln p=kh+\ln 15$

$\ln p-\ln 15=kh$

$\ln\frac{p}{15}=kh$

$\frac{p}{15}=e^{kh}$

$p=15e^{kh}$

Substitute $p=10, h=9800$:

$10=15e^{9800k}$

$\frac{2}{3}=e^{9800k}$

$e^k=(2/3)^{1/9800}$

$p=15(2/3)^{h/9800}$

49. $\frac{dx}{dt}=1-0.25x$

$\frac{dx}{0.25x-1}=-dt$

$4\ln(0.25x-1)=-t+c$

Substitute $x=12\text{ ft}^3$ when $t=0$:

$c=4\ln 2=2.77$

$4\ln(0.25x-1)=-t+4\ln 2$

$\ln\frac{0.25x-1}{2}=-0.25t$

$0.25x-1=2e^{-0.25t}$

$x=4+8e^{-0.25t}$

31.7 Higher-order Homogeneous Equations

1. $D^2y - 5Dy = 0$

$m^2 - 5m = 0$

$m(m-5) = 0$

$m = 0, m = 5$

$y = c_1 + c_2e^{5x}$

5. $3\dfrac{d^2y}{dx^2} + 4\dfrac{dy}{dx} + y = 0$

$3D^2y + 4Dy + y = 0$

$3m^2 + 4m + 1 = 0$

$(3m+1)(m+1) = 0$

$m_1 = -\dfrac{1}{3}, m_2 = -1$

$y = c_1e^{-(1/3)x} + c_2e^{-x}$

9. $2D^2y - 3y = Dy$

$2D^2y - Dy - 3y = 0$

$2m^2 - m - 3 = 0$

$(2m-3)(m+1) = 0$

$m = \dfrac{3}{2}, m = -1$

$y = c_1e^{\frac{3x}{2}} + c_2e^{-x}$

13. $3D^2y + 8Dy - 3y = 0$

$3m^2 + 8m - 3 = 0$

$(3m-1)(m+3) = 0$

$m_1 = \dfrac{1}{3}$ and $m_2 = -3$

$y = c_1e^{x/3} + c_2e^{-3x}$

17. $2\dfrac{d^2y}{dx^2} - 4\dfrac{dy}{dx} + y = 0$

$2D^2y - 4Dy + y = 0$

$2m^2 - 4m + 1 = 0$

Quadratic formula: $m = \dfrac{4 \pm \sqrt{16-8}}{4}$;

$m_1 = 1 + \dfrac{\sqrt{2}}{2},\ m_2 = 1 - \dfrac{\sqrt{2}}{2}$

$y = c_1e^{\left(1+\left(\sqrt{2}/2\right)\right)x} + c_2e^{\left(1-\left(\sqrt{2}/2\right)\right)x}$

21. $y'' = 3y' + y$

$D^2y - 3Dy - y = 0$

$m^2 - 3m - 1 = 0$

Quadratic formula:

$m = \dfrac{3 \pm \sqrt{9+4}}{2}; m_1 = \dfrac{3}{2} + \dfrac{\sqrt{13}}{3}; m_2 \dfrac{3}{2} - \dfrac{\sqrt{13}}{2}$

$y = c_1e^{\left((3/2)+\left(\sqrt{13}/2\right)\right)x} + c_2e^{\left((3/2)-\left(\sqrt{13}/2\right)\right)x}$

$y = e^{3x/2}\left(c_1e^{x\sqrt{13}/2} + c_2e^{-x\left(\sqrt{13}/2\right)}\right)$

25. $2D^2y + 5aDy - 12a^2y = 0,\ a > 0$

$2m^2 + 5am - 12a^2 = 0$

$(2m-3a)(m+4a) = 0$

$m = \dfrac{3a}{2}, \quad m = -4a$

$y = c_1e^{\frac{3a}{2}x} + c_2e^{-4ax}$

29. $D^2y - Dy = 12y$

$D^2y - Dy - 12y = 0$

$y = 0$ when $x = 0$; $y = 1$ when $x = 1$

$m^2 - m - 12 = 0$

$(m-4)(m+3) = 0$

$m_1 = 4; m_2 = -3$

$y = c_1e^{4x} + c_2e^{-3x}$

Substituting given values:

$0 = c_1 + c_2$; therefore, $c_1 = -c_2$;

$1 = c_1e^4 + c_2e^{-3}$

$1 = c_1e^4 + c_2e^{-3} = -c_2e^4 + \dfrac{c_2}{e^3} = \dfrac{-c_2e^7 + c_2}{e^3}$

$e^3 = c_1\left(1-e^7\right)$; therefore,

$$c_2 = -\frac{e^3}{\left(1-e^7\right)}, c_1 = \frac{e^3}{\left(1-e^7\right)}$$

$$y = -\frac{e^3}{\left(1-e^7\right)}e^{4x} + \frac{e^3}{\left(1-e^7\right)}e^{-3x}$$

$$= \frac{e^3}{e^7-1}e^{4x} - \frac{e^3}{e^7-1}e^{-3x}$$

$$y = \frac{e^3}{e^7-1}\left(e^{4x}-e^{-3x}\right)$$

33. $D^4y - 5D^2y + 4y = 0$

$m^4 - 5m^2 + 4 = 0$

$(m^2-4)(m^2-1) = 0$

$m_1 = 2, m_2 = -2, m_3 = 1, m_4 = -1$

$y = c_1e^x + c_2e^{-2x} + c_3e^x + c_4e^{-x}$

37. $D^2y + 4Dy = 0$

$D(D+4)y = 0$

Let $z = Dy$

$(D+4)z = 0$

$\frac{dz}{dx} = -4z$

$\frac{dz}{z} = -4\,dx$

$\ln z = -4x + \ln c_0$

$z = c_0e^{-4x}$

$Dy = c_0e^{-4x}$

$dy = c_0e^{-4x}\,dx$

$y = -\frac{c_0}{4}e^{-4x} + c_1$

$y = c_0'e^{-4x} + c_1; c_0' = -\frac{c_0}{4}$

31.8 Auxiliary Equation with Repeated or Complex Roots

1. $\frac{d^2y}{dx^2} + 10\frac{dy}{dx} + 25y = 0$

$D^2y + 10Dy + 25y = 0$

$m^2 + 10m + 25 = 0$

$(m+5)^2 = 0$

$m = -5, -5$

$y = e^{-5x}(c_1 + c_2x)$

5. $D^2y - 2Dy + y = 0$

$m^2 - 2m + 1 = 0$

$(m-1)^2 = 0$

$m = 1, 1$

$y = e^x(c_1 + c_2x)$

9. $D^2y + 9y = 0$

$m^2 + 9 = 0$

$m_1 = 3j$ and $m_2 = -3j$

$\alpha = 0, \beta = 3$

$y = e^{0x}(c_1\sin 3x + c_2\cos 3x)$

$y = c_1\sin 3x + c_2\cos 3x$

13. $D^4y - y = 0$

$m^4 - 1 = 0$

$\left(m^2-1\right)\left(m^2+1\right) = 0$

$(m-1)(m+1)\left(m^2+1\right) = 0$

$m = \pm 1, m = \pm j$ $(\alpha=0, \beta=1)$

$y = c_1e^x + c_2e^{-x} + c_3\sin x + c_4\cos x$

17. $16D^2y - 24Dy + 9y = 0$

$16m^2 - 24m + 9 = 0$

$(4m-3)^2 = 0$

$m = \frac{3}{4}, \frac{3}{4}$

$y = e^{3x/4}(c_1 + c_2x)$

21. $2D^2y + 5y = 4Dy$

$2D^2y + 5y - 4Dy = 0$

$2m^2 - 4m + 5 = 0$

Quadratic formula:

$$m = \frac{4 \pm \sqrt{16-40}}{4} = \frac{4 \pm 2\sqrt{-6}}{4}$$

$$m_1 = 1 + \frac{\sqrt{6}}{2}j; m_2 = 1 - \frac{\sqrt{6}}{2}j; \alpha = 1, \beta = \frac{1}{2}\sqrt{6}$$

$$y = e^x \quad c_1 \cos\frac{1}{2}\sqrt{6}x + c_2 \sin\frac{1}{2}\sqrt{6}x$$

25. $2D^2y - 3Dy - y = 0$

$2m^2 - 3m - 1 = 0$

By the quadratic formula,

$$m = \frac{3 \pm \sqrt{9+8}}{4}$$

$$m_1 = \frac{3}{4} + \frac{\sqrt{17}}{4}, m_2 = \frac{3}{4} - \frac{\sqrt{17}}{4}$$

$$y = c_1 e^{((3/4)+(\sqrt{17}/4))x} + c_2 e^{((3/4)+(\sqrt{17}/4))x};$$

$$y = e^{(3/4)x}\left(c_1 e^{x(\sqrt{17}/4)} + c_2 e^{-x(\sqrt{17}/4)}\right)$$

29. $D^3y - 6D^2y + 12Dy - 8y = 0$

$m^3 - 6m^2 + 12m - 8 = 0$

$(m-2)(m^2 - 4m + 4) = 0$

$(m-2)(m-2)(m-2) = 0$

$m = 2, 2, 2$ repeated root

$y = e^{2x}(c_1 + c_2 x + c_3 x^2)$

33. $D^2y + 2Dy + 10y = 0$

$m^2 + 2m + 10 = 0$

By the quadratic formula,

$$m = \frac{-2 \pm \sqrt{4-40}}{2}$$

$m_1 = -1 + 3j; m_2 = -1 - 3j, \alpha = -1, \beta = 3;$

$y = e^{-x}(c_1 \sin 3x + c_2 \cos 3x)$

Substituting $y = 0$ when $x = 0$:

$0 = e^0(c_1 \sin 0 + c_2 \cos 0); c_2 = 0$

Substituting $y = e^{-\pi/6}, x = \frac{\pi}{6}$:

$$e^{-\pi/6} = e^{-\pi/6} \quad c_1 \sin\frac{\pi}{2}$$

$e^{-\pi/6} = e^{-\pi/6} c_1$

$c_1 = 1$

$y = e^{-x} \sin 3x$

37. $y = c_1 e^{3x} + c_2 e^{-3x}$

The roots are $m = 3, m = -3$, so the auxiliary equation is

$(m-3)(m+3) = 0$

$m^2 - 9 = 0$

Therefore,

$D^2y - 9y = 0$

41. $D^2y + 4Dy + 4 = 0$

$(D+2)^2 y = 0$

$(m+2)^2 = 0 \to m = -2$ is a double root

$y = c_1 e^{-2t} + c_2 t e^{-2t}$

$t = 0 \to 0 = c_1$

$t = 1 \to 0.5 = c_2 e^{-2} \to c_2 = 0.5e^2$

$y = 0.5e^2 t e^{-2t}$

31.9 Solutions of Nonhomogeneous Equations

1. $b: x^2 + 2x + e^{-x}$

We choose the particular solution to be of the form

$y_p = A + Bx + Cx^2 + Ee^{-x}$

5. $D^2y - Dy - 2y = 4$

$m^2 - m - 2 = 0$

$(m-2)(m+1) = 0$

$m_1 = 2, m_2 = -1$

$y_c = c_1e^{2x} + c_2e^{-x}$

Assume a particular solution of the form $y_p = A$

$Dy_p = 0; D^2y_p = 0$

Substituting in the differential equation:

$0 - 0 - 2A = 4$

$A = -2$

Therefore, $y_p = -2$ and the complete solution is

$y = c_1e^{2x} + c_2e^{-x} - 2$

9. $y'' - 3y' = 2e^x + xe^x$

$D^2y - 3Dy = 2e^x + xe^x$

$m^2 - 3m = 0$

$m(m-3) = 0; m_1 = 0, m_2 = 3$

$y_c = c_1e^0 + c_2e^{3x} = c_1 + c_2e^{3x}$

Let $y_p = Ae^x + Bxe^x$

Then $Dy_p = Ae^x + B(xe^x + e^x) = Ae^x + Bxe^x + Be^x$

$D^2y_p = Ae^x + Be^x + B(xe^x + e^x)$

$= Ae^x + Be^x + Bxe^x + Be^x$

$D^2y_p = Ae^x + 2Be^x + Bxe^x$

Substituting in the differential equation:

$Ae^x + 2Be^x + Bxe^x - 3(Ae^x + Bxe^x + Be^x)$

$= Ae^x + 2Be^x + Bxe^x - 3Ae^x - 3Bxe^x - 3Be^x$

$= -2Ae^x - Be^x - 2Bxe^x$

$= e^x(-2A - B) + xe^x(-2B)$

$= 2e^x + xe^x$

Equating coefficients of similar terms:

$-2A - B = 2$ and $-2B = 1$

$B = -\frac{1}{2}$ and $A = -\frac{3}{4}$

The complete solution is

$y = c_1 + c_2e^{3x} - \frac{3}{4}e^x - \frac{1}{2}xe^x$

13. $\frac{d^2y}{dx^2} - 2\frac{dy}{dx} + y = 2x + x^2 + \sin 3x$

$D^2y - 2Dy + y = 2x + x^2 + \sin 3x$

$m^2 - 2m + 1 =$

$(m-1)^2 = 0$

$m = 1$ (double root)

$y_c = e^x(c_1 + c_2x)$

Let $y_p = A + Bx + Cx^2 + E\sin 3x + F\cos 3x$

Then $Dy_p = B + 2Cx + 3E\cos 3x - 3F\sin 3x$

$D^2y_p = 2C - 9E\sin 3x - 9F\cos 3x$

Substituting in the differential equation:

$2C - 9E\sin 3x - 9F\cos 3x$

$-2(B + 2Cx + 3E\cos 3x - 3F\sin x)$

$+A + Bx + Cx^2 + E\sin 3x + F\cos 3x$

$= 2C - 9E\sin 3x - 9F\cos 3x - 2B - 4Cx - 6E\cos 3x$

$+6F\sin x + A + Bx + Cx^2 + E\sin 3x + F\cos 3x$

$= (2C - 2B + A) + \sin 3x(6F - 8E)$

$+\cos 3x(-8F - 6E) + x(B - 4) + Cx^2$

$= 2x + x^2 + \sin 3x$

Equating the coefficients of similar terms:

$2C - 2B + A = 0; 6F - 8E = 1; -8F - 6E = 0;$

$B - 4 = 2; C = 1$

Therefore,

$C = 1, B = 6, A = 10, E = -\frac{2}{25}, F = \frac{3}{5}$

The complete solution is

$y = e^x(c_1 + c_2x) + 10 + 6x + x^2 - \frac{2}{25}\sin 3x + \frac{3}{50}\cos 3x$

17. $D^2y - Dy - 30y = 10$

$m^2 - m - 30 = 0$

$(m-6)(m+5) = 0$

$m_1 = -5, m_2 = 6$

$y_c = c_1e^{-5x} + c_2e^{6x}$

Let $y_p = A$

Then $Dy_p = 0, D^2y_p = 0$

Substituting in the differential equation:

$0 - 0 - 30A = 10$

$A = -\frac{1}{3}$

The complete solution is

$y = c_1e^{-5x} + c_2e^{6x} - \frac{1}{3}$

21. $D^2y-4y=\sin x+2\cos x$

$D^2y-4y=0$

$m^2-4=0$

$m_1=2; m_2=-2$

$y_c=c_1e^{2x}+c_2e^{-2x}$

Let $y_p=A\sin x+B\cos x$

Then $Dy_p=A\cos x-B\sin x$

$D^2y_p=-A\sin x-B\cos x$

Substituting into the differential equation:

$-A\sin x-B\cos x-4A\sin x-4B\cos x$

$=-5A\sin x-5B\cos x$

$=\sin x+2\cos x$

Equating the coefficients of similar terms:

$-5A=1$, so $A=-\frac{1}{5}$; $-5B=2$, so $B=-\frac{2}{5}$

The complete solution is

$y=c_1e^{2x}+c_2e^{-2x}-\frac{1}{5}\sin x-\frac{2}{5}\cos x$

25. $D^2y+5Dy+4y=xe^x+4$

$D^2y+5Dy+4y=0$

$m^2+5m+4=0$

$(m+1)(m+4)=0$

$m_1=-1, m_2=-4$

$y_c=c_1e^{-x}+c_2e^{-4x}$

Let $y_p=Ae^x+Bxe^x+C$

$Dy_p=Ae^x+Bxe^x+Be^x$

$D^2y_p=Ae^x+B(xe^x+e^x)+Be^x$

$=Ae^x+2Be^x+Bxe^x$

Substituting in the differential equation:

$Ae^x+2Be^x+Bxe^x+5(Ae^x+Bxe^x+Be^x)$

$+4(Ae^x+Bxe^x+C)$

$=(10A+7B)e^x+10Bxe^x+4C$

$=xe^x+4$

Equating the coefficients of similar terms:

$10A+7B=0; 10B=1; 4C=4$

Therefore, $B=\frac{1}{10}, C=1, A=-\frac{7}{100}$

The complete solution is

$y=c_1e^{-x}+c_2e^{-4x}-\frac{7}{100}e^x+\frac{1}{10}xe^x+1$

29. $D^2y+y=\cos x$

$D^2y+y=0$

$m^2+1=0$

$m=\pm i$

$y_c=c_1\sin x+c_2\cos x$

Let $y_p=x(A\sin x+B\cos x)$

Then $Dy_p=x(A\cos x-B\sin x)+A\sin x+B\cos x$

$D^2y_p=x(-A\sin x-B\cos x)+A\cos x$

$-B\sin x+A\cos x-B\sin x$

Substituting in the differential equation:

$-x(A\sin x+B\cos x)+2A\cos x-2B\sin x$

$+x(A\sin x+B\cos x)=\cos x$

$2A\cos x-2B\sin x=\cos x$

$2A=1, B=0$

$A=\frac{1}{2}$

$y_p=\frac{1}{2}x\sin x$

The complete solution is

$y=c_1\sin x+c_2\cos x+\frac{1}{2}x\sin x$

33. $D^2y-Dy-6y=5-e^x$

$D^2y-Dy-6y=0$

$m^2-m-6=0$

$(m-3)(m+2)=0$

$m_1=3, m_2=-2$

$y_c=c_1e^{3x}+c_2e^{-2x}$

Let $y_p=A+Be^x$

Then $Dy_p=Be^x$ and $D^2y_p=Be^x$

Substituting in the differential equation:

$Be^x-Be^x-6(A+Be^x)=5-e^x$

$-6A-6Be^x=5-e^x$

$-6A=5$, so $A=-\frac{5}{6}$

$-6B=-1$ so $B=\frac{1}{6}$

The complete solution is

$y=c_1e^{3x}+c_2e^{-2x}-\frac{5}{6}+\frac{1}{6}e^x$

Substituting $x = 0, y = 2$:

$2 = c_1 + c_2 - \frac{5}{6} + \frac{1}{6}$

$c_1 = \frac{8}{3} - c_2$

$D_y = 3c_1e^{3x} - 2c_2e^{-2x} + \frac{1}{6}e^x$

Substituting $Dy = 4, x = 0$:

$4 = 3c_1 - 2c_2 + \frac{1}{6}$

Substituting c_1:

$4 = 8 - 3c_2 - 2c_2 + \frac{1}{6}$

$c_2 = \frac{5}{6}, c_1 = \frac{11}{6}$

The solution is

$y = \frac{11}{6}e^{3x} + \frac{5}{6}e^{-2x} + \frac{1}{6}e^x - \frac{5}{6}$

$= \frac{1}{6}\left(11e^{3x} + 5e^{-2x} + e^x - 5\right)$

37. $Dy - y = x^2$

$m - 1 = 0$

$m = 1$

$y_c = c_1e^x$

Let $y_p = A + Bx + Cx^2$

Then $Dy_p = B + 2Cx$

Substituting in the differential equation:

$B + 2Cx - A - Bx - Cx^2 = x^2$

Equating the coefficients of similar terms:

$-C = 1$, so $C = -1$

$2C - B = 0$, so $B = -2$

$A - B = 0$, so $A = -2$

The solution is

$y = c_1e^x - 2 - 2x - x^2$

Solving with an integrating factor would require integration by parts, so this method is simpler.

31.10 Applications of Higher-order Equations

1. $x = c_1 \sin 4t + c_2 \cos 4t; x = 2, Dx = 4$ for $t = 0$

Substituting $x = 2, t = 0$:

$2 = c_1 \sin(4(0)) + c_2 \cos(4(0))$

$c_2 = 2$

$x = c_1 \sin 4t + 2\cos 4t$

$Dx = 4c_1 \cos 4t - 8\sin 4t$

Substituting $Dx = 4, t = 0$:

$4 = 4c_1 \cos(4(0)) - 8\sin(4(0))$

$c_1 = 1$

The solution is

$x = \sin 4t + 2\cos 4t$

5. **(a)** (a)$D^2x + bDx + 100x = 0$

$k^2 = 100$, so $k = 10$

For the motion to be underdamped,

$b^2 < 4k^2$

$b^2 < 400$

$b < 20$

(b) $D^2x + bDx + 100x = 0$

$k^2 = 100$, so $k = 10$

For the motion to be overdamped,

$b^2 > 4k^2$

$b^2 > 400$

$b > 20$

9. $D^2y + bDy + 25y = 0$

$m^2 + bm + 25 = 0$

$m = \frac{-b \pm \sqrt{b^2 - 4(25)}}{2} = \frac{-b \pm \sqrt{b^2 - 100}}{2}$

For critical damping: $b^2 - 100 = 0$;

$b = 10$ $(b > 0)$

13. To get spring constant: $F = kx$

$4.00 = k(0.125)$

$k = 32.0$ lb/ft

To get mass of object:

$F = ma; 4.00 = m(32)$

$m = 0.125$ slugs

Using Newton's second law:

mass × accel. = restoring force

$$0.125\frac{d^2x}{dt^2} = -4x$$

$$D^2x + 32x = 0$$

$$m^2 + 32 = 0$$

$$m = \pm 4\sqrt{2}j$$

$$x = c_1 \sin 4\sqrt{2}t + c_2 \cos 4\sqrt{2}t$$

$$D_x = 4\sqrt{2}c_1 \cos 4\sqrt{2}t - 4\sqrt{2}c_2 \sin 4\sqrt{2}t$$

Let $x = 0.25$ ft when $t = 0$:

$$0.25 = c_1 \sin 0 + c_2 \cos 0$$

$$c_2 = 0.25$$

Let $Dx = 0$ when $t = 0$:

$$0 = 4\sqrt{2}c_1 \cos 4\sqrt{2}t - \sqrt{2}c_2 \sin 4\sqrt{2}t$$

$$c_1 = 0$$

$$x = 0.25 \cos 4\sqrt{2}t$$

17. $$L\frac{d^2q}{dt^2} + R\frac{dq}{dt} + \frac{q}{C} = E$$

$$0.200\frac{d^2q}{dt^2} + 8.00\frac{dq}{dt} + 10^6 q = 0$$

$$0.200m^2 + 8.00m + 10^6 = 0$$

$$m = \frac{-8.00 \pm \sqrt{64.0 - 0.800 \times 10^6}}{0.400};$$

$$m = -20.0 \pm 2240j; \alpha = -20.0, \beta = 2240$$

$$q = e^{-20.0t}(c_1 \sin 2240t + c_2 \cos 2240t)$$

$$t = 0, q = 0$$

$$0 = e^{-20.0(0)}(c_1 \sin 0 + c_2 \cos 0)$$

Therefore, $c_2 = 0$

$$\frac{dq}{dt} = e^{-20.0t}(2240c_1 \cos 2240t - 2240c_2 \sin 2240t) + (c_1 \sin 2240t + c_2 \cos 2240t)(-20.0e^{-20.0t})$$

$t = 0, i = 0.500$; therefore, $c_1 = 2.24 \times 10^{-4}$;

$$q = e^{-20.0t}(2.24 \times 10^{-4}) \sin 2240t$$

$$q = 2.24 \times 10^{-4} e^{-20.0t} \sin 2240t$$

21. $$L\frac{d^2q}{dt^2} + R\frac{dq}{dt} + \frac{q}{C} = E$$

$$0.500D^2q + 10.0Dq + \frac{q}{200 \times 10^{-6}} = 120 \sin 120\pi t$$

$$1.00D^2q + 20.0Dq + 10^4 q = 240 \sin 120\pi t$$

$$1.00D^2q + 20.0Dq + 10^4 q = 0$$

$$1.00m^2 + 20.0m + 10^4 = 0$$

$$m = \frac{-20.0 \pm \sqrt{400 - 4 \times 10^4}}{2.00} = -10.0 \pm 99.5j$$

$$\alpha = -10, \beta = 99.5$$

$$q_c = e^{-10.0t}(c_1 \sin 99.5t + c_2 \cos 99.5t)$$

Let $q_p = A \sin 120\pi t + B \cos 120\pi t$, then

$$Dq_p = 120\pi A \cos 120\pi t - 120\pi B \sin 120\pi t$$

$$D^2 q_p = -120^2 \pi^2 A \sin 120\pi t - 120^2 \pi^2 B \cos 120\pi t$$

Substituting in the differential equation:

$$-120^2\pi^2 A \sin 120\pi t - 120^2 \pi^2 B \cos 120\pi t + 20(120\pi A \cos 120\pi t - 120\pi B \sin 120\pi t) + 10^4(A \sin 120\pi t + B \cos 120\pi t) = 240 \sin 120\pi t$$

Equating the coefficients of similar terms:

$$-132000A - 7540B = 240$$

$$-132000B + 7540A = 0$$

Therefore,

$$A = -1.81 \times 10^{-3}, B = -1.03 \times 10^{-4}$$

$$q = e^{-10.0t}(c_1 \sin 99.5t + c_2 \cos 99.5t) - 1.81 \times 10^{-3} \sin 120\pi t - 1.03 \times 10^{-4} \cos 120\pi t$$

25. $L\frac{d^2q}{dt^2}+R\frac{dq}{dt}+\frac{q}{C}=E$

$1.00D^2q+5.00Dq+\frac{q}{150\times10^{-6}}=120\sin100t$

$1.00D^2q+5.00Dq+6670q=120\sin100t$

To find the steady-state current, we only need the particular solution. Let

$q_p=A\sin100t+B\cos100t$

Then $Dq_p=100A\cos100t-100B\sin100t$

$D^2q_p=-10^4A\sin100t-10^4B\cos100t$

Substituting in the differential equation:

$-10^4A\sin100t-10^4B\cos100t$
$+5.00(100A\cos100t-100B\sin100t)$
$+6670(A\sin100t+B\cos100t)$
$=120\sin100t$

$(-10^4A-500B+6670A)\sin100t$
$+(-10^4B+500A+6670B)\cos100t$
$=120\sin100t$

Equating the coefficients of similar terms:

$3330A+500B=-120$

$500A-3330B=0$

$A=-0.0352; B=0.005\ 28$

The particular solution for charge is

$q_p=-0.0352\sin100t+0.005\ 28\cos100t$

The particular solution for current (the derivative of charge) is

$i_p=-3.52\cos100t+0.528\sin100t$

31.11 Laplace Transforms

1.

$f(t)=1, t>0$

$L(f)=\int_0^\infty e^{-st}\cdot1\,dt$

$=\lim_{c\to\infty}\frac{-1}{s}\int_0^c e^{-st}(-s\,dt)$

$=-\frac{1}{s}\lim_{c\to\infty}e^{-st}\Big|_0^c$

$=-\frac{1}{s}\lim_{c\to\infty}\left(e^{-sc}-1\right)$

$=-\frac{1}{s}(0-1)$

$L(f)=\frac{1}{s}$

5. $f(t)=e^{3t}$; from Transform 3 of the table, with $a=-3$:

$L(e^{3t})=\frac{1}{s-3}$

9. $f(t)=\cos2t-\sin2t$

$L(f)=L(\cos2t)-L(\sin2t)$

By Transforms 5 and 6 with $a=2$:

$L(f)=\frac{s}{s^2+4}-\frac{2}{s^2+4}$

$L(f)=\frac{s-2}{s^2+4}$

13. $f''+f'; f(0)=0; f'(0)=0$

$L[f''+f']$
$=L(f'')+L(f')$
$=s^2L(f)-sf(0)-f'(0)+sL(f)-f(0)$
$=s^2L(f)-s\cdot0-0+sL(f)-0$
$=s^2L(f)+sL(f)$

17. $F(s)=\frac{2}{s^3}$

$L^{-1}(F)=L^{-1}\ \frac{2}{s^3}\ =2L^{-1}\ \frac{1}{s^3}$

$L^{-1}(F)=\frac{2t^2}{2}=t^2$ (Transform 2)

21. $F(s)=\frac{1}{s^3+3s^2+3s+1}$

$L^{-1}(F)=L^{-1}\ \frac{1}{(s+1)^3}$

$=\frac{1}{2}L^{-1}\ \frac{2!}{(s+1)^3}$

$=\frac{1}{2}t^2e^{-t}$

(Transform 12)

25. $F(s)=\frac{4s^2-8}{(s+1)(s-2)(s-3)}=\frac{-\frac{1}{3}}{s+1}+\frac{-\frac{8}{3}}{s-2}+\frac{7}{s-3}$

$L^{-1}(F)$

$=-\frac{1}{3}L^{-1}\ \frac{1}{s+1}\ -\frac{8}{3}L^{-1}\ \frac{1}{s-2}\ +7L^{-1}\ \frac{1}{s-3}$

$=-\frac{1}{3}e^{-t}-\frac{8}{3}e^{2t}+7e^{3t}$

29. Take $f(t) = e^{-at}$. Then

$$F(s) = L(f) = \frac{1}{s+a}$$

$$-\frac{d}{ds}F(s) = -\frac{d}{ds}\frac{1}{s+a}$$

$$= -\frac{-1}{(s+a)^2}$$

$$= \frac{1}{(s+a)^2}$$

From Transform 11,

$$L\{tf(t)\} = L\left(te^{-at}\right) = \frac{1}{(s+a)^2}$$

Therefore,

$$L\{tf(t)\} = -\frac{d}{ds}F(s)$$

31.12 Solving Differential Equations by Laplace Transforms

1. $2y' - y = 0; y(0) = 2$

$$L(2y') - L(y) = L(0)$$

$$2L(y') - L(y) = 0$$

$$2sL(y) - 2(2) - L(y) = 0$$

$$L(y) = \frac{4}{2s-1} = 2\frac{1}{s-\frac{1}{2}}$$

From Transform 3,

$$y = 2e^{\frac{1}{2}t}$$

5. $y' + y = 0; y(0) = 1$

$$L(y') + L(y) = L(0)$$

$$L(y') + L(y) = 0$$

$$sL(y) - y(0) + L(y) = 0$$

$$sL(y) - 1 + L(y) = 0$$

$$(s+1)L(y) = 1$$

$$L(y) = \frac{1}{s+1}$$

Transform 3 with $a = -1$

$$y = e^{-t}$$

9. $y' + 3y = e^{-3t}; y(0) = 1$

$$L(y') + L(3y) = L(e^{-3t})$$

$$[sL(y) - 1)] + 3L(y) = \frac{1}{s+3}$$

$$(s+3)L(y) = \frac{1}{s+3} + 1$$

$$L(y) = \frac{1}{(s+3)^2} + \frac{1}{s+3}$$

The inverse is found from Transforms 11 and 3

$$y = te^{-3t} + e^{-3t} = (1+t)e^{-3t}$$

13. $4y'' + 4y' + 5y = 0,\ y(0) = 1,\ y'(0) = -\frac{1}{2}$

$$4L(y'') + 4L(y') + 5L(y) = 0$$

$$4\left(s^2L(y) - sy(0) - y'(0)\right) + 4\left(sL(y) - y(0)\right)$$

$$+5L(y) = 0$$

$$4s^2L(y) - 4s + 2 + 4sL(y) - 4 + 5L(y) = 0$$

$$\left(4s^2 + 4s + 5\right)L(y) = 4s + 2$$

$$L(y) = \frac{4s+2}{4s^2+4s+5} = \frac{4\left(s+\frac{1}{2}\right)}{4\left(s^2+s+\frac{1}{4}\right)+4}$$

$$L(y) = \frac{s+\frac{1}{2}}{\left(s+\frac{1}{2}\right)^2+1}$$

Use Transform 20, $a = \frac{1}{2}, b = 1$

$$y = e^{-1/2t}\cos t$$

17. $y'' + y = 1; y(0) = 1; y'(0) = 1$

$$L(y'') + L(y) = L(1)$$

$$s^2L(y) - s - 1 + L(y) = \frac{1}{s}$$

$$(s^2+1)L(y) = \frac{1}{s} + s + 1$$

$$L(y) = \frac{1}{s(s^2+1^2)} + \frac{s}{s^2+1^2} + \frac{1}{s^2+1^2}$$

By Transforms 7, 5, and 6:

$$y = 1 - \cos t + \cos t + \sin t$$

$$y = 1 + \sin t$$

21. $y''-4y=10e^{3t}, y(0)=5, y'(0)=0$

$L(y'')-4L(y)=10\cdot L(e^{3y})$

Use Transform 3 with $a=-3$:

$$s^2L(y)-sy(0)-y'(0)-4L(y)=\frac{10}{s-3}$$

$$s^2L(y)-5s-0-4L(y)=\frac{10}{s-3}$$

$$(s^2-4)L(y)=5s+\frac{10}{s-3}$$

$$L(y)=\frac{5s}{(s+2)(s-2)}+\frac{10}{(s+2)(s-2)(s-3)}$$

Write in partial fractions,

$$L(y)=\frac{\frac{5}{2}}{s+2}+\frac{\frac{5}{2}}{s-2}+\frac{\frac{1}{2}}{s+2}+\frac{-\frac{5}{2}}{s-2}+\frac{2}{s-3}$$

$$L(y)=\frac{3}{s+2}+\frac{2}{s-3}$$

Use Transform 3 twice:

$y=3e^{-2t}+2e^{3t}$

25. $2v'=6-v$; since the object starts at rest, $v(0)=0,\ v'(0)=0$

$2L(v')+L(v)=6L(1)$

Use Transform 1:

$$2sL(v)-0+L(v)=\frac{6}{s}$$

$$(2s+1)L(v)=\frac{6}{s}$$

$$L(v)=\frac{6}{s(2s+1)}=6\ \frac{\frac{1}{2}}{s(s+\frac{1}{2})}$$

By Transform 4 with $a=\frac{1}{2}$:

$v=6(1-e^{-t/2})$

29. $L\dfrac{d^2q}{dt^2}+R\dfrac{dq}{dt}+\dfrac{q}{C}=E; q(0)=0$

$$0(q'')+50q'+\frac{q}{4\times10^{-6}}=40$$

$$50q'+\frac{10^6}{4}q=40$$

$$50L(q')+\frac{10^6}{4}L(q)=L(40)$$

$$50[sL(q)-q(0)]+\frac{10^6}{4}L(q)=\frac{40}{s}$$

$$L(q)\ \ 50s+\frac{10^6}{4}\ \ =\frac{40}{s}$$

$$L(q)=\frac{40}{s\left(50s+\frac{10^6}{4}\right)}$$

$$L(q)=\frac{40}{50s(s+5000)}$$

$$=\frac{40}{50(5000)}\cdot\frac{5000}{s(s+5000)}$$

$$=0.00016\cdot\frac{5000}{s(s+5000)}$$

$$q=L^{-1}\ \ 0.00016\cdot\frac{5000}{s(s+5000)}$$

$q=1.60\times10^{-4}(1-e^{-5000t})$

(from Transform 4, with $a=5000$)

33. $D^2y+9y=18\sin 3t;\ y=0,\ Dy=0, t=0$

$y''+9y=18\sin 3t$

$L(y'')+9L(y)=18L(\sin 3t)$

Use Transform 6 with $a=3$:

$$s^2L(y)-sy(0)-y'(0)+9L(y)=18\cdot\frac{3}{s^2+9}$$

$$L(y)(s^2+9)=\frac{54}{s^2+9}$$

$$L(y)=\frac{54}{(s^2+9)^2}$$

Use Transform 15, with $a=3, 2a^3=54$

$$y=L^{-1}\ \frac{54}{(s^2+9)^2}$$

$y=\sin 3t-3t\cos 3t$

37. $L\frac{d^2q}{dt^2}+R\frac{dq}{dt}+\frac{q}{C}=E;\ q(0)=0, q'(0)=0$

$$60\times10^{-3}\frac{d^2q}{dt^2}+8\frac{dq}{dt}+\frac{q}{300\times10^{-6}}=60$$

$$6\frac{d^2q}{dt^2}+800\frac{dq}{dt}+\frac{10^6q}{3}=6000$$

$$6D^2q+800Dq+\frac{10^6}{3}q=0$$

$$6m^2+800m+\frac{10^6}{3}m=0$$

$$m=\frac{-800\pm\sqrt{800^2-4(6)(\frac{10^6}{3})}}{12}$$

$$=-66.7\pm226j$$

$$q_c=e^{-66.7t}(c_1\sin(226t)+c_2\cos(226t))$$

Let $q_p=A$, then $\frac{dq_p}{dt}$ and $\frac{d^2q_p}{dt^2}=0.$

Substituting in the differential equation,

$\frac{10^6A}{3}=6000$, so $A=0.018$

$$q=e^{-66.7t}(c_1\sin(226t)+c_2\cos(226t))+0.018$$

Substituting $q=0, t=0:$

$$0=c_2+0.018$$

$$c_2=-0.018$$

$$q=e^{-66.7t}(c_1\sin(226t)-0.018\cos(226t))+0.018$$

$$i=\frac{dq}{dt}=e^{-66.7t}(226c_1\cos(226t)+(0.018)(226)\sin(226t))$$
$$-66.7(c_1\sin(226t)-0.018\cos(226t))e^{-66.7t}$$

$$i=\frac{dq}{dt}=e^{-66.7t}(226c_1\cos(226t)+(0.018)(226)\sin(226t))$$
$$-66.7(c_1\sin(226t)-0.018\cos(226t))e^{-66.7t}$$

Substituting $t=0, i=0:$

$$0=226c_1-66.7(-0.018)$$

$$c_1=-\frac{66.7(0.018)}{226}$$

$$i=e^{-66.7t}[(-66.7)(0.018)\cos(226t)+(0.018)(226)\sin(226t)+\frac{66.7^2(0.018)}{226}\sin(226t)+66.7(0.018)\cos(226t)]$$

$$i=4.42e^{-66.7t}\sin(226t)$$

Review Exercises

1. This is true.

$y = ce^{-x} + 2e^{2x}$

$Dy = -ce^{-x} + 4e^{2x}$

$D^2y = ce^{-x} + 8e^{2x}$

$$D^2y - Dy - 2y = \left(ce^{-x} + 8e^{2x}\right) - \left(-ce^{-x} + 4e^{2x}\right) - 2\left(ce^{-x} + 2e^{2x}\right)$$
$$= 0$$

5. This is false.

The general solution is $c_1e^{3x} + c_2xe^{3x}$.

9. $4xy^3dx + \left(x^2 + 1\right)dy = 0$

The equation is separable; divide by $y^3(x^2 + 1)$.

$$\frac{4x}{x^2 + 1}dx + \frac{dy}{y^3} = 0$$

Integrating ($u = x^2 + 1, du = 2xdx$),

$$2\ln\left(x^2 + 1\right) - \frac{1}{2y^2} = c$$

13. $2D^2y + Dy = 0$

The auxiliary equation is

$2m^2 + m = 0$

$m(2m + 1) = 0$

$m_1 = 0$ and $m_2 = -\frac{1}{2}$

$y = c_1e^0 + c_2e^{-x/2}$

$y = c_1 + c_2e^{-x/2}$

17. $(x + y)dx + \left(x + y^3\right)dy = 0$

$x\,dx + y\,dx + x\,dy + y^3\,dy = 0$

$x\,dx + d(xy) + y^3\,dy = 0$

Integrating,

$$\frac{1}{2}x^2 + xy + \frac{1}{4}y^4 = c_1$$

$$2x^2 + 4xy + y^4 = c$$

21. $dy = (2y + y^2)dx$

By separation of variables:

$$\frac{dy}{y(2+y)} = dx$$

Integrating by formula 2:

$$-\frac{1}{2}\ln\ \frac{2+y}{y} = x + \ln c_1$$

$$\ln\frac{2+y}{y} = -2x - \ln c_1^2$$

$$c_1^2\ \frac{2+y}{y} = e^{-2x}$$

$$y = c_1^2(y+2)e^{2x}$$

$$y = c(y+2)e^{2x}$$

$$y(1-ce^{2x}) = 2c$$

$$y = \frac{2c}{1-ce^{2x}}$$

25. $y' + 4y = 2e^{-2x}$

$$\frac{dy}{dx} + 4y = 2e^{-2x}$$

$$dy + 4\cdot y\,dx = 2e^{-2x}dx$$

$$e^{\int 4dx} = e^{4x}$$

$$ye^{4x} = \int 2e^{-2x}e^{4x} + c$$

$$= e^{2x} + c$$

$$y = e^{-2x} + ce^{-4x}$$

29. $2D^2s + Ds - 3s = 6$

$$2m^2 + m - 3 = 0$$

$$(m-1)(2m+3) = 0$$

$$m_1 = 1,\ m_2 = -\frac{3}{2}$$

$$s_c = c_1e^t + c_2e^{-3t/2}$$

Let $s_p = A$, then $s_p' = 0$; $s_p'' = 0$

Substituting into the differential equation,

$$2(0) + 0 - 3A = 6$$

$$A = -2$$

$$s_p = -2$$

The complete solution is

$$s = c_1e^t + c_2e^{-3t/2} - 2$$

33. $9D^2y - 18Dy + 8y = 16 + 4x$

$$D^2y - 2Dy + \frac{8}{9}y = \frac{16}{9} + \frac{4}{9}x$$

$$m^2 - 2m + \frac{8}{9} = 0$$

$$m = \frac{2 \pm \sqrt{4 - 4\left(\frac{8}{9}\right)}}{2};$$

$$m_1 = \frac{2}{3};\ m_2 = \frac{4}{3}$$

$$y_c = c_1e^{2x/3} + c_2e^{4x/3}$$

Let $y_p = A + Bx$. Then $y_p' = B$; $y_p'' = 0$

Substituting into the differential equation.

$$0 - 2B + \frac{8}{9}(A + Bx) = \frac{16}{9} + \frac{4}{9}x$$

$$-2B + \frac{8}{9}A\ + \frac{8}{9}Bx = \frac{16}{9} + \frac{4}{9}x$$

$$-2B + \frac{8}{9}A = \frac{16}{9};\ \frac{8}{9}B = \frac{4}{9}$$

$$B = \frac{1}{2}$$

$$-2\ \frac{1}{2}\ + \frac{8}{9}A = \frac{16}{9}$$

$$A = \frac{25}{8}$$

$$y_p = \frac{1}{2}x + \frac{25}{8}$$

The complete solution is

$$y = c_1e^{2x/3} + c_2e^{4x/3} + \frac{1}{2}x + \frac{25}{8}$$

37. $y''-7y'-8y=2e^{-x}$

$D^2y-7Dy-8y=0$

$m^2-7m-8=0$

$(m+1)(m-8)=0$

$m=-1,\ m=8$

$y_c=c_1e^{-x}+c_2e^{8x}$

Let $y_p=Axe^{-x}$

Then $y_p'=Ae^{-x}(1-x)$,

$y_p''=Ae^{-x}(x-2)$

Substituting in the differential equation:

$Ae^{-x}(x-2)-7Ae^{-x}(1-x)-8Axe^{-x}=2e^{-x}$

$A(x-2)-7A(1-x)-8Ax=2$

$-9A=2$

$A=\frac{-2}{9}$

$y_p=\frac{-2}{9}xe^{-x}$

$y=y_c+y_p$

$y=c_1e^{-x}+c_2e^{8x}+\frac{-2}{9}xe^{-x}$

41. $3y'=2y\cot x$

$\frac{dy}{dx}=\frac{2}{3}y\cot x$

$\frac{dy}{y}=\frac{2}{3}\cot x\,dx$

$\ln y=\frac{2}{3}\ln\sin x+\ln c$

$\ln y-\ln\sin^{2/3}x=\ln c;$

$\ln\frac{y}{\sin^{2/3}x}=\ln c$

$\frac{y}{\sin^{2/3}x}=c$

$y=c\sin^{2/3}x$

Substituting $y=2, x=\frac{\pi}{2}$:

$2=c\sin^{2/3}\frac{\pi}{2}$

$c=2$

Therefore,

$y=2\sin^{2/3}x=2\sqrt[3]{\sin^2 x}$

$y^3=8\sin^2 x$

45. $D^2v+Dv+4v=0$

$m^2+m+4=0$

$m=\frac{-1\pm\sqrt{-15}}{2}$

$m=-\frac{1}{2}\pm\frac{\sqrt{15}}{2}j;\ \alpha=-\frac{1}{2},\ \beta=\frac{\sqrt{15}}{2};$

$v=e^{-t/2}\ \ c_1\sin\frac{\sqrt{15}}{2}t+c_2\cos\frac{\sqrt{15}}{2}t$

$Dv=e^{-t/2}\ \ \frac{\sqrt{15}}{2}c_1\cos\frac{\sqrt{15}}{2}t-\frac{\sqrt{15}}{2}c_2\sin\frac{\sqrt{15}}{2}t$

$+\ \ c_1\sin\frac{\sqrt{15}t}{2}c_2\cos\frac{\sqrt{15}}{2}t\ \ \ -\frac{1}{2}e^{-t}$

Substituting $v=0$ when $t=0$:

$0=e^0(c_1\cdot 0+c_2)$

$c_2=0$

Substituting $Dv=\sqrt{15}, t=0$:

$\sqrt{15}=e^0\ \ \frac{\sqrt{15}}{2}c_1$

$c_1=2$

Therefore,

$y=2e^{-t/2}\sin\ \ \frac{\sqrt{15}}{2}t$

49. $4y'-y=0; y(0)=1$

$L(4y')-L(y)=0$

$4L(y')-L(y)=0$

$4sL(y)-1-L(y)=0$

by Eq. (30.24)

$(4s-1)L(y)=1$

$L(y)=\frac{1}{4s-1}=\frac{1}{4(s-\frac{1}{4})}$

By Transform 3 with $a=-\frac{1}{4}$,

$y=\frac{1}{4}e^{t/4}.$

53. $y'' - 6y' + 9y = t, y(0) = 0. y'(0) = 1$

$$L(y'' - 6y' + 9y) = L(t)$$

$$s^2L(y) - sy(0) - y'(0) - 6sL(y) - 6y(0) + 9L(y) = \frac{1}{s+2}$$

$$(s-3)^2 L(y) - 1 = \frac{1}{s+2}$$

$$(s-3)^2 L(y) = 1 + \frac{1}{s+2}$$

$$L(y) = \frac{1}{(s-3)^2} + \frac{1}{(s-3)^2(s+2)}$$

$$L(y) = \frac{6}{5}\cdot\frac{1}{(s-3)^2} + \frac{1}{25}\cdot\frac{1}{s+2} - \frac{1}{25}\cdot\frac{1}{s-3}$$

$$y = \frac{6}{5}te^{3t} + \frac{1}{25}e^{-2t} - \frac{1}{25}e^{3t}$$

57. $L(y) = \frac{1}{16}\frac{3}{\left(s^2 + \frac{9}{16}\right)(s-1)}$

Writing in partial fractions,

$$L(y) = \frac{3}{25}\ \frac{1}{(s-1)} - \frac{s}{\left(s^2 + \frac{9}{16}\right)} - \frac{4}{3}\frac{\frac{3}{4}}{s^2 + \frac{9}{16}}$$

Using Transforms 3,5 and 6:

$$y = \frac{3}{25}\ e^x - \cos\frac{3x}{4} - \frac{4}{3}\sin\frac{3x}{4}$$

61. $\frac{dy}{dx} - 2y = e^x; y(0) = 1$

(a) As a linear equation:

$dy - 2y\,dx = e^{3x}dx$

$ye^{-2dx} = e^{-2dx}e^{3x}dx$

$ye^{-2x} = e^x dx = e^x + c_1$

$y = e^{3x} + c_1e^{2x}$

Substituting $y = 1,\ x = 0$:

$1 = 1 + c_1$

$c_1 = 0$

The solution is $y = e^{3x}$.

(b) Using Laplace transforms:

$L(y') - 2L(y) = L(e^{3x})$

Use Transform 3:

$$sL(y) - y(0) - 2L(y) = \frac{1}{s-3}$$

$$sL(y) - 1 - 2L(y) = \frac{1}{s-3}$$

$$L(y) = \frac{1}{(s-3)(s-2)} + \frac{1}{s-2}$$

Use Transform 9 with $a = -3$, $b = -2$
and Transform 3 with $a = -2$:

$$y = e^{3x} - e^{2x} + e^{2x} = e^{3x}$$

65. $\frac{dx}{dt} = 2t$

$x = t^2 + c$

Substitute $x = 1, t = 0$:

$1 = 0^2 + c$

$c = 1$

Therefore, x in terms of t is given by

$x = t^2 + 1$

Taking derivatives with respect to t on both sides of $xy = 1$:

$$x\frac{dy}{dt} + y\frac{dx}{dt} = 0$$

Substituting $\frac{dx}{dt} = 2t$ and $x = \frac{1}{y}$:

$$\frac{1}{y}\frac{dy}{dt} + y(2t) = 0$$

Multiplying by dt / y,

$$\frac{1}{y^2}dy + 2t\,dt = 0$$

$$-\frac{1}{y} + t^2 = c$$

Substituting $y = 1, t = 0$:

$$-\frac{1}{1} + 0^2 = c$$

$$c = -1$$

$$-\frac{1}{y} + t^2 = -1$$

$$-1 + yt^2 = -y$$

$$y\left(t^2 + 1\right) = 1$$

$$y = \frac{1}{t^2 + 1}$$

In terms of t, $x = t^2 + 1$ and $y = \frac{1}{t^2 + 1}$.

69. $\frac{dm}{dt} = km$

Solve by separation of variables:

$\frac{dm}{m} = kdt$

$\ln m = kt + c$

Substituting $t = 0, m = m_0$:

$\ln m_0 = k(0) + c$

$c = \ln m_0$

$\ln m = kt + \ln m_0$

$\ln \frac{m}{m_0} = kt$

$\frac{m}{m_0} = e^{kt}$

$m = m_0 e^{kt}$

73. $\frac{dy}{dx} = \frac{y}{y - x}, (-1, 2), y > 0$

$ydy = ydx + xdy = d(xy)$

$\frac{y^2}{2} = xy + C$

Substituting $x = -1, y = 2$:

$\frac{2^2}{2} = (-1)(2) + C$

$C = 4$

$\frac{y^2}{2} = xy + 4$

$y^2 = 2xy + 8$

$y^2 - 2xy - 8 = 0$

To find y explicitly, we use the quadratic formula with $a = 1, b = -2x, c = -8$:

$y = \frac{2x \pm \sqrt{4x^2 + 32}}{2}$

$= x \pm \sqrt{x^2 + 8}$

The path is a hyperbola that consists of the two functions:

$y = x + \sqrt{x^2 + 8}$ and

$y = x - \sqrt{x^2 + 8}$

77. $N = N_0 e^{kt}$

Substituting $N = N_0/2, t = 1.28\times10^9$:

$$\frac{N_0}{2} = N_0 e^{k(1.28\times10^9)}$$

$$k = \frac{\ln 0.5}{1.28\times10^9}$$

$$N = N_0 e^{-\frac{\ln 0.5}{1.28\times10^9}t}$$

Find t for $N = 0.75N_0$:

$$0.75N_0 = N_0 e^{\frac{\ln 0.5}{1.28\times10^9}t}$$

$$t = \frac{(\ln 0.75)(1.28\times10^9)}{\ln 0.5}$$

$t = 5.31\times10^8$ years

81. See Example 2, Section 31.6.

$$y = kx^5,\ k = \frac{y}{x^5}$$

Take the derivative with respect to x :

$y' = 5kx^4$

Substitute the expresion for k :

$$y' = 5\ \frac{y}{x^5}\ x^4 = \frac{5y}{x}$$

The slope of the orthogonal trajectories will be the negative reciprocal of y', therefore,

$$\left.\frac{dy}{dx}\right|_{OT} = -\frac{x}{5y}$$

Solve by separating variables:

$5y\,dy = -x\,dx$

$$\frac{5}{2}y^2 = -\frac{1}{2}x^2 + \frac{1}{2}c$$

$5y^2 + x^2 = c$

The orthogonal trajectories are ellipses.

85. $L\frac{di}{dt} + Ri = E$

$$2\frac{di}{dt} + 40i = 20$$

$$\frac{di}{dt} + 20i = 10$$

$di + 20i\,dt = 10\,dt$

$P = 20, Q = 10$

$$ie^{\int 20\,dt} = 10e^{\int 20\,dt} + c$$

$$ie^{20t} = 10e^{20t} + c$$

$$= 0.5e^{20t} + c$$

Substituting $i = 0, t = 0$:

$$0 = 0.5 + c$$

$$c = -0.5$$

$$ie^{20t} = 0.5e^{20t} - 0.5$$

$$= 0.5\left(e^{20t} - 1\right)$$

$$i = 0.5\left(1 - e^{-20t}\right)$$

89. $LD^2q + RDq + \dfrac{q}{C} = E$

$L = 0.5$ H, $R = 6\Omega$, $C = .02$ F, $E = 24\sin 10t$

$0.5D^2q + 6Dq + 50q = 24\sin 10t$

$D^2q + 12Dq + 100q = 48\sin 10t$

$D^2q + 12Dq + 100q = 0$

$m^2 + 12m + 100 = 0$

$m = \dfrac{-12 \pm \sqrt{144 - 400}}{2} = -6 \pm 8j$

$q_c = e^{-6t}\left(c_1 \sin 8t + c_2 \cos 8t\right)$

Let $q_p = A\sin 10t + B\cos 10t$, then

$Dq_p = 10A\cos 10t - 10B\sin 10t$

$D^2q_p = -100A\sin 10t - 100B\cos 10t$

Substituting in the differential equation:

$-100A\sin 10t - 100B\cos 10t + 12\left(10A\cos 10t - 10B\sin 10t\right)$

$\quad + 100\left(A\sin 10t + B\cos 10t\right) = 48\sin 10t - 120B\sin 10t + 120A\cos 10t = 48\sin 10t$

Equating the coefficients of similar terms:

$-120B = 48; 120A = 0$

Therefore, $B = -0.4$, $A = 0$

The complete solution is

$q = e^{-6t}\left(c_1 \sin 8t + c_2 \cos 8t\right) - 0.4\cos 10t$

Substituting $q = 0$, $t = 0$:

$0 = c_2 - 0.4$, so $c_2 = 0.4$

$q = e^{-6t}\left(c_1 \sin 8t + 0.4\cos 8t\right) - 0.4\cos 10t$

$Dq = e^{-6t}\left(-6c_1 \sin 8t - 2.4\cos 8t + 8c_1 \cos 8t - 3.2\sin 8t\right) + 4\sin 10t$

Substituting $Dq = 0, t = 0$:

$0 = -2.4 + 8c_1$, so $c_1 = 0.3$

The solution is

$q = e^{-6t}\left(0.3\sin 8t + 0.4\cos 8t\right) - 0.4\cos 10t$

93. $2\dfrac{di}{dt}+i=12,\ i(0)=0$

$2L(i')+L(i)=12L(1)$

Use Transform 1:

$2 \quad sL(i) \quad i(0) \quad L(i) \quad \dfrac{12}{s}$

$(2s+1)L(i)=\dfrac{12}{s}$

$$L(i)=\frac{12}{s(2s+1)}$$

$$=12\frac{\frac{1}{2}}{s(s+1/2)}$$

Use Transform 4, $a=\frac{1}{2}$:

$$i=12\left(1-e^{-t/2}\right)$$

Evaluate i for $t=0.3$:

$$i=12\left(1-e^{-0.3/2}\right)$$

$$i=1.67 \text{ A}$$

97. R = rate in L / t at which mixtures flow in/out of container

$V(t)$ = volume of O_2 in L in container at time t

$V(0)=5.00$ L. When 5.00 L of air passed into the container,

$Rt=5,\ t=\dfrac{R}{5}$. Find $V \quad \dfrac{R}{5}$.

$\dfrac{dV}{dt}=0.20R-\dfrac{V}{5.00}R=-0.20R(V-1)$

$\dfrac{dV}{V-1}=-0.20\ R\ dt$

$\ln(V-1)=-0.20\ Rt+C$

Substituting $t=0, V=5.00$:

$\ln(5.00-1)=-0.20R\cdot 0+C$

$C=\ln 4.00$

$\ln(V-1)=-0.20\ Rt+\ln 4.00$

Find V for $t=\frac{5}{R}$:

$\ln(V-1)=-0.20R\left(\dfrac{5}{R}\right)+4.00$

$\ln\dfrac{V-1}{4.00}=-1.00$

$\dfrac{V-1}{4.00}=e^{-1}$

$V=4.00e^{-1}+1$

$V=2.47$ L

101. $EI\frac{d^2y}{dx^2}=M,\ M=2000x-40x^2$

$$EID^2y=2000x-40x^2$$

$$D^2y=\frac{1}{EI}\left(2000x-40x^2\right)$$

$$Dy=\frac{1}{EI}\left(1000x^2-\frac{40}{3}x^3\right)+c_1$$

$$y=\frac{1}{EI}\left(\frac{1000}{3}x^3-\frac{10}{3}x^4\right)+c_1x+c_2$$

$y=0$ for $x=0$ and $x=L$

$$0=\frac{1}{EI}(0-0)+0+c_2,\ c_2=0$$

$$0=\frac{1}{EI}\left(\frac{1000}{3}L^3-\frac{10}{3}L^4\right)+c_1L$$

$$c_1=\frac{1}{EI}\left(\frac{10}{3}L^3-\frac{1000}{3}L^2\right)$$

$$y=\frac{1}{EI}\left(\frac{1000}{3}x^3-\frac{10}{3}x^4\right)+\frac{1}{EI}\left(\frac{1000}{3}L^3-\frac{1000}{3}L^2\right)x$$

$$=\frac{10}{3EI}\left(100x^3-x^4+L^3x-100L^2x\right)$$